Urban Watersheds

Geology, Contamination, and Sustainable Development

Martin M. Kaufman • Daniel T. Rogers
Kent S. Murray

CRC Press
Taylor & Francis Group
Boca Raton London New York

CRC Press is an imprint of the
Taylor & Francis Group, an **informa** business

CRC Press
Taylor & Francis Group
6000 Broken Sound Parkway NW, Suite 300
Boca Raton, FL 33487-2742

Printed in the United States of America on acid-free paper
10 9 8 7 6 5 4 3 2 1

International Standard Book Number: 978-1-4398-5274-3 (Hardback)

Library of Congress Cataloging-in-Publication Data

Kaufman, Martin M.
Urban watersheds : geology, contamination, and sustainable development / by Martin M. Kaufman, Daniel T. Rogers, Kent S. Murray.
p. cm.
Summary: "This handbook presents 10 years of research on contaminants affecting lands and waters within heavily urbanized. It integrates research and professional practice from the fields of environmental geology, geochemistry, risk analysis, hydrology, and urban planning and offers knowledge on improving watershed management practices and urban development. It explains different aspects of urban land and water contamination; including sources, extents, and risks. The text includes case studies on successful and unsuccessful approaches to contaminant remediation, as well as practical methods for environmental risk assessment"-- Provided by publisher.
Includes bibliographical references and index.
ISBN 978-1-4398-5274-3 (hardback)
1. Urban runoff--Management. 2. Urban watersheds--Protection. 3. Water--Pollution. 4. Sustainable development. 5. Cities and towns--Environmental conditions. 6. Urban pollution. I. Rogers, Daniel T. (Daniel Timothy), 1959- II. Murray, Kent S. III. Title.

TD657.K38 2011
363.72'84--dc22 2010045339

Visit the Taylor & Francis Web site at
http://www.taylorandfrancis.com

and the CRC Press Web site at
http://www.crcpress.com

To my brothers, Alan and Bob.

Martin M. Kaufman

To my family, my wife Vivian, and our two children, Hannah and Matt, for their encouragement, patience, understanding, and love for the many years I have pursued my passion.

Daniel T. Rogers

To my family, my wife Lynda, my son Sean, and stepsons Brett and Steven.

Kent S. Murray

Contents

Preface ... xv
Acknowledgments ... xvii
Authors ... xix

1. The Watershed Approach ... 1
1.1 Introduction ... 1
1.2 Historical Transformation of Urbanized Watersheds by Industrial Development ... 1
1.3 The Evolution of Industry and the Rouge Watershed ... 3
1.4 Major Themes of the Watershed Approach Used in This Book ... 7
1.4.1 Theme #1: Near-Surface Geology, Hydrogeology, and Surface Hydrology ... 7
1.4.1.1 Geology ... 8
1.4.1.2 Hydrogeology ... 8
1.4.1.3 Surface Hydrology ... 9
1.4.2 Theme #2: Science and Planning—The Movement of Water ... 10
1.4.3 Theme #3: Industrial Property Abandonment, Contamination, and Risk ... 11
1.4.4 Theme #4: Remediation of Contaminated Sites ... 13
1.4.5 Theme #5: Science-Based Landscape Planning ... 14
1.5 Organization of This Book ... 15
References ... 16

Part I Geology

2. Geology of Urban Watersheds ... 21
2.1 Introduction ... 21
2.2 Geological Processes Affecting Urban Watersheds ... 22
2.2.1 Formation of Sediments and Rocks ... 23
2.2.2 Principle of Uniformity ... 26
2.2.3 Principles of Stratigraphy ... 26
2.2.4 Unconformities ... 28
2.3 Sedimentary Processes: From Removal to Rock Formation ... 29
2.3.1 Clastic Processes ... 30
2.3.1.1 Weathering ... 30
2.3.1.2 Mass Wasting ... 31
2.3.1.3 Stream Transport ... 33
2.3.2 Non-Clastic Chemical Sedimentary Processes ... 38
2.3.2.1 Limestone and Dolostone ... 39
2.3.2.2 Evaporites ... 39
2.3.3 Formation of Sediments ... 40
2.3.4 Rock Formation: Diagenesis ... 41

2.4 Identification and Classification of Sedimentary Deposits 41
2.4.1 Identification and Classification Methods 42
2.4.1.1 Sorting 42
2.4.1.2 Grain Shape 43
2.4.1.3 Grain Size 43
2.4.1.4 Composition 43
2.4.1.5 Thickness 45
2.4.2 Bedded or Stratified Depositional Features 45
2.4.3 Other Depositional Features 47
2.5 Sedimentary Depositional Environments 49
2.5.1 Alluvial 50
2.5.2 Fluvial 50
2.5.3 Lacustrine 53
2.5.4 Eolian 54
2.5.5 Wetlands 54
2.5.6 Glacial 54
2.5.7 Non-Clastic Chemical Depositional Environments 57
2.6 Disturbance 57
2.6.1 Anthropogenic Disturbance 57
2.6.2 Naturally Occurring Disturbance 58
2.7 Summary and Conclusion 59
References 60

3. Water and the Hydrogeology of Watersheds 65
3.1 Introduction 65
3.2 Water Basics 66
3.2.1 Basic Water Chemistry 66
3.2.2 Water Cycle 67
3.2.3 Distribution of Water on Earth 67
3.2.4 Potable Water Use 68
3.3 Surface Water in Watersheds 71
3.3.1 Watersheds 71
3.3.2 Surface Water Drainage Development and Patterns 74
3.4 Groundwater in Watersheds 76
3.5 Fundamental Concepts of Groundwater 78
3.5.1 Aquifer Types and Basic Structure 79
3.5.2 Heterogeneity and Homogeneity 80
3.5.3 Hydraulic Conductivity 83
3.5.4 Groundwater Movement 84
3.5.5 Primary Porosity 86
3.5.6 Secondary Porosity 88
3.5.7 Water Table and Hydraulic Gradient 90
3.5.8 Determining the Direction of Groundwater Flow 90
3.5.9 Estimating the Rate of Groundwater Flow 92
3.5.10 Groundwater Withdrawal 94
3.6 Surface Water–Groundwater Interaction 96
3.6.1 Stream Interaction with Groundwater 96
3.6.2 Lake Interaction with Groundwater 98

3.6.3 Wetland Interaction with Groundwater 100
3.6.4 Ocean Water, Groundwater, and Surface Water Interaction........ 100
3.7 Anthropogenic Influences on Surface Water and Groundwater........ 103
3.7.1 Anthropogenic Surface Water Influences........ 103
3.7.1.1 Destruction of Wetlands 104
3.7.1.2 Dam Construction........ 105
3.7.2 Anthropogenic Groundwater Influences 107
3.7.2.1 Saltwater Intrusion........ 107
3.7.2.2 Changes in Groundwater Flow 107
3.7.2.3 Artificial Recharge 109
3.7.2.4 Land Subsidence........ 111
3.8 Summary and Conclusion 112
References 113

4. Conducting Subsurface Environmental Investigations in Urban Watersheds 117
4.1 Introduction 117
4.2 Types and Description of Subsurface Environmental Investigations, Studies, Plans, or Reports 119
4.2.1 Phase I Environmental Site Assessment 120
4.2.2 Health and Safety 124
4.2.3 Phase II Investigation 127
4.2.4 Geophysical Investigation........ 130
4.2.5 Ecological and Human Health Risk Assessment........ 131
4.2.6 Feasibility Study........ 134
4.2.7 Remedial Investigation 134
4.2.8 Natural Resource Damage Assessment........ 135
4.2.9 Environmental Impact Statement........ 135
4.2.10 Remedial Action Plan........ 136
4.2.11 Completion Evaluation Study 136
4.2.12 Summary of Environmental Investigations........ 137
4.3 Common Environmental Sampling Methods........ 137
4.3.1 Soil Sampling and Description 138
4.3.1.1 Surface Sampling........ 138
4.3.1.2 Excavating Test Pits........ 139
4.3.1.3 Drilling Using a Hand Auger........ 139
4.3.1.4 Mechanical Drilling Methods 139
4.3.1.5 Sample Containers 143
4.3.1.6 Documentation 143
4.3.2 Groundwater Sampling Methods........ 150
4.3.2.1 Installing a Monitoring Well 150
4.3.2.2 Groundwater Sampling........ 154
4.3.3 Sediment Sampling 157
4.3.4 Surface Water Sampling........ 157
4.3.5 Air Sampling 158
4.3.5.1 Indoor Air Sampling........ 158
4.3.5.2 Outdoor Air Sampling........ 160
4.4 Summary and Conclusion 161
References 162

5. Urban Geologic Mapping 165
5.1 Introduction 165
5.2 Geologic Mapping in an Urban Environment 166
5.2.1 Information and Mapping Challenges 166
5.2.2 Information Sources, Integration, and Evaluating Usability 167
5.2.3 Building the Conceptual Geologic Model 171
5.2.4 Fieldwork 172
5.2.5 Facies Analysis 177
5.2.6 Developing and Constructing Cross Sections 179
5.2.7 Creating the Base Map 181
5.2.8 Final Draft Map Preparation 182
5.2.8.1 Succession of Maps 183
5.2.8.2 Three-Dimensional Maps 185
5.2.8.3 Maps Using GIS 185
5.3 Summary and Conclusion 190
References 193

6. Geologic Vulnerability 195
6.1 Introduction 195
6.2 Subsurface Vulnerability and Vulnerability Map Development 196
6.3 Methods 197
6.4 Map-Building Example 202
6.5 Demonstrating the Significance of Vulnerability Mapping 205
6.5.1 Site 1: Low Vulnerability Site 207
6.5.2 Site 2: High Vulnerability Site 210
6.5.3 Site Comparison Analysis 217
6.6 Summary and Conclusion 219
References 219

Part II Contamination

7. Common Contaminants in Urban Watersheds 225
7.1 Introduction 225
7.2 Contaminant Toxicity 227
7.3 Volatile Organic Compounds 230
7.3.1 Light Nonaqueous Phase Liquids 230
7.3.2 Dense Nonaqueous Phase Liquids 233
7.3.3 Trihalomethanes 235
7.4 Polynuclear Aromatic Hydrocarbons 236
7.5 Polychlorinated Biphenyls 237
7.6 Semivolatile Organic Compounds 240
7.6.1 Phthalates 240
7.6.2 Phenol 242
7.6.3 Amines 243
7.6.4 Esters 243
7.7 Heavy Metals 244
7.8 Other Contaminants 247

7.8.1 Pesticides and Herbicides ... 247
7.8.2 Dioxins ... 249
7.8.3 Fertilizers ... 250
7.8.4 Cyanide ... 250
7.8.5 Asbestos ... 251
7.8.6 Acids and Bases ... 253
7.8.7 Radioactive Compounds ... 254
7.8.8 Greenhouse Gases ... 255
7.8.9 Carbon Monoxide ... 256
7.8.10 Ozone ... 258
7.8.11 Sulfur Dioxide ... 260
7.8.12 Particulate Matter ... 260
7.8.13 Bacteria, Parasites, and Viruses ... 263
7.8.14 Emerging Contaminants ... 264
7.9 Summary and Conclusion ... 266
References ... 266

8. Contaminant Fate and Transport ... 273
8.1 Introduction ... 273
8.2 Contaminant Releases into the Environment ... 274
8.3 Principles of Contaminant Fate and Transport ... 276
8.3.1 Basic Contaminant Transport Concepts ... 276
8.3.2 Basic Contaminant Degradation Concepts ... 281
8.3.2.1 Biotic Degradation ... 282
8.3.2.2 Abiotic Degradation ... 283
8.3.3 Fate and Transport of Contaminants in Soil ... 284
8.3.4 Fate and Transport of Contaminants in Surface Water ... 285
8.3.4.1 Contaminant Persistence and Bioaccumulation ... 287
8.3.5 Fate and Transport of Contaminants in Groundwater ... 288
8.3.5.1 Karst Topography ... 293
8.3.6 Fate and Transport of Contaminants in the Atmosphere ... 293
8.4 Fate and Transport of Contaminants ... 297
8.4.1 VOCs ... 297
8.4.2 PAHs ... 301
8.4.3 PCBs ... 301
8.4.4 SVOCs ... 302
8.4.5 Heavy Metals ... 302
8.4.6 Other Contaminants ... 303
8.4.6.1 Pesticides and Herbicides ... 303
8.4.6.2 Dioxins ... 304
8.4.6.3 Fertilizers ... 304
8.4.6.4 Cyanide ... 305
8.4.6.5 Asbestos ... 305
8.4.6.6 Acids and Bases ... 305
8.4.6.7 Radioactive Compounds ... 306
8.4.6.8 Greenhouse Gases ... 306
8.4.6.9 Carbon Monoxide ... 306
8.4.6.10 Ozone ... 306

8.4.6.11 Sulfur Dioxide ... 307
8.4.6.12 Particulate Matter ... 307
8.4.6.13 Bacteria, Parasites, and Viruses ... 308
8.4.6.14 Emerging Contaminants ... 308
8.5 Summary and Conclusion ... 308
References ... 309

9. Metal Contamination in Urban Watersheds ... 315
9.1 Introduction ... 315
9.2 Heavy Metals in Soil ... 315
9.3 Case Study: Metals in Soil, Rouge River Watershed ... 316
9.3.1 Study Objectives, Study Area, and Methods ... 316
9.3.2 Results ... 321
9.4 Case Study: Dissolved Metals in Shallow Groundwater, Rouge River Watershed ... 326
9.4.1 Study Objectives, Study Area, and Methods ... 327
9.4.2 Results ... 329
9.5 Conclusions ... 336
References ... 337

10. Contaminant Risk Factors ... 341
10.1 Introduction ... 341
10.2 Surface Risk Factor ... 341
10.3 Groundwater Contaminant Risk Factor ... 346
10.4 Soil Contaminant Risk Factor ... 355
10.5 Air Contaminant Risk Factor ... 363
10.6 Discussion and Implications ... 370
10.7 Summary and Conclusion ... 374
References ... 374

11. Remediation: Techniques and Cost ... 377
11.1 Introduction ... 377
11.2 Remediation Overview ... 377
11.3 Common Soil Remediation Technologies ... 379
11.3.1 Soil Excavation ... 379
11.3.2 Active and Passive Bioremediation ... 380
11.3.3 Monitored Natural Attenuation ... 381
11.3.4 Mechanical Soil Aeration ... 381
11.3.5 Capping ... 382
11.3.6 Land Use Restrictions ... 382
11.3.7 Soil Vapor Extraction ... 383
11.3.8 Phytoremediation ... 383
11.3.9 Soil Washing ... 384
11.3.10 In Situ Thermal Treatment ... 384
11.3.11 Ex Situ Thermal Treatment ... 385
11.3.12 Electrokinetics ... 386
11.3.13 Solidification/Stabilization ... 386
11.3.14 Fracturing ... 386

11.3.15 Vitrification ... 386
11.3.16 Chemical Oxidation ... 387
11.3.17 Chemical Dehalogenation ... 387
11.4 Summary of Soil Remediation Technologies ... 388
11.5 Common Groundwater Remediation Technologies ... 390
11.5.1 In Situ Soil Flushing ... 392
11.5.2 Air Sparging ... 392
11.5.3 Permeable Reactive Barriers ... 393
11.5.4 Injection of Biological Agents ... 393
11.5.5 Injection of Chemical Agents ... 393
11.5.6 Monitored Natural Attenuation ... 394
11.5.7 Institutional Controls ... 394
11.5.8 Multiphase Extraction ... 394
11.6 Summary of Groundwater Remediation Technologies ... 394
11.7 Sediment Remediation ... 397
11.7.1 Source Control ... 398
11.7.2 Excavation ... 398
11.7.3 Dredging ... 398
11.7.4 Bioremediation ... 399
11.7.5 Capping ... 399
11.7.6 Natural Attenuation ... 399
11.8 Surface Water Remediation ... 399
11.8.1 Spills ... 401
11.9 Air Remediation ... 403
11.10 Cost of Remediation ... 404
11.10.1 Results ... 407
11.11 Summary and Conclusion ... 410
References ... 411

Part III Sustainable Development

12. Urbanization and the Disruption of Matter and Energy Flows within Watersheds ... 417
12.1 Introduction ... 417
12.2 Watershed Structure ... 418
12.2.1 The Atmosphere ... 418
12.2.2 The Surface and Subsurface ... 420
12.2.3 Zonation of Surface Watersheds ... 421
12.3 Watershed Function ... 422
12.4 Urbanization and Watersheds: Disruption of Matter and Energy Flows ... 424
12.4.1 Water Infrastructure and Ecosystems ... 425
12.4.2 Stormwater ... 428
12.4.2.1 Case Study ... 430
12.4.3 Erosion and Sedimentation ... 432
12.4.4 The Urban Heat Island ... 432
12.5 Summary and Conclusion ... 433
References ... 434

13. Pollution Prevention 437
13.1 Introduction 437
13.2 Pollution Prevention in the United States 438
13.3 Implementing Pollution Prevention Techniques: Industrial Point Sources 439
13.3.1 Step 1: Establishing Objectives and Gathering Background Data 440
13.3.2 Step 2: Inventory of Hazardous Substances 441
13.3.3 Step 3: Assessing CRFs of Substances Present 442
13.3.4 Step 4: Preliminary Assessment of Geologic Vulnerability 442
13.3.5 Step 5: Preventing Pollution through ESPM Methods 443
13.3.5.1 Elimination 443
13.3.5.2 Substitution 443
13.3.5.3 Prevention 444
13.3.5.4 Minimization 448
13.3.6 Step 6: Assessing Results 449
13.3.7 Financial Incentives for Industrial Pollution Prevention 450
13.4 Implementing Pollution Prevention Techniques for Nonpoint Sources in Urbanized Watersheds 450
13.4.1 Applying the Source Control Framework to Nonpoint Sources 451
13.4.1.1 Step 1: Establishing Objectives and Gathering Background Data 451
13.4.1.2 Step 2: Inventory of Hazardous Substances 453
13.4.1.3 Step 3: Assessing CRFs of Substances Present 454
13.4.1.4 Step 4: Preliminary Assessment of Geologic Vulnerability 454
13.4.1.5 Step 5: Preventing Pollution through ESPM Methods 455
13.4.1.6 Step 6: Assessing Results 459
13.5 Summary and Conclusion 459
References 460

14. Case Studies: Successful and Unsuccessful Redevelopment of Contaminated Sites 463
14.1 Introduction 463
14.2 Case Study 1: Michigan 463
14.3 Case Study 2: Illinois 470
14.4 Case Study 3: Indiana 473
14.5 Case Study 4: Wisconsin 477
14.6 Case Study 5: California 481
14.7 Case Study 6: New Jersey 485
14.8 Summary and Conclusion 488
References 490

15. Science-Based Landscape Planning in Urban Watersheds 491
15.1 Introduction 491
15.2 The Science–Policy Gap 492
15.2.1 Land Development at the Community Level 493
15.2.2 Narrowing the Science–Policy Gap 494
15.3 Science-Based Landscape Planning 496
15.3.1 How Landscapes Form and Change 497
15.3.2 The Special Role of Water in Landscape Change 497
15.3.3 The Dimensions of Landscape 498

15.3.4 Site and Situation 498
15.3.5 On-Site Interactions and Thresholds 499
15.3.5.1 Interaction 1: Soil Type with Soil Water 500
15.3.5.2 Interaction 2: Soil Type with Vegetation 500
15.4 Watershed Management 501
15.5 Summary and Conclusion 503
References 503

16. Attaining Sustainable Urban Watersheds 505
16.1 Introduction 505
16.2 Sustainability and Sustainable Development 506
16.3 A Framework for Beginning Urban Watershed Management 506
16.3.1 Prevent Further Land Contamination and Degradation 507
16.3.2 Preserve Undamaged and Minimally Damaged Land 510
16.3.3 Remediate and Rehabilitate Degraded Brownfields 512
16.4 Linking the First Step to the Big Picture 514
16.4.1 Boundaries 515
16.4.2 Structural and Nonstructural Controls 515
16.4.3 Land and Water Management 516
16.4.4 Proactive and Reactive Response 517
16.4.5 Lack of Education 517
16.5 Summary and Conclusion 518
References 518

Glossary 521

Index 537

Preface

We now hear almost daily about how billions, and sometimes trillions of dollars are spent to bail out banks, failing companies, and to provide economic stimulus. Lost among these highly publicized events are the chronic problems we face in the United States and throughout the world with environmental contamination. Here, too, tremendous sums of money are being spent. In fact, over 2% of the U.S. gross domestic product is allocated annually for environmental remediation.

Much of our environmental contamination affects our land, water, and air, and originates from manufacturing and other activities. A large percentage of the affected land lies within urban areas, where it poses significant threats to the public health through water and soil contamination. Many of the contaminants such as hexavalent chromium and chlorinated hydrocarbons are mobile and persistent; others, like lead and arsenic, lie still, waiting in soil to be disturbed by children playing in a park.

We have been studying the contaminants affecting our lands and waters for more than two decades within the heavily urbanized Rouge River watershed of southeast Michigan. We have published over 30 peer-reviewed scientific papers on a variety of related urban contamination topics, including the role of geology in the assessment and management of urban development and redevelopment strategies; analytical (but accessible) methods for assessing the risks associated with groundwater, soil, and air contamination; the behavior and characterization of heavy metals in urban soils and groundwater; the ecological and public health implications associated with the contamination in the Great Lakes Basin; and the observed cost patterns for remediating different chemicals in specific geologic environments. We have come to the conclusion that there is an urgent need to collate the findings from this research and present it in a framework usable by those attempting to improve the urban environment.

This book logically and carefully tells the story from an environmental perspective of the fundamental challenges faced by humans in sustaining our urban watersheds. It highlights important pieces of critical scientific information that are missing from the literature, and explores the plethora of existing contaminants, most of which are caused by anthropogenic activities and do not naturally occur in the environment. The book specifies evaluation criteria and priorities for combining geologic theory, field methods, and contaminant characteristics, and uses these as a guide to conduct assessments of the environmental risks at the geographic scales of a single parcel of property or an entire watershed. Information obtained from these efforts is integrated into a framework for sustaining our urban watersheds into the future.

An interdisciplinary watershed-based approach is used throughout the book. Research and professional practice is incorporated from the fields of environmental geology, geochemistry, risk analysis, hydrology, and urban planning so practicing professionals and students can (1) become familiar with the different aspects of urban land and water contamination—their sources, extents, and risks; (2) use scientific knowledge to improve watershed management practices and urban development and redevelopment practices; and (3) improve the effectiveness of environmental restoration by learning about successful and unsuccessful approaches to contaminant remediation.

There are two broad audiences who can benefit from this book: The first consists of the students in college courses related to watershed management, environmental science, and environmental geology. Most of these courses exist at the undergraduate level, and to accommodate this reality, only basic math and statistical operations are used. The second audience includes the members of organizations involved with practical watershed management; for example, local and regional government officials; streamkeeper activists; urban and regional planners, environmental compliance managers working in industry, environmental consultants, state and federal regulators, and professionals involved with brownfield remediation.

PowerPoint® presentations of selected portions of the book are available with qualifying course adoption. Key terms appear in bold and are defined in the glossary.

Acknowledgments

This book is the result of more than 20 years of research in geology and contamination of urban areas, and information contained in this book is the result of collaboration with many dedicated professionals. We wish to thank the following contributors for their assistance, cooperation, ideas, and spirited debate: Tony Anthony, Michelle Arsenault, Greg Barrows, Larry and Cathy Bean, Richard Berak, Rhonda Berger, Gary Blinkiewicz, Ed Brosius, Tom Buggey, Kristine Casper, Russ Chadwick, Bob Cigale, Tom Cok, Caroline (Piper) Copeland, Jim Depa, Sheryl Doxtader, Walter Eifert, Rob Ellis, Rose Ellison, Rob Ferree, Jennifer Formoso, Orin Gelderloos, Beth Gotthelf, Mary Hall, Jon Hanson, Jack Heintz, James Heller, Steve Hoin, Mao Huang, Cheryl Kaspzyk, George Karalus, Kevin Kincare, Keith King, Steve Kitler, Steve Kulpanowski, Betty Locey, Daniel J. Lombardi, Bill Looney, Doug McVey, Mary Metcalf, Mark Mikesell, Kim Myers, Ernie Nimister, Tom O'Hara, Ray Ostrowski, Paul Owens, Bruce Patterson, Fred Payne, Mark Penzkover, Jeff Rienke, Eric Ross, Jeanne Schlaufman, Dave Slayton, Dave Smith, Stephen Smith, Rebecca Spearot, Ed Stewart, Keith Stigall, Michele Strickland, Patricia Thornton, Mary Vanderlaan, Kerby Vulgamott, Chris Venezia, Brad Venman, Nick Welty, Cheryl Wilson, Tom Wenzel, Kellie Wing, Derek Wong, and Brian Young.

We wish to thank several environmental consulting and remediation firms, including AECOM; AKTPeerless; Arcadis; Atwell-Hicks; Bureau Veritas; Endpoint Solutions; Environmental Consulting and Technology, Inc.; Roux Associates, Inc.; and St. John–Mittelhauser & Associates; as well as several state agencies, especially the Michigan Department of Natural Resources and Environment, for assisting in providing input or access to environmental data for research purposes.

This book would not have been possible without the dedicated assistance and technical expertise in the development of many of the figures from Jim Depa, Omar Sanchez, Kurt Kamholz, Matt Lyter, and Daniel J. Lombardi. We would like to especially single out the efforts, technical expertise, and artistic expression of Brian Peterson for many of the figures, including a majority of the color plates. A special note of additional thanks also goes to Daniel J. Lombardi and Matt Lyter for technical review and input in the development of the figures.

Finally, we would like to thank those who assisted in reviewing portions of the manuscript, and the professionals who believed in this book's concept and helped bring it to fruition: Irma Shagla-Britton, Kathryn Younce, Arlene Kopeloff, and Richard Tressider at CRC Press, and Remya Divakaran at SPI Technologies India.

Authors

Martin M. Kaufman is a professor of earth science at the University of Michigan–Flint, Flint, Michigan. He has over 25 years of experience in geographic information systems, hydrologic investigations, and watershed management. He is the author of *Critical Thinker's Guide to the Environment* (1996) and over 20 peer-reviewed publications in the areas of watershed hydrology, applied GIS, environmental risk assessment, science-based planning methods, and groundwater vulnerability. Throughout his career, he has received numerous research grants covering a diverse set of topics, including brownfield redevelopment (U.S. EPA), stormwater management (Genesee County Drain Commission), and the risks to children from hazardous materials at home (Ruth Mott Foundation). In 2000, he was the recipient of an American Water Works Association Best Paper Award and in 2009 was named the David M. French Distinguished Professor at the University of Michigan–Flint, awarded for continuous scholarly achievement, teaching excellence, and service. He has a BA in geography, an MUP (urban planning), and a PhD in environmental planning all from the University of Michigan in Ann Arbor.

Daniel T. Rogers is the director of environmental affairs at Amsted Industries Incorporated in Chicago, Illinois. Amsted Industries is a diversified manufacturing company of industrial components serving primarily railroad, vehicular, and construction and building markets. It has more than 50 manufacturing locations in 11 different countries. Throughout his career, Rogers has managed and conducted hundreds of geologic and hydrogeologic investigations and remediated and successfully closed industrial sites in the United States and internationally. He has published over 75 research papers in professional and academic publications and peer-reviewed journals on subjects such as environmental geology, hydrogeology, geologic vulnerability and mapping, contaminant fate and transport, urban geology, conducting environmental investigations, contaminant risk, brownfield redevelopment, and remediation. He is also the author of *Environmental Geology of Metropolitan Detroit* (1996) and has published surficial geologic maps of the Rouge River watershed in southeastern Michigan. He has taught geology and environmental chemistry at Eastern Michigan University and the University of Michigan and has presented guest lectures at several colleges and universities both in the United States and internationally. He has a BS and an MS in geology from Utah State University.

Kent S. Murray is a professor of geology at the University of Michigan–Dearborn, Dearborn, Michigan. He has over 35 years experience in environmental geology, groundwater hydrology, and low-temperature geochemistry. He has authored more than 40 publications in peer-reviewed journals, generated over 100 professional papers, and has made over 90 presentations at national and international scientific meetings. He has received numerous awards over the years, including the University of Michigan–Dearborn Distinguished Researcher award and the NCID (National Center for Institutional Diversity) Fellowship for his work in southwest Detroit. Professor Murray has received over $2 million in funding to support his research from foundations, state, and federal agencies, including the National Science Foundation, the Environmental Protection Agency, and the Department of Housing and Urban Development. Professor Murray has also worked with the U.S.G.S., Center of Astrogeology, and the California Energy Commission, and has taught at California State University and Oregon State University. He serves as an expert witness for several environmental law and environmental consulting firms and is also an adjunct faculty member in the Department of Water Resources at the China Geosciences University in Beijing, People's Republic of China.

1

The Watershed Approach

1.1 Introduction

John Wesley Powell, scientist, explorer, and an early director of the U.S. Geological Survey, put it best when he said that a watershed is

> ... that area of land, a bounded hydrologic system, within which all living things are inextricably linked by their common water course and where, as humans settled, simple logic demanded that they become part of a community.

Powell spoke these words in the late 1800s. Since then, population has increased dramatically and more people have become part of the watershed community. This increase in population within a finite hydrologic system has created new and increasingly complex human–environmental interactions.

Watersheds come in all shapes and sizes. They cross county, state, and national boundaries. In the continental United States, there are 2110 watersheds; including Hawaii, Alaska, and Puerto Rico, there are 2267 watersheds. Many of these watersheds contain homogenous land uses and land cover, such as agriculture, forest, or mountains. Others are more densely populated and consist of political units each pursuing their own self-interest, making it even more difficult to achieve consensus on how to solve a particular problem. A watershed also contains a great diversity of people. Among this diversity are people who live a rural lifestyle, suburbanites, and those living in an urban setting.

Urban watersheds are unique in the sense they typically contain a full palette of land uses and land cover types. Headwater regions of urban streams may contain endangered species of fish and local farms, while older manufacturing sites often sit idle within the original urban core. Many of the fundamental processes that contribute to the occurrence and distribution of contamination in the urban environment such as the movement of water and the location of industry are closely linked to drainage patterns. For this reason, an interdisciplinary scientific *watershed approach* is required to account for their full natural variability and the resulting implications for society. The watershed approach is thus a framework that uses science from various disciplines to characterize the physical and social processes occurring within a watershed in order to help solve some of its environmental problems.

1.2 Historical Transformation of Urbanized Watersheds by Industrial Development

In the process of urban development, both technology and environment have profoundly influenced the relationships between site and geographic situation. Early settlements in North America tended to be dependent on water for power, transportation, and supply, which led to the development of major industrial centers located near large trunk streams (Rugg 1972). As cities grew, they became increasingly chaotic: factory smoke smothered neighborhoods where workers lived, cars demanded an increasing share of the land, and the geographic order of land uses in and around cities became disrupted and fragmented. The massive rise in automobile use after World War II led to the frenetic and unplanned growth of suburbs.

After World War II, the United States began the largest public works project in history—the Interstate Highway System. Over 72,000 km (45,000 mi) of new highway were built between 1956 and 1992. Virtually every major city was connected, and many central business districts were bisected by these new and wide expressways. Aided by these new routes, urban populations moved to the suburbs, where urban **sprawl** not only ate up more land but also changed the character of land use. A separation of land uses arose where subdivisions housed the large hordes of urban commuters and other locations supported the new variety of retail and commercial zones that included strip malls, outlet stores, large shopping malls, "big box" retail stores, and corporate parks. Meanwhile, the older cities saw their tax bases decline, and increasing abandonments of industrial properties as the general manufacturing capacity of the United States declined due to foreign competition and a shift to service sector jobs.

The historical development of the heavily urbanized Rouge River watershed in southeastern Michigan (Figure 1.1) provides many compelling reasons for selecting it as a base case for this book:

- There is a lot of water in the region (the Great Lakes).
- A major urban industrial center lies within it (Detroit).

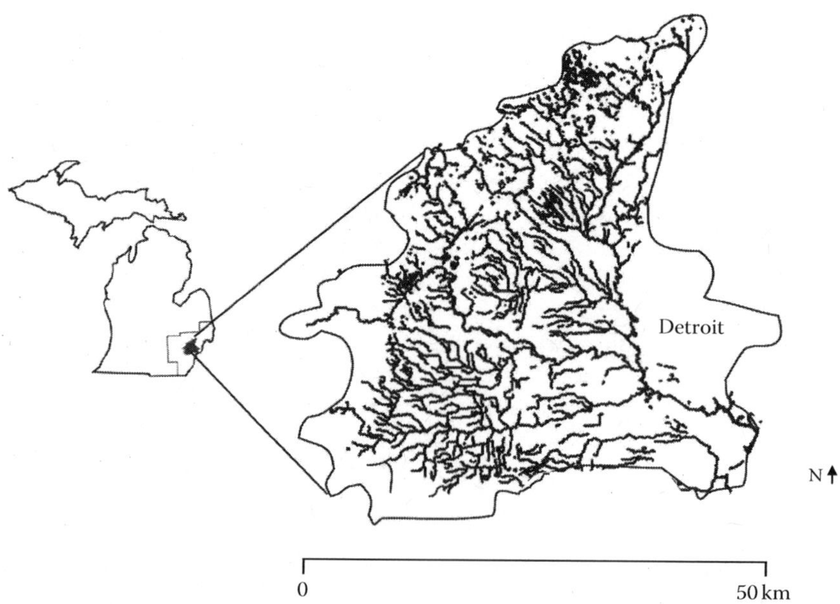

FIGURE 1.1
Rouge River watershed in southeast Michigan. (From USEPA, U.S. Environmental Protection Agency, 2004 Rouge River Remedial Action Plan Revision, http://www.epa.gov/greatlakes/aoc/rougriv/2004_Rouge-River-RAP-Revision.pdf (accessed August 1, 2009), 2004b.)

- A long industrial history exists (e.g., metal processing and automotive).
- **Brownfields** are plentiful.
- Varied geology.
- An abundance of contamination to study.
- Heavily populated (currently over 1.5 million in the watershed; but historically the population was much greater).
- Varied land use (industrial, residential, commercial, agricultural, open space, parkland).
- Urban sprawl is widespread.
- High stream density.
- A relatively small watershed that is easy to study (approximately 1200 km^2).
- High volumes of data about the watershed exist from State of Michigan Remedial Action Plans, USEPA, and contaminated site investigations by the authors.

The watershed's 175 years of industrial development include a 100 year legacy of automobile manufacturing. What began as an area of postglacial swampland, beach deposits, and forests has evolved into a landscape now littered with brownfields and broadly contaminated water resources from the **groundwater** to the surface waters, and into the water vapor in the lower atmosphere. If we follow this principle "it is important to understand how things were, so we can understand how they are today," then it becomes necessary to provide some historical context for this book by highlighting the key aspects of the industrial development within the southeast Michigan region and its effects on the Rouge watershed.

1.3 The Evolution of Industry and the Rouge Watershed

The growth of industry in Detroit and its surrounding region can be attributed to the basic geographical factors of site and situation, and its geology. Detroit's site (the actual area within the original settlement of 1701) consisted largely of marshes, was heavily forested, and underlain by thick wet clay—a remnant of past glaciations. In 1815, Edward Tiffin, the surveyor general for the northwest reported to the National Government his assessment of the Detroit area's agricultural potential: "the streams were narrow and deep…the intermediate spaces between the inland lakes are a poor and barren sandy land," and concluded: "the balance is bad, there could not be more than one acre out of a hundred, if there would be one out of a thousand, that would in any case admit of cultivation"* (American State Papers 1834).

With respect to its growth potential, this agricultural deficiency at the **site** was more than compensated for by Detroit's **situation**—the quantity and quality of its linkages and interactions (e.g., physical, social, and economic) with the surrounding region. Detroit's early economic development was spurred by a combination of situational factors: the opening of the Erie Canal in 1826, the city's Great Lakes location, the increasing use of rail

* The Congressional impetus for the Tiffin survey was to locate 2,000,000 acres of bounty land (land given as a reward for military service) for veterans of the War of 1812.

transport, the growing lumber and flour-milling industries, abundant copper and iron-ore deposits in the Lake Superior region, the availability of a skilled labor force, and the role Detroit played during the Civil War by providing supplies and provisions to the Union cause (Parkins 1918).

Another factor aiding Detroit's growth was the accelerating westward migration of population during the 1800s, as many of these people ended their journey in Michigan (Brown 1948). More population meant altering the land and creating a higher diversity of land uses. While most of the new land uses were commercial in nature, these were soon followed by industrial varieties. For example, a new commercial land use emerged in the early 1800s at the confluence of the Rouge and Detroit Rivers as large docking structures for holding ferries and steamboats were constructed. By the late 1800s, docks lined 8 km (5 mi) of riverfront (Kerr et al. 2003). By 1837, a sawmill and gristmill were built on the Rouge River and led to the development of Southfield, a suburb where a little more than a century later the nation's first regional shopping center arose (Northland Shopping Center). During the 1850s, the Rouge River also housed a site for the growing railroad manufacturing industry (Parkins 1918).

While migration, commerce, and small-scale industry were important catalysts for Detroit's growth, it was the mining of copper and iron ore in the Upper Peninsula that led to the transformation of Detroit into a major industrial center. Mining of both metals began in earnest during the 1840s, but the lack of labor in the Lake Superior region and the ability to ship its product on the nearby Great Lakes led to the development of the smelting and iron industries in Detroit. Paleo-Indian cultures (5000–3000 YBP) were the first to obtain copper in the region for use as tools, and the French explorers of the Lake Superior region were aware of Indians trading copper by the mid-1600s (Dorr and Eschman 1970). After some technological advances in smelting occurred in Europe during the 1700s, they eventually found their way into the United States and commercial mining in the sparsely settled Upper Peninsula began in the 1840s. The mined ore was shipped on the Great Lakes to Detroit for smelting where labor was plentiful. When it was soon learned that workers could be lured to the Upper Peninsula and the costs reduced by smelting the ore closer to the mines, Detroit's role in smelting copper declined (Parkins 1918).

Detroit's industrial growth was then jump-started by iron ore. There were large quantities of iron ore deposited as sediments in the Lake Superior region of **Precambrian** origin. Out of these sediments, highly concentrated natural ore bodies formed where locally concentrated groundwater flows leached or dissolved away the silica from a portion of the large iron formation, leaving behind enriched iron minerals as residues (Dorr and Eschman 1970). This physical process enabled a direct connection between the regional geology and Detroit's industrial growth, since the iron residue was *direct shippable ore*, meaning it contained more than 60% iron and could be used in a blast furnace with minimal preparation and cost.

The iron ore and Great Lakes connection was good for a while, but Detroit's manufacturing future was still not out of the woods. Since the early blast furnaces used charcoal to make iron, the local forests were soon depleted. As a result, iron production became dependant upon coal, and the blast furnaces moved closer to the major sources of coal in Pennsylvania. After 1880, the iron industry declined in southeast Michigan (Parkins 1918).

By 1900, Detroit was the thirteenth largest city in the United States with 285,000 residents, most living within a few kilometer radius of its downtown adjacent to the Detroit River. As was the case with most nineteenth-century industrial cities, manufacturing was concentrated along the river for easy access to power and transportation. There was no dominant industry at the time. The leading Detroit industries included stove manufacture, tobacco goods, drugs and chemicals, metal working, and food production. In 1896, Charles B. King determined Detroit's destiny when he drove a horseless carriage on the city streets. Soon Henry Ford introduced his own version of this conveyance, and Detroit was on its way to becoming the automobile capital of the world. Along with Ford, such automotive pioneers as W.C. Durant, Walter P. Chrysler, Ransom Olds, Henry Leland, and the Dodge brothers laid the foundation for the companies that emerged as the Big Three automakers—Ford, General Motors, and Chrysler—by the latter half of the twentieth century (Sugrue 2005).

The characteristics of site and situation present in Detroit during the 1800s had changed, largely due to the regionally evolving locations of population, coal, and steel-making. By 1900, Detroit was in the center of America's industrial heartland—a region that extended from lower New England down to Pennsylvania and across the Appalachians westward through Ohio, Indiana, and Illinois. All of the raw materials needed for automobile production were easily accessible to the city by the Great Lakes waterways and by rail. The raw materials within easy reach included the coal in Pennsylvania and West Virginia; the steel in the mills of Pittsburgh, Youngstown, Cleveland, Gary, and Chicago; the limestone and gypsum near Cleveland; and the iron and copper ore embedded in the regions of northern Michigan and Minnesota. In addition, as shown by Figure 1.2, Detroit's central location at the confluence of East and Midwest gave its auto producers easy access to the capital and markets necessary for its phenomenal growth (Sugrue 2005).

Detroit's first auto plants were small operations, but they contributed significant modifications to the Rouge watershed. Between 1910 and 1920, Henry Ford dammed the Rouge River at six locations to supply power to some of the small factories producing parts for

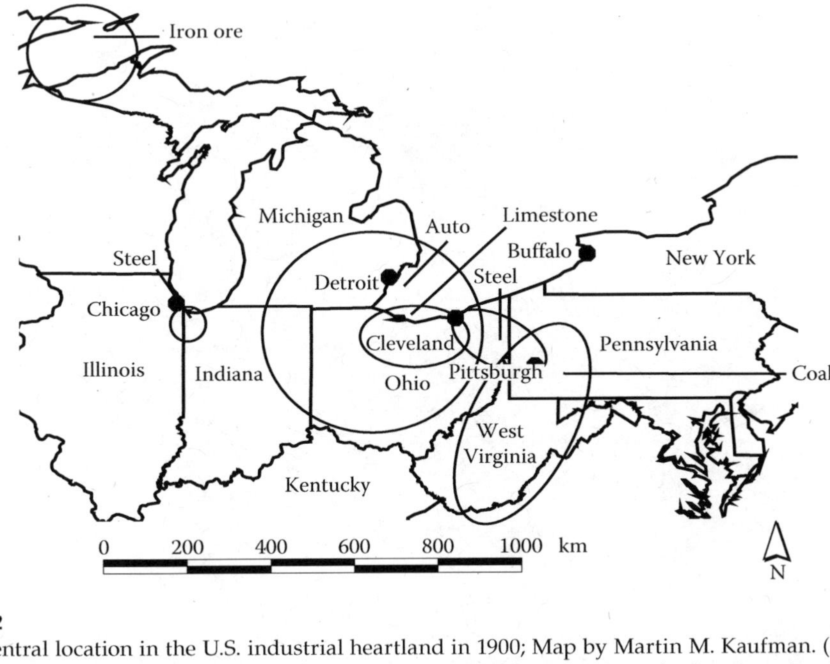

FIGURE 1.2
Detroit's central location in the U.S. industrial heartland in 1900; Map by Martin M. Kaufman. (From Sugrue, T.J, From motor city to motor metropolis: How the automobile industry reshaped urban America, Automobile in American Life and Society website Dearborn: Henry Ford Museum and University of Michigan, http://www.autolife.umd.umich.edu (accessed May 17, 2010), 2005.)

his auto plants, and built another dam to supply power to his mansion. After Ford bought 810 ha (2000 ac) along the lower Rouge west of Detroit in 1915, the River was dredged near its mouth in 1918 in the area that was to become the Ford Rouge Complex (USEPA 2004b). Finished in 1927, the River Rouge plant consisted of nineteen separate buildings in an industrial complex sprawling over more than 5 km^2 (2 mi^2). The complex included the aforementioned deep sea harbor; the world's largest steel foundry; 150 km (94 mi) of railroad track; and stamping, glass making, and auto assembly buildings.

At its peak, over 90,000 people worked at the Rouge, and the other members of the Big Three automakers followed the format of building large facilities with large workforces. By the 1920s, Dodge had built the 27 ha (67 ac) Dodge Main Plant in Hamtramck—a city enclosed by Detroit and just a short distance from Ford's Highland Park facility where the production line originated. Also during this period, General Motors constructed a large plant on 19 ha (47 ac) in southwest Detroit to produce the Cadillac. After a downturn in the industry during the Great Depression, the auto industry rebounded in the 1940s by turning its energies to the war effort. Ford opened a bomber factory and Chrysler a tank plant, leading to a new nickname for Detroit—"the arsenal of democracy"—a phrase made famous in one of President Franklin Roosevelt's fireside chats (Davis 2007).

The centralization of auto production in Detroit (and Michigan) began to decline in the 1950s, as automakers sought ways to cut costs by moving factories to the suburbs and non-union states, and avoiding situations where a strike at one plant could cripple the whole industry (Sugrue 2005). These actions corresponded with the expansion of the national highway network and accelerated the growth of suburbs and urban sprawl. Figure 1.3 shows the history of developed land in southeast Michigan from 1905 to 1992. The shaded areas represent area covered by "urban or built up land," according to Anderson et al. (1976).

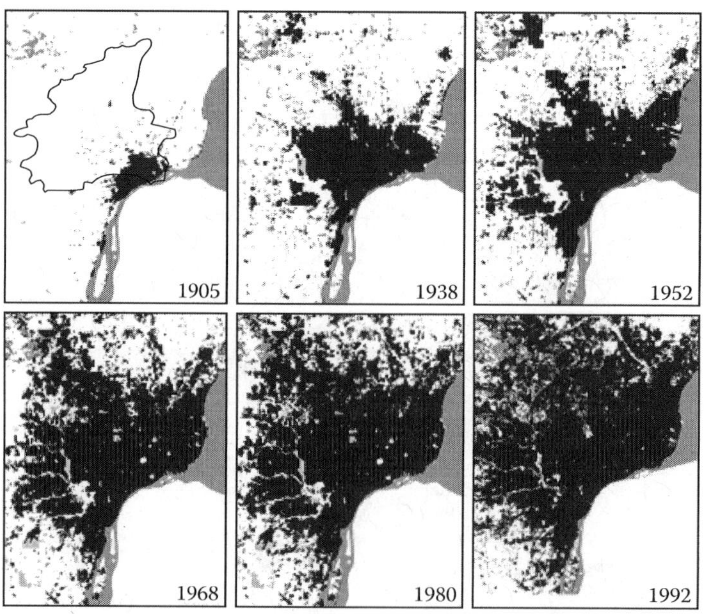

FIGURE 1.3
Developed land in southeast Michigan (1905–1992); Rouge watershed boundary is shown in the 1905 image at the top left. (From Richards, L., *Detroit River Corridor, Preliminary Assessment of Land Use Change,* Urban Dynamics Research Program, USGS Ames Research Center, Moffett Field, CA, 2003.)

The urban sprawl shown in Figure 1.3 completely engulfs the Rouge watershed by the 1970s. This land use boom, in concert with unregulated industrial discharges, had detrimental consequences for the water quality of the river. By 1968, the Rouge was flowing orange from pickling liquor waste and was one of the rivers in the Great Lakes Basin to catch on fire. In 1985, the Rouge River was identified as one of 42 Areas of Concern in the Great Lakes Basin. Today, the elimination of some of the **Combined Sewer Overflows** and the reduction of industrial discharges have improved the water quality within some reaches of the river. Yet, a major challenge remains in this watershed related to the amount of developed land cover and its relationship to surface runoff. Most watershed surface areas in the United States consist of over 90% land (as does the Rouge's), so taking good care of the land and reducing its inputs of pollution are therefore required to ensure adequate water quality. Recent trends, however, indicate that each house being built in southeast Michigan is consuming more land. For example, between 1990 and 2000, the amount of land used for homes increased by 19%, while the number of households only grew by 9%. Prior to 1990, there were 1.14 housing units per ha (2.84 per ac), but this has decreased to an average of 0.5 per ha (1.26 per ac) after 1990 (SEMCOG 2003). This increase in the amount of land used for each house is significant because it accounts for 43% more land developed than would have been with the higher-density construction before 1990. In most instances, more runoff is produced when more land is developed, and much of this runoff contains contaminants that threaten surface water and groundwater.

1.4 Major Themes of the Watershed Approach Used in This Book

A watershed approach is consistent with a geoscience perspective for the solving of environmental problems. Watersheds consist of surface water and their drainage systems, the groundwater that supplies the surface water in gaining streams, the surface and subsurface geology including landforms and soil, the flora and fauna, the air, the social and land use history of the region, and the current political decision making processes within them. Within this context, we identify five major themes, each having the common denominators of geology, contamination, and sustainable urban development.

1.4.1 Theme #1: Near-Surface Geology, Hydrogeology, and Surface Hydrology

An accurate physical characterization of a watershed, including its geology, hydrogeology, and surface hydrology, is a key prerequisite for a successful watershed restoration effort. This assertion is corroborated by the successful watershed management and restoration efforts that have been accomplished in North America. Consider the Don River, a stream flowing through the urbanized area of Toronto, Canada. This watershed is now in the process of a major restoration effort, which has been largely successful. Here, the watershed restoration effort began with a scientific foundation that accounted for the glacial origins of the topography. This science was translated into the need to preserve certain natural features so as to maintain natural drainage patterns and minimize investment in new infrastructure.

The following annotated outline summarizes some of the essential elements for characterizing the geology, hydrogeology, and surface hydrology of a watershed in order to perform contamination analysis and sustainable remediation:

1.4.1.1 Geology

- Historical geology of the watershed
- Local and regional structural formations (e.g., faults, basins, etc.)
- Types of rock groups present in the study area
- **Stratigraphy** of the rock types
- A geologic cross-section
- Near-surface geology, including descriptions of the major soil units; glacial and/or depositional history

Example:

The study area is located on the southeastern edge of the Michigan Basin, a large-scale structural feature comprised primarily of limestones, shales, and sandstones. These sedimentary rocks are Paleozoic in age and directly overlie Precambrian crystalline rock that comprises the crystalline core of the continent. The Paleozoic rocks rarely exist as natural outcrops because of the presence of a thick deposit of glacial drift in the region. Beneath the study area, the Paleozoic rocks range from 425 to 730 m thick and gently dip toward the center of the basin to the northwest. The depth to the Paleozoic bedrock ranges from more than 110 m in the northwest of the study area to less than 15 m in the southeast (Rieck 1981a).

The surficial geology in Michigan is dominated by glacial sediments that are typically over 60 m thick and, at some locations, more than 300 m thick (Rieck 1981a). These sediments were deposited during the Pleistocene Epoch by the Wisconsinan stage of glaciation and consist of **outwash, moraine**, and beach, bar and lake deposits (Farrand 1988). Varied and complex lithologies are exhibited within these deposits, including coarse gravels, fine-grained sands, and clays (Bergquist and MacLachlan 1951; Mozola 1969; Rieck 1981a, b).

Five distinct near-surface geologic units have been identified within the study area (Leverett 1911; Sherzer 1916; Farrand 1982; Rogers 1996). The units are classified by their composition and include moraine, sandy clay, sand, sandy and silty clay, and upper clay. With the exception of the moraine unit, all the surficial geologic units within the study area are of **glacial lacustrine** origin (Farrand 1982; Rogers 1996) (Figure 1.4).

1.4.1.2 Hydrogeology

- Location and quantity of groundwater
- Factors influencing the **hydraulic conductivity** of the groundwater
- Groundwater flow direction

Example:

Shallow groundwater within the watershed is typically encountered within 3 m of the surface. Within the moraine and sand units, the shallow groundwater has been present at sufficient quantities to enable continuous pumping at volumes exceeding 100 L/min and occasionally at more than 250 L/min (Rogers 1996; Rogers and Murray 1997). Shallow groundwater within the predominantly clay units is not present in the volumes encountered within the moraine and sand units, but is still widely encountered at the geologic contact between the upper glacial-lacustrine units and the lower unconsolidated unit of glacial origin (i.e., lower clay unit) (Rogers 1996).

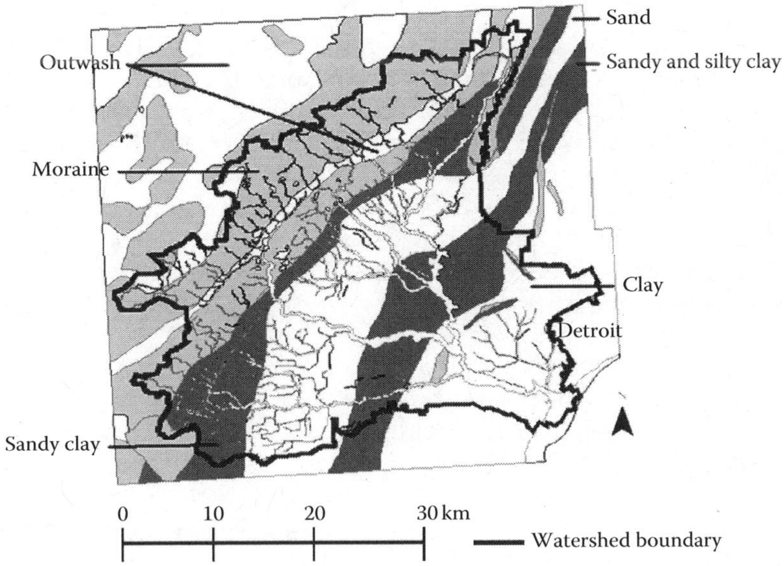

FIGURE 1.4
Simplified map of near surface geology, Rouge River watershed. (From Murray et al., *J. Environ. Qual*, 33, 163, 2004. With permission, American Society of Agronomy.)

A thick lower clay unit of glacial origin (labeled as the lower clay unit) is located directly beneath the surficial units. The lower clay unit ranges in thickness from 30 to 60 m and exhibits a consistent hydraulic conductivity horizontally and vertically on the order of 1×10^{-8} centimeters per second (cm/s), which essentially makes the unit an effective **confining layer** (Rogers 1996). The thickness, horizontal distribution, and low and uniform **hydraulic conductivity** of the lower clay unit have influenced near-surface groundwater within the entire study area. The presence of the lower clay unit has essentially prevented groundwater from migrating to deeper aquifers. Instead, groundwater in near-surface aquifers discharges to surface water (Rogers and Murray 1997). Inspection of the groundwater flow direction at each contaminated site revealed a good correspondence with the direction of surface water flow.

1.4.1.3 Surface Hydrology

- Regional climate and stream type relationships (e.g., **influent/effluent stream types**, drainage density)
- Watershed area
- Surface water characterization: streams, wetlands, lakes
- Floodplain structure and flood frequency
- Water quality issues

Example:

Situated within the **humid microthermal climate** zone of the Midwestern United States, the Rouge River is effluent, meaning it is fed by groundwater entering as baseflow. Thus, any significant contamination of groundwater that does not readily degrade or attenuate will ultimately be delivered to the surface water drainage network (Murray et al. 2000).

The Rouge River watershed consists of over 200 km of streams, tributaries, wetlands, lakes, and ponds within a 1200 km^2 area. Stream density is highest in the sand and moraine units. The glacial topography consisting of outwash plains, eskers, and kames has created irregular drainage patterns, and the wetlands are often associated with these features. Most of the inland lakes are kettles or impoundments formed by damming the river at different locations.

The Rouge floodplain is not uniform in height due to the erratic deposition during floods. Historically, flooding was common during heavy rains, and floodplains of river silt were built at a height of several meters above the ordinary stage of the water. Today, flooding of certain low-lying sections occurs during heavy summer thunderstorms and prolonged precipitation events yielding over 38 mm (1.5 in.) of rain.

The fan-shaped watershed includes all or part of 47 different municipalities and is currently the focus of intense scientific study and restoration. It has been identified as an Area of Concern by the International Joint Commission (Hartig and Zarull 1991) and cited as a significant source of pollution to the lower Great Lakes (Murray and Bona 1993).

1.4.2 Theme #2: Science and Planning—The Movement of Water

There is a fundamental relationship that exists between basic science and the ability to make informed environmental planning decisions. Regarding the movement of water on earth, the relevant basic science is the **hydrologic cycle**, the solar-initiated and gravity-sustained renewable flow of water between five major reservoirs: the oceans, atmosphere, ice caps, surface water, and groundwater. Two of these water reservoirs—groundwater and atmospheric water vapor—have the unique attribute of being invisible. Being out of sight often makes groundwater out of mind to policy makers and the general public, and this creates special management challenges. Problems often arise when basic hydrologic processes linked to groundwater such as surface water flow are not included, such as the incorrect identification of the extent of a contamination event.

Let us consider groundwater contamination in the Great Lakes region, where the flows of many contaminants will be largely dictated by the pathways of the hydrologic cycle within that given region, as shown below:

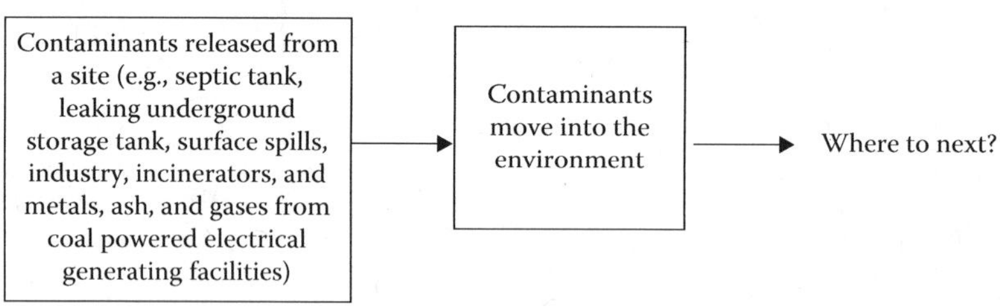

Different soil types and the near-surface geologic environment will play key roles in the migration of the contaminant. Releases are typically mixtures of contaminants (e.g., gasoline), which contain over 100 individual chemical compounds. Each specific compound behaves and migrates in the near surface geologic environment differently. Geologic environments consisting largely of clay materials tend to inhibit the migration of contaminants downward, whereas in a sandy geological environment, contaminants often reach the water table. As the next sequence of boxes demonstrates, the surface environment and groundwater flow direction will help determine the extent of the contamination.

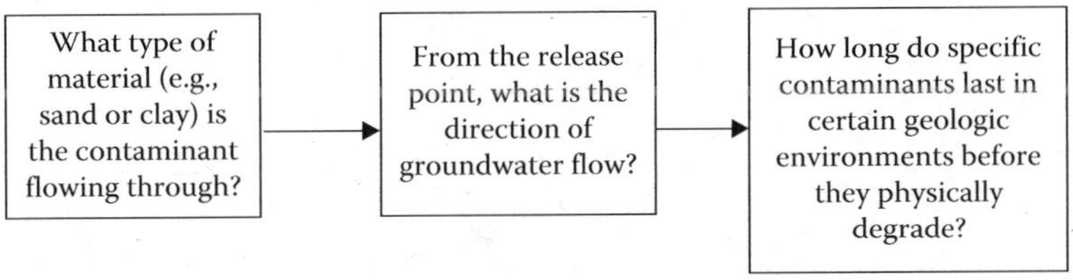

In humid climates, groundwater flows into surface water, and this process creates the need to know more about the local geography of the water resources, particularly the locations of surface streams, lakes, and wetlands. The locations of streams and their flow patterns are required to determine the potential for contaminant transport and the larger-scale environmental problems that may result. And, since the water within lakes and wetlands consists of groundwater above the water table, knowledge of their flow patterns is also necessary to obtain a complete picture of the contamination potential.

The three maps in Figure 1.5 show the outcome for generalized categories of contaminants spilled from selected surface locations within the Rouge River watershed in the Great Lakes region. There are five contaminant categories that have been selectively investigated. These categories include **dense nonaqueous phase liquids** (DNAPLs—chlorinated solvents), **light nonaqueous phase liquids** (LNAPLs—gasoline compounds), **polycyclic aromatic hydrocarbons** (PAHs—oil compounds), **polychlorinated biphenyls** (PCBs) and a group of several heavy metals, including chromium and lead. Only those locations where the near-surface geologic environments have a high migration potential are shown (sand and moraine). At the top left, the map shows buffers around some of these locations, with the buffer sizes representing the average distance these contaminants traveled based upon measurements taken at their points of release These buffered sites were overlaid (symbolized by the "+") on the surface stream network map to the right (including all first-order streams and higher) within the watershed. The composite map shown below the arrow indicates the areas where there are intersections between the average contaminant extent and a stream channel. To avoid map clutter, the circles represent a sample of some of the locations where highly toxic DNAPL compounds contaminated groundwater in the sand and moraine geological units.

Inspection of the contaminant sites and surface stream maps indicates there are numerous other sites where contamination within groundwater has the potential to migrate to surface water within a few days. This outcome is likely due to the high higher flow rates within the sand and moraine units in this watershed and the high drainage densities within these same units. At this geographic scale, the hydrologic cycle flows follow these pathways: soil > groundwater > low order surface streams > higher order surface streams > Great Lakes. Some persistent contaminants such as tetrachloroethene (a common dry cleaning chemical) and chromium VI (the compound used in chrome plating) will travel along with the water. The lesson demonstrated here is that by omitting the linkage between groundwater and flowing surface water, the consideration of ecological impacts to the larger region of the Great Lakes may not occur.

1.4.3 Theme #3: Industrial Property Abandonment, Contamination, and Risk

We spoke earlier of the decentralization of the automobile industry that began in the 1950s. This trend has accelerated in recent years, as a geographic shift of production from Michigan and other parts of the upper Midwest continues southward and to other countries. In addition, so-called "transplants"—foreign nameplate companies producing

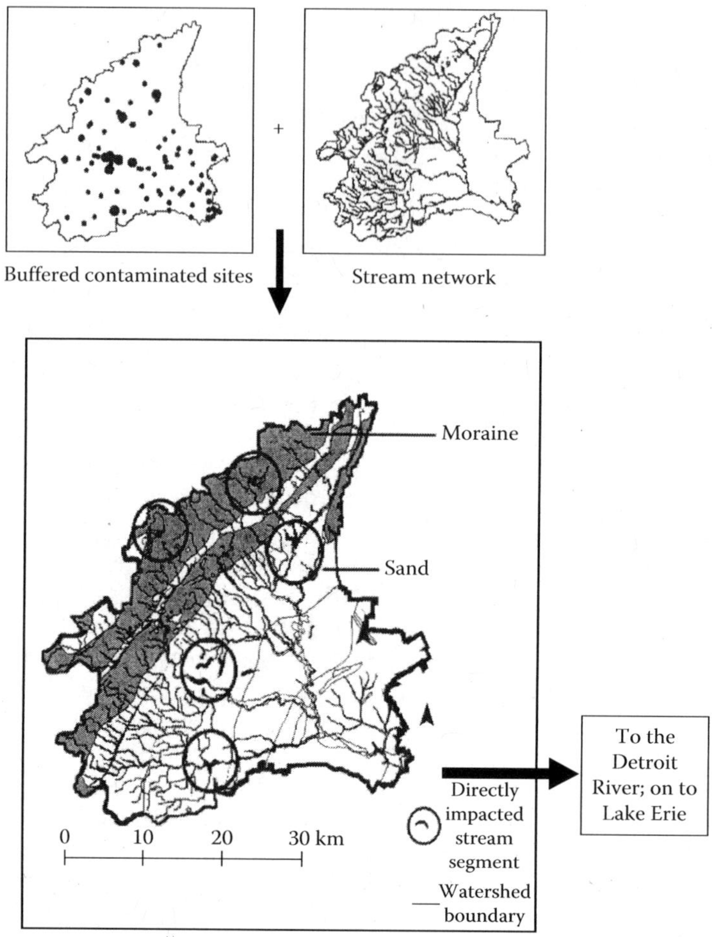

FIGURE 1.5
Contaminant pathways via groundwater discharging to local surface streams. (With kind permission from Springer Science+Business Media: *Environ. Geol.*, Using soil and contaminant properties to assess the potential for groundwater contamination to the lower Great Lakes, USA, 56, 2009, p. 1019, Kaufman, M.M. et al., Figure 5, © Springer-Verlag 2008.)

vehicles in the U.S.—have captured significant market share from the Big Three automakers. Transplant production also largely takes place in the South. The statistics of this shift are staggering: from 2000 through July, 2005 year-to-date, Michigan lost 42% of its auto assembly jobs versus a 14% loss in the United States located outside of the three Midwest auto-intensive states (Ohio and Indiana are the other two). Auto parts tell an even larger part of the story, since there are four times as many jobs in parts as assembly operations. Parts makers tend to be located near the assembly plants for historical reasons and, more recently, because "just-in-time" production requires proximity for many parts such as seats and subassemblies. Michigan's parts employment is down 34% since year 2000, versus 19% in the rest of the United States (Federal Reserve Bank of Chicago 2005). On the ground, the results are clear: auto plants have closed, suppliers have gone bankrupt, and many manufacturing and warehouse facilities lie vacant. This sequence of industrial–commercial decline and property dormancy is not just confined to Detroit—it has occurred in many urban areas around the country.

The sites where industrial or commercial facilities once operated but are now inactive were named "brownfields" (USEPA 2007). At these locations, expansion or redevelopment is complicated by real or perceived environmental contamination. There are perhaps over 1,000,000 brownfield sites in the United States alone, with an estimated remediation cost of several hundred billion dollars. Although the concern over brownfields has traditionally been the toxicity level of the contamination present, a second factor—the geologic environment—is perhaps more important and is generally not taken into account. This science angle exists because the contamination at a brownfield site results from the substances released on and just below the surface by human activity and their interactions with the near surface soils and shallow aquifer (Foster 1987; Daly and Warren 1994).

The type of substance released, specifically its toxicity, mobility, and persistence all work together to determine the potential risks involved with the contamination at a given site. An area's geology plays an important role in determining the risk posed by any contaminant released into the environment, since risks also vary depending on the final destination of certain contaminants. While certain metals such as lead are typically immobile because soils with clay tend to **adsorb** it so it does not reach groundwater, lead still poses a significant risk to children who might put a bit of soil in their mouths. Other contaminants such as **hexavalent chromium** if released into a larger-grained soil type like sand are highly mobile and will often migrate into groundwater. When this type of event occurs, there are significant potential risks to the public health and ecology at a larger geographical scale (Murray et al. 2006).

1.4.4 Theme #4: Remediation of Contaminated Sites

Environmental management at industrial facilities has traditionally focused on maintaining compliance with environmental regulations, especially those regulations affecting production. The emphasis on production-specific regulations has resulted in a higher proportion of inspections targeted at the by-products of the production process (Owen 1995; Daugherty 1996).

This pattern of industrial compliance and regulatory enforcement has not provided an effective mechanism for reducing environmental spending. Overall, environmental spending in the United States continues to increase and represents approximately 2% of the gross national product or about 150 billion dollars per year (Morgenstern et al. 1998; USEPA 2004a, 2004b). Of this total, only 25% is associated with compliance-related expenditures, while the remaining 75% is related to other expenses such as investigation and remediation of existing or historical contamination (Owen 1995; USEPA 2004a, 2004b). In 2004, USEPA estimated there were 350,000 sites of environmental contamination in the United States requiring remediation within the next 30 years. Furthermore, USEPA estimates that the cost to remediate those sites will exceed $250 billion. Excluded from this list of 350,000 sites are those sites where remediation is in progress or has been completed. In addition, there are an estimated 500,000–1 million abandoned industrial facilities or brownfield sites that are also excluded from the list because a cost estimate to investigate and remediate these sites has not been conducted. Most of these abandoned sites are located in urban regions in the United States, possess some degree of contamination, and will require investigation and remediation (USEPA 2004a, 2004b). Thus, it is likely that the environmental costs to investigate and remediate sites of environmental contamination in the future will exceed the $250 billion dollar estimate.

The business community has implemented pollution prevention initiatives through what has been termed environmental management systems (Fletcher and Paleogos 2000) and environmental risk management systems (Telego 1998). Environmental management and risk systems are a series of standards that are used to develop a business model for an integrated management system to identify, control, and monitor environmental risks (Voorhees and Woellner 1998). Both initiatives focus efforts on what has essentially been identified as life-cycle analysis. Life-cycle analysis involves the evaluation of the raw materials, by-products, wastes, and the final product using a decision matrix to reduce costs and environmental liabilities (Curran 1996). The purpose of conducting such a life-cycle analysis on environmental matters is to reduce risk and liabilities predominantly through awareness (USEPA 1996).

The effectiveness of these approaches is limited. As we will demonstrate, an aggressive pollution prevention program is required to eliminate the use of those contaminants that are especially expensive to remediate. There is also the need to develop stronger and more effective engineering controls at facilities located in sensitive ecological areas. These environmental management and risk reduction methods need to be more efficient and cost effective, and, most importantly, company-specific. Developing company-specific objectives based on past quantified experience has been a critical factor in obtaining company cooperation and commitment at all levels of the organization.

1.4.5 Theme #5: Science-Based Landscape Planning

It took nature millions of years to develop the most efficient ways to evolve species, create landscapes and erode rocks, transport sediment and water, and form soil. Any time humans alter a landscape and fail to mimic the processes nature used to form that landscape, we create an imbalance, primarily in the flows of energy and materials. **Ecosystems** are nature's organizers of energy and material transport, and this is why human activity—especially the large inputs of energy involved with large-scale earth moving for new subdivisions or mining—causes negative environmental impacts.

Organisms within ecosystems take in energy from an outside source (usually the sun), convert and use that energy to produce more biomass (growth), and then release some of the energy back into the environment through respiration. At this basic level, and without going into more complex ecosystem dynamics such as **trophic levels** or **speciation**, an analogy between ecosystems and urban areas can be constructed (Figure 1.6).

Referring to Figure 1.6, please note that the primary processes—input, energy conversion/use, production, and energy/material release—are similar. The differences occur in

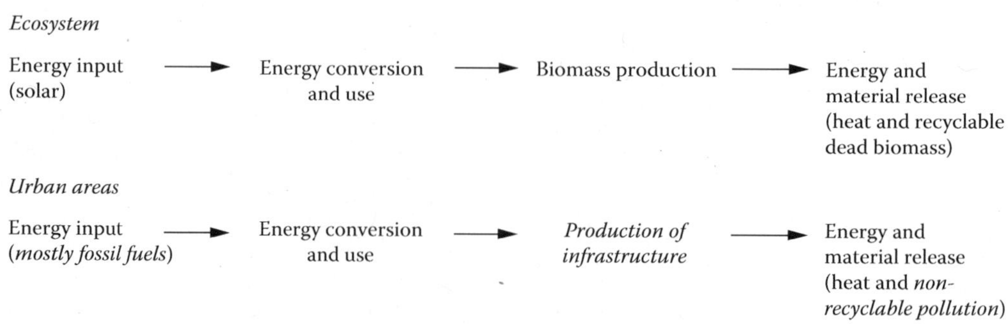

FIGURE 1.6
Urban areas as ecosystems (differences are italicized).

the urban areas through *the character* of their corresponding processes: (1) their energy inputs come from a source that yields pollution (fossil fuels); (2) the production process produces growth, but it is in the form of infrastructure, which is likely to impact existing natural ecosystems through their removal; and (3) the energy and material releases differ in their magnitudes and impacts. Urban areas release tremendous amounts of heat in the form of reradiated long-wave radiation from darker surfaces such as asphalt. Substantial heat is also emitted from automobiles and power plant discharges. This excessive heat often creates an "urban heat island," which affects the local microclimate, specifically the magnitude and frequency of summer thunderstorms (Changnon 1978). In terms of the material released, in natural ecosystems, the dead biomass is part of the **carbon cycle** and eventually becomes a part of another living entity. For example, the biomass from decomposed trees adds organic materials to the soil, which when they decompose will help resupply the carbon dioxide in the atmosphere to be taken up later by other plants through photosynthesis. The biomass in this process is therefore recyclable.

In urban areas, the production of infrastructure does not produce recyclable biomass, and this reality creates the critical challenge for urban **sustainability**. How can we continue to produce "things" in concentrated urban areas, and yet not wreck the surrounding ecosystems supporting our existence? Part of the answer lies in reducing pollution at its source—before it enters the environment. Here, the elimination of certain toxic and persistent chemicals from the manufacturing process can help. Another piece involves landscape-based urban planning, so our waters are more effectively protected. The planning process is required because watersheds are ecosystem units encompassing urban areas, and their surface area is typically composed of 5% water and 95% land. A third component to a sustainable urban environment must address the reality that large areas within them are already contaminated, and these contaminated sites need to be brought back "on line." Successful remediation of damaged soil and groundwater must make the land and water recyclable again.

To succeed, all of these efforts must proceed from a scientific foundation, one which includes environmental geology, geochemistry, risk analysis, hydrology, and science-based landscape planning. The following chapters follow this route.

1.5 Organization of This Book

The three parts of this book, Geology, Contamination, and Sustainable Development address the five themes just mentioned within the context of a watershed approach. The Geology part (Chapters 2 through 6) addresses the role of geology in watershed investigations; water and hydrogeology of watersheds; preparing a geological analysis of a watershed, including case studies focusing on the surface geology in various terrains; and developing geological and vulnerability maps of an urban watershed. Historically, the majority of geologic investigations have either emphasized identification of what was present within a few inches of the surface of the ground or characterizing deeper bedrock. Detailed geologic investigations of the unconsolidated materials between these two points have only recently become of interest. This topic forms a critical piece of the linkage between urban geology, contamination, and urban redevelopment, by focusing on the surface and near-surface geologies and their implications for contaminant sources and sinks, contaminant migration, hydraulic conductivities, and potable water availability.

The Contamination part (Chapters 7 through 11) is dedicated to learning about contaminants; specifically their chemical properties, their behavior once released into the environment, and the risks they pose to humans and the environment. The final chapter in this part examines remediation techniques and their costs under different contamination scenarios. Here we present the results from actual contamination events and demonstrate how different contaminants and near-surface geological environments influence remediation techniques and cost. Special attention is given to heavy metals, since they are a major concern in urban watersheds.

Chapters in the Sustainable Development part (Chapters 12 through 16) will analyze the disruption brought on by urbanization, present case studies of environmental damage, and offer strategies for source control of contaminants at the site scale (i.e., specific properties). Case studies of successful and unsuccessful redevelopment at contaminated sites are used to point out some of the broader issues related to attaining scientifically-informed landscape planning at the watershed scale. We conclude by reviewing the current methods used for landscape planning at the community level and present a science-based framework for improving watershed management efforts that can lead to sustainable urban watersheds.

As noted in the preface, the figures and tables from this book are available at the CRC Press website: http://www.crcpress.com/product/isbn/9781439852743

References

American State Papers, 1834. *Public Lands*, Vol. III. Washington, DC: Duff Green.

Anderson, J.R., E.E. Hardy, J.T. Roach et al. 1976. *A Land Use and Land Cover Classification System for Use with Remote Sensor Data*. Geological Survey Professional Paper 964. Washington, DC: United States Government Printing Office.

Bergquist S.G. and D.C. MacLachlan. 1951. *Guidebook to the Study of Pleistocene Features of the Huron-Saginaw Ice Lobes in Michigan*. The Glacial Field Trip of the Geological Society of America, Detroit Meeting.

Brown, R.H. 1948. *Historical Geography of the United States*. New York: Harcourt, Brace & World.

Changnon, S. 1978. Urban effects on severe local storms at St. Louis. *Journal of Applied Meteorology* 17:578–586.

Curran, M.A. 1996. *Environmental Life-Cycle Assessment*. New York: McGraw Hill.

Daly, D. and W.P. Warren. 1994. *Vulnerability Mapping*. Newsletter 25. Dublin, U.K.: Geological Survey of Ireland.

Daugherty, J.E. 1996. *Industrial Environmental Management*. Rockville, MD: Government Institutes.

Davis, M.W.R. 2007. *Detroit's Wartime Industry: Arsenal of Democracy*. Charleston, SC: Arcadia Publishing.

Dorr, J.A and D.F. Eschman. 1970. *Geology of Michigan*. Ann Arbor, MI: University of Michigan Press.

Farrand, W.R. 1982. *Quaternary Geology of Southern (& Northern) Michigan*. Michigan Department of Natural Resources, MI Department of Natural Resources, Geological Survey Division, Lansing, MI. 1:500,000.

Farrand, W.R. 1988. *The Glacial Lakes around Michigan*. Bulletin 4. Michigan Department of Natural Resources, Lansing, MI.

Federal Reserve Bank of Chicago. 2005. Michigan Auto Woes, October, 4, 2005, http://midwest.chicagofedblogs.org/archives/auto_industry/ (accessed August 2, 2009).

Fletcher, C.D. and E.K. Paleologos. 2000. *Environmental Risk and Liability Management*. Westminster, CO: American Institute of Professional Geologists.

Foster, S.S.D. 1987. Fundamental concepts in aquifer vulnerability: Pollution risk and protection strategy. In *Vulnerability of Soil and Groundwater to Pollutants*, Number 38, ed. W. van Duijvenbooden and H.G. van Waegeningh, pp. 69–87. The Hague, the Netherlands: TNO Committee on Hydrological Research.

Hartig, J.H. and M.A. Zarull. 1991. Methods of restoring degraded areas in the Great Lakes. *Review of Environmental Contamination and Toxicology* 117:127–154.

Kaufman, M.M., D.T. Rogers, and K.S. Murray. 2009. Using soil and contaminant properties to assess the potential for groundwater contamination to the lower Great Lakes, USA. *Environmental Geology* 56:1009–1021.

Kerr, J.K., W.S. Olinek, and J.H. Hartig. 2003. The Detroit River as an artery of trade and commerce. In *Honoring Our Detroit River: Caring for Our Home*, ed. J.H. Hartig, pp. 35–47. Bloomfield Hills, MI: Cranbrook Institute of Science.

Leverett, F.B. 1911. *Map of the Surface Formations of the Southern Peninsula of Michigan*. Geological Survey of Michigan, Publication 25, Lansing, MI, 1:1,000,000.

Morgenstern, R.D., W.A. Pizer, and J.S. Shih. 1998. *The Cost of Environmental Protection*. Washington, DC: Resources for the Future.

Mozola, A.J. 1969. *Geology for Land and Groundwater Development in Wayne County*, Michigan. Michigan Geologic Survey, Lansing, MI.

Murray, J.E. and J.M. Bona. 1993. *Rouge River National Wet Weather Demonstration Project*. Detroit, MI: Wayne County.

Murray, K.S., D.T. Rogers, and M.M. Kaufman. 2004. Heavy metals in an urban watershed in southeastern Michigan. *Journal of Environmental Quality* 33:163–172.

Murray, K.S., D.T. Rogers, and M.M. Kaufman. 2006. Dissolved heavy metals in shallow groundwater in an urban watershed in southeastern Michigan. *Journal of the American Water Resources Association* 42:777–792.

Murray, K.S., X. Zhou, M. McNulty et al. 2000. Relationship between land use, near surface geology and water quality in an urban watershed, Southeast Michigan, USA. *Proceedings of the International Symposium on Hydrology and the Environment*, Wuhan, China, pp. 67–71.

Owen, J. 1995. Environmental compliance: managing the mandate. *Manufacturing Engineering* 113:59–66.

Parkins, A.E. 1918. *The Historical Geography of Detroit*. Lansing, MI: Michigan Historical Commission.

Richards, L. 2003. *Detroit River Corridor, Preliminary Assessment of Land Use Change*. Urban Dynamics Research Program. Moffett Field, CA: USGS Ames Research Center.

Rieck, R.L. 1981a. *Glacial Drift Thickness, Southern Peninsula. Plate 15*. Hydrogeologic Atlas of Michigan. Kalamazoo, MI: Western Michigan University.

Rieck, R.L. 1981b. *Community Public Water Supplies. Plate 22*. Hydrogeologic Atlas of Michigan. Kalamazoo, MI: Western Michigan University.

Rogers, D.T. 1996. *Environmental Geology of Metropolitan Detroit*. Novi, MI: Clayton Environmental Consultants, Inc.

Rogers, D.T. and K.S. Murray. 1997. Occurrence of groundwater in metropolitan Detroit, Michigan, USA. In *Groundwater in the Urban Environment*, ed. J. Chilton et al., pp. 155–160. Rotterdam, the Netherlands: Balkema.

Rugg, D.S. 1972. *Spatial Foundations of Urbanism*. Dubuque, IA: W.C. Brown.

SEMCOG, Southeast Michigan Council of Governments. 2003. *Land Use Change in Southeast Michigan: Causes and Consequences*. Detroit, MI: SEMCOG.

Sherzer, W.H. 1916. *Geologic Atlas of the United States, Detroit folio Number. 205*. United States Geological Survey. Reston, VA. 1:62,500.

Sugrue, T.J. 2005. From motor city to motor metropolis: How the automobile industry reshaped urban America. Automobile in American Life and Society website Dearborn: Henry Ford Museum and University of Michigan, http://www.autolife.umd.umich.edu (accessed May 17, 2010).

Telego, J.D. 1998. A growing role: Environmental risk management. *Risk Management* 17:19–21.

USEPA, U.S. Environmental Protection Agency. 1996. *Valuating Potential Environmental Liabilities for Managerial Decision Making: A Review of Available Techniques*; EPA-742/R-96/003. Washington, DC: U.S. Government Printing Office.

USEPA, U.S. Environmental Protection Agency. 2004a. *Cleaning Up the Nation's Waste Sites: Markets and Technology Trends, 2004 Edition*; EPA-542/R-04/101. Washington, DC: U.S. Government Printing Office.

USEPA, U.S. Environmental Protection Agency. 2004b. 2004 Rouge River Remedial Action Plan Revision. http://www.epa.gov/greatlakes/aoc/rougriv/2004_Rouge-River-RAP-Revision.pdf (accessed August 1, 2009).

USEPA, U.S. Environmental Protection Agency. 2007. *Expediting cleanup and Redevelopment of Brownfields: Addressing the Major Barriers to Private Sector Involvement—Real or Perceived*. Washington, DC.

Voorhees, T.H. and R.A. Woellner. 1998. *International Environmental Risk Management*. Boca Raton, FL: Lewis Publishers.

Part I

Geology

2

Geology of Urban Watersheds

2.1 Introduction

We walk on a history book every day, yet most of us are not fully aware of its significance. The arrangement, thickness, and composition of the sediment layers beneath our feet have a profound influence on where cities are located, how buildings are constructed, where roads are built, and perhaps most important to the development and redevelopment of our urban centers—how contaminants behave and how they affect the environment and people.

Geology is the science dedicated to the study of the history, structure, and composition of the earth (Flint and Skinner 1974). The study of anthropogenic or human impacts on the geologic environment is termed environmental geology (Bennett and Doyle 1997). For the most part, the study of geology has been conducted for purely scientific and academic purposes, natural resource exploitation, and geologic hazard evaluation and avoidance. Only recently has the geologic environment in urban areas become of interest to geologists and other scientists. The extreme upper portion of the lithosphere is affected by human activities, and this near-surface geologic environment frequently acts as the migration pathway for contaminants to travel from a specific point of release to a point of ecologic or human exposure. In many cases, this pathway goes undetected because the contaminants migrate beneath the surface of the ground, and cannot be observed or detected until it is perhaps too late.

The study of the near-surface geologic environment will become increasingly important in the future. The world's human population is increasing in urban areas at a rate greater than the human population as a whole. Nearly 3 billion people currently live in urban areas. By the year 2030, the human population in urban areas is projected to be nearly 5 billion or 60% of the 8.6 billion people on the planet. The current level of population has placed enormous pressure on our environment—including the geologic environment—and the projected increases are likely to cause additional stress. Therefore, a thorough understanding of the subsurface geologic environment is critical in order to understand and protect the cities and places we live.

The environmental problems we face in the typical urban setting can in part be attributed to the poor understanding of the subsurface geology. Lack of attention to this area of inquiry by the scientific community is one factor contributing to this knowledge gap. Geologists generally have not spent very much effort or time studying shallow unconsolidated geologic deposits beneath urban areas for environmental reasons because they

- Do not typically contain natural resources of high economic value.
- Have only become important in the last 30 years with the onset of significant environmental regulations.
- Are covered by buildings and roads and other human-constructed surfaces.

- Are difficult to access because many desirable study locations are on private property.
- Have been disturbed and modified due to development.
- Are typically located on unconsolidated sediments of Quaternary age and were historically considered overburden and not as scientifically valuable as the bedrock located beneath the unconsolidated deposits.
- Are very complex and time consuming to study.
- Involve issues that may take decades to be noticed by the general community as compared with other more sudden types of geologic hazards such as earthquakes, tsunamis, and volcanic eruptions.

In addition, the geology of most urban regions generally goes unnoticed by most residents and even the scientific community because it is not visually spectacular like the geology seen in our national parks. Most residents would be more inclined to learn about a region's geology and its underlying processes if their area was prone to earthquakes, such as parts of California, or volcanic eruptions, such as the Pacific Northwest or Hawaii. The geology beneath most urban regions affects our day-to-day lives in more subtle ways than an earthquake or volcanic eruption; however, its ultimate impact on the quality of our lives could be just as significant.

Watersheds functioned well before humans began to alter their natural flows of water; and now many are in a distressed state. Understanding what was once at a given location (site) and how it came into existence are therefore prerequisites for correctly analyzing the current problem, and prescribing an effective and sustainable remediation strategy. As more and more sites of human interference are analyzed, the overall picture of the linkages between the watershed's geology, biology, geography, and human interference becomes clearer. As more of these linkages are made, the geographic extent and impacts of contamination in different geological environments become more evident, appropriate remediation efforts can be more effectively implemented, and the methods for sustainable redevelopment and new development refined.

Our efforts in this chapter concentrate on the sedimentary processes influencing urban areas and their watersheds. We begin with the formation of rocks and the geologic principles governing sedimentary formations and their arrangements (uniformity, stratigraphy, and unconformities), then move on to the processes involved in determining sedimentary rock composition and transport, and the environments created from the combination of the rock characteristics and their depositional processes. The chapter concludes with a discussion of natural and anthropogenic disturbance.

2.2 Geological Processes Affecting Urban Watersheds

Geological processes affecting urban watersheds generally take place at or very near the surface and are primarily sedimentary-type processes, which include processes involving liquid and solid water (ice), gravity, and tectonic activity. **Sedimentary deposits** are defined as the accumulation of natural materials and sediments formed at or near the surface of the Earth at ordinary temperatures and pressures (Flint and Skinner 1974). Geological processes of sedimentary origin usually occur gradually over long periods of time. On rare occasions, however, sudden events affect sedimentary processes and urban environments. These short-duration events are typically labeled as geologic hazards and may include tectonic processes such as earthquakes, tsunamis, volcanic eruptions, and landslides. Some short-term geologic processes are also triggered by a weather event such as a flood caused by extensive rains, sudden melting of snowpacks, or a hurricane.

2.2.1 Formation of Sediments and Rocks

The fundamental building blocks of the Earth's crust are rocks. Understanding and evaluating the rocks beneath urban areas requires knowledge of their composition, and how they were deposited at a given location. On Earth, the dominant physical force that builds and destroys rock is plate tectonics. **Plate tectonics** is based on the concept that the Earth's lithosphere (crust and upper mantle) is composed of moving plates (Seyfert and Sirkin 1973; Oreskes 2003). In 1912, a German meteorologist, Alfred Wegener, noticed that some sedimentary rocks known only to form in tropical environments were located in Arctic and Antarctic regions. Wegener then compared these sedimentary deposits with their approximate relative ages and distances from the equator and concluded that the continents had "drifted" to their present locations.

Wegener's hypothesis of continental drift was ridiculed because it lacked a mechanism to forcibly move the continental land masses. Oceanographic research in the decades following World War II shed a great amount of detail on the structure of the earth and a potential mechanism to move the continental crust. Today, we recognize that the continental and oceanographic crust ride passively along on lithospheric plates. Energy to drive the movement of these tectonic plates is believed to be caused by the heat rising in the form of convection currents generated at the core–mantle boundary. These concepts became the theory of plate tectonics.

Figure 2.1 presents a world map showing each of the major tectonic plates. These lithospheric plates (crust and uppermost mantle) move on top of the upper mantle called the **asthenosphere** (Figure 2.2). Their movement is responsible for most earthquakes, volcanoes, seafloor spreading, mountain building, and ocean trench formation (Stanley 1999). The energy source driving the movement of the plates is believed to be heat from convection currents originating deep within the interior of the Earth. Figure 2.3 shows the concepts of plate tectonics.

Plate tectonics is responsible for, or plays a significant role in the formation, deformation, and weathering of most rocks on Earth. The primary classification of rocks recognizes three categories (igneous, metamorphic, and sedimentary) based on their method of formation.

Igneous rocks are formed by the solidification of molten material either beneath Earth's surface (plutonic igneous rocks) or at the surface (volcanic igneous rocks) (AGI 1962). Examples of plutonic (intrusive) igneous rocks include granite, diorite, gabbro, and peridotite; volcanic (extrusive) igneous rocks include andesite, basalt, and rhyolite. Each type of igneous rock is named based on its mineralogic and chemical composition.

Metamorphic rocks are formed by applying extreme temperature and pressure to existing rocks that rearrange their structure and mineralogy to create a new rock. Examples of metamorphic rocks include gneiss (formed from the metamorphism of granite); quartzite (metamorphosed from quartz sandstone); and marble (metamorphosed limestone).

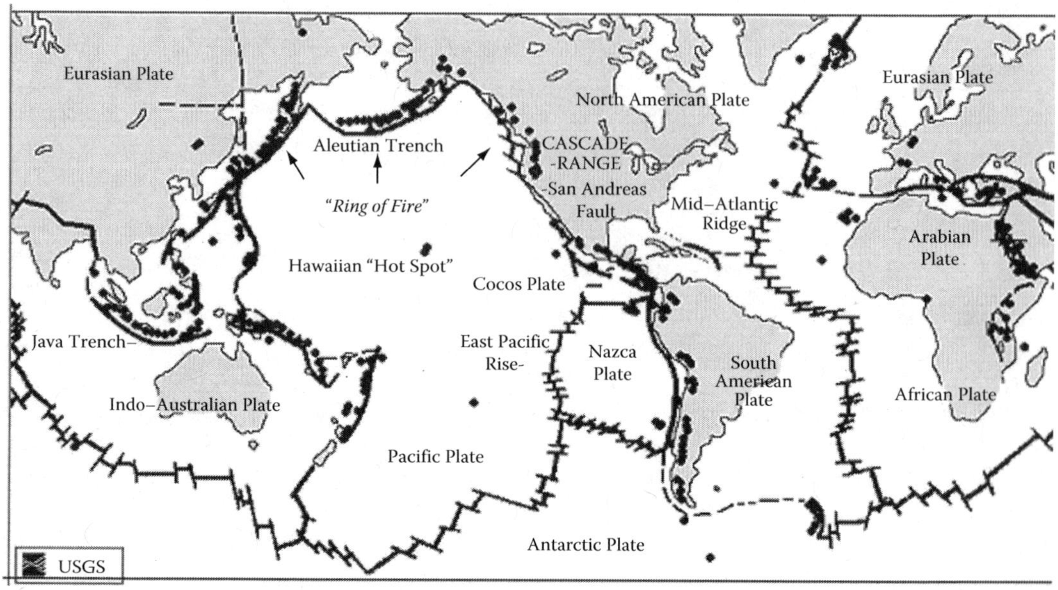

FIGURE 2.1
The major tectonic plates of the world. (From United States Geological Survey, World map showing plate boundaries and active volcanoes, http://vulcan.wr.usgs.gov/Glossary/PlateTectonics/Maps/map_plate_tectonics_world.html (accessed June 18, 2010), 2010a.)

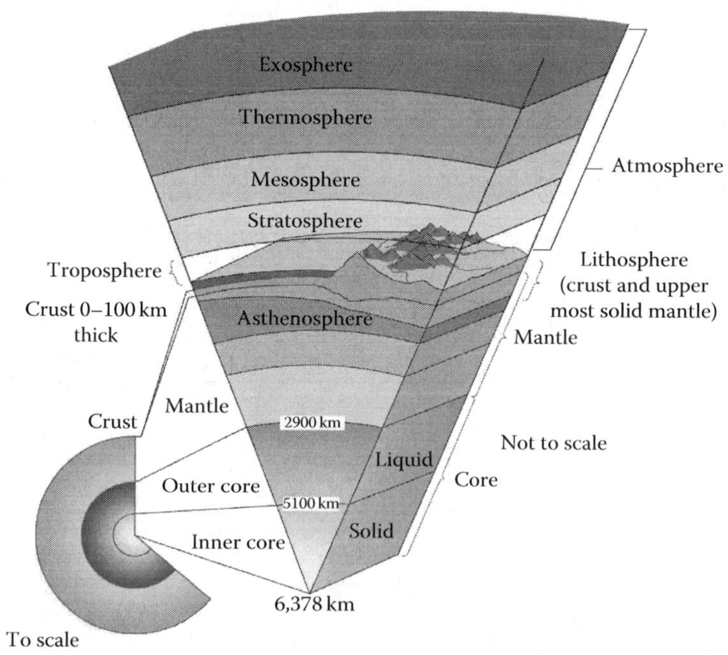

FIGURE 2.2
(See color insert.) Cross section of the interior portions of the Earth showing the asthenosphere. (Modified from Kious, W.J. and Tilling, R.I., *The Dynamic Earth: The Story of Plate Tectonics,* United States Geological Survey, Governmental Printing Office, Washington, DC, 1996.)

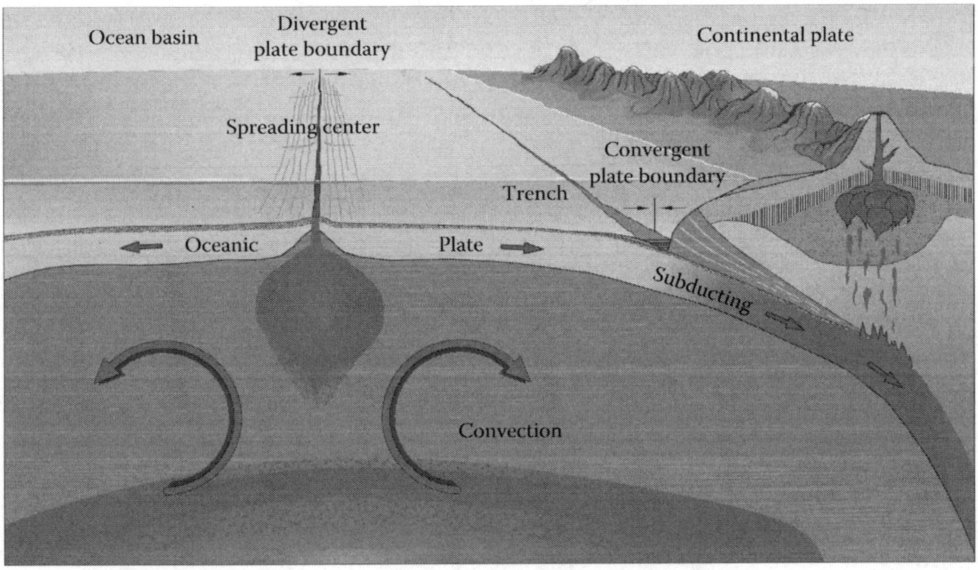

FIGURE 2.3
(See color insert.) Concepts of plate tectonics and continental drift.

Sedimentary rocks are formed from the mechanical weathering or erosion of preexisting rock or from dissolved material that precipitates from solution. Therefore, there are two types of sedimentary rocks (Pettijohn 1975):

- *Clastic*: Composed of grains of fragments of rocks and minerals that have been eroded, transported, deposited, and eventually become cemented together to form a rock.
- *Chemical or biochemical*: Composed of minerals that have been precipitated from water by inorganic processes or biological processes. For example, coal is formed from plant debris.

Common examples of clastic sedimentary rocks include conglomerate, sandstone, siltstone, and shale.

Some of the more common chemical and biochemical sedimentary rocks include limestone, dolostone, coal, and the evaporites, halite and gypsum. When coal undergoes significant changes through exposure to high temperatures and pressures, it forms a progressively higher-grade of coal, eventually becoming anthracite, considered to be a metamorphic rock (Pettijohn 1975).

A **rock** is simply composed of minerals (Flint and Skinner 1974). A mineral is a naturally occurring solid formed through geological processes, which has a characteristic chemical composition, a highly ordered atomic structure, and specific physical properties (Hurlbut and Klein 1985). Minerals range in composition from pure elements to very complex silicates, oxides, sulfates, and others with thousands of known forms. Rocks do not require a specific chemical composition but have ranges of compositions of typically two or more different minerals. For instance, igneous rocks differ in composition based on the chemical makeup of their parent magma and their rate of cooling. The composition of sedimentary rocks differs based on the parent material and the methods of transport and deposition.

Before we focus on sedimentary rocks, a brief overview of some basic geological principles is required. These principles (uniformity, stratigraphy, and unconformity) provide the necessary foundation for understanding the general characteristics of sedimentary rocks in urban environments.

2.2.2 Principle of Uniformity

During the late 1700s, James Hutton, a Scottish geologist, doctor, and farmer was studying the rock outcrops around Scotland. Hutton, influenced by the Age of Enlightenment, cast aside the existing religious dogma of a very young Earth and developed the **principle of uniformity**, which stated that the external and internal processes on and within the Earth observed today have been operating unchanged, and at the same approximate rates, for a very long time throughout most of the Earth's history (Hutton 1788). Essentially, the meaning of the principle of uniformity is that the present is the key to the past. The significance of this principle cannot be understated, because the recognition of this principle presents the geologist with a great tool for exploring the history of the Earth. Geologists today look at features within observable rocks and search for insights into how they formed long ago. For example, in some urban areas, the presence of sand found near the surface today might indicate the presence of a beach thousands or millions of years ago. As we will soon learn, the shape of the sand grains and their orientation can reveal even more information about the past and the current physical and anthropogenic impacts.

2.2.3 Principles of Stratigraphy

Stratigraphy is the study of rock layers or unconsolidated sediment and strata; particularly their ages, composition, and relationship with other layers (AGI 1962; Christopherson 2008). In geology, the term **strata** refers to the layers of rock or sediment with internally consistent characteristics that distinguish it from contiguous layers (Krumbein and Sloss 1963). Each individual layer or stratum is generally one of a number of parallel and originally horizontal layers that were deposited by natural forces and lay one upon the other. Therefore, the study of stratigraphy of a particular area provides important clues concerning its geologic history.

Stratigraphy relies on four basic principles (Anderton 1985):

1. The *principle of original horizontality* states that the sediments formed from sedimentary deposits are generally deposited as horizontal sheets. An example of very fine horizontal layering of a sedimentary deposit composed of layers of clay and fine silt is shown in Figure 2.4.

   This principle is important because it allows the geologist to distinguish between disturbed or deformed layers. In urban areas, this principle is especially important because so much of the near-surface sediments and soil layers have undergone significant disturbance due to development or deformation from tectonic activity. Some examples of near-surface disturbance include farming, constructing buildings and foundations, road building, mining, and landscaping. The term deformation generally refers to both brittle and ductile processes such as faulting or folding caused by tectonic activity. This type of deformation explains why, for example, sedimentary rock deposited thousands of feet below sea level may now be found thousands of feet above sea level. This type of uplift is almost always the result of plate interaction. Erosion or removal of a portion of a sedimentary layer through the action of water

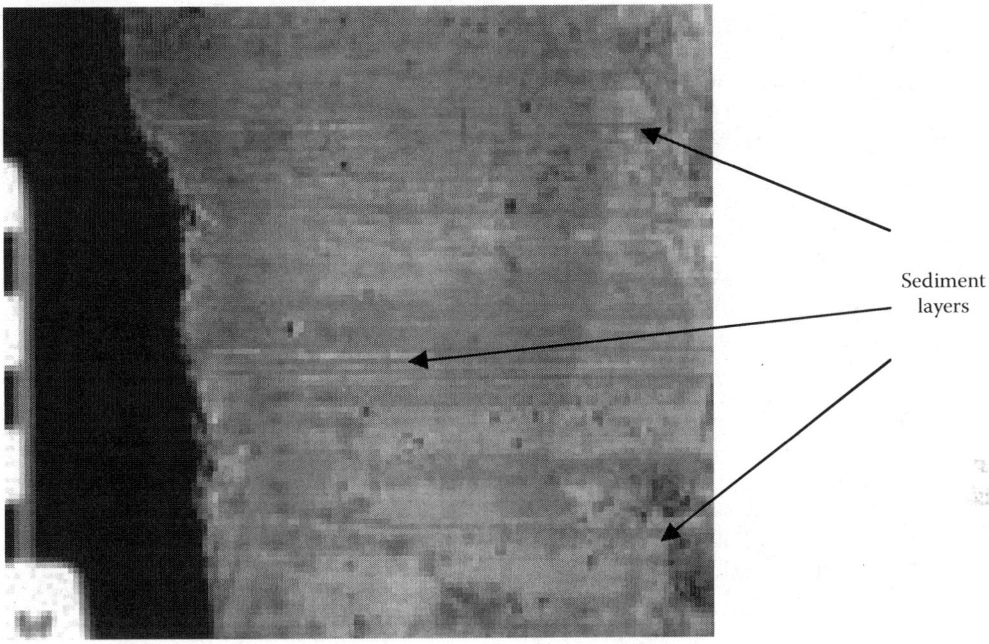

FIGURE 2.4
Example of very fine horizontal layering within a clastic sedimentary deposit. Sample is approximately 8 cm vertical by 6 cm horizontal. (Photo by Daniel T. Rogers.)

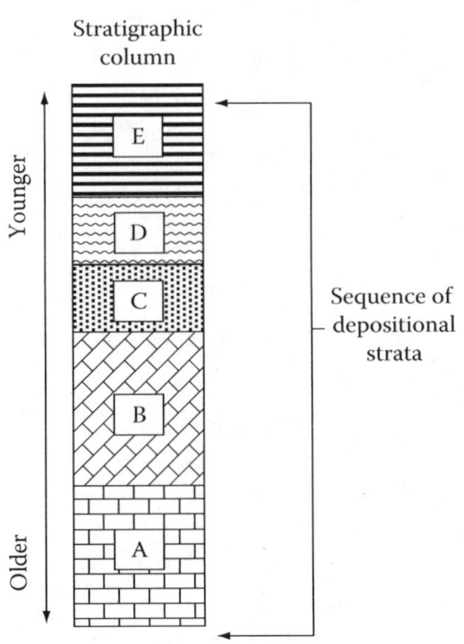

FIGURE 2.5
Example of a stratigraphic sequence demonstrating the principle of superposition.

and wind can also significantly erode or rework a layer of sediment after it has been deposited.

2. The *principle of superposition* defines the depositional sequence in undeformed strata of rock or sediment. Strata are sheets or individual layers of sedimentary deposits, and as a group, are visibly distinct from those above and below. Figure 2.5

demonstrates the principle of superposition. The principle of superposition results from the simple observation that new layers of sediment are deposited on top of older layers of sediment. Therefore, strata are deposited sequentially, layer after layer. It is important in areas where tectonic activity has occurred to establish the correct stratigraphic sequence to ensure that the entire sequence of layers has not been overturned.

3. The *principle of crosscutting relations* states when a layer of sediment is cut, perhaps by faulting, or the intrusion of an igneous dike or sill; the sedimentary layer is older than the event or feature that cuts the layer. In urban environments, where frequent developments have occurred, it is common, for example, to have a glacial moraine or a river terrace cut by development that obscures the correct interpretation of age and stratigraphic relationships.
4. The *principle of faunal and floral succession* explains the relationship between biological evolution and sedimentary deposits. Because plants and animals evolve over long time periods into new species, sedimentary deposits of different ages will contain fossils of different species. Knowing the age of a fossil will assist in dating a particular sedimentary deposit containing the fossil. This relationship may help when evaluating two sedimentary deposits at different locations.

2.2.4 Unconformities

When examining the vertical succession of sedimentary layers, sometimes the layering may not represent a complete timeline of deposition. There may be present within the succession of layers of strata a period of time when there was no deposition, or evidence that erosion removed some preexisting layers. When this occurs and is documented in a depositional succession, it is termed an unconformity, and the missing period of time is called a hiatus. An **unconformity** is defined as a buried erosional surface separating two rock masses or sedimentary strata of different ages, indicating that sediment deposition was not continuous (Flint and Skinner 1974).

There are several types of unconformities (Wicander and Monroe 2007):

- **Disconformity**—An unconformity that exists between parallel layers of sedimentary rocks.
- **Nonconformity**—An unconformity where horizontal strata or layers of sedimentary rock overlie crystalline rock, either igneous or metamorphic.
- **Angular unconformity**—An unconformity where horizontal strata or layers of sedimentary rock are deposited on tilted and eroded strata. The result produces an angular discordance with the overlying horizontal layers.

Figure 2.6 is a figure of several types of unconformities reproduced from Hutton (1788), who was the first geologist to describe them (modified from Hutton 1788).

Examination of Figure 2.6 reveals the sequence of major events creating the unconformities:

- The layers identified as 1 at the bottom portion of the figure were deposited and then metamorphosed, uplifted, folded, and tilted vertically; they represent the oldest layers.
- After folding and tilting of the rocks, they were exposed and eroded.

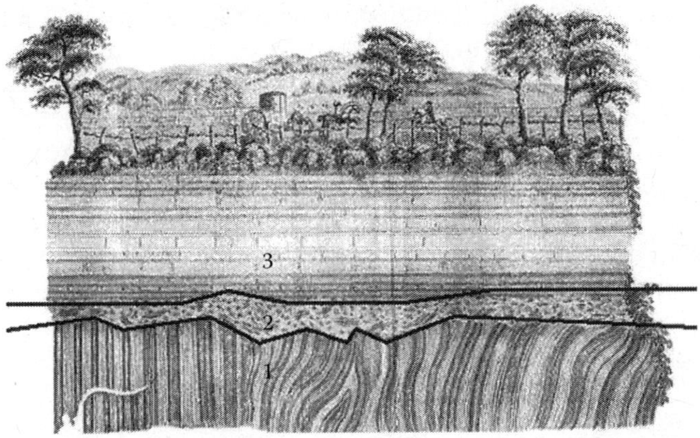

FIGURE 2.6
Concept of unconformities. Unconformities identified by black line. (From Hutton, J., *Trans. R. Soc. Edinb.*, 1, 209, 1788.)

- Following erosion, sediments identified within the zone labeled as 2 were deposited on top of the folded and tilted rock layers, creating a nonconformity (bold line between zones 1 and 2).
- After the layers identified as 2 were deposited, they were uplifted, exposed, and eroded to create a second erosion surface.
- Following a rise in sea level, sequence 3, a marine limestone was deposited, creating a disconformity (bold line between zones 2 and 3).

2.3 Sedimentary Processes: From Removal to Rock Formation

Sedimentary deposits make up only 5% of the Earth's crust (Figure 2.7a), but account for approximately 75% of the rocks exposed at the earth's surface (Figure 2.7b) (Flint and

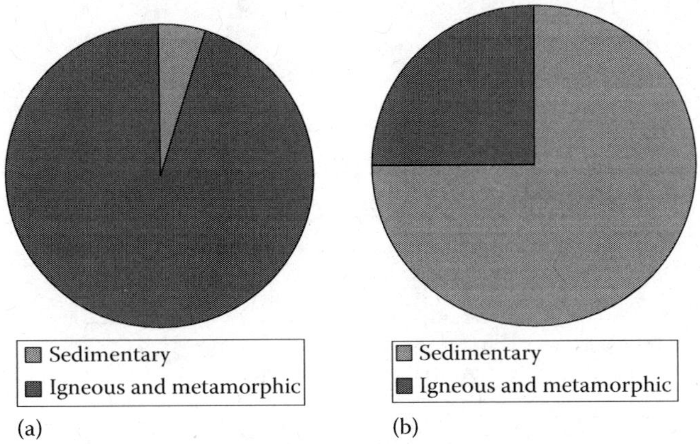

FIGURE 2.7
(a) Sedimentary deposits within Earth's crust. (b) Sedimentary deposits at Earth's surface.

Skinner 1974; Pettijohn 1975). The study of sedimentary deposits involves surficial processes that include the accumulation, consolidation, deformation, origin, age, content, and stratigraphy of sediments. Sedimentary deposits result from a sequence of processes characterized by their removal from one location, some form of transport (e.g., down-gradient by water or aloft by wind), and then final deposition in a **sink**. In sinks, sediments are not at peace. Some may be subjected to heat and pressure and become metamorphic rocks; others may move slowly with the ocean floor and become part of a subducting tectonic plate and end up as molten magma in the mantle; and many will coevolve with the regional and local biota to create unique ecosystems such as beaches, river deltas, and dunes. The vast majority of sedimentary rocks were created in a marine environment as vast shallow seas transgressed and regressed across the continental land masses.

2.3.1 Clastic Processes

There are three broad categories of removal processes involved with the formation of clastic sedimentary rocks: weathering, mass wasting, and erosion (Figure 2.8). As shown in Figure 2.8, the first process involved in the formation of a sedimentary clastic deposit is the weathering (breakdown) of bedrock, which provides the sediment or clastic material necessary to form a sedimentary deposit. **Bedrock** is defined as the continuous solid rock of the continental crust (Flint and Skinner 1974). Once bedrock begins to erode, it forms a material known as **regolith**—or weathered rock devoid of organic material.

2.3.1.1 Weathering

Weathering (process #1, Figure 2.8) is a destructive process by which rocks and minerals are broken down through exposure to atmospheric agents such as air, wind, water, and ice (Christopherson 2008). Weathering processes fall into two general categories: physical and chemical. Physical weathering involves the fragmentation of larger rock into smaller pieces by a mechanical process, without changing the chemical composition of the rock. Examples of physical weathering include (1) abrasion, where rock is dislodged from the impact of grains carried by wind, water, or ice; (2) fragmentation, from the downslope movement of

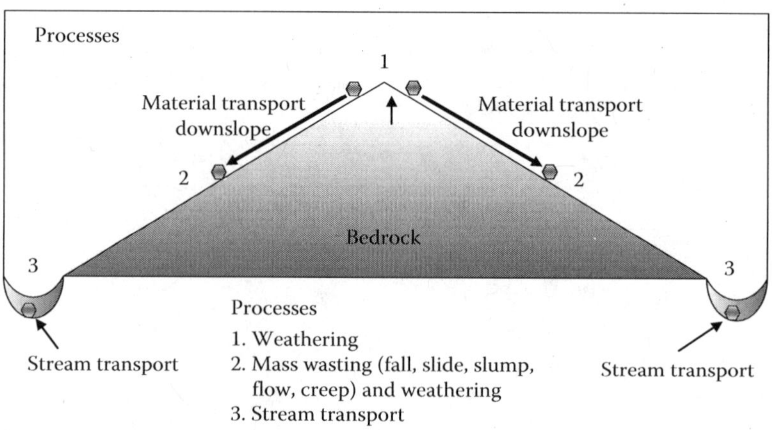

FIGURE 2.8
Diagram showing the processes of clastic sedimentary rock formation: weathering, mass wasting, and sediment transport.

rock in landslides and rockfalls; (3) frost wedging, caused by wetting and freezing cycles; and (4) thermal expansion, from the contraction of rock by heating and cooling.

Chemical weathering is the breakdown of rocks and minerals through reactions between the rocks and minerals and atmospheric constituents including water, oxygen, and carbon dioxide. This process alters the chemical composition of the rocks involved. The most common types of reactions include (1) solution, where molecules and elements in rocks and minerals dissolve directly into water; (2) oxidation and hydration, consisting of reactions between oxygen, water, and iron-bearing minerals, which assist in weakening and breaking down rocks; and (3) hydrolysis, a complex weathering reaction forming clays. A more detailed treatment of chemical weathering is presented later in this section and Section 2.3.2 within the context of sediment composition and karst topography.

The processes of weathering are the first step in soil formation.

2.3.1.2 Mass Wasting

Once weathering occurs and a portion of the bedrock is separated from itself and becomes regolith, it becomes prey for other forces that move the freed materials downslope under the influence of gravity. The several types of movement acting on regolith to transport the material downslope are collectively termed mass wasting. **Mass wasting** (process #2, Figure 2.8) is the movement of rock or soil downslope by gravity without the aid of moving water, glaciers, or wind (Flint and Skinner 1974). On any slope, mass wasting processes depend on the interplay between gravity, slope angle, moisture content, cohesion, and friction. When the gravitational force acting on a slope exceeds its resisting force, slope failure or mass wasting occurs. The slope materials' strength and cohesion and the amount of internal friction between the materials contribute to maintaining slope stability and are collectively known as the slope's shear strength. The steepest angle that a cohensionless slope can maintain without losing its stability is known as its **angle of repose** (Pudasaini and Hutter 2007).

Mass wasting can affect large bodies of regolith or just one small rock particle. In addition, mass wasting is not confined to continental regions; it also takes place in marine environments where submarine landslides occur on steep slopes and material subsequently spreads out onto the seafloor. In areas of low relief, mass wasting may still take place but at a very slow rate. By contrast, in mountainous areas, mass wasting processes can be very fast, as in a rockfall, rockslide, or debris flow.

Factors affecting the potential or rate of mass wasting include (1) a change in slope angle, (2) weakening of material through weathering, (3) increased water content, (4) changes in vegetation cover, and (5) overloading (Wicander 2005).

Anthropogenic sources increase the rate of mass wasting commonly through the building of roads requiring the construction of road cuts through uneven terrain. This practice creates situations where the cutback slope is constructed beyond the angle of repose. Overloading commonly occurs from anthropogenic sources such as heavy loading on roadways and building structures. In addition, irrigation may significantly increase the rate of mass wasting in urban areas of varied terrain.

Types of mass wasting include

- Rockfall or debris fall—the rapid descent of a rock mass, vertically from a cliff or by leaps down a very steep slope (Figure 2.9).
- Rockslide or debris slide—the rapid, sliding descent of a rock mass down a slope (Figure 2.9).

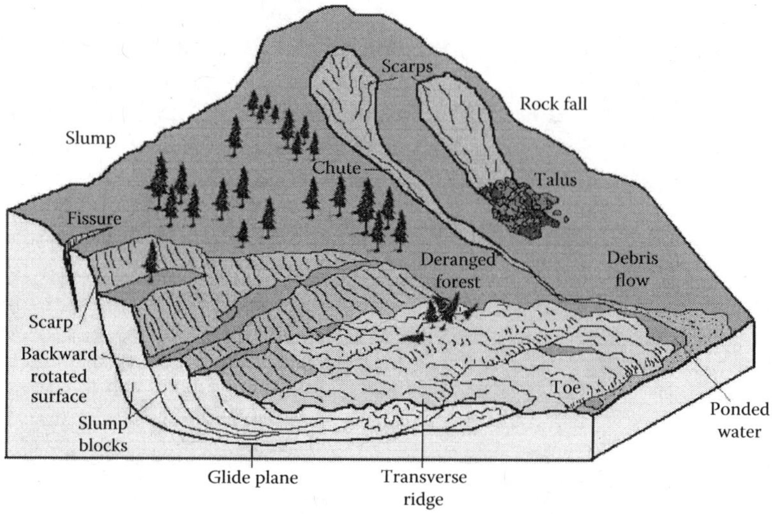

FIGURE 2.9
Examples of mass wasting. (From United States Geological Survey, USGS glossary, http://3dparks.wr.usgs.gov/haywardfault/images/slump.jpg, (accessed June 28, 2010), 2010b.)

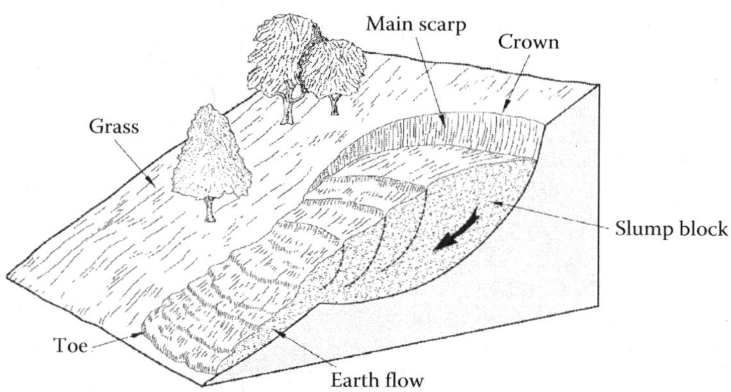

FIGURE 2.10
Example of slumping. (Reprinted from Bloom, A.L., *Geomorphology: A Systematic Analysis of Late Cenozoic Landforms*, 3rd edn., Waveland Press, Long Grove, IL, 2004. With permission.)

- Slump—The downward slipping of a coherent body of rock, regolith, or soil along a curved surface or rupture. The original surface of the slumped mass and any flat-lying planes within it become rotated as the mass slides downward. Figures 2.9 and 2.10 show a more detailed example of slumping.
- Debris flow—The rapid downslope plastic flow of a mass of regolith, soil, or debris. Commonly forms as an apron-like or tongue-like area at its terminus with a very irregular surface (Figure 2.9).
- Mudflow—The rapid downslope movement of a very plastic and almost fluid mass of regolith, soil, or debris. Very similar to a debris flow but with a higher water content (Figure 2.9).

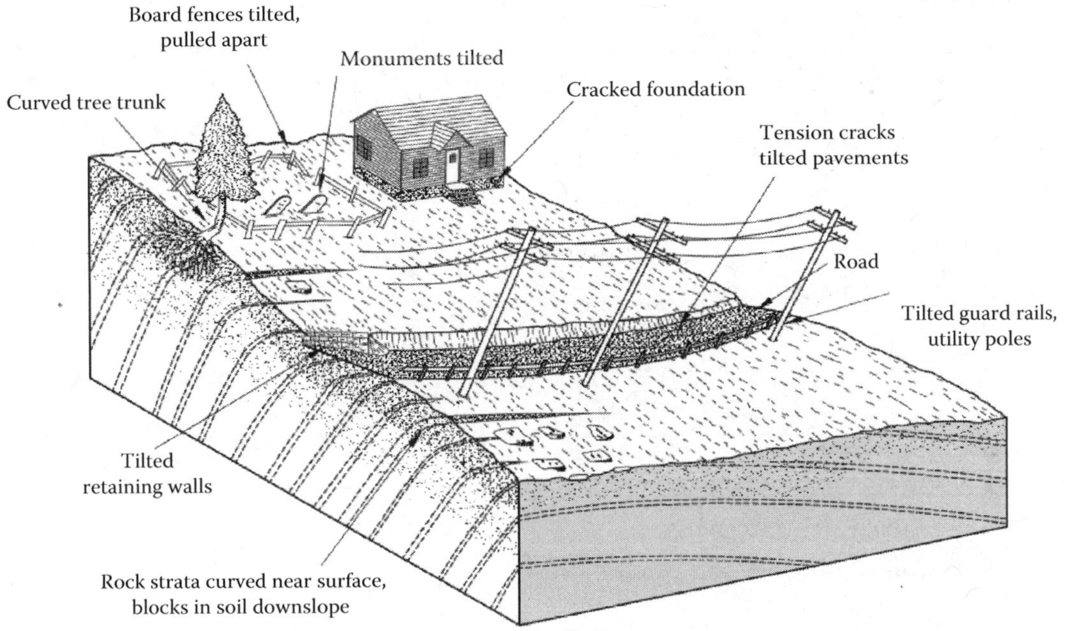

FIGURE 2.11
Example of soil creep. (Reprinted from Bloom, A.L., *Geomorphology: A Systematic Analysis of Late Cenozoic Landforms*, 3rd edn., Waveland Press, Long Grove, IL, 2004. With permission.)

- Soil creep—Slow movement of slope. Soil creep is the most common type of mass wasting and is always present on any slope where there is soil. This process tends to go unnoticed because the process is slow. Some of the effects from soil creep are shown in Figure 2.11.

Gravity is not the sole force in operation during mass wasting. Other transport mediums such as wind, flowing water, ice, or molten lava may be involved. Although flowing water is excluded from the process by definition, water nevertheless plays an important role in mass wasting. Flowing water can erode the base of a hillslope and weaken the mid-slope region, leading to an increase in the effects of gravity. In a process known as piping, water migrating near the soil surface from seepage fronts of groundwater may subsequently produce slope instability and increase the "lubrication" of a preexisting plane of weakness. More frequently, however, water entering the soil simply increases the pore pressure, reducing grain-to-grain contact of the soil particles and results in slope failures such as debris flows.

2.3.1.3 Stream Transport

After the downslope transport of material, streams carry a large portion of these solids away (process #3, Figure 2.8). The larger and heavier rocks become part of the channel bottom, while a small fraction of the solid material reacts chemically with water and becomes dissolved load that takes the same journey as the stream. Most of the downslope material, however, becomes suspended load and eventually ends up in the continental shelves as a sedimentary deposit.

2.3.1.3.1 Erosion

Before the deposition of clastic sedimentary material occurs, the material must be eroded. **Erosion** is defined as the complex group of related processes by which rock is broken down physically and chemically and its products removed (Christopherson 2008). The rate of erosion depends on many factors and includes

- Climatic factors
  - Amount and intensity of precipitation
  - Average temperature
  - Temperature range
  - Wind speed and changes in wind direction
  - Storm intensity and frequency
- Geologic factors
  - Sediment composition
  - Sediment size and shape
  - Sediment porosity and permeability
  - Slope of surface
  - Orientation of sediment surface (tilted, faulted, or folded)
- Biological factors
  - Vegetative cover type and amount
  - Animal habitation behaviors
  - Land use

Estimates of the natural erosion rate in North America ranges between 1 and 3 cm every 1000 years (Flint and Skinner 1974). This estimate, however, does not include anthropogenic sources of erosion that can double its rate (U.S. Soil Conservation Service 1992). In addition, the rate of erosion is not uniform. Some areas may experience significant amounts of erosion over a relatively short period of time, whereas in other areas, erosion may not be apparent.

Figure 2.12 shows the different types of water erosion on the land surface. Gravity is responsible for the erosion associated with mass wasting; while precipitation initiates a sequence of splash erosion (where a soil particle is dislodged), leading to sheet erosion from overland flow, and in extreme cases, rills (small indentations) or gullies may form. Strong downslope winds on unvegetated slopes can increase the potential for rills and gully formation; this occurs in southern California during the Santa Ana wind season.

In urban environments, the creation of an impervious surface during development activities increases the volume and velocity of surface runoff, and erosion is generally accelerated (Figure 2.13a and b). Other types of anthropogenic processes that enhance or accelerate erosion include

- Removal of vegetative cover through logging
- Overgrazing of livestock
- Changing native vegetation
- Overwatering of vegetation
- Slash and burn agriculture and forestry practices, especially on steep slopes

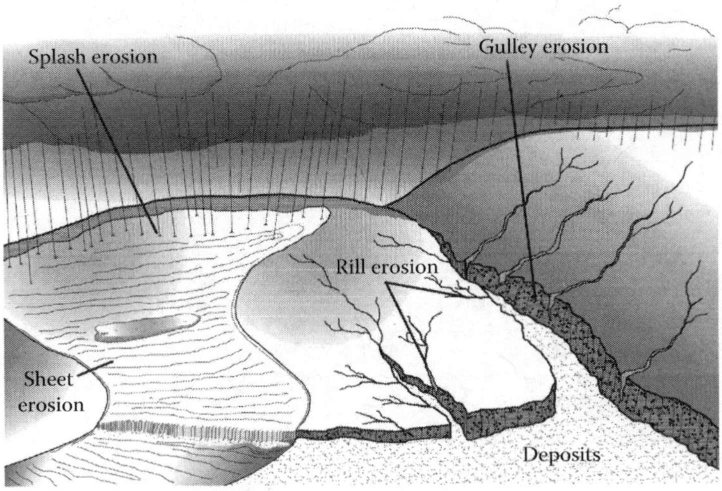

FIGURE 2.12
Water erosion on the land surface.

FIGURE 2.13
(a) Rill erosion in an urban area. (Photo by Martin M. Kaufman.). (b) Gullying around a stormdrain. (Photo by Martin M. Kaufman.)

- Cultivation or farming
- Removal of leaf litter, small shrubs, and grasses

On coastlines, wind, waves, and storms promote the most significant erosion. Wind produces waves and storms produce high wind and big waves. Large storm waves have been known to produce 907 kg (2000 lb) of pressure per square foot. The pure energy of waves and the chemical content of the water combine to erode sediment and rock along a coastline (Bird 1985). Other factors, such as rising sea levels, tides, tidal currents, and drainage patterns from land naturally influence coastal erosion.

Waves erode shorelines in three main ways:

- Hydraulic action—waves strike a steep slope and dislodge material
- Abrasion—solid materials such as sand or rocks grind together causing them to become rounded and smaller in size
- Corrosion or dissolution—sea water reacts with the sediments along the shoreline and slowly dissolves portions of the solid material

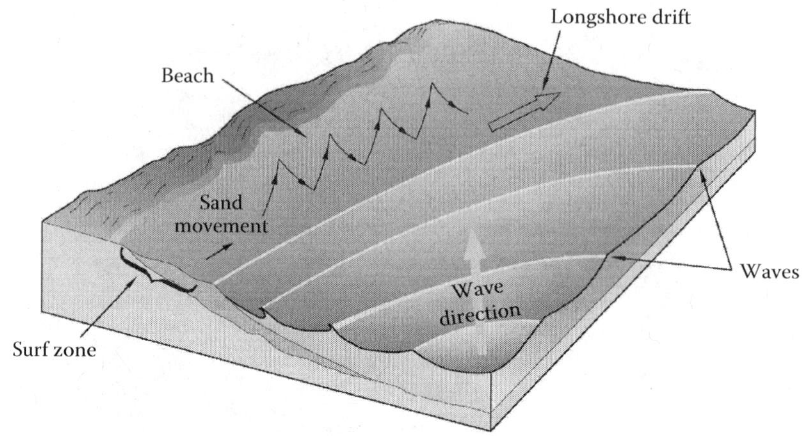

FIGURE 2.14
Processes of beach erosion by longshore current or drift.

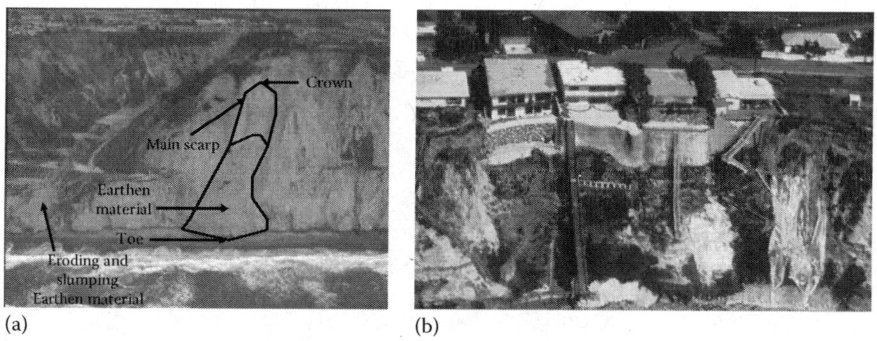

FIGURE 2.15
(a) Coastal erosion near San Francisco, CA. (USGS 1998) U.S. Department of the Interior, U.S. Geological Survey, Center for Coastal Geology http://coastal.er.usgs.gov/cgi-bin/response.pl?site=cc&loc=9. (b) Homes in coastal California in jeopardy of falling into the Pacific Ocean. (Photo by Richard McCarthy.)

The sand along a coastline is much easier for the waves to erode than rock. Along some beaches, there is an annual cycle where sand is removed from a beach one season, only to be returned by waves during the successive season. In addition, a steady or seasonal wind direction at an oblique angle toward the shoreline produces waves and crash along the beach at an angle. The release of energy produces a **longshore current**. Longshore currents are responsible for erosion and migration of beach sediments along the beach front from the action of waves. When a longshore current is present, sediments in the surf zone move parallel to the beach in the direction of the wind.

Figure 2.14 shows the processes of beach erosion. Seasonal winds drive the longshore current and transport large quantities of sand along a beach.

In the United States, coastal erosion is significant since roughly 15% of the land area in the United States is coastal and 40% of the population lives near the coastline. This distribution of population places many of our urban environments under significantly more stress than other less urbanized locations and is illustrated in Figure 2.15a and b.

Within coastal zones, some anthropogenic activities influence—and in some cases—make coastal erosion more severe (NOAA 2009):

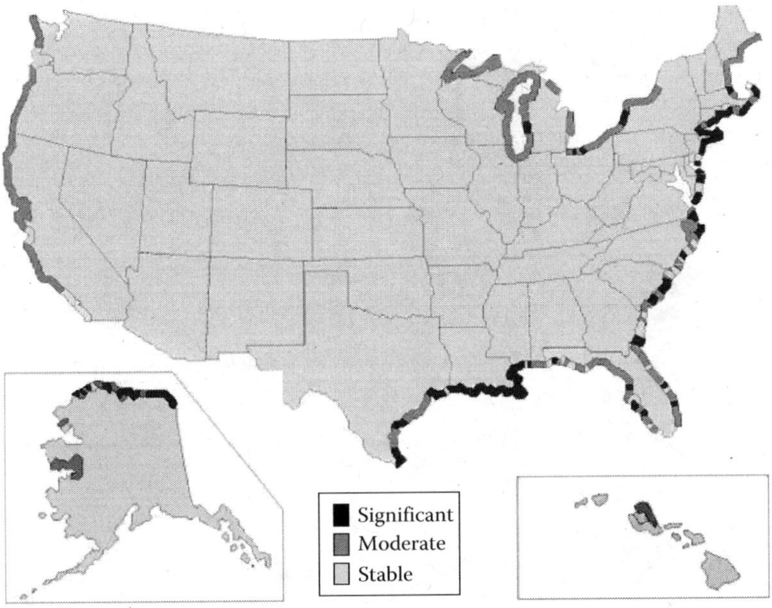

FIGURE 2.16
Severity of coastal erosion in the United States. (From United States Global Change Research Program, Climate change impacts on the United States: The potential consequences of climate variability and change, Overview: Coastal Areas and Marine Resources, http://www.usgcrp.gov/usgcrp/Library/nationalassessment/overview-coastal.htm, (accessed July 4, 2010), 2000.)

- Dredging channels may change the bathymetry offshore, leading to an increase in the energy of waves striking the shoreline
- Mining beach sand
- Construction of sea walls and spits
- Rising sea levels caused by global warming

Figure 2.16 is a map of the United States showing the significance of coastal erosion. The generally higher severity of erosion along the eastern and southern coastlines can be attributed to sandy beaches; the western beaches tend to have higher rock content.

2.3.1.3.2 Groundwater Erosion

A significant type of erosion often overlooked results from the movement of groundwater. Groundwater erosion frequently occurs in areas of significant deposits of non-clastic chemical sedimentary rocks such as limestone, dolostone, and evaporite deposits.

Groundwater erosion can eventually lead to a type of landform called karst topography. **Karst topography** is a landscape created by groundwater slowly dissolving sedimentary rock such as limestone. The resulting landforms from this type of erosion include sinkholes and caves. A **sinkhole** is a rounded depression of the ground and typically forms when the surface rock and soil above a cave collapses. A **cave** is a natural feature beneath the surface formed by groundwater dissolving away the rock. At many locations throughout the United States, extensive cave systems have formed (Davies and Legrand 1972). Figure 2.17 shows the most common features of a karst area, and Figure 2.18 shows areas of the United States with sedimentary-type deposits prone to karst feature formation.

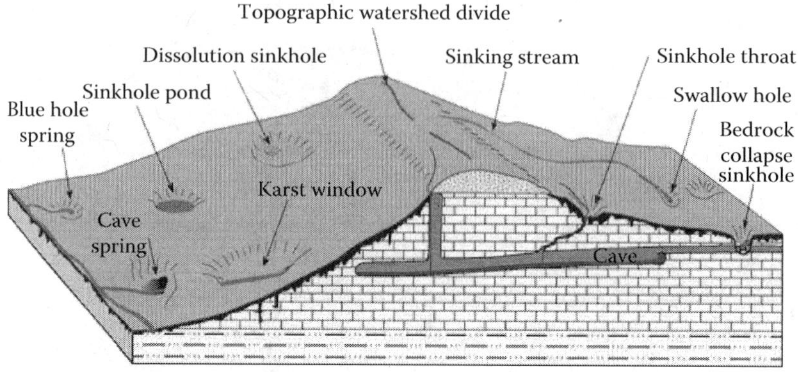

FIGURE 2.17
Common features of karst topography. (From Kentucky Geological Survey, Karst land in Kentucky, http://www.uky.edu/KGS/water/general/karst/index.htm, (accessed August 29, 2009), 2009.)

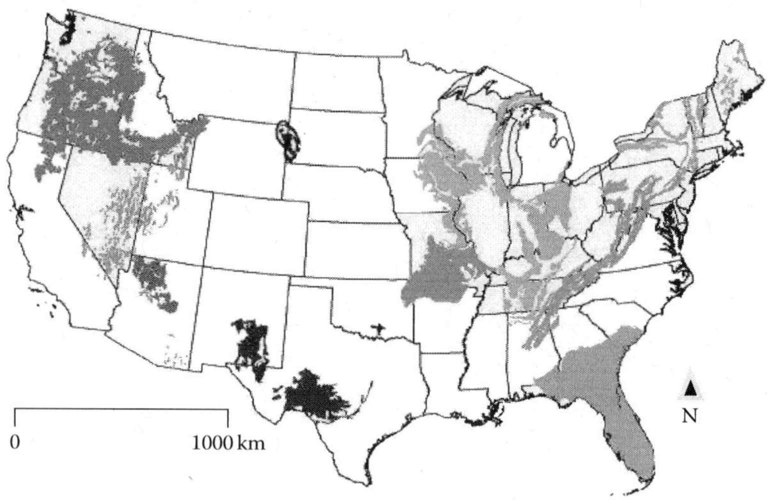

FIGURE 2.18
Extent of karst in the continental United States. (Differences in shading represent different regions and formations) (From United States Geological Survey, Eolian history of the United States, http://esp.cr.usgs.gov/info/eolian/task1.html (accessed February 26, 2010), 2009c.)

2.3.2 Non-Clastic Chemical Sedimentary Processes

Non-clastic sedimentary processes involve the precipitation of minerals from water by inorganic or biological processes. Limestone, dolostone, chert, and evaporites are the most common sedimentary rocks formed. Proper identification of sedimentary deposits from chemical processes requires the identification of the primary mineral.

As noted by Deer et al. (1966), common minerals of chemical sedimentary rocks include

- Limestone—composed of calcium carbonate ($CaCO_3$) and predominantly forms in a marine environment from the accumulation of shells, shell fragments, or from the direct precipitation from water (usually seawater).
- Dolostone—composed of magnesium–calcium carbonate ($CaMg[CO_3]_2$) and associated with a marine-type environment.

- Chert—composed of silica (SiO_2) and may form from microscopic organisms that secrete silica shells around their bodies or as a secondary precipitant.
- Halite—a mineral composed of sodium chloride (NaCl), and is formed from the evaporation of concentrated seawater in a restricted basin.
- Gypsum—a sulfate mineral ($CaSO_4 \cdot 2H_2O$), and is an evaporite deposit formed in a similar manner as halite.

2.3.2.1 Limestone and Dolostone

Limestone and dolostone are the dominant chemical and biochemical sedimentary deposits formed in marine environments. Limestone found on the continental landmasses was deposited in shallow seas that transgressed and regressed across the continents during times of rising and falling sea levels. Limestone may also form as reef deposits, such as the 440 million year old Niagara Escarpment that surrounds parts of the Great Lakes Basin. In general, significant limestone deposition requires a marine environment with three characteristics: clear water, warm water, and shallow water (Bathhurst 1975; Wilson 1975). It should be noted that karst is a landscape formed by groundwater solution—it is not a deposit.

Deposition of limestone sediments now occurs within the equatorial belt of warm ocean currents between latitudes of 30° north and south, but deposits as far as 40° north and south have been observed (Wilson 1975). As noted above, warm tropical waters are only one of three criteria for limestone deposition—clear water is also required, particularly for limestone reefs. As a result, significant limestone production does not occur in environments located near significant sources of clastic sediments such as a river delta (e.g., the Mississippi) because significant clastic sediment influx tends to choke limestone sedimentation. In addition, severe storms such as hurricanes may interrupt limestone sediment formation, or in some cases destroy the entire production of limestone sediment in local and regional areas. Large storms and hurricanes have much higher wave action and potentially stronger water currents. The waves from these events erode and destroy portions of near-surface reefs and also transport significant amounts of suspended sediments. As the water turbidity increases from these suspended sediments, interruption of limestone formation may occur. With respect to limestone formation in shallow water, the controlling factors are the availability of sunlight and warmer water located near the surface. Limestone will not form in the deeper parts of the ocean because $CaCO_3$ dissolves in cold water.

Throughout geologic time, limestone production has been significant. Plate tectonics coupled with the rise and fall of sea levels throughout the Paleozoic and Mesozoic have resulted in thick sequences of limestone deposits throughout the world, including a significant portion in the United States as a result of the transgressing seas. In fact, more than 10% of bedrock in the United States is composed of limestone, with significant deposits located within the eastern United States (USGS 2003).

Formation of dolostone typically occurs as a slow replacement process and generally follows limestone deposition (Wilson 1975); although primary dolomite formation has also been documented. About half of the calcium in the $CaCO_3$ molecule is replaced with magnesium forming dolostone ($CaMg[CO_3]_2$). Because of the addition of magnesium in its molecular structure, dolostone is more resistant to chemical erosion than limestone.

2.3.2.2 Evaporites

An **evaporite** forms as a result of the evaporation of water (Blatt and Tracey 1996). As the list at the beginning of the section shows, common evaporite deposits include the minerals halite and gypsum. During the evaporation of water, the concentration of constituents dissolved in the water increases until the time when the minerals begin to precipitate. Complete evaporation leaves behind mineral deposits such as halite and gypsum. Evaporite deposits generally form when water accumulates in closed basins in arid and semi-arid environments (Section 2.5).

2.3.3 Formation of Sediments

Once bedrock is weathered to produce sediment, it is transported by the agents of erosion, particularly water, and also by wind and ice. During transport, the sediment will be sorted—a process that reduces the grain size and causes them to become progressively more rounded. Minerals such as feldspar and micas will weather to clay minerals and may be separated from the rest of the sediment at the margins of lakes and oceans. Once sediment is deposited, it may be composed of tightly compacted individual mineral grains of varying size, composition, thickness, and geometric shapes. Sediments form in three basic sedimentary environments:

1. Continental or terrestrial environments—all those occurring on the land surface
2. Marine—deposited by or within the ocean
3. Transitional environmental—the boundary between continental or terrestrial and marine environments

Common examples of the geological processes forming sediments are listed in Table 2.1.

TABLE 2.1

Common Sedimentary Geological Processes That Form Unconsolidated Sediments

Location	Type of Geological Process
Continental	• Fluvial—deposited by a river such as a sand bar or floodplain deposits • Lacustrine—deposited by a lake • Glacial—deposited by a glacier • Alluvial fan—deposited by water flowing from a steep slope onto a level plain • Certain volcanic-related deposits—deposited by mud and debris flows caused by a volcanic eruption • Ash—deposited from a volcanic eruption • Eolian—deposited by the wind • Dune—deposited by the wind • Landslides—generally forms under the force of gravity • Rockslides and rockfalls—generally form on steep unvegetated slopes in mountainous regions • Soil creep—forms on slopes and moves very slowly • Mudflows—deposited by water typically during or shortly after a severe rain or flood event; may be associated with a sloped surface
Marine	• Reef—typically non-clastic chemical sedimentary deposits such as limestone • Layers of fine-grained sediments deposited from settling • Submarine landslides and slumping
Transitional	• Delta—deposited by a river discharging to a lake or ocean • Beach—deposited along the shoreline

2.3.4 Rock Formation: Diagenesis

After sediments are formed, the processes of diagenesis turn the sediment into rock. **Diagenesis** describes all the physical, chemical, or biological changes sediment undergoes after it has been deposited.

The types of diagenesis a sedimentary deposit may experience include (Pettijohn 1975)

- Compaction
- Deformation
- Dissolution
- Cementation
- Authigenesis
- Replacement
- Recrystallization
- Hydration
- Bacterial action
- Concretion development

The most important diagenetic processes acting upon a sedimentary deposit are compaction and cementation (lithification). **Compaction** occurs over time within a sedimentary deposit where individual grains are rearranged to form a more tightly packed sediment. The process of compaction forces out water that may be trapped in pore spaces between mineral grains. Compaction is a function of burial depth, and is the first step to a sediment's journey of becoming a rock.

Cementation involves the precipitation of minerals at grain-to-grain contacts and within the deposit's pore spaces. Only after cementation has occurred can a sedimentary deposit be considered a rock. Although compaction may have significantly reduced the amount of pore space within a sedimentary deposit, some pore space usually remains. Fluids containing minerals migrate through the remaining pore spaces and precipitate cementing materials within the remaining pore spaces and finalize the conversion of a sediment to a rock. The most common cements in sedimentary deposits are calcium carbonate ($CaCO_3$), silica (SiO_2), and iron oxides (Fe_xO_x) (Flint and Skinner 1974).

2.4 Identification and Classification of Sedimentary Deposits

To this point, sedimentary rocks have been subjected to significant stress. They have been torn from their original home, broken apart and perhaps changed chemically, and then put into nature's equivalent of the trash compactor and cement mixer. We now have them right where we want them—in an observable form near the Earth's surface—so we can proceed with their identification and classification. The knowledge gained from these procedures allows the investigator in urban areas to identify sedimentary deposits of potential commercial value; including the sand and gravel used in construction and consumer products, and the clay used in the manufacture of bricks. The most critical knowledge, however, gained from these procedures is the understanding of how contaminants behave and migrate in these geological environments. We cover this topic beginning with Chapter 7.

2.4.1 Identification and Classification Methods

The successful identification and classification of sedimentary deposits requires an evaluation of several features, including the degree of sorting, grain shape, grain size, composition, thickness, and other characteristics.

2.4.1.1 Sorting

Sorting is the spatial arrangement of particles during their transport and deposition by size. Sedimentary deposits range from very well sorted to very poorly sorted. The degree of sorting in a sedimentary deposit or layer is controlled by the method of transport and deposition. Sedimentary layers formed from wind transport are generally considered well sorted because the wind can only transport fine-grained materials. Sedimentary deposits formed from the action of glaciers are generally poorly sorted because glaciers are not selective—they move everything.

Figure 2.19 shows gradational examples of individual grains ranging from well rounded to angular. Figure 2.20 shows an example of a poorly sorted material.

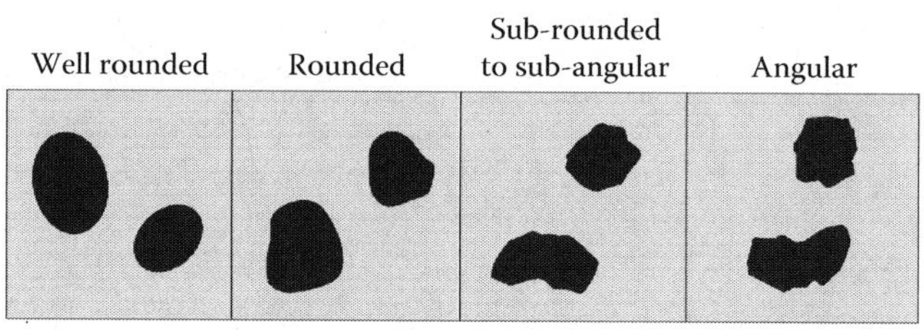

FIGURE 2.19
Examples of well-sorted, well-rounded, and angular sedimentary materials.

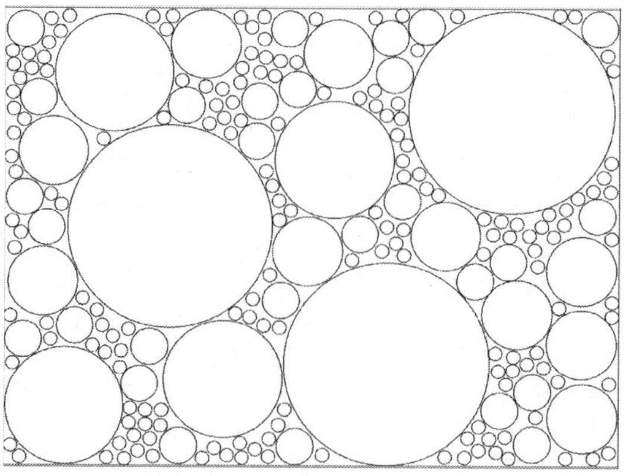

FIGURE 2.20
Example of a poorly sorted sedimentary material.

Individual grains or particles can be sorted by grain size—where the largest grains tend to settle first, and by composition; where those grains with the highest specific gravity minerals tend to settle first.

2.4.1.2 Grain Shape

Individual grains also vary in shape. The shapes of individual grains are characterized by their degree of roundness; the degree of edge and corner removal; and their sphericity, the degree to which the grain approaches the shape of a sphere. Roundness and sphericity vary significantly and depend upon several factors, including composition, the method of transport, and their distance traveled.

In general, roundness and sphericity increase the farther the material is transported before it is deposited. Knowing (1) the shape of the grains and (2) that the largest grains and those with the highest specific gravity settle out first during their deposition enables geologists to make inferences about these deposits. For example, when examining a sedimentary deposit, it may be possible to determine the relative distance sediment was transported, potential energy of the transport system, and method of transport.

2.4.1.3 Grain Size

Individual grains are also classified according to size and given specific names depending on the grain size. For instance, a grain that has a size between 1/16 and 2 mm is classified as a sand grain and a grain having a size between 4 and 64 mm is termed a pebble. The Wentworth (1922) scale is the common method for classifying grain size, and the ranges are listed in Table 2.2.

2.4.1.4 Composition

The grains in a sedimentary deposit tend to be resistant to mechanical or chemical weathering. The composition of sand deposits is highly variable and depends upon the source of the material. One of the most resistant minerals common in sedimentary deposits is quartz, a dominant mineral of beach, river, and dune deposits (Boggs 2000). The quartz content in modern-day sand deposits from major rivers and beaches in the United States ranges between 67% and 99%, with the second most common mineral being feldspar. By contrast, in tropical regions, a dominant component in sand deposits is limestone. Sand deposits comprise approximately 25% of all clastic sedimentary deposits (Pettijohn 1975). Figure 2.21 is an example of a quartz sand that is medium grain, well-sorted, with sub- to well-rounded grains.

Composition is also used to help evaluate the degree of sorting in many sedimentary deposits. Clay minerals are the predominate minerals constituting clay deposits, followed by quartz, feldspar, carbonates, and iron oxide minerals, respectively (Pettijohn 1975; Hillier 2003). Clay and silt deposits tend to be more erodible than quartz because they contain higher amounts of organic material in addition to their mineral components of micas and feldspars. Types of clay minerals include kaolinite; montmorillonite-smectite; illite; and chlorite.

Clay minerals typically form over long periods of time from the gradual chemical weathering of rocks, usually silicate-bearing, by low concentrations of carbonic acid and other diluted solvents. The formation of carbonic acid involves the reaction between water, such as rain and carbon dioxide in the atmosphere. As rain falls through the atmosphere,

TABLE 2.2
Grain Size Classification

Wentworth Size Class	Millimeters	Inches
Boulder	256 or greater	10.08 or greater
Cobble	64–256	2.52–10.08
Pebble	4–64	0.16–2.52
Very coarse gravel	32–64	1.26–2.52
Coarse gravel	16–32	0.63–1.26
Medium gravel	8–16	0.31–0.63
Fine gravel	4–8	0.16–0.31
Granule (very fine gravel)	2–3	0.08–0.16
Very coarse sand	1–2	0.04–0.08
Coarse sand	0.5–1	0.02–0.04
Medium sand	0.25–0.5	0.01–0.02
Fine sand	0.125–0.25	0.005–0.01
Very fine sand	0.0625–0.125	0.002–0.005
Silt	0.004–0.0625	0.0002–0.002
Coarse silt	0.031–0.0625	0.001–0.005
Medium silt	0.0156–0.031	0.0005–0.001
Fine silt	0.0078–0.0156	0.0003–0.0005
Very fine silt	0.0039–0.0078	0.0002–0.0003
Clay	Less than 00039	Less than 0.0002

Source: New Jersey Department of Environmental Protection (NJDEP). 2005. *Field Sampling Procedures Manual.* Trenton, NJ: NJDEP.

FIGURE 2.21
Example of a sub- to well-rounded, well-sorted, medium-grained, quartz sand. The light colored mineral grains are quartz. (Photo by Daniel T. Rogers.)

rainwater dissolves small quantities of carbon dioxide, so when the rain reaches the ground, it is a weak carbonic acid. As water migrates through the soils and sediments, the strength of the acid may increase significantly through the addition of carbon dioxide created by the bacterial decay of vegetation. Once carbonic acid forms, it ionizes to form hydrogen and bicarbonate ions. Figure 2.22 shows the chemical reactions leading to the formation of carbonic acid, hydrogen, and bicarbonate ions.

$$\underset{\text{(Water)}}{H_2O} + \underset{\text{(Carbon dioxide)}}{CO_2} \longrightarrow \underset{\text{(Carbonic acid)}}{H_2CO_3} \longrightarrow \underset{\text{(Hydrogen ion)}}{H^{+1}} + \underset{\text{(Bicarbonate ion)}}{(H_2CO_3)^-}$$

FIGURE 2.22
Formation of carbonic acid and hydrogen and bicarbonate ions.

$$\underset{\text{(Potassium feldspar)}}{2KAlSi_3O_8} + \underset{\text{(Hydrogen ions)}}{2H^{+1}} + \underset{\text{(Water)}}{H_2O} \longrightarrow \underset{\text{(Potassium ions)}}{2K^{+1}} + \underset{\text{(Kaolinite)}}{Al_2Si2O_5(OH)_4} + \underset{\text{(Silica)}}{4SiO_2}$$

FIGURE 2.23
Formation of kaolinite through hydrolysis of potassium feldspar.

Hydrogen ions are extremely effective in decomposing minerals through a process of hydrolysis. **Hydrolysis** is a chemical process in which a molecule is cleaved into two parts by the addition of a molecule of water (Greenwood and Earnshaw 1997). The formation of the clay mineral kaolinite from potassium feldspar is shown in the hydrolysis reaction in Figure 2.23.

Clay deposits may be formed in place but usually are eroded and transported from their original location to generally low-energy depositional environments such as large lakes and marine deposits.

2.4.1.5 Thickness

Thicknesses of unconsolidated sedimentary deposits can vary enormously over short distances. Just a few feet of separation can result in significant variation. The thickness of unconsolidated sediments beneath many urban regions ranges from merely a few meters to more than 305 m (1000 ft) (USGS 2009). The deposition of these materials forms a complex interconnected web that may drastically change both vertically and horizontally. It is not uncommon to observe changes in sediment type and origin of formation in only a few centimeters in the vertical or horizontal directions. Figure 2.24 is an example of how sediment layers can change over very short distances in both grain size and composition.

2.4.2 Bedded or Stratified Depositional Features

Depositional features provide clues about the life cycle of sedimentary deposits (Wicander and Monroe 2007), and provide valuable information about the environment where the deposits are formed. Some of the more commonly observed features include

- Bedding or stratification—defined as the layering of sediments as they are formed and deposited (Wicander and Monroe 2007). Layering of sediments is the result of vertical differences in composition, texture, and grain fabric. The most widespread deposition structure of sedimentary deposits is their bedding or stratification. Bedding layers may be very thin, on the order of less than a millimeter, to very thick, on the order of several meters. One variety of bedding layers is termed **laminae**, which are defined as very thinly bedded layers on the order of a millimeter (Potter et al. 1980). Careful inspection of stratified sediments must be conducted because compaction may enhance and distort bedding irregularities of hydraulic origin. There are several different types of stratification, including (Potter et al. 1980; Pettijohn et al. 1987)

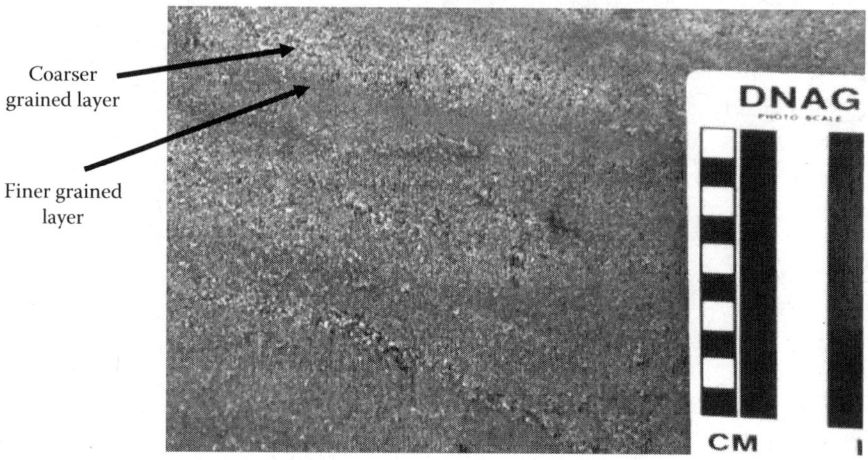

FIGURE 2.24
An example of sediment layer variability over short distances. (Photo by Daniel T. Rogers.)

- Even parallel
- Even discontinuous parallel
- Wavy parallel
- Wavy nonparallel
- Discontinuous wavy parallel
- Discontinuous wavy nonparallel
- Discontinuous curved parallel
- Curved nonparallel
- Structureless
- Mottled
- Brecciated
- Algal
- Graded
- Cross

Parallel horizontal bedding is characterized by distinct beds or laminae, with either sharp or gradational boundaries that are continuous and parallel. This type of stratification is the most common in fine-grained sediments such as silts and clays. Figure 2.24 is an example of this type of bedding. *Discontinuous, but even parallel stratification* is closely related and represents a stratification of slightly less uniformity.

Wavy stratification may be parallel, nonparallel, or discontinuous and is predominantly formed by small ripples of silt and sand. When these particles are interbedded with fine-grained material, they are called **flaser** bedding. Wavy bedding of all types is an indication of higher energy and implies a small but discontinuous micro-relief on the bottom where the sediment accumulates. *Discontinuous wavy, nonparallel bedding* is a less orderly arrangement and probably represents a higher level of current flow. *Curved parallel or curved nonparallel bedding* are locally present in many fine-grained sediments such as clays, and probably indicates lateral accretion in channel scours.

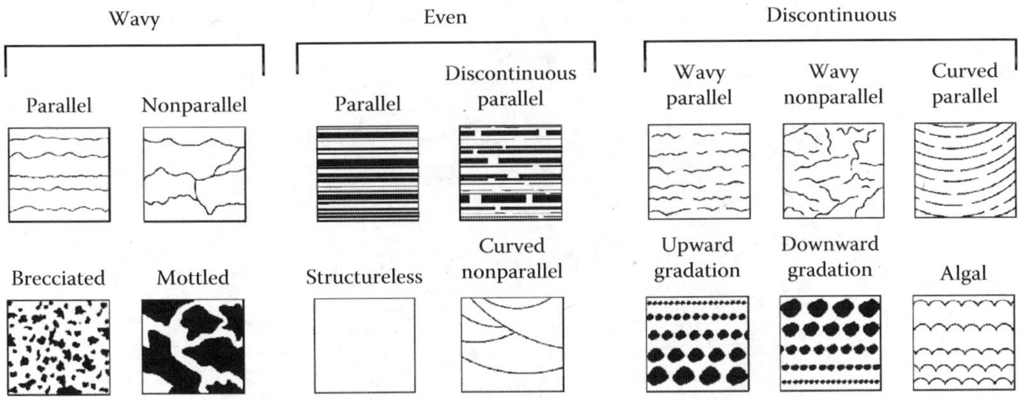

FIGURE 2.25
Stratification types in fine-grained sediments.

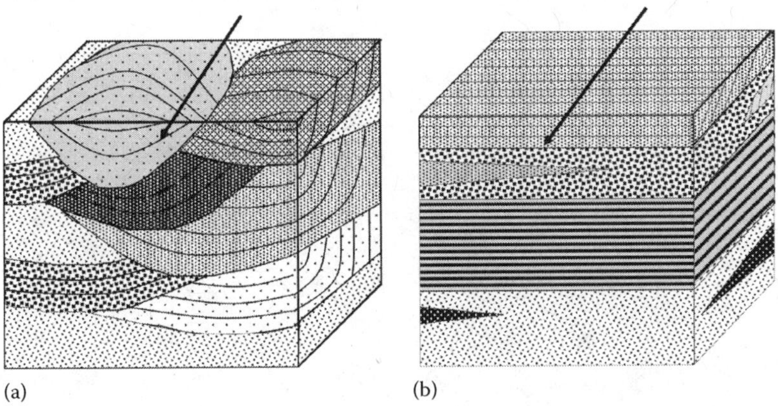

FIGURE 2.26
Planar tabular (a) and trough or festoon (b) cross-bedding.

Structureless bedding represents rapid sedimentation with little or no change or variation in the source material. Figure 2.25 shows many types of stratification in fine-grained sediments.

Cross-bedding are sets of bedded sediments at various, on-lapping, and inclined angles. Cross-bedding indicates changing directions and perhaps energy in the direction of origin of the sediments being deposited. In sedimentary deposits, cross-bedding typically indicates the deposit in question was affected by the forces of wind or changing water currents. Cross-bedding occurs most often within coarse-grained sediments such as sand. Figure 2.26 is a diagram showing the concept of cross-bedding (a is planar tabular and b is trough or festoon cross-bedding). Figure 2.27 is a photograph of a sand deposit exhibiting a festoon type of cross-bedding, and Figure 2.28 is a photograph of a sand deposit exhibiting a planar tabular type of cross-bedding.

2.4.3 Other Depositional Features

Other depositional features may include remnants or identification marks useful for identifying the environment and method creating the deposit. In addition to bedded and stratified features, these other depositional features provide significant clues to geologists

FIGURE 2.27
Sand deposit with festoon-type cross-bedding in Michigan. Hard hat in upper left for scale. (Photo by Daniel T. Rogers.)

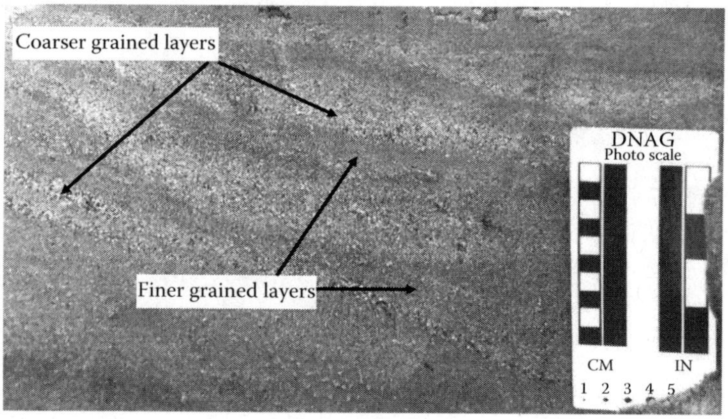

FIGURE 2.28
Sand deposit with planar tabular type of cross-bedding. (Photo by Daniel T. Rogers.)

investigating urban sites and help them distinguish whether the deposits being examined originated from a natural process or formed by anthropogenic activity.

- Ripple marks—formed within wind-blown dune deposits or along beach deposits generally in shallow water. Ripple marks are very small dune-like deposits whose long axes are perpendicular to the air or water current forming them. Ripple marks commonly form either as asymmetrical ripple marks or as oscillation ripple marks. Asymmetrical ripple marks form in shallow water or as dune deposits where there is a dominant direction of water current or wind. Oscillation ripple marks form in shallow water within the surf zone where there are competing and opposite directions of water flow.
- Mudcracks—are also commonly termed desiccation cracks. These features are formed when a fine-grained deposit—often clay-sized—undergoes drying and then is buried, preserving the feature.

- Raindrop prints—are small circular impressions formed from raindrops hitting the soft surface of exposed sediment.
- Fossils—observing fossils within a sedimentary deposit offer geologists many insights as to how the deposit formed, the age of the deposit, and many other factors. When placed in a stratigraphic sequence with deposits from both above and below, fossils are used to reconstruct the history of the region. For instance, observing fossils or remnants of fish indicate the deposit formed under water, and observing terrestrial fossils, such as a Mammoth or Mastodon indicates the land surface was nearby.
- Bioturbation and tracks—bioturbation is the physical rearrangement of the soil profile by soil organisms. The resulting features are also described as trace fossils and can range in scale from worm burrows to animal tracks.
- Varves—form in a variety of marine and lacustrine depositional environments from seasonal variations in clastic, biological, and chemical layering of sediments. Varves are a special variety of stratification. The classic varve deposit appears as alternative light and dark layering of sediment in a glacial lake. The light layers usually are composed of silt and fine-grained sand, and are higher energy deposits because they traveled farther, originating as glacial meltwater. These light layers are also typically coarse-grained than the darker clay layers. In lakes, the layering pattern is a function of their seasonal freezing and thawing, and/or the seasonal fluctuations in sediment loading.

2.5 Sedimentary Depositional Environments

Sedimentary depositional environments represent a combination of physical, chemical, and biological processes associated with the deposition of a specific type of sediment. Grain size, shape, sorting, composition, thickness, anthropogenic disturbance, natural disturbance, unconformities, and other features present clues as to how, where, and when both clastic and chemical sedimentary deposits formed. Since sedimentary deposits form at or very near the surface of the ground, they also communicate to geologists what the climatic conditions were at the time of deposition (Wicander and Monroe 2007). This demonstrates the interdisciplinary nature of geology since a significant portion of the formation and preservation of sedimentary deposits is atmospheric and weather-related. Figure 2.29 shows several sedimentary depositional environments.

An example of a depositional structure containing direct evidence about the environment of its deposition is an extensive cross-bedded sand deposit, indicating there were either dunes or a beach nearby. Shell fragments embedded in this sand would further indicate the sand was deposited in shallow water. This pattern may have significant implications because it could signal the historical climatic shifts and changes, and perhaps present clues related to atmospheric-related hazards (e.g., floods) affecting the site and its surrounding area. Table 2.3 lists the dominant types of clastic sedimentary deposits and corresponding depositional environments in which they form.

Most urban areas are not located directly upon bedrock. Urban landscapes tend to be located in continental sedimentary depositional environments including alluvial, beach, eolian, fluvial, glacial, lacustrine, and volcanic-type deposits. Many cities are also located

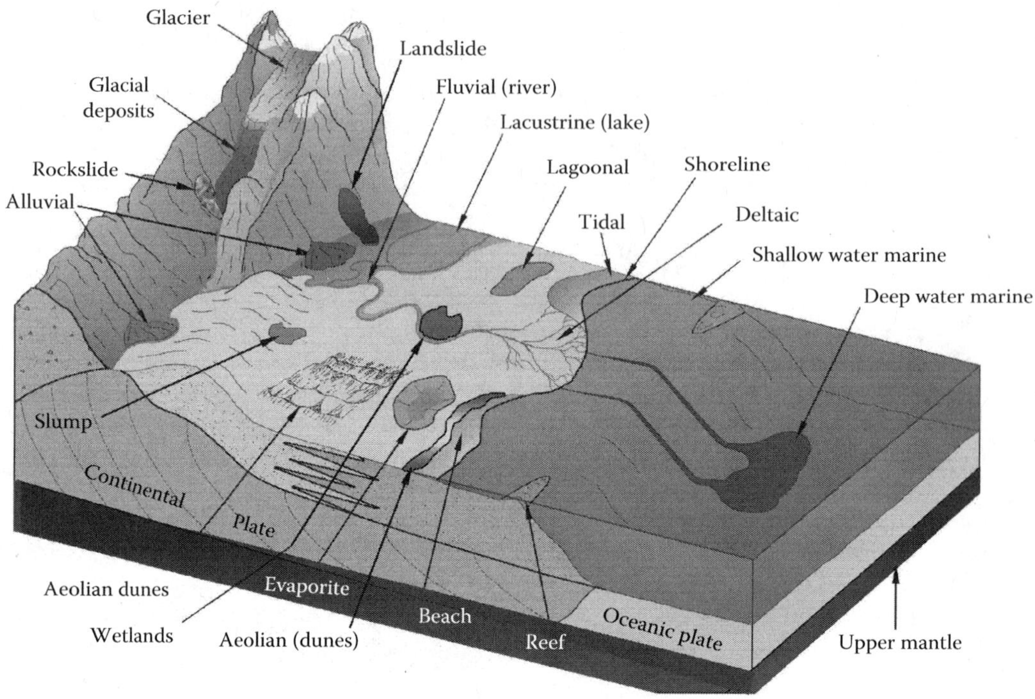

FIGURE 2.29
(See color insert.) Sedimentary depositional environments.

on sedimentary deposits of marine origin. Table 2.4 lists 29 urban areas in the United States and identifies the dominant geology beneath each area by type of depositional environment. All of the sedimentary depositional environments are represented.

2.5.1 Alluvial

Alluvial deposits are formed by fast-flowing stream water. They range in size from fine silt to boulders, and also include debris flows and rock slides from adjacent mountain slopes. One example of an alluvial deposit is an alluvial fan, which is a flat to gently sloping mass of sediment and loose rock material shaped like a fan. These deposits are created at the base of a mountain slope where fast-moving steams coming out of the mountains meet the relatively flat surfaces of a basin floor or broad valley. At this junction, the stream gradient is significantly decreased, which removes much of the energy available for sediment transport by the flowing water (Bloom 2004). The result is the deposition of sediment in a fan shape as depicted in Figure 2.30. Alluvial fans are common in the western United States.

2.5.2 Fluvial

Fluvial deposits are sediments deposited by a river or stream. Deposition results from fast- and slow-moving water currents on channel beds and bars. Low-velocity deposition occurs on floodplains during floods. Here, the deposits are typically fine-grained silts and clays, and may extend over large areas. According to Walker and Coleman (1987), the Atlantic and Gulf coastal plains, in large part, owe their origins to fluvial deposition over

TABLE 2.3

General Depositional Environments of Select Clastic Sedimentary Deposits

Location	Depositional Environment	Sedimentary Process	Type of Deposit
Continental	Alluvial Fan	Stream deposits from fast-flowing water. Also includes debris flows and rock slides from adjacent mountain slopes	Conglomerate and coarse sands that are poorly to moderately sorted
	Rivers (fluvial)	Deposition from fast- and slow-moving water current on channel beds and bars. Low-velocity deposition on floodplains during overbank flow generally during floods	Sand, moderately to well-sorted and potentially cross-bedded in higher energy type deposits, to clay and silt in floodplain deposits
	Deserts (eolian or aeolian)	Sediment transport by wind (eolian or sometimes referred to as aeolian). Leads to formation and migration of dune deposits	Sand, very well-sorted and rounded, quartz-rich
	Glaciers	Wide-range of deposits from unsorted till in moraines to well sorted, coarse- to fine-grained deposits near unique glacial type deposits termed kames and kettles	Fine-grained conglomerate-type deposits; poorly sorted to moderately well-sorted kame deposits
	Lakes (lacustrine)	Deposition of fine-grained sediments in deep water to coarse-grained in beach deposits. Also evaporation of lake water causing chemical precipitation, especially in arid environments	Finely bedded silt and clay deposits indicating deeper water formation to coarse-grained sand indicating beach deposits
	Wetland	Mainly composed of decayed vegetative material and very fine-grained sediments	Buried peat; when buried and exposed to heat and pressure may develop into coal
Transitional	Deltas	Deposition of fast- to slow-moving water at the mouth of a river or stream entering a large body of water. Re-working of sediments by wave action. Deposition of organic debris in swamps	Sand moderately- to well-sorted to silt and clay. May be fossiliferous and may also contain layers of peat
	Beach	Reworking of sediment layers by waves. May also include formation of dunes	Sand fine- to coarse-grained, moderately- to well-sorted, quartz-rich
	Tidal flats	High- to low-velocity flow in channels and may also have sheet flow. Periodic wetting and drying	Sand to clay deposits, fine-grained and well sorted, may be organic and fossiliferous, mudcracks and trace fossils
Marine	Reefs and Lagoons	High to low energy flow through submarine locations and deposition of increasingly finer material with waning flow. May exhibit submarine disturbance indicative of sliding and debris flows	Conglomerate, sand, to clay. Moderately well-sorted, typically exhibits graded bedding (coarse- to fine-grained upward sequence). Also unsorted sliding or debris flow deposits
	Deeper water	Abyssal plain-type deposits that typically exhibit very fine non-descript dark muddy-type deposits	Very fine-grained clay

TABLE 2.4

Major U.S. Urban Centers and Their Sedimentary Environments

	Geologic Environment								
Urban Area	**Alluvial**	**Beach**	**Eolian**	**Fluvial**	**Glacial**	**Lacustrine**	**Marine**	**Volcanic**	**Bedrock**
Atlanta[1]	X			X		X			
Baltimore[2]	X			X			X		
Boston[3]	X			X	X		X		X
Chicago[4]		X		X	X	X			
Cincinnati[5]	X			X	X				X
Cleveland[5]		X		X	X	X			
Dallas[6,7]	X		X	X		X			X
Detroit[8, 9]		X		X	X	X			
Denver[10]	X		X	X	X				X
Houston[6,7]	X			X			X		
Indianapolis[30]				X	X	X			
Kansas City[11,12]	X			X	X				
Las Vegas[13]	X		X	X		X			X
Los Angeles[14]	X	X	X	X		X	X		X
Miami[15]	X			X			X		
Minneapolis[16]	X			X	X	X			X
New Orleans[17]		X		X		X	X		
New York[18]	X	X		X	X	X	X		X
Philadelphia[19, 20]		X		X		X	X		X
Phoenix[21, 22]	X		X	X		X			X
Pittsburgh[19]	X			X	X	X			X
Portland[23]	X			X	X	X	X	X	X
San Antonio[6,7]	X		X	X		X			X
Saint Louis[24]	X	X		X	X				X
Salt Lake City[25]	X	X		X	X	X			X
San Francisco[26]	X	X		X			X		X
Seattle[28]	X	X		X	X	X	X	X	X
Tampa[15, 27]	X	X		X		X	X		X
Washington, DC[29]	X			X		X			X

Sources: King and Beikman (1974) and others

[1]Abra ms (1984); [2]USGS (2009a); [3]Kaye (1976); [4]Alder (1902); [5]Coogan (1996); [6]USGS (2009b); [7]Spearing (1991); [8]Rogers (1996); [9]Rogers (1997); [10]Halka (1980); [11]Aber (1984); [12]Gentle (1984); [13] Page et al. (2005); [14]Bilodeau et al. (2007); [15]Florida Geological Survey (1994); [16]Ojakangas and Matsch (1982); [17]Snowden (1980); [18]USGS (2003); [19]Barnes and Sevon (1996); [20]Goodwin (1964); [21]Allen (2003); [22]Halka (1983); [23]Trimble (1963); [24]Missouri Department of Natural Resource, Geological Survey Division; [25]Stokes (1989); [26]Stoffer and Gordon (2001); [27]USGS (2005); [28]Stewart (1992); [29]USGS (2007); [30]Indiana Geological Survey.

the last 65 million or more years by the numerous rivers that drain off the Appalachian Mountains. Fluvial sediments occur as four main types (Leopold et al. 1992):

1. Channel-floor sediments deposited on the bottom of the river or stream; sizes range from fine sand to gravel depending on flow velocity.
2. Bar sediments deposited along the river and stream banks, especially along the inside portions of river or stream channel bends (Figure 2.31). Deposits are sand and gravel with some silt.

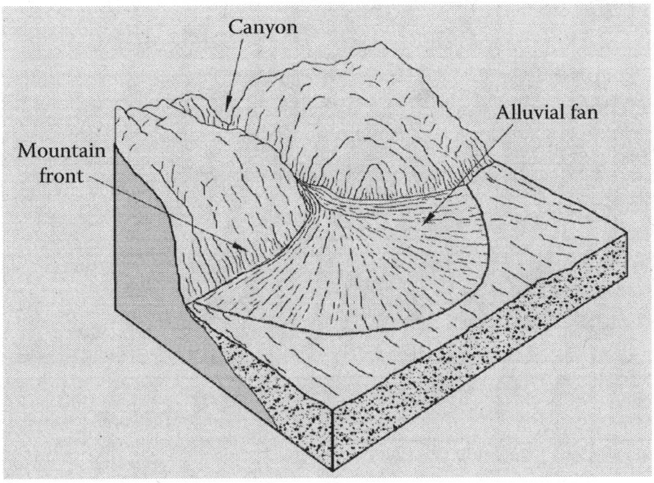

FIGURE 2.30
Alluvial fan.

FIGURE 2.31
Example of fluvial bar sediment formation in a fully mature stream. (Photo by Daniel T. Rogers.)

3. Channel-top and bar-top sediments deposited on top of the bar sediments are formed in the shallow water regions in channels and abandoned channels. These types of deposits are generally sand and silt-sized deposits.
4. Floodplain sediments deposited when the water level rises above the confines of the river or stream bank and overflow into the surrounding area. These deposits tend to be fine-grained and are composed of sand, silts, and clays.

2.5.3 Lacustrine

Lacustrine sediments are lake deposits characterized by the presence of a range of fine-grained sediments in deep water to coarse-grained sediments in shallow water near the shoreline. Large lakes, such as the Great Lakes, may have well-defined and significant beach deposits composed of sand and gravel. In some arid regions, such as the southwestern

FIGURE 2.32
Fine-grained lacustrine deposit. (Photo by Daniel T. Rogers.)

portion of the United States, evaporation of lake water leaves behind evaporite deposits. Fine-grained lacustrine deposits tend to be thinly layered or stratified. Figure 2.32 is an example of a thinly layered lacustrine deposit.

2.5.4 Eolian

Eolian (sometimes referred to as Aeolian) deposits result from the wind, and in the United States occur primarily in arid or semiarid climates such as the southwest because these areas often have a large supply of unconsolidated sediments. Many other regions, however, also exhibit eolian deposition generated during the last 2 million years. Some of the locations include parts of Nebraska, Kansas, Texas, South Dakota, Colorado, Oklahoma, Illinois, Iowa, sections of the east coast, southwest, and many other areas (Gustavson et al. 1991; Gibbard and Kolfschoten 2004). **Loess** is a typical wind-blown deposit consisting primarily of silt-sized fragments. Most loess deposits were derived originally from glacial till, and then transported by the wind.

2.5.5 Wetlands

A **wetland** is an area whose soil is saturated with water permanently or seasonally but is saturated long enough to support aquatic plants. Wetlands also include areas partially or totally covered by shallow pools of water. Examples of wetlands are swamps, marshes, and bogs. Deposits in wetlands are primarily composed of decaying vegetative matter. Sedimentary deposits derived from wetlands are typically peat deposits and some very fine-grained sediments such as clay (Mitsch et al. 2009). Extensive wetland deposits are located in the Midwest and portions of the southeastern United States.

2.5.6 Glacial

Glacial sediments are deposits associated directly or indirectly with glaciers (Benn and Evans 1999). Although most of us do not live near active glaciers, the northern half of North America experienced several episodes of glacial advance and retreat during the last 2 million years (Imbrie and Imbrie 1979; Wright 1989). In 1894, Thomas Chamberlin was the first to attempt to draw a map of North America depicting the extent of glaciation during the Pleistocene (Figure 2.33).

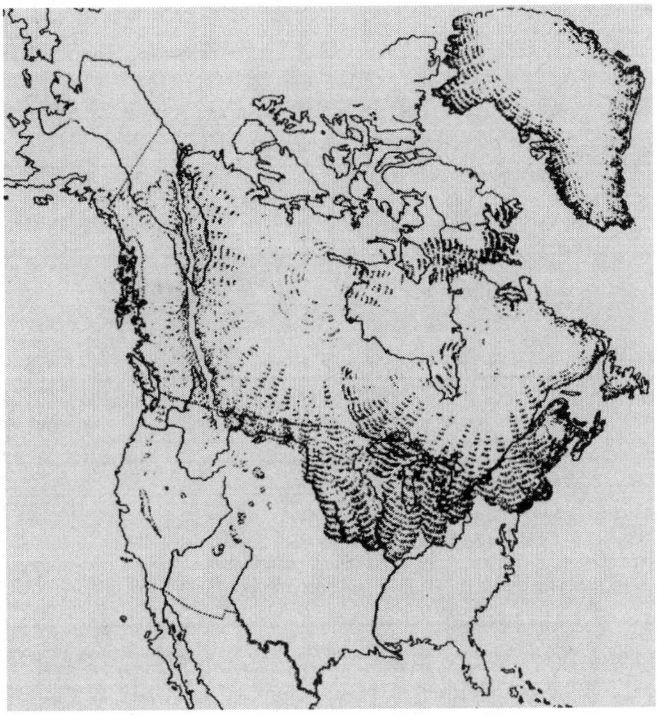

FIGURE 2.33
Extent of Pleistocene Age glaciers (From Chamberlin, T.C. and Salisbury, R.D., *A College Textbook of Geology*, The Lakeside Press, Chicago, IL, 1909.)

Figures 2.34a and b show the variety of glacial deposits as they are formed in the glacial sedimentary environment (modified from Farrand 1988).

The following discussion refers to Figures 2.34a and b and is based on the description of glacial deposits found in Bennett and Glasser (1996) and Benn and Evans (1999):

- Moraines—are formed by the deposition of material from a glacier and are exposed after the glacier has receded. These features usually are referred to as glacial till. **Glacial till** is material directly deposited from glacial ice, and usually appears as linear mounds of unsorted material, or as a very poorly sorted mixture of rock, gravel, and boulders within a matrix of a fine-grained silt- and clay-sized material. There are three main types of moraines:
  - Lateral moraine: formed on the sides of glaciers.
  - End moraine: formed at the end of a glacier or at locations where the glacier had been stagnant during a retreating or regressive phase.
  - Ground moraine: formed under the ice between the lateral moraine. Deposits of the ground moraine variety tend to be fine-grained than lateral or end moraines. Ground moraines may also be termed lodgment tills.
- Kames—Ice contact features shaped by glacial meltwater; these form conical-shaped masses of course-grained sand and gravel with little or no fine-grained material.

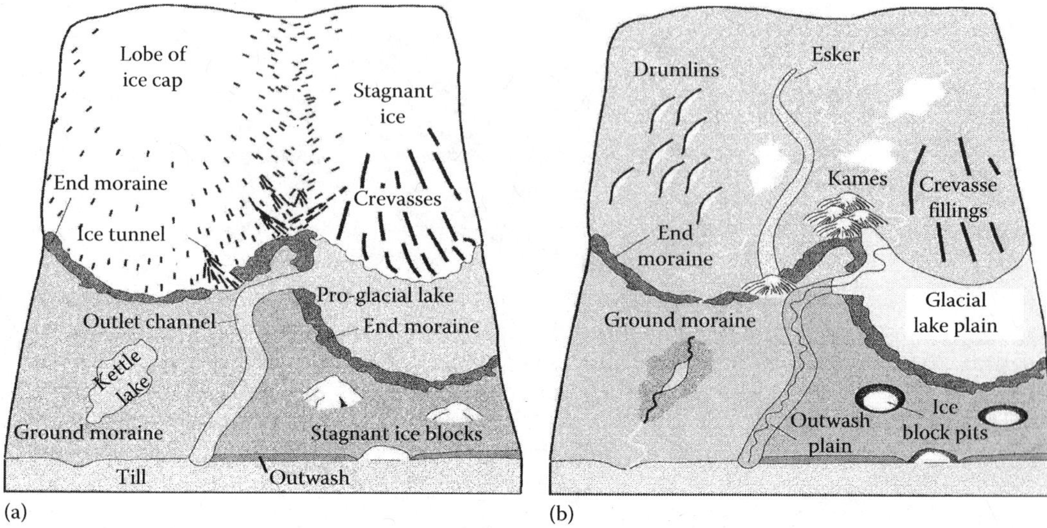

FIGURE 2.34
Glacial deposits formed in the glacial sedimentary environment.

- Kettles—Depressions formed when portions of ice from the glacial front become partially or wholly buried with glacial outwash. Groundwater fills the depressions to form kettle lakes. Many of the inland lakes in Michigan are kettle lakes.
- Eskers—Formed beneath glaciers in ice-walled tunnels of outwash; they typically deposit coarse-grained gravel deposits in an irregular pattern, deposited by streams flowing under the ice.
- Outwash—Glacial sediments deposited by the action of meltwater transporting glacial till away from the glacier front. Glacial meltwater streams from a glacial front may form a type of pattern called braided streams consisting of a maze of interconnected channels with high sediment loads.
- Drumlins—Elongated hills composed of glacial till; they are much steeper on one side than the other. The steeper slope indicates the direction of ice movement (Figure 2.34b).
- Glacial lakes—Many glacial lakes form between the ice front and end moraines of a retreating glacier. The end moraines form an effective barrier for glacial meltwater to form. Deposits in glacial lakes tend to be fine-grained and may contain varves.

Glacial lacustrine lakes or pro-glacial lakes as shown in Figure 2.34a were very common throughout North America during the Late Pleistocene (Benn and Evans 1998). Many urban areas such as Chicago, Detroit, Cleveland, and Buffalo are located along or near the shoreline of these glacial lakes (Farrand 1988; Rogers 1996; Benn and Evans 1998). Other notable large Pleistocene glacial lakes include Lake Bonneville, which occupied the area where Salt Lake City in Utah is located (Green and Currey 1988; Morrison 1991); Lake Missoula and others that were located in Montana, Idaho, and eastern Washington (Alt 2001); and Lake Agassiz in the northern United States and southern Canada (Imbrie and Imbrie 1979). Many of these Pleistocene glacial lakes were as large as the current Great Lakes and they form, or have influenced, many of the landscapes of urban areas in the northern United States.

2.5.7 Non-Clastic Chemical Depositional Environments

As noted earlier, chemical sedimentary deposits form as minerals precipitated from water through organic or inorganic processes. Identification of sedimentary deposits from chemical processes requires the identification of the primary mineral.

Common evaporite deposits include halite and gypsum. All water bodies on the surface and in aquifers contain dissolved constituents including salts. For an evaporite deposit to form, the water must evaporate into the atmosphere, causing the minerals dissolved in the water to precipitate. Evaporites generally form when water accumulates in closed basins in arid and semiarid environments. Enclosed basins in arid environments are favored because they provide high evaporation rates due to low humidity and lack an outlet for the water to drain, which essentially traps the water. Examples of extensive evaporite deposits include areas in and around the Great Salt Lake of Utah and parts of Nevada.

Many limestone deposits form from the accumulation of shells or precipitants from microorganisms. The deposition of limestone strata is thus a byproduct and indicator of biological activity in the geological record (Blatt and Tracey 1996; Taylor and Wilson 2003). Over time, these deposits build and sometimes become hundreds of feet thick.

2.6 Disturbance

Disturbance of sedimentary deposits occurs constantly from the onset of sediment formation. There are several anthropogenic and natural types of disturbance, and when conducting studies in an urban environment, these should be described and evaluated.

2.6.1 Anthropogenic Disturbance

One common concern for any investigator evaluating shallow unconsolidated deposits is whether any disturbance has occurred since deposition. This concern is ratcheted up in urban and developed areas since anthropogenic disturbance often includes excavation or filling.

Disturbance can also occur naturally, although at times it may be difficult to determine whether the disturbance is natural or anthropogenic. For instance, spoils from dredging historically were spread over the land surface in low areas to enable the development of these areas. Dredging a river bottom and spreading the material over the land surface is accomplished by pumping a combination of river water and bottom sediment to the desired area of disposal. Using water as the transport agent for the sediment may create some difficulty in identifying the source of the disturbance because the water and this mixture of sediments and other materials can leave depositional structures similar to natural deposits. Therefore, special care should be undertaken when evaluating sediment layers in urban settings.

Often, this type of activity can be readily observed—but this is not always the case. Typically, to evaluate the existence of anthropogenic disturbance in a specific area, geologists rely on a few clues for assistance. Some of these clues include

- Evidence of historical development
- Evidence of landfilling
- Evidence of grading or land surface disturbance

FIGURE 2.35
Example of anthropogenically deposited material. (Photo by Daniel T. Rogers.)

- Nonnaturally occurring debris located within sediments
- Evidence of excavation
- Lack of original deposition structures within the deposits in question
- Inconsistency or evidence of stressed or absent vegetative cover
- Topography inconsistent with surrounding area or region

Figure 2.35 is a photograph showing landfill material. In this case, several metal containers were excavated indicating anthropogenic disturbance.

2.6.2 Naturally Occurring Disturbance

Examples of naturally occurring types of disturbance include the following (Pettijohn 1975):

- Diagenesis, typically involving compaction and lithification.
- Erosional removal of a portion or an entire deposit (causing an unconformity).
- Bioturbation involving the disturbance of very near-surface soil layers, typically by worms and other macro invertebrates.
- Vegetative disturbance from trees with large and deep root systems.
- Differential compaction or subsidence; these events can obscure, bend, or offset original depositional deposits and sequences and make interpretation of deposition challenging.
- Redistribution of sand deposits by wind along a beach, forming dune deposits often difficult to recognize because many dune deposits do not contain easily recognizable depositional structures.
- Offsets in bedding planes or sediment layers from faulting or other tectonic activity. The results of tectonic offsets may make the matching of depositional layers impossible depending on the distance of the offset.
- Landslides along steep slopes or unvegetated hillslopes resulting from fire or flooding.

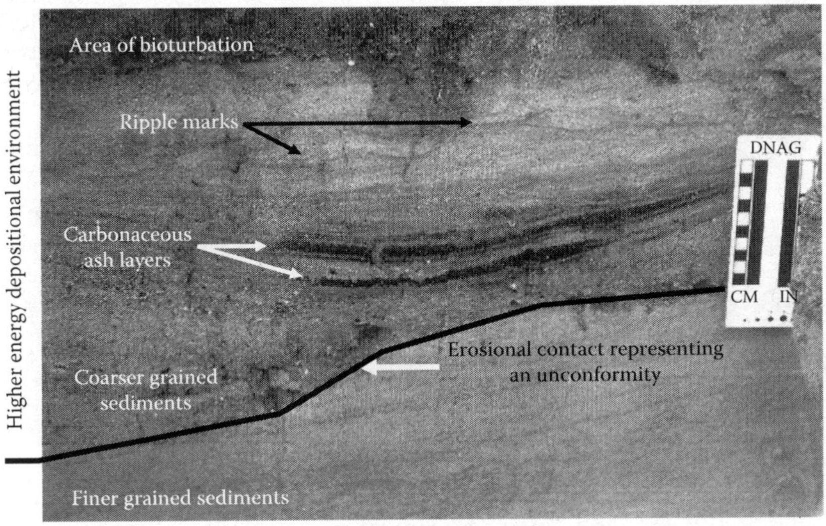

FIGURE 2.36
Late Pleistocene unconsolidated sedimentary deposit in the Rouge River watershed composed of sand (lighter color) and ash (black). (Photo by Daniel T. Rogers.)

- Chemical disturbance from the migration of water from the surface through the sedimentary layers; leaving behind a chemical precipitant or dissolving a portion of the original constituents of the sediments. Given the correct circumstances, this type of disturbance may lead to a karst topography.

The upper portion of Figure 2.36 shows an example of a disturbance likely originating from a naturally occurring source because the line separating the disturbed portion from the non-disturbed portion is not sharp, as would be characteristic of an anthropogenic source. Instead, the boundary is transitional and gradual in most areas, suggesting the disturbance was probably a combination of bioturbation or differential subsidence.

2.7 Summary and Conclusion

The geology of a watershed literally forms the foundation of where we live. Moreover, the geology of any urban region shapes the manner in which we live: our lifestyles; food production; natural resource availability; economic activity, and ultimately determines our standard of living and our sustainability. We have explored the different types of geological environments where urban developments are located, and discovered that most urban areas exist in geological settings of sedimentary origin. Along the way, it was also discovered why the geology of urban areas is very complex and changes rapidly. In addition, we learned every urban area of the United States has a unique set of geologic deposits formed from more than two—and in some cases up to eight—distinct geologic processes.

A representative sample of the many types of geological environments of urban areas throughout the United States has been presented. What is abundantly apparent from this investigation is that many urban areas are located in geological settings dominated by sedimentary deposits located near water. There exists a complex, dynamic, codependant, and sometimes delicate relationship between the presence of water and geology. When urban areas are placed into the mix, these complexities and delicate relationships may be thrown out of balance. Although not always intentional, the resulting human impacts may alter our lives or the environment negatively, and in some cases, catastrophically. What determines our sustainability is only in small part determined by urban location; most critical are the specific actions we undertake to understand and protect the urban environment.

Finally, we have briefly introduced the most important factor determined by geology at any location, and especially in urban areas: the presence and availability of water—the most precious resource required for living organisms and their continued sustainability. Nace (1969), a noted global hydrologist, once stated, "the story of the growth of civilization and science could be written largely in terms of human concern with water." The next chapter explores water in the urban environment from a scientific perspective and encompasses a wide range of processes from its basic molecular structure and chemical behavior to its actions on the surface and below the ground.

References

Aber, J.S. 1984. History of Kansas glacial geology. *Journal of Earth Science History* 3:134–142.

Abrams, C.F. 1984. *Geology of Greater Atlanta*. Geological Survey Bulletin 96. Atlanta, GA: USGS.

Alder, W.C. 1902. *Geology of Chicago: The 1902 Chicago Folio*. Geologic Atlas of the United States Number 81. Washington, DC: USGS.

Allen, R. 2003. *The Rocks of the Valley of the Sun*. Phoenix, AZ. http://www.Gemland.com (accessed August 27, 2009).

Alt, D. 2001. *Glacial Lake Missoula and its Humongous Flood*. Missoula, MT: Mountain Press Publishing.

American Geological Institute. 1962. *Dictionary of Geological Terms*. New York: Dolphin Books.

Anderton, R. 1985. *Clastic Facies Models and Facies Analysis*. Geological Society of London, Special Publication. Vol. 18. London, England.

Barnes, J.H. and W.D. Sevon. 1996. *The Geological Story of Pennsylvania*; Educational series book Number. 4. Harrisburg, PA: Pennsylvania Geological Survey.

Bathurst, R.G.C. 1975. *Carbonate Sediments and Their Diagenesis*. Developments in sedimentology, 12. New York: Elsevier Scientific.

Benn, D.I. and D.J. Evans. 1998. *Glaciers and Glaciation*. New York: John Wiley & Sons.

Bennett, M.R. and P. Doyle. 1997. *Environmental Geology: Geology and the Human Environment*. New York: John Wiley & Sons.

Bennett, M.R. and N.F. Glasser. 1996. *Glacier Geology: Ice Sheets and Landforms*. New York: John Wiley & Sons.

Bilodeau, W.C., S.W. Bilodeau, E.M. Gath et al. 2007. Geology of Los Angeles. *Environmental and Engineering Geoscience Journal* 13:99–160.

Bird, E.C. 1985. *Coastal Changes: A Global Review*. New York: John Wiley & Sons.

Blatt, H. and R.J. Tracey. 1996. *Petrology: Igneous, Sedimentary, and Metamorphic*, 2nd edn. New York: W. H. Freeman & Company.

Bloom, A.L. 2004. *Geomorphology: A Systematic Analysis of Late Cenozoic Landforms*, 3rd edn. Long Grove, IL: Waveland Press.

Boggs, J.R. 2000. *Principles of Sedimentology and Stratigraphy*, 3rd edn. Toronto, Canada: Merrill Publishing Company.

Chamberlin, T.C. and R.D. Salisbury. 1909. *A College Textbook of Geology*. Chicago, IL: The Lakeside Press.

Christopherson, R.W. 2008. *Geosystems: An Introduction to Physical Geography*, 7th edn. Upper Saddle River, NJ: Prentice Hall.

Coogan, A.H. 1996. *Ohio's Surface Rocks and Sediments*. Ohio Geological Survey Bulletin 70. Columbus, OH.

Davies, W.E. and H.E. Legrand. 1972. Karst of the United States. In: *Karst*, ed. M. Herak and M.T. Springfield, pp. 467–505. United States Geological Survey.

Deer, W., R.A. Howie, and J. Zussman.1966. *An Introduction to Rock Forming Minerals*. Essex, England: Longman Group Limited.

Farrand W.R. 1988. *The Glacial Lakes Around Michigan*. Michigan Department of Natural Resources, Geological Survey Division Bulletin Number 4. Lansing, MI.

Flint, F.F. and B.J. Skinner. 1974. *Physical Geology*. New York: John Wiley & Sons.

Florida Geological Survey. 1994. *Florida's Geological History and Geological Resources*. Tallahassee, FL: Florida Geological Survey.

Gentile, R.J. 1984. *Geology of the Circumferential Highway System at Kansas City, Missouri*. National Association of Geology Teachers—Central Section Guidebook. Kansas City, MO.

Gibbard, P. and T. Van Kolfschoten. 2004. The pleistocene and holocene epochs. In *A Geologic Time Scale*, ed. F.M. Gradstein, J.G. Ogg, and A.G. Smith, pp. 441–452. Cambridge, U.K.: Cambridge University Press.

Goodwin, B.K. 1964. *Guidebook to the Geology of the Philadelphia Area*. Philadelphia, PA: Philadelphia Geological Society.

Green, S.A. and D.R. Currey. 1988. The Stansbury shoreline and other transgressive deposits of the Lake Bonneville cycle. In *In the Footsteps of G. K. Gilbert—Lake Bonneville and Neotectonics of the Eastern Basin and Range Province*, Guidebook for Field Trip Number Twelve. Boulder, CO: Geological Society of America.

Greenwood, N.N. and A. Earnshaw. 1997. *Chemistry of the Elements*, 2nd edn. Oxford: Butterworth-Heinemann.

Gustavson, T. C., R.W. Baumgardner, S.C. Caran et al. 1991. Quaternary geology of the southern Great Plains and an adjacent segment of the rolling hills. In *The Decade of North American Geology. Volume. K-2. Quaternary Nonglacial Geology: Conterminous United States*, ed. R.B. Morrison, pp. 477–501. Boulder, CO: The Geological Society of America.

Halka, C. 1980. *Roadside Geology of Colorado*. Denver, CO: Mountain Press Publishing Company.

Halke, C. 1983. *Roadside Geology of Arizona*. Denver, CO: Mountain Press Publishing Company.

Hillier, S. 2003. Clay Mineralogy. In *Encyclopedia of Sediments and Sedimentary Rocks*, ed. G.V. Middleton, M.J. Church, M. Coniglio, L.A. Hardie, and F.J. Longstaffe, pp. 139–142. Dordrecht, The Netherlands: Kluwer Academic Publishers.

Hurlbut, C.S. and C. Klein. 1985. *Dana Manual of Mineralogy*, 20th edn. New York: John Wiley & Sons.

Hutton, J. 1788. Theory of the Earth; or an investigation of the laws observable in the composition, dissolution, and restoration of the land upon the globe. *Transactions of the Royal Society of Edinburgh*, 1: 209–304.

Imbrie, J. and K.P. Imbrie. 1979. *Ice Ages; Solving the Mystery*. Cambridge, MA: Harvard University Press.

Indiana Geological Survey. 2010. Indiana geology. Indiana Geological Survey. Bloomington, Indiana. http://igs.indiana.edu/Geology/index.cfm (accessed April 12, 2010).

Kaye, C.A. 1976. *The Geology and Early History of the Boston Area of Massachusetts: A Bicentennial Approach*. USGS Special Report 1476. Washington, DC.

Kentucky Geological Survey. 2009. Karst land in Kentucky. http://www.uky.edu/KGS/water/general/karst/index.htm (accessed August 29, 2009).

King, P.B. and H.M. Beikman. 1974. *Geologic Map of the United States; Excluding Alaska and Hawaii*. United States Geological Survey, Washington, DC: Governmental Printing Office. 1:2,500,000, 2 Sheets.

Kious, W.J. and R.I. Tilling. 1996. *The Dynamic Earth: The Story of Plate Tectonics*. United States Geological Survey. Washington, DC: Governmental Printing Office.

Krumbein, W.C. and L.L. Sloss. 1963. *Stratigraphy and Sedimentation*, 2nd edn. San Francisco, CA: W. H. Freeman.

Leopold, L.B., M.G. Wolman, and J.P. Miller. 1992. *Fluvial Processes in Geomorphology*. New York: Dover.

Missouri Department of Natural Resources. 2004. *Geologic Map of Missouri*. Geological Survey Division. Rolla, MO.

Mitsch, W.J., J.G. Gosselink, C.J. Anderson et al. 2009. *Wetlands Ecosystems*. New York: John Wiley & Sons.

Morrison, R.B. 1991. Quaternary stratigraphic, hydrologic, and climatic history of the Great Basin, with emphasis on Lake Lohontan, Bonneville, and Tecopa. In *Quaternary Nonglacial Geology of Conterminous United States, The Geology of North America*, ed. R.B. Morrison, pp. 283–320. Boulder, CO: Geological Association of America.

Nace, R.L. 1969. *Water and Man—A World View*. Paris, France: UNESCO.

National Oceanic and Atmospheric Administration (NOAA). 2009. Our changing planet. http://www.umac.org/images/ocp/coastal_pie.jpg (accessed August 27, 2009).

New Jersey Department of Environmental Protection (NJDEP). 2005. *Field Sampling Procedures Manual*. Trenton, NJ: NJDEP.

Ojakangas, R.W. and C.L. Matsch. 1982. *Minnesota's Geology*. Minneapolis, MN: University of Minnesota Press.

Oreskes, N. 2003. *Plate Tectonics: An Insider's History to the Modern Theory of the Earth*. Boulder, CO: Westview Press.

Page, W.R., S.C. Lundstrom, A.G. Harris et al. 2005. *Geologic and Geophysical Map of the Las Vegas 30′×60′ Quadrangle, Clark and Nye Counties, Nevada, and Inyo County, California*. United States Geological Survey Pamphlet to Accompany Scientific Investigations Map 2814. Washington DC.

Potter, P.E., J.B. Maynard, and W.A. Pryor. 1980. *Sedimentology of Shale*. New York: Springer–Verlag.

Pettijohn, F.J. 1975. *Sedimentary Rocks*, 3rd edn. New York: Harper & Row.

Pettijohn, F.J., P.E. Potter, and R. Siever. 1987. *Sand and Sandstone*, 2nd edn. New York: Springer-Verlag.

Pudasaini, S.P. and K. Hutter. 2007. *Avalanche Dynamics: Dynamics of Rapid Flows of Dense Granular Avalanches*. New York: Springer-Verlag.

Rogers, D.T. 1996. *Environmental Geology of Metropolitan Detroit*. Novi, MI: Clayton Environmental Consultants.

Rogers, D.T. 1997. *Surface Geologic Map of the Rouge River Watershed, Southeastern Michigan*. Wayne County, Michigan. 1:63,360. 2 Sheets.

Seyfert, C.K. and L.A. Sirkin. 1973. Earth *History and Plate Tectonics: An Introduction to Historical Geology*. New York: Harper & Row.

Snowden, J.O. 1980. *Geology of Greater New Orleans*. New Orleans, LA: New Orleans Geological Society.

Spearing, D. 1991. *Roadside Geology of Texas*. Denver, CO: Mountain Press Publishing Company.

Stanley, S.M. 1999. *Earth System History*. New York: W. H. Freeman.

Stewart, M. 1992. *Karst Characteristics of the USF Geology Park*. Tampa, FL: Department of Geology, University of Southern Florida.

Stoffer, P.W. and L. Gordon. 2001. *Geology and Natural History of the San Francisco Bay Area—A Field Trip Guidebook*. USGS Bulletin 2188. Washington, DC.

Stokes, W.L. 1989. *Geology of Utah*. Utah museum of natural history occasional paper Number. 6. Salt Lake City, UT.

Taylor, P.D. and M.A. Wilson. 2003. Paleoecology and evolution of marine hard substrate communities. *Earth-Science Reviews* 62:1–103.

Trimble, D.E. 1963. *Geology of Portland, Oregon and Adjacent Areas*. USGS Bulletin 1119. Washington, DC.

United States Geological Survey. 1998. U.S. Department of the Interior, U.S. Geological Survey, Center for Coastal Geology. http://coastal.er.usgs.gov/cgi-bin/response.pl?site=cc&loc=9 (accessed November 26, 2010).

United States Geological Survey. 2003. Geology of the New York City region. http://3dparks.wr.usgs.gov/nyc/index.html (accessed August 27, 2009).

United States Geological Survey. 2005. The new geologic map and geodatabase of Seattle. University of Washington. Seattle, Washington. http://geomapnw.ess.washington.edu/docs/Sea_Map_Fact_Sheet.pdf (accessed August 27, 2009).

United States Geological Survey. 2007. *Geology of the National Capital Region—A Field Trip Guidebook*. USGS Circular 1264. Washington, DC.

United States Geological Survey. 2009a. *Geology of the Mid-Atlantic Urban Corridor (GOMAC)*. USGS Circular 1148. Washington, DC.

United States Geological Survey. 2009b. *Geology of Texas*. http://mrdata.usgs.gov/sgmc/tx.html (accessed August 27, 2009).

United States Geological Survey. 2009c. *Eolian History of the United States*. http://esp.cr.usgs.gov/info/eolian/task1.html (accessed August 27, 2009).

United States Geological Survey. 2009c. *Karst and the United States Geological Survey*. http://water.usgs/gov/ogw/karst/index (accessed February 26, 2010).

United States Geological Survey. 2010a. World map showing plate boundaries and active volcanoes. http://vulcan.wr.usgs.gov/Glossary/PlateTectonics/Maps/map_plate_tectonics_world.html (accessed June 18, 2010).

United States Geological Survey. 2010b. USGS glossary. http://3dparks.wr.usgs.gov/haywardfault/images/slump.jpg (accessed June 28, 2010).

United States Global Change Research Program. 2000. Climate change impacts on the United States: The potential consequences of climate variability and change. Overview: Coastal Areas and Marine Resources. http://www.usgcrp.gov/usgcrp/Library/nationalassessment/overview-coastal.htm (accessed July 4, 2010).

United States Soil Conservation Service. 1992. *Soil Erosion by Water*. Washington, DC: United States Department of Agriculture.

Walker, H.J. and J.M. Coleman. 1987. Atlantic and Gulf Coast province. In *Geomorphic Systems of North America. Volume. 2. Decade of North American Geology*, ed. W.L. Graf, pp. 51–110. Boulder, CO: Geological Society of America.

Wentworth, C.K. 1922. A scale of grade and class terms for clastic sediments. *Journal of Geology* 30:377–392.

Wicander, R. 2005. *The Changing Earth: Exploring Geology and Evolution*. New York: Thompson Brooks/Cole.

Wicander, R. and J. Monroe. 2007. *Historical Geology*. New York: Thompson Brooks/Cole.

Wilson, J.L. 1975. *Carbonate Facies in Geologic History*. Heidelberg, Germany: Springer-Verlag.

Wright, H.E. 1989. The quaternary. In *The Decade of North American Geology. Volume. A. The Geology of North America—An Overview*, ed. A.W. Bally and A.R. Palmer, pp. 523–536. Boulder, CO: The Geological Society of America.

3

Water and the Hydrogeology of Watersheds

3.1 Introduction

The importance of water on Earth cannot be overstated. Water, along with energy and organic molecules, was the prerequisite for the origin of life here billions of years ago. Today, water plays a central role in maintaining our survival as a species, preservation of the natural environment, and therefore achieving a sustainable planet (United Nations 2003). Yet, there are immense challenges facing humans with respect to securing water for their basic needs and long-term quality of life. Although almost three-fourths of Earth's surface is covered by water, most of this water is not potable, and a high percentage of the fresh stuff is either frozen, underground, or in a gaseous phase.

Wait—there are other complications. Because of the spherical nature of the Earth and its axial tilt, the arrangement of land and water, and differences in surface elevation, it is not possible to achieve a uniform distribution of the incoming solar energy we receive. Surpluses and deficits of energy arise at different locations and create uneven atmospheric pressures and densities. In a thermodynamic system, these inequalities try to even themselves out, so fluids move in a quest to achieve overall energy equilibrium. Air (via wind) transports part of the solar energy it has absorbed and moves it from zones of higher pressure to zones of lower pressure. Water moves the excess energy it has absorbed from the equatorial region in a general poleward direction via ocean currents. All of this movement results in Earth's topside having a peculiar and unpredictable precipitation pattern: some areas receive precipitation almost daily; others may not get any for years; and many places lie between these extremes.

Precipitation from the atmosphere recharges the underground water reservoirs and plays a central role in shaping the geology of urban areas through weathering, sediment transport and deposition, chemical precipitation and dissolution, and erosion. Given the haphazard nature of precipitation, it is a good thing that nature is so organized. Surface flows of water are controlled hierarchically by a system of topographically bounded spatial units called watersheds, which are really just water collection bowls. Small watersheds consisting of small streams feed larger watersheds with larger streams; for example, the Scioto River watershed in Ohio delivers its flows into the Ohio River, which then flows into the larger Mississippi River. Some of the precipitation is intercepted by vegetation, some pools on the surface, and a good part of it moves downward through the soil where it recharges the groundwater reservoir.

This chapter opens with a brief discussion of basic water chemistry and the geography of Earth's water, and how it moves cyclically between the atmosphere and the oceans. Understanding basic water chemistry and refreshing ourselves about the global water cycle provide the knowledge base for the more detailed material later in the chapter and later sections of this book. Key objectives of this chapter include an investigation of the multifaceted nature of surface water and groundwater as discrete entities, while also considering their interactions. The chapter concludes with a discussion of the anthropogenic effects on surface water and groundwater.

3.2 Water Basics

This section begins with the basic chemistry and properties of water. The focus then turns to water's abundance and distribution on Earth, with special attention given to the work water accomplishes as it moves through its global cycle. We finish with a discussion of potable water.

3.2.1 Basic Water Chemistry

A water molecule is composed of two hydrogen atoms bonded with one atom of oxygen. This bonding ratio gives water a chemical formula of H_2O. The spatial arrangement of the atoms within each water molecule is shown in Figure 3.1 (Hill et al. 1993). The hydrogen atoms are attached to each side of the oxygen atom, and this configuration results in a **polarity** or directional charge. The water molecule is positively charged at the hydrogen end and negatively charged on the opposite side where the oxygen atom is located. Since opposite electrical charges attract, water molecules have an affinity for each other, and, in its liquid state, water forms the linear configuration as shown in Figure 3.2 (Hill et al. 1993).

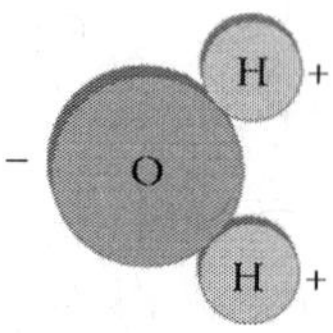

FIGURE 3.1
Water molecule.

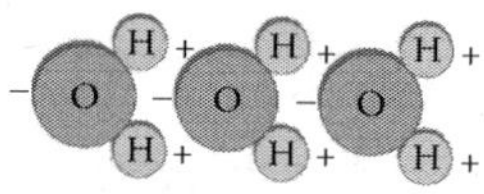

FIGURE 3.2
Electrical attraction of water molecules.

The angular arrangement of the water molecule shown in Figure 3.1 results from the properties of its chemical bonds. Hydrogen atoms are attached to the oxygen atom in a **covalent bond** (Figure 3.3), with each black dot representing an electron. Note the circled pairs of electrons on the two sides opposite the hydrogen atoms. The pairing of electrons creates a slightly greater negative charge, which acts to push the two hydrogen atoms closer. This push results in a slightly bent tetrahedral molecule with an angle of approximately 107.5° (Figure 3.4).

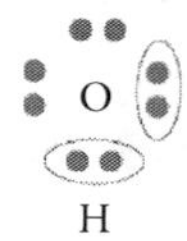

FIGURE 3.3
Water molecule showing electron arrangement.

The polarity of a water molecule is significant, because it permits water to

- Stick together with other water molecules, as in a drop of water.
- Interact with other polar molecules—this is why other substances are able to dissolve in water. Water is known as the "universal solvent" because it dissolves more substances than any other liquid.
- Interact with electricity.

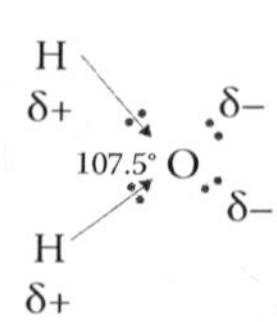

FIGURE 3.4
Angular configuration and polarity of the water molecule.

Some important chemical properties of water include

- It freezes at 32°F or 0°C.
- It boils at 212°F or 100°C at mean sea level.

- Liquid water is denser than ice.
- It has a high **specific heat capacity**.
- It has a very high surface tension.

Water is unique, because it is the only substance on Earth found in all three states (liquid, solid, and gas) within our planet's temperature ranges. Liquid water is essential to life, as it comprises approximately 60% of the human body by weight and 70% of the human brain. Some organisms are 90% liquid water by weight.

Without water, humans simply would not exist. The ability of water to dissolve so many different substances allows cells to use the nutrients, minerals, and chemicals in biological processes. In natural systems, wherever water goes—the air, the ground, streams and lakes, biota, or through our bodies—it takes valuable chemicals, minerals, and nutrients picked up along the way. This transport ability also means that water can carry substances harmful to humans and the environment. If these contaminants are present at a sufficient concentration and the exposure is long enough, potentially harmful effects can occur. The damage can be immediate and obvious, as when oil is washed up on a beach and kills waterfowl—or slow developing and silent—a scenario represented all to often by the benzene that originated from a leaking gasoline tank and carried by groundwater into a drinking water well.

This is why we spend so much time and effort studying water.

3.2.2 Water Cycle

Water is constantly on the move through the four spheres of the geosphere—the atmosphere, biosphere, hydrosphere, and lithosphere. This movement of water is cyclical and is called the hydrologic cycle or water cycle. In this cycle, water changes phase many times and exists as a liquid (surface waters and groundwater), solid (snow and ice), or gas (water vapor). The global movement of water is initiated by solar energy, which evaporates surface water into the atmosphere. Much of this water vapor condenses and falls as some form of precipitation on a distant land surface where it either evaporates, flows back into the oceans through rivers and streams, is taken up by vegetation and slowly released into the atmosphere as evapotranspiration, or infiltrates into the ground. Groundwater also migrates back to the oceans (Jones 1997; Alley et al. 1999). Figure 3.5 depicts the water cycle on Earth.

As with all physical systems, the water cycle performs some work. These three main chores are accomplished: (1) evaporation desalinates large quantities of ocean water and creates freshwater that is potentially potable; (2) nutrients are transported—a process called the biogeochemical cycles—which moves carbon, phosphorus, and other nutrients between their reservoirs. For example, the carbon cycle transports enormous quantities of carbon between the atmosphere and the oceans by precipitation, with much of this carbon held within the carbonic acid (H_2CO_3) formed in the atmosphere by the combination of CO_2 and water vapor; and (3) landscapes are altered by flowing water (erosion and sedimentation) and water's phase changes between ice and liquid (ice wedging and frost heaving). Water, with its higher viscosity and weight per unit volume than air, is the key agent of landscape change on Earth.

3.2.3 Distribution of Water on Earth

The vast majority of water on the Earth is saline and resides in the oceans. Although the water in the oceans is important for life on Earth, it is the freshwater contained in ice and

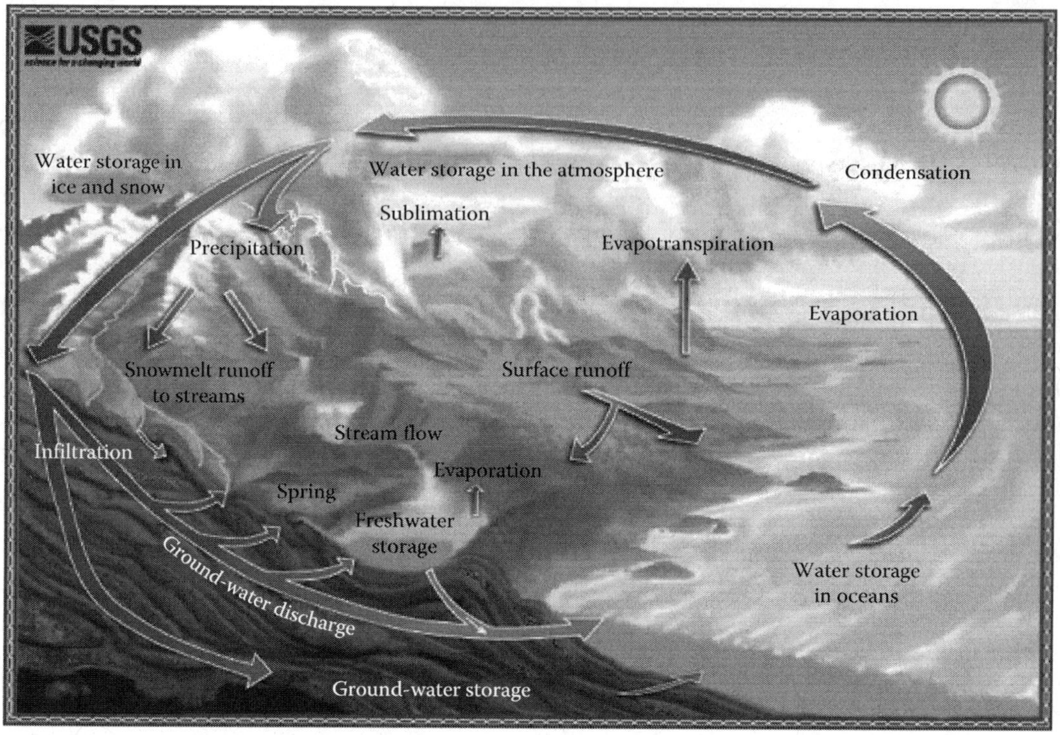

FIGURE 3.5
(See color insert.) The water cycle. (From United States Geological Survey, Summary of the water cycle, http://ga/water.usgs.gov/edu/watercyclesummary.html, 2010a.)

snow, rivers and streams, lakes, swamps, and groundwater that provides people with the water needed to live. Figure 3.6 shows the percent distribution of Earth's water, with the left column indicating 97% of this water is saline and only 3% is freshwater. Moving to the middle column, we see the 3% of freshwater breaking down into a majority of ice (68.7%) and a smaller percentage (30.1%) as groundwater. This leaves only 0.3% of Earth's total water on the surface, and as shown by the right column, a small percentage is held in primarily in lakes (87%), and the rest in rivers, streams, and swamps.

Table 3.1 presents a more detailed accounting of Earth's water reserves; specifically where the water resides, the estimated quantities within these reservoirs, and the amounts of freshwater as a percentage of total water (Shiklomanov 1993; Jones 1997).

The major lessons learned from the characteristics of Earth's water are (1) most of our water is not potable and the accessible portion of potable water is small and (2) contamination of a significant amount of our potable water resources can and does create regional shortages, despite the existence of an adequate supply of freshwater per capita on Earth.

3.2.4 Potable Water Use

The term **potable water** means water used for any human purpose (Fuhrer et al. 1999) and includes uses other than drinking water. From a water quality standpoint, potable water contains less than 10,000 μg/L of total dissolved solids. In the United States, potable water use includes: (modified from Fuhrer et al. 1999)

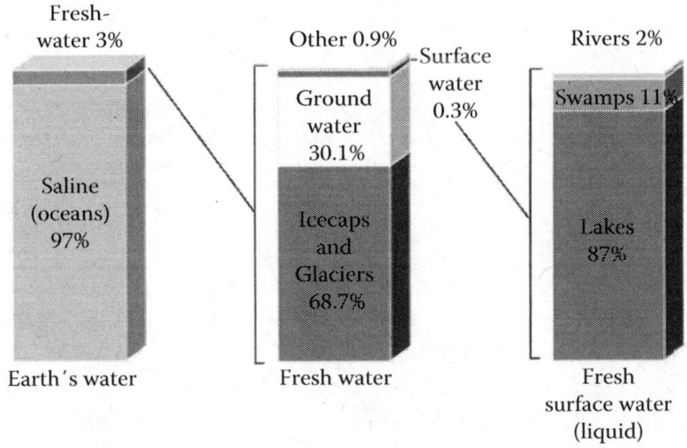

FIGURE 3.6
Distribution of Earth's water. (From United States Geological Survey, Water properties, http://ga.water.usgs.gov/edu/waterproperties.html, 2009a.)

TABLE 3.1
Earth's Water by Source, Volume, and Type

Water Source	Water Volume (km^3)	Percent Freshwater	Percent Total Water
Oceans, seas, and bays	1,337,986,366	—	96.5
Ice caps, glaciers, and permanent snow	32,399,277	68.7	1.74
Groundwater	23,400,173	—	1.7
• Fresh	10,528,827	30.1	0.76
• Saline	12,871,345	—	0.94
Soil moisture	16,502	0.05	0.001
Ground ice and permafrost	299,984	0.86	0.022
Lakes	176,397	—	0.013
• Fresh	90,991	0.26	0.007
• Saline	85,406	—	0.006
Atmosphere	12,901	0.04	0.001
Swamps	11,471	0.03	0.0008
Rivers	2122	0.006	0.0002
Biological water	1121	0.003	0.0001
Total	1,385,920,457	—	100.0

- Drinking and bottling (including beverages such as soft drinks)
- Washing
- Recreation
- Heating and cooling
- Fire fighting
- Irrigation
- Industrial use

- Commercial use
- Livestock
- Mining
- Electricity production
- Wastewater treatment

In 2000, people in the United States used an average of 407.5 billion gallons of water per day. Of this total amount, 79% or 323 billion gallons was from a surface water source and 21% or 84.5 billion gallons originated from groundwater (USGS 2004). Over the last 50 years, the usage of water for electricity production increased by more than 500% and, since 1965, has been the category of highest water use. Water used for irrigation purposes has also risen 50% between 1950 and 2000—another significant increase.

From 1970 to 2000, the use of fresh surface water in the United States has remained relatively constant at a usage rate of approximately 275 billion gallons of water per day. During this same time period, the use of fresh groundwater has increased from approximately 65 billion to 85 billion gallons of water per day (USGS 2009b). In many states, groundwater is the key source of potable water for irrigation; Nebraska, Texas, and California are examples. Some states, such as California, have designated specific areas for **groundwater recharge**. Here, during the short rainy season, surface water seeps into the ground and is stored until it is needed for human use during dryer periods.

Figure 3.7 shows the percent of population in each state using groundwater as a source of drinking water. Over 90% of the people in Idaho, New Mexico, Mississippi, and Florida use groundwater for this purpose. By contrast, the populations of Colorado, Rhode Island, and Kentucky obtain less than 30% of their drinking water from groundwater (USGS 2004).

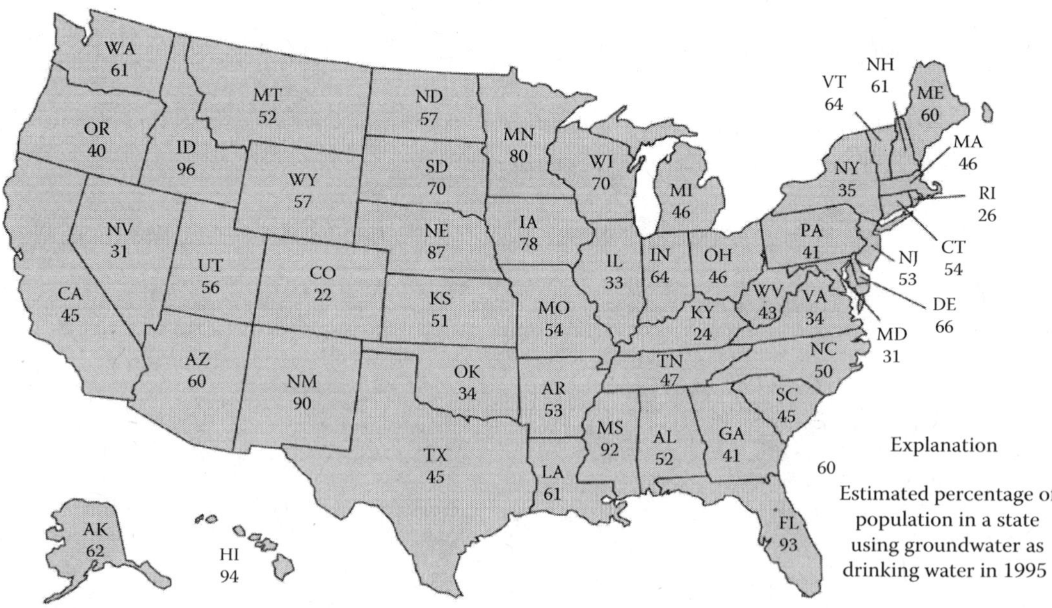

FIGURE 3.7
Groundwater use by state. (From Alley, W.M. et al., *Sustainability of Groundwater Resources*, United States Geological Survey Circular 1186, Denver, CO, 1999.)

The data on water use indicate that humans require and use enormous quantities of water every day to provide energy, food, and to sustain their lives. But the geography of water often dictates specific usage patterns. Many regions of the United States lack abundant freshwater supplies and are required to use water more wisely because of its relative scarcity compared to other areas. The arid southwest and the humid region of the Great Lakes demonstrate this contrast (Meyer 1989). Because of this geo-climatic factor and other considerations (one being the water law doctrines existing in eastern and western states), water is not equally protected throughout the United States, and the methods of conservation and recycling of water vary widely depending on location.

We now turn our attention to watersheds—the organizers of freshwater supplies. The focus here will be on urban watersheds and the specific implications their surface waters and groundwater have for regional geology, contaminant sources and sinks, and the long-term sustainability of the landscape.

3.3 Surface Water in Watersheds

The importance of surface water to urban centers cannot be overstated. This water acts as a source of potable water and supports agriculture, commerce, and transportation. From an environmental perspective, surface water is responsible for the erosion and deposition of a significant portion of the sediments lying beneath urban areas, and when surface water is in the form of a stream, it carries suspended and dissolved materials into other bodies of water.

Nature organizes streamflow by adopting the most energy efficient method of transport within a physical framework consisting largely of the local geology and climate. In areas of fractured rock, the streams follow the fractures; whereas in sedimentary environments consisting of surficial unconsolidated sediments, a dendritic or treelike pattern of drainage may develop. If it rains more, then a denser stream network evolves. The primary objective of a stream is to transport the water it receives and the materials it carries without delay, so the next time a precipitation event occurs, the transport can continue. The organizational structure nature came up with to perform these tasks is called a watershed.

3.3.1 Watersheds

A more modern definition of a watershed sees them as an extent of land where water as rain or snowmelt drains downhill into a body of water such as a river, lake, swamp, estuary, reservoir, wetland, bay, sea, or ocean (Winter et al. 1998). In some cases, the drainage areas for water at the surface and below within a watershed may differ. When the areas are different, the groundwater watershed is usually larger than the surface watershed (DeBarry 2004). Watersheds act as a conduit for the transport of water and materials at the surface and below the surface (groundwater). Figure 3.8 shows the functional regions of a surface watershed and their geofluvial characteristics; i.e., their flow volumes and erosion/deposition balance.

The basic moniker used for watersheds varies. Other terms describing a watershed include catchment, catchment area, catchment basin, drainage area, drainage basin, river

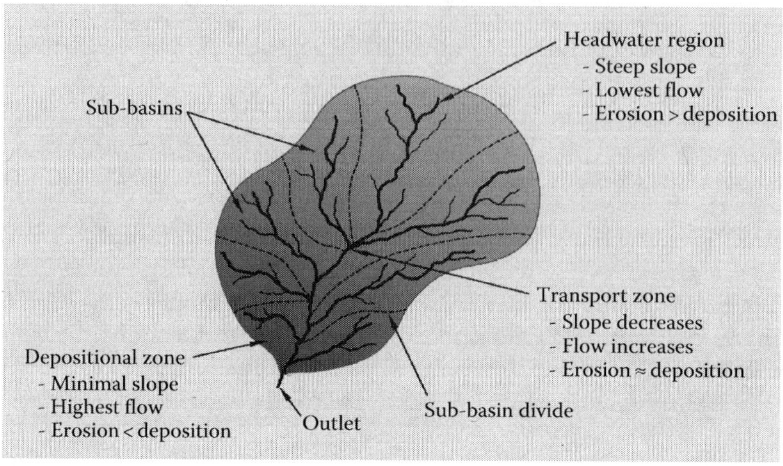

FIGURE 3.8
Watershed regions and transport functions.

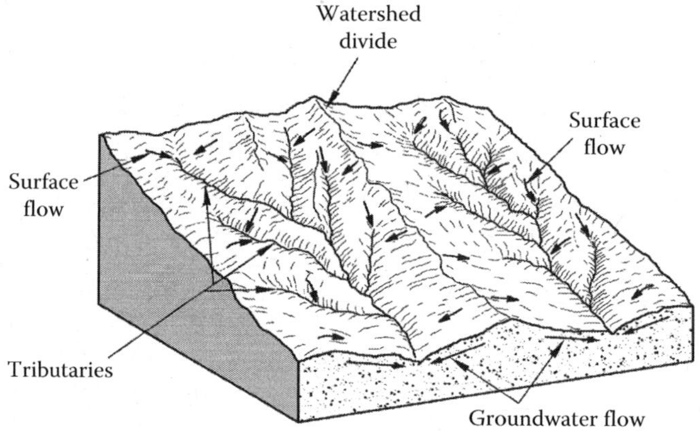

FIGURE 3.9
Two adjacent watersheds.

basin, or water basin. For consistency and to avoid the use of a thesaurus, we use the term watershed throughout the book. Figure 3.9 shows two adjacent watersheds to illustrate their structure and flow patterns at the surface and below (Alley 1999).

The entire land surface of the Earth can be subdivided into watersheds. Some of the more notable watersheds in North America include those drained by the rivers listed in Table 3.2 (Kammerer 1990). Figure 3.10 shows the major watersheds of the United States.

Whether water drains off the land into successively larger watersheds or drains from a watershed like the Mississippi, it ultimately discharges into an ocean. The Atlantic Ocean receives surface water from 47% of the land surface of the world and the Pacific and Indian Oceans each drain 13% of the land surface. The remaining surface drainage either ends up in the Arctic Ocean or does not reach an ocean. Non-ocean bound surface water drains into enclosed basins, such as the Great Basin located in the western portion of the United States (Kammerer 1990).

TABLE 3.2
Some Major Rivers of the United States

River	Length (km)	Drainage Area (km²)	Discharge (m³/s)	Major Tributaries
Missouri	4,023	1,370,104	2158	Yellowstone River
				Platte River
Mississippi	3,764	2,978,486	16,792	Illinois River
				Ohio River
				Missouri River
				Arkansas River
Rio Grande	3,034	870,236	283	Pecos River
Arkansas	2,350	416,988	1,161	Canadian River
				Cimarron River
Colorado	2,334	637,137	425	Green River
Columbia	1,854	668,217	7,504	Snake River
Ohio	1,569	525,768	7,957	Allegheny River
				Monongahela River
St. Lawrence	1,223	1,025,635	9,854	Great Lakes
				Ottawa River

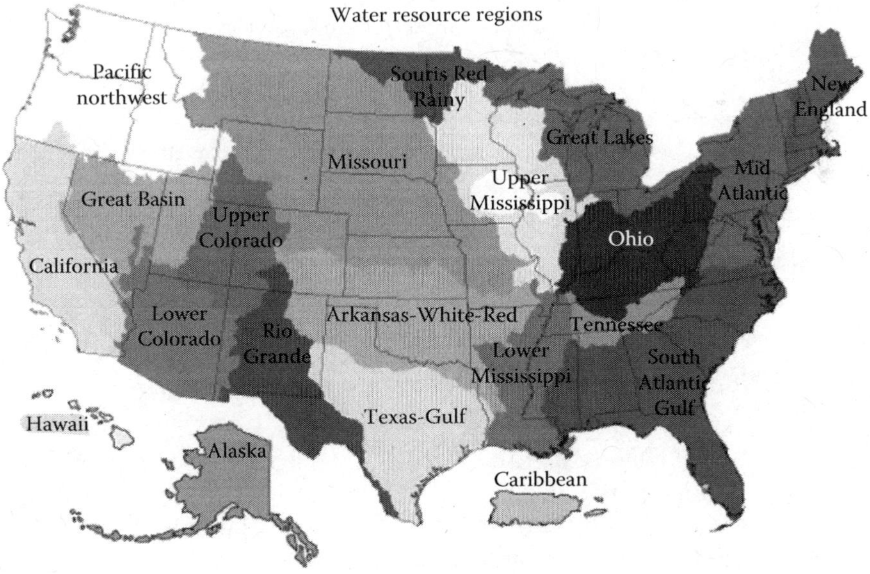

FIGURE 3.10
Major watershed regions of the United States. (From United States Geological Survey, Science in your watershed, http://water.usgs.gov/wsc/map_index.html, 2010b.)

Table 2.4 lists the major urban areas of the United States and their associated geologic environments. Table 3.3 contains the same urban areas as Table 2.4, but matches them with their closest major surface water feature.

Of the 29 urban centers listed in Table 3.3, 12 are located near the ocean and 8 more are located near bodies of water deep and wide enough to support large-scale shipping. The

TABLE 3.3

Major Urban Areas of the United States and Associated Surface Water Features

Urban Area	Surface Water Features
Atlanta	Chattahoochee River
Baltimore	Atlantic Ocean
Boston	Atlantic Ocean
Chicago	Lake Michigan and Chicago River
Cincinnati	Ohio River
Cleveland	Lake Erie
Dallas	Trinity River and White Rock River
Detroit	Detroit River
Denver	South Platte River
Houston	Gulf of Mexico
Indianapolis	White River
Kansas City	Missouri River
Las Vegas	Colorado River
Los Angeles	Pacific Ocean
Miami	Atlantic Ocean
Minneapolis	Mississippi River
New Orleans	Mississippi River and Gulf of Mexico
New York	Atlantic Ocean and Hudson River
Philadelphia	Atlantic Ocean and Delaware River
Phoenix	Gila River and Salt River
Pittsburgh	Ohio River
Portland	Pacific Ocean, Columbia River and Willamette River
Saint Louis	Mississippi River and Missouri River
Salt Lake City	Great Salt Lake and Jordan River
San Antonio	Salado River, San Antonio River, and Olmos River
San Francisco	Pacific Ocean
Seattle	Pacific Ocean
Tampa	Gulf of Mexico
Washington, DC	Potomac River

12 located near an ocean include Baltimore, Boston, Houston, Los Angeles, Miami, New Orleans, New York, Philadelphia, Portland, San Francisco, Seattle, and Tampa. The eight cites located near bodies of water large enough to support large-scale shipping include Chicago, Cincinnati, Cleveland, Detroit, Kansas City, Minneapolis, Pittsburgh, and Saint Louis. The remaining seven urban locations are located near water bodies that are either used for energy sources (hydroelectric), recreational functions, and sources of drinking water. These seven remaining urban areas include Atlanta, Dallas, Denver, Indianapolis, Las Vegas, Phoenix, Salt Lake City, and San Antonio (USGS 1984).

3.3.2 Surface Water Drainage Development and Patterns

Climate and geology were mentioned earlier as important factors influencing the development of surface water flow patterns and drainage density. A more comprehensive list of factors involved in these processes includes (Leopold et al. 1992)

- Climate
- Vegetation
- Topography
- Rainfall
- Geology

Rivers and streams go through an evolutionary process as they age and follow the general sequence depicted in Figure 3.11a through d (Dutch 2009).

A young stream is characterized by a V-shaped valley, rapids and waterfalls, and no floodplain (Figure 3.11a). Over time, a young stream deepens its valley (Figure 3.11b). As the stream matures, it begins to develop a floodplain and sand and gravel bars. The valley deepening slows (Figure 3.11c). A fully mature stream exhibits a very wide floodplain with pronounced meanders and cut-off meanders called **ox-bow lakes** (Figure 3.11d) Young streams can also be characterized as predominantly eroding streams and fully mature streams as depositional streams (Leopold 1992).

The drainage development pattern of a watershed may yield some clues about its subsurface geology. For instance, Figure 3.12 shows a dendritic drainage pattern with a higher drainage density in the headwater region than in the depositional zone. In the headwater region, silty clay underlies the streams; whereas in the depositional zone, the sediments are porous with a high-infiltration capacity. Since the clay sediments in the headwater region are not very permeable and restrict infiltration, most of the precipitation drains as surface water. The result is a denser stream drainage pattern required to drain the same amount of land. This same drainage pattern can also be formed by a steeper slope gradient in the headwater region than in the depositional zone.

There are other processes at work that determine a watershed's drainage density, so we must consider other factors in addition to soil infiltration capacity. For example, groundwater at the surface shapes the Earth's topography through a process known as seepage

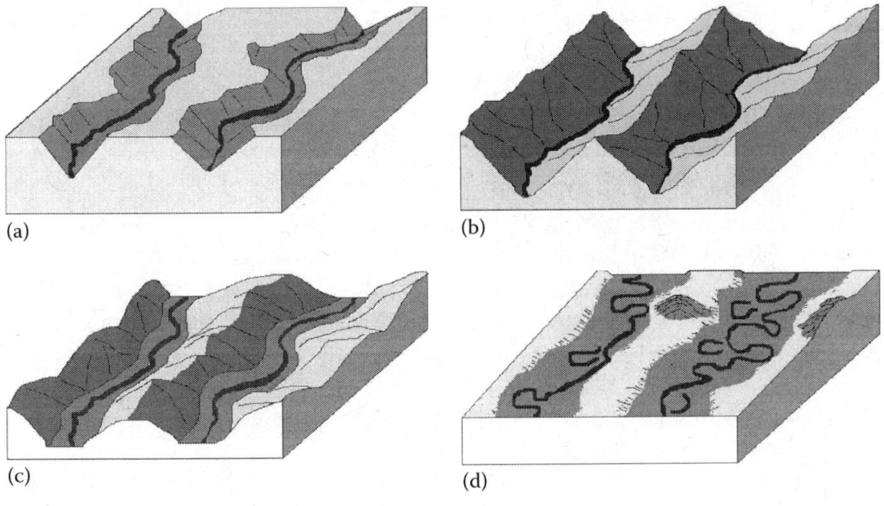

FIGURE 3.11
(a) Young stream, (b) mature (early) stream, (c) mature (late) stream, and (d) old age stream. (From Dutch, S., Erosion and landscape evolution, University of Wisconsin Green Bay, http://www.uwgb.edu/dutchs/EarthSC202Notes/erosion.htm, 2009. With permission.)

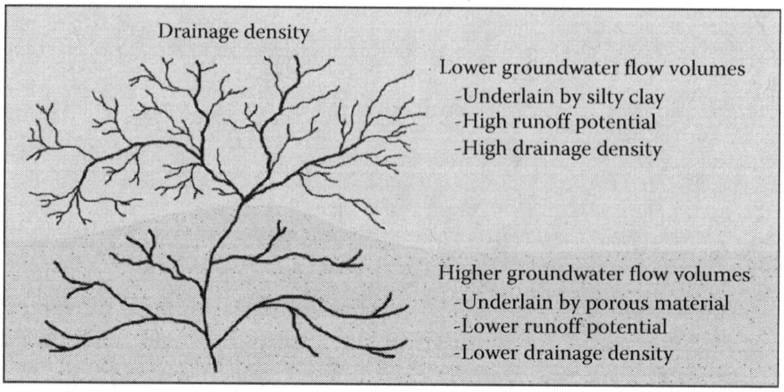

FIGURE 3.12
Stream drainage densities based on differences in geology.

erosion. In combination with overland flow, seepage erosion contributes to the initiation and growth of channel networks (Dunne 1990; Abrams et al. 2009). Many of the United States' urban watersheds of glacial–lacustrine origin have higher drainage densities in sandy soils, and this pattern could be related to the initiation of channels by groundwater. Although the exact causes of channel density have not been resolved (Abrams et al. 2009), the regional geology provides a good starting point for gaining important information about the surface streams.

As this discussion reveals, there is a close relationship between surface water and groundwater. At many locations, surface water recharges groundwater and, in turn, groundwater discharges to surface water, but this is only one part of a bigger picture. We now turn our attention to groundwater.

3.4 Groundwater in Watersheds

Groundwater is defined as any water beneath the surface of the ground (Heath 1983) and is often described in terms of its occurrence in an aquifer. An **aquifer** is a naturally occurring mappable geologic unit(s) composed of water-saturated porous media capable of storing and transmitting significant quantities of water under ordinary conditions (modified from Freeze and Cherry 1979). For example, the unconsolidated geologic material composing many aquifers consists of sands and gravels, and these formations are deemed aquifers because they meet the requirements just defined. Unconsolidated geologic material composed of silt and clay can also make the grade for aquifer status, but sometimes these finer-grained materials do not transmit water quickly enough to fulfill the definition. In fact, clay layers have such a low permeability that we will refer to them as confining layers that retard the downward flow of groundwater.

Early civilizations such as the Persians and Romans recognized the importance of groundwater as a source of drinking water and as a supplement to precipitation for crop irrigation. The presence of water in the subsurface is influenced by fundamental geologic processes, including the subsurface rock type and the character of the tectonic activity influencing the region (e.g., folding or faulting). Today, the distribution and abundance of groundwater is a primary concern facing the growing human population on Earth and especially those within arid and semiarid watersheds. Although the use of groundwater dates back thousands of years, it has only begun to be understood in the last couple of centuries—and major gaps in our knowledge still persist today. Scientists now understand the role of groundwater in the hydrologic cycle but still lack detailed information about groundwater in urban areas. This information includes groundwater's geographic distribution and migration pathways, anthropogenic influence, contaminant behavior and transport, and its connection with and influence on surface water.

At the watershed scale, more knowledge is required to understand the geology of groundwater and nongroundwater-bearing strata as well as its quality, flow, residence time, age, and the anthropogenic contributions and impacts to this resource. In addition, we need a deeper understanding of the fundamental hydrologic and hydrogeologic processes where groundwater plays a critical role, such as infiltration, surface runoff, and the recharge of surface waters.

The many reasons underlying the lack of understanding concerning groundwater can be organized into a few broad categories:

1. Hydrogeological
    - An absence of readily available groundwater in a particular area. There is a tendency to study only things we can see.
    - An abundance of groundwater in a particular area. In this case, groundwater may be taken for granted and accepted without critical inquiry.
    - An in-depth knowledge and understanding of geological and hydrological principles is required. These skills are limited to a relatively small segment of the population.
2. Geographical
    - Other readily available and exploitable sources of water such as a large river or lake are nearby. As a result, groundwater is ignored.
3. Financial
    - Groundwater is expensive to investigate, study, and develop as a water source. Money woes can limit curiosity.

Groundwater has only been studied outside of academic circles when something negative occurs. This includes events where contaminated groundwater is consumed from an impacted aquifer, or when the water supply runs out or is significantly reduced. This is particularly disturbing because approximately 50% of the population in United States obtains its drinking water from groundwater, and more than 40% of irrigation water comes from groundwater.

The importance of groundwater cannot be overstated, because groundwater accounts for over 95% of all the freshwater on the planet—excluding the hard to reach water stored in ice caps and glaciers. The estimated total world reserve of groundwater is estimated to be approximately 10,528,827 km^3 (2.5 million mi^3) (Table 3.1), which is about 100 times more than all the freshwater in lakes, streams, rivers, and swamps on Earth. To envision this amount of water is difficult, but suppose all of the groundwater in the world was spread out on the surface of the United States. The water depth would be close to 0.8 km (0.5 mi). Let us go global. If all the groundwater was placed on the land surface of the Earth, it would be 46 m (150 ft) deep.

The generally poor understanding of groundwater is compounded in urban areas, because the near-surface unconsolidated sediments here are also poorly understood. This is somewhat alarming, since a significant portion of the human population currently resides in urban areas and this population is projected to increase rapidly in the future (United Nations 2008).

Groundwater plays an important part in the water cycle on the planet; a cycle involving a complex set of interactions between water and the surface, subsurface, and atmosphere (Figure 3.5). The next section builds the foundation for understanding groundwater within this context.

3.5 Fundamental Concepts of Groundwater

Describing the occurrence, distribution, behavior, and migration of groundwater beneath the surface is best accomplished from the perspective of geographic scale. At the macro or regional scale, groundwater exists in well-defined geologic formations (aquifers) and, at the microscale, groundwater assumes many forms, including any water unavailable for flow such as the water surrounding minerals grains due to surface tension.

The term "significant quantities" used to describe water in an aquifer is a relative term and depends on whether the water is intended for human exploitation. The outcome is largely dependent upon location and geography. For instance, an aquifer in a desert region may not yield as much water as an aquifer located elsewhere, but it still may be considered significant due to its geographical location. The majority of aquifers in the world considered high-yielding are composed of sand and gravel deposits; for example, the high-yielding Ogallala aquifer stretching from the Texas panhandle to North Dakota in the Great Plains region is sand and gravel. Figure 3.13 shows the distribution of aquifers in the

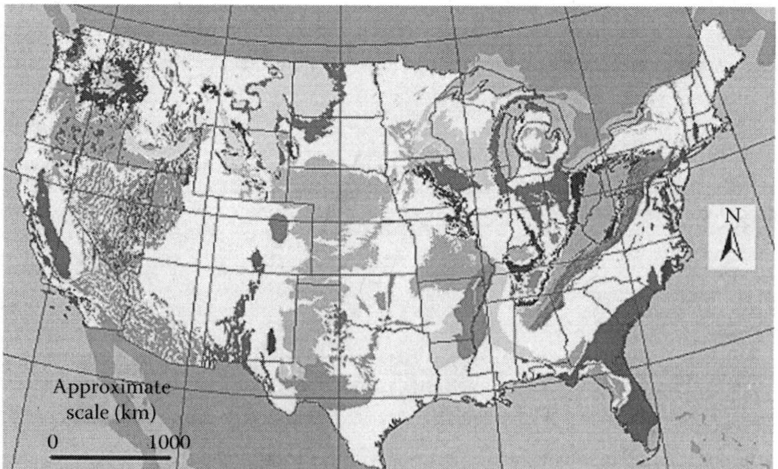

FIGURE 3.13
Distribution of major U.S. aquifers. (From United States Geological Survey, Principal aquifers of the 48 conterminous United States, http://water.usgs.gov/GIS/browse/aquifers_us.jpg, 2010c.)

conterminous United States. Most of these aquifers are composed of sedimentary rock and underlie a majority of the country's major urban areas.

3.5.1 Aquifer Types and Basic Structure

There are two types of aquifers, confined and unconfined, based on the presence or absence of a water table. A **water table or phreatic surface** is the upper extent of the zone of saturation in direct contact with atmospheric air pressure through void spaces in the overlying geologic material. An **unconfined aquifer**, therefore, is defined as a saturated geologic material where the surface of the saturated layer is equal to atmospheric pressure (Heath 1983). The depth to the water table surface can range from just a few meters or less in humid regions to more than 305 m (1000 ft) in dry desert regions. Unconfined aquifers are usually the first saturated waters encountered beneath the surface. If this water is of limited areal extent, it is commonly known as a perched aquifer or perched water depending on the amount and extent of saturated materials.

Perched water or perched aquifers are commonly encountered in urban environments. Many building foundations, sewers, and roads are backfilled with materials such as sand and gravel with higher **permeability** (capable of being penetrated by water) than the indigenous soils often composed of clay. In these cases, water eventually saturates the more permeable materials surrounding the foundation and perched water develops. In a pure geologic sense, this underground configuration of water does not fit the definition of an aquifer because it is not a mappable geologic unit, does not occur in a naturally occurring formation, and exists because of anthropogenic activities.

Confined aquifers are permeable geologic formations or strata bounded above and below by relatively impervious geologic material consisting of clay layers or other similar materials and contain water at pressure greater than atmospheric (Heath 1983). These confining layers or strata separate the higher permeability aquifer material from direct contact with atmospheric pressure and impede the upward and downward movement of water.

As shown in Figure 3.14, the confining layer above the confined aquifer creates a column of water with very limited exposure to the ground surface. This overburden of water creates additional pressure within these formations, so any well installed within a confined aquifer will have a water level higher than the elevation where the water was encountered. If the water level in the well rises but does not reach the surface, the well is termed an **artesian well**. When the pressure in a confined aquifer is sufficient to allow the water level

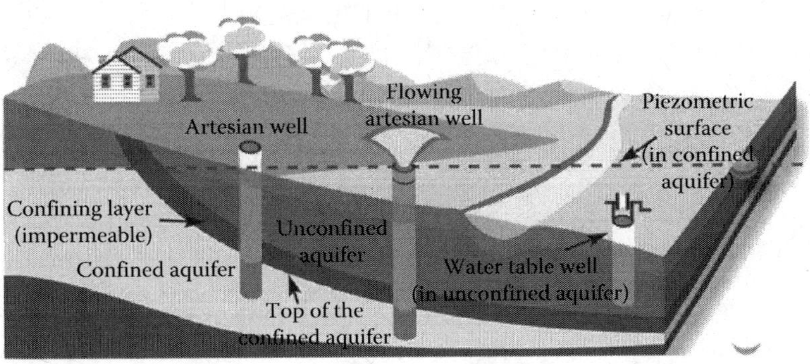

FIGURE 3.14
Example of an unconfined and confined aquifer and artesian and flowing artesian wells. (From Environment Canada, Aquifers and wells, http://www.ec.gc.ca/water/images/nature/gedwtr/a5f3e.htm, 2009.)

to reach or exceed the ground surface, the well is called a **flowing artesian well** (Driscoll 1986). Figure 3.14 shows a schematic of an unconfined and a confined aquifer and also shows artesian and flowing artesian wells. The elevation water can reach in wells drilled into a confined aquifer is called the piezometric surface and is indicated by the dashed line in the figure.

Many types of aquifers can exist in certain geological environments where thick sequences of variable unconsolidated deposits are present. In these cases, the term commonly applied is an **aquifer system**.

In areas of karst topography, groundwater flows through bedrock and initially concentrates between bedding layers, fractures, or faults within the sedimentary deposits. Rock types prone to the development of karst topography contain a high amount of calcite ($CaCO_3$), which is the primary mineral in limestone. Through time, chemical weathering along some of the groundwater migration pathways begins to dissolve away the rock and enlarges the pathway. If this enlargement continues, a cavity or void space called a **cavern** is formed. As the cavern grows with continued groundwater migration, it may connect with other caverns and lead to increased groundwater flow as void space increases. If water levels drop due to natural groundwater fluctuation or increased rates of groundwater flow, the caverns fill with air instead of water and become a cave. A **cave** is a natural cavity beneath the surface created by groundwater dissolving rock (Burger and Dubertret 1984; Ford and Williams 2007).

Karst formations are cavernous and have high rates of permeability. At many locations with well-developed karst topography, surface streams disappear and flow underground through networks of caves only to reappear at the surface sometimes several kilometers away. Flow rates under these circumstances do not fit into the range of usual groundwater velocities; they approach those of surface streams. Overall, groundwater flow conditions in karst topography become difficult to evaluate not only because the flow rates vary widely but also because flow directions can be unpredictable. These flow properties of karst-resident groundwater are a function of the fracture and dissolution patterns of the host rock (Ford and Williams 2007). Figure 3.15 is a diagram of karst topography.

Evaluating an aquifer in karst topography requires mapping the fracture pattern of the host rock. The mapping process begins with tracer testing. A tracer die is placed at an upgradient location, and the monitoring devices or observation points are set up at several downgradient locations. The elapsed time between the upgradient introduction of the die and the first observation of the die at downgradient locations allows for an estimation of the rate of groundwater flow, general groundwater flow direction, and provides information on the fracture pattern of the aquifer (Ford and Williams 2007). Figure 3.16 shows fractured bedrock on the left and a solution enhanced fractured bedrock on the right (Heath 1983).

3.5.2 Heterogeneity and Homogeneity

Heterogeneity and homogeneity are measures of variability used to classify geological diversity. In geological terms, **heterogeneity** is a term applied to highly variable or poorly sorted geologic materials, whereas **homogeneity** refers to geologic formations without much variability and characterized as well-sorted (Heath 1983).

Aquifers are typically composed of coarse-grained sediments derived from processes characterized as high energy. The types of geological processes leading to the formation of aquifers in unconsolidated sedimentary deposits are fluvial, glacial, lacustrine, beach, or eolian (USGS 1999). These sediments consist of layered sands and gravels and may also be cross-bedded (Figure 2.28). As a rule, no two aquifers—or for that matter—no two

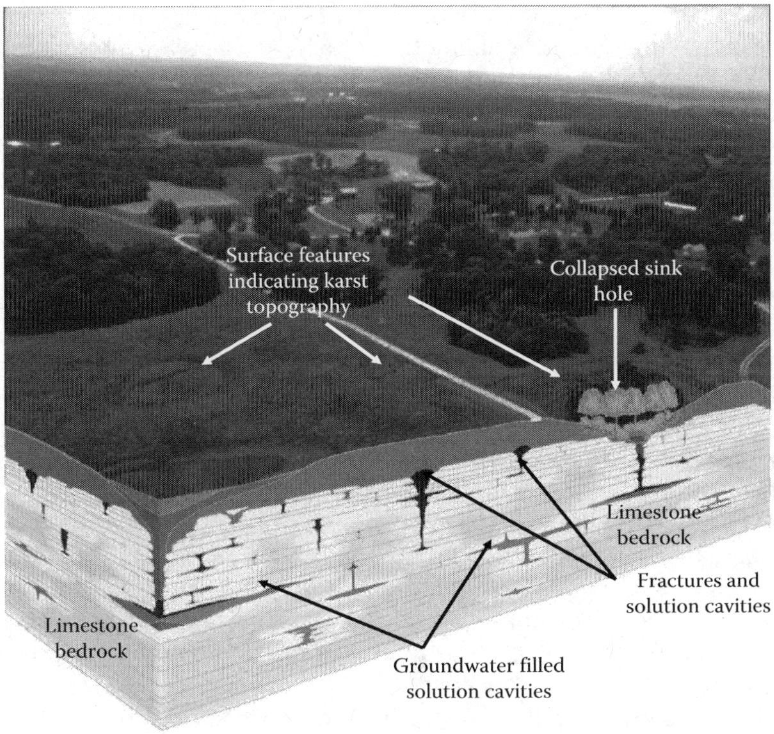

FIGURE 3.15
Karst topography. (From Illinois State Geological Survey, Geobit 7 karst, http://www.isgs.edu/maps-datapub/publications/geobits7.shtml, 2010.)

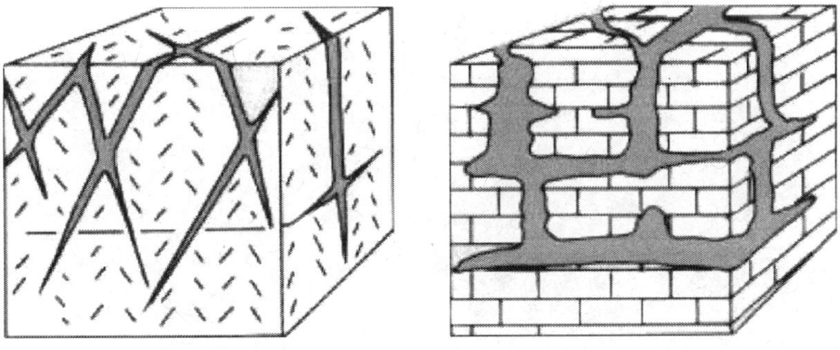

FIGURE 3.16
Fractured and solution enhanced bedrock.

locations within an aquifer are exactly alike. Since aquifers are a member of a larger set of geologic units (e.g., glacial deposits), some of those principles used to evaluate the heterogeneity and homogeneity of the larger set also apply to aquifers:

- The size and shape of any geologic unit is inherently different when evaluated from a three-dimensional perspective. Confined and unconfined aquifers range in thickness and areal distribution and composition of matrix materials (e.g., sand, gravel and fractured limestone).

- Each geologic unit has a unique arrangement and network of interconnecting features created by the depositional method. Karst aquifers are demonstrably different in this respect than sand and gravel aquifers.
- Comparison of heterogeneity and homogeneity is possible at several different geographical scales. The mega scale is a geologically dependent formational scale and is regional in extent. An example here is the heterogeneity of large river deposits compared to a glacial moraine deposit. The macroscale refers to the variability within the formation itself. In this case, you could compare the variability of two sand bars deposited by the same river. The microscale refers to the variability and arrangement of depositional layers and individual grains within a specific deposit.

When an aquifer has high variability, it is called heterogeneous. An aquifer of low variability may be labeled as a homogeneous or relatively homogeneous aquifer. The probability of encountering a truly homogeneous aquifer is effectively zero, since all aquifers demonstrate some degree of variability, with most displaying a high degree of variability.

The terms **isotropic** and **anisotropic** are used in describing the degree of variability related to a given direction for a specified geologic feature. For instance, a geologic formation may exhibit a more homogeneous texture within a specific bedding layer trending in a certain direction. This phenomenon then qualifies as a layer displaying either more of an isotropic tendency or anisotropic tendency in a given direction from a defined location. Figure 3.17 has examples showing the different combinations of homogeneous, heterogeneous, isotropic, and anisotropic concepts.

The concept of heterogeneity describing the variability within an aquifer is significant and requires a sound scientific understanding. Heterogeneity has a profound effect on the migration and fate of contaminants present in groundwater and influences how those contaminants are removed, destroyed, or chemically altered through specially designed remedial technologies (Payne et al. 2008).

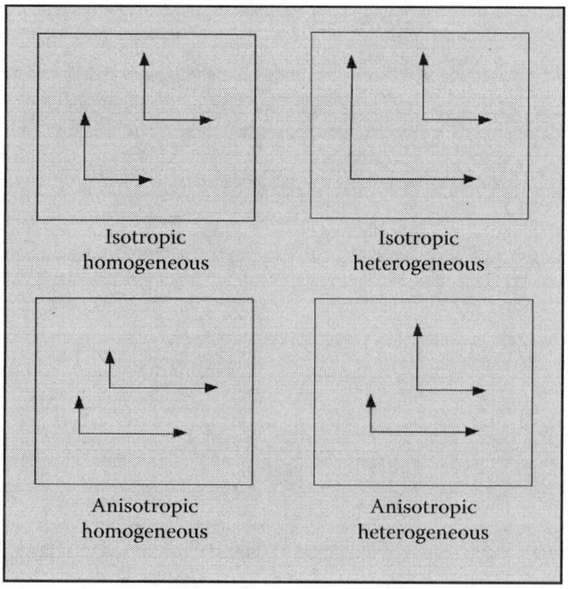

FIGURE 3.17
Visualization of variability concepts.

3.5.3 Hydraulic Conductivity

Hydraulic conductivity (K) is a parameter used to measure the relative ease water flows through an aquifer or other geologic formations such as a confining unit. **Hydraulic conductivity** (K) is defined as the ability of saturated geologic media to conduct water under an induced hydraulic or pressure gradient (Payne et al. 2008). Geologic formations composed of sands and gravels typically have a high hydraulic conductivity or K value (e.g., 10^{-2}–10^{+1} cm/s). Geologic formations composed of clay that might be considered a confining unit frequently have very low hydraulic conductivities (e.g., 10^{-6}–10^{-8} cm/s). Figure 3.18 shows the average range of hydraulic conductivities for different geologic materials.

In 1856, a researcher from France named Henry Darcy conducted an experiment to evaluate the flow of water through sand beds. His experiments resulted in the development of Darcy's law. **Darcy's law** predicts the flow of water (Q [cm^3/s]) through a sand bed is *proportional* to its hydraulic conductivity (K), area (A [cm^2]) and the water level difference between each end of the sand bed (dH [cm]), and *inversely proportional* to the thickness (L) of the sand bed.

Equation 3.1 presents Darcy's law.

$$Q = K \times A \times \frac{dH}{dL} \tag{3.1}$$

where
K is the hydraulic conductivity (cm/s)
A is the area (cm^2)
dH is the change in height of the bed of sand
dL is the length of the bed of sand

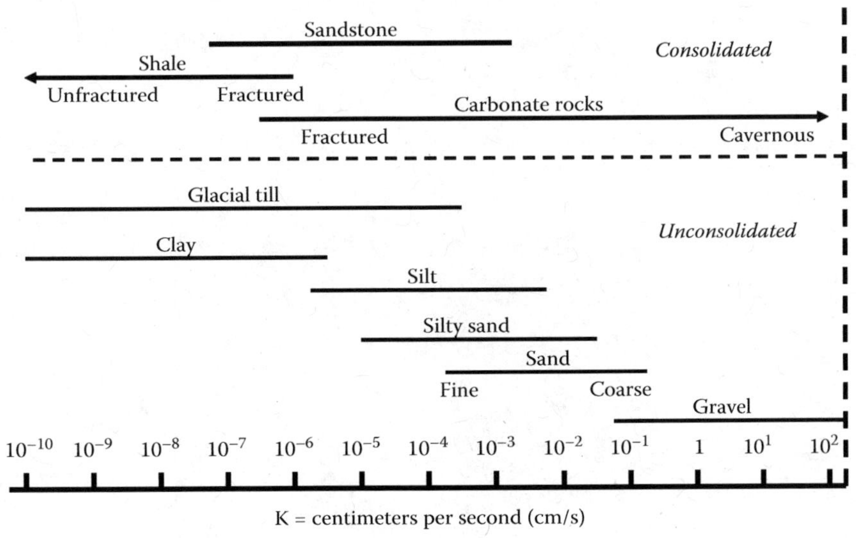

FIGURE 3.18
Range of hydraulic conductivities for different geologic materials. (Modified from Heath, R.C., *Basic Ground-Water Hydrology*, United States Geological Survey, Water Supply Paper 2220, United States Government Printing Office, Alexandria, VA, 1983.)

3.5.4 Groundwater Movement

Groundwater is in constant motion, although groundwater flow rates are usually much slower than those of surface streams. Unlike surface streams, groundwater is pulled downward by gravity through an intricate network of passageways between the pore spaces in the soil and sediment. This area is called the **vadose zone** (Heath 1983). The vadose zone (also called the zone or aeration or unsaturated zone) is the subsurface area where water is present, but air still occupies part of the overall pore space. Figure 3.19 shows the location of the vadose zone relative to the water table and saturated zone.

A thin layer of water—called **hygroscopic water**—will always be attracted to the surface of mineral grains. This attraction is a result of unsatisfied ionic charges on the surface of the mineral grains combined with the polar nature of water molecules. This is termed the force of **molecular attraction** and is shown in Figure 3.20.

The thickness of the vadose zone varies by location. In desert locations, the vadose zone may be 100 m to more than 305 m (1000 ft) thick; in humid regions, where there is significant precipitation, the thickness may be just a few meters or even less. There are also seasonal

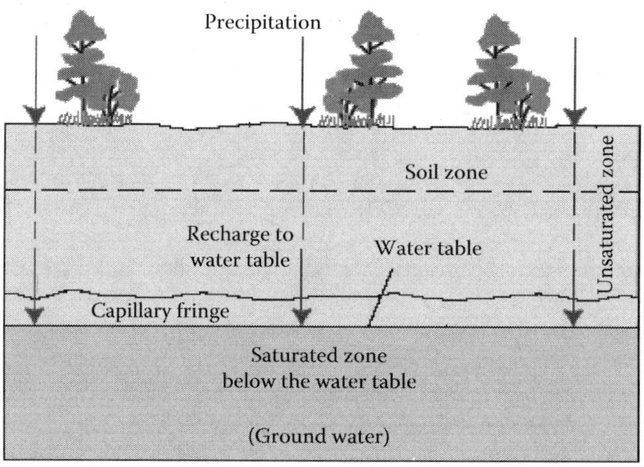

FIGURE 3.19
Diagram of the vadose zone. (From United States Geological Survey, Summary of the water cycle, http://ga/water.usgs.gov/edu/watercyclesummary.html, 2010a.)

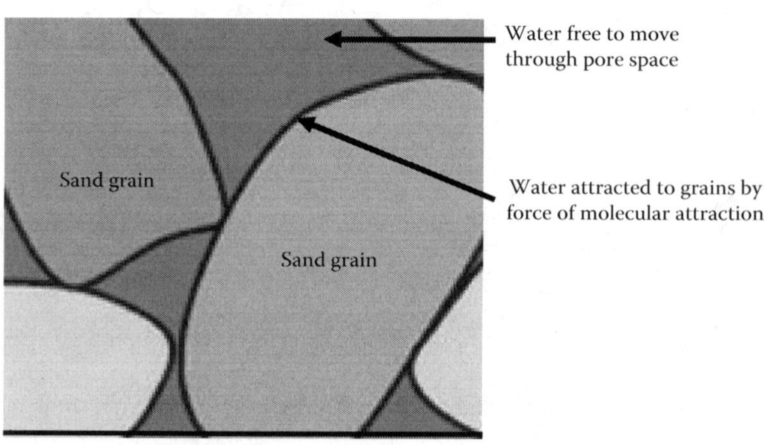

FIGURE 3.20
Force of molecular attraction.

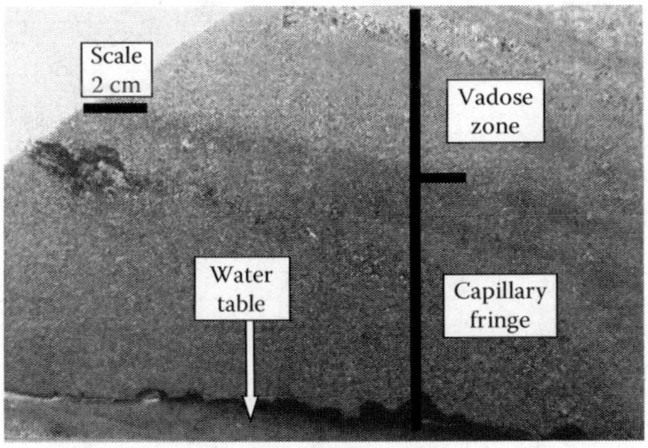

FIGURE 3.21
Photograph showing vadose zone, capillary fringe, and water table surface. (Photo by Daniel T. Rogers.)

variations in the thickness of the vadose zone. During prolonged periods of drought, the vadose zone may become much thicker as opposed to periods of prolonged rainfall when the thickness of the vadose zone may shrink significantly.

As surface water continues its journey downward through the unsaturated zone, it encounters the capillary fringe. The **capillary fringe** is the subsurface layer where groundwater seeps up from the water table by capillary action to fill pore spaces (USGS 1999). At the base of the capillary fringe, the pore spaces are filled with water due to tension saturation. The thickness of the capillary fringe is dependent upon the balance between the soil/sediment's adhesion binding capability and surface tension, and the gravitational force of the lifted water mass. What factors prevail here? In most cases, the thickness of the capillary fringe is less in course-grained materials than finer-grained materials (Payne et al. 2008). Figure 3.21 shows the vertical arrangement of the subsurface zones just described.

Once water reaches the water table, it migrates downward through the available interconnected pore spaces under the force of gravity and from zones of higher pressure to areas of lower pressure. Groundwater can also move upward from areas of higher pressure to areas of lower pressure. Imagine a U-shaped tube partially filled with water. If pressure is exerted on one side of the tube, the water on the other side rises proportionally in response to the pressure exerted on the other side. This one-sided pressure is provided by the cork in the image on the right side of Figure 3.22.

Similar variations in pressure occur beneath the surface of the Earth. Pressure is higher beneath hills and lower in valleys (Heath 1983). Elevated land contains more mass than valleys and therefore exerts more force per unit area—the definition of pressure. The lower pressure below valleys creates a pressure gradient toward the surface, allowing deeper

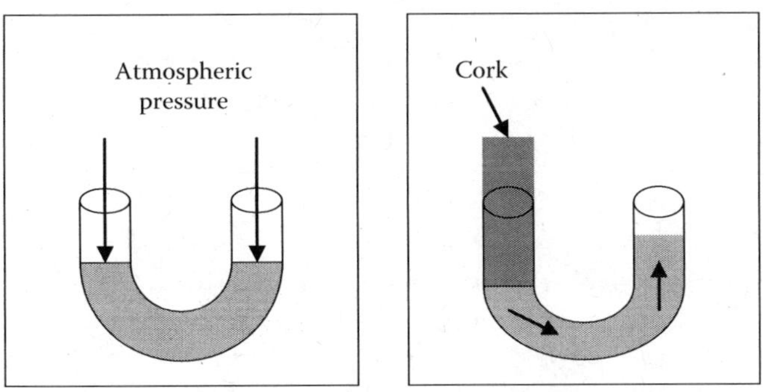

FIGURE 3.22
Water movement under pressure.

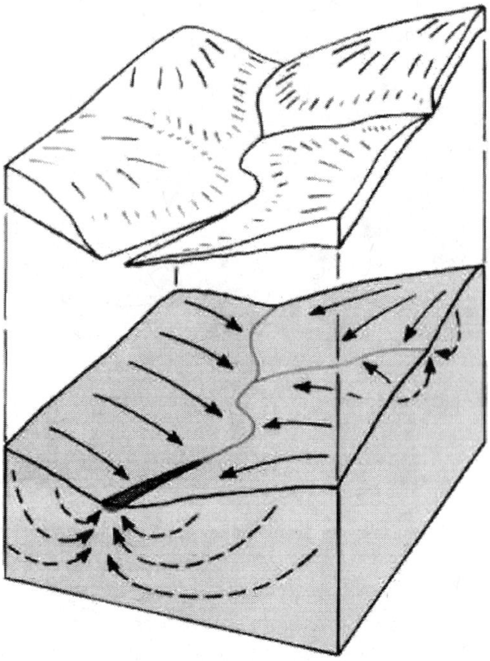

FIGURE 3.23
Groundwater flow along topographically generated pressure gradients. (Modified from Heath, R.C., *Basic Ground-Water Hydrology*, United States Geological Survey, Water Supply Paper 2220, United States Government Printing Office, Alexandria, VA, 1983.)

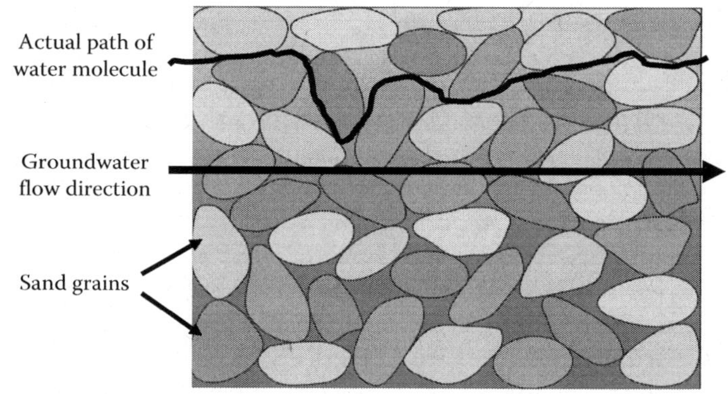

FIGURE 3.24
Groundwater flow on the micro scale.

groundwater to flow upward and reach surface water. The arrows in Figure 3.23 show the directions of groundwater flow in response to these pressure differentials.

At the microscale within the saturated zone, groundwater does not travel in a straight line. Figure 3.24 shows one pathway a particle of water may take when moving within the saturated zone.

3.5.5 Primary Porosity

To reiterate, groundwater travels through the subsurface through the interconnected pores between the individual minerals grains. **Porosity** is the percentage of the volume of the sediment that is open space (pore space) and is depicted in Figure 3.25.

Well-sorted and well-rounded sediments have a higher degree of porosity compared to poorly sorted sediments, because there are no fine-grained particles to fill the pore space between the well sorted and rounded grains. Figure 3.25 depicts well-sorted materials and

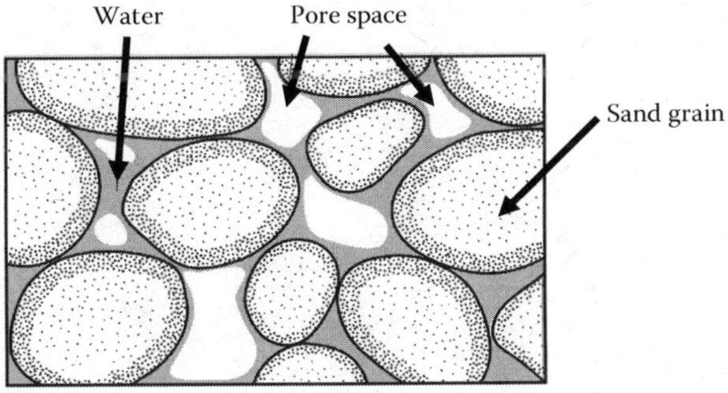

FIGURE 3.25
Example of pore space in sediments. (Modified from Alley et al., *Sustainability of Groundwater Resources*, USGS Circular 1186, Denver, CO, 1999.)

also qualitatively demonstrates the difference between a well-sorted, well-rounded, large-grained material compared to a smaller-grained, less well-rounded, well-sorted material.

Porosity (Ø) is defined by Equation 3.2:

$$\varnothing = \frac{V_V}{V_T} \tag{3.2}$$

where

V_V is the volume of void space
V_T is the total bulk volume of the material including the solid and void components

Porosity is composed of the interstitial space between soil materials; the water here consists of the hygroscopic water bound tightly to the material surface and the other water capable of draining freely under the influence of gravity. Water capable of draining freely under the influence of gravity is termed **specific yield (Sy)** or **effective porosity (n_e)** and represents the approximate measure of the interconnected pore space groundwater flows through within geologic strata or layers. The remaining water is termed **specific retention (Sr)**. **Total porosity** is the sum of effective porosity and specific retention and is commonly expressed as Equation 3.3 (Heath 1983).

$$N = Sy + Sr \tag{3.3}$$

where

N is the total porosity
Sy is the specific yield
Sr is the specific retention

As stated earlier, a poorly sorted sediment is not as porous as a well-sorted sediment because the fine-grained particles occupy spaces between the larger grains. This arrangement of particles is depicted in Figure 3.26.

An additional type of primary porosity not often discussed and more difficult to evaluate occurs between nonconformable contacts or unconformities. As noted in Chapter 2, an unconformity represents a buried erosional surface separating two different rock masses or deposits. This surface represents a plane or surface of discontinuity often exhibiting a different and highly localized porosity. As a result, this surface is often saturated with water, especially if the geologic units above and below are low permeable formations. These types of contacts do not contain large quantities of groundwater, but they often contain enough water to enable their flows to be measured.

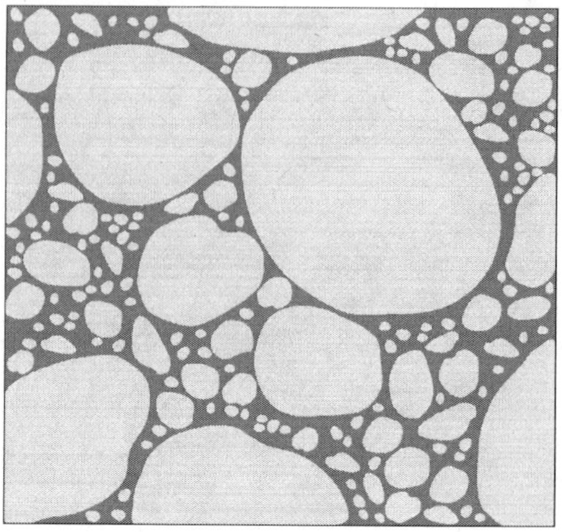

FIGURE 3.26
Example of a poorly sorted sedimentary material.

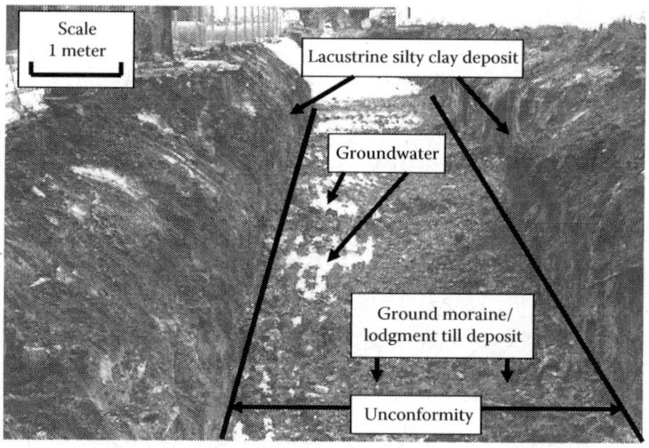

FIGURE 3.27
Presence of groundwater due to presence of an unconformity. (Photo by Daniel T. Rogers.)

An example of this type of porosity is presented in Figure 3.27. This figure is a photograph of an excavation through a fine-grained lacustrine deposit composed of silty clay overlying a ground moraine or lodgment till deposit of glacial origin. The unconformable contact between the two different and distinct geologic units is represented at the floor of the excavation. Groundwater is not encountered in the upper lacustrine deposit nor is it encountered within the ground moraine deposit. Instead, groundwater is encountered at the unconformable contact between the two geologic units.

This type of porosity is significant because it represents an apparent anomaly when evaluating the porosity and permeability of the two geologic units individually. By themselves, the geologic units may not exhibit significant porosity, but the contact between the two units creates enough porosity for groundwater to be present and for flow to occur.

3.5.6 Secondary Porosity

Secondary porosity refers to the porosities formed within a geological unit after the material has been deposited (Freeze and Cherry 1979). Types of secondary porosity include

- Fractures caused by differential compaction
- Fractures caused by freeze-thaw cycles
- Fractures caused by vegetative root fragments
- Bioturbation
- Vugs (small to medium-sized cavities inside rocks), fissures, or solution cavities caused by dissolution from migrating waters or other chemicals that dissolve or precipitate minerals

Freeze and Cherry (1979) note the extent of freeze-thaw within some soils in colder climates can reach up to 9 m (30 ft) beneath the surface. This may be particularly significant when examining the integrity of a confining unit at shallow depths. Any of the above listed types of secondary porosity can also influence the integrity of an aquiclude from shallow to sometimes significant depths beneath the ground surface. Figures 3.28 and 3.29 show a root fragment in a clayey soil and a vertical fracture in a clayey soil, respectively.

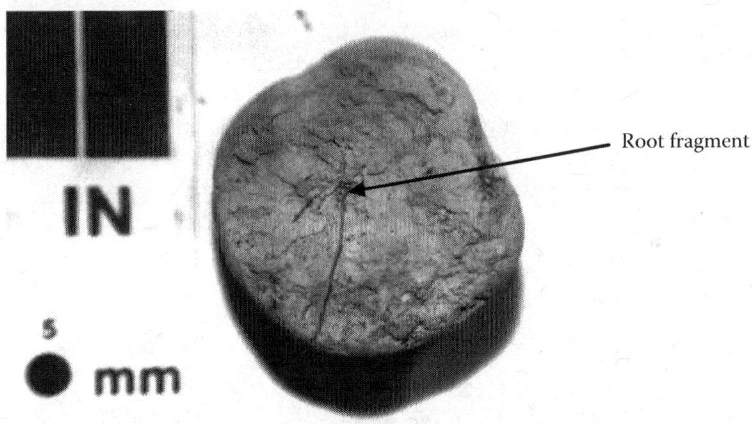

FIGURE 3.28
Root fragment in a deposit composed of clay. (Photo by Daniel T. Rogers.)

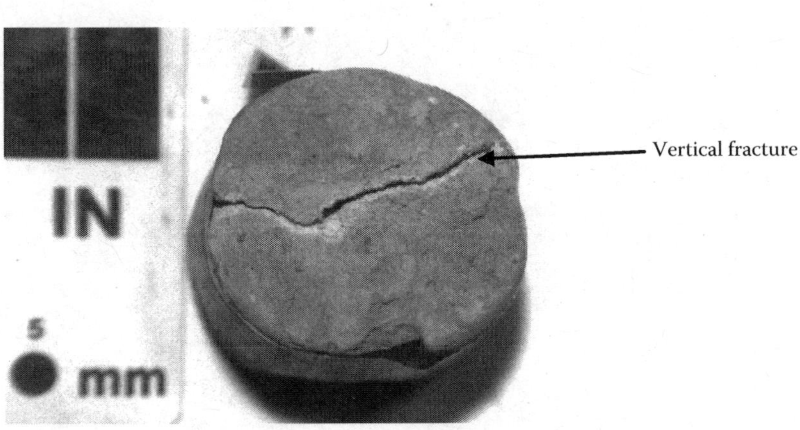

FIGURE 3.29
Vertical fracture in a sediment composed of clay. (Photo by Daniel T. Rogers.)

Secondary porosity in the form of vertical fractures and root fragments combined with the porosity observed at unconformable contacts may assist in explaining why groundwater is observed in some locations. Geologic units previously characterized as impermeable or not having enough groundwater to permit flow may indeed yield measurable water. The awareness of this possibility may help when evaluating anthropogenic influences on the geology and hydrogeology of urban areas for groundwater and surface water protection purposes (Howard and Gerber 1997; Murray et al. 2000).

3.5.7 Water Table and Hydraulic Gradient

The water level in a well penetrating an unconfined aquifer undergoing horizontal flow will be equal to the level of the water table. When these water levels are joined across multiple wells in two-dimensional space, a water table plane or **potentiometric surface** is defined (Freeze and Cherry 1979). This constantly changing surface is a measure of the upper physical boundary of the water table. Fluids flow between the differences in pressure or elevations existing at distinct locations. For example, groundwater flows from areas of high pressure to areas of lower pressure or in hydrologic terms groundwater flows from areas of higher potential to areas of lower potential. A *pressure* differential is called a pressure gradient, whereas the *elevation* difference between wells is called a hydraulic gradient. The hydraulic gradient between the water levels in Well 2 and Well 1 depicted in Figure 3.30 is often referred to as hydraulic head. To calculate the hydraulic head between two wells, take the ratio of the rise or height difference between the two wells (dH) to the run or horizontal distance between the two wells (dL). This ratio is shown in Figure 3.31.

3.5.8 Determining the Direction of Groundwater Flow

Calculating the direction of groundwater flow requires at least three data points that accurately measure the level of groundwater under atmospheric pressure. The vertical accuracy required is usually within 3 mm (0.01 ft). Three wells is the minimum number necessary to determine the direction of groundwater flow, and if only three wells are used, their location, pattern, or placement should be in a triangular pattern.

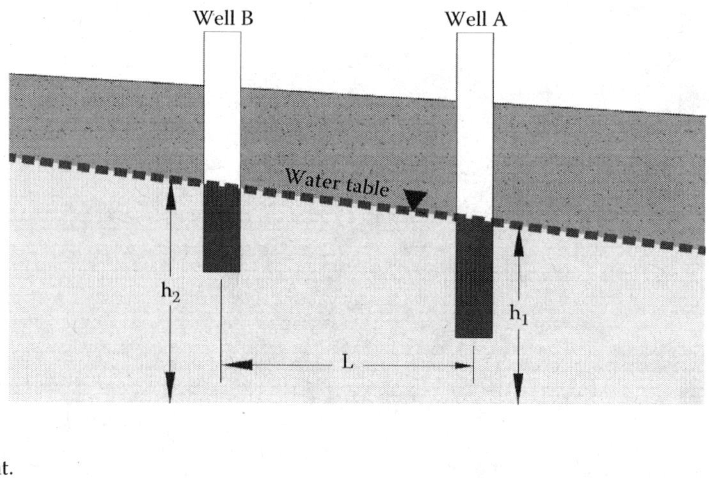

FIGURE 3.30
Hydraulic gradient.

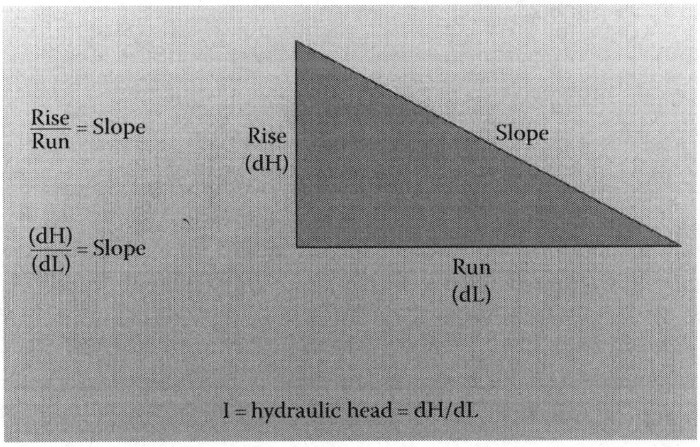

FIGURE 3.31
Calculating the hydraulic head.

This information is needed to determine the general direction of groundwater flow between the three wells:

- Relative geographic position of the wells
- Distance between the wells
- Elevation of water in each well

Calculating the direction of groundwater flow involves the following steps (Heath 1983), and an example of the procedure is shown in Figure 3.32:

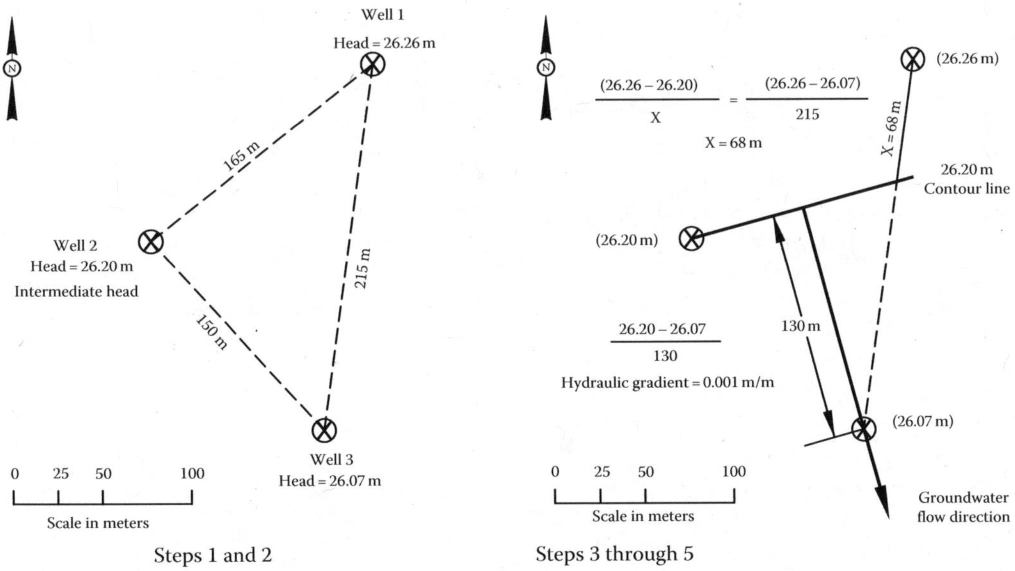

FIGURE 3.32
Determining the direction of groundwater flow. (Modified from Heath, R.C., *Basic Ground-Water Hydrology*, United States Geological Survey, Water Supply Paper 2220, United States Government Printing Office, Alexandria, VA, 1983.)

1. Identify the well with the intermediate water level.
2. Interpolate the location of the intermediate water level well along a straight line between the wells having the highest and lowest relative water levels.
3. Create a line segment of equal potential to the intermediate well by drawing a straight line from the well with the intermediate water level to its water equivalent elevation plotted along the straight line between the wells having the highest and lowest relative water elevations.
4. Draw a perpendicular line to the water-level contour just plotted with the well of highest or lowest water elevation. This line parallels the general direction of groundwater flow.
5. Calculate the difference between the water level of the well and the intermediate well-level contour by using the distance between the well and the contour to reveal the hydraulic gradient.

3.5.9 Estimating the Rate of Groundwater Flow

Calculating the rate or velocity of groundwater flow requires site-specific knowledge or estimation of the following parameters:

- Hydraulic conductivity (K) of the saturated geologic strata
- Effective porosity (n_e) of the saturated geologic strata
- Hydraulic gradient (dH/dL) or slope of the groundwater surface

Once the above parameters are measured or estimated, the rate or velocity of groundwater flow can be calculated using the stream economy equation and Darcy's law. The stream economy equation is

$$Q = Av \tag{3.4}$$

where
Q is the discharge (volume of water per unit time (cm^3/s)
A is the cross sectional area (at right angle to the water flow direction) cm^2
v is the velocity (cm/s)

$$Q = K \times A \times \frac{dH}{dL} \tag{3.5}$$

where
K is the hydraulic conductivity (cm/s)
A is the area (cm^2)
dH/dL is the hydraulic gradient

Step 1 of Equation 3.6 demonstrates how groundwater velocity can be computed by substituting Av (which equals Q) in Equation 3.4 for the Q in Equation 3.5. After the area (A) is cancelled on both sides of the equation, the velocity (v) is equal to hydraulic conductivity

(K) multiplied by the hydraulic gradient (step 2). Groundwater, however, does not migrate in open space; it travels through aquifer materials impeding its velocity. To account for this, the effective porosity of the saturated geologic strata (specific yield) is required to accurately estimate the velocity of groundwater flow (step 3).

$$\begin{array}{ccc} \text{Step 1} & \text{Step 2} & \text{Step 3} \\ Av = k \times A \times \frac{dH}{dL} & \rightarrow v = K \times \frac{dH}{dL} & \rightarrow v = \frac{K}{n_e} \times \frac{dH}{dL} \end{array} \tag{3.6}$$

To illustrate the difference in the velocity of groundwater flow through two different materials, aquifers composed of sand and clay with the same hydraulic gradients are compared:

Sand aquifer example:

$K = 0.1$ cm/s

$dH/dL = 1$ cm/1000 cm

$n_e = 0.22$

$$v = \frac{K}{n_e} \times \frac{dH}{dL}$$

$$v = 0.1\,\text{cm/s} \times 1/0.22 \times 1/1000 = 2.2^{-5}\,\text{cm/s}$$

Clayey aquifer example:

$K = 0.000001$ cm/s

$dH/dL = 1$ cm/1000 cm

$n_e = 0.20$

$$v = \frac{K}{n_e} \times \frac{dH}{dL}$$

$$v = 0.000001\,\text{cm/s} \times 1/0.20 \times 1/1{,}000 = 5.0^{-9}\,\text{cm/s}$$

Given the same hydraulic gradient, the velocity of groundwater flow in the sand aquifer is approximately 10,000 times greater than the velocity of groundwater flow in the clay aquifer.

The residence time of water as groundwater can range from days to thousands of years and is a function of several variables. The dominant influence is depth, manifested by the strong correlation between increasing groundwater depth and residence time. Other factors influencing residence time include

- The composition of the aquifer material
- Effective porosity
- Hydraulic gradient
- Distance to point of discharge

The span of residence times for groundwater is depicted in Figure 3.33, and the residence times for water in other environments are listed in Table 3.4.

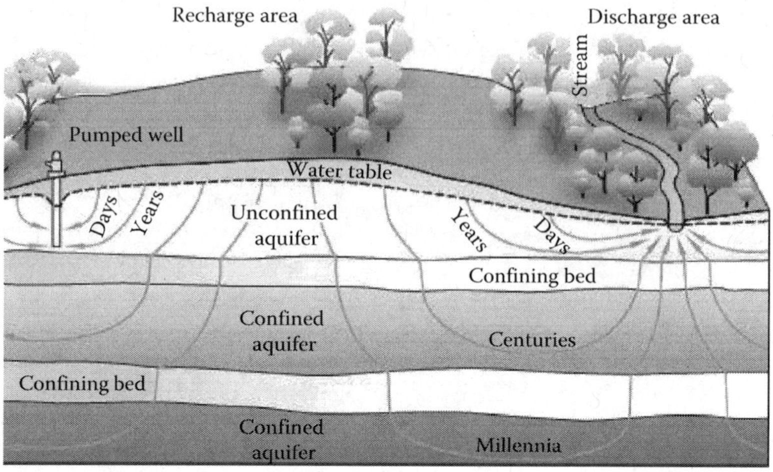

FIGURE 3.33
Range in groundwater residence time. (From Focazio, M.J. et al., *Assessing Ground-Water Vulnerability to Contamination: Providing Scientifically Defensible Information for Decision Makers*, United States Geological Survey Circular 1224, Denver, CO, 2001.)

TABLE 3.4
Estimated Depth and Residence Time of the World's Water Supply

Parameter	Equivalent Depth (m)	Approximate Residence Time
Oceans and Seas	2,500	4,000 years
Lakes and Reservoirs	0.250	10 years
Swamps	0.007	1–10 years
River channels	0.003	2 weeks
Soil moisture	0.130	2 weeks to 1 year
Groundwater	120	2 weeks to 10,000 years
Ice caps and Glaciers	60	10–10,000 years
Atmospheric water	0.025	10 days
Biospheric water	0.001	1 week

Source: Environment Canada, Aquifers and wells, http://www.ec.gc.ca/water/images/nature/gedwtr/a5f3e.htm, 2009.

3.5.10 Groundwater Withdrawal

Withdrawal of groundwater is accomplished by pumping a well installed within the aquifer. As groundwater is pumped from the aquifer, the water level in the aquifer begins to decline as water is removed from storage. A hydraulic gradient is created as the water level falls below the level of the surrounding aquifer. Water then begins to migrate from the aquifer into the well. The water level will continue to decline within the aquifer, and the flow rate of water into the well will increase until the inflow rate is equal to the withdrawal rate. This process occurs only when the pumping rate does not exceed the capacity of the aquifer formation to transmit groundwater to the well. In this case, the diameter of the well may be too small, and the pump should be placed lower in the well, or the withdrawal rate should be decreased.

Water from the aquifer converges on the well from all directions and the hydraulic gradient gets steeper near the well. For these reasons, the three-dimensional shape of water withdrawal from a well is termed a **cone of depression**. The cones of depression shown in Figure 3.34 result from a pumping well withdrawing water from unconfined (a) and confined aquifers (b).

Transmissivity (T) is a measure of the amount of water that can be transmitted horizontally through a unit width by the full saturated thickness of the aquifer under a hydraulic gradient of 1. Taking into account the definition of hydraulic conductivity, transmissivity (T) is actually equal to the hydraulic conductivity (K) multiplied by the thickness of the aquifer (b) (Equation 3.7).

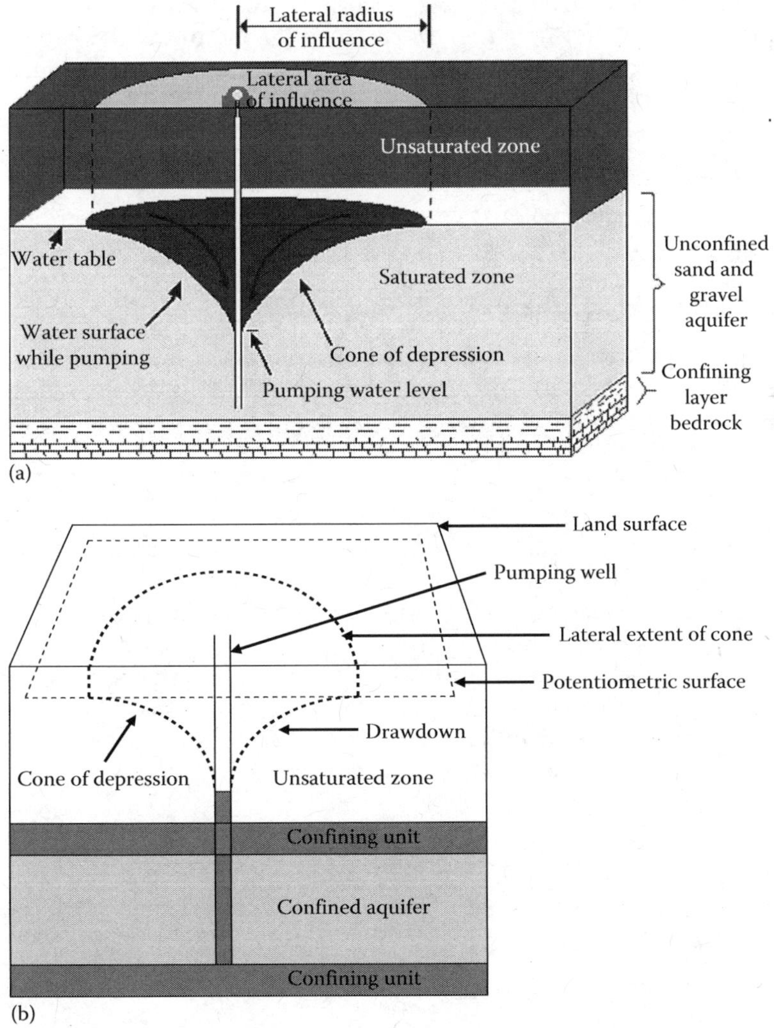

FIGURE 3.34
(a) Cone of depression from a pumping well in an unconfined aquifer (IEPA 2010). (b) Cone of depression from a pumping well in a confined aquifer. (Modified from Heath, R.C., *Basic Ground-Water Hydrology*, United States Geological Survey, Water Supply Paper 2220, United States Government Printing Office, Alexandria, VA, 1983.)

$$T = Kb \quad (3.7)$$

where
T is the transmissivity
K the hydraulic conductivity
b the aquifer thickness

The aquifer parameters of transmissivity (T) and storage coefficient (S) are variables that can dictate the shape of the cone of depression. The storage coefficient is the volume of water that a permeable unit will absorb or expel from storage per unit surface area per unit change in head. For a confined aquifer, the expansion of the cone of depression in response to depressurizing (pumping) is very small, but the compression pressures can be significant. This permits the cone of depression to expand and deepen rapidly when pumped, so a lower storage coefficient (S) value will create a deeper and wider cone than a higher storage coefficient. An aquifer with a low transmissivity will develop a deep and narrow cone of depression, and an aquifer with a high transmissivity will develop a shallow and wide cone of depression.

3.6 Surface Water–Groundwater Interaction

Surface water and groundwater are fundamentally connected, and their interaction occurs everywhere in all watersheds. It is difficult to separate the two because they "feed" off of each other and are interconnected through the water cycle.

3.6.1 Stream Interaction with Groundwater

Groundwater accounts for most and often all of baseflow in rivers and streams. **Baseflow** is defined as groundwater seepage into a stream channel (Freeze and Cherry 1979). This influx of groundwater is very important to streams and rivers because it provides flow during periods of low precipitation or drought and helps sustain aquatic life. On hillslopes, many streams originate as groundwater as shown in Figure 3.35 (Winter et al. 1998).

When groundwater flows toward and discharges to a surface stream, the stream is termed a **gaining stream** as depicted in Figure 3.36. When a stream loses water to groundwater, the stream is termed a **losing stream** (Figure 3.37).

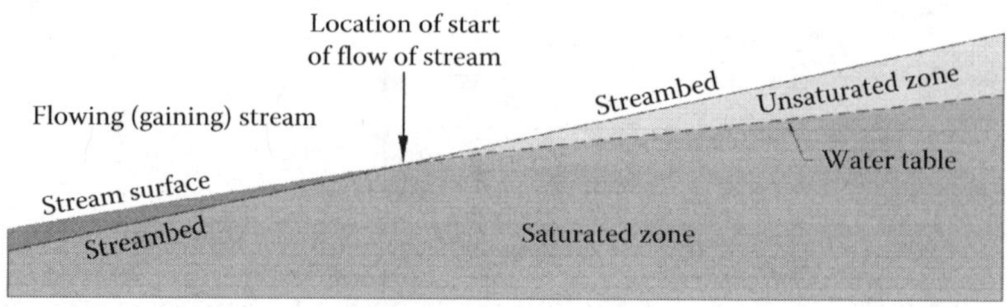

FIGURE 3.35
Beginning of stream flow from groundwater.

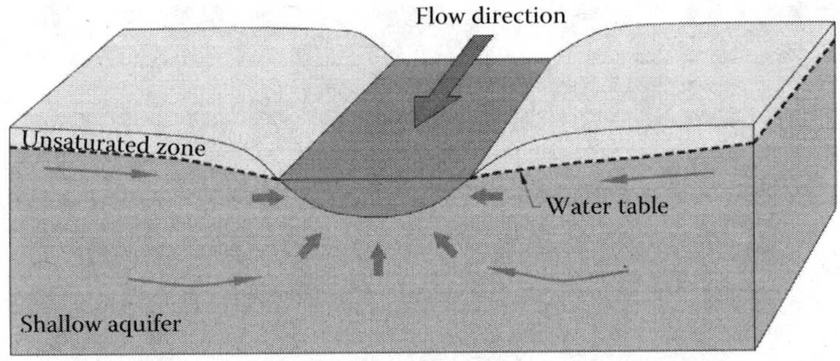

FIGURE 3.36
Example of gaining stream. (From Alley, W.M. et al., *Sustainability of Groundwater Resources*, United States Geological Survey Circular 1186, Denver, CO, 1999.)

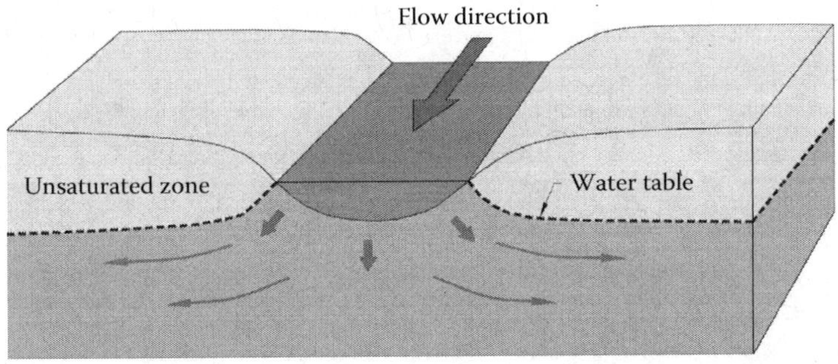

FIGURE 3.37
Example of a losing stream. (From Alley, W.M. et al., *Sustainability of Groundwater Resources*, United States Geological Survey Circular 1186, Denver, CO, 1999.)

Another stream interaction with groundwater is termed disconnected and occurs when the surface water stream and groundwater are separated by the unsaturated or vadose zone. This usually occurs when the sediments within the unsaturated zone beneath the stream have a low hydraulic conductivity and impair significant communication between the two. Another cause of this interaction is a lowering of the water table due to drought or the local geology. Figure 3.38 shows an example of a disconnected scenario

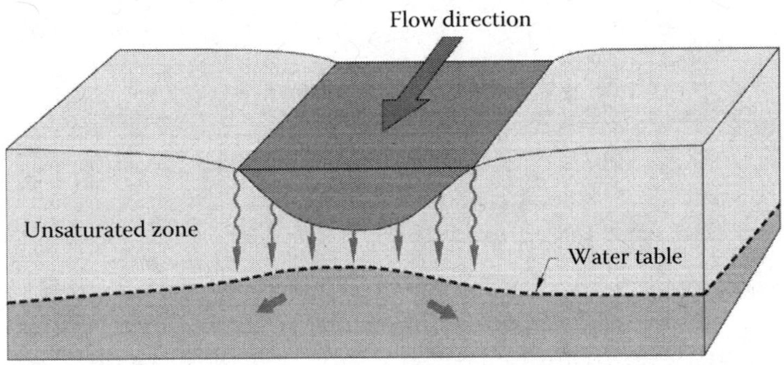

FIGURE 3.38
Example of a disconnection between surface water and groundwater.

FIGURE 3.39
Groundwater seep in the Rouge River watershed where groundwater contamination is discharging to surface water. (Photo by Daniel T. Rogers.)

(Alley et al. 1999). Since streams and rivers flow through varied terrain and geology, it is possible that one stream may exhibit the three types of connections depicted in Figures 3.36 through 3.38.

At any specific location, a stream is either gaining, losing, or disconnected depending on geographic, geologic, hydrologic, and anthropogenic factors.

During one study of the Rouge River watershed in southeastern Michigan, the direction of groundwater flow was evaluated at several hundred sites within the watershed (Rogers 1993, 1994). Groundwater flow direction was established using a minimum of 6 and a maximum of 80 monitoring wells per site, and each monitoring well was surveyed by a licensed surveyor. In a more detailed study of the lower branch of the Rouge River, Murray et al. (2000) installed 54 groundwater monitoring wells along the river along with 18 well points installed into the river's bed sediment.

The results of these studies conclusively show that groundwater flow is discharging through the bed sediment into the river and in a downgradient direction, consistent with the direction of surface water flow.

This is significant for an urban watershed like the Rouge for these reasons: (1) the direct connection between groundwater and surface water means that contaminants released within the watershed may impact shallow groundwater and (2) this contamination could also reach surface water and the lower Great Lakes (Rogers 1996, 1997; Rogers and Murray 1997; Murray et al. 2000). Figure 3.39 is a photograph showing one of numerous groundwater seeps within the Rouge River watershed, and Figure 3.40 provides a close-up view of this seeping water.

3.6.2 Lake Interaction with Groundwater

Lakes interact with groundwater in three basic ways. Some lakes receive groundwater through the entire bed of the lake, some lakes have water loss through the entire bed of the lake, and perhaps most commonly, lakes receive groundwater at some locations of the bed and have water loss to groundwater at other locations. These three scenarios are depicted in Figure 3.41 (Winter et al. 1998). Scenario A represents water loss from a lake to

FIGURE 3.40
Surface water from groundwater seep in the Rouge watershed. (Photo by Daniel T. Rogers.)

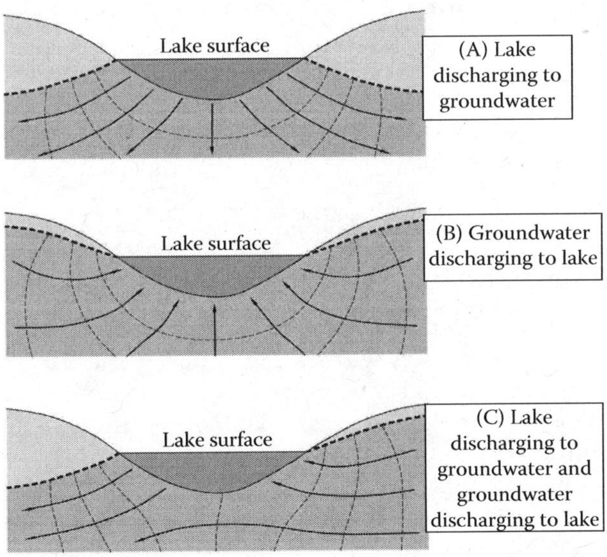

FIGURE 3.41
Lake–groundwater interaction.

groundwater, scenario B represents a lake receiving groundwater, and scenario C represents both water loss and water gain from groundwater to a lake.

In natural lakes not controlled by dams, the water level does not change as rapidly as the water level in streams and rivers, where it may fluctuate widely in response to large rain events. Evaporation affects the water levels of lakes more than streams and rivers because

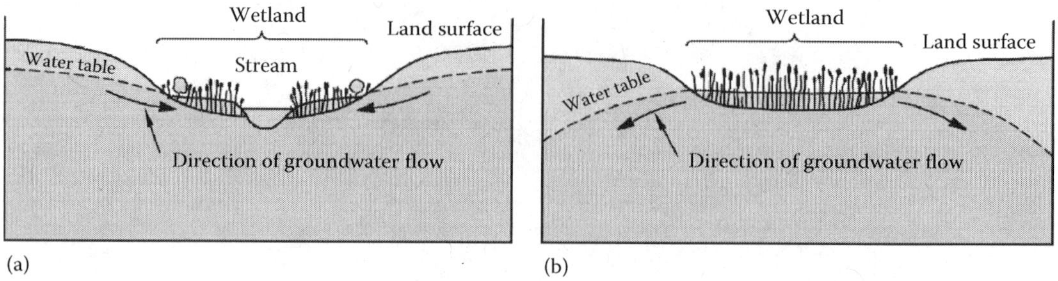

FIGURE 3.42
(a) Wetland receiving water. (b) Wetland discharging water.

(1) lakes have larger surface areas and (2) there is less shading from the vegetation around a lake than from the dense riparian vegetation often present along a stream or riverbank (Winter et al. 1998).

3.6.3 Wetland Interaction with Groundwater

Wetlands prevent rapid drainage of water from the land surface, and similar to lakes and streams, wetlands can receive groundwater inflow, recharge groundwater, or do both. The wetlands situated in topographic depressions of the land surface exhibit interactions similar to lakes and streams. Figure 3.42a depicts a wetland receiving groundwater inflow, and Figure 3.42b shows a wetland discharging water to groundwater (Winter et al. 1998).

The United States has just over 40,468,564 ha (100 million ac) of wetlands remaining (Dahl and Allord 1994). The remaining wetland areas in the United States are concentrated along the southeast coast, the Mississippi River Valley, and the formerly glaciated regions of the upper Midwest. Figure 3.43 shows where the country's wetlands remain (Winter et al. 1998).

3.6.4 Ocean Water, Groundwater, and Surface Water Interaction

Coastal regions may exhibit more complex hydrogeology than other regions, because surface water and groundwater interact with saline water at the ocean surface and in the subsurface. In addition, other forces such as ocean currents, storms, and tidal fluctuations may significantly influence these interactions. Therefore, understanding the geology of coastal regions is the first step toward understanding this complex system (Barlow 2003). Figure 3.44 shows a generalized view of surface water, groundwater, and ocean water interaction (Heath 1983).

Under normal conditions, the seaward movement of freshwater prevents saline water from encroaching coastal aquifers as depicted in Figure 3.45. The **zone of dispersion or transition zone** shown in Figure 3.45 is a diffuse boundary where freshwater and saline water mix (Barlow 2003).

One method used to estimate the depth to the zone of dispersion is called the Ghyben–Herzberg relation named after the two scientists who developed it independently in the late 1800s (Barlow 2003). The thicknesses of freshwater and seawater is related by Equation 3.8:

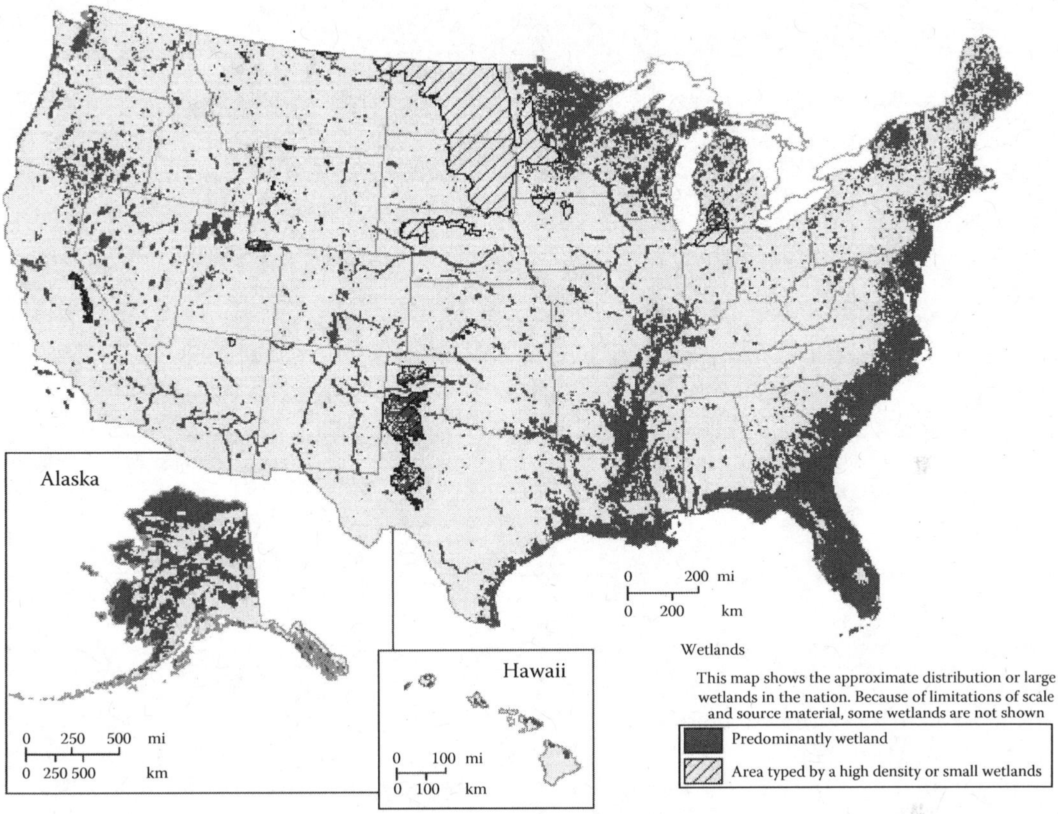

FIGURE 3.43
Location of wetlands in the United States.

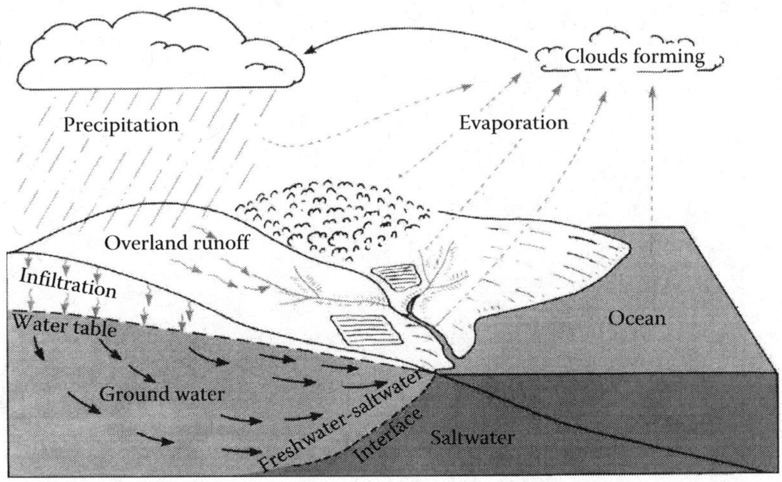

FIGURE 3.44
Generalized diagram showing groundwater, surface water, and ocean water interaction.

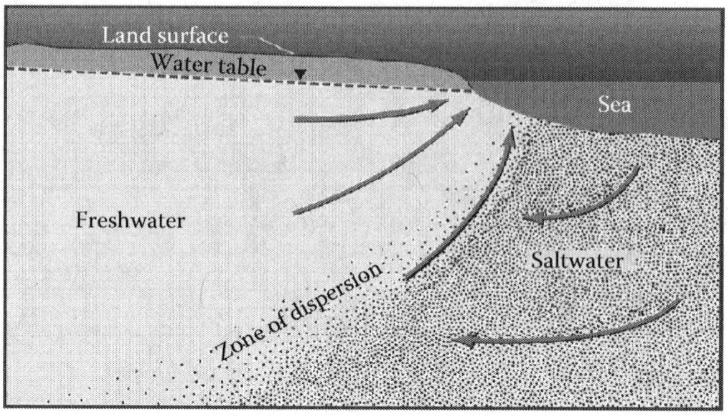

FIGURE 3.45
Zone of dispersion. (Modified from Cooper, H.H., *A Hypothesis Concerning the Dynamic Balance of Fresh Water and Salt Water in a Coastal Aquifer*, United States Geological Survey Water—Supply Paper 1613, Washington, DC, 1964.)

$$z = \frac{\rho_f}{\rho_s - \rho_f} h \quad (3.8)$$

where
ρ_f is the density of freshwater (kg/m^3)
ρ_s the density of seawater (kg/m^3)
h is the thickness of the freshwater zone above sea level (m)
z is the thickness of the freshwater zone below sea level (m)

These parameters are shown in Figure 3.46.

Freshwater has a density of 1.000 g/cm^3 at 20°C, whereas seawater has a density of approximately 1.025 g/cm^3. Although the density differential between seawater and freshwater is small, solving Equation 3.8 demonstrates that 12 m (40 ft) of freshwater exist below sea level for every 0.3 m (1 ft) of freshwater above sea level.

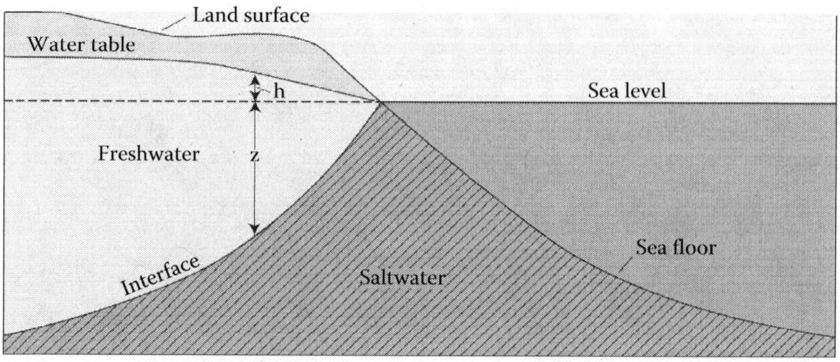

FIGURE 3.46
Estimating the Ghyben–Herzberg freshwater-seawater relation (From Barlow, P.M., *Ground Water in Freshwater—Saltwater Environments on the Atlantic Coast*, United States Environmental Protection Agency Circular 1262, Washington, DC, 2003.)

3.7 Anthropogenic Influences on Surface Water and Groundwater

Urbanization has a profound influence on the hydrogeology of an urban area because urbanization creates an anthropogenic infrastructure. This infrastructure is intended to mimic the natural systems it replaces, but ends up changing many or all of the characteristics of the surface water and groundwater in developed areas (Vuono and Hallenbeck 1995; Zaporozec and Eaton 1996; Kibel 1998). For example, surface water drainage patterns are changed, runoff velocity increases, groundwater levels are raised and lowered, and groundwater recharge areas can expand or disappear (Burn et al. 2007; Garcia-Fresca 2007; Howard et al. 2007; Mohorlok et al. 2007). While many of us are aware of anthropogenic influences on surface water, it is much more difficult to recognize, evaluate, and monitor anthropogenic influences on groundwater because they are not readily observable, and a limited amount of research on urban groundwater exists.

3.7.1 Anthropogenic Surface Water Influences

A more detailed accounting of the anthropogenic influences on surface water can be found in Kaufman and Marsh (1997) and Winter et al. (2008), and includes

- Changing drainage patterns, most notably the creation of ephemeral drainage networks (e.g., driveways and streets) in the headwater zones of streams
- Increasing runoff volume and velocity
- "First flush" pollution; occurring when road oil and other auto-related liquids are washed off the pavement during intense thunderstorms
- Increased erosion and sediment loading
- Wetland destruction
- Dam construction for hydroelectric power and reservoirs
- Rising groundwater elevations in areas near dams and reservoirs
- Contaminated groundwater discharging to surface water in urban areas
- Using surface water for irrigation
- Discharge of industrial and sanitary wastewater to surface water
- Diversion of polluted stormwater into surface water bodies
- **Combined sewer overflows**
- Infiltration/inflow problems within the sewer and water supply pipe networks of older urban areas
- Decreased baseflow resulting from the construction of impervious surfaces such as roads and buildings
- Pruning (removal) of streams due to new development
- Direct and indirect discharges of contaminants to surface water

Some of the motive forces behind drainage pattern modification include:

- Flood control through (1) construction of levees, (2) deepening of streams and rivers through dredging, (3) construction of stormwater retention ponds, and

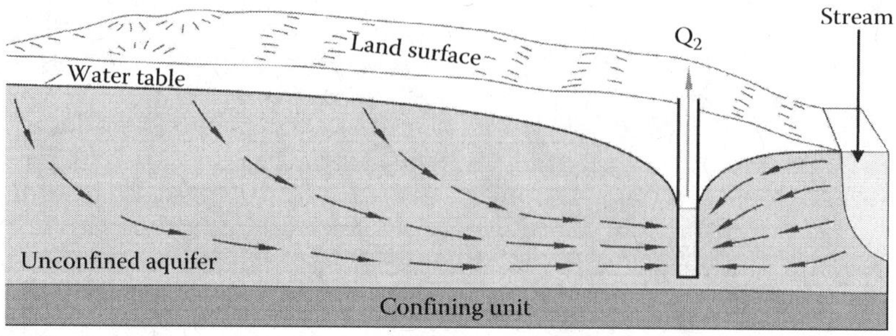

FIGURE 3.47
Potential effects of groundwater withdrawal near a stream.

(4) straightening meanders and bends in streams and rivers, and the removal of bankside vegetation; processes collectively termed **channelization**.

- Canal construction.
- Road construction.
- Landscaping.
- Urban sprawl; the practice of creating an urban landscape equivalent to monocrop agriculture. Mixed residential, retail, and commercial land use within neighborhoods is replaced by regional malls and large big box retail stores with massive parking lots.

One of the more notable anthropogenic changes to surface water flow was the reversal of the flow of the Chicago River in 1900 after completion of the Chicago Sanitary and Ship Canal. This flow modification made the transport of goods and shipping between the Great Lakes and Mississippi River possible. Reversing the flow of the river also provided a clean source of water and changed the local nickname of the river, which had been known as the "Stinking River" (Friends of the Chicago River 2009).

Withdrawing groundwater close to a surface water body provides another example of the human influences on surface water flow. In this scenario, pumping large quantities of groundwater near a surface water source may result in some depletion of water from the surface water source (Figure 3.47) (Alley et al. 1999). In one case known to the authors, contaminated surface water was drawn into drinking water wells as a result of overpumping. This recently happened in the town of Franklin, Michigan, a rural community located in the northwestern part of the Rouge River watershed. Drinking water wells located adjacent to the Rouge River were drawing river water into the wells and consumed by residents for several months before the situation was detected and rectified. The consumption of contaminated drinking water led to a variety of serious intestinal illnesses of several Franklin residents.

3.7.1.1 Destruction of Wetlands

Wetlands are nature's sponge and filter. They perform the following functions: adsorbing excess water; cleaning and biologically degrading harmful chemicals such as oil and other petroleum compounds; assisting in keeping lakes and rivers clean and free of excess sediment; slowing the erosion process; decreasing flooding; and are important areas of groundwater recharge.

In the early 1600s, the United States had an estimated 89,435,527 ha (221 million ac) of wetlands, but by the mid-1980s their extent had been reduced to approximately 41,682,621 ha (103 million ac)—a 54% reduction (Dahl and Allord 1994). Seven states including Indiana, Illinois, Missouri, Kentucky, Iowa, California, and Ohio have lost more than 80% of their wetlands. Although much of this wetland loss is attributed to agricultural land conversion, other anthropogenic causes include

- Construction of dikes and levees along rivers to divert or minimize flooding
- Logging
- Mining
- Road construction
- Building construction
- Urban expansion
- Non-native plant invasion
- Overgrazing

As recently as the 1970s, destruction of wetlands in the United States was encouraged. It was not until the passage of the Emergency Wetland Resources Act of 1986 that wetlands became protected as a vital ecologic and hydrological resource.

Approximately 150 species of birds including the bald eagle, great blue heron, and belted kingfisher and more than 200 species of fish depend on wetlands for their survival. Mammals such as the muskrat, white-tailed deer, and numerous species of reptiles, amphibians, and invertebrates also rely on wetlands. Wetlands exceed all other land types in wildlife productivity and diversity because they are ecotones—a region where two ecosystems overlap and exhibit a synergistic (nonlinear) effect with respect to habitat quality. In the case of wetlands, the overlapping ecosystems are an aquatic lowland area and the adjacent drier upland terrain.

3.7.1.2 Dam Construction

Dams are constructed for hydroelectric power generation, flood control, improvement of navigation, and water supply to provide a water source for irrigation and recreation (Figure 3.48). Dams cause the water table to rise and may also increase head pressures in confined aquifers if conditions are favorable. Two major concerns about dams are whether the rise in groundwater levels will weaken the surrounding geologic material, and if this rise will impact the natural groundwater flow patterns. These concerns were realized during the Teton Dam collapse on June 5, 1976, when a leak appeared in the earthen dam as it was being filled. The leak first appeared approximately 40 m (130 ft) below the crest of the dam indicated by the arrow in Figure 3.49. The dam collapsed a couple hours later (Figure 3.50) (Randle et al. 2000).

The arrow in Figure 3.49 shows the hole caused by seepage of water through the dam on the left portion of the figure. A D-9 bulldozer is at work above the seepage point attempting to plug the hole. Figure 3.50 was taken a couple hours later and shows the nearly complete collapse of the dam. Eleven people lost their lives in the collapse of the Teton Dam. The collapse also caused more than 30 landslides, killed more than 13,000 cattle, incurred more than $300 million in property damage, and flooded portions of Rexburg, Wilford, Sugar City, Salem, and Hibbard, Idaho (Randle et al. 2000).

FIGURE 3.48
Glen Canyon dam and Lake Powell in southern Utah and northern Arizona. (Photo by Daniel T. Rogers.)

FIGURE 3.49
Leak prior to Teton Dam collapse. (Photo by Mrs. Eunice Olsen, June 5, 1976.)

FIGURE 3.50
Teton Dam failure. (Photo by Mrs. Eunice Olsen, June 5, 1976.)

3.7.2 Anthropogenic Groundwater Influences

Documentation of anthropogenic influences on groundwater in urban areas is a recent phenomenon (Barber 1997). The late arrival of this technique occurred because the connection between anthropogenic processes at the surface and shallow groundwater in urban areas is complex and difficult to investigate—or was simply not thought to be important.

Some anthropogenic influences on groundwater include (Howard 1997; Sharp 1997)

- Excessive groundwater withdrawal resulting in saltwater intrusion along coastal areas.
- Significant decreases in groundwater surface elevations in arid regions or any region where a consistently high volume of groundwater withdrawal occurs.
- Decreasing natural groundwater recharge resulting from urban development and the construction of impervious surfaces.
- Decreasing residence time of water in shallow aquifers in urban areas through construction and de-watering efforts.
- Decreasing residence time of water in deeper aquifers through excessive withdrawal.
- Increasing anthropogenic groundwater recharge resulting from leaking sewers, septic tanks, and injection wells.
- Ground surface subsidence resulting from excessive groundwater withdrawal; an event frequently evident in urban regions where groundwater withdrawal is excessive.
- Flooding in urban areas due to the increase in impervious surfaces as a by-product of urban development.
- Excessive sewer leakage resulting in groundwater quality degradation and potential flooding.
- Increased potential for land subsidence and landslides in areas near dams and reservoirs due to rising groundwater levels.
- Direct and indirect discharges of contaminants to groundwater through accidental releases, leaking sewers, septic systems, injection wells, and residential, commercial, and industrial discharges.

3.7.2.1 Saltwater Intrusion

Fresh groundwater in the United States is surrounded by saline groundwater both vertically and laterally. This is especially evident along the coastlines as noted in Section 3.6.4, although much of the interior of the country is underlain by deep saline aquifers (Alley et al. 1999). This juxtaposition of freshwater and saltwater creates the potential for saltwater intrusion and rendering the freshwater nonpotable. This scenario can occur from installing and pumping a water well placed too close to a saltwater source and is shown in Figure 3.51 (Alley et al. 1999).

3.7.2.2 Changes in Groundwater Flow

As established in Section 3.5.8, a minimum of three monitoring wells are necessary to estimate the direction of groundwater flow. Because urban areas are complex, using the

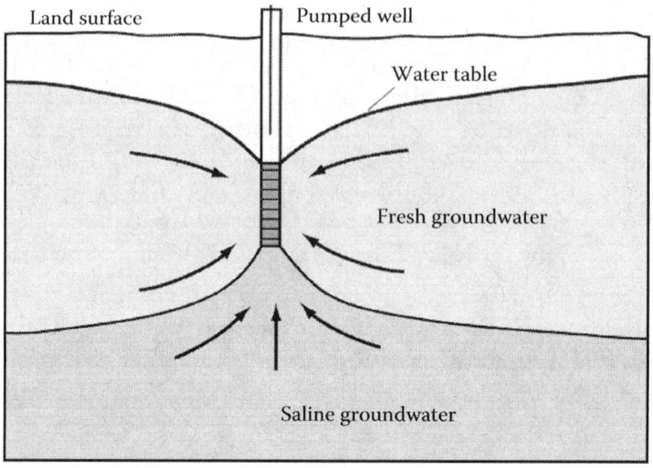

FIGURE 3.51
Saltwater intrusion by pumping groundwater.

minimum of three monitoring wells is usually not sufficient. The flow patterns of groundwater in urban areas vary considerably due to anthropogenic influence, especially if the water table is shallow (less than 6 m or 20 ft beneath the surface), so using more than three monitoring wells is necessary. Although no precise number of monitoring wells necessary to adequately characterize the groundwater flow pattern beneath a specific location has been established, the authors recommend installing a monitoring well network of six or more monitoring wells to begin the process.

The following considerations can help with the evaluation:

- Size and shape of a site: large or irregularly shaped sites require more monitoring wells to evaluate groundwater flow patterns.
- Surface water features: if a stream is located along the boundary or dissects the site.
- Potential presence of multiple aquifers or perched aquifers. The presence of multiple aquifers may indicate different groundwater flow directions.
- Seasonal fluctuations in surface water discharge and precipitation patterns.
- Anthropogenic influences including the following:
    - Utility corridors or sewers. Utility corridors or sewers potentially influence groundwater flow in two basic ways: (1) leaking water supplies act to recharge groundwater and (2) storm sewers and other utilities corridors act as preferential groundwater flow pathways when the water table intersects the utility corridor.
    - Extraction or pumping wells.
    - Deep building foundations.
    - Land disturbance.
    - Artificial recharge areas.
    - Pipelines.
    - Wastewater discharge points.

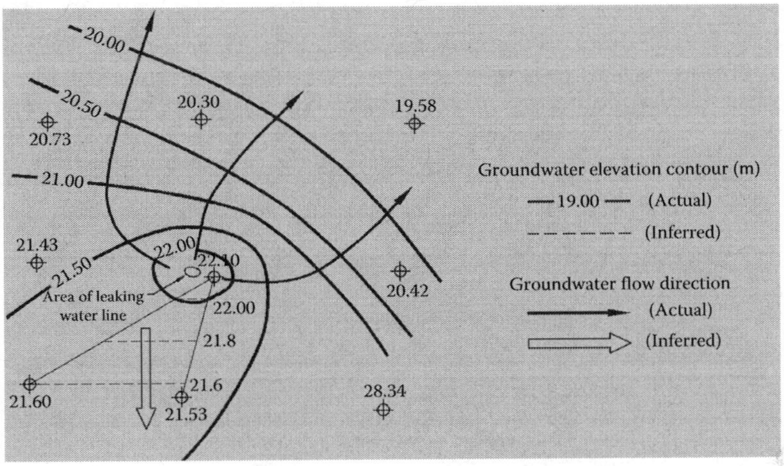

FIGURE 3.52
Anthropogenic influence on groundwater flow.

Figure 3.52 presents a situation where three monitoring wells do not provide a large enough well network to evaluate the direction of groundwater flow. This number was not enough because the presence of a leaking pipe from a potable water source modified the natural groundwater flow pattern (OEPA 1995).

An additional example of anthropogenic influence on groundwater is presented in Figure 3.53. In this case, an industrial site in the Midwest overlies groundwater within a fine- to medium-grained lacustrine sand beach deposit. Groundwater is encountered at a depth of less than 3 m (10 ft) beneath the surface and is anthropogenically influenced in two ways: (1) storm sewers were installed at a depth below the water table and (2) foundations for the buildings were constructed to depths below the water table. Equal potential lines or groundwater contours are identified on the figure with a contour interval of 0.3 m (1 ft). Groundwater flow on the western portion of the figure is toward the east but as groundwater flows across the site, it is interrupted.

Flow interruption occurred where the groundwater migrated along the out-casing of the storm sewers because they were backfilled with a coarse gravel. Gravel is much more permeable than the naturally occurring geologic formation and groundwater leaked into the storm sewers through joints and cracks. The contaminants dissolved in groundwater followed the anthropogenic pathway to migrate to a potentially sensitive ecological habitat—in this case, the storm sewers discharge to a lake containing a trout population. As identified in Figure 3.53, there are three locations where groundwater enters the storm drain, each identified as 1, 2, and 3. In addition, the groundwater flow lines are spaced relatively close together along the western (left side of figure) site boundary and are disrupted toward the eastern (right side of figure) indicating that the building foundations have influenced groundwater flow across the site.

3.7.2.3 Artificial Recharge

Artificial groundwater recharge is defined as the practice of increasing by artificial means the amount of water that enters an aquifer and has been performed in the United States for over 50 years (Todd 1959). This definition has been expanded to include direct recharge of groundwater for potable sources through spreading basins, pits, and injection

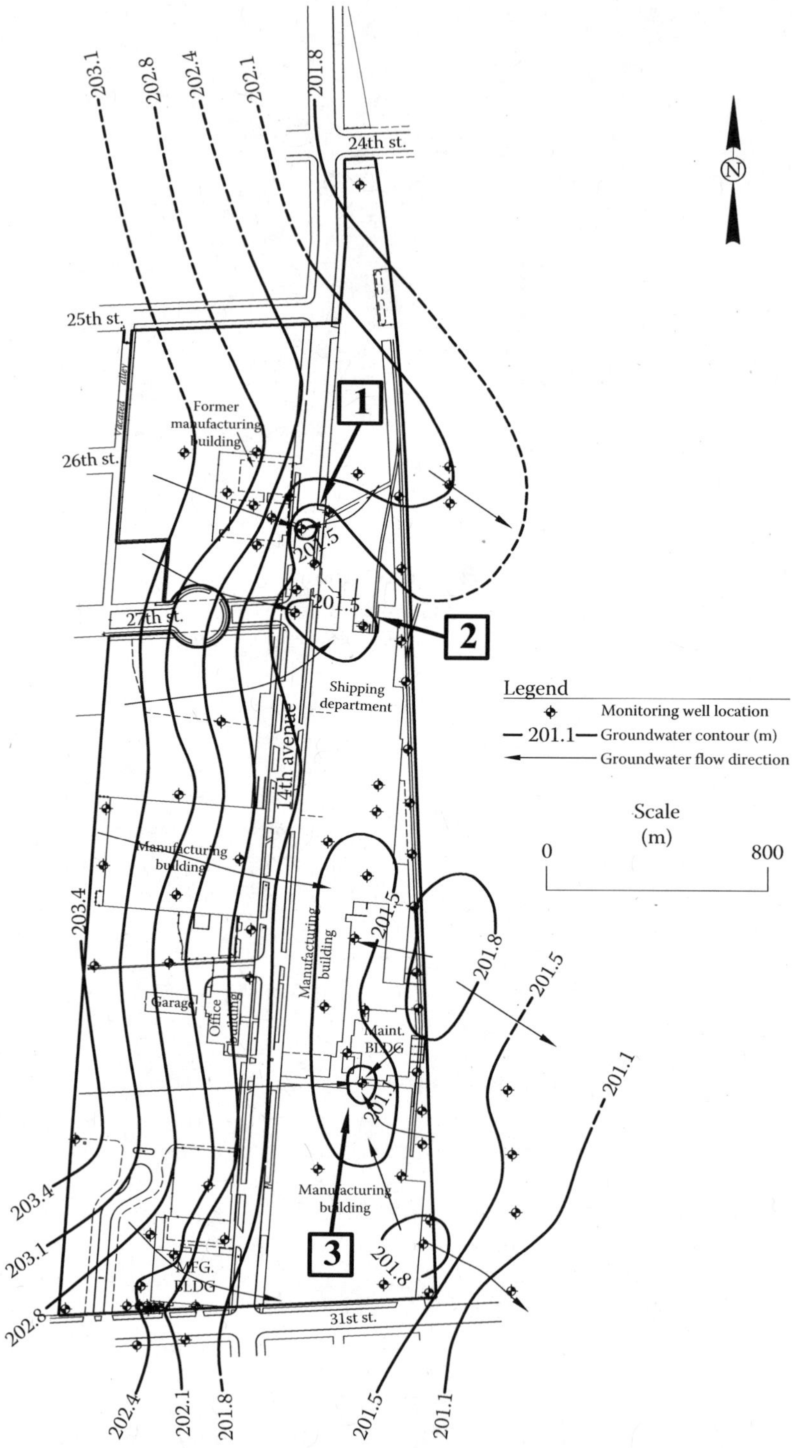

FIGURE 3.53
Figure showing anthropogenic influence of groundwater flow.

through wells. Not included is seepage of sewage or deep well injection of wastes (Aiken and Kuniansky 2002).

Many states practice artificial recharge techniques to replenish groundwater reserves for potable purposes when other sources of potable are either not available or too expensive. Florida, New York, and California are examples (Aiken and Kuniansky 2002), and other states in the high plains region of the United States including Kansas, Nebraska, Colorado, South Dakota, and North Dakota are also using artificial recharge techniques.

The artificial recharge technique most commonly employed in a **spreading basin**. Here, water is spread over a large surface area composed of permeable sediments such as sand and gravel and permitted to seep into the ground and migrate to the aquifer below.

Artificial recharge of groundwater offers several advantages:

- Limits loss of water through evaporation.
- Lower cost than dam construction or construction of water towers.
- Can store very large quantities of water.
- Lessons impact of subsidence.
- Permits the **conjunctive use** of groundwater and surface water. During wet periods, groundwater is stored for later use during dry weather and reduces the stress placed on surface water reservoirs for supply.

The disadvantages for artificial recharge include

- Ineffective recovery leading to water loss
- Storing water beneath urban areas may lead to degraded groundwater quality and pollution
- May increase dissolution if chemistry of infiltrating water is significantly different than native groundwater chemistry
- Changes natural groundwater flow paths and directions

Some countries, especially those in more arid regions, are evaluating and experimenting with techniques of artificial recharge using stormwater and reclaimed water (Dillon et al. 2007; Howard et al. 2007; Limaye 2007). These techniques may become commonplace in the future as our climate changes and population increases generate a greater need for freshwater. Urban areas where large populations use large quantities of water can improve groundwater recharge through an understanding of the hydrogeological environment. This understanding can help avoid aquifer contamination and groundwater mounding (Howard et al. 2007).

3.7.2.4 Land Subsidence

Land subsidence is defined as the gradual settling or sudden sinking of the Earth's surface caused by the movement of subsurface materials (Galloway et al. 2001). Subsidence is a global issue. In the United States, more than 49,030 km^2 (17,000 mi^2) of land surface in 45 States has experienced some form of land subsidence. This is equivalent to a land area the size of Vermont and New Hampshire combined.

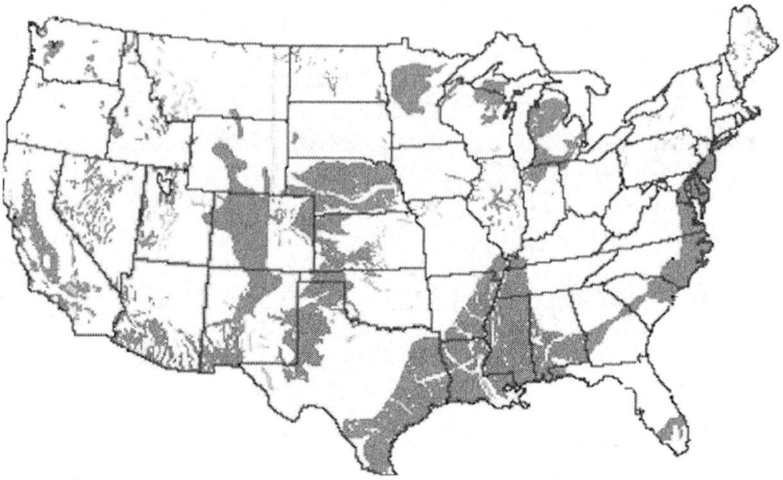

FIGURE 3.54
United States land areas affected by subsidence from pumping groundwater.

The principal causes of subsidence include (National Research Council 1991):

- Excessive groundwater withdrawal
- Aquifer-system compactions
- Drainage of organic soils
- Underground mining
- Hydrocompaction
- Natural compaction
- Sinkholes
- Thawing permafrost

More than 80% of the identified subsidence in the United States is a result of exploitation of groundwater resources (Galloway et al. 2001). In the future, land subsidence may become exacerbated by the increasing rates of withdrawal and exploitation of groundwater. Regions especially prone to subsidence are the southwest portion of the United States; areas overlain by more soluble geological deposits such as limestone, gypsum, or halite; and areas where groundwater withdrawal is significant or a necessity because no dependable surface water source is available. Figure 3.54 shows these areas (Galloway et al. 2001).

3.8 Summary and Conclusion

Geology and water work together to shape our urban environment. Human impact on our environment has also been significant, and, within every urban area, there exists a unique set of environmental concerns from geologic, hydrogeologic, and anthropocentric perspectives. This will become clearer when we move on to explore the presence, migration, and persistence of contaminants in urban areas.

Water is a very good solvent and has the ability to dissolve and erode substances and materials at the surface and below. Water can transport these substances and materials great distances. An appropriate quote heard in geological circles states, "Water destroys everything that the Earth creates and puts it in the oceans (reference source unknown)."

The next chapter applies the principles learned from Chapters 2 and 3 to the geologic investigations performed at specific urban sites. These investigations are called Environmental Investigations and are conducted across a broad array of urban regions in the United States at thousands of properties every year.

References

Abrams, D.M., A.E. Lobkovsky, A.P. Petroff et al. 2009. Growth laws for channel networks incised by groundwater flow. *Nature Geoscience* 2:193–196.

Aiken, G.R. and E.L. Kuniansky. 2002. *United States Geological Survey Artificial Recharge Workshop Proceedings, Sacramento, California*. United States Geological Survey Open-File Report 02-89. Washington, DC.

Alley, W.M., T.E. Reilly, and O.E. Franke. 1999. *Sustainability of Groundwater Resources*. United States Geological Survey Circular 1186. Denver, CO.

Barber, C. 1997. Perspectives on innovations and solutions for groundwater management. In *Groundwater in the Urban Environment*, ed. J. Chilton et al., pp. 3–10. Rotterdam, the Netherlands: Balkema.

Barlow, P.M. 2003. *Ground Water in Freshwater—Saltwater Environments on the Atlantic Coast*. United States Environmental Protection Agency Circular 1262. Washington, DC.

Burger, A.C. and L. Dubertret. 1984. *Hydrogeology of Karstic Terrains: Case Histories*. International Association of Hydrogeologists, International Contributions to Hydrogeology, Vol. 1. Rotterdam, the Netherlands: Balkema.

Burn, S., M. Eiswirth, R. Correll et al. 2007. Urban infrastructure and its impact on groundwater contamination. In *Urban Groundwater—Meeting the Challenge*, ed. K.W.F. Howard, pp. 29–40. London, U.K.: Taylor & Francis.

Cooper, H.H. 1964. *A Hypothesis Concerning the Dynamic Balance of Fresh Water and Salt Water in a Coastal Aquifer*. United States Geological Survey Water-Supply Paper 1613. Washington, DC.

Dahl, T.E. and G.J. Allord. 1994. *Technical Aspects of Wetlands; History of Wetlands in the Conterminous United States*. United States Geological Survey Water Supply Paper 2425. Washington, DC.

De Barry, P.A. 2004. *Watersheds: Processes, Assessments, and Management*. New York: John Wiley & Sons.

Dunne, T. 1990. Hydrology, mechanics, and geomorphic implications of erosion of subsurface flow. In *Groundwater Geomorphology:* The role of subsurface water in earth-surface processes and landforms, ed. C.G. Higgins and D.R. Coates, pp. 1–28. Special paper 252. Boulder, CO: Geological Society of America.

Dillon, P., S. Toze, and P. Pavelic. 2007. Conjunctive use of urban surface water and groundwater for improved urban water security. In *Urban Groundwater—Meeting the Challenge*, ed. K.W.F. Howard, pp. 163–174. London, U.K.: Taylor & Francis.

Driscoll, F.G. 1986, *Groundwater and Wells*, 2nd edn. St. Paul, MN: Johnson Filtration Systems.

Dutch, S. 2009. Erosion and landscape evolution. University of Wisconsin Green Bay. http://www.uwgb.edu/dutchs/EarthSC202Notes/erosion.htm (accessed July 14, 2010).

Environment Canada. 2009. Aquifers and wells. http://www.ec.gc.ca/water/images/nature/gedwtr/a5f3e.htm (accessed August 29, 2009).

Focazio, M.J., T.E. Reilly, M.G. Rupert et al. 2001. *Assessing Ground-Water Vulnerability to Contamination: Providing Scientifically Defensible Information for Decision Makers*. United States Geological Survey Circular 1224. Denver, CO.

Ford, D.C. and P. Williams. 2007. *Karst Hydrogeology and Geomorphology*, 2nd edn. New York: John Wiley & Sons.

Freeze, R.A. and J. Cherry. 1979. *Groundwater*. Englewood Cliffs, NJ: Prentice-Hall.

Friends of the Chicago River. 2009. Brief history of the Chicago River. http://chicagoriver.org/upload/Chicago%20History.pdf (accessed September 1, 2009).

Fuhrer, G.J., R.J. Gilliom, P.A. Hamilton et al. 1999. *The Quality of Our Nation's Water—Nutrients and Pesticides*. United States Geological Survey Circular 1225. Reston, VA.

Galloway, D., D.R. Jones, and S.E. Ingebritsen 2001. *Land Subsidence in the United States*. United States Geological Survey Circular 1182. Denver, CO.

Garcia-Fresca, B. 2007. Urban-enhanced groundwater recharge: Review and case study of Austin, Texas, USA. In *Urban Groundwater—Meeting the Challenge*, ed. K.W.F. Howard, 3–18. London, U.K.: Taylor & Francis.

Heath, R.C. 1983. *Basic Ground-Water Hydrology*. United States Geological Survey, Water Supply Paper 2220. Alexandria, VA: United States Government Printing Office.

Hill, J.W., D.M. Feigl, and S.J. Baum. 1993. *Chemistry and Life*. New York: Macmillan.

Howard, K.W.F. 1997. Incorporating policies for groundwater protection in urban planning. In *Groundwater in the Urban Environment*, ed. J. Chilton et al., pp. 31–40. Rotterdam, the Netherlands: Balkema.

Howard, K.W.F. and R.E. Gerber 1997. Do tills beneath urban Toronto provide adequate groundwater protection? In *Groundwater in the Urban Environment*, ed. J. Chilton et al., pp. 433–438. Rotterdam, the Netherlands: Balkema.

Howard, K.W.F., S. Di Biase, J. Thompson et al. 2007. Stormwater infiltration technologies for augmenting groundwater recharge in urban areas. In *Urban Groundwater—Meeting the Challenge*, ed. K.W.F. Howard, pp. 175–188. London, U.K.: Taylor & Francis.

Illinois Environmental Protection Agency (IEPA), 2010. Wellhead protection for new community water supply wells. http://www.epa.state.il.us/water/wellhead-protection/ (accessed July 15, 2010).

Illinois State Geological Survey. 2010. Geobit 7 karst. http://www.isgs.edu/maps-datapub/publications/geobits7.shtml (accessed July 2, 2010).

Jones, J.A. 1997. *Global Hydrology: Processes, Resources, and Environmental Management*. London, U.K.: Longman.

Kammerer, J.C. 1990. *Largest Rivers in the United States*. United States Geological Survey Open-File Report 87-242. Washington, DC.

Kaufman, M.M. and W.M. Marsh. 1997. Hydro-ecological implications of edge cities. *Landscape and Urban Planning* 36:277–290.

Kibel, P.S. 1998. The urban nexus: Open space, brownfields, and justice. *Boston College Environmental Affairs Law Review* 25:589–618.

Leopold, L.B., M.G. Wolman, and J.P. Miller. 1992. *Fluvial Processes in Geomorphology*. New York: Dover.

Limaye, S.D. 2007. Aspects of urban groundwater management and use in India. In *Urban Groundwater—Meeting the Challenge*, ed. K.W.F. Howard, pp. 189–194. London, U.K.: Taylor & Francis.

Meyer, G. 1989. Fresh water of the North American continent. In *The Decade of North American Geology. Volume. A. The Geology of North America—An Overview*, ed. A.W. Bally and A.R. Palmer, pp. 537–554. Boulder, CO: The Geological Society of America.

Mohrlok, U., C. Cata, and M. Bucker-Gittel. 2007. Impact on urban groundwater by wastewater infiltration into soils. In *Urban Groundwater—Meeting the Challenge*, ed. K.W.F. Howard, pp. 57–63. London, U.K.: Taylor & Francis.

Murray, K.S., X. Zhou, M. McNulty et al. 2000. Relationship between groundwater, surface water quality and land use in an urban watershed. In *Proceedings of the International Symposium on Hydrogeology and the Environment*, pp. 67–71. Wuhan, P.R. China.

National Research Council. 1991. *Mitigating Losses from Land Subsidence in the United States.* Washington, DC: National Academic Press.

Ohio Environmental Protection Agency (OEPA). 1995. *Technical Guidance Manual for Hydrogeologic Investigations and Ground Water Monitoring*. Columbus: OEPA.

Payne, F.C., J.A. Quinnan, and S.T. Potter. 2008. *Remediation Hydraulics*. Boca Raton, FL: CRC Press.

Randle, T.L., J.A. Bountry, R. Klinger et al. 2000. *Geomorphology and River Hydraulics of the Teton River Upstream of the Teton Dam, Teton River, Idaho*. United States Department of the Interior, Denver, CO: Bureau of Reclamation.

Rogers, D.T. 1993. The accuracy of slug tests in estimating hydraulic conductivities in sandy unconfined aquifers. In *Groundwater Management Book No. 16*. pp. 669–682. Dublin, OH: National Groundwater Association.

Rogers, D.T. 1994. *Hydrogeology of Pleistocene Glacial Deposits of Southeastern Michigan*. Geological Society of America. Abstracts with Programs. Seattle, WA.

Rogers, D.T. 1996. *Environmental Geology of Metropolitan Detroit*. Novi, MI: Clayton Environmental Consultants.

Rogers, D.T. 1997. The influence of groundwater on surface water in Michigan's Rouge River watershed. *Proceedings of the American Water Resources Association, Conjunctive Use of Water Resources; Aquifer Storage and Recovery*, pp. 173–180. Long Beach, CA.

Rogers, D.T. and K.S. Murray 1997. Occurrence of groundwater in Metropolitan Detroit. In *Groundwater in the Urban Environment*, ed. J. Chilton et al., pp. 155–161. Rotterdam, the Netherlands: Balkema.

Sharp, J.M. 1997. Ground-water supply issues in urban and urbanizing areas. In *Groundwater in the Urban Environment*, ed. J. Chilton et al., pp. 67–74. Rotterdam, the Netherlands: Balkema.

Shiklomanov, I. 1993. World fresh water resources. In *Water in Crisis: A Guide to the World's Fresh Water Resources*, ed. P.T. Gleick, pp. 13–24. New York: Oxford University Press.

Todd, D.K. 1959. *Annotated Bibliography on Artificial Recharge of Groundwater in the United States through 1954*. United States Geological Survey Water Supply Paper 1477. Washington, DC.

United Nations. 2003. International decade for action; water for life 2005–2015. New York. http://www.un.org/waterforlifedecade/background.html (accessed August 28, 2009).

United Nations. 2008. World population prospects: The 2008 revision population database. United Nations Population division. http://esa.un.org/unpp.p2k0data.asp (accessed December 4, 2009).

United States Geological Survey. 1984. *National Water Summary 1983—Hydrologic Events and Issues.* United States Geological Survey Water-Supply Paper 2250. Washington, DC.

United States Geological Survey. 2004. *Estimated Use of Water in the United States.* USGS Circular 1268. Washington, DC.

United States Geological Survey. 2009a. Water properties. http://ga.water.usgs.gov/edu/waterproperties.html (accessed September 1, 2009).

United States Geological Survey. 2009b. Trends in water use. http://ga.water.usgs.gov/edu/totrendbar/html (accessed August 29, 2009).

United States Geological Survey. 2010a. Summary of the water cycle. http://ga/water.usgs.gov/edu/watercyclesummary.html (accessed June 17, 2010).

United States Geological Survey. 2010b. Science in your watershed. http://water.usgs.gov/wsc/map_index.html (accessed June 19, 2010).

United States Geological Survey. 2010c. Principal aquifers of the 48 conterminous United States. http://water.usgs.gov/GIS/browse/aquifers_us.jpg (accessed July 2, 2010).

Vuono, M. and R.P. Hallenbeck. 1995. Redeveloping contaminated properties. *Journal of Risk Management* 42:58–69.

Winter T.C., J.W. Harvey, O.E. Franke et al. 1998. *Ground Water and Surface Water: A Single Resource.* United States Geological Survey Circular 1139. Denver, CO.

Zaporozec, A. and T.T. Eaton. 1996. Ground-water resource inventory in urbanized areas. In *Hydrology and Hydrogeology of Urban and Urbanizing Areas.* American Institute of Hydrology (AIH) Annual Meeting Proceedings, St. Paul, MN.

4

Conducting Subsurface Environmental Investigations in Urban Watersheds

4.1 Introduction

This chapter presents the sequence of science-based procedures used to evaluate a specific land parcel or larger area for environmental contamination and assess its environmental risk. We conclude with specific methods available to environmental professionals for collecting samples from the soil, groundwater, sediments, surface water, and air. First, we define and present some basic concepts necessary for understanding and framing the concept of a subsurface environmental investigation.

The **environment** is a broad term encompassing all living and nonliving things on the Earth or some region thereof (United Nations 1987). This definition identifies the natural world existing within the atmosphere, hydrosphere, lithosphere, and biosphere, and also includes the built or developed world (anthrosphere). **Environmental risk** is defined as the probability of an event resulting in an adverse impact upon the environment or humans (Fletcher and Paleologos 2000). Environmental risk increases when a completed human pathway occurs or a sensitive ecological system is impacted (USEPA 2002).

Environmental *assessment* contributes to our understanding of human impacts on the environment and is considered the first step in evaluating and ensuring the long-term viability of Earth as a habitable planet. For hundreds of years, the practice of conducting assessments to determine adverse environmental impacts was never a significant priority. Since the last quarter of the twentieth century, however, significant amounts of human and financial resources have been mobilized to evaluate the actual or potential environmental damage incurred from human activities of the past, present, or those proposed for the future.

Environmental *investigations* are conducted when there is the likelihood that certain environmental contaminants exist at a specific property or site having the potential to cause material harm to human health or the environment historically, presently, or in the foreseeable future. Some subjectivity is inherent in the decision to proceed with an environmental investigation at any given property. This uncertainty occurs because most properties contain, store, or at some time have used substances that could pose potential harm to human health and/or the environment. For instance, most households store and use potentially harmful substances, including

- Cleaners (ammonia)
- Solvents (acetone, mineral spirits)
- Gasoline for lawn maintenance, snow removal, and other purposes
- Pesticides (e.g., Permethrin, used for flea and tick control)
- Herbicides (e.g., glyphosate)

- Lawn fertilizers (organophosphates)
- Some paints (**volatile organic compounds** [VOCs])
- Oil
- Grease

In addition, some household products contain substances posing contamination risks to the environment if discarded improperly. Some of the most common or recognizable products include

- Computer equipment (lead, cadmium, mercury)
- Televisions (lead, cadmium, mercury)
- Some electrical equipment (lead, copper)
- Most batteries (cadmium)
- Some building products (VOCs, e.g., formaldehyde and benzene)
- Wood with certain applied preservatives or coatings (arsenic)

The purpose for conducting any environmental investigation is to evaluate whether there is an unacceptable environmental risk to a specific site or property being studied (Jain et al. 1993).

Because we live in the midst of so many potentially harmful chemicals, some well-informed and educated interpretation is necessary to evaluate the actual risk potential when environmental contamination is identified at a site. Martin (2003) notes the triggering mechanisms for conducting an environmental investigation at a property occur when

- A property is involved in a real estate transaction.
- Contamination of some kind is released or suspected to have been released, such as a tanker spill.
- A property is developed or redeveloped.
- A change in land use is proposed.
- A lending institution is involved with a property and requires an environmental investigation before committing any financial resources.
- Something is observed at a property indicating contamination may be present.

Subsurface environmental investigations often begin at the ground surface—the intersection of air, land, water, and human activity, and frequently involve hazardous materials and remnants of contamination from the past. These characteristics of subsurface investigations impart a distinct set of rigorous requirements:

- Investigations focus on evaluating the potential existence of hazardous substances. Therefore, the investigation must be conducted with great care to minimize the potential risks of chemical exposure and other site hazards to investigators.
- The results of investigations may be used in a court of law, so the expertise level of the investigator must be unimpeachable, and any method used—including those for site investigation, quality control, and documentation must be expertly

performed. Interpretations and conclusions must rely upon a firm foundation of science.

- Evaluating the possible presence of chemicals in the environment often requires searching for specific compounds at extremely low concentrations, routinely in the microgram per liter (or kilogram) range. A microgram per liter is roughly equivalent to a part per billion (ppb), which in time units equates to 1 s per 31 years. This analogy demonstrates why these low quantities of evidence require great care to ensure the integrity of the investigation.
- Well-defined activities characterize subsurface environmental investigations, including: scientific inquiry; historical review; data collection; interpretation and assessment; and in some cases, remedial action to lower or eliminate any environmental risk considered unacceptable. During these activities, special emphasis is placed on evaluating the presence or potential likelihood of uncontrolled releases or migration of contaminants.
- Once discovered, the key to defining the nature and extent of contamination is through a detailed understanding of the site geology and hydrogeology, and the behavior of every contaminant detected. Geological and hydrogeological expertise and experience are crucial for achieving a successful outcome for the site/area being investigated.

Subsurface environmental investigations exist as a subset of a larger set of environmental investigation types. Throughout the chapter when we refer to an environmental investigation, the reference is to a subsurface environmental investigation, unless otherwise noted.

4.2 Types and Description of Subsurface Environmental Investigations, Studies, Plans, or Reports

Environmental investigations are conducted for many different purposes and objectives but generally follow a step-wise, logic- and progressive-based approach, where the results of one investigation are used to evaluate the need and scope of subsequent investigations if necessary. The most common types of environmental investigations include

- Phase I environmental site assessment (ESA)
- Phase II investigation
- Geophysical investigation
- Remedial investigation
- Feasibility study
- Ecological and human health risk assessment
- Natural resource damage assessment (NRDA)
- Environmental impact statement (EIS)
- Remedial action plan (RAP)
- Confirmatory investigation

Each type of investigation is discussed in greater detail in the following sections.

4.2.1 Phase I Environmental Site Assessment

The Phase I ESA is typically the first environmental investigation conducted at a specific property. The Phase I ESA is conducted with the objective of qualitatively evaluating the environmental condition and potential environmental risk of a property or site. A **property** is defined here as a parcel of land with a specific and unique legal description. A **site** is defined here as a parcel of land including more than one property or easement and refers to an area of contamination potentially affecting more than one property.

As the first environmental investigation, the Phase I ESA is often regarded as the most important activity because all subsequent decisions concerning the property are, in part, based on the results of the Phase I ESA (Rogers 1992). Therefore, great care, scrutiny, scientific inquiry, and objectivity should be exercised while conducting the Phase I ESA.

Standards for conducting Phase I ESAs were published by the American Society for Testing Materials (ASTM) in 1993 and were revised in 1997, 2000, and 2005 (ASTM 2005a). On November 1, 2006, the United States Environmental Protection Agency (USEPA) established federal standards for conducting Phase I ESAs termed "Standards and Practices for All Appropriate Inquiries" under the Comprehensive Environmental Response Compensation and Liability Act (CERCLA) of 1980, commonly known as Superfund (USEPA 2005).

Under CERCLA, United States courts have ruled that a buyer, lessor, or lender may be held responsible for remediation of hazardous substance residues, even if a prior owner caused the contamination. However, the performance of a Phase I ESA may create a safe harbor or protection from liability, known as the "Innocent Landowner Defense" for new purchasers or lenders. Therefore, Phase I ESAs have become the standard type of environmental investigation employed when initially investigating a property.

According to USEPA (2005) requirements, an environmental professional, such as a geologist, or environmental scientist must conduct the Phase I ESA. General requirements for conducting a Phase I ESA include

- Extensive review of current and historical written records, operations, and reports
- Extensive site inspection
- Interviews with knowledgeable and key on-site personnel
- Assessment of potential environmental risks from off-site properties
- Data gap or data failure analysis

A Phase I ESA is typically a noninvasive assessment, performed without sampling or analysis. In some instances, limited sampling may be conducted on a case-by-case basis, if in the professional judgment of the person conducting the assessment—or due to other requests or mitigating factors—sampling is justified. In most cases, collecting and analyzing samples is usually deferred to the Phase II investigation, but if it does occur, the sampling conducted during a Phase I ESA may include the following:

- Sediment
- Drinking water
- Surface water
- Groundwater
- Waste material
- Soil

- Suspected lead-based paint
- Suspected asbestos-containing materials
- Mold
- Radon

The environmental professional conducting the Phase I ESA must perform extensive research and review all available written records. These researching activities apply not only to the property in question, but also to the surrounding properties within a radius of up to 1.6 km (1 mile) of the investigated property or site. The research/review process typically encompasses the following items and actions:

- Title search and environmental liens
- Historical operations documents
- Historical chemical ordering documents
- Historical photographs
- Historical aerial photographs
- Engineering reports and diagrams
- Previous environmental reports
- Previous accident reports indicating whether a spill or release of a hazardous substance occurred
- Material safety data sheets (MSDS)
- Historical and current environmental compliance documents
- Hazardous substance inventories and mass balance evaluations
- Fire insurance maps
- Soil Conservation Service, soil maps, and reports
- USGS topographic maps
- USGS investigation reports
- USGS geologic maps
- Environmental agency records
- Native American Tribal records
- Local governmental records including sewer and utility departments
- Building permits
- Constructing diagrams and blueprints

The site inspection consists of a walk-through of the property or site. Items to evaluate and document during the site inspection include (modified from Rogers 1992)

- Interviews with knowledgeable and key facility personnel
- Stressed vegetation
- Areas absent of vegetation
- Stained soil or pavement
- Aboveground storage tanks

- Underground storage tanks
- Chemical storage areas
- Back doors
- The "back 40," meaning the rear of a property
- Signage
- Storage sheds
- Refuse storage and containers
- Special labeling
- Evidence of fill or mounding
- Recent excavations or land disturbance
- Depressions in land surface
- General topography
- Wetlands
- Mold
- Animal droppings
- Insects
- Pits
- Trenches
- Floor and roof drains
- Broken concrete
- Areas not inspected or inaccessible
- Weather conditions
- Recent precipitation events
- Nearest water body
- On-site and off-site drainage patterns
- On-site or off-site evidence of wells or borings
- Potential asbestos-containing materials
- Potential septic systems
- Utilities
- On-site or off-site dumping
- Off-site inspection
- Soil type(s)
- Potential contaminant migration pathways
- Potential ecological and human receptor pathways

Once the environmental professional has completed the data collection and site inspection portion of the Phase I ESA, an evaluation of whether there is evidence of an existing release, a past release, or a material threat of a release of any hazardous substance or petroleum is made. If a product was released and then migrated into structures on the property or into the ground, groundwater, or surface water of the property, then this situation is termed a *Recognized Environmental Condition (REC).*

FIGURE 4.1
Recognized environmental condition. (Photo courtesy of Daniel T. Rogers.)

If a REC is discovered, further investigation will likely be recommended to evaluate its potential significance. Many sites have more than one REC, and many of these RECs may require further evaluation. An example of a REC with significant amounts of oil-stained soil, stressed, and dead vegetation is presented in Figure 4.1.

RECs are intended to exclude de minimis conditions generally not presenting a threat to human health or the environment and typically would not be the subject of an enforcement action if brought to the attention of appropriate governmental agencies. However, there may be impacts encountered whose perceived severity falls between a de minimis condition and a REC. In these situations, the term **environmental concern** is applied. Items listed as environmental concerns become RECs if left unattended or lead to a release.

Figure 4.2 is a photograph of a paved parking space with a residual amount of what appears to be a petroleum product discharged from an automobile. In the opinion of the environmental professional who conducted a Phase I ESA at this property, this condition was not evaluated to be a REC and was characterized as de minimis.

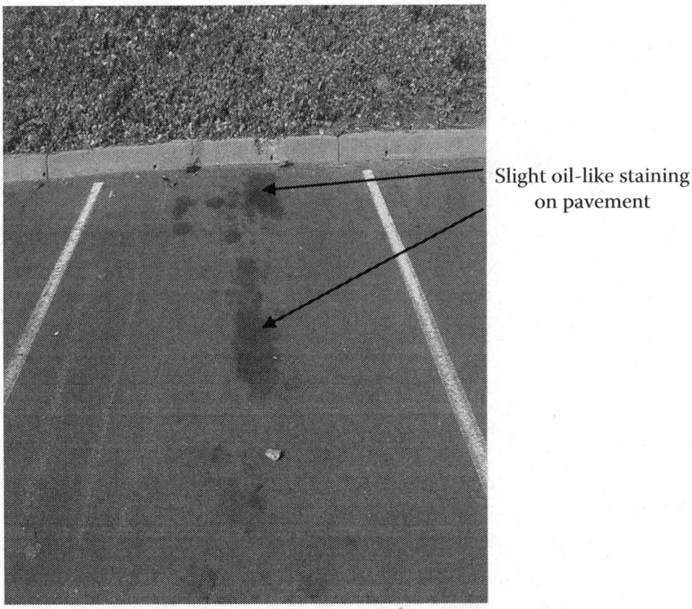

FIGURE 4.2
Example of a de minimis release. (Photo courtesy of Daniel T. Rogers.)

FIGURE 4.3
(a) Aerial photograph showing farmland; (b) Earlier aerial photograph showing landfilling activities. (Photos courtesy of Daniel T. Rogers.)

Historical aerial photographs are effective sources of information and often help with environmental investigations. In urban areas especially, aerial photographs from several sources are readily available. These sources include

- Private local companies specializing in aerial photography
- Private national companies specializing in aerial photography
- Local and state historical societies
- Local and state agencies
- Utility companies
- Local companies
- Federal agencies such as USEPA, USGS, Soil Conservation Service, National Forest Service, Bureau of Land Management, and National Park Service.

Figure 4.3a and b demonstrate how historical aerial photographs can help identify RECs. On the left, Figure 4.3a shows a particular property as farmland with no identifiable REC. Analysis of an aerial photograph taken a few years earlier of the same property (Figure 4.3b to the right) indicates the property was used as a landfill, which is almost always considered a REC.

4.2.2 Health and Safety

If the result of the Phase I ESA indicates further investigation is needed, a health and safety plan must be prepared before the Phase II investigation can begin (Occupational Safety and Health Administration [OSHA] 1989, as revised). This plan can only be prepared by a qualified professional because onsite Phase II investigations are conducted at locations containing physical and chemical hazards. The potential for exposure to these hazards requires specialized health and safety training for personnel going into the field—an absolute necessity for anybody doing this work—and especially critical in urban settings with their additional array of hazards.

The following items represent the minimum level of health and safety planning required before fieldwork begins:

- List of activities or scope of work to be conducted.
- List of emergency contact names, titles, and contact information.
- Identification of potential risks (excluding chemical exposure) including
  - Potential physical hazards
  - Confined space entry
  - Overhead hazards
  - Traffic hazards
  - Weather
  - Other natural and anthropogenic hazards
  - List of chemicals that may be present.
- MSDS for every chemical potentially encountered during on-site activities.
- Review medical procedures for each chemical potentially encountered during on-site activities to ensure that proper safety equipment is onsite and readily available.
- Map showing nearest medical treatment facility with directions.
- List of appropriate personal protective clothing.
- List of other clothing requirements such as hard hat, steel toed shoes, gloves, reflective safety vests, and hearing protection.
- Review safety procedures for nonroutine activities.
- Review buddy system requirements, if appropriate.
- Review emergency hand signals, if required.
- Check safety equipment to ensure it has been properly decontaminated and in working order.
- Contingency plans in case an incident occurs.
- Update the health and safety plan as appropriate, as new information is obtained.

Specific health and safety equipment may be required when conducting Phase II investigations. The following list also contains quality control and quality assurance considerations:

- Hard hat
- Steel toed boots
- Specialized gloves
- Hearing protection
- Safety glasses
- Reflective vests, if appropriate
- Face protection, when appropriate
- Boot covers, when appropriate

- Vapor monitors and detectors for compounds or conditions including
  - Explosive conditions
  - Oxygen levels
  - Hydrogen sulfide
  - Carbon monoxide
  - Volatile organic compounds
  - Chlorine gas
  - Temperature
  - Other potential hazards, as appropriate

Depending on the specific physical and chemical hazards encountered, increased levels of health and safety procedures and personal protective equipment may be required to ensure the health and safety of personnel conducting the investigation. The highest level of protection would include a specialized suit and breathing apparatus supplying purified air, along with several layers of specialized gloves and boots. Figure 4.4 shows a mid-level personal protection configuration consisting of specialized gloves, boots, body suit, full-face respirator, and hard hat.

In many cases, personal protective equipment is promptly removed and disposed of properly after field activities to prevent the inadvertent spread of any contamination beyond the impacted area.

FIGURE 4.4
Mid-level personal protective equipment. (Photo by Daniel T. Rogers.)

4.2.3 Phase II Investigation

When a Phase I ESA has been completed and there are one or more RECs identified, a Phase II investigation will likely be required to confirm the existence of the REC(s). The scope of Phase II investigations vary widely and largely depend on the RECs identified in the Phase I ESA. Under normal circumstances, a Phase II investigation will involve the collection and analysis of samples collected from the area identified as a REC in the Phase I ESA. These samples are collected and analyzed to evaluate the specific contaminants present, and if they exceed federal- or state-mandated concentrations for protection of human health and the environment.

An example of the sampling procedure is shown in Figure 4.5a. This map shows an industrial site where a Phase I ESA was conducted; the 25 RECs and 5 environmental concerns discovered are represented by circles enclosing the letters A through DD. Figure 4.5b shows the location of soil borings drilled during the Phase II investigation at the same site used to evaluate each of the RECs and environmental concerns. A comparison of the two figures indicates the soil borings were drilled in close proximity to each of the RECs and environmental concerns. Figure 4.5b also identified areas where chemicals were stored or used; this information assisted with the selection of analytical testing parameters and associated field sampling procedures.

Further examination of Figure 4.5a and b reveals soil borings drilled in areas not associated with any REC. These borings were drilled primarily along the western and eastern boundaries and in isolated areas in the central portion of the site for three reasons:

1. To evaluate background conditions in areas not believed to be impacted. This assessment is routine because heavy metals naturally occur in the environment, and are likely to be present at large properties where there are numerous RECs.
2. To evaluate contamination migration potential onto the site from an off-site source. The results can determine if an off-site source has impacted the property.
3. To more fully characterize the site geology and hydrogeology, since these site attributes can, and typically do change over short distances. Accurately characterizing the geology and hydrogeology of any site is the key to evaluating the nature and extent of contamination.

Phase II investigations are designed to use the scientific data gathered from each step to evaluate the need for, or to plot the direction of the subsequent step(s). The USEPA has published guidance for conducting investigations under the CERCLA (1988) and under the Resource Conservation and Recovery Act (RCRA) 1986. In addition, many States have published guidelines for conducting investigations; e.g., New Jersey (2005), Ohio (1995, 2006), and California Environmental Protection Agency (1995).

The investigative process follows three steps consisting of find, define, and refine. Each step in the sequence is described in greater detail as follows (Rogers et al. 2009):

- **Find**: This type of investigation has the dual purpose of finding and identifying specific contaminants and is accomplished by collecting and analyzing biased or "worst-case" samples. Examples of "worst-case" samples are those: nearest the source of the release, directly beneath an outfall, appearing the most heavily-stained, exhibiting the strongest odor, or those registering the highest readings on field-screening equipment.

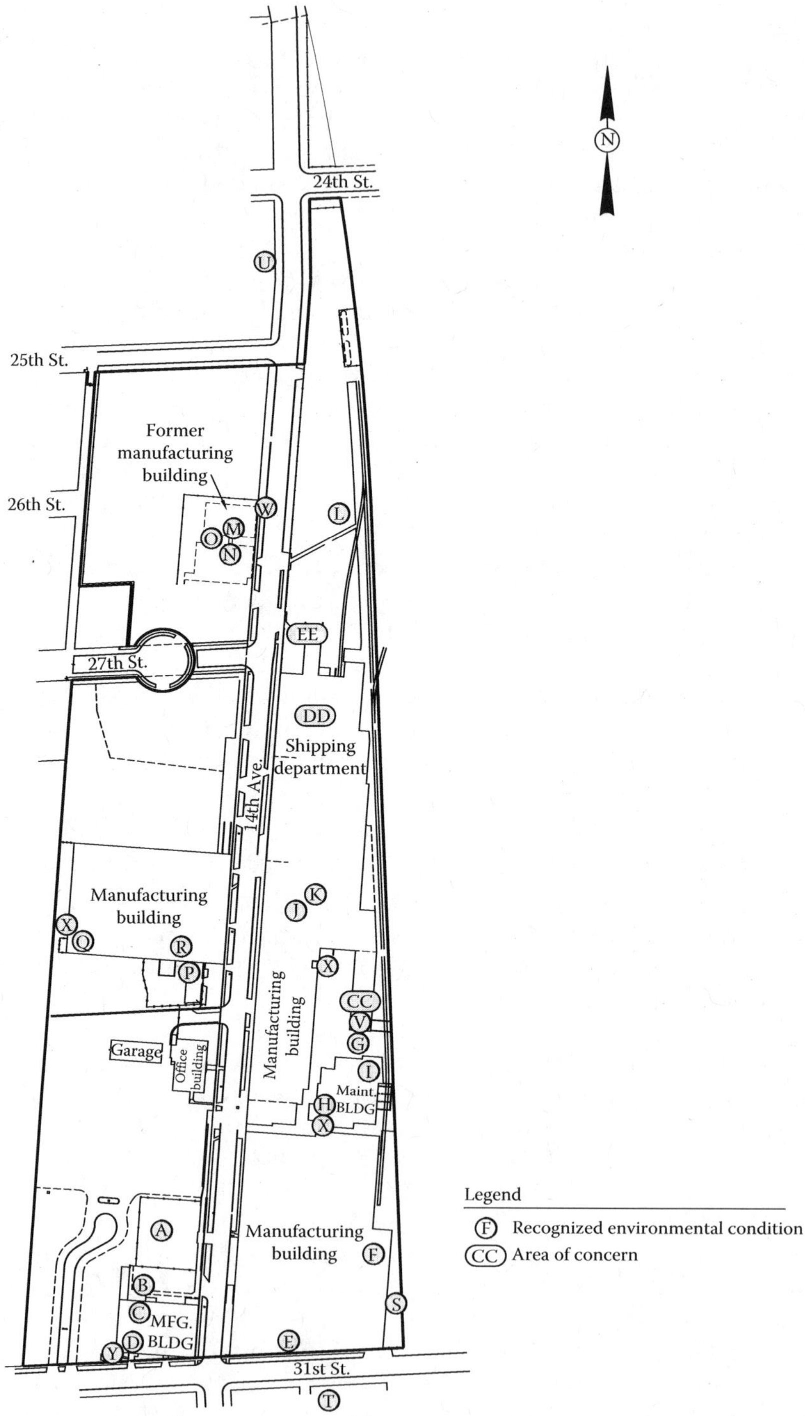

FIGURE 4.5
(a) Map of a site showing location of RECs and environmental concerns.

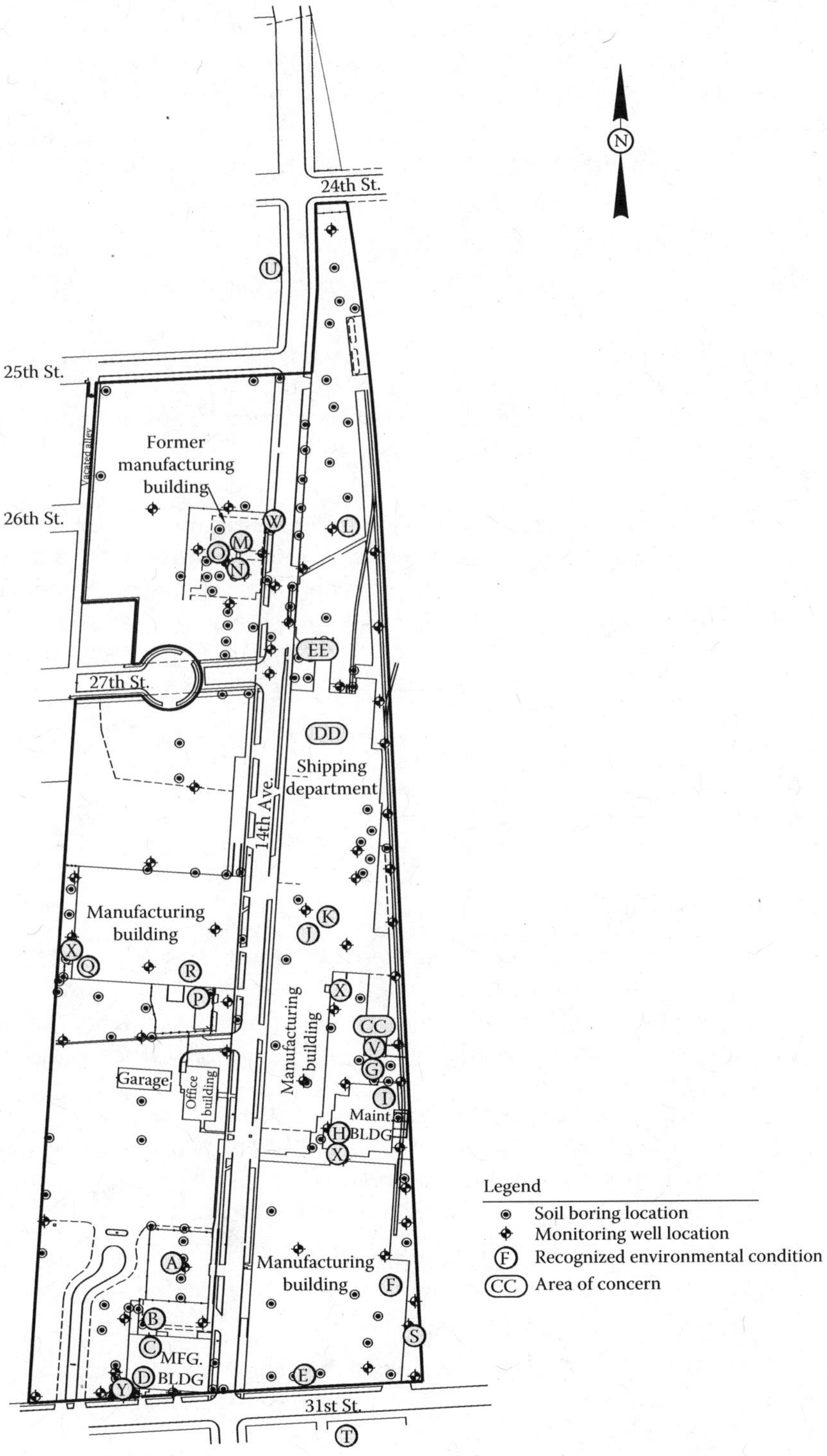

FIGURE 4.5 (continued)
(b) Map showing locations of soil boring, relative to locations of RECs and environmental concerns.

- **Define**: After contamination is found, the purpose of the define-type of investigation is to evaluate the nature and extent of contamination. Define-type investigations may consist of several phases requiring a greatly expanded scope compared to the find-type of investigation. The resulting scope depends upon the following factors:
    - Federal- or state-specific requirements for each contaminant of concern
    - The media being evaluated (e.g., soil, groundwater, sediment, surface water)
    - The pathways being considered (e.g., ingestion, inhalation, dermal contact)
    - Background concentrations of heavy metals such as lead or arsenic

The extent is derived laterally and vertically by these two equivalencies: (1) to the concentration equal to or less than the lowest applicable cleanup criteria for nonnaturally occurring compounds; and (2) to background concentrations for those compounds naturally occurring in the environment, such as heavy metals.

- **Refine**: The purpose of the "refine" type of investigation is to gather additional information about the contamination and determine if an unacceptable risk to human health and the environment exists and requires remedial action. Refine-type investigations include collecting and analyzing samples from locations of highest risk for specialized analytical parameters to evaluate certain remedial options, or to gather more specific data on the nature and extent of contamination.

Phase II environmental investigations are conducted separately for soil and groundwater and are discussed in Section 4.3.

4.2.4 Geophysical Investigation

There are occasions when direct sampling under a traditional Phase II investigation approach does not adequately characterize a specific site. Examples may include areas not accessible to sampling equipment, extremely large sites, or locations where buried objects are of great concern or are under investigation. In these cases, surface geophysical techniques may be used to fill in the data gaps.

Since geophysical investigations provide supplemental data, they rarely are the only type of environmental investigation conducted at a site. There are a wide range of geophysical techniques to accommodate different objectives, investigative requirements, and site limitations, and these techniques offer certain advantages, including

- Increasing the accuracy and area of coverage of subsurface investigations
- Increasing data density
- Collecting data on subsurface geology and hydrogeology at locations inaccessible to other investigative techniques, or areas considered too dangerous to investigate using other methods
- Decreasing the cost and time necessary to characterize certain sites, and are generally noninvasive

Geologists conducting environmental investigations apply geophysical techniques for the following reasons:

- To collect geologic and hydrogeologic data.
- To detect and map contaminant plumes.
- To detect and map subsurface features such as buried drums, utilities, tanks, trenches, landfills, buried wastes, or other buried objects.
- To evaluate soil properties.
- To evaluate anthropogenic subsurface structures.

Geophysical investigations include the following techniques and benefits:

- Seismic refraction and reflection. Seismic refraction and reflection measurements involve the measurement of seismic waves traveling through the subsurface (ASTM 2000b). Stratigraphy, structure, and material properties can be evaluated using seismic methods.
- Electrical resistivity. Electrical resistivity measures the apparent resistivity averaged over a volume of material (ASTM 1999a). These measurements can be used to evaluate the depth and thickness of geologic units, map saltwater intrusion and contaminant plumes, locate buried wastes, and evaluate lateral changes in geologic units.
- Electromagnetics. Electromagnetic methods measure the conductivity of subsurface materials and are frequently used to detect buried metal objects (ASTM 2001).
- Gravity survey. Gravity or microgravity surveys measure changes in subsurface density (ASTM 1999b). Natural variations in the density of subsurface materials such as soil and rock types, cavities, buried channels, faults, fractures, and joints can be detected and evaluated using gravity surveys.
- Ground penetrating radar (GPR). GPR uses high-frequency electromagnetic waves to evaluate subsurface strata (ASTM 2005b). GPR is one of the more popular geophysical techniques used in environmental investigations because it is easy to use and economical, and employable in a variety of challenging accessibility situations. GPR can determine objects having a dielectric contrast with surrounding areas, i.e., buried drums, disturbed soil, etc.
- Borehole geophysics. Borehole geophysics uses instruments to measure and record different properties outside of a well or borehole as the instrument is lowered down the borehole (Keys 1990; ASTM 2005c). Borehole geophysics is used to evaluate well construction, lithology, permeability and porosity, and water quality.

Figure 4.6 contains seismic survey output from a site used to evaluate and map sedimentary layers.

4.2.5 Ecological and Human Health Risk Assessment

After the investigation phase has been completed and the nature and extent of contamination have been defined, an environmental **risk assessment** may be conducted. A common

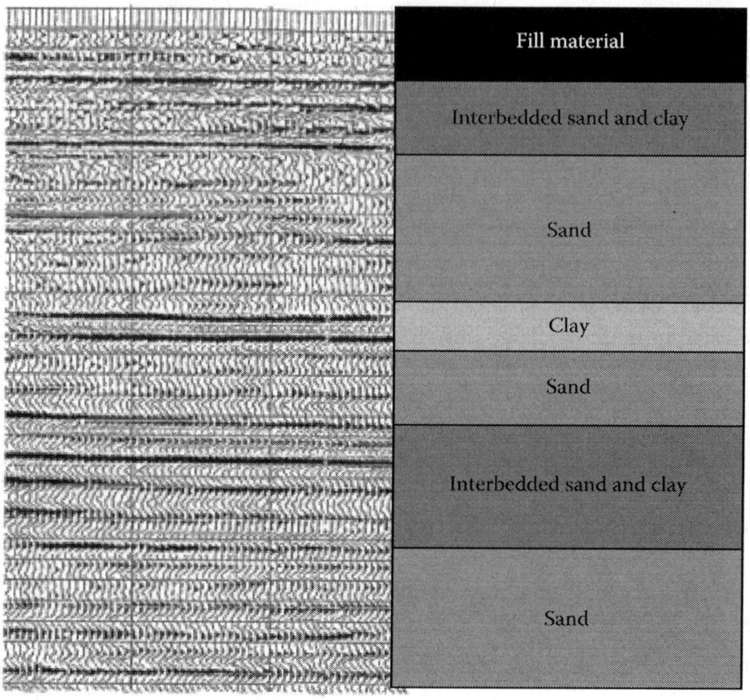

FIGURE 4.6
Seismic profile showing subsurface geology.

objective for conducting a risk assessment is to evaluate whether there is an unacceptable risk posed by the presence of contamination at a given site for the future intended use of the property (USEPA 2002). Intended land use usually fits into the three broad categories of industrial, commercial, and residential. In general, if the intended land use is industrial or commercial, higher concentrations of contaminants are allowed if certain site conditions are satisfied to minimize the exposure potential to the contamination.

An environmental risk assessment follows scientific procedures for evaluating the risk to humans and the environment posed by the contamination present at a specific site. Three basic components are used to evaluate risk: (1) toxicity or potency of contaminants present, (2) exposure pathways, and (3) receptors (California Environmental Protection Agency 1996; Oregon Department of Environmental Quality [ODEQ] 1998; USEPA 2002).

Toxicity: Toxic substances or toxicants are health threats because of the resultant effects on biotic receptors when exposure to these toxicants occur. The following chemical hazards found commonly at hazardous waste sites may be classified as toxic: chemicals fatal to humans in low doses or to over 50% of test animals at certain concentrations; Hazardous substances: chemicals that can cause harm because they are flammable, explosive, strongly acidic or basic, or can induce allergic reactions; Carcinogens: chemicals along with radiation and viruses that can cause or promote the growth of malignant tumors or cancer; Mutagens: chemicals, including ionizing radiation and viruses that cause mutations—inheritable changes in DNA molecules in the genes found in chromosomes; Teratogens: chemicals, ionizing radiation and viruses that cause birth defects. Toxicants include a wide variety of substances such as metals, most commonly Sb, As, Ba, Be, Cd, Cr, Co, Cu, Hg,

Mn, Ni, Pb, and Zn, asbestos, nitrate and nitrite, or organic chemicals. The toxicities of organic chemicals are difficult to fully assess. We need to take into account not only the exposure level, but also the duration of exposure (Yong 2002).

Exposure pathways fall into three broad categories: inhalation, dermal adsorption, and ingestion (USEPA 1989). Any chemical can affect the human body, with the outcome depending on several factors:

- The nature and toxicity of the chemical itself
- The concentration of the chemical
- The distribution of the chemical at a particular site
- The chemical's route of entry into the human body (inhalation, dermal adsorption, or ingestion)
- The length of the exposure
- Individual human susceptibility

Receptors implies biotic receptors, but can also pertain to the physical land environment, including surface water and groundwater (see Section 4.2.8, Natural Resource Damage Assessment). Consequently, the nature and extent of the threat of a particular pollutant will not only depend on the nature and distribution of the pollutant but also the target that is threatened.

Risk assessment is accomplished using a four step process:

1. Hazard identification—answers this question: Does exposure to a chemical or agent cause an increase in the incidence of an adverse health effect (e.g., cancer or birth defects)?
2. Dose–response assessment—quantitative characterization of the relationship between the dose of a chemical or agent and the incidence of an adverse health effect.
3. Exposure assessment—evaluation of the intensity, frequency, duration, and routes of exposure to the chemical or agent.
4. Risk characterization—estimation of the potential incidence of a health effect, calculated by obtaining information from the dose–response assessment, along with information from the exposure assessment.

The outcome of a risk assessment usually produces a hazard quotient, which is the result of evaluating each chemical of concern for potential carcinogenic risk and chronic health risk. USEPA (2002) currently has an acceptable carcinogenic risk of 1 in a 100,000 (1×10^{-5}) cancer incidence, with values under 1 being an acceptable noncarcinogenic chronic health risk. However, many risk assessments evaluate cumulative risk. For instance, there may be 10 different chemicals present, and one chemical alone may not result in an unacceptable risk, but exposure to more than one may result in an unacceptable exposure. In most circumstances, this is an acceptable methodology, especially if contaminants overlap.

USEPA and most states have published guidelines for conducting ecological and human risk assessments, and these address whether a cumulative risk evaluation is warranted. A review of federal, local, and state requirements is recommended before conducting an ecological or human health risk assessment.

4.2.6 Feasibility Study

A feasibility study is conducted after the nature and extent of contamination have been defined. The timing of this study corresponds with the risk assessment if contamination is present at sufficient levels to justify the possibility of remedial action. Selecting the most appropriate remedial technology or technologies—if more than one is necessary—is a key objective of the feasibility study. Technology selection can be evaluated within this framework: (1) the nature and extent of contamination at a specific site; (2) the future intended land use or uses; (3) the results of the risk assessment; and (4) other sites conditions, if present (USEPA 1988; New York State Department of Environmental Conservation 2009).

A feasibility study evaluates many different methods and technologies to reduce contaminant risk. Some cases may not involve a specific technology, but include alternatives such as institutional controls and deed restrictions. The purposes of institutional controls are to restrict or prevent access to areas posing exposure risk, or to reduce the risk of accidentally spreading any of the remaining onsite contamination. Examples of institutional controls include

- Vapor barriers
- Pavement
- Maintaining building structures
- Drainage control
- Access restrictions
- Fencing
- Constructing other physical barriers

Common examples of deed restrictions include special handling of materials if they are disturbed and restricting the use of surface water or groundwater.

As part of a feasibility study, a bench scale and/or pilot level test is conducted using a particular remedial technology to evaluate its effectiveness as applied to real site conditions, contaminants, and on-site geologic and hydrogeologic conditions. A bench scale test involves collecting samples of contaminated material from a site and conducting tests in the laboratory under controlled conditions. The samples are then used to assess the efficacy of the particular technology, and if successful, additional engineering and design data for the pilot test are collected.

A pilot test is larger in scale than a bench scale test and is conducted in situ (at the site of the contamination). Results from the pilot test are used for final engineering design, dosing (if used), and evaluating the time required to complete remediation. Information obtained from the bench scale and pilot tests assist in selecting and designing a remedial technology.

4.2.7 Remedial Investigation

Remedial investigations represent the range of actions taken at a site of environmental concern, and typically represent a combination of the Phase I ESA, Phase II Investigation, Risk Assessment, and Feasibility Assessment outcomes. Remedial investigations generate large amounts of geological and hydrogeological information about the area, and often the region being investigated, but often require two or more years to complete and may cost several million dollars. These time and cost factors dictate that these investigations are conducted only when it is generally accepted that the site poses an obvious and significant risk to human health and the environment. Combining the investigation and evaluation process can shorten the time required to investigate a site and save financial resources, provided the necessary amount of environmental risk is removed to protect human health (USEPA 1988).

4.2.8 Natural Resource Damage Assessment

An NRDA is a study that evaluates the damage or injuries to the environment (USEPA 2009). The term injury refers to an actual adverse impact to, or loss of, a natural resource. Damage refers to the monetary cost of restoration or replacement of the natural resource. The USEPA (2009) defines a natural resource as

- Land
- Air
- Water
- Fish
- Biota
- Wildlife
- Groundwater
- Drinking water supplies
- Other potentially identifiable natural resources

NRDAs are usually conducted after the extent of contamination has been defined and before any final remedy is implemented. Typically, an NRDA is conducted at large sites where there is clear evidence a human-induced environmental impact has significantly degraded the ecological and natural resources of a specific region.

4.2.9 Environmental Impact Statement

An EIS is conducted to identify and evaluate the positive and negative biophysical, social, and other environmental effects that a proposed development action may have on the environment (Glasson et al. 2005).

An EIS includes

1. Describing the proposed action or activity to be conducted.
2. Describing the extent to which the proposed action or activity will potentially affect the environment.
3. Analyzing each potential environmental impact.
4. Presenting a range of development alternatives for the public to consider regarding the potential environmental impacts of the proposed action. Included in this range of alternatives is a "no action" option, meaning the project should not be undertaken.

If the EIS process concludes that the proposed action has a high potential to cause significant environmental degradation, other options may be required to reduce the environmental impact. The process thus does not prohibit the proposed action from causing any harm to the environment; it merely requires disclosure and understanding of the potential impacts prior to taking any action (Glasson et al. 2005).

4.2.10 Remedial Action Plan

A RAP describes the actions implemented to lower the risk posed by the presence of contamination at a specific site to an acceptable level. RAPs include (USEPA 1988; New Hampshire Department of Environmental Services 2008)

- A summary of each investigation completed
- A list of the contaminants of concern (COC)
- Cleanup objectives and goals for each COC
- The technology selected to reduce concentrations of each COC
- Methodology and justification for technology selection
- A time frame to complete each activity
- A plan outlining methods to confirm the remedial action has achieved the desired result and has met the cleanup standard
- Contingency plan in case the selected remedial action or actions do not achieve the desired result

A RAP includes detailed engineering and planning drawings to precisely depict the methods and systems proposed for lowering the contaminant concentrations at a particular site to an acceptable level.

4.2.11 Completion Evaluation Study

A Completion Evaluation Study is conducted after the site is remediated. The purpose of the Completion Evaluation Study is to evaluate the effectiveness of the remediation and to confirm the desired result of permanently reducing contaminant concentrations to acceptable levels has been achieved. Samples of previously affected media, including soil, groundwater, sediment, surface water, and air are collected to help confirm the remediation was successful. To avoid bias, requirements for the Completion Evaluation Study are usually outlined in the RAP. In addition, many individual states have general guidelines for determining when remediation has been completed. A Completion Evaluation Study may require several years to finish, especially if groundwater is a media of concern and has undergone an active remedial action. The requirements for a Completion Evaluation Study include

- Confirmatory sample locations
- Number of samples
- Media to be sampled
- Frequency of sampling
- Quality control and quality assurance requirements
- Analysis and analytical procedures

TABLE 4.1

Types and Purposes of Environmental Investigations

Investigation Type	Purpose
Phase I ESA	Qualitative review of a specific site to evaluate if there is any scientific basis the site poses any environmental risk
Phase II	Quantitative study that finds, defines, and refines the nature and extent of contamination
Geophysical Investigation	Used to supplement and add geologic and hydrogeologic information to Phase II investigations; helps identify and locate buried objects, utilities or other buried structures
Risk Assessment	Evaluate whether the presence of contamination poses an unacceptable risk to human health and the environment
Feasibility Study	Evaluate potential remedial alternatives with the objective of selecting the most appropriate technology(s) to lower risk to an acceptable level
Remedial Investigation	Defines the nature and extent of contamination at large sites likely posing significant risk to human health and the environment
NRDA	Assess damages to natural resources
EIS	Evaluate potential impacts of future development on natural resources and the environment
RAP	Plan outlining what actions will be conducted to lower risk to an acceptable level
Completion Study	Quantitative investigation conducted to confirm that the remediation at a site is sufficient and complete

- Statistical methodology, if statistical methods are used
- Cleanup criteria for each COC
- Contingency plan in case one or more samples indicate remediation has not been completed

4.2.12 Summary of Environmental Investigations

Table 4.1 lists each type of environmental investigation and its major purpose. The Phase II Investigation, Remedial Investigation, and the Geophysical Investigation typically generate the most geological and hydrogeological information—and in some instances—generate a large volume of detailed information. In addition, the Phase I ESA often has valuable information useful for geological purposes, especially if a Phase II or other investigation is subsequently conducted.

4.3 Common Environmental Sampling Methods

Sampled items during an environmental investigation typically include soil, groundwater, sediment, surface water, or air. The primary objectives when collecting any sample during an environmental investigation are (1) collecting a representative sample and (2) ensuring the integrity of samples collected by taking measures not to contaminate or cross-contaminate the samples, or altering any of the original properties of the sample (USEPA 1988).

With these objectives in mind, acceptable sampling techniques have been developed by the USEPA. In addition, most states have published protocols for the collection of samples for environmental purposes. An overview of common sampling methods for soil, groundwater, sediment, and surface water are described in the following sections. In addition to sample collection requirements and guidelines, transportation or shipment of samples may be subject to Department of Transportation (DOT) rules described in Title 49 of the Code of Federal Regulation (CFR), Parts 171 through 179. Therefore, applicable local, state, and federal guidelines should be followed when conducting an environmental investigation at a specific site.

4.3.1 Soil Sampling and Description

Soil sampling is usually initiated during the Phase II investigation and involves collecting surface and subsurface samples from borings drilled within and perhaps adjacent to RECs identified in the Phase I ESA. The number and depth of the borings is dependent upon the source and nature of contamination suspected to have been released. Soil samples are collected using a wide variety of methods, including

- Collecting surface samples using a stainless steel trowel or other similar device
- Using manual hand augers to depths of approximately 1.5–3 m (5–10 ft)
- Hydraulically driving a probe called a Geoprobe® to depths of 6 m (20 ft)
- Excavating test pits using a backhoe or a similar type of equipment to depths approaching 6 m
- Boring with hollow-stem augers to depths greater than 15 m (50 ft)
- Using Roto-sonic boring to depths of 30 m (100 ft) or more
- Using a Cone Penetrometer for depths greater than 30 m

4.3.1.1 Surface Sampling

Many investigations begin simply by collecting surface soil samples in each area of concern. Figure 4.7 shows a location where a surface sample was collected. This location was selected to represent a "worst-case" sample and to fingerprint the contaminant's release area depicted by the area of heavy staining. The soil sample was collected using a stainless steel trowel, then placed into a stainless steel bowl. The sample was then transferred into appropriate sample containers before transport to the laboratory for analysis.

FIGURE 4.7
Surface soil sampling using a stainless steel trowel and bowl in an area of concern. (Photo by Daniel T. Rogers.)

FIGURE 4.8
Example test pit. (Photo by Daniel T. Rogers.)

4.3.1.2 Excavating Test Pits

When possible, excavating test pits using a backhoe is a preferred method of collecting and characterizing shallow subsurface geology. Test pits expose much more of the subsurface than other methods and give the geologist greater opportunity to observe and gather geologic information. An added advantage with test pits is the ability to select optimal locations for collecting soil samples for analysis (Figure 4.8).

While test pits may be the preferred method of characterizing subsurface geology and providing optimal sample collection points, excavating test pits in an urban setting is uncommon. In many cases, test pits cannot be excavated because they disturb much larger areas than drilling a soil boring using a Geoprobe or other similar method. In addition, many sampling locations are located near utilities, buildings, or beneath paved areas making test pits very difficult, dangerous, or impossible to excavate.

4.3.1.3 Drilling Using a Hand Auger

Using a hand auger to characterize subsurface geology has severe limitations. The twisting required to advance the auger deeper into the subsurface causes the loss of most, if not all depositional structures and features in the recovered samples. For this reason, a hand auger is used to characterize soil or sediment type and is not used to interpret most complex geological features or structures. A hand auger may also be used to drill the first few feet to explore and avoid buried utilities before a drill rig completes the rest of the boring. Figure 4.9 shows a typical hand auger.

4.3.1.4 Mechanical Drilling Methods

For drilling in unconsolidated materials, preferred mechanical drilling methods include a Geoprobe, hollow-stem auger, or Roto-sonic methods. Characterizing subsurface geology from soil borings obtained by mechanical methods is much more difficult than test pits, because these methods create a long and very narrow sample (4–10 ft long, and only 5–10 cm [2–4 in.] in diameter). These dimensions of sample size increase the difficulty of the analysis. In addition (1) the outer edges of soil samples collected from a mechanical

FIGURE 4.9
Hand auger. (Photo by Daniel T. Rogers.)

method may be smeared from the sample collection method and (2) the sample probe must be reinserted into the borehole after each sample is collected, and this action may lead to cross-contamination or result in a gap in the depositional history of the area. Additional care and scrutiny must be undertaken when characterizing subsurface geologic features from samples collected using these methods.

Using a Geoprobe for soil sampling is the preferred technology in urbanized areas. A Geoprobe is typically faster than other methods, can penetrate hard surfaces such as pavement and concrete, and can be used inside buildings and other locations with limited access and space. Some Geoprobe drill rigs are capable of directional drilling, providing advantages in some unique situations. Soil samples collected using a Geoprobe are typically 5 cm (2 in.) in diameter and either 2 or 4 ft long. The sample is collected inside a plastic sleeve inserted into the bottom of a metal rod hydraulically pushed and vibrated into the ground. There are drawbacks to the use of Geoprobe devices however. First, great care must be taken not to "push" contaminated soil downward to lower levels resulting in misidentification of the vertical extent of contamination. For this reason, Geoprobe devices are not allowed in California when conducting remedial investigations. Second, the usefulness of a Geoprobe is limited to the vertical extent of its capability, generally to within 6–9 m (20–30 ft). However, under favorable conditions, or using a specially outfitted Geoprobe, depths of greater than 23 m (75 ft) have been achieved. Finally, if a soil boring is to be converted to a semi-permanent monitoring well, a hollow stem, truck-mounted drill rig would need to be brought onto the site, resulting in additional costs.

Soil samples collected from a hollow-stem auger drilling rig are typically 5 cm (2 in.) in diameter and are 61 cm (2 ft) long. The sample is collected inside a plastic sleeve inserted into a tool called a split-spoon. A split-spoon is a geotechnical soil sampling tool comprised of two halves of a hollow metal tube. The tube is threaded on each end where end caps are screwed onto the split spoon to hold it in place while the sample is collected. The split spoon is lowered down the borehole to the desired sampling depth through the center of the hollow-stem auger and then is pounded into the ground using a 64 kg (140 lb) hammer. The number of hammer blows required to drive the split spoon into the ground 2 ft is recorded to provide geotechnical information concerning soil type and resistance.

Figure 4.10 is a photograph of a Geoprobe, and Figure 4.11 is a photograph of a hollow stem auger drilling rig. Figure 4.12 shows a 5 cm diameter by 1.2 m (4 ft) long soil sample collected using a Geoprobe being inspected by a geologist. The plastic sleeve has been sliced in half longitudinally to provide proper access to the sample for inspection. Figure 4.13 shows a Roto-sonic drilling rig, and Figure 4.14 shows a sample being extracted from a Roto-sonic drilling rig.

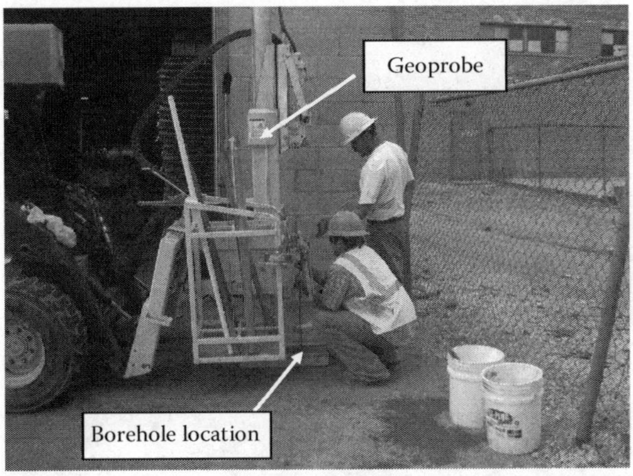

FIGURE 4.10
Geoprobe. (Photo by Daniel T. Rogers.)

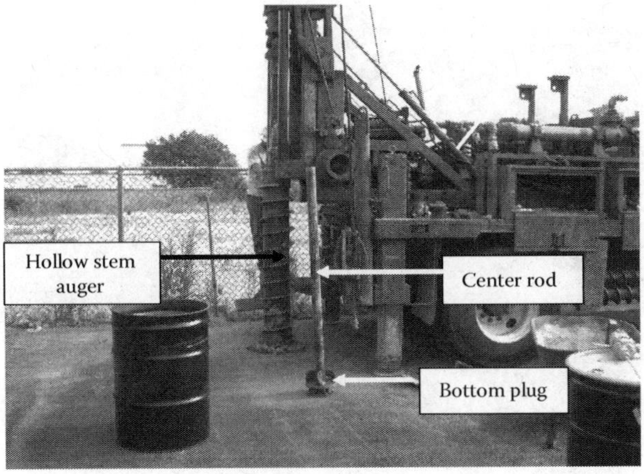

FIGURE 4.11
Hollow-stem auger drilling rig. (Photo by Daniel T. Rogers.)

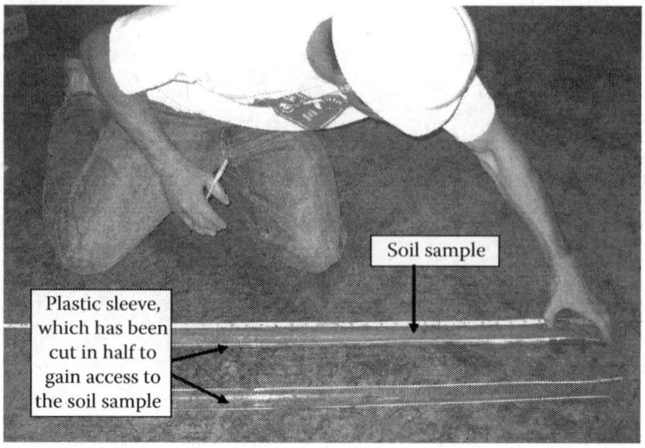

FIGURE 4.12
Soil sample collected from a Geoprobe. (Photo by Daniel T. Rogers.)

FIGURE 4.13
Roto-sonic drilling rig. (Photo by Daniel T. Rogers.)

FIGURE 4.14
Extracting a sample from a Roto-sonic drilling rod. (Photo by Daniel T. Rogers.)

The Roto-sonic method is similar to the Geoprobe method of advancing a boring into unconsolidated subsurface deposits. One advantage of the Roto-sonic method when compared to the Geoprobe is its ability to reach greater depths with a larger diameter borehole. This capability enables a larger volume of sample to be collected and also results in improved monitoring well installation success rates. Roto-sonic also produces less waste and has a faster drilling rate than the hollow-stem auger.

Other drilling methods include direct mud rotary, directional drilling, solid flight and bucket augers, jetting, direct air rotary, cable tool, and rotary diamond. These methods are used when drilling through solid rock, and not preferred for unconsolidated materials.

4.3.1.5 Sample Containers

Soil samples using any of the methods described above are transferred into appropriate sample containers following their collection. Most containers consist of glass having a Teflon inner lid, designed to minimize the loss of vapors, and are filled to capacity as shown in Figure 4.15. Specification of sample containers, and the methods, protocols, and equipment required for sample transfer vary with the specific contaminants being evaluated. In addition, sampling procedures and methodology differ from state to state for certain analytes. To avoid confusion, a work plan identifying the appropriate sampling procedures should be prepared prior to initiation of sampling activities.

4.3.1.6 Documentation

Proper description of each soil sample by a qualified scientist is crucial. A field log book houses this information.

Information recorded in a field log book when each soil boring is being advanced should include

- Drilling method
- Size of auger or drill bit
- Drilling or advance rates

FIGURE 4.15
Filled 250 mL glass sample container of soil. (Photo by Daniel T. Rogers.)

- Drilling or advance difficulties
- Equipment failures
- Drilling difficulties
- Weather conditions

After extraction, each soil sample is inspected and information related to the site's geology and soil properties is recorded in the field log book, and shown in the following:

- Soil or sediment type (i.e., sand, clay, etc.) using the Unified Soil Classification System
- Grain size
- Grain shape
- Grain sorting
- Grain composition
- Consistency
- Plasticity
- Stratigraphy
- Observable depositional features and bedding
- Presence of fill or anthropogenic materials or substances
- Color using the Munsell Color Chart developed by Munsell (1905)
- Color changes
- Degree of saturation
- Field screening results
- Odors
- Location of sample collection for laboratory analysis, if any

Table 4.2 lists the grain size dimensions used to differentiate between clay, silt, and sand. Figure 4.16 is a diagram that shows the relative size difference between a particle of sand, silt, and clay. Figure 4.17 is a soil textural chart used in the field as a guide to identify and

TABLE 4.2

Soil Classification by Grain Size

Soil Type	Diameter (Range in mm or Sieve Size)
Very Coarse Sand	1.0–2.0
Coarse Sand	0.5–1.0
Medium Sand	0.25–0.50
Fine Sand	0.125–0.25
Very Fine Sand	0.0625–0.125
Silt	0.0039–0.0625
Clay	<0.0039

Source: Modified from New Jersey Department of Environmental Protection, *Field Sampling Procedures Manual*, Trenton, NJ, 2005.

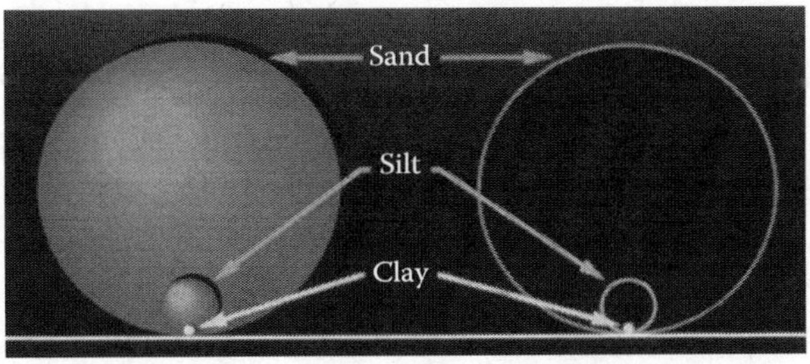

FIGURE 4.16
Relative size difference between a sand, silt, and clay particle. (From United States Department of Agriculture, From the surface down. An introduction to soil surveys for agronomic use, Washington, DC, http://ftp-fc.egov.usda.gov/NSSC/Educational_Resources/surdown.pdf (accessed July 4, 2010), 2010a.)

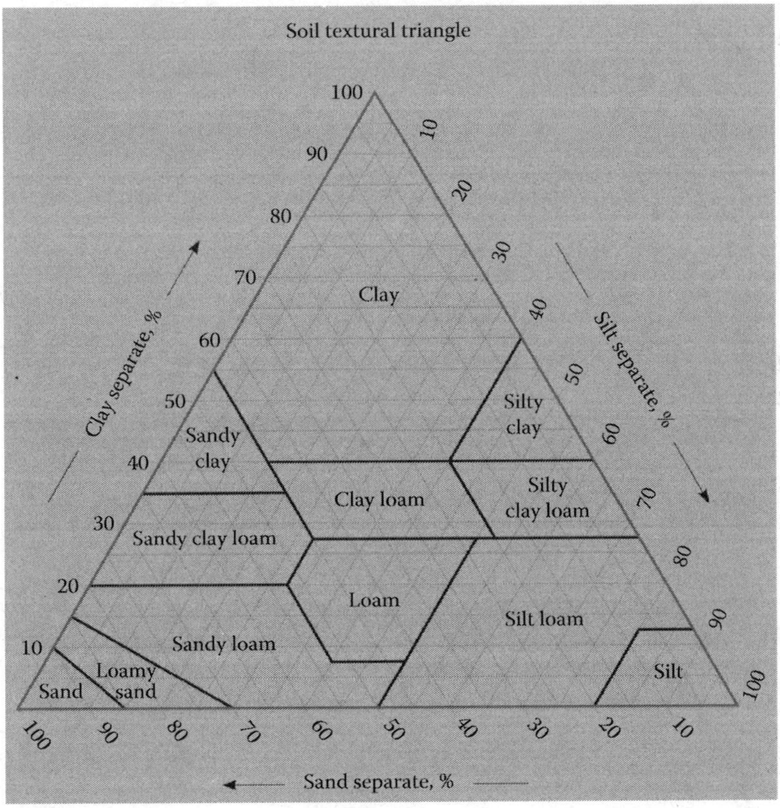

FIGURE 4.17
Soil textural chart. (From United States Department of Agriculture, Guide to texture by feel, http://soils.gov/education/resources/lessons/texture (accessed July 4, 2010), 2010b.)

describe soils on a consistent basis (USDA 2010b). Figure 4.18 is a useful guide when using the textural chart shown in Figure 4.17, and Figure 4.19 contains the guidelines used to describe many properties of soil (Virginia Department of Transportation 2003).

Consistency refers to the relative ease a sediment can be deformed. There are various methods for determining consistency, but in the field, consistency can be estimated using

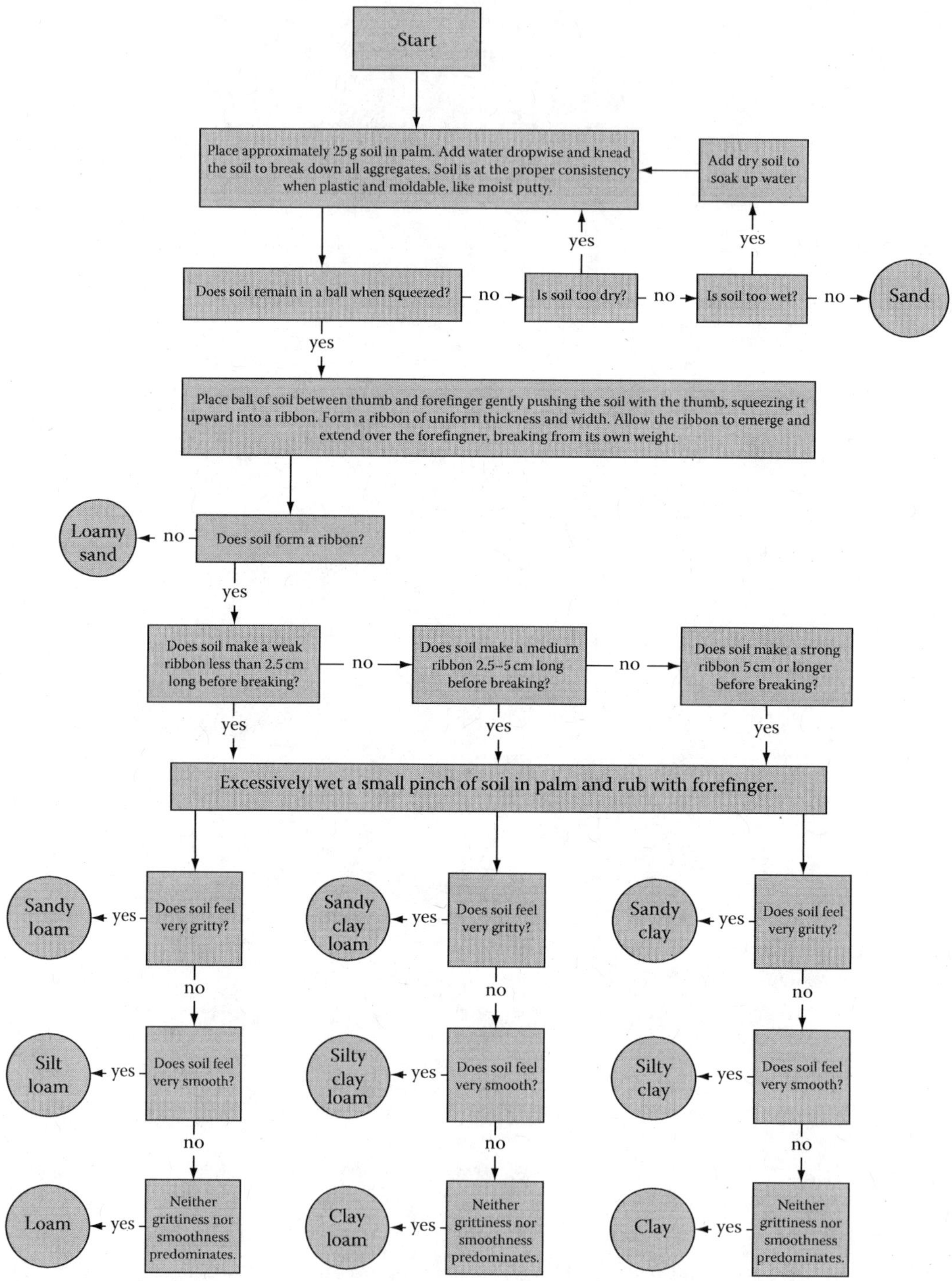

FIGURE 4.18
Guidance on soil textural determination. (From United States Department of Agriculture, Guide to texture by feel, http://soils.gov/education/resources/lessons/texture (accessed July 4, 2010), 2010b.)

Unified Soil Classification and Symbol Chart		
Coarse-grained soils (more than 50% of material is larger than No. 200 sieve size.)		
Gravels More than 50% of coarse fraction larger than No. 4 sieve size		Clean gravels (less than 5% fines)
	GW	Well-graded gravels, gravel–sand mixtures, little or no fines
	GP	Poorly-graded gravels, gravel–sand mixtures, little or no fines
		Gravels with fines (more than 12% fines)
	GM	Silty gravels, gravel–sand–silt mixtures
	GC	Clayey gravels, gravel–sand–clay mixtures
Sands 50% or more of coarse fraction smaller than No. 4 sieve size		Clean sands (less than 5% fines)
	SW	Well-graded sands, gravelly sands, little or no fines
	SP	Poorly graded sands, gravelly sands, little or no fines
		Sands with fines (more than 12% fines)
	SM	Silty sands, sand–silt mixtures
	SC	Clayey sands, sand–clay mixtures
Fine-grained soils (50% or more of material is smaller than No. 200 sieve size)		
Silts and clays liquid limit less than 50%	ML	Inorganic silts and very fine sands, rock flour, silty of clayey fine sands or clayey silts with slight plasticity
	CL	Inorganic clays of low to medium plasticity, gravelly clays, sandy clays, silty clays, lean clays
	OL	Organic silts and organic silty clays of low plasticity
Silts and clays liquid limit 50% or greater	MH	Inorganic silts, micaceous or diatomaceous fine sandy or silty soils, elastic silts
	CH	Inorganic clays of high plasticity, fat clays
	OH	Organic clays of medium to high plasticity, organic silts
Highly organic soils	PT	Peat and other highly organic soils

FIGURE 4.19
Unified Soil Classification System. (From Virginia Department of Transportation, Unified soil classification system. http://www.occ.state.ok.us/Divisions/PST/Forms/Technical%20Forms/Orbca%20Guidance/unified%20soil%classification.pdf (accessed July 4, 2010), 2003.)

the methodology described in Table 4.3. To determine consistency using Table 4.3, a representative sample of the soil or sediment is necessary; the recommended sample size is approximately 5–10 cm^3.

Moisture content should be estimated in the field using the criteria listed in Table 4.4.

Plasticity is the property of soil or rock allowing it to be deformed beyond the point of recovery without cracking or exhibiting appreciable change in volume—its "plastic limit." Criteria for estimating plasticity of soil samples in the field are presented in Table 4.5 (OEPA 2006).

Pertinent information on each soil boring is recorded onto a **boring log**. Figure 4.20 shows a sample boring log (Iowa Department of Natural Resources 2002).

Additional physical parameters may be used to characterize subsurface solid materials, but these cannot be reliably measured in the field and most often require determination at a laboratory. These additional parameters include (ASTM 1996)

TABLE 4.3

Criteria for Describing Consistency

Description	Criteria
Very soft	Thumb will penetrate easily, typically more than 25 mm (1 in.)
Soft	Thumb will penetrate approximately 25 mm
Firm	Thumb will indent soil approximately 6 mm (0.25 in.)
Hard	Thumb will not indent soil, but readily indented with thumbnail
Very hard	Thumbnail will not indent soil

Source: Ohio Environmental Protection Agency, *Technical Guidance for Ground Water Investigations*, Columbus, OH, 2006.

TABLE 4.4

Criteria for Describing Soil Moisture

Description	Criteria
Dry	Absence of observable moisture, dry to the touch
Moist	Damp but no visible water
Wet	Water is observed when sample is squeezed
Saturated	Free water present without squeezing the sample

Source: Ohio Environmental Protection Agency, *Technical Guidance for Ground Water Investigations*, Columbus, OH, 2006.

TABLE 4.5

Criteria for Describing Plasticity

Description	Criteria
Nonplastic	A sample cannot be rolled to an approximate diameter of 12 mm (0.5 in.)
Low plasticity	A sample can be rolled, but is lumpy and difficult, and cannot be rolled when sample is drier than the plastic limit
Medium plasticity	A sample is easily rolled and not much time is required to reach the plastic limit
High plasticity	A sample is very easily rolled and it takes considerable time to reach the plastic limit

Source: Ohio Environmental Protection Agency, *Technical Guidance for Ground Water Investigations*, Columbus, OH, 2006.

Soil Boring Log and Monitoring Well Construction Diagram			
Boring/well number: MW–6	Facility name: ABC facility	Facility street address: 123 West avenue-somewhere, Iowa	
Boring depth (ft) × diameter (in): 15.0 × 6		Drilling method: HSA	
Well contractor name: BEST consultants Registration number: C98765		Logged by: A. bore	
Ground surface Elevation (ASL): 1234.55		Top of casing Elevation (ASL): 1235.43	
Date: 1/2/02 Start time: 8.00 am	Date: 1/3/02 End time: 12.30 pm	UST number: 9901234	LUST number: 9ZYXW00

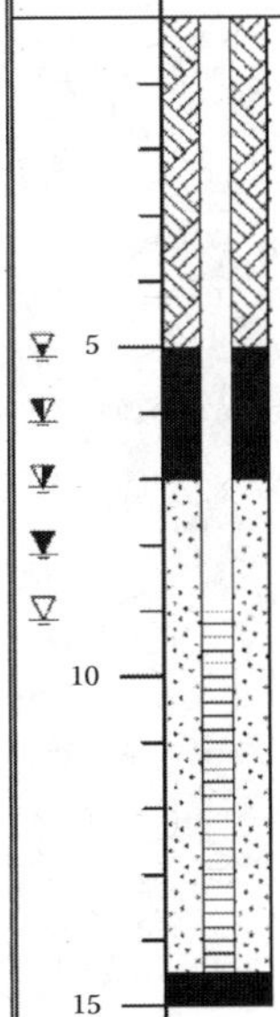

Depth (ft)	Well construction details	Blow count if applicable	Sample No.	Sample Type*	PID/FID Reading	Rock formations, soil, color and classifications, observations (moisture, odor, etc.) first column for USCS	
0	Cement grout	20	1	SS	0	CL	Brown sandy clay, medium stiff, moist (fill)
5	Bentonite pellets	25	2	SS	50	SM	Light brown to gray silty sand to clayey sand, medium dense to dense moist to wet
	Sand filter						
10	Slotted PVC casing	30	3	SS	100	ML	Dark brown to black clayey silt, stiff to very stiff, saturated
15	Bentonite bottom seal	100/6″	4	SS	ND		Bottom of MW-6 @ 15 feet

* SS (split spoon) HS (hollow stem auger)

Observations	Date:	1/3/02	1/3/02	1/3/02	1/3/02	1/3/02
Water levels (ASL)	Level:	1,225.43	1,226.43	1,227.43	1,228.43	1,229.43
Static water level symbol	Time:	8.00 am	9.00 am	10.00 am	11.00 am	12.30 pm

DNR Form 542-1392

FIGURE 4.20
Sample boring log. (From Iowa Department of Natural Resources. 2002. *Sample Boring Log DNR Form 542-1392*. Des Moines, IA: Iowa Geological Survey.)

- Bulk density
- Cation exchange capacity
- Organic carbon content
- pH
- Specific gravity
- Mineral content

4.3.2 Groundwater Sampling Methods

If groundwater is encountered during the drilling of soil borings or excavating test pits, then in most cases, the environmental investigation must also evaluate the potential for groundwater contamination. Sampling groundwater usually requires the installation of a monitoring well, but this is not always the case. Occasionally, a water sample is collected directly from the borehole or excavation pit, or a temporary monitoring well is installed and removed after a groundwater sample is collected.

A groundwater sample collected directly from an open borehole is called a grab sample. Groundwater samples collected from a temporary monitoring well may also be called grab samples, depending on the construction of the temporary well, how long the well had been in place before sampling, and how the well was developed. In general, groundwater samples collected from open boreholes or temporary monitoring wells installed without great care are used as a rough screening of groundwater to evaluate the need for a more permanent monitoring well. Collecting and analyzing groundwater samples from an open borehole or temporary monitoring well is not always recommended because the analytical results may not be representative of the actual groundwater in the aquifer.

Figure 4.21a through e show the process of collecting a groundwater grab sample from a borehole advanced into the saturated zone using a Geoprobe. Figure 4.21a shows a stainless steel well screen which is then placed inside the lead drill rod. Once the desired depth is reached, the outer rod is pulled up to expose the screen. A polyvinyl chloride (PVC) flexible hose is inserted into the inside of the drill rods as shown in Figure 4.21b, and groundwater is pumped from the subsurface using a pump (Figure 4.21c). Groundwater is then pumped into a drum as depicted in Figure 4.21d. After the pumping continues for several minutes, a groundwater sample is collected in an appropriate container (Figure 4.21e).

4.3.2.1 Installing a Monitoring Well

Collecting a representative groundwater sample requires the installation of a monitoring well. The objectives of installing a monitoring well include

- Collecting a representative sample of groundwater for analytical testing to evaluate groundwater quality
- Evaluating the direction of groundwater flow
- Estimating the hydraulic conductivity
- Evaluating specific yield
- Monitoring seasonal fluctuations in flow direction
- Future monitoring of groundwater quality

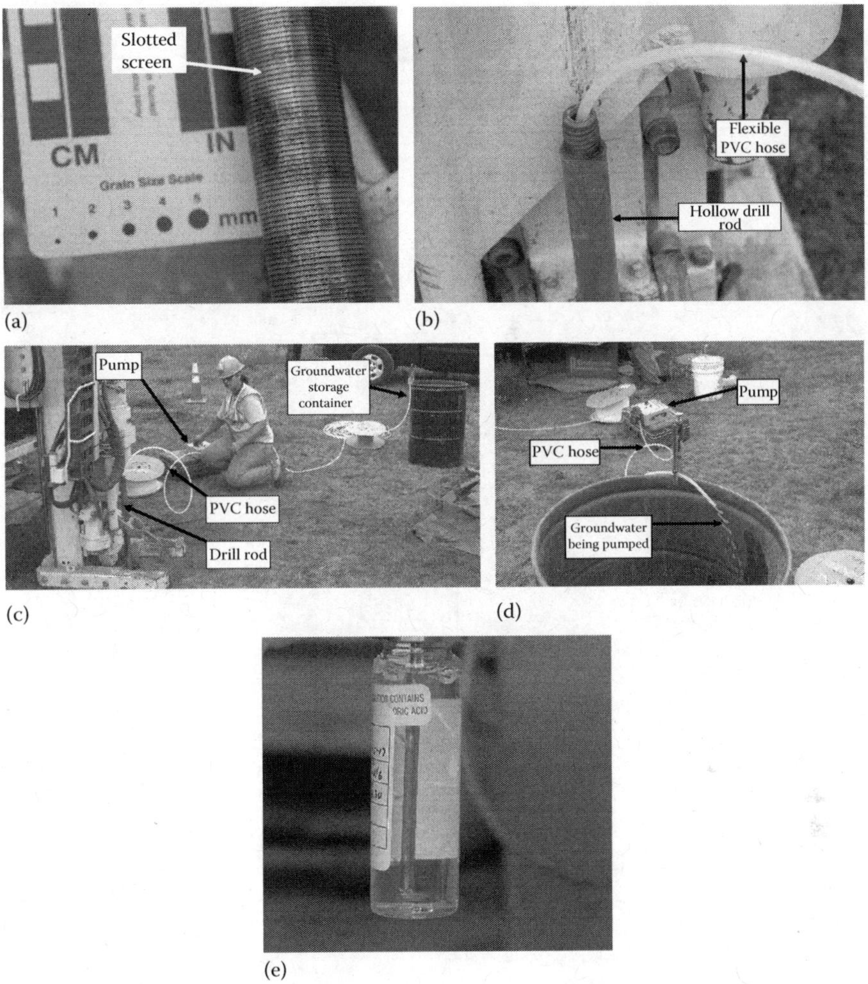

FIGURE 4.21
(a) Stainless steel well screen; (b) Hollow drill rod and PVC hose; (c) Pumping groundwater; (d) Pumping groundwater; (e) Groundwater sample in an appropriate 40 mL glass container. (Photos by Daniel T. Rogers.)

Groundwater samples are typically analyzed for specific chemical compounds having detection limits in the microgram per liter (µg/L) or ppb range. These detection limits mandate a representative sample not only because of these low detection limits—cleanup standards are also very low for certain chemical compounds.

Monitoring wells are typically constructed of PVC, but may also be constructed of stainless steel (ss), Teflon, or steel. Figure 4.22 has a construction diagram of a typical monitoring well.

The **casing** in a monitoring well is the hollow tube placed inside the borehole and provides access to the subsurface. The **screen** at the bottom of the well allows water to seep into the well from the aquifer. Figure 4.23 shows the slots in the PVC screen that permit groundwater to enter the well casing. The slits have a very small diameter so the water enters the well but not the filter pack around the outside of the well screen. A filter pack is placed around the well screen and serves to prevent any fine-grain sediment in the aquifer

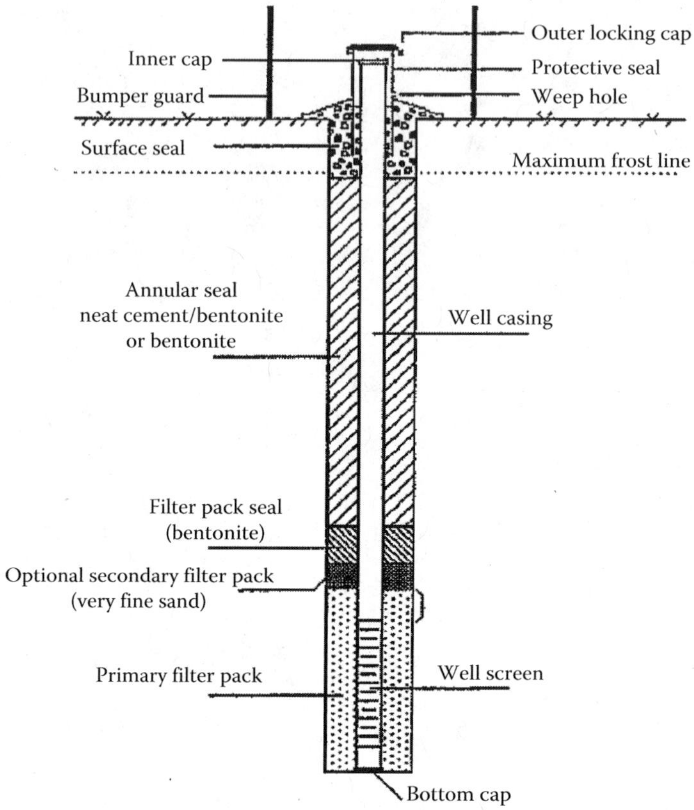

FIGURE 4.22
Monitoring well construction diagram. (From Ohio Environmental Protection Agency, *Technical Guidance Manual for Hydrogeologic Investigations and Ground Water Monitoring*, Division of Drinking and Ground Waters, Columbus, OH, 1995.)

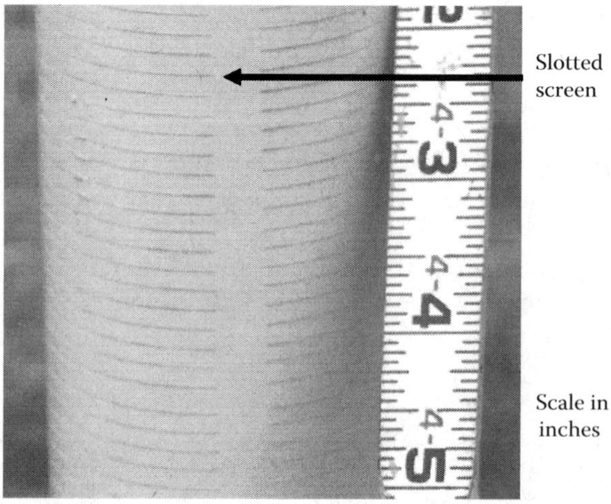

FIGURE 4.23
PVC well screen. (Photo by Daniel T. Rogers.)

formation from entering the well. The **filter pack** contains granular sand or gravel of uniform size and is placed between the aquifer formation and the well screen. The filter pack aids in screening out the fine particles within the aquifer mobilized by well pumping, so this device is important in aquifers composed of unconsolidated sediments that tend to be poorly sorted. Removing solid particles before they enter the well keeps the groundwater sample free from analytes of concern that may be sorbed onto sediment particles.

Many states require that the filter pack and the slot size of the well-screen be "engineered," based on the textural characteristics of the aquifer material. This is accomplished by performing a sediment size analysis of the aquifer material and creating a grain-size distribution curve. The desirable percentage of open slots in a well screen should be at least equal to the porosity of filter pack. Although desirable, this is not always achievable, particularly if the aquifer material consists of fine sand. Moreover, the cost of performing a grain size analysis of aquifer material to optimize the filter pack and the well screen is frequently too great, especially considering the cost of having the drill rig and crew standing by while this analysis is being performed. As a result, we cannot overstate the importance of contracting an experienced drilling company that has already successfully installed monitoring wells in the area where you are working.

An **annular or bentonite** seal composed of bentonite clay is placed above the filter pack and above the well screen surrounding the casing. This placement of the seal helps to minimize potential infiltration from surface water and other fluids via the outside casing of the well.

Most monitoring well screens are 2 or 3.5 m (5–10 ft) long and 5 cm (2 in.) in diameter. Many local regulatory agencies require well screens to be no more than 2 m (5 ft) in length. Monitoring wells installed in an unconfined aquifer generally straddle the water table, and these usually have a 3.5 m (10 ft) long screen (if allowed by the regulatory agency) to allow for the seasonal fluctuation of the water table. Half the well screen (2 m) is installed below the water table, with the other half above the water table. Monitoring wells installed entirely below the water table (termed fully submerged) have well screens 2 m in length.

Installing more than one monitoring well per location may be required in cases of significant aquifer thickness of over 6–8 m (20–25 ft). Multiple monitoring wells at a single location are installed to evaluate the groundwater quality vertically through the length of the aquifer. Figure 4.24 shows a cross section of an aquifer composed of sand with a saturated thickness of approximately 25 m (≈75 ft). In addition to the monitoring wells installed straddling the water table, three monitoring wells were installed as a set at three depths within the aquifer. One set straddles the water table, one set was placed at the approximate mid-point of the aquifer, and the last set was located at the bottom of the aquifer. However, in some cases, even installing three monitoring wells at one location may not be sufficient for gathering enough data to adequately understand the inner dynamics of a contaminant plume. An example of this is where multiple plumes are present. In this scenario, a technique termed **High Resolution Aquifer Profiling** is used and involves collecting numerous groundwater samples vertically at several locations to develop a detailed three-dimensional depiction of a groundwater plume or plumes (Welty et al. 2010).

Requirements for the design and installation of monitoring wells exist and vary state by state and often county to county. The differences occur for items such as obtaining a permit to install, location, depth, seals, certification, length of well screen, diameter of well screen, placement of screen, construction materials, and installation procedures. Before installing a monitoring well, state and local requirements should be examined.

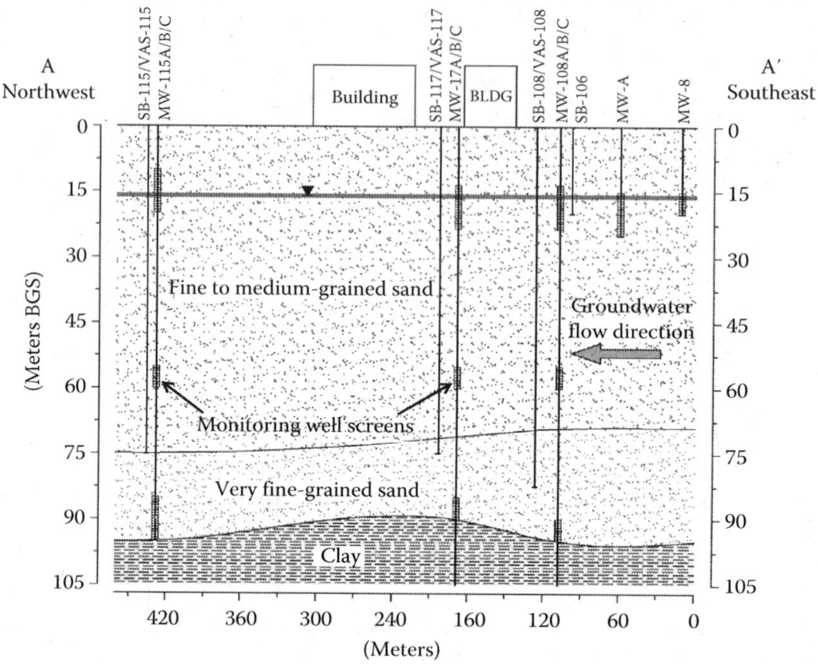

FIGURE 4.24
Cross section showing multiple monitoring wells, horizontally and vertically (in m).

4.3.2.2 Groundwater Sampling

Collecting groundwater samples consists of a process involving the following nine steps (modified from Harter 2003):

1. Preparation
2. Inspecting and accessing the monitoring well
3. Measuring the water level
4. Purging the well
5. Letting the well recover
6. Measuring the water level
7. Collecting the water sample
8. Decontamination
9. Securing the well after sampling

The preparation step involves the following:

- Establishing the list of parameters for analysis.
- Ensuring the purging and sample collection equipment including bailers and pumps are in working order; these items should be new or properly decontaminated.
- Verifying the correct type and quantity of gloves, splash protection, and other potentially required health and safety equipment are working properly; these items should be new or properly decontaminated.

- Checking the proper function of the water quality measuring equipment; these items should be new or properly decontaminated. Equipment may include: temperature, pH, conductivity, dissolved oxygen, reduction-oxidation potential, hydrogen sulfide, photoionization, explosivity, and lower explosive limit detection devices or meters.
- Ordering appropriate sample containers.
- Establishing appropriate protocol for sample labels.
- Making sure chain of custody and laboratory log-in sheets have been prepared in advance to the extent possible.
- Ensuring the sampling preservation, transport, and packing materials are appropriate, including: coolers, ice, sample preservation chemicals, filters, and any specialized laboratory instructions.

Figure 4.25a and b show accessing a monitoring well and lowering a Teflon bailer into a well. Figure 4.26 shows a water level indicator; it consists of a graduated tape measure

(a) (b)

FIGURE 4.25
(a) Monitoring well cover; (b) Lowering a Teflon bailer down into a monitoring well. (Photos by Daniel T. Rogers.)

FIGURE 4.26
Water level indicator. (Photo by Daniel T. Rogers.)

attached to an electronic sounding device at the tip that emits an audio alarm when water is encountered.

The depth to groundwater using the water level indicator should be measured from the top of the monitoring well casing to the water table surface. Usually a reference mark is placed at some common point on the top of the monitoring well casing, because accurate measurements to within 3 mm (1/100 of a foot) are routinely required when measuring the depth to groundwater. This mark should be consulted every time a water level measurement reading is conducted to ensure consistency and accuracy because the top of the monitoring well casing is not always level.

As explained in Chapter 3, the elevation at the top of the casing of the monitoring well must also be measured accurately to determine the direction of groundwater flow. The geographic position and elevation at the top of the casing of each monitoring well should be measured by a licensed surveyor.

If the monitoring wells will be used for any length of time, usually more than 1 year, consideration should be given to periodically checking the elevation of each monitoring well to ensure the elevation of the casing has not changed. The elevation can change from natural causes, such as frost heave, and from anthropogenic causes, such as damage caused if snow removal or other equipment accidentally run into the monitoring well.

Figure 4.27 is a photograph of a Teflon bailer containing groundwater with relatively high turbidity (cloudy appearance) extracted from the monitoring well located near the bottom of the picture. Teflon bailers are routinely used for collecting groundwater samples. The bailer pictured in Figure 4.27 has a check valve at the bottom that allows water to fill the bailer when groundwater is encountered as the bailer is slowly lowered into the well. After the check valve is engaged upon lifting of the bailer, water is prevented from escaping.

FIGURE 4.27
Teflon bailer containing groundwater. (Photo by Daniel T. Rogers.)

4.3.3 Sediment Sampling

Sediment sampling is conducted with several types of specialized equipment. One instrument commonly used is called a Ponar Dredge. A Ponar dredge is equipped with a set of jaws and is deployed in the open position while being lowered through the water column until it becomes embedded into the bottom sediment. Once the instrument is pulled upward with minimal force to avoid pulling it off the bottom, the jaws are engaged and come together as the instrument is raised. This sequence of events traps sediment in the instrument bucket. These types of samplers allow the collection of small or large samples depending on the size of the bucket (USEPA 2007).

Another popular sediment sampler is called a tube or grab sampler; it consists of a hollow tube typically constructed of stainless steel, PVC, or Teflon (Capri et al. 2005). These types of samplers are intended for collecting shallow sediment samples within water columns of less than 1.5–3 m (5–10 ft). Proper function of these devices depends upon the presence of a strong water current or other associated factors that may cause the sampler to drift and not collect an appropriate sample size, or sample the wrong location.

4.3.4 Surface Water Sampling

Sampling surface water uses many of the same procedures as groundwater sampling. The first step in sampling surface water is evaluating the morphology and hydrology of the surface water body so appropriate sampling location(s) can be identified. Collecting an appropriate and representative sample of surface water also depends upon the type of surface water body; e.g., a lagoon, stream, lake, river, or spring.

Factors affecting sampling location include

- Chemical compounds of concern
- Depth of water body
- Flow rate
- Size
- Whether there is a specific location of concern
- Topography and composition of the bottom of the water body

Each surface water sample location should be marked and surveyed in case a future sampling event is required. Placements of stakes, flags, or buoys are common ways to mark surface water sampling locations. The elevation of a surface water body is also measured during every sampling event to assess the current hydrologic conditions; e.g., drought or flood. Placing a benchmark on a stationary object, such as the bottom or side of a bridge, at the edge of a paved roadway, or building corner usually works.

In complex surface water settings, when measuring water quality data such as temperature, pH, and dissolved oxygen it helps to know if any stratification exists either horizontally or vertically within the water column. Monitoring these factors improves the chances you will select appropriate sampling locations and an effective number of samples.

Surface water samples can be collected using a glass tube, pumps, or bailers, or obtained directly using a sample container as depicted in Figure 4.28. Specialized surface water instruments include the Kemmerer Bottle, Bacon Bomb Sampler, or Dip Sampler. These types of instruments are used from a boat or a bridge and are lowered into the water column and can collect a water sample from any depth. The exception

FIGURE 4.28
Collecting a surface water sample using the dip method. (Photo by Daniel T. Rogers.)

here is the Dip Sampler, which only collects samples from the surface and is essentially a sample container dipped into the water. Dip samplers are useful where access is limited (Figure 4.28).

4.3.5 Air Sampling

Conducting subsurface investigations sometimes requires sampling air. Some contaminants migrate through subsurface soil and groundwater and may come into contact with subsurface structures such as basements. When this event occurs, certain contaminants have the potential to accumulate in these subsurface void spaces and contaminant the air. Additionally, contaminants are released directly into the air and may become deposited on the ground or water surfaces.

Under certain conditions, air sampling is justified when conducting a subsurface environmental investigation for three purposes:

- Evaluating whether the source of soil or water contamination originated from air deposition
- Protecting the health and safety of people during an investigation
- As part of conducting a risk assessment

Sampling for contaminants in air falls into two broad categories: contaminants in the vapor or gas phase, and contaminants as particulates or sorbed to particulates (USEPA 1991). The objectives of air sampling include the assessment of exposure for health reasons, and for normal or ambient air quality. Exposure monitoring usually is performed with indoor air, and ambient air monitoring is typically done with outdoor air.

4.3.5.1 Indoor Air Sampling

Before conducting any air sampling, a site inspection should be completed—especially if the sampling will be conducted indoors. The inspection seeks to identify two conditions:

anything affecting or interfering with the proposed testing, and obvious sources of air contaminants.

The site inspection should include (New York Department of Health 2005)

- Type of structure
- Physical conditions
- Odors such as solvents, mold, sweet, sour
- Airflow and ventilation engineering design of the building, structure, or residence
- An analysis of potential contaminants sources including
    - Chemicals used
    - Chemical storage areas
    - Chemical use areas
    - Chemical disposal methods
    - Chemical disposal areas
- Weather conditions

Conducting a chemical inventory is recommended prior to any indoor sampling. This inventory will greatly assist in providing valuable information concerning the identification of the COC, and whether air sampling is necessary. Material Safety Data Sheets (MSDS) for each chemical at the site should be reviewed.

At least 24 h prior to conducting any indoor air sampling, the measures listed in the following should be avoided since they may skew the sample results (New York Department of Health 2005):

- Opening windows
- Operating ventilation fans
- Smoking
- Painting
- Using wood stoves or fireplaces
- Using kerosene heaters
- Operating or storing automobiles in a structure connected to the location where indoor sampling will be conducted (i.e., attached garage)
- Cleaning, waxing, or polishing furniture or other surfaces with petroleum- or oil-based products
- Operating any lawn maintenance equipment including lawn mowers or other mechanical instruments using liquid fuels
- Using cosmetics, e.g., hairspray, deodorant, nail polish, nail polish removers, perfume, and cologne
- Applying pesticides and herbicides

Air samples should be collected from an adequate number of locations to understand the likely sources of potential chemical exposure, and to evaluate their potential exposure to occupants at various locations. In private residences, air samples should be collected from all the floors, including the basement, and outdoors.

For comparison purposes, indoor samples collected from commercial or industrial settings should be collected during normal operating hours and again during times when the facility or locations are not in operation and occupied. To ensure a robust sample—one that is representative of the conditions being tested and minimizes other influences—samples should be collected for at least 1 h, and not more than 8 h.

Sample collection techniques vary depending on the analytical method(s) being used, and must conform to the specific sample collection method. Some methods also specify that duplicate samples must be collected. Indoor air sampling is either conducted on individuals or at stationary locations. Sampling conducted on individuals is either passive or by a sampler attached to a pump that pumps air through a filter or sampling media. This process captures potential contaminants and stores them until analysis can be conducted. To be as representative as possible, indoor air samples are usually collected within the breathing zone between 1 and 2 m (3 and 6 ft) above the floor.

4.3.5.2 Outdoor Air Sampling

Outdoor air sampling for particulates is conducted using high volume air sampling equipment as depicted in Figure 4.29 (USEPA 1998). The ambient air sampling device shown pumps air from all directions across a 18 × 23 cm (7 × 9 in.) exposed filter inside the shelter at flow rates ranging from 1.1 to 1.7 m^3 (39–60 ft^3) of air per minute. The roof design depicted in Figure 4.30 of the shelter is a standard design acting as a plenum above the filter to permit the free flow of air into the plenum space. The sizes of particles collected using the sampler range from 0.3 μ to as large as 25–50 μ in aerodynamic diameter, and are dependent upon wind direction and speed (USEPA 2008).

Filters used for collecting outdoor air sampling for particulates are composed of either glass or cellulose fibers. Glass fiber samples have been extensively used, and cellulose fiber filters are commonly used when sampling for heavy metals. One drawback to cellulose fibers is their potential to increase the potential for adsorption of water and enhance artifact formation of nitrates and sulfates on the filter. The filters are kept in a clean and sterile environment prior to sampling at a constant temperature and a relative humidity of approximately 3%, and are precisely weighed (USEPA 1998). After the sample is collected, several parameters must be recorded, including

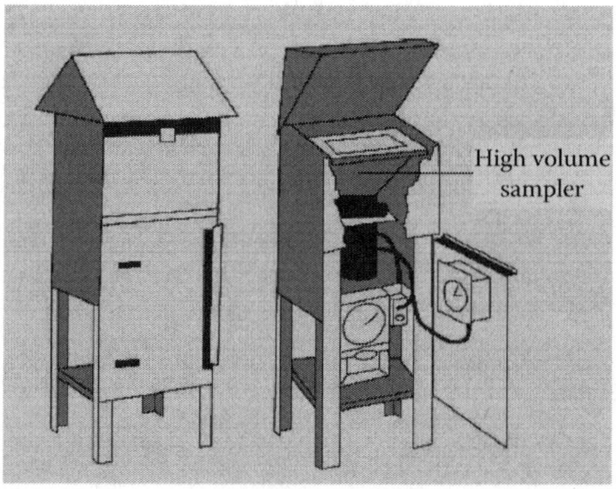

FIGURE 4.29
Outdoor particulate air sampler. (From United States Environmental Protection Agency, *List of Designated Reference and Equivalent Methods for Air Sampling*, United States Environmental Protection Agency National Exposure Research Laboratory, Washington, DC, 2008.)

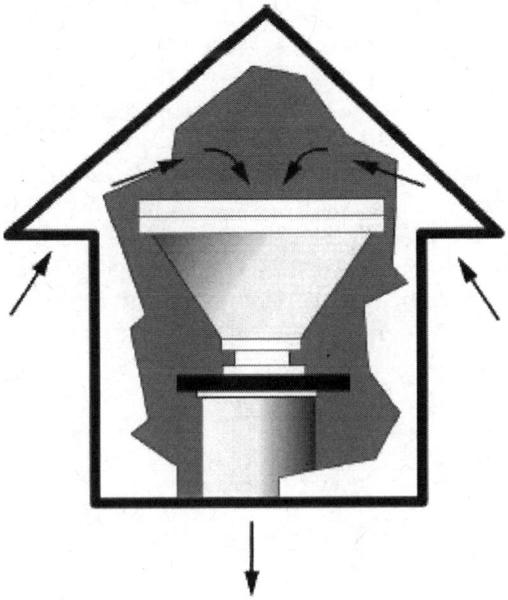

FIGURE 4.30
Airflow pattern through outdoor particulate sample. (From United States Environmental Protection Agency, *List of Designated Reference and Equivalent Methods for Air Sampling*, United States Environmental Protection Agency National Exposure Research Laboratory, Washington, DC, 2008.)

- Starting time and date
- Ending time and date
- Airflow rate
- Temperature ranges during sampling
- Relative humidity difference during sampling
- Summary of conditions that may affect results (construction activities in the area, etc.)
- Barometric pressure at start and end

4.4 Summary and Conclusion

Environmental subsurface investigations are very detailed scientific investigations. These investigations collect enormous amounts of geologic and hydrogeologic information, and significant amounts of other information. All of this information allows the professionals conducting the investigation to determine the existence of contamination at a site or property, and if the level of any contamination present requires a remedial action because it presents an unacceptable risk to human health or the environment.

Environmental investigations must be conducted with a high level of scientific care for several reasons: (1) because contaminants requiring a remedial action may be present at very low levels and precautions must be taken to minimize the chances of any cross-contamination; (2) because the investigation and potential remediation must be conducted in such a manner as to enable their admission in a court of law; and (3) to protect the investigators from any exposure to contaminants while collecting and handling samples taken during field activities.

Over the last 25 years, environmental subsurface investigations have been the most common type of environmental investigation conducted. In urban areas, they have provided extensive geologic and hydrogeologic information of the highest quality. Therefore, environmental subsurface investigations can form the cornerstone of mapping the geology and hydrogeology of an urban watershed.

The next chapter uses geologic and hydrogeologic information obtained from urban environmental investigations along with other sources of geologic information and fieldwork to map the geology of an urban watershed. This chapter provides the insight and tools necessary to take geologic information from numerous subsurface investigation sites within an urban watershed, and then generate a geologic map capable of telling the story of the near-surface depositional history of that same watershed.

References

American Society for Testing Materials (ASTM). 1996. *Standard Practice for Description and Identification of Soils*. D2488-90. West Philadelphia, PA: ASTM.

American Society for Testing Materials (ASTM). 1999a. *Standard Guide for Using the Direct Current Resistivity Method for Subsurface Investigation*. D6431-99. West Philadelphia, PA: ASTM.

American Society for Testing Materials (ASTM). 1999b. *Standard Guide for Using the Gravity Method for Subsurface Investigation*. D6430-99. West Philadelphia, PA: ASTM.

American Society for Testing Materials (ASTM). 2000a. *Standard Guide for Selection of Methods for Assessing Ground Water or Aquifer Sensitivity and Vulnerability*. D6030-96. West Philadelphia, PA: ASTM.

American Society for Testing Materials (ASTM). 2000b. *Standard Guide for Using the Seismic Refraction Method for Subsurface Investigation*. D5777-00. West Philadelphia, PA: ASTM.

American Society for Testing Materials (ASTM). 2001. *Standard Guide for Using Electromagnetic Methods for Subsurface Investigations*. D6639-01. West Philadelphia, PA: ASTM.

American Society for Testing Materials (ASTM). 2005a. *Standard Practice for Environmental Site Assessments: Phase I Environmental Site Assessment Process*. E1527-05. West Philadelphia, PA: ASTM.

American Society for Testing Materials (ASTM). 2005b. *Standard Guide for Using the Surface Ground Penetrating Radar Method for Subsurface Investigation*. D6432-99. West Philadelphia, PA: ASTM.

American Society for Testing Materials (ASTM). 2005c. *Standard Guide for Planning and Conducting Borehole Geophysical Logging*. D5753-05. West Philadelphia, PA: ASTM.

California Environmental Protection Agency. 1995. *Guidelines for Hydrogeologic Characterization of Hazardous Substance Release Sites. Volume 1: Field Investigation Manual*. Sacramento, CA.

California Environmental Protection Agency. 1996. *Guidance for Ecological Risk Assessment at Hazardous Waste Sites and Permitted Facilities*. Department of Toxic Substances and Control. Sacramento, CA.

Capri, J., B.A. Schumacher, S. Wanning et al. 2005. *Collecting Undisturbed Surface Sediments* United States Environmental Protection Agency. EPA/600/R-05/076. Washington, DC.

Fletcher, C.D. and E.K. Paleologos, 2000. *Environmental Risk and Liability Management*. Westminster, CO: American Institute of Professional Geologists.

Glasson, J., R. Therivel, and A. Chadwick. 2005. *An Introduction to Environmental Impact Assessments*. London, U.K.: Routledge.

Harter, T. 2003. *Groundwater Sampling and Monitoring*. University of California ANR Publication 8085. Oakland, CA: University of California, Division of Agriculture and Natural Resources.

Iowa Department of Natural Resources. 2002. *Sample Boring Log DNR form 542–1392*. Des Moines, IA: Iowa Geological Survey.

Jain, R.K., L.V. Urban, G.G. Stacey et al. 1993. *Environmental Assessment*. New York: McGraw-Hill.

Keys, W.S., 1990. Borehole geophysics applied to ground-water investigations. In *United States Geological Survey Techniques of Water-Resources Investigation, Book 2*, ed. A.A.R. Zhody et al. Washington, DC: U.S. Government Printing Office.

Martin, S.N. 2003. *Site Assessment and Remediation Handbook*. Boca Raton, FL: CRC Press.

Munsell, A.H. 1905. *A Color Notation*. Boston, MA: Geo. H. Ellis and Company.

New Hampshire Department of Environmental Services. 2008. *Contaminated Site Management Remedial Action Plan Checklist*. Concord, NH.

New Jersey Department of Environmental Protection. 2005. *Field Sampling Procedures Manual*. Trenton, NJ.

New York Department of Environmental Conservation. 2009. Remedial investigation and feasibility study fact sheet. http://www.dec.ny.gov/chemical/8658.html (accessed September 6, 2009).

New York Department of Health. 2005. *Indoor Air Sampling and Analysis Guidance*. Albany, NY: New York Department of Health. Division of Environmental Health Assessment. Center for Environmental Health.

Occupational Safety and Health Administration (OSHA). 1989. *Hazardous Waste Operations and Emergency Response (HAZWOPER)*. Code of Federal Regulations 40 (CFR) 1910.120. Washington, DC.

Ohio Environmental Protection Agency. 1995. *Technical Guidance Manual for Hydrogeologic Investigations and Ground Water Monitoring*. Columbus, OH: Division of Drinking and Ground Waters.

Ohio Environmental Protection Agency. 2006. *Technical Guidance for Ground Water Investigations*. Columbus, OH.

Oregon Department of Environmental Quality. 1998. *Guidance for Ecological Risk Assessment: Levels I, II, III and IV*. Portland, OR: Oregon Department of Environmental Quality.

Rogers, D.T. 1992. The importance of site observation and follow-up environmental site assessment—A case study. In *Groundwater Management Book 12*, pp. 563–573. Dublin, OH: National Groundwater Association.

Rogers, D.T., M.M. Kaufman, and K.S. Murray. 2009. *An Analysis of Remedial Technology Effectiveness at Brownfield Sites*, abstract, p. 525. Association of American Geographers Annual Meeting. Las Vegas, NV.

United Nations. 1987. Report on the world commission on environment and development. http://www.un.org/documents/ga/res/42/ares42-187.htm (accessed September 15, 2009).

United States Department of Agriculture. 2010a. From the surface down. An introduction to soil surveys for agronomic use. Washington, DC. http://ftp-fc.egov.usda.gov/NSSC/Educational_Resources/surdown.pdf (accessed July 4, 2010).

United States Department of Agriculture. 2010b. Guide to texture by feel. http://soils.gov/education/resources/lessons/texture (accessed July 4, 2010).

United States Environmental Protection Agency (USEPA). 1976. *Resource Conservation and Recovery Act* (RCRA). U.S.C. 6901. Washington, DC.: U.S. Government Printing Office.

United States Environmental Protection Agency (USEPA). 1980. *Comprehensive, Environmental Response, Compensation and Liability Act (CERCLA)*. U.S.C. 9601–9675. Washington, DC.: U.S. Government Printing Office.

United States Environmental Protection Agency (USEPA). 1988. *Guidance for Conducting Remedial Investigations and Feasibility Studies under CERCLA*. Office of Emergency and Remedial Response. EPA 540/G-89-004. Washington, DC: U.S. Government Printing Office.

United States Environmental Protection Agency (USEPA). 1989. *Risk Assessment Guidance for Superfund. Volume 1: Human Health Evaluation Manual (Part A)*. United States Environmental Protection Agency, Office of Emergency and Remedial Response. OSWER Directive 9285.7 Ola. Washington, DC.

United States Environmental Protection Agency (USEPA). 1991. *Evaluating Exposures to Toxic Air Pollutants*. United States Environmental Protection Agency EPA Document 450/3-90-23. Washington, DC.

United States Environmental Protection Agency (USEPA). 1994. *Standard Operating Procedure for Surface Water Sampling*. S.O.P. 2013. Washington, DC.

United States Environmental Protection Agency (USEPA). 1998. *Guidance for Using Continuous Air Monitors*. United States Environmental Protection Agency Office of Air Quality Planning and Standards. Washington, DC.

United States Environmental Protection Agency (USEPA). 2002. *Risk Assessment Guidance for Superfund*. EPA-540/R-02/002. Washington, DC: U.S. Government Printing Office.

United States Environmental Protection Agency (USEPA). 2005. *Standards and Practices for All Appropriate Inquiries*. 40 Code of Federal Regulation (CFR), Part 312. Washington, DC: U.S. Government Printing Office.

United States Environmental Protection Agency (USEPA). 2007. *Sediment Sampling*. USEPA Region 4, Science and Ecosystem Support Division. Athens, GA.

United States Environmental Protection Agency (USEPA). 2008. *List of Designated Reference and Equivalent Methods for Air Sampling*. United States Environmental Protection Agency National Exposure Research Laboratory. Washington, DC.

United States Environmental Protection Agency (USEPA). 2009. Natural resource damage assessment. http://www.epa.gov/superfund/programs/nrd/nrda2.htm#pagetop (accessed September 6, 2009).

Virginia Department of Transportation. 2003. Unified soil classification system. http://www.virginiadot.org/business/resources/Materials/MCS_Study_Guides/bu-mat-Chapter1Soils.pdf (accessed July 4, 2010).

Welty, N., D.T. Rogers, and J. Quinnan. 2010. *High Resolution Profiling of Aquifer Permeability and Contaminant Mass*. Battelle Chlorinated Remediation Conference. Monterey, CA, p. 13.

Yong, R.N. 2002. *Geoenvironmental Engineering, Contaminated Soils, Pollution Fate, and Mitigation*. Boca Raton, FL: CRC Press.

5

Urban Geologic Mapping

5.1 Introduction

The methods presented in this chapter provide the basic building block for understanding the geology and hydrogeology beneath urban regions. This chapter begins by explaining the process of interpreting and integrating information from the environmental investigations presented in Chapter 4 with other sources. The chapter concludes by combining this information with field mapping techniques to construct a geologic map of a watershed.

Geologic maps tell a story of the history of a particular area, and their intended users are geoscience professionals and the general public. Urban geologic maps present many new challenges to geologists and are typically complex to construct. One of the key challenges is to create an easily understandable and usable geological map for members of the nonscientific community. This should not come as a surprise to the geoscience community, since it has long been one of the major goals to construct geologic maps with the ability to clearly communicate scientific information to a diverse audience. John Wesley Powell, who as director of the USGS stated in 1888, "maps are designed not so much for the specialist as for the people, who justly look to the official geologist for a classification, nomenclature, and system of convention so simple and expressive as to render his work immediately available" (Powell 1888).

Effectively communicating complex information is joined by other unique challenges when developing urban geologic maps. As many geologists studying the geology of urban areas know, access and disturbance of the natural landscape present some of the greatest problems when trying to develop a map or while conducting an environmental subsurface investigation. The geologist must be clever, tenacious, and opportunistic because a large geological map base covering urban areas does not exist. Only recently has interest in geologic mapping of urban areas become a focus in the United States (USGS 1998, 2009; Great Lakes Geologic Mapping Coalition 2009). The unconsolidated near-surface geology of urban regions has been largely ignored because urban areas are not in the spotlight like areas with valuable natural resources or risk zones such as those prone to earthquakes or volcanoes. Despite this lack of high-profile attention, characterizing, understanding, and evaluating the geology at urban locations have been quietly conducted at tens of thousands of sites throughout the United States, and these efforts collectively represent a massive information database of fundamental geologic knowledge.

The complexity associated with urban geologic maps is directly related to the characteristics of the urban environment. As shown in Table 2.4, the vast majority of large urban areas in the United States are located on unconsolidated sedimentary deposits. Most of these areas are less than 2 million years old—young by geological standards—but complex in nature by virtue of them having natural- and anthropogenic-induced unconformities as a common occurrence. These urban areas are also proximate to large amounts of surface water and groundwater, and this water plays a fundamental role in functioning as the prime transport mechanism for contaminants. This combination of complex sedimentary deposits and flowing water is why the shallow geologic environment beneath urban areas must be understood at a level of detail at least equal to, or even greater than other areas of the Earth, because these places are where we live and form the tightest bonds between nature and society.

Geologic maps thus form the foundation for ecosystem management (USGS 1998). The natural environment and the humans within it acquire an imposed organization resulting from the geologic framework at any given location. Geologic maps provide us with knowledge of that framework. Specifically, a geologic map facilitates the basic means for evaluating resources, evaluating hazards, and environmental data relative to human activities (Thomas 2004). As noted in Chapter 2, the arrangement, thickness, and composition of the sediment layers beneath our feet have a profound influence on where cities are located, how buildings are constructed, where roads are built, and how contaminants behave and affect people. Constructing a geologic map of a region or a specific site is critical and should be the first step in evaluating the development or redevelopment potential of an urban region or site (USGS 1998; Berg et al. 1999; Berg 2002). It is therefore not an exaggeration to say that sustainable urban environments cannot be achieved without accurate and usable geological maps.

5.2 Geologic Mapping in an Urban Environment

Mapping the geology of urban areas increases our understanding of their fundamental natural processes and identifies the types and locations of anthropogenic impact to consider during the sustainable development and redevelopment of city cores and outlying areas. Standards exist for geologic mapping (USGS 2010a) and always require that extensive field work be completed when constructing a geologic map of a desired area (Maley 2005). However, many urban locations are not accessible, often making it imperative to gather and evaluate as much reliable historical data and information as possible from multiple sources. The next nine sections present a sequence beginning with the collection of relevant information and ending with the preparation of the final urban geologic map.

5.2.1 Information and Mapping Challenges

Mapping the geology of an urban region presents many challenges to geologists and can be more difficult than mapping other regions (USGS 1998). Some of these challenges include

- Maintaining safety. Conducting field activities in urban areas poses many safety concerns, including physical hazards from equipment, machinery, automobiles, and other mechanical equipment. In addition, there may be other hazards such as heat stress and related hazards imposed by weather conditions and the requirement for wearing specialized personal protective clothing to prevent exposure to contaminants.
- Inaccessibility due to legal aspects of trespass and private property. Many locations of interest are not on public property and can only be accessed by receiving permission from private property owners.
- Areas covered with buildings and other anthropogenic structures obscure original features and eliminate the possibility of physical inspection.

- Developments such as parks, landfills, or other landscaped areas may cover, destroy, or remove original deposition features.
- Fieldwork—such as excavating test pits in a historic park—may significantly disturb an area and create the need for expensive measures to restore the land to an acceptable standard.
- Influencing and altering the natural environmental balance through large construction projects affect the subsurface environment. For example, the construction of a dam may raise the regional groundwater levels or flood areas with water before a sufficiently detailed geologic study can be conducted.
- Vegetative cover in urban areas occupies large amounts of surface area, especially in regions with humid climates. This extensive vegetative cover contributes to the difficulty in observing natural deposits and also increases the potential for disturbing the natural deposits through either physical or chemical means.
- Introduction of chemicals and anthropogenic sources of water may destroy, disturb, or dissolve original depositional features and deposits. Sources of these external change agents include
    - Stormwater runoff
    - Leaking sewers
    - Landfills
    - Old dumps
    - Underground tanks
    - Underground utilities
    - Pipelines
    - Septic systems
    - Wastewater treatment plant discharges
    - Industrial sites
    - Fertilizer application
    - Pesticides and herbicide application
    - Vehicle exhaust deposition
    - Old cemeteries
    - Golf courses
    - Fire-fighting efforts
    - Tanker spills
    - Brownfield sites
    - Abandoned buildings
    - Power plant particulate and water discharges (nuclear and coal)
    - Acidic precipitation

5.2.2 Information Sources, Integration, and Evaluating Usability

To counter the challenges faced by mapping the geology of urban regions, geologists have to be more creative and diligent in pursuing the information necessary to adequately understand the geology of the near-surface urban environment.

There are several sources of information in urban regions useful for geological purposes, which, in some cases, may be more extensive than what is available for mapping undeveloped regions. Often, to be successful in gathering information on the geology of urban areas, the geologist must become a forensic geologist, because the near-surface unconsolidated deposits in many urban areas of the United States have experienced significant disturbance. Being a forensic geologist means knowing and evaluating whether the information collected is reliable and valid and whether it should be included in the final urban geologic map.

Mapping the geology of an urban region begins with a literature review of existing data and publications using the following sources as a guide:

- Current and historical maps:
  - Geologic maps
  - Planning maps
  - Historical highway maps
  - Fire protection maps (often referred to as "Sanborn maps")
  - Topographic maps (7.5 min scale is best)
  - Trail maps
  - Historical interest maps
  - Park and recreational maps
  - Maps prepared for site-planning purposes; these often contain accurate locations of existing underground infrastructure and above-ground water management structures, such as detention basins
  - Others
- United States Geological Survey studies and reports
- State Geological Survey studies and reports
- State and local water well records
- Army Corps of Engineers reports and studies
- Oil and gas records, studies, and reports
- Department of transportation studies and reports
- Building construction reports and permits
- Geotechnical reports and studies
- State and local environmental investigation reports and studies at sites of contamination
- State and local historical societies studies; publications, and photographs
- State and local universities and colleges
- Sources of historical aerial photographs
- Previously published scientific literature

You may think historical highway maps cannot provide valuable information to help understand the geology of a particular area or region—but this is simply not true. Old and even new highway maps provide valuable information on mapping the geology of urban areas (Bennison 1974).

A few examples include

- Names of highways and roads. The names of highways and roads sometimes provide insight into a region's geology. For instance, a highway named "Ridge Road" creates a potential association between topography and geology. Others performing this task include Muddy Road, Sandy Flats, Rocky Road, Sandy Hill Road, Wash Road, Winding Road, Coastal Highway, Beach Highway, Marsh Creek Road, Peat Bog Road, etc.
- Location of cemeteries. Cemeteries provide information on a region's geology, because cemeteries during the nineteenth and early twentieth centuries were usually located in areas of well-drained soil such as sand or gravel. This siting prevented the occasional casket from popping out of the ground after burial; an event possible in more poorly drained soil such as a clay.
- Location of parks and recreational areas. Parks and recreational areas are often found in areas with a scenic or educational theme related to the geology of a region.
- Historical markers and historical sites. These locations often exist over or near significant natural disasters, battlefields, and trails.
- Locations of historical settlements and buildings. Before urbanization occurred, many historical settlements used foundation materials of native stone collected from nearby locations, and these items are useful when assessing a region's geology.

These sources of information must be carefully inspected for integrity and usefulness. Perhaps, the information was not initially collected for geological purposes and may not have been collected by a qualified geologist. What if you are consulting historical geological maps? Unfortunately, most historical mapping of urban regions was produced in two dimensions and did not map the subsurface geology to a depth of more than a few meters. Relying upon a two-dimensional map is problematic because effective and useful urban geologic maps produced for environmental purposes must represent three dimensions. This spatial requirement means these maps are much more complex and require significantly more information (Barnes 1993). Some forensic methods described throughout this chapter can help evaluate the accuracy and ultimate usability of questionable historical geologic information.

Just when things are starting to sound bleak, we get some relief in the form of geologic and hydrogeologic information obtained from environmental investigations. These activities can supply much of the baseline information to evaluate whether questionable geologic information is usable and also supply the necessary geologic information to enable detailed three-dimensional mapping. In many urban areas, environmental investigations have been conducted at thousands of locations. A large percentage of these environmental investigations yield very detailed geologic and hydrogeologic information collected by qualified and certified professionals. This information is readily accessible through federal, state, and local environmental regulatory agencies. For example, over 3000 reports of the subsurface geologic investigations performed in the Rouge River watershed are available for review in one state agency, the Michigan Department of Environmental Quality (now the Michigan Department of Natural Resources and Environment) (MDEQ 2008). It is therefore quite possible the best sources of geologic and hydrogeologic information are within easy reach.

Figure 5.1 shows the location of nearly 500 sites where detailed and large-scale environmental investigations have been conducted within the Rouge River watershed (Rogers 1996). In terms of the geologic and hydrogeologic data collected, each site had a minimum

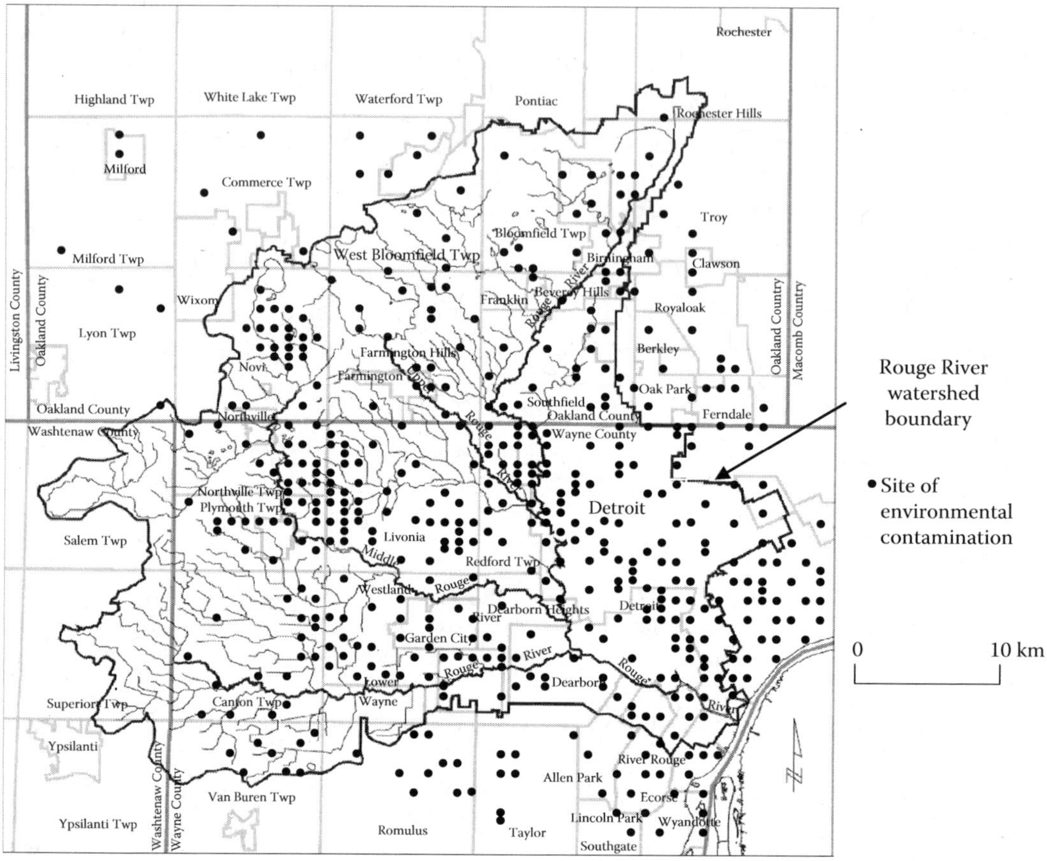

FIGURE 5.1
Sites of extensive environmental subsurface investigation in an urban watershed.

of 20 to sometimes more than 300 individual soil borings drilled and up to 80 monitoring wells installed. Investigative depths reached as deep as 30–45 m (100–150 ft) beneath the ground surface.

Environmental investigations in urban areas usually do not extend beyond 15 m (50 ft) beneath the surface. However, unconsolidated sediments in many urban areas extend to depths greater than 15 m and may even approach 305 m (1000 ft) or beyond as in Salt Lake City, Los Angeles, Phoenix, and Las Vegas to name a few (Halka 1983; Stokes 1989; Allen 2003; Page et al. 2005; Bilodeau et al. 2007). The evaluation of data from these investigations can be supplemented and perhaps confirmed by the information supplied from well drillers and well logs.

Water supply and exploratory wells are routinely drilled to depths exceeding 30 m beneath the surface and some may extend to over 45 m. Drillers installing water supply wells are often required to register the well with the appropriate local, county, or state officials and submit a written well log similar in many cases to the example shown in Figure 4.20. Water supply well logs often contain valuable and abundant sources of geologic and hydrogeologic information. However, as stated earlier, care should be taken before accepting the accuracy of data from a water supply well or other similar sources providing geologic or hydrogeologic data because they may not have been conducted by a qualified professional. Therefore, some forensics are required. Cross-referencing the geologic information obtained during an environmental investigation with a water supply well log that was installed very close to the investigation site can provide information on the accuracy and ultimately acceptability of the water supply well log. For instance, assume a water supply well was installed to a depth of 100 m on the same property where an environmental subsurface investigation was conducted to a depth of 15 m. If the water well log correctly interpreted the geology in the upper 15 m, a reasonable conclusion is that the water well driller correctly interpreted the geology at depths greater than 15 m.

There is another point regarding the installation of water wells worth mentioning here from a forensic point of view. Historically, it was not uncommon for a qualified geologist to accompany water supply well drillers during drilling operations, because this was the only source of affordable geologic information. On many occasions, the geologist's name would appear on the water supply well log (Sherzer 1913; Mozola 1954, 1969). There is some benefit then to consult with geologists in the area who were historically employed by the United States Geological Survey, State Geological Survey, or who teach at local colleges and universities. This action will help evaluate and cross-reference the information when reviewing historical water supply well logs.

Geophysical information in various forms is capable of providing valuable geologic information to assist in geologic mapping both horizontally and vertically. Geophysical information is collectable in most urban areas without causing any disturbance and can be extremely useful in interpreting data between boreholes where the geology is known. Geophysical information is most helpful when the areas are inaccessible to other methods of collecting geologic or hydrogeologic information and within areas of complex geology.

5.2.3 Building the Conceptual Geologic Model

Following the review of existing information, a conceptual geologic model can generally be constructed. The conceptual model is a working draft geologic map of the desired region that identifies areas by their basic geologic processes. For instance, most urban areas are composed of unconsolidated sediments resulting from more than one type of geologic process; e.g., fluvial, glacial, alluvial, and marine. The conceptual model simply identifies geologic processes that formed the deposits along with their general location and horizontal and vertical boundaries.

In many cases, enough information has been gathered to allow the initially identified geologic processes to be subdivided into individual geologic units or formations. It is during this step when data gaps and difficulties in interpreting geologic information usually become apparent. These data gaps then form the basis for concentrating the field work activities described in the next section. Listing every data gap, irregularity, or area of interest is recommended until each has been resolved with adequate certainty. After each data gap, irregularity, or area of interest has been addressed, reinterpretation of the conceptual geologic model is recommended to evaluate if any changes are necessary.

Another important decision to make while building the conceptual geologic model is whether anthropogenic deposits such as fill material and natural deposits, such as top soil or peat, should be included in the mapping process. Fill material is usually noted, but not included on a geologic map unless the areal extent is significant. Top soil may be noted at specific locations, but is generally not mapped unless the deposit is either (1) significant in areal and vertical extent and/or (2) represents a peat or other similar material resulting from deposition in a current or former swamp, bog, or wetland as described in

Chapter 2. The investigator should pay attention to these deposits during the information gathering phase of the mapping process. Then, he or she can make a determination whether the deposits are sufficiently large in scale compared to the map being developed, and if they contribute to the history and understanding of the area being mapped.

5.2.4 Fieldwork

Information gathered and evaluated from the sources listed in Sections 5.2.2 and 5.23 will assist the geologist in understanding the area to be mapped before extensive field work begins. The geologist must be skilled in understanding sedimentary processes and in identifying the difference between a natural and anthropogenic deposit. This is sometimes difficult even when there is a detailed understanding of the geology of a site or region. An example is the anthropogenic deposition of dredge materials noted in Section 2.6.2, which can take on the appearance of a natural deposit due to the common presence of bedding layers indicating a stratified sequence of deposition. Complications such as this in many urban areas may require several rounds of fieldwork, literature search, and historical forensic review before sufficient understanding can be achieved to construct a map.

Based on the authors' review of large volumes of historical information and the construction of numerous conceptual geologic models, fieldwork in urban areas is usually concentrated at the following locations:

- Areas with limited or no information, or areas with gaps or missing information in the existing data set
- Locations with geologic irregularities or complexities
- Areas with extensive anthropogenic disturbance
- Areas of geologic significance such as formation contacts and unconformities
- Areas where geologic processes of formation overlap or are located in close proximity
- Locations requiring confirmation of previous interpretations
- Areas not easily accessible
- Areas of ecological significance
- Areas with a high potential of encountering contamination

Geologic mapping in an urban area often requires the geologist to excavate through potentially disturbed soil profiles to reach material representing naturally occurring deposits. This fieldwork is often a dangerous activity due in part to broken glass, metal fragments and other debris, underground pipes, and the existence of potentially harmful contamination. There may also be other hazards present, such as high vehicular traffic. Therefore, extreme care and proper planning should be employed, and the geologist should perform the following actions before conducting geologic mapping activities in the field:

- Conduct a reconnaissance of each area or location of interest
- Prepare a written scope of work for each desired location, including
  - Purpose and objective
  - Activities to be conducted
  - Equipment list

    - Methods of investigation
    - Sampling collection procedures
    - Chain of custody (if required)
    - Photographic documentation
- Obtain written permission of private property owner(s) if any type of invasive fieldwork may be conducted
- Obtain written permission from the local municipality or appropriate governmental agency if any type of invasive fieldwork has the potential of being conducted on public property
- Contact local utilities to ensure the proper marking of all utilities in the area where fieldwork is to be conducted
- Prepare an appropriate written health and safety plan before beginning fieldwork using the guidelines in Section 4.2.2

Fieldwork may simply consist of digging a test pit with a shovel or drilling a soil boring by hand using a hand auger. In some cases, test pits may be excavated using a backhoe or other type of excavator. If you are fortunate, many opportunities exist for conducting fieldwork without much effort. Areas under construction often offer the urban geologist ample opportunity to conduct field activities and mapping because the subsurface geology is already exposed for other purposes. Some of these locations or activities include

- Road building
- Subsurface utility work
- Excavation of foundations for new buildings or for repairing existing buildings
- Demolition activities where the footings and foundation of existing buildings are excavated and removed
- Sewer repair
- Stormwater control construction activities

Conducting inspections and fieldwork at construction sites provide the geologist with many opportunities to collect valuable data at a much lower cost, with less effort, and in a much shorter time. Contact local municipalities to assist fieldwork planning and to obtain the necessary permissions before construction activities begin. Review the building permits before the start of construction and contact the appropriate Department of Transportation and utility company personnel in areas where fieldwork is to be performed. Then secure the necessary permissions before conducting any activities and employ the appropriate health and safety measures.

Figures 5.2 through 5.4 document the exposure of the near-surface geology in an urban area and show the opportunities a construction site offers the geologist. Each example represents a good opportunity to gather an abundance of geologic information in a short amount of time and at no cost. Next to each figure is a stratigraphic column used later in this chapter to assist in building a geologic map and evaluating the recent geologic history of the region.

After examining the depositional layers from excavation pits as shown in Figure 5.4, evaluating and recording the strike and dip of the depositional units is required before a three-dimensional map can be developed (Lahee 1961). Strike refers to the attitude or

FIGURE 5.2
Geologic strata exposed at a construction site. (Photo by Daniel T. Rogers.)

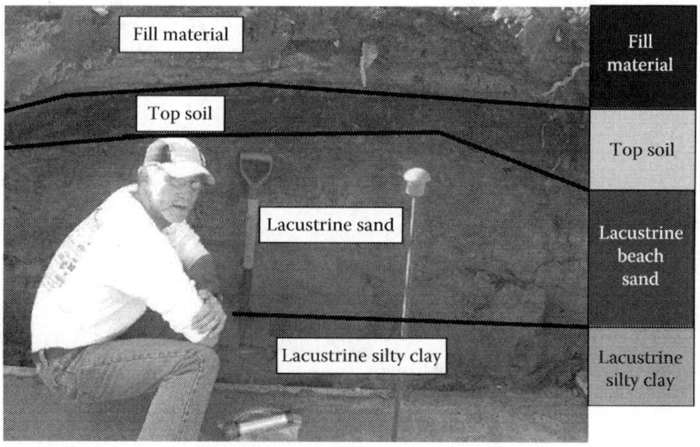

FIGURE 5.3
Construction site with exposed geologic strata. (Photo by Daniel T. Rogers.)

FIGURE 5.4
Excavation at a construction site exposing geologic strata. (Photo by Daniel T. Rogers.)

trend of a particular deposit. For instance, if a geologist were mapping a sandy beach deposit from an ancient lake and recorded the trend of the deposit at several locations, the geologist would be able to determine the size of the lake. The strike of a deposit, outcrop, or other planar feature is represented on a geologic map as a short straight line segment oriented parallel to the compass direction of the strike. The Dip is defined as the angle at which the geologic deposit, feature, or structure is tilted relative to the horizontal plane. The dip is represented on map as a line segment perpendicularly attached to the strike symbol.

For unconsolidated deposits, evaluating strike and dip is much more difficult to measure since the sediments being evaluated may be loose. However, measuring the strike and dip is still important and may provide valuable information concerning origin of the formation and whether the unit has experienced any type of deformation since being deposited. Evaluating a beach deposit from a former glacial lake provides an example of natural deformation. Many former beach deposits from former glacial lakes are not horizontal because the strike of many glacial lake deposits dip very slightly away from the former glacial front. This dip provides an indication of the amount of isostatic rebound that has occurred since the glacier retreated and provides a basis for estimating the thickness of the glacier. Measuring the slight angle produced by this process requires high accuracy and precision since many deposits only dip 0.3–0.6 m (1–2 ft) per 1.6 km (mi) at most locations. In addition, measuring and evaluating strike and dip of unconsolidated deposits can also be used to evaluate whether the unit has experienced any anthropogenic disturbance. Figure 5.5 shows strike and dip of a set of layered deposits. Note the perpendicular dip direction with respect to the strike direction. Figure 5.6 shows the strike and dip of several geologic units and the corresponding information provided by strike and dip when constructing a three-dimensional representation of the data collected from a site. The numerical values on the figure represent the angle at which the units plunge into the subsurface.

Producing detailed intermediate maps showing the geology of an individual site is a crucial step in the process of constructing a three-dimensional geologic map of an entire region and watershed. Figure 5.7 shows a basic geologic map beneath a site where each number represents a distinct deposit; Figure 5.8 illustrates a planar representation of the

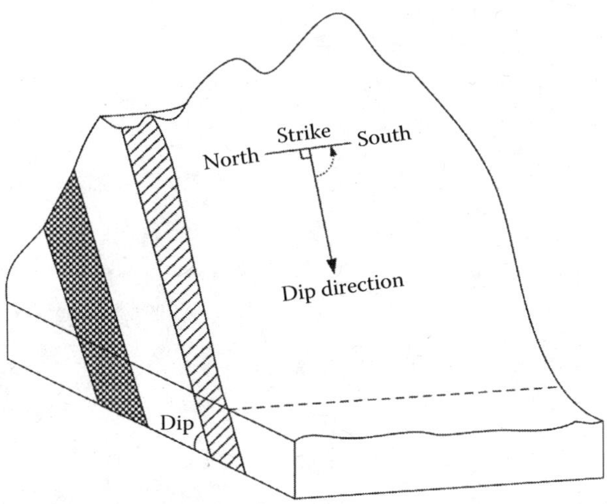

FIGURE 5.5
Concept of strike and dip.

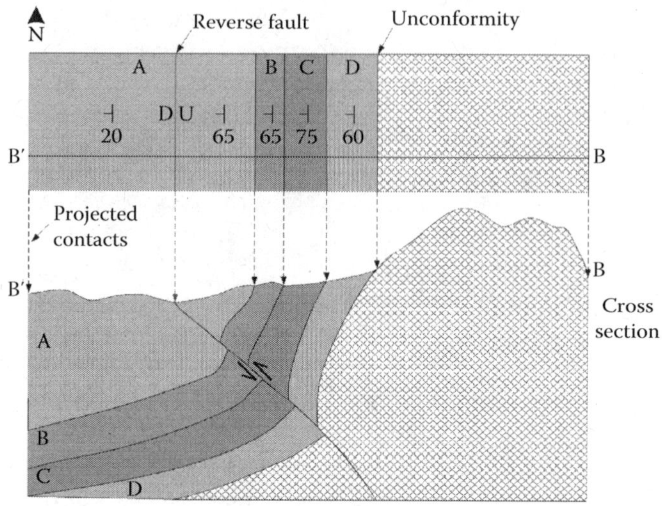

FIGURE 5.6
Concept of strike and dip as depicted on a map.

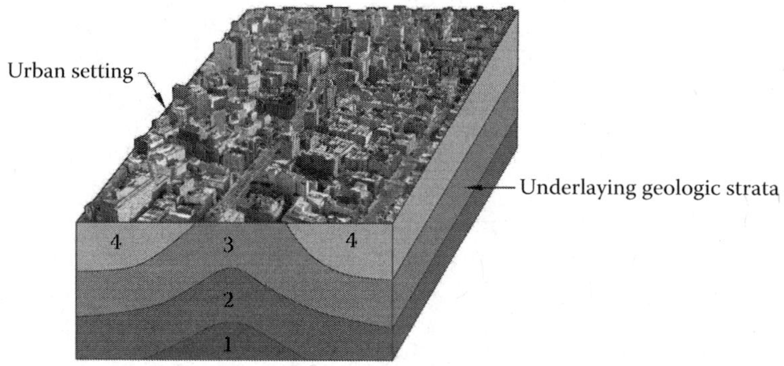

FIGURE 5.7
Urban setting showing underlying geologic units at a site.

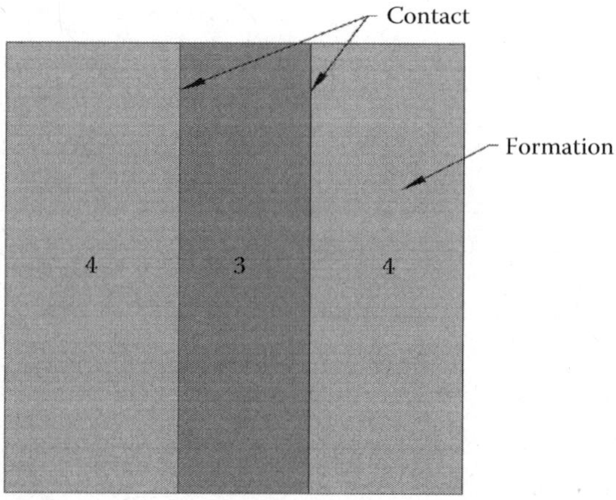

FIGURE 5.8
Geologic site map of area shown in Figure 5.7.

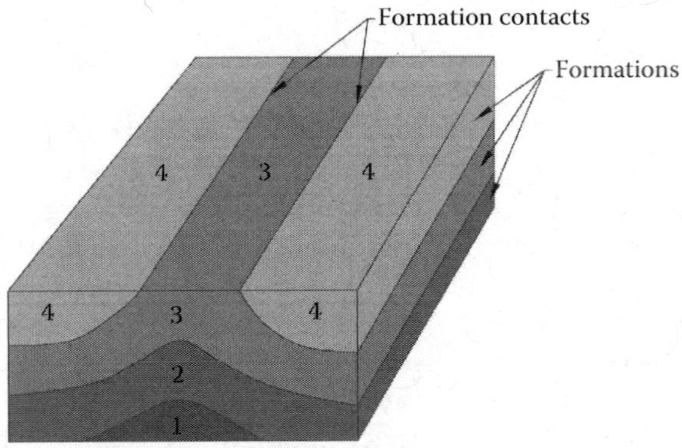

FIGURE 5.9
Composite block map developed from Figures 5.7 and 5.8.

individual strata present at a given region; and Figure 5.9 is a composite map of the underlying geology shown in Figure 5.7 and the individual strata depicted in Figure 5.8.

Combining geologic information from many individual sites using facies analysis is the next step in constructing a three-dimensional geologic map for a region.

5.2.5 Facies Analysis

Facies are the characteristics of a sedimentary rock mass that reflects its depositional environment. Facies analysis or cross-dating is a technique used to help understand sedimentary environments and the processes leading to the origin of their sediments (Anderton, 1985). Facies analysis can be performed when two or more stratigraphic columns are developed for different locations within a given area. Facies analysis is conducted to

- Evaluate the completeness of sedimentary deposits at any one location
- Identify any gaps in the depositional record
- Reconstruct the geologic history of a region
- Identify or confirm the historical occurrence of geological processes within a given region
- Reconstruct a succession of events
- Create a vertical succession of depositional record

The type of facies analysis most often employed to reconstruct the geologic history of a particular region is termed relative dating. This technique uses the four principles of stratigraphy introduced and described in Chapter 2 and repeated here for reference: (1) principle of original horizontality, (2) principle of superposition, (3) principle of cross-cutting relations, and (4) principle of faunal and floral succession. Through the use of these four principles, facies analysis can evaluate and establish the chronology of depositional events within the area being mapped. For example, during the course of reviewing historical literature and conducting fieldwork, it usually becomes apparent that a few locations contain a nearly complete and undisturbed vertical geologic section. These accessible columns essentially have preserved the depositional history so it can be studied in detail and are called type sections. Once the chronology of deposition has been firmly established, a stratigraphic column can be constructed for the entire area being mapped (USGS 2007).

As shown in Table 2.4, many types of geologic processes are responsible for building the geologic history of the large urban areas within the United States. We will demonstrate the process of facies analysis using the stratigraphic columns from Figures 5.2 through 5.4 because they contain a good record of the depositional history of the Rouge watershed. The stratigraphic columns from each site are presented in Figure 5.10.

The distance separating the three sites and represented by each column shown in Figure 5.10 is approximately 1.5 km. Each site is perpendicularly aligned (north to south) to the strike of the geologic units identified at each location (west to east). The lines drawn between each stratigraphic column connect geologic units appearing in more than one stratigraphic column such as the Lacustrine Beach Sand Deposit and the Lacustrine Silty Clay Deposit. The units of Fill Material and the Beach Dune Sand Deposit exhibit converging lines because they are not present in the other stratigraphic columns within the study area.

A simple analysis of the facies represented by each depositional unit in Figure 5.10 indicates each site was once part of a former lake, with the beach and associated dunes located toward the west and the deeper water. The small amount of dip present between columns I and II (approximately 10 m per km) is also characteristic of these glacial lacustrine features. The presence of a lake between columns I and III is further confirmed by the absence of

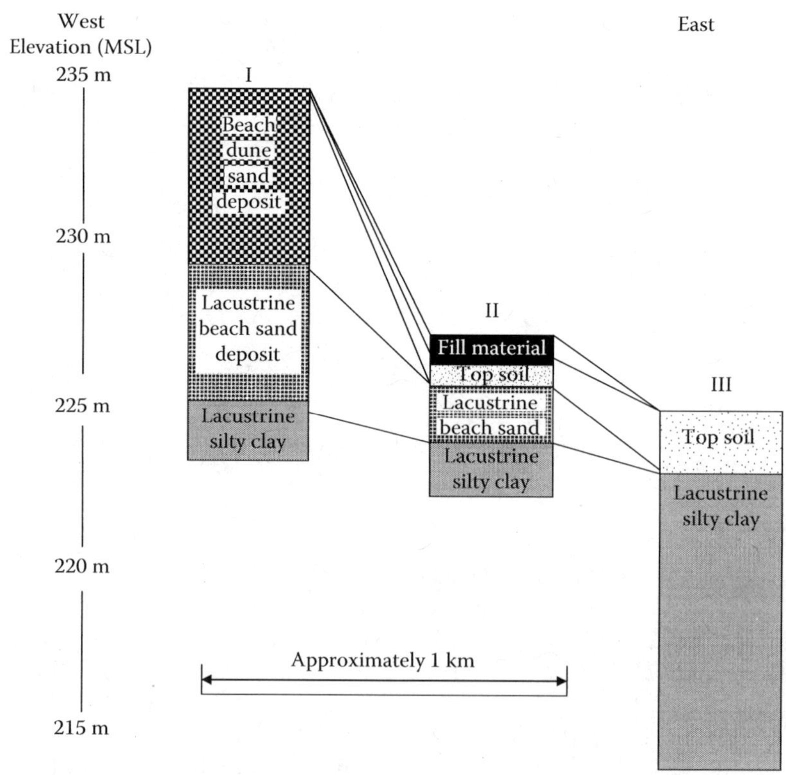

FIGURE 5.10
Example of facies analysis or cross dating using stratigraphic columns from Figures 5.2 through 5.4.

the sand and dune deposits in column III toward the east and the continued presence of the lacustrine silty clay deposit.

It should be noted that the presence of sedimentary deposits dipping in this example represents the slope or contour of the original land surface and does not represent a structural cause such as faulting. However, some isostatic adjustment has occurred since the materials were deposited and is evident from an elevation rise in the beach elevation of approximately 1 m every 2 km toward the northeast. This rise is likely a result of the historical glacial ice retreat from the region.

Additional analysis of facies also suggests the lake was once larger because a deposit of lacustrine silty clay is present beneath the beach sand deposit in Stratigraphic Column I. This suggests the lake was getting smaller and in a regressive stage before it disappeared. The absence of top soil in Stratigraphic Column I and the presence of fill material in Stratigraphic Column II also indicate anthropogenic disturbance.

Please do not get too comfortable—facies analysis is not always this straightforward. Gradational or interpolated boundaries between geologic units require separate and careful treatment. Gradational boundaries occur both laterally and vertically and are fairly common in sedimentary deposits. Gradational boundaries should be identified and explained in the legend and accompanying text according to the following guidelines:

- Nature of the gradational boundary
- Cause of the gradational boundary
- Extent of boundary

A sufficient facies analysis contains numerous stratigraphic columns with enough areal coverage to enable an understanding of the region's depositional history. This information allows for the construction of cross sections. Cross sections display the geologic deposits without any breaks or gaps between the stratigraphic sections and help facilitate the construction of a three-dimensional geological map (USGS 2002). Some degree of interpretation is usually necessary to construct a cross section, and these finer points are explained in greater detail in the next section.

5.2.6 Developing and Constructing Cross Sections

A cross section is similar to facies analysis, but cross sections add levels of complexity and additional information. A cross section represents a view of the subsurface as though it was cut open and viewed from the side. Geologic cross sections are intended to be interpretive and demonstrate the arrangement, relationship, structural composition, and depositional history of geologic units displayed in the cross section.

General guidelines for constructing a geologic cross section include

- Choosing appropriate horizontal and vertical scales
- Properly locating control points or points along the cross section where stratigraphic sequences are known
- Ensuring the legend incorporates and explains each geologic material and feature
- Using appropriate symbols to identify each geologic unit
- Using appropriate orientations and landmark features
- Including vertical and horizontal scales
- Including a statement of vertical exaggeration

When constructing a cross section, the amount of geologic interpretation between stratigraphic sections depends upon the

- Number of stratigraphic sections
- Distribution of stratigraphic sections
- Number of data gaps requiring interpretation
- Complexity of each stratigraphic section

Selecting the line or slice direction is perhaps the most important operation when constructing a cross section. Usually, several cross sections are constructed to create a more complete picture of the vertical dimension and depositional sequence. Generally, the cross section containing the most significant information is one constructed as a perpendicular slice to the strike or trend of the geologic units being evaluated. These types of cross sections are most effective because more geologic units are usually represented, and the relationship, structure, and depositional history are more easily revealed and understood. This is especially relevant to urban watershed restoration efforts because the majority of the deposits existing below urban areas are sedimentary in nature.

An example cross section constructed perpendicular to the strike is shown for the Rouge River watershed in southeastern Michigan in Figure 5.11 (Rogers 1996, 1997; Kaufman et al. 2005). This cross section includes several types of sedimentary deposits of glacial, lacustrine, and fluvial origin commonly found in most urban areas of the United States. Figure 5.12 is a three-dimensional representation of several intersecting cross sections

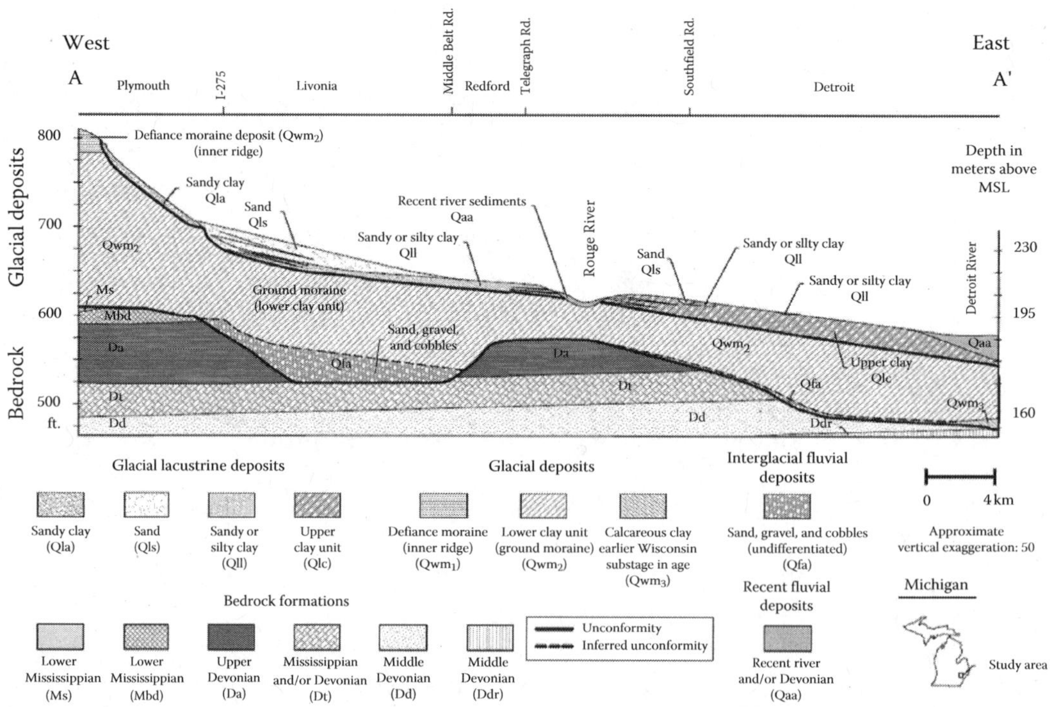

FIGURE 5.11
Detailed cross section of Rouge River watershed in southeastern Michigan.

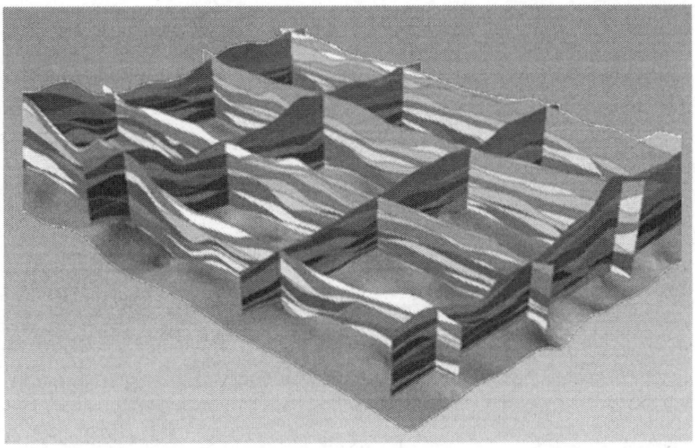

FIGURE 5.12
Several interconnecting cross sections. (From United States Geological Survey, National cooperative geologic mapping program, http://ncgmp.usgs.gov, 2009.)

(Great Lakes Mapping Coalition 2009). Cross sections of this type aid in interpreting the depositional history of the area being mapped.

5.2.7 Creating the Base Map

After the review of existing data, conducting field work, facies analysis, and constructing cross sections, a base map can be prepared. The steps required to prepare a base map include (1) using appropriate nomenclature, (2) selecting appropriate symbols, (3) selecting the scale for the map, and (4) determining what type of map is most appropriate.

Using recognized and acceptable nomenclature and symbols ensures the geologic map being prepared maintains consistency and minimizes ambiguity. The standard nomenclature for geologic maps in North America is The North American Stratigraphic Code (American Association of Petroleum Geologists 2005). This code is used for naming, describing, establishing, redefining, and abandoning geologic units. For symbols, the standard is the Federal Geographic Data Committee Digital Cartographic Standard for Geologic Map Symbolization (USGS 2006). Items such as line symbols, point symbols, and colors and patterns used to portray the different features on a geologic map are covered by this standard.

The first step in preparing a geologic map is to start small and then add adjacent locations to build the entire map. One way to accomplish this process is to begin with a single geologic unit that is relatively simple or easy to identify in the field and then add adjacent geologic units both horizontally and vertically through the entire unconsolidated strata until bedrock is encountered. For reference, the applicable 7.5 min series USGS Quadrangle map ("quad") is recommended to construct the field base map. The 7.5 min series quad is a 1:24,000 scale map—a scale suitable for many environmental applications. These maps also have valuable information to greatly assist in mapping any given area, including

- Each map is very accurate and reliable.
- Most urban areas in the United States have readily available 7.5 min series Quadrangle maps.

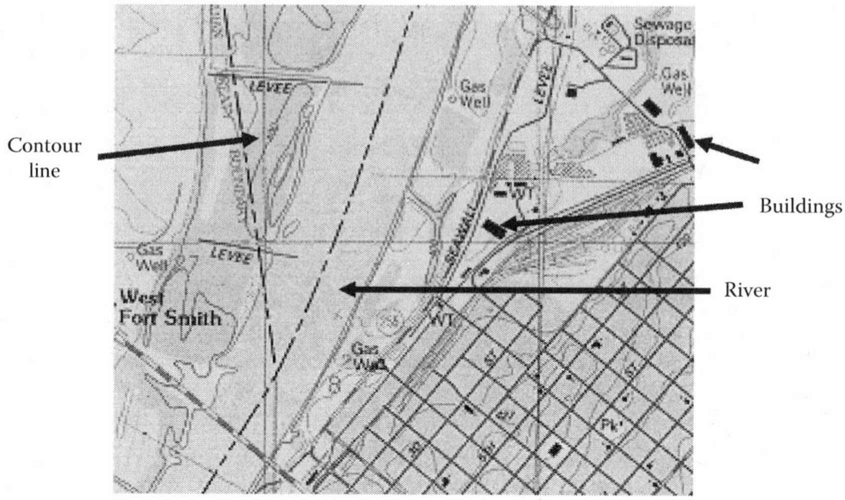

FIGURE 5.13
Portion of 7.5 min USGS series map of Fort Smith, Arkansas. (From United States Geological Survey, Geographic information system, USGS, Washington, DC, http://egsc.usgs.gov/ish/pubs/gis_poster, 2010b.)

- Elevation contours are present to assist with the identification of geologic units and help establish formation boundaries.
- Many land features of interest to geologists are identified.
- Anthropogenic features such as developed areas, roads, and parks are identified.
- Natural features such as lakes, rivers, streams, forested areas, and points of interest are identified.
- Each map is well labeled.
- Coordinates of longitude and latitude are present on each map.
- Magnetic declination is included on each map.
- The location of springs and wells are located on many maps.
- Information can be easily transferred to other mapping technologies such as geographic information systems (GIS) and three-dimensional maps.
- Many maps are available electronically, so areas with complex geology or areas requiring finer resolution can be mapped using larger cartographic scales.

Figure 5.13 presents a portion of a 7.5 min USGS series map.

5.2.8 Final Draft Map Preparation

Over 100 years ago, John Wesley Powell stated that "maps are designed not so much for the specialist as for the people." He elaborates more on this topic (and we paraphrase here) by saying the system of convention selected should be simple and expressive, so the concepts and processes depicted can be easily understood. Conveying geologic information with the simplest of terms is difficult and challenging, because the information is often very complex. Do not despair—there are several methods available to allow an effective presentation of the information to intended users. Selecting an appropriate method or combination of methods depends on (1) the level of complexity, detail, and scale of the information and area being mapped; (2) the intended users, such as scientific or academic, land-use planners, or the general public; and (3) cost.

Some common methodologies used to prepare the map include

1. Preparing a series of maps showing the regional development and evolution of the landscape of a particular region. These maps tend to lack great detail because the geographic scale is generally very large. The purpose of these maps is to convey major concepts of regional significance.
2. Preparing several maps beginning with the deposition of the bottom or oldest layer of sediment, and then continuing with the display of distinctive intervals of equal time or significant geologic events of deposition. These maps have advantages because they can be very detailed and focus on significant geological events.
3. Adding geological processes that formed the sedimentary deposits such as stream or river succession, lake elevations, or glacial advances or retreats. This type of map combines the areal distribution of geological deposits with the methods of their formation. Although this type of map is very effective at conveying processes with geological deposits, it may lack desired detail depending on the scale of the map.
4. Producing maps in three dimensions. Constructing maps in three dimensions generally requires more time and effort, but has the benefit of providing better data gap analysis and integrating landscapes into the geologic map.
5. Using GIS to combine geological maps with other digital data layers such as roads, buildings, parks, utilities, drainage patterns, proposed developments, and population. A composite map representing the overlay of some or all of these layers can produce new information about the study area.

The most common methods are #3, #4, and #5, or some combination of each.

5.2.8.1 Succession of Maps

A succession of maps is one method used to decrease the time necessary to understand the geological processes responsible for the development of sedimentary deposits in urban areas. This technique presents complex processes gradually through the use of several maps that incrementally add complexities and depositional processes. At a regional scale, building layers showing different geological processes occurring at specified intervals and/or depicting the sequences forming deposits can simplify the communication of geologic information. These types of maps can also incorporate three-dimensional and GIS formats to present the information.

An example of a succession of regional maps of the Great Lakes region is presented in Figure 5.14a through h (modified from Hough 1958, 1963). This map succession shows the regional development and evolution of the landscape for a particular region at a specific time—the Great Lakes during retreat of the last glacial advance of the late Pleistocene. These maps are general in nature and lack detail because their geographic scale is very large.

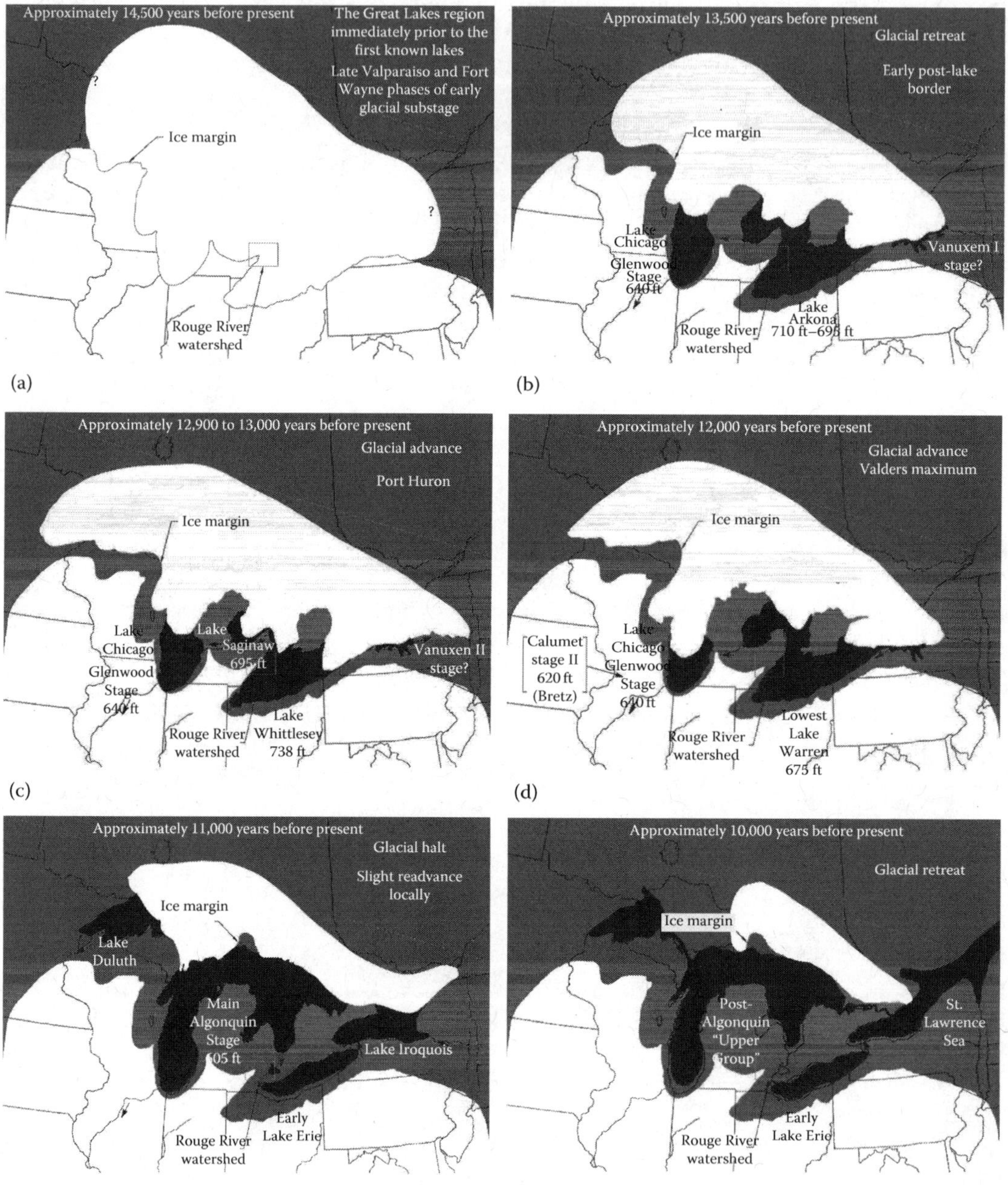

FIGURE 5.14
(a) Great Lakes region, approximately 14,500 years before the present. (b) Approximately 13,500 years before the present. (c) Approximately 13,000 years before the present. (d) Approximately 12,000 years before the present. (e) Approximately 11,000 years before the present. (f) Approximately 10,000 years before the present.

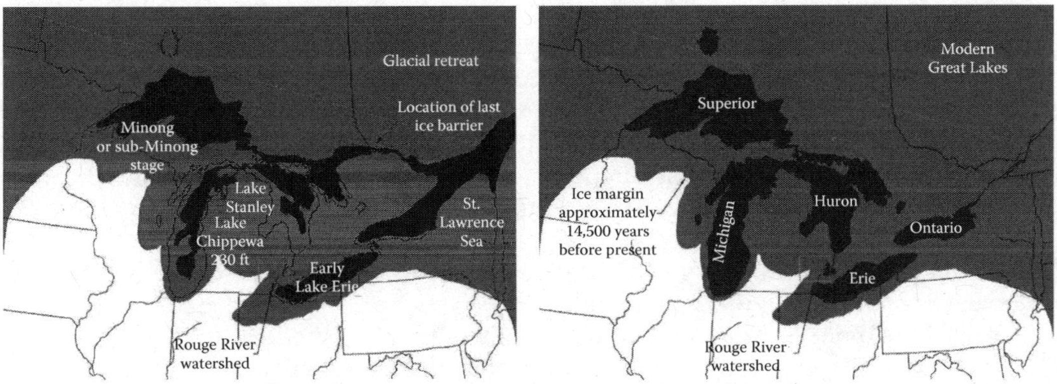

(g) (h)

FIGURE 5.14 (continued)
(g) Approximately 9,000 years before the present. (h) Modern time.

The purpose of these maps is to (1) convey the concept of regional geological significance; i.e., the glacial origin of unconsolidated sediments; and (2) provide a reference point for the next set of more-detailed maps presented as the figure sequence 5.15a through h. By starting at the general regional scale and then moving to the more detailed watershed scale, more information of a specific nature can be included. This progression illustrates the educational capability of this method.

The map succession shown in Figures 5.15a through h provides an example of how the use of progressive complexity at the smaller scale of an urban watershed can depict the processes forming the sedimentary deposits. For example, this sequence shows the glacial retreat and the associated sedimentary deposits in three dimensions. Figure 5.16 represents the current geology of the watershed in an enlarged format with a complete legend describing the stratigraphic column.

5.2.8.2 Three-Dimensional Maps

In addition to map sequences showing increasing amounts of information, three-dimensional digital techniques are also available to develop geologic maps (USGS 2005).

Examples of three-dimensional maps are shown in Figures 5.17 and 5.18 (Great Lakes Mapping Coalition 2009). Figure 5.17 shows an area being mapped in three dimensions with the data derived from borings drilled within the site. Using this approach assists with mapping and data gap analysis while developing the final geologic map. Figure 5.18 shows a completed draft of a three-dimensional map.

5.2.8.3 Maps Using GIS

GIS are useful for displaying and integrating many sources of information, including geological maps of urban areas (Chang 2007). One advantage of using GIS is its popularity in urban areas, and most major urban areas in the United States have significant GIS capabilities. Since the systems already exist, integrating geological information with other types of readily available layers of digital data provides a distinct cost advantage.

A common operation of a GIS is to combine specific digital data layers of similar geographic extent to produce new maps containing the input layers' spatial and descriptive attributes. Figure 5.19 is an example of this layering process.

The use of multiple layers is recommended when constructing the base map using geologic data of similar types of formation. For example, you can use a layer for glacial deposits, fluvial deposits, lacustrine deposits, and marine deposits. Subdividing the input of information into layers also assists with the editing process and may also be used to develop multiple maps.

An example two-dimensional geologic map constructed using GIS of the Rouge River watershed is shown in Figure 5.20 (Rogers 1997).

Figure 5.21 is the stratigraphic column for the Rouge watershed. Stratigraphic columns list all geological formations depicted on a map in sequential order by age. Because

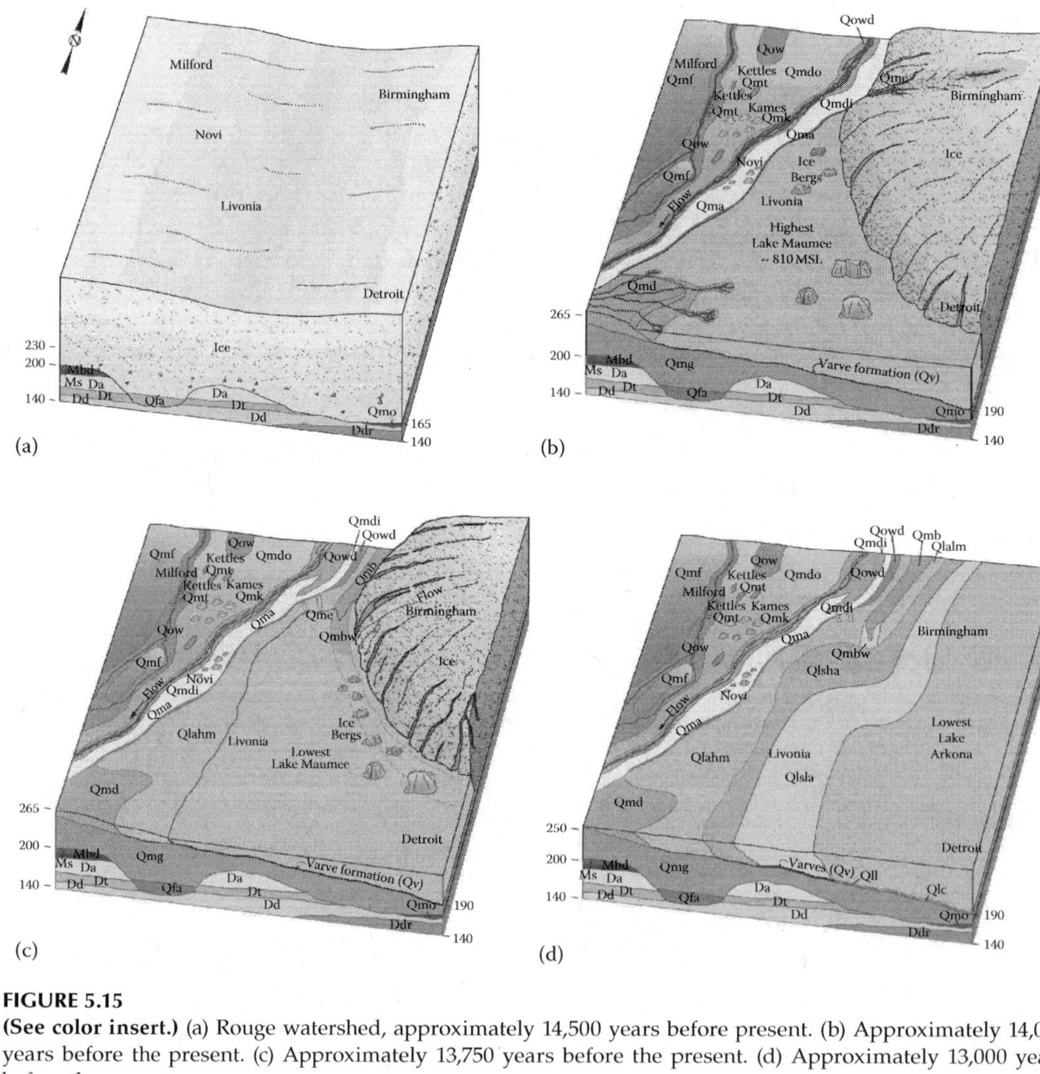

FIGURE 5.15
(See color insert.) (a) Rouge watershed, approximately 14,500 years before present. (b) Approximately 14,000 years before the present. (c) Approximately 13,750 years before the present. (d) Approximately 13,000 years before the present.

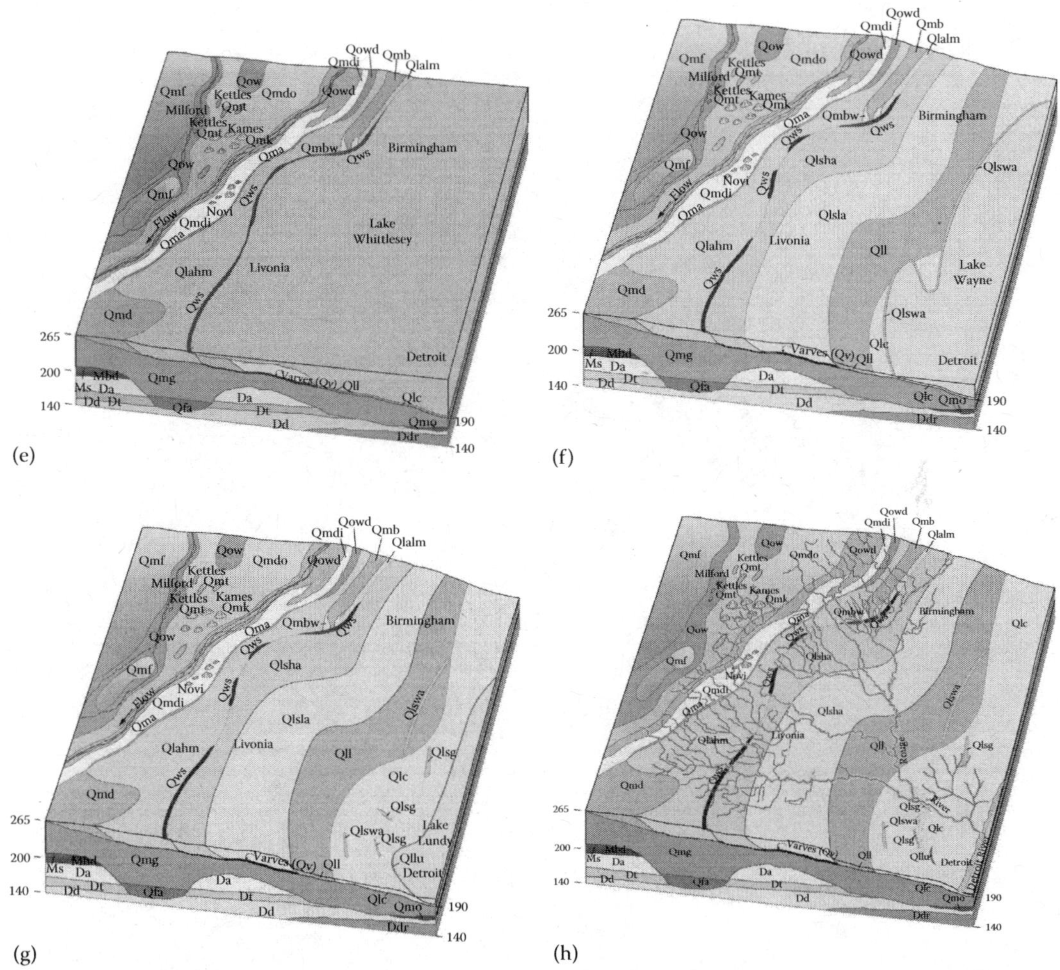

FIGURE 5.15 (continued)
(e) Approximately 12,900 years before the present. (f) Approximately 12,500 years before the present. (g) Approximately 11,000 years before the present. (h) Rouge River watershed geologic map, present day.

they list each deposit, stratigraphic columns are a necessary part of constructing a geologic map.

Many of the hydrogeologic principles demonstrated through the mapping of the Rouge watershed in southeastern Michigan are applicable to many other areas in the United States and the world. Some of these principles include

1. Sediments are composed of geologic materials originating from several different sedimentary processes including fluvial, lacustrine, and glacial; this is very typical of most watersheds in the United States.
2. The drainage pattern within the watershed is reflective of the geology depicted in Figure 5.16. Specifically, the drainage pattern changes density once the sand unit is encountered.

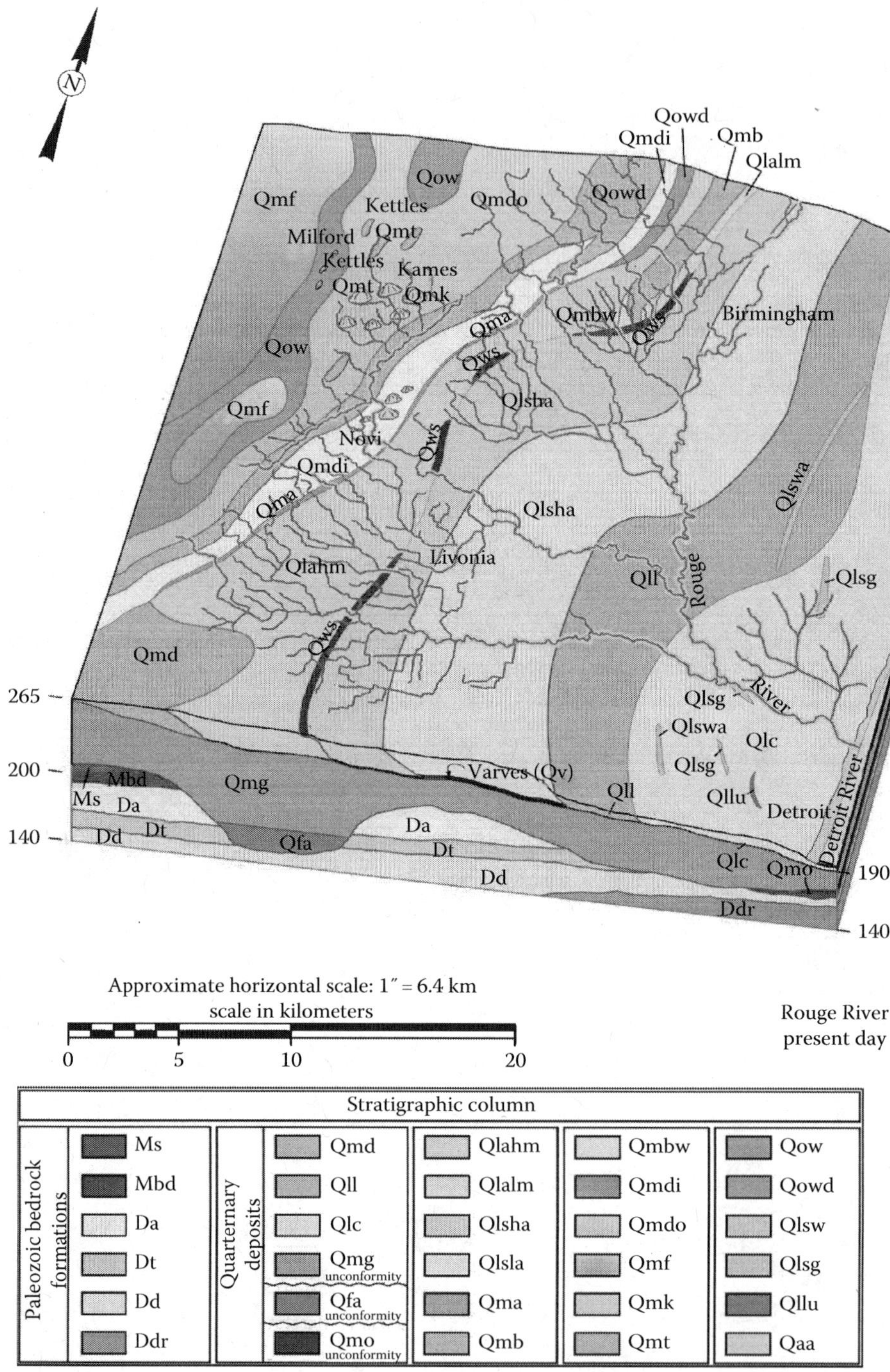

FIGURE 5.16
(See color insert.) Enlarged and annotated current geologic map of the Rouge River watershed in southeastern Michigan. (From Rogers, D.T. et al., *Quaternary Geology of the Rouge River Watershed in Southeastern Michigan, USA*, 33rd International Geological Congress, Abstracts with Programs, Oslo, Norway, 2008.)

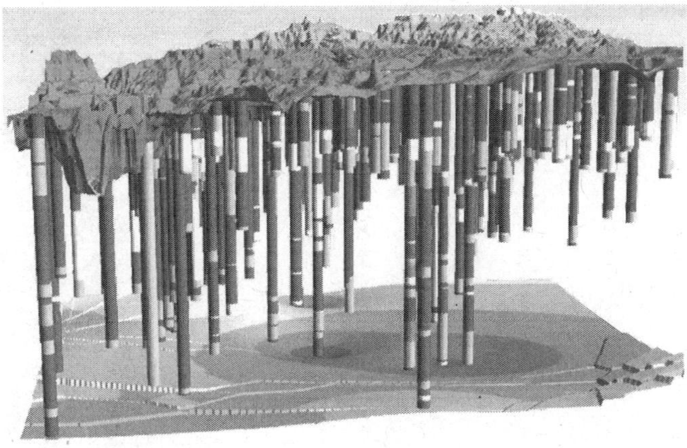

FIGURE 5.17
Mapping in three dimensions using boring logs. (From Great Lakes Geologic Mapping Coalition, Geologic mapping introduction, Indiana Geologic Survey, http://igs.indiana.edu/GreatLakesGeology/introduction.html, 2009.)

FIGURE 5.18
Three-dimensional geologic map. (From Great Lakes Geologic Mapping Coalition, Geologic mapping introduction, Indiana Geologic Survey, http://igs.indiana.edu/GreatLakesGeology/introduction.html, 2009.)

3. The dendritic drainage pattern of the Rouge watershed is similar to most inland urban watersheds lying over sedimentary rock anywhere in the world (even those not glaciated); coastal zones may be different because of regional tectonic influences.
4. The trend or strike of the glacial and lacustrine deposits is generally from the southwest to the northeast and follows the retreating ice as the glaciers melted.
5. The types of deposits change—sometimes significantly—over short distances. This is evident from the map as the deposits vary from glacial to lacustrine to fluvial.

The advantages of a three-dimensional map compared to a two-dimensional map become apparent by comparing Figure 5.16—a simplified three-dimensional map of the Rouge River watershed—with the two-dimensional geologic map of the Rouge River shown in Figure 5.20. These advantages include (1) showing the thickness of the sediment layers; (2) showing formations, such as the Lower Clay Unit, which is not even represented in the two-dimensional map; and (3) showing the dip, if any, of the geologic units.

FIGURE 5.19
Example of GIS layered mapping capability. (From United States Geological Survey, Topographic mapping, Online edition, http://egsc.usgs.gov/ish/pubs/booklets/topo/topo.html, 2010c.)

5.3 Summary and Conclusion

In this chapter, we have introduced the concept of "forensic geology" and discussed many unique circumstances that offer the geologist valuable opportunities to observe and collect geologic information in urban settings. This information can supplement and confirm geologic findings gathered onsite and on file from local and state regulatory agencies. These data can be used to develop urban environmental investigations in a short period of time and at low cost. By combining fieldwork with other sources of geological information, a detailed geologic map of any urban watershed in the United States can be produced.

Geologic mapping in urban areas requires a special skill set: a thorough geologic knowledge of near-surface sedimentary environments; the ability to identify the differences between natural and anthropogenic deposits; knowledge of the urban landscape; awareness and expertise in the methods used for obtaining access to sites of interest and information sources; and the facility to use a simple and expressive system to display the concepts and processes. Of these skills, the most critical, other than getting the geology right, is the presentation of data. Even the most detailed information database is rendered useless if it cannot be effectively and easily communicated. The use of multiple maps designed to add geologic knowledge and complexity through time, used alone, or coupled with GIS and three-dimensional mapping techniques, also helps simplify the communication of complex information. In addition, these tools can integrate geological information with other useful forms of information for the intended audience.

After obtaining a detailed understanding of urban geology and hydrogeology, the next step in evaluating the sustainability of a watershed can begin. This involves organizing the geologic and hydrogeologic information and bringing in other important data sources to construct a geologic vulnerability map—the subject of the next chapter.

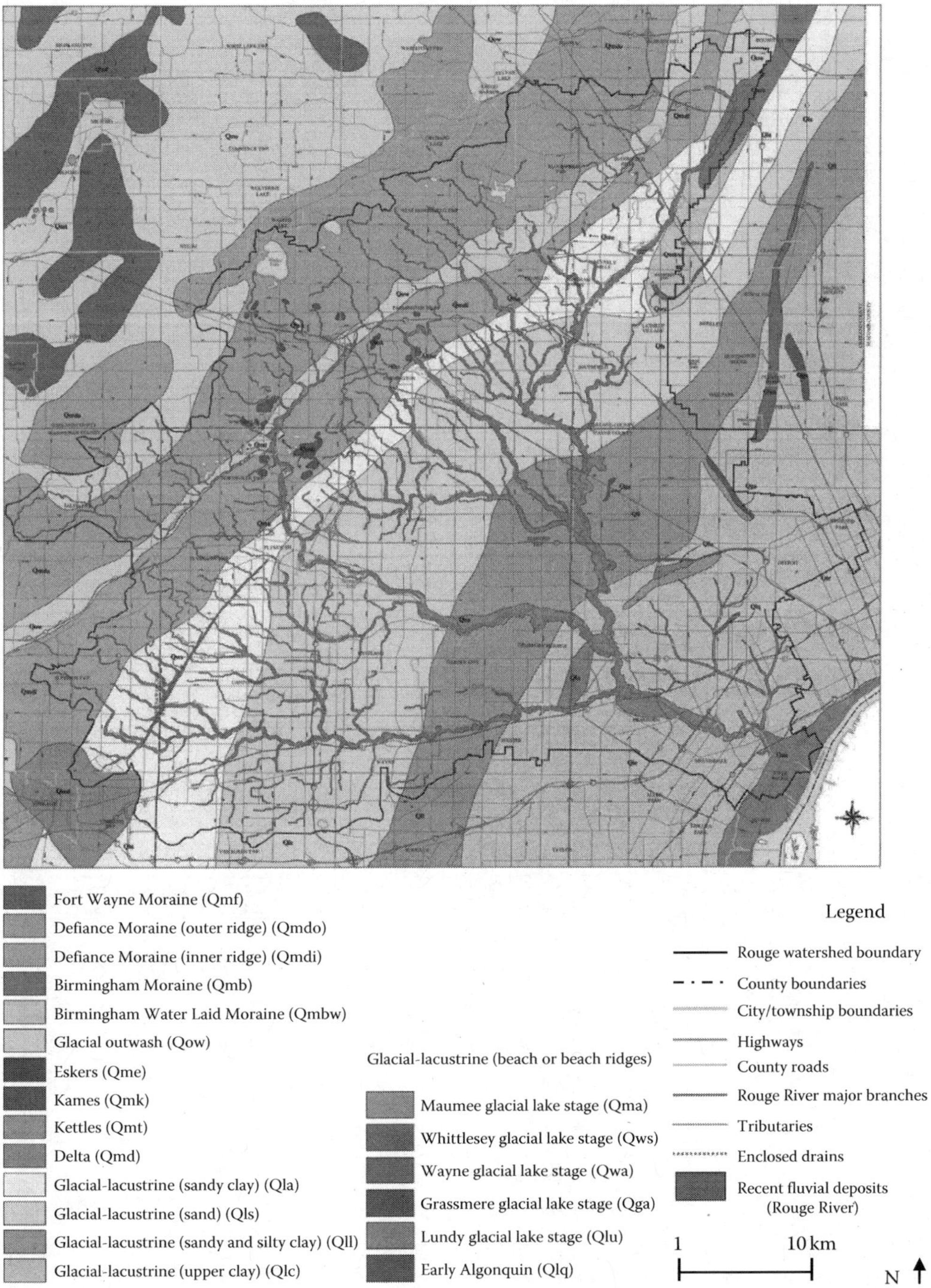

FIGURE 5.20
(See color insert.) Geologic map of the Rouge River watershed produced using GIS. (From Rogers, D.T., Surficial geologic map of the Rouge River watershed in southeastern Michigan, Wayne County, MI, Map Scale 1:62,500, 2 Sheets, 1997.)

Era	Period	Epoch	Stage		Age (approximate years before present)	Symbol	Unit name
Cenozoic	Quaternary	Holocene	Wisconsinan	Fluvial	<10,000		Qaa — Recent Rouge River fluvial deposits
		Pleistocene		Glacial lucustrine	<14,500		Qllu — Glacial lacustrine beach deposit Lake Lundy
							Qlsg — Glacial lacustrine beach deposit Lake Grassmere
							Qlswa — Glacial lacustrine beach deposit Lake Wayne
							Qlc — Glacial lacustrine clay
							Qll — Glacial lacustrine silty clay
							Qws — Glacial lacustrine beach ridge Lake Whittlesey
							Qlsla — Glacial lacustrine sand, lowest Lake Arkona
							Qlsha — Glacial lacustrine sand, highest Lake Arkona
							Qv — Varves
							Qlahm — Glacial lacustrine sandy clay Maumee Lake deposit
							Qma — Glacial lacustrine beach highest Lake Maumee
				Moraine	<14,500		Qmbw — Birmingham waterlaid moraine
							Qmb — Birmingham Moraine
				Delta	<14,500		Qmd — Delta
				Outwash	<14,500		Qmd — Outwash defiance
				Esker	<14,500		Qme — Esker
				Moraine	<14,500		Qmdi — Inner Defiance Moraine
							Qmdo — Outer Defiance Moraine
				Glacial outwash	<14,500		Qow — Outwash Fort Wayne
				Kame	<14,500		Qmk — Kame
				Kettle	<14,500		Qmt — Kettle
				Moraine	<14,500		Qmf — Fort Wayne Moraine
							Qmg — Ground moraine / lodgement till
					Unconformity		
				Fluvial	22,000		Qfa — Fluvial (interglacial) >22,000 ybp
					Unconformity		
				Moraine	22,000		Qmo — Pre-late Wisconsinan glacial till depost > 22,000 ybp
					Unconformity		
Paleozoic	Mississippian	Lower	Kinderhookian	Sunbury shale	345,000,000		Ms — Sunbury shale
				Bedford shale	345,000,000		Mbd — Bedford shale
	Devonian	Upper	Senecan	Antim shale	350,000,000		Da — Antrim shale
			Erian	Traverse group	355,000,000		Dt — Traverse group
		Middle	Erian	Dundee limestone	360,000,000		Dd — Dundee limestone
			Ultsterian	Detroit River dolomite	370,000,000		Ddr — Detroit River formation

FIGURE 5.21
(See color insert.) Stratigraphic column of Rouge River watershed in southeastern Michigan. (From Rogers, D.T., Surficial geologic map of the Rouge River watershed in southeastern Michigan, Wayne County, MI, Map Scale 1:62,500, 2 Sheets, 1997.)

References

Allen, R. 2003. The rocks of the Valley of the Sun. Phoenix, AZ. http://www.Gemland.com (accessed August 27, 2009).

American Association of Petroleum Geologists. 2005. The North American stratigraphic code. The North American Commission of Stratigraphic Nomenclature. *The American Association of Petroleum Geologists Bulletin* 89:1547–1591.

Anderton, R. 1985. *Clastic Facies Models and Facies Analysis*, Vol. 18. London, U.K.: Geological Society of London, Special Publication.

Barnes, J. 1993. *Basic Geologic Mapping*, 2nd edn. Geological Society of London Handbook. London, U.K.: John Wiley & Sons.

Bennison, A.P. 1974. *Geological Highway Map of the Great Lakes Region*. Tulsa, OK: The American Association of Petroleum Geologists and United States Geological Survey. 1 Sheet.

Berg, R.C., N.K. Bleuer, B.E. Jones et al. 1999. *Mapping the Glacial Geology of the Central Great Lakes Region in Three-Dimensions—A Model for State-Federal Cooperation*. United States Geological Survey Open File Report 99-349. Washington, DC.

Berg, R.C. 2002. Geoenvironmental mapping for groundwater protection in Illinois, USA. In *Geoenvironmental Mapping; Methods, Theory, and Practice*, ed. P.T. Bobrowsky, pp. 273–294. Abingdon, U.K.: Balkema.

Bilodeau, W.C., S.W. Bilodeau, E.M. Gath et al. 2007. Geology of Los Angeles. *Environmental and Engineering Geoscience Journal* 13:99–160.

Chang, K. 2007. *Introduction to Geographic Information Systems*, 4th edn. New York: McGraw Hill.

Great Lakes Geologic Mapping Coalition. 2009. Geologic mapping introduction. Indiana Geologic Survey. http://igs.indiana.edu/GreatLakesGeology/introduction.html (accessed September 8, 2009).

Halke, C. 1983. *Roadside Geology of Arizona*. Denver, CO: Mountain Press.

Hough, J.L. 1958. *Geology of the Great Lakes*. Champaign, IL: University of Illinois Press.

Hough. 1963. The prehistoric Great Lakes of North America. *American Scientist* 51:84–109.

Kaufman, M.M., D.T. Rogers, and K.S. Murray. 2005. An empirical model for estimating remediation costs at contaminated sites. *Journal of Water, Air and Soil Pollution* 167:365–386.

Lahee, F.H. 1961. *Manual of Field Geology*, 6th edn. New York: McGraw Hill.

Maley, T.S. 2005. *Field Geology*. Ann Arbor, MI: Sheridan Books.

Michigan Department of Environmental Quality. 2008. *Michigan Sites of Environmental Contamination*. Lansing, MI.

Mozola, A.J. 1954. *A Survey of Groundwater Resources of Oakland County Michigan*. Michigan Geologic Survey. Publication Number 48. Lansing, MI.

Mozola, A.J. 1969. *Geology for Land and Groundwater Development in Wayne County, Michigan*. Michigan Geologic Survey. Lansing, MI.

Page, W.R., S.C. Lundstrom, A.G. Harris et al. 2005. *Geologic and Geophysical Map of the Las Vegas 30′ x 60′ Quadrangle, Clark and Nye Counties, Nevada, and Inyo County, California*. United States Geological Survey Pamphlet to Accompany Scientific Investigations Map 2814. Washington DC.

Powell, J.W. 1888. *Methods of Geologic Cartography in Use by the United States Geological Survey*. Washington, DC: USGS.

Rogers, D.T. 1996. *Environmental Geology of Metropolitan Detroit*. Novi, MI: Clayton Environmental Consultants.

Rogers, D.T. 1997, Surficial geologic map of the Rouge River watershed in southeastern Michigan. Wayne County, MI. Map Scale 1:62,500. 2 Sheets.

Rogers, D.T., M.M. Kaufman, and K.S. Murray. 2008. *Quaternary Geology of the Rouge River Watershed in Southeastern Michigan, USA*. 33rd International Geological Congress. Abstracts with Programs. Oslo, Norway.

Sherzer, W.H. 1913. *Geology of Wayne County*. Geological Survey of Michigan. Lansing, MI.

Stokes, W.L. 1989. *Geology of Utah*. Utah Museum of Natural History Occasional Paper No. 6. Salt Lake City, UT.

Thomas, W.A. 2004. *Meeting the Challenges with Geologic Maps*. Alexandria, VA: American Geological Institute (AGI).

United States Geological Survey. 1998. *United States Geological Survey Circular 1148*. Forum on geologic mapping applications in the Washington-Baltimore urban area. Washington, DC.

United States Geological Survey. 2002. *3D Geologic Maps and Visualization: A New Approach to the Geology of the Santa Clara (Silicon) Valley, California*. United States Geological Survey Open-File Report 01-223.

United States Geological Survey. 2005. *3D Modeling Techniques for Geological and Environmental Visualization and Analysis*. United States Geological Survey Open-File Report 2005-1428. Washington, DC.

United States Geological Survey. 2006. *Federal Geographic Data Committee (FGDC) Digital Cartographic Standard for Geologic Map Symbolization*. FGDC Document Number FGDC-STD-013-2006. Washington, DC.

United States Geological Survey. 2007. *Divisions of Geologic Time—Major Chronostratigraphic and Geochronologic Units*. United States Geological Survey Fact Sheet 2007-3015. Washington, DC.

United States Geological Survey. 2009. National cooperative geologic mapping program. http://ncgmp.usgs.gov (accessed September 8, 2009).

United States Geological Survey. 2010a. Geologic mapping standards. USGS. National Cooperative Geologic Mapping Program. http://ncgmp.usgs.gov (accessed July 5, 2010).

United States Geological Survey. 2010b. Geographic information systems. USGS. Washington, DC. http://egsc.usgs.gov/ish/pubs/gis_poster (accessed July 5, 2010).

United States Geological Survey. 2010c. Topographic mapping. Online edition. http://egsc.usgs.gov/ish/pubs/booklets/topo/topo.html (accessed July 5, 2010).

6

Geologic Vulnerability

6.1 Introduction

We have control over the chemicals we use, and where and how we use them. Control over the geologic environment, however, is beyond our means. Therefore, we must understand the geologic environment where our urban areas are located and develop methods to minimize or eliminate the potential harmful effects of contaminants upon human health and the environment. A logical first step to this end is through an *understanding* of urban geology, followed by an *evaluation* of the extent that a given urban area's geology influences the migration of contaminants. Since water plays a critical role in assessing a region's vulnerability to contamination, the analyses performed during the evaluation step require an understanding of water occurring at the Earth's surface and beneath.

As demonstrated earlier, urban areas within the United States and throughout the world share a geologic environment dominated by sedimentary deposits. Many of those sedimentary deposits are saturated with water and function as sources of drinking water and/or as hydraulic connections to surface water and sensitive ecosystems. Given these relationships, using a watershed approach during the assessment process is scientifically sound because the geology of a region greatly influences the surface drainage patterns and often contributes significantly to the baseflow of surface water. Moreover, as described in Chapter 3, water is considered the universal solvent, so any contamination released into the environment from anthropogenic or natural sources has the potential to migrate and impair surface water or groundwater.

The factors controlling the severity of the damage are (1) the hydrogeologic environment, (2) the physical chemistry of the contaminants and amounts released, and (3) the mechanism in which the release occurs (Rogers 1996; Murray and Rogers 1999a; Kaufman et al. 2005). As noted in Chapters 4 and 5, there are several techniques available for investigating and managing the complexity of urban water pollution. For example, a significant amount of detailed geologic and hydrogeologic information for many urban areas is currently available to support environmental assessments. In addition, there are several methods that enable access to contaminated locations for conducting fieldwork, and procedures exist to collect the relevant data without causing disturbance to any natural or anthropogenic surface features.

Despite the availability of specific methods and procedures, the environmental assessment of many urban areas can become a daunting task. This situation arises because the near-surface geologic deposits in urban areas are poorly understood, difficult to study, complex, have been anthropogenically disturbed, and exhibit high variability over short distances. Therefore, to achieve any level of success in mitigating environmental contamination, it becomes a prerequisite to understand the watershed's hydrology, and the fate and migration of contaminants within its specific geology.

The focus of this chapter is to identify urban regions vulnerable to contamination and those areas where widespread contamination is less likely. To accomplish this task, an additional level of interpretation consisting of a comprehensive vulnerability analysis is added to the near-surface geologic maps of urban areas presented in Chapter 5. This chapter also explains why certain types of geology may be especially susceptible to contamination—a topic explored and discussed in greater detail later in this book.

6.2 Subsurface Vulnerability and Vulnerability Map Development

The concept of vulnerability of the subsurface to contamination originated in France during the 1960s and was introduced into the scientific literature by Albinet and Margat (1970). Since then, the concept of subsurface vulnerability has evolved to include both a distinction between and combination of vulnerability and risk assessment. Groundwater vulnerability is currently interpreted as a function of the natural properties of the overlying soil or sediments of the unsaturated zone, aquifer properties (e.g., effective porosity and recharge area), and aquifer material (Foster and Hirata 1988; Robins et al. 1994; Rogers et al. 2007).

Geologic vulnerability mapping can be divided into two groups: subjective rating methods and statistical and process-based methods (Focazio et al. 2001). The subjective rating methods are characterized by numerical scales representing low to high vulnerabilities. Typically the results are applied to large areas and used for policy and management objectives. By contrast, the statistical process-based methods produce finite values, such as areas exceeding specific water quality values. With these methods, the results are usually not applied to large areas due to data gaps and variable geology. In addition, the results are generally obtained under more detailed site-specific assessments and used for purely scientific purposes (Focazio et al. 2001). In practice, the subjective rating methods are preferred for conducting vulnerability assessments on a watershed scale (Murray and Rogers 1999a).

The concept of geologic vulnerability relies on the assessment and representation of various hydrogeologic parameters such as vadose zone characteristics (e.g., thickness and infiltration capacity), depth to water, and amount of recharge (Zaporozec and Eaton 1996; Eaton and Zaporozec 1997). The utility of this concept, however, becomes more important when the geologic data are supplemented with environmental, economical, and political insight gained through past environmental cleanup efforts (Foster et al. 1993; Loague et al. 1998). A specific example of this data augmentation is provided at the end of the chapter.

Successful development of geologic vulnerability maps can be difficult to achieve in areas experiencing rapid growth. As noted in Chapters 2 and 3, urbanization and the artificial infrastructure it produces (e.g., sewers and detention ponds) can have a profound influence on the regional hydrogeology (Vuono and Hallenbeck 1995; Zaporozec and Eaton 1996; Kibel 1998). Basic processes affecting surface water and groundwater are modified, including surface water drainage patterns and velocities, evaporation rates, infiltration, and aquifer recharge (Burn et al. 2007; Garcia-Fresca 2007; Howard et al. 2007; Mohrlok et al. 2007). The difficulties in vulnerability map development are also compounded by the differences in the amount and type of geologic and hydrogeologic information available in urban areas and rural settings. For these reasons, a uniform assessment of data while conducting vulnerability studies in urbanizing areas is difficult to achieve.

Geologic vulnerability mapping provides a starting point for quantifying anticipated environmental risk at a particular site and can also highlight locations where additional information is warranted. In lieu of specific site information, this risk assessment can serve, if necessary, as a proxy for anticipated future cleanup costs. Additionally, this method can be used by other interested parties and stakeholders during the recycling of industrial sites to estimate the liability of sites (Stiber et al. 1995; Murray and Rogers 1999b). With respect to water resource allocation, since surface and groundwater interact, mapping groundwater also provides valuable information concerning their respective distributions (Rogers and Murray 1997; Pierce et al. 2007).

We demonstrate later in the book how it is often impossible or prohibitively expensive to clean up contaminated groundwater effectively. This reality is why a better approach for ensuring groundwater quality is to map actual groundwater contamination, and groundwater potentially vulnerable to contamination, and then to use this information as an integral part of land use planning and management (Zaporozec and Eaton 1996). Then, the information gathered through the process of geologic vulnerability assessment would allow decision makers to assess the current and future environmental risks associated with any particular site as long as it was contained within the area mapped.

6.3 Methods

The evaluation of geologic vulnerability in an urban watershed using a subjective rating method requires a combination of geologic and hydrogeologic data, identification of the potential receptor sites, and political and economic information. The first and most crucial step is mapping the near-surface geology. Once the geologic map is created, the process of developing a geologic vulnerability map can be initiated.

The authors (Murray and Rogers 1999a; Kaufman et al. 2003, 2005) have developed a method for geologic vulnerability mapping using a modified DRASTIC model (Aller et al. 1987). This method contains a subjective numerical rating system and uses different weighting coefficients for various geologic and hydrogeologic parameters of concern, and incorporates potential receptors and political data into the model. The geologic and hydrogeologic factors used to develop the conceptual geologic vulnerability model were introduced and discussed in detail in Chapters 2 and 3. Listed below are those geologic and hydrogeologic factors, along with the political, ecological, and environmental factors covered in Chapter 4 that are routinely collected during environmental subsurface investigations at known or suspected sites of contamination (Rogers 1996; Murray and Rogers 1997; Rogers 2002; Kaufman et al. 2003, 2005):

- Soil or sediment type, composition, color, texture, thickness, and relative soil moisture
- Stratigraphy of geologic units
- Horizontal and vertical extent of geologic units
- Variation within geologic units (heterogeneity and anisotropism)
- Type of primary and secondary features
- Presence, extent, and structure of unconformities
- Presence of groundwater
- Depth to groundwater

- Relative abundance of groundwater
- Groundwater flow direction
- Areas of groundwater recharge
- Areas of groundwater discharge
- Groundwater–surface water interaction features
- Anthropogenic features including
  - Storm sewer placement relative to groundwater elevations
  - Utility corridors
  - Building footings and other subsurface structures
  - Landfills
  - Artificial recharge basins
  - Surface water confinement features
  - Dams
  - Stormwater retention or detention basins
  - Large paved areas
  - Artificial surface drainage pattern alteration
  - Roads and road cuts
- Potential receptors or points of potential exposure:
  - Water supply wells
  - Irrigation wells
  - Surface water bodies such as lakes, streams, rivers, swamps, bogs, springs, groundwater seeps, etc.
  - Buildings and building foundations
  - Parks, schools, playgrounds, day care facilities, retirement communities, hospitals, long-term care facilities, etc.
- Potential sources of contamination:
  - State and federal lists of environmental contamination
  - Hazardous waste facilities
  - Abandoned dumps
  - Brownfield sites
  - Historical industrial manufacturing sites
  - Electrical generating facilities
  - Gasoline service stations
  - Dry cleaning facilities
  - Refineries
  - Other known or identified sources

Taken together, these factors provide a framework for the evaluation of the geologic vulnerability within any urban area. This framework is constructed through the development of a vulnerability matrix and scoring system as shown in Table 6.1 (Murray and Rogers 1999a; Rogers and Murray 2002).

TABLE 6.1

Geologic Vulnerability Matrix and Scoring System

Parameter Identification Number	Parameter Description	Rating Strength
1	*Depth to groundwater*	
	<3 m (10 ft) below the ground surface	10
	3–10 m (10–30 ft)	5
	>10 m	1
2	*Composition, areal extent, and thickness of soil units in the unsaturated zone*	
	Thick and extensive sequence of sand and gravel	10
	Interbedded sands and clay deposits	5
	Thick and extensive sequence of clay	1
3	*Composition, areal extent, and thickness of saturated zone*	
	Thick and extensive sequence of sand and gravel	10
	Interbedded sands and clay deposits	5
	Thick and extensive sequence of clay	1
4	*Occurrence and relative abundance of groundwater*	
	25% or less likelihood before encountering an confining layer	10
	25%–74% likelihood	5
	>75% likelihood	1
5	*Area of groundwater recharge*	
	Significant area of recharge	10
	Moderate area of recharge	5
	Not a significant area of recharge	1
6	*Areas of groundwater discharge*	
	Significant area of recharge	10
	Moderate area of recharge	5
	Not a significant area of recharge	1
7	*Travel time and distance to point of potential exposure*	
	<10 years	10
	10–25 years	5
	>25 years	1
8	*Source of potable water*	
	Current source of potable water	10
	Potential source of potable water	5
	Not a potential source of potable water	1

A sequential description of the vulnerability matrix parameters presented in Table 6.1 follows:

- *Depth to groundwater.* The closer to the surface groundwater is encountered the higher the geologic vulnerability because contaminants have a shorter distance to migrate vertically before encountering saturated conditions.

- Depth to groundwater varies widely across the United States depending on geology, climate, and anthropogenic influence. For example, groundwater is routinely present at depths very near the surface along the east coast and southeast; from a few meters to tens of meters in the Midwest; to perhaps several hundred meters in the southwest. With respect to anthropogenic influence, the landscape in urban areas is characterized by extensive modifications. The natural hydrology is altered by the impervious surface, the construction of stormwater retention basins, and the use of more permeable areas for artificial recharge or groundwater storage. In addition, leaks from sanitary sewers, storm sewers, potable water lines, and surface watering (especially golf courses) may profoundly impact the depth to groundwater and the direction of flow locally or over larger areas.

- *Composition, areal extent, and thickness of soil units in the unsaturated zone.* Composition of soil or sediment above the water table is important to evaluate because it helps to determine whether the soil or sediments above the water table will impede the vertical migration of contaminants through the soil or sediment column. For instance, clay soils are fine grained and generally impede or slow migration, whereas sand and gravel deposits tend to facilitate contaminant migration. Interbedding of geologic units is also considered.
  - Other significant factors include the presence of unconformities, primary and secondary porosity and the potential for unsaturated soils located above aquifers to be subject to anthropogenic disturbance. For instance, a clay layer that is 6 m thick may appear to significantly impede the vertical migration of contaminants. However, if portions are excavated, or if the integrity of the clay has been compromised by vertical fractures, root fragments, or anthropogenic activities, these events increase the potential for contaminant migration through the overlying clay into groundwater (Murray et al. 1997).
  - Scoring this value therefore involves a consideration of multiple factors. With the 10-point scoring system, sands and gravels typically receive 10 points. Clay deposits of adequate integrity, thickness (generally more than 3–6 m thick), and areal extent receive a score of 3 or less.

- *Composition, areal extent, and thickness of saturated zone.* Composition, areal extent, and thickness of aquifer materials are important factors because they influence the rate and direction of groundwater flow. Risk may be higher in a scenario where the saturated thickness is 23 m (75 ft) and the material is composed of sand, as opposed to a saturated thickness of 1 m composed of silt. The areal extent and thickness of the aquifer must also be evaluated to establish whether multiple aquifers exist and are hydraulically connected. For instance, using the example above, if the 1 m thick saturated silt layer was hydraulically connected to the 23 m thick saturated sand layer; this would change the risk evaluation.
  - In an urban environment, perched saturated zones often result from anthropogenic influences. Construction activities such as roads, building foundations, and utility corridors are backfilled with porous materials, and then become more porous than natural soils and sediments. Over time these porous backfilled materials become saturated with water. Therefore, investigative activities must be very detailed in nature to evaluate the significance of potential anthropogenic influences on the hydrology of the area being investigated and mapped.

  - Scoring this parameter is similar to scoring the composition of the unsaturated zone (parameter #2). In practice, however, the factors determining the score are the areal extent of the saturated zone and whether the aquifer is hydraulically connected to additional or larger aquifers. If a saturated zone is encountered and evaluated to be perched or discontinuous and therefore isolated, the scoring value of 1 may be assigned. However, if the saturated thickness is 23 m and lies within a mappable geologic formation, the scoring value of 10 may be assigned.
- *Occurrence and relative abundance of groundwater.* This parameter is similar to the depth of groundwater (parameter #1) but focuses on whether groundwater is present at relatively shallow depths, and of a sufficient quantity to sustain a rate of withdrawal. A key aspect of this measure is whether any groundwater is encountered before a confining layer. The abundance of groundwater relates to the transmissivity of the formation and is not intended to focus on whether groundwater is encountered.
  - With this parameter, there is high scoring variability depending on the type of materials in the deposit. Consider an extensive deposit composed of saturated sand and gravel. This deposit would tend to have a much higher transmissivity value, and would likely receive a high vulnerability score of 8 or more. When this deposit is compared to a similarly extensive and saturated silty clay deposit with a low relative transmissivity, the silty clay deposit would receive a score of 5 or less.
- *Area of groundwater recharge.* This parameter focuses on the source of groundwater. If the area being evaluated is a source for groundwater recharge, its risk will be higher than an area not considered a significant source of groundwater discharge.
  - Special care should be taken in urban areas to evaluate the anthropogenic sources of groundwater recharge. Significant anthropogenic influences include storm water infiltration, sanitary sewer leakage, and water supply leakage. Other human impacts to consider are detention basins and wetland modification.
- *Areas of groundwater discharge.* This parameter requires an understanding of groundwater flow at different geographic scales; including specific sites, off-site areas of larger extent, and the entire watershed.
  - The migration and final discharge locations for groundwater includes an understanding of the relationship between groundwater and surface water, and permits an evaluation of the potential for interconnecting aquifer systems and discharge to deeper aquifers. It is also essential to understand the geology of a region and the potential influence of unconformities associated with the depositional units within the watershed.
- *Travel time and distance to point of potential exposure.* In general, the longer the time and distance required for a contaminant to reach its potential point of exposure, the lower the risk. This is true for most contaminants that degrade naturally in the environment.
  - Evaluating anthropogenic influence with this parameter is also important. For instance, if groundwater is used as a source of potable water and pumped from the ground using extraction wells, the residence time of potential contaminants in the aquifer may be greatly reduced.

- *Source of potable water.* If the aquifer encountered beneath a particular site is used as a source of potable water or is connected to a source of potable water, the risk increases.

Evaluating this parameter must also include other potential uses besides the provision of drinking water. Further information may be necessary to identify those uses and accurately score this parameter.

6.4 Map-Building Example

Because of its diversity, a good example of building a geologic vulnerability map is provided by the Rouge River watershed located in southeastern Michigan (Figure 6.1).

As noted in Section 1.2, several key features making the Rouge River watershed an ideal watershed to study include

- It is an urban watershed
- It has varied geology and hydrogeology
- There are different stream patterns and densities throughout
- The population is high: approximately 1.5 million people
- The watershed has been studied extensively

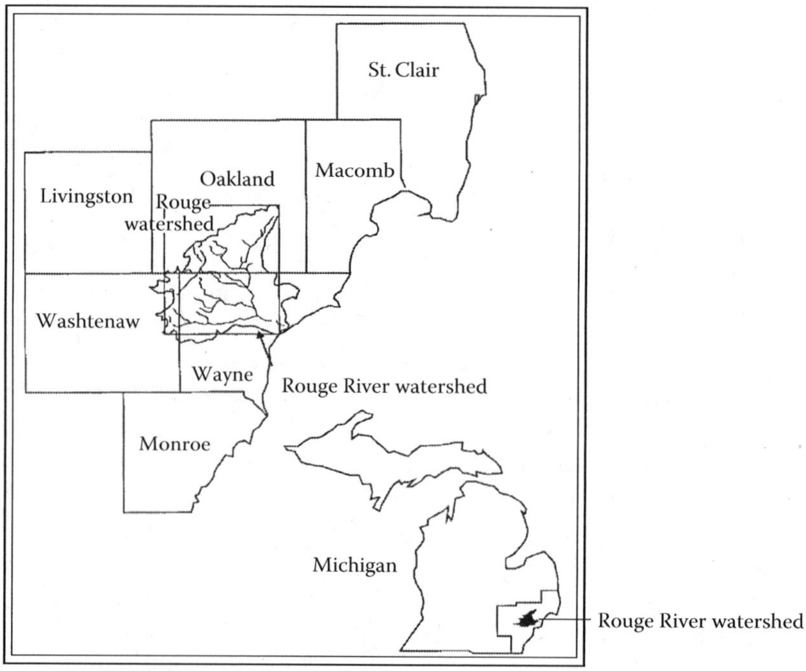

FIGURE 6.1
Rouge River watershed.

- There is an abundance of geologic and hydrogeologic data
- It has a long history of industrial output
- It has significant environmental impacts
- It has similar geology to many other cities in the United States

The Rouge River watershed has over 200 km of streams, tributaries, lakes, and ponds, and encompasses an area of approximately 1200 km^2 (460 mi^2) (Murray and Rogers 1999a). The fan-shaped watershed includes all or part of 47 municipalities in three counties. Over 66% of the watershed has been developed, and in the year 2000, 99% of the watershed's population lived within the U.S. Census-defined urbanized areas—making it the watershed with the highest population density in the eastern United States (Kaufman et al. 2003).

The near-surface geology of the watershed is dominated by glacial deposits, glacial lacustrine deposits, and the recent fluvial deposits from the Rouge River itself. The glacial deposits are generally greater than 61 m (200 ft) in thickness, the glacial lacustrine deposits are rarely more than 9 m (30 ft) thick, and the recent fluvial deposits are generally less than 3 m (10 ft) thick (Farrand 1982, 1988; Rogers 1996, 1997c). Distinctive geologic units identified and mapped within the watershed include

- Four different surface moraine units that include the Fort Wayne Moraine, the Outer and Inner Defiance Moraine, and the Birmingham Moraine
- Glacial outwash deposits
- Several beach deposits composed of sand from historical glacial lakes
- Silt and clay deposits from historical glacial lakes
- Recent fluvial deposits from the Rouge River
- A ground moraine or lodgment till composed of clay underlies the entire watershed beneath the near surface deposits described above
- A sand and gravel deposit located beneath the ground moraine in the center portion of the watershed. This feature represents a fluvial deposit from a large river that was present before the last glacial advance approximately 22,000 YBP
- Weathered bedrock encountered beneath the ground moraine at other locations
- Bedrock composed of shale, limestone, and sandstone of Paleozoic age present beneath the unconsolidated units

The geologic map and stratigraphic column for the watershed are shown as Figures 5.20 and 5.21, respectively. The geologic map of the Rouge River watershed was produced using the methods described in Chapters 4 and 5 and included information from more than 3000 subsurface environmental investigations conducted within or adjacent to the watershed. Production of this geologic map also relied on the information gathered from numerous sources of historical literature.

Anthropogenic effects within the Rouge River watershed have been numerous and significant. Surface water and groundwater within the watershed have been severely degraded. In recent assessments, the Rouge River was ranked as one of the most toxic sites in Michigan (Michigan Department of Environmental Quality 2008). The watershed is also the focus of ongoing intense scientific study and restoration. It has been identified as an Area of Concern by the International Joint Commission (Hartig and Zarull 1991) and cited as a significant source of contamination to the lower Great Lakes (Murray and Bona 1993).

Rouge River sediments have shown a significant presence of heavy metals and polychlorinated biphenyls (PCBs) (Michigan Department of Natural Resources 1988; Murray 1996; Murray et al. 1999).

Shallow groundwater within the Rouge River discharges to surface water and accounts for most of the baseflow (Rogers and Murray 1997). Average annual precipitation amounts and the geology of the watershed determine this flow pattern. Climatically, the Rouge watershed is situated within the humid microthermal zone of the Midwestern United States, meaning the Rouge River is effluent and fed by groundwater entering as baseflow. With respect to its geology, the lower clay unit (ground moraine or lodgment till) that completely underlies the watershed is very thick (ranging from 9 m to more than 55 m thick), has a very low hydraulic conductivity (less than 1×10^{-8} cm/s), and does not show signs of unconformities or features suggesting any significant secondary porosity (i.e., vertical fractures or root fragments). Therefore, the lower clay unit is a very effective confining layer. This type of formation is not uncommon. Most other areas in North America glaciated during the Pleistocene have similar ground moraine or lodgment till deposits of similarly low hydraulic conductivity (Keller et al. 1989). This is a significant finding because any contamination that does not degrade and reaches groundwater will eventually migrate and discharge to the surface waters of the Rouge River. From there the contamination subsequently discharges into the lower Great Lakes (Rogers and Murray 1997; Rogers 1997a). Figure 6.2 shows this discharge sequence based on the flow patterns of the watershed's surface water and groundwater.

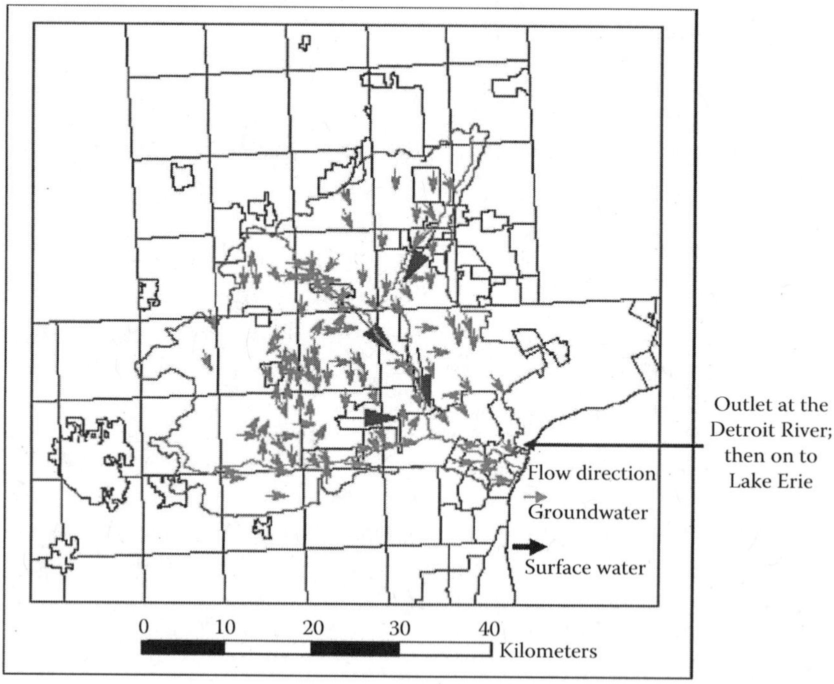

FIGURE 6.2
Surface water and groundwater flow patterns in the Rouge River watershed. (Murray, K. et al., Dissolved heavy metals in shallow ground water in a southeastern Michigan urban watershed, *Journal of the American Water Resources Association*, 2006, 42(3), 777–792. Copyright Wiley-VCH Verlag GmbH & Co. KGaA. Reproduced with permission.)

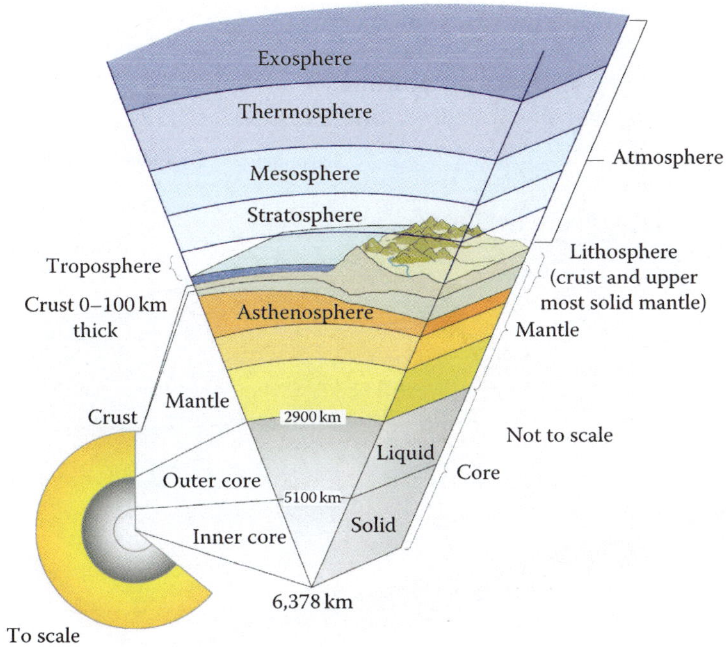

FIGURE 2.2
Cross section of the interior portions of the Earth showing the asthenosphere. (Modified from Kious, W.J. and Tilling, R.I., *The Dynamic Earth: The Story of Plate Tectonics,* United States Geological Survey, Governmental Printing Office, Washington, DC, 1996.)

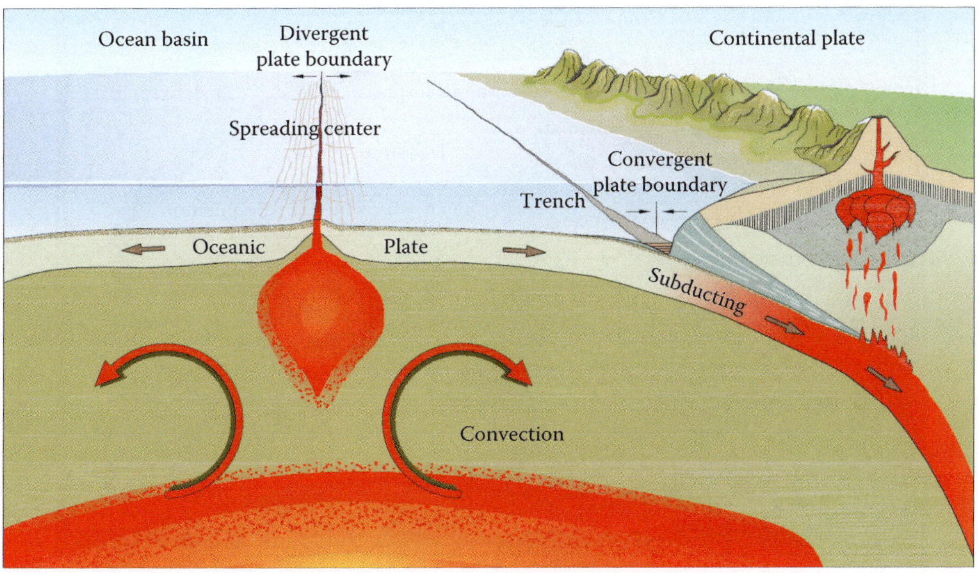

FIGURE 2.3
Concepts of plate tectonics and continental drift.

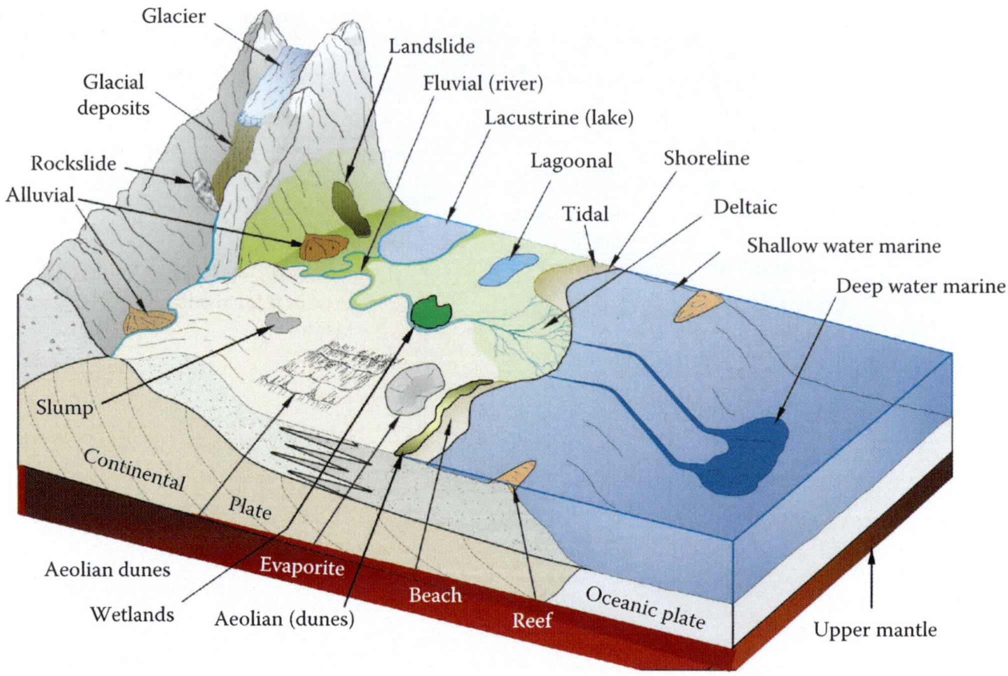

FIGURE 2.29
Sedimentary depositional environments.

FIGURE 3.5
The water cycle. (From United States Geological Survey, Summary of the water cycle, http://ga/water.usgs.gov/edu/watercyclesummary.html, 2010a.)

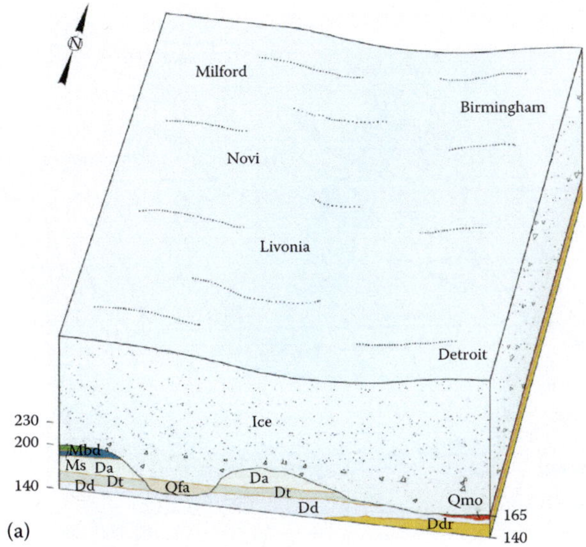

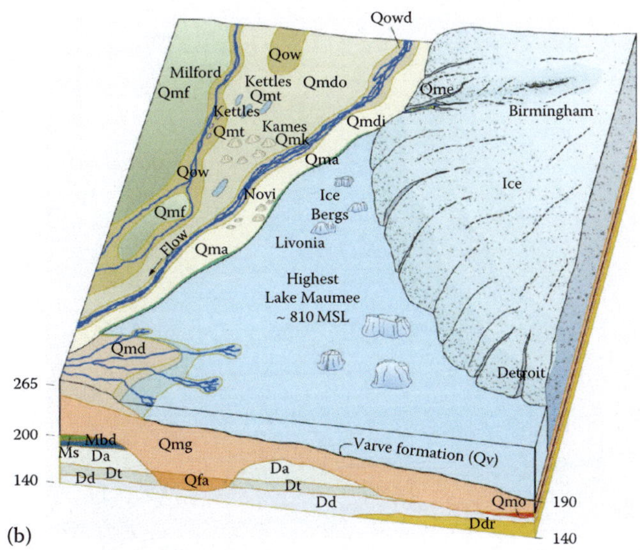

FIGURE 5.15
(a) Rouge watershed, approximately 14,500 years before present. (b) Approximately 14,000 years before the present.

(continued)

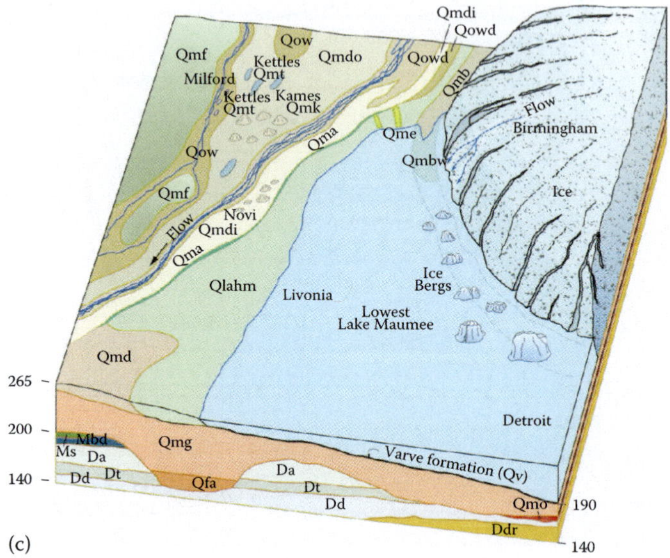

(c)

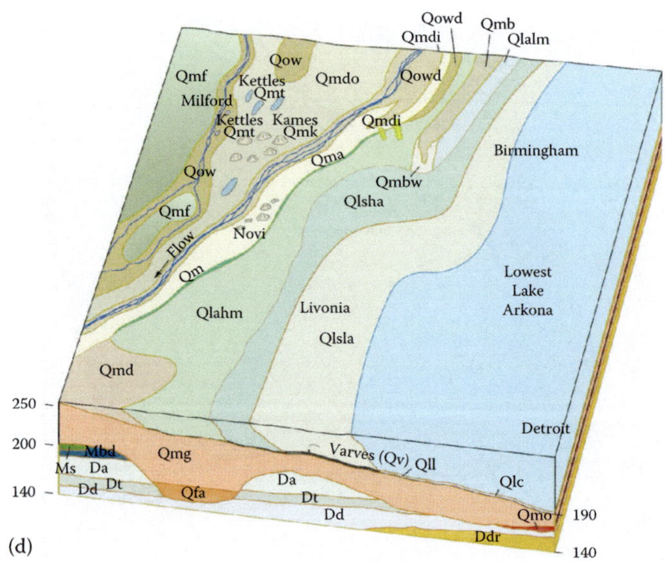

(d)

FIGURE 5.15 (continued)
(c) Approximately 13,750 years before the present. (d) Approximately 13,000 years before the present.

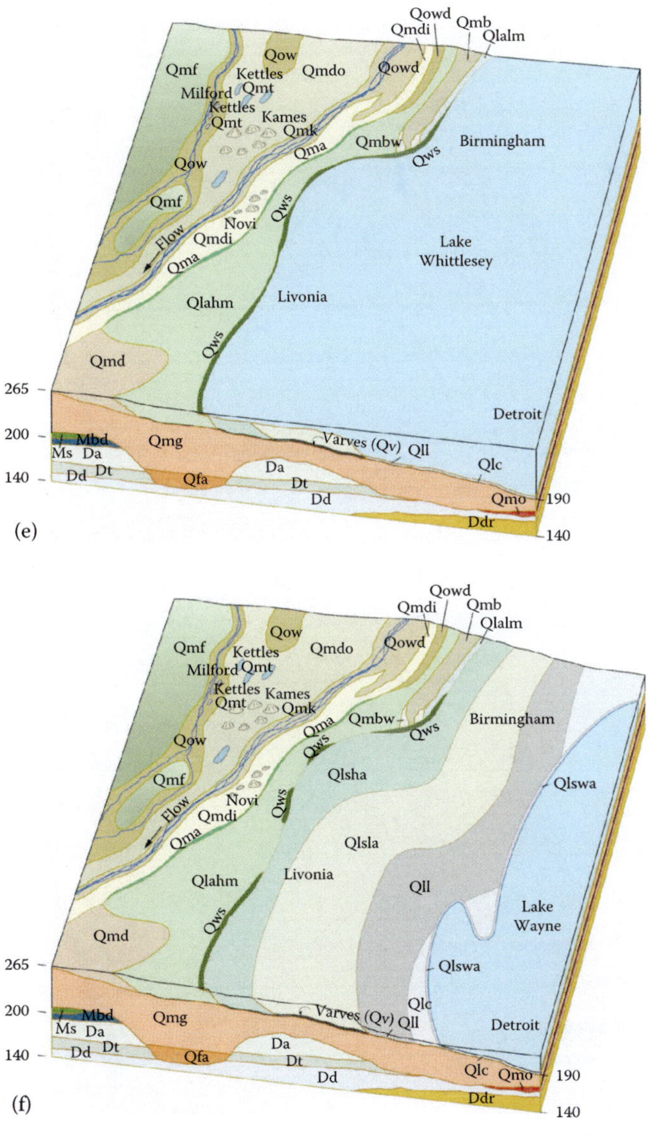

FIGURE 5.15 (continued)
(e) Approximately 12,900 years before the present. (f) Approximately 12,500 years before the present.

(continued)

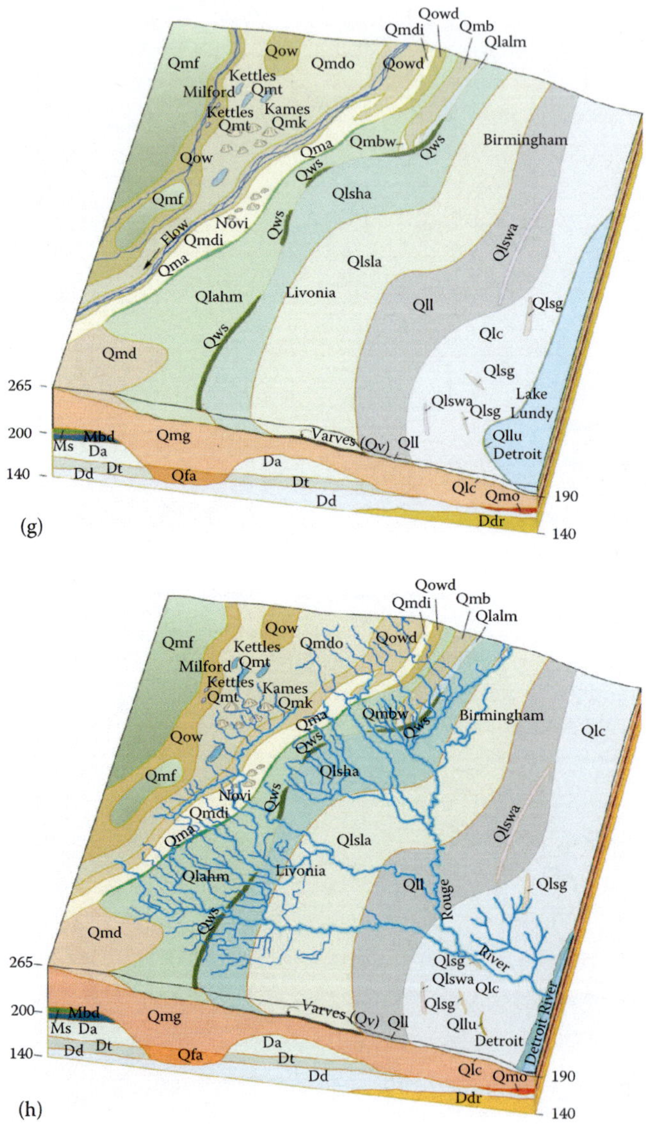

FIGURE 5.15 (continued)
(g) Approximately 11,000 years before the present. (h) Rouge River watershed geologic map, present day.

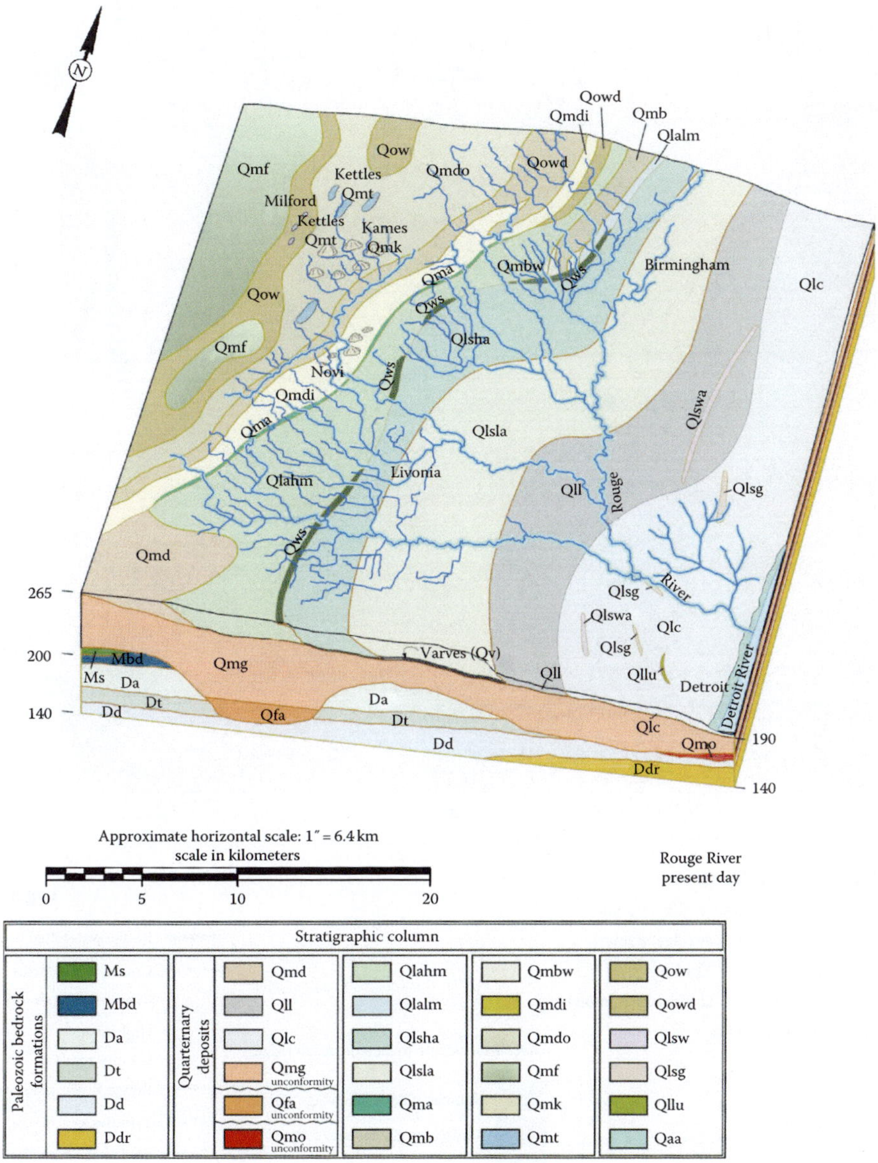

FIGURE 5.16
Enlarged and annotated current geologic map of the Rouge River watershed in southeastern Michigan. (From Rogers, D.T. et al., *Quaternary Geology of the Rouge River Watershed in Southeastern Michigan, USA*, 33rd International Geological Congress, Abstracts with Programs, Oslo, Norway, 2008.)

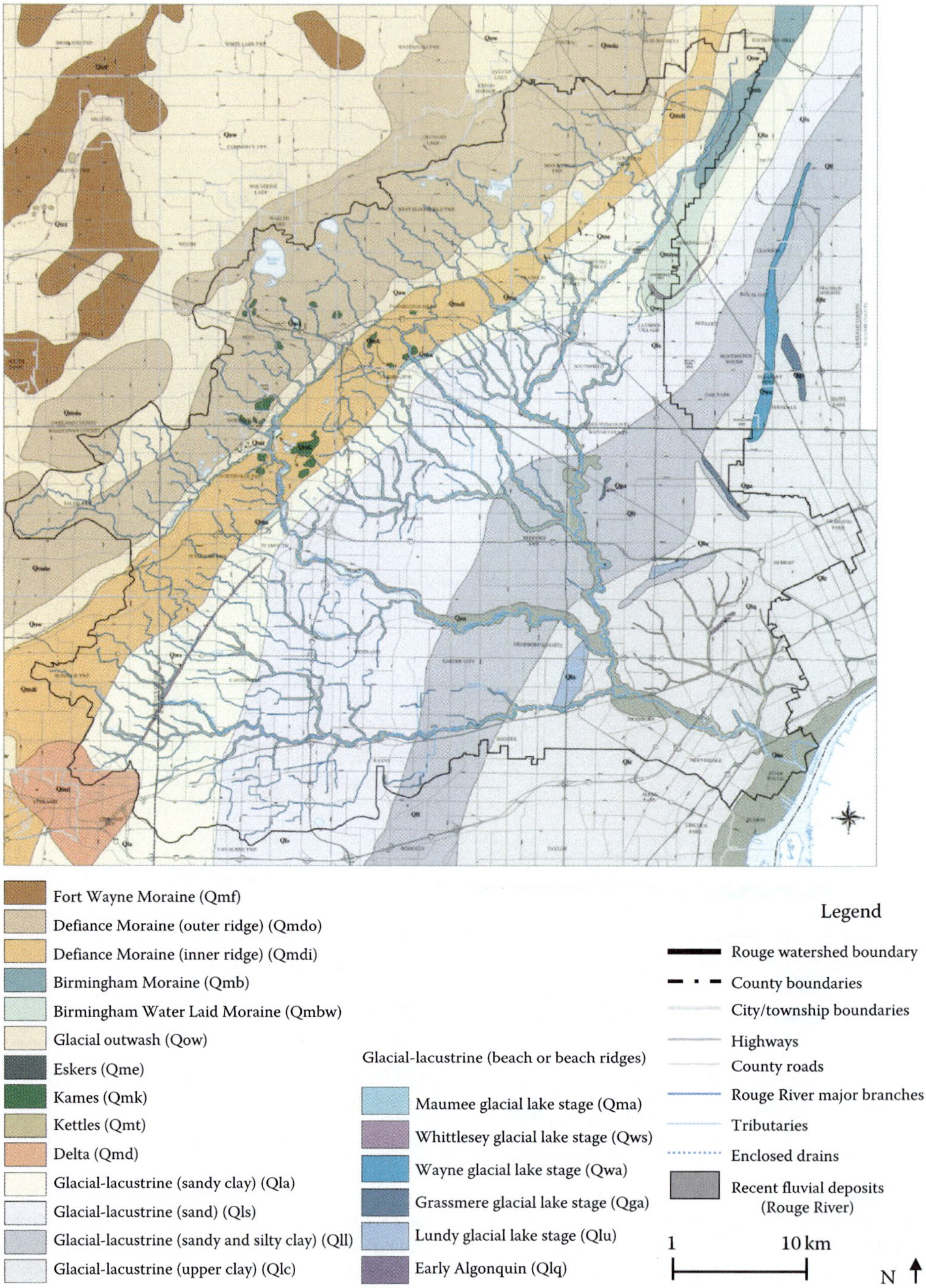

FIGURE 5.20
Geologic map of the Rouge River watershed produced using GIS. (From Rogers, D.T., Surficial geologic map of the Rouge River watershed in southeastern Michigan, Wayne County, MI, Map Scale 1:62,500, 2 Sheets, 1997.)

Era	Period	Epoch	Stage		Age (approximate years before present)	Symbol	Unit name
Cenozoic	Quaternary	Holocene	Wisconsinan	Fluvial	<10,000		Qaa — Recent Rouge River fluvial deposits
		Pleistocene		Glacial lucustrine	<14,500		Qllu — Glacial lacustrine beach deposit Lake Lundy
							Qlsg — Glacial lacustrine beach deposit Lake Grassmere
							Qlswa — Glacial lacustrine beach deposit Lake Wayne
							Qlc — Glacial lacustrine clay
							Qll — Glacial lacustrine silty clay
							Qws — Glacial lacustrine beach ridge Lake Whittlesey
							Qlsla — Glacial lacustrine sand, lowest Lake Arkona
							Qlsha — Glacial lacustrine sand, highest Lake Arkona
							Qv — Varves
							Qlahm — Glacial lacustrine sandy clay Maumee Lake deposit
							Qma — Glacial lacustrine beach highest Lake Maumee
				Moraine	<14,500		Qmbw — Birmingham waterlaid moraine
							Qmb — Birmingham Moraine
				Delta	<14,500		Qmd — Delta
				Outwash	<14,500		Qmd — Outwash defiance
				Esker	<14,500		Qme — Esker
				Moraine	<14,500		Qmdi — Inner Defiance Moraine
							Qmdo — Outer Defiance Moraine
				Glacial outwash	<14,500		Qow — Outwash Fort Wayne
				Kame	<14,500		Qmk — Kame
				Kettle	<14,500		Qmt — Kettle
				Moraine	<14,500		Qmf — Fort Wayne Moraine
							Qmg — Ground moraine / lodgement till
					Unconformity		
				Fluvial	22,000		Qfa — Fluvial (interglacial) >22,000 ybp
					Unconformity		
				Moraine	22,000		Qmo — Pre-late Wisconsinan glacial till depost >22,000 ybp
					Unconformity		
Paleozoic	Mississipian	Lower	Kinderhookian	Sunbury shale	345,000,000		Ms — Sunbury shale
				Bedford shale	345,000,000		Mbd — Bedford shale
	Devonian	Upper	Senecan	Antim shale	350,000,000		Da — Antrim shale
			Erian	Traverse group	355,000,000		Dt — Traverse group
		Middle	Erian	Dundee limestone	360,000,000		Dd — Dundee limestone
			Ultsterian	Detroit River dolomite	370,000,000		Ddr — Detroit River formation

FIGURE 5.21
Stratigraphic column of Rouge River Watershed in southeastern Michigan. (From Rogers, D.T., Surficial geologic map of the Rouge River watershed in southeastern Michigan, Wayne County, MI, Map Scale 1:62,500, 2 Sheets, 1997.)

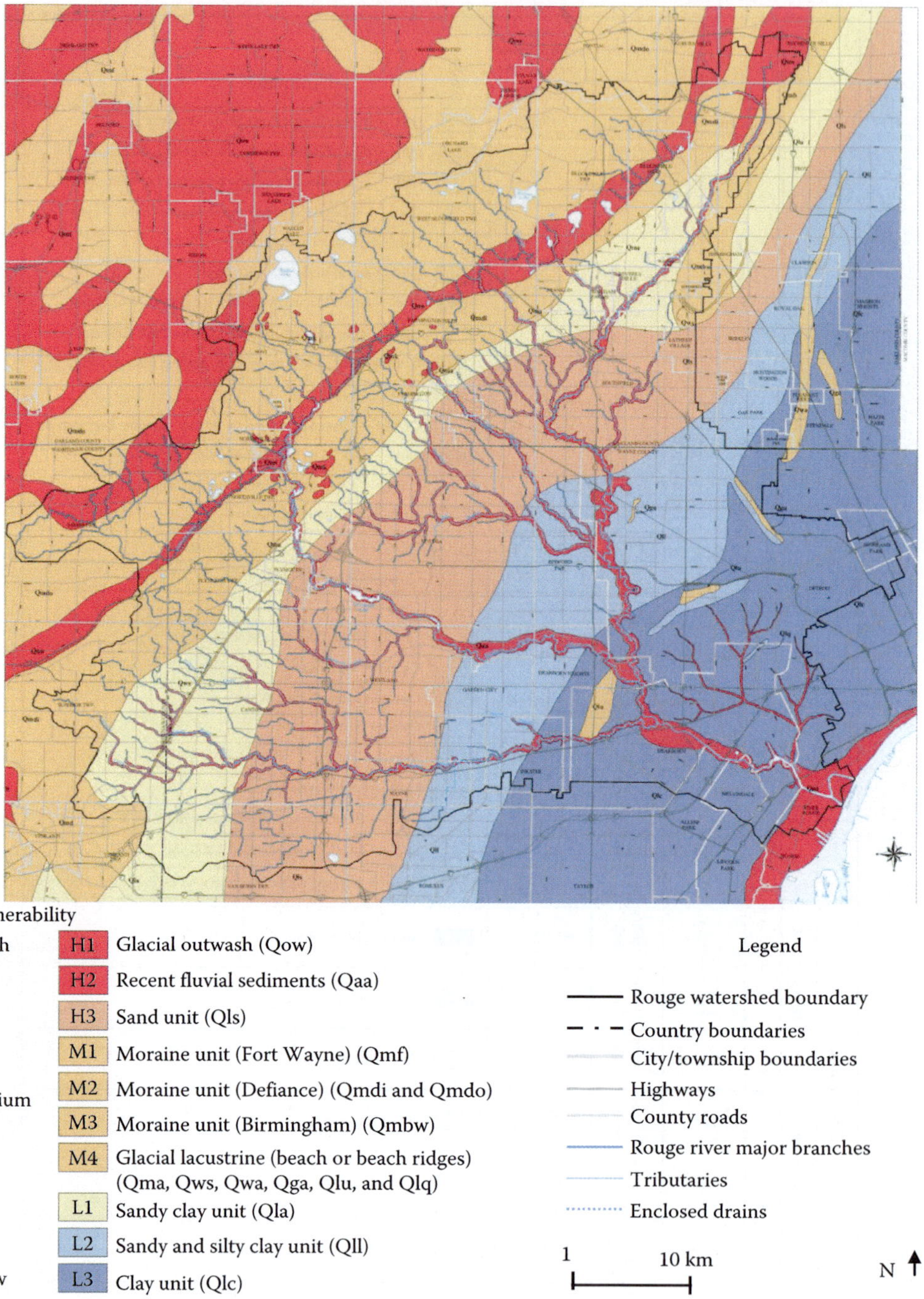

FIGURE 6.3
Geologic vulnerability map: Rouge River watershed. (From Rogers, D.T., Geologic sensitivity map of the Rouge River in southeastern Michigan, Wayne County, MI, 1:62,500. 1 Sheet, 1997c.)

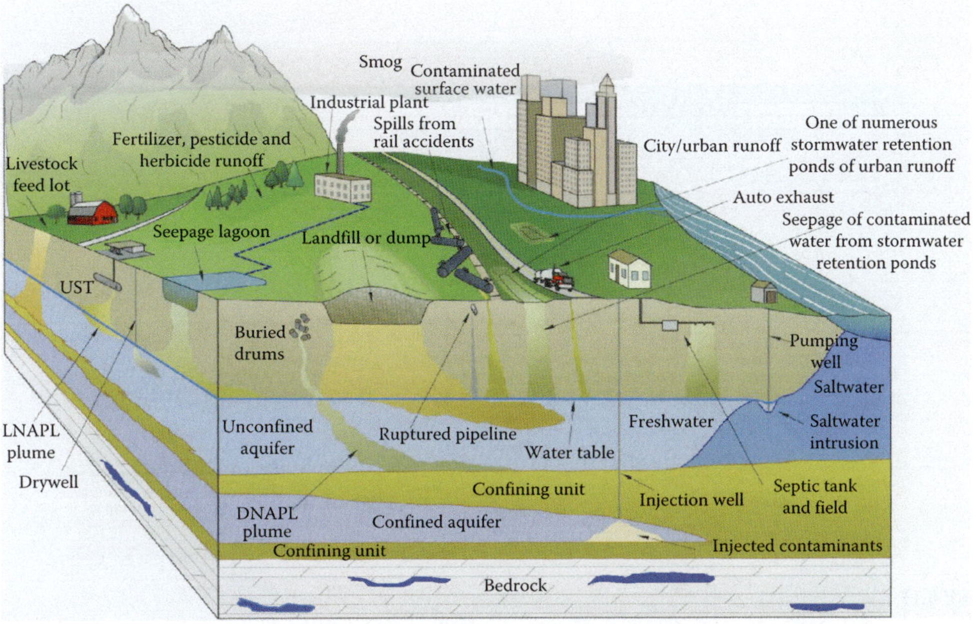

FIGURE 8.1
Sources and locations of contaminant release.

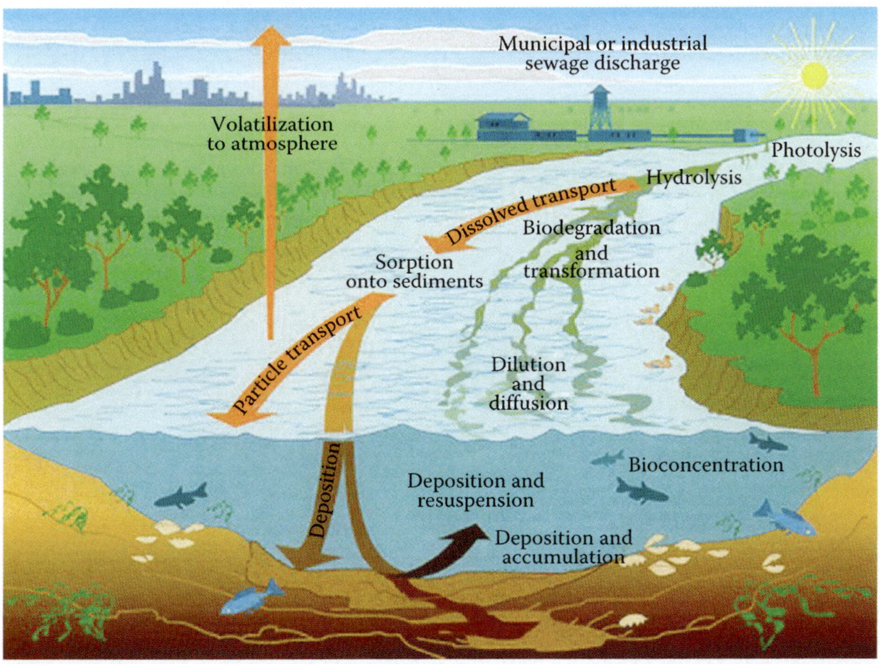

FIGURE 8.5
Fate and transport effects in surface water. (From United States Geological Survey (USGS), Contaminants in the Mississippi River, USGS Circular 1133, Washington, DC, 1995a.)

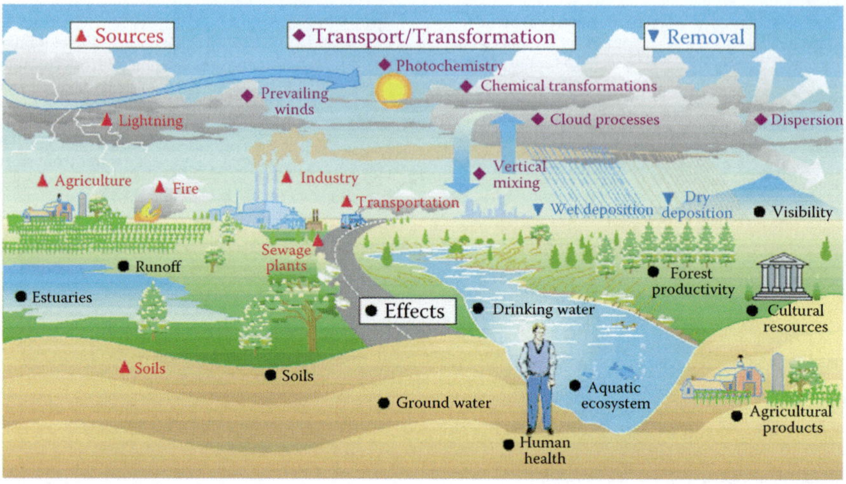

FIGURE 8.15
Sources, transport methods, and removal of air pollution. (From United States Environmental Protection Agency (USEPA), *Latest Findings on National Air Quality: Status and Trends through 2006*, EPA454/R-07-007, Research Triangle Park, NC, 2008a.)

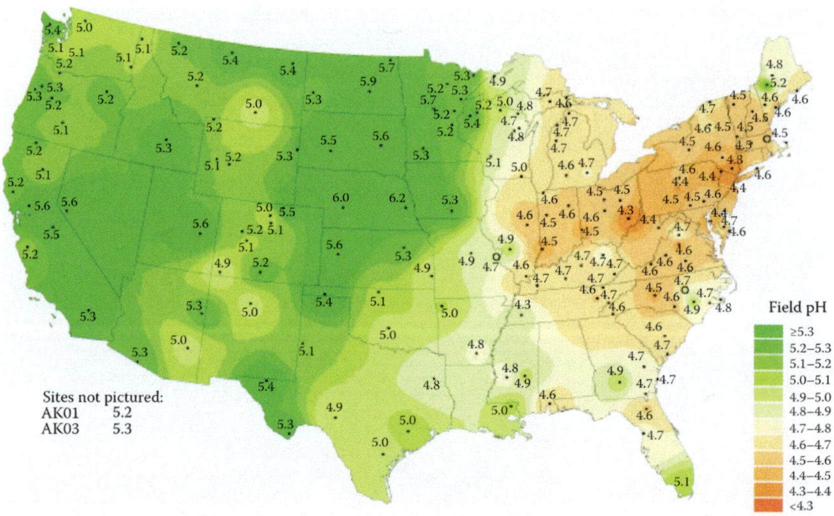

FIGURE 8.25
pH of precipitation in the United States in 2004. (From National Oceanic and Atmospheric Administration (NOAA), Atmospheric CO2 at Mauna Loa observatory, http://www.esrl.noaa.gov/gmd/ccgg/trends/co2_data_mlo.html (accessed November 11, 2009), 2009.)

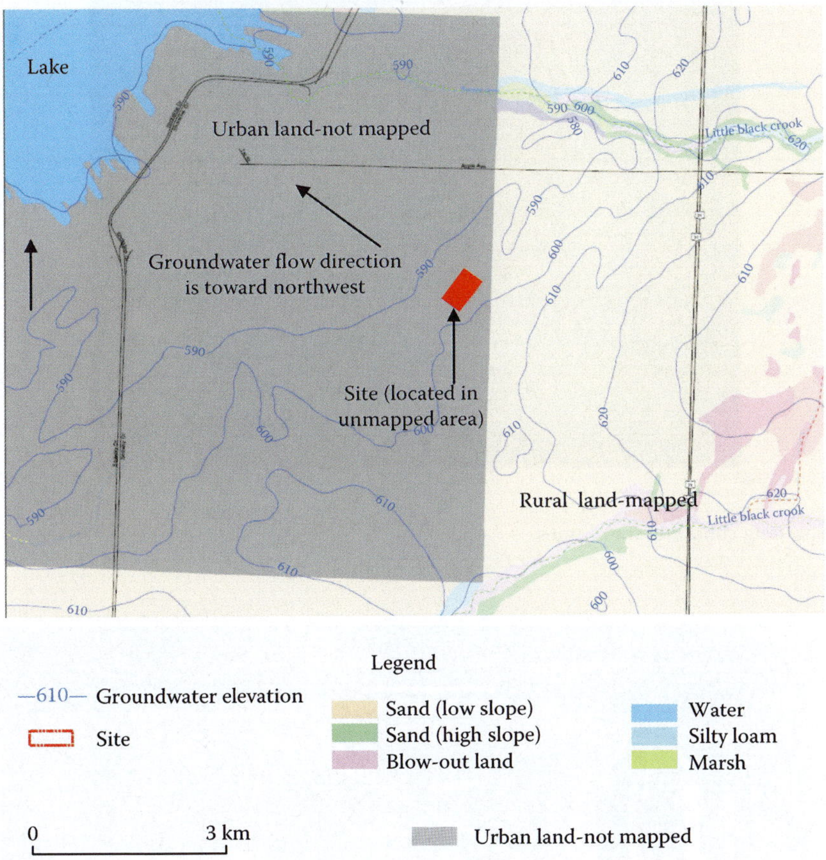

FIGURE 14.5
Available geologic map showing lack of available data in urban area shaded gray.

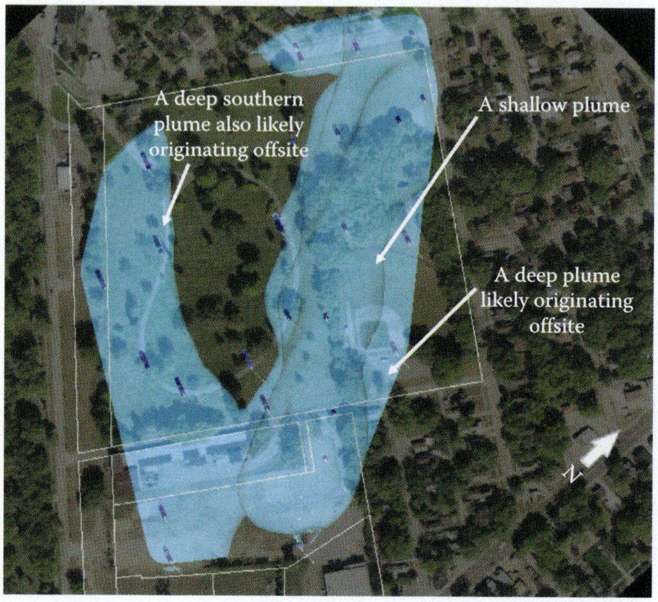

FIGURE 14.6
Surface map showing contaminant plumes.

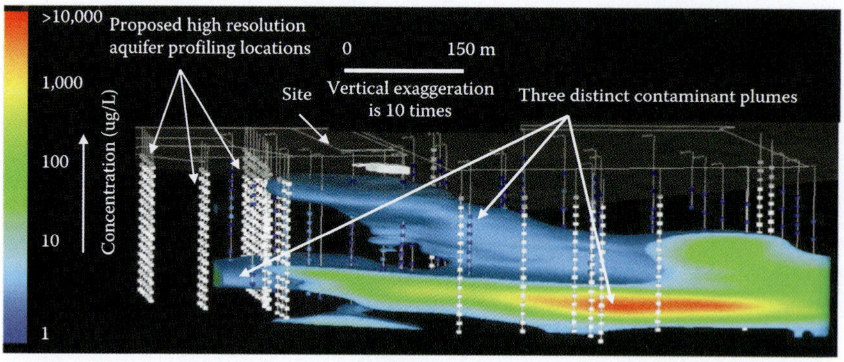

FIGURE 14.7
Three-dimensional view of DNAPL VOC contaminant plumes.

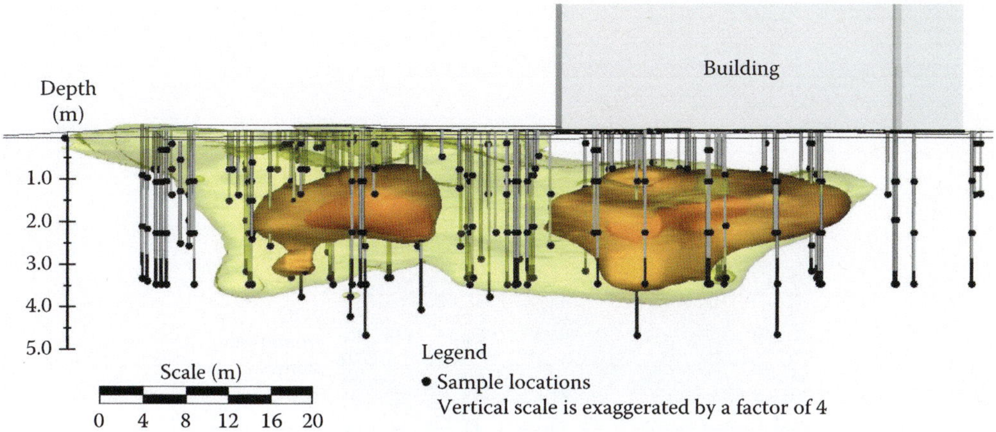

FIGURE 14.10
Cross section showing soil boring locations and extent and magnitude of CrVI impacts (darker color indicates higher concentration).

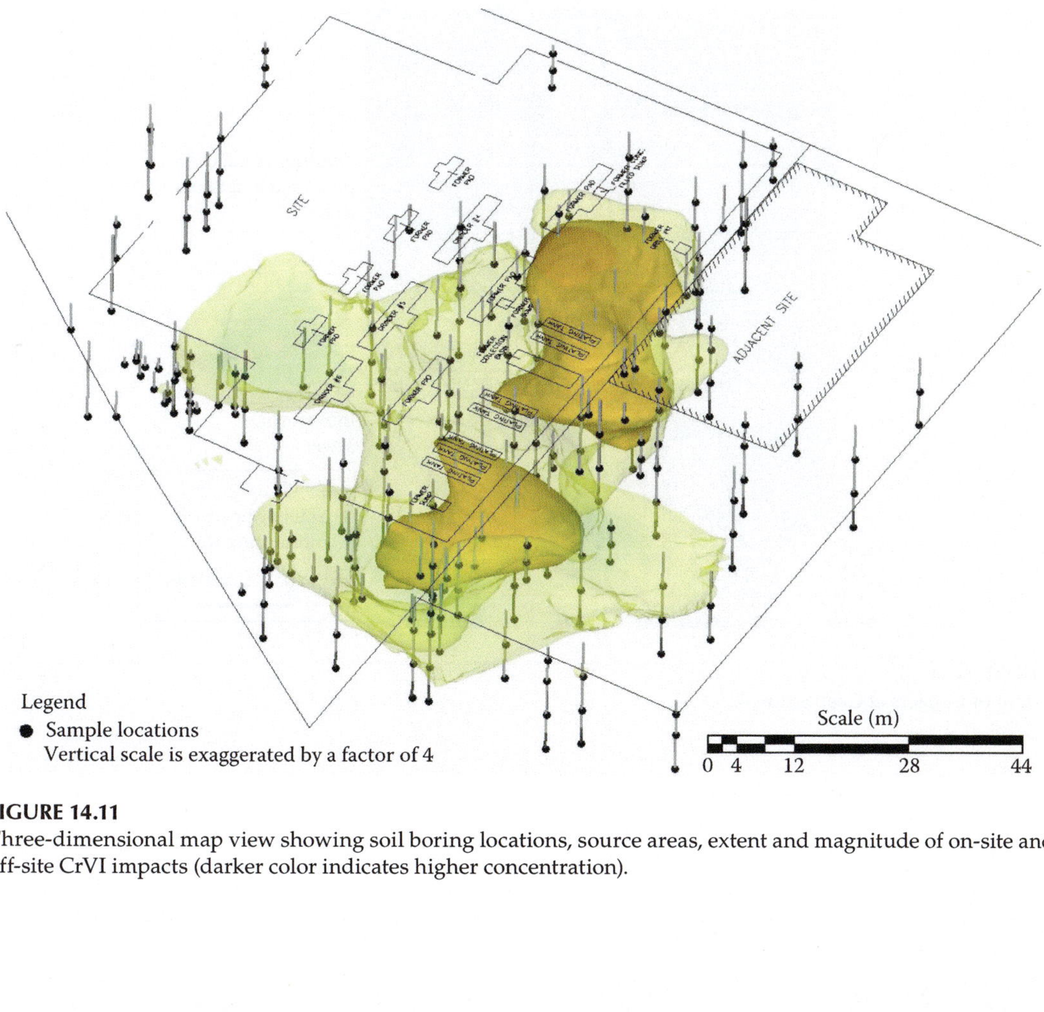

FIGURE 14.11
Three-dimensional map view showing soil boring locations, source areas, extent and magnitude of on-site and off-site CrVI impacts (darker color indicates higher concentration).

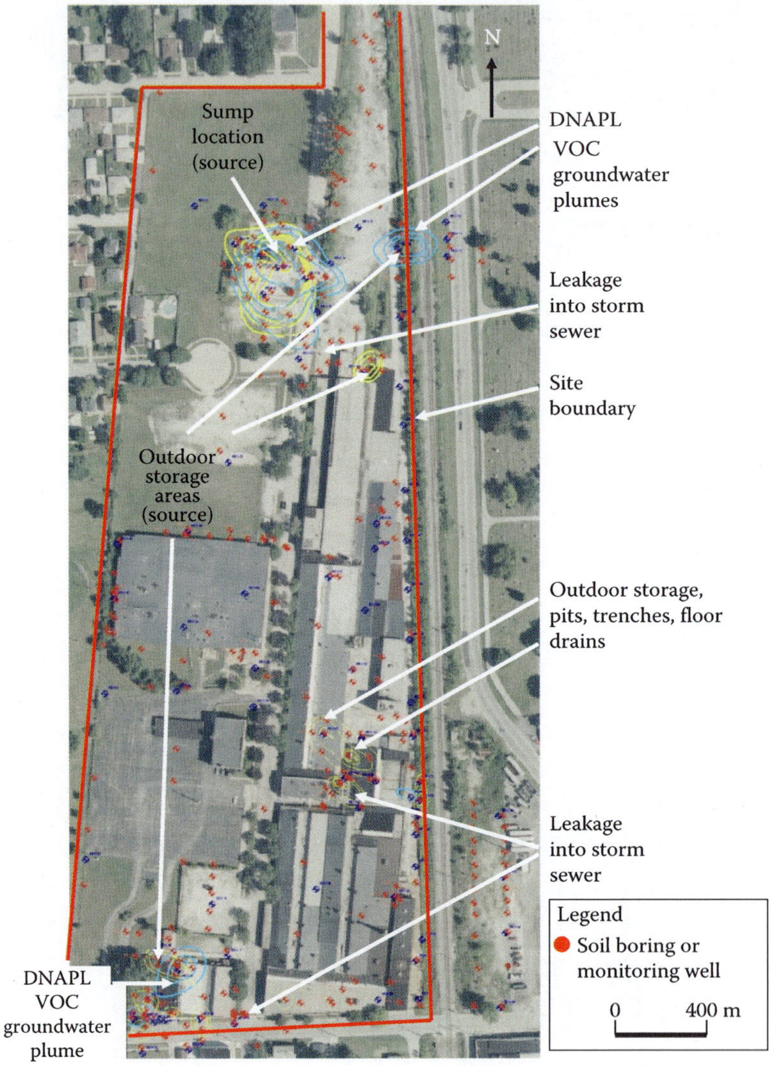

FIGURE 14.16
Extent of impacts at Case Study 4.

TABLE 6.2

Rouge River Geologic Vulnerability Scoring

	Parameters and Vulnerability Scoring									
Geologic Unit	**1**	**2**	**3**	**4**	**5**	**6**	**7**	**8**	**Total Score**	**Rank**
Outwash unit	10	9	9	10	10	10	10	10	79	1
Recent fluvial	10	6	6	8	10	10	10	10	70	2
Main sand unit	10	9	9	10	10	10	6	1	65	3
Other sand units	10	8	8	10	10	10	6	1	63	4
Moraine unit 1	5	7	7	8	8	10	5	10	60	5
Moraine unit 2	5	6	6	7	7	10	5	10	56	6
Moraine unit 3	5	5	5	7	7	10	5	10	54	7
Moraine unit 4	5	4	4	6	6	10	5	10	50	8
Sandy clay unit	4	3	3	4	1	1	3	1	20	9
Sandy and silty clay unit	4	2	2	4	1	1	3	1	18	10
Upper clay unit	3	1	1	3	1	1	2	1	13	11
Lower clay unit (ground moraine unit)	1	1	1	1	1	1	1	1	8	12

Using this geologic, hydrogeologic, and anthropogenic-impact information, a vulnerability matrix was developed and is presented in Table 6.2 (modified from Murray and Rogers 1999b; Rogers and Murray 2002). The geologic units with the highest geological vulnerability (represented by the total score column) were the outwash unit, recent fluvial deposits associated with the Rouge River, and the sand unit of glacial lacustrine beach origin. The next group of units with moderate risk were the four surface moraine units. The geological units with the lowest risk were a sandy clay unit, silty clay unit, and an upper clay unit, all associated with the glacial lacustrine deposits.

The outwash, recent fluvial, and sand units were evaluated to be the most geologically vulnerable geologic units in the watershed because they are generally composed of coarse-grained sediments, have an abundance of groundwater, and account for the majority of the base flow of surface water to the Rouge River. The four moraine units were of moderate risk because they are composed of finer-grained deposits, have less abundance of groundwater, and do not serve to recharge groundwater or discharge to surface water as significantly as the outwash, recent fluvial, or sand units. The low vulnerability units are the sandy clay unit, sandy and silty clay unit, upper clay unit, and the lower clay unit. These geological units generally do not have an abundance of groundwater, are very fine-grained units, and generally impede the migration of groundwater.

Using the surface geological map of the Rouge River watershed (Rogers 1997b) as a base map, and the information contributing to the construction of Table 6.2, Figure 6.3 was developed as the geologic vulnerability map for the Rouge River watershed (Rogers 1997c).

6.5 Demonstrating the Significance of Vulnerability Mapping

As noted by Foster et al. (1993) and Loague et al. (1998), the significance of geologic vulnerability is not appreciated until it can be put into environmental, economic, or political

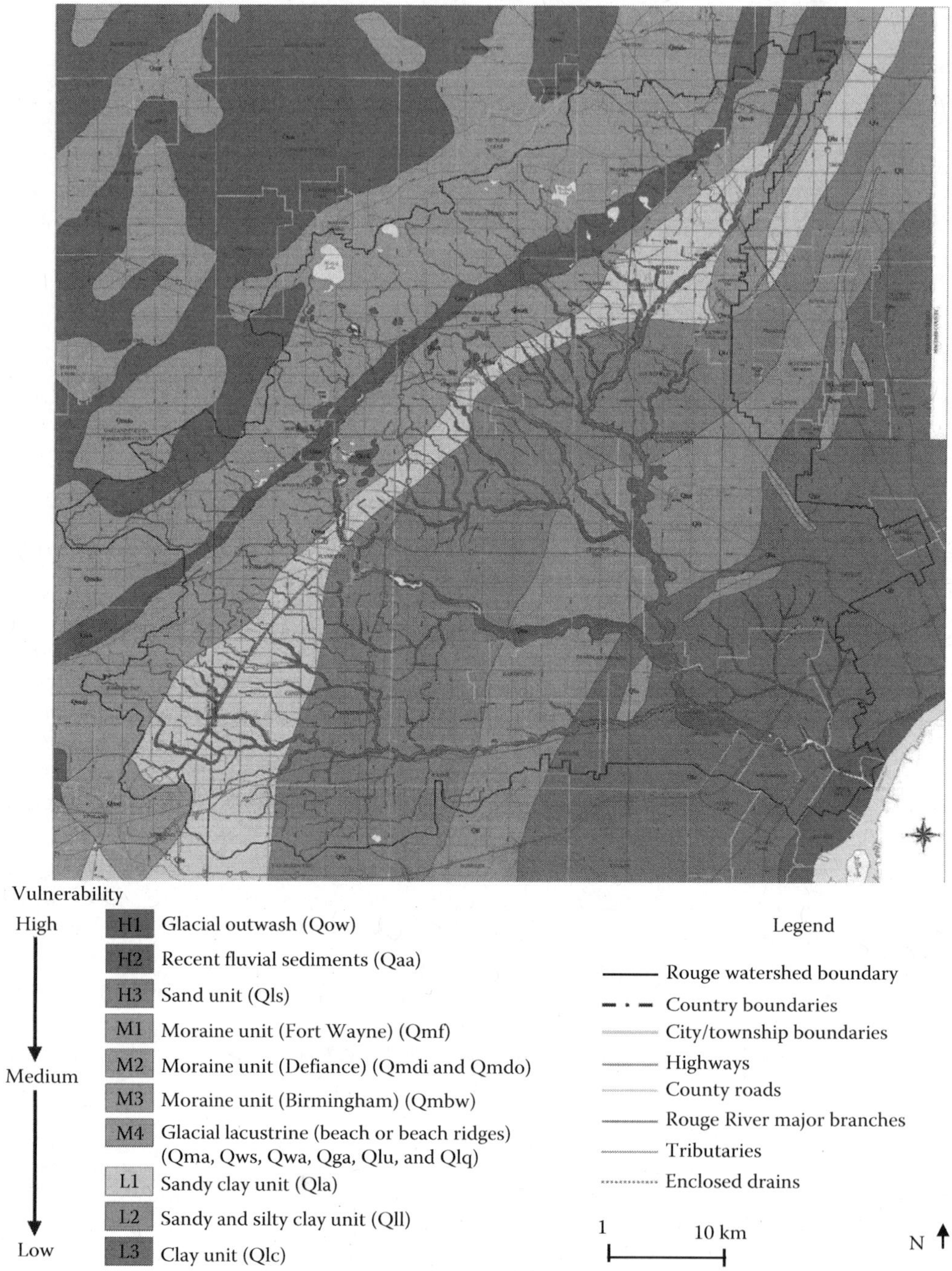

FIGURE 6.3
(See color insert.) Geologic vulnerability map: Rouge River watershed. (From Rogers, D.T., Geologic sensitivity map of the Rouge River in southeastern Michigan, Wayne County, MI, 1:62,500. 1 Sheet, 1997c.)

perspective by actually cleaning up sites of environmental contamination. Therefore, to evaluate whether certain geologic units are vulnerable to contamination, a comparison between specific sites located in low geologic vulnerability areas to sites located in high geologic vulnerability areas must be conducted. If valid, the geologic vulnerability mapping should confirm that the sites situated above high vulnerability locations pose greater risk of exposure than sites located above low vulnerability locations.

For this analysis, a site of low vulnerability located in a geological environment predominantly composed of clay sediments (Site 1) is compared to a high vulnerability site located in a geological environment predominantly composed of sand (Site 2). Both sites are located in the Rouge River watershed and are separated by only 11 km (7 mi). However, Site 1 (1) is significantly larger than Site 2 (approximately twice the size), (2) has a much longer heavy industrial operational history (operated approximately 40 years longer), and (3) has significantly more contamination and types of contaminants (nearly 10 times the mass and three times as many contaminants) released into the environment.

Without considering the geology of each site, it would be logical to assume the environmental risks were higher at Site 1, and the associated clean up costs would be higher and reflect its contamination history. We now determine if the vulnerability map predicts these outcomes.

6.5.1 Site 1: Low Vulnerability Site

Site 1 is a former heavy manufacturing facility located on approximately 16 ac of land that operated for approximately 70 years. A Phase I environmental site assessment was required by the lending institution and conducted due to a real estate transaction involving the property. This initial assessment identified six recognized environmental conditions (RECs).

During the next investigational period (Phase II), several subsurface investigations were conducted at the facility and four main sources of contaminant release were identified that required remediation. These sources are shown as areas of impact on Figure 6.4 and include (1) surface spills, (2) an aboveground storage tank, and (3) spills and leaks of hazardous liquids located in waste storage areas. Other sources or releases were identified during the course of evaluating the site but were not severe enough to warrant further action.

The six RECs identified during the Phase I investigation included

- *Former chemical storage areas.* Evidence of surface staining indicating some spillage of liquids was observed on bare ground near the two former storage areas. No staining was observed near the current storage area.
- *Current storage area.* The current storage area was located inside the main manufacturing building (northern building). The concrete flooring was heavily cracked, providing a potential pathway for spills and leaks to contaminate the ground beneath the building.
- *A former above-ground storage tank that stored gasoline.* A limited amount of surface staining was observed at the general locations of the tank.
- *Surface soil staining and stressed vegetation.* These conditions were observed near a backdoor of the facility and close to an inside location where maintenance activities were conducted and dense nonaqueous phase liquid (DNAPL) solvents were used.

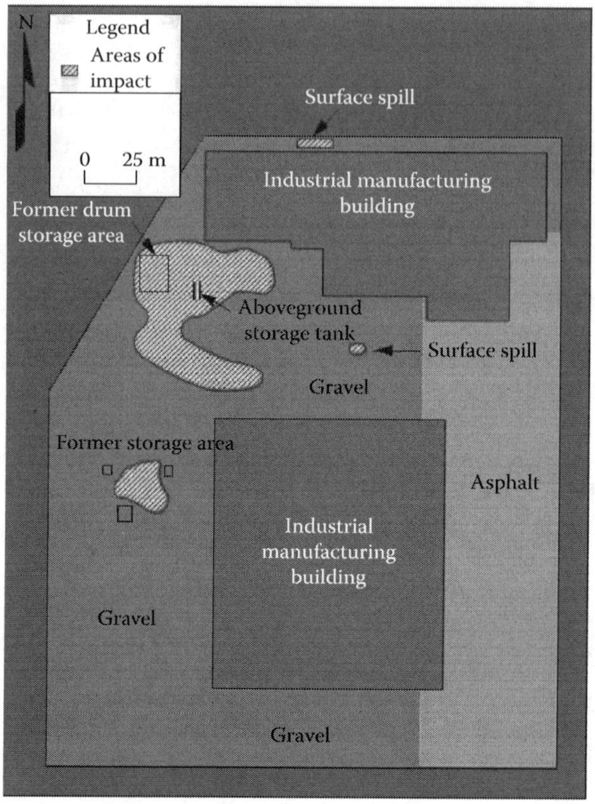

FIGURE 6.4
Generalized site map of Site 1 located in a low vulnerability area.

- Stained soil and stressed vegetation were observed at a location of bare ground where deliveries to the facilities were conducted, and where materials including liquids were offloaded from trucks.

The initial Phase II investigation involved drilling 15 soil borings in areas with the highest likelihood of detecting contamination. This initial subsurface investigation had the following objectives: (1) begin to characterize the subsurface geology of the site, (2) identify the contaminants that were suspected to have been released, and (3) evaluate the magnitude of contaminants present by collecting and analyzing "worst-case" samples from each area of suspected contamination.

The results confirmed the presence of contamination at five of the six RECs identified during the Phase I environmental site assessment. Those compounds detected were at significant concentrations and required further evaluation. The current storage area was eliminated as an area of concern because contamination could not be confirmed through the drilling of three soil borings into the most visibly vulnerable areas of the concrete and the analysis of six soil samples.

Three additional investigation phases were conducted to define the nature and extent of contamination. During the subsequent phases, a total of 40 additional soil borings were drilled and six monitoring wells were installed to evaluate whether there was enough groundwater present for analysis and to establish the direction of groundwater flow. In

TABLE 6.3

Description of Geology for Site 1

Geologic Unit	Depth (m)	Color	Soil Class	Moisture Content	Description
Fill	0–0.6	Gray to brown	Fill	Dry	Fill material consisting of sand and silt, and some construction debris including brick and wood fragments. Fill material ranges from 0.3 to 0.6 m (1–2 ft) thick throughout the site and is not present beneath the buildings.
Clay with some silt	0.6–2.4	Light brown	CL	Dry to moist	Light brown clay with occasional very thin, discontinuous silt layers indicating layered deposition. Silt layers range in thickness from less than a millimeter to not more than two millimeters.
Clay	2.4–12	Light olive gray to blue gray	CH	Damp	Blue to gray colored ground moraine clay. Upper portions very plastic. No visible signs of any silt or original depositional structures to indicate depositional layering of any sort. Very consistent in lithology and color with depth. No signs of larger grained materials such as pebbles. Groundwater was not observed at the contact between the upper lacustrine clay and the ground moraine lower blue clay.

many of the soil borings, more than two soil samples were analyzed to gather data on the vertical extent of contamination. The maximum depth of the soil borings was 4.6 m (15 ft) beneath the surface in impacted areas, yet the vertical extent of contamination did not exceed a depth of 1.5 m (5 ft). The monitoring wells indicated damp to moist soils existed at some locations within very thin layers of silt. These silty layers were just a few millimeters thick and were also observed in soil samples collected from some of the soil borings drilled during investigative activities. After the monitoring wells failed to detect any groundwater seepage, they were pulled from the ground and the boreholes sealed with a bentonite clay grout.

Characterization of the geology at the site was accomplished by drilling a soil boring to a depth of 12 m (40 ft) in a nonimpacted area. In general, the site was underlain by a clay deposit from a Pleistocene age glacial lake that occupied the region more than 12,000 years before the present. Historical geological literature of the region indicates a ground moraine or lodgment till deposit extended to depths of approximately 55 m (180 ft) beneath the ground at the site. Table 6.3 details the geology between the ground surface and 12 m (40 ft) beneath the site.

Other pertinent technical and geological information concerning the site included

- Storm sewers in the immediate vicinity did not intersect any of the contamination
- Surface water drainage was controlled by storm sewers
- No buried utilities intersected contaminated areas

- Potable water was supplied by the municipality and the source was more than 16 km (10 mi) away
- The contamination did not extend beyond the property boundary

The types of contaminants detected at the facility included

- Volatile organic compounds (VOCs) including
  - DNAPLs commonly referred to as chlorinated solvents used to degrease and clean metal surfaces
  - Light nonaqueous phase liquids (LNAPLs) used as solvents, paint thinners, and cleaning products, and are common constituents in fuels such as gasoline
- Polynuclear aromatic hydrocarbons (PNAs or PAHs) commonly used as lubricants and motor oils and cutting fluids
- PCBs used in electrical equipment
- Heavy metals (arsenic, chromium, and lead) commonly used in paints and pigments, batteries, and metal plating

A list of specific chemical compounds, highest concentrations detected, and estimated contaminant mass remediated are listed in Table 6.4.

Soil excavation and disposal of the contaminated soils at a licensed landfill was the remedial method of choice for the contamination at this site. The overriding considerations for selecting this method were: the contaminated areas were less than 1.5 m deep and were not located beneath any buildings; the activities could be conducted quickly without disturbing ongoing facility operations, and it represented the lowest cost alternative. Approximately 5,400 m^3 (7,000 cubic yards)—equivalent to 9,072 ton of soil (10,000 ton)—was excavated and transported to a local landfill for disposal. The total cost for investigation and remediation was approximately $400,000. This translates into a remediation cost per kilogram of contaminant of $362 ($800/lb).

After the remediation was verified by the regulatory authority through the collection and analysis of the soil samples taken from each area remediated, closure was granted and a "No Further Action Required" letter was issued for the site. The closure was deemed unrestricted, meaning the site had been remediated to comply with residential land use requirements. Eighteen months had elapsed since the Phase I environmental site assessment.

6.5.2 Site 2: High Vulnerability Site

Site 2 is a former heavy manufacturing facility approximately 3.2 ha (8 ac) in size that operated for 30 years. Figure 6.5 shows the layout of the site.

The Phase I environmental site assessment identified four RECs:

- *Two former chemical storage areas.* Evidence of surface staining indicated some spillage of liquids at one storage area located on bare ground near an area storing waste paints. An additional waste storage area was located on stained asphalt pavement that was heavily cracked and broken.
- *A former underground storage tank.* This underground storage tank once stored gasoline and was identified as a REC because during its removal (1) it was not

TABLE 6.4

Contaminant Types, Concentration, Mass Remediated for Site 1

Contaminant	Maximum Concentration (μg/kg)[a]	Estimated Contaminant Mass Remediated kg/(lb)
VOCs		91/(200)
DNAPL compounds		
Tetrachloroethene	60,100	
Trichloroethene	45,000	
Cis-1,2-dichoroethene	20,000	
Trans-1,2-dichloroethene	3,000	
Methylene chloride	800	
LNAPL compounds		
Ethyl benzene	10,000	
Xylenes	10,000	
Acetone	280	
Carbon disulfide	200	
PNAs		113/(250)
Naphthalene	339,000	
Acenaphthalene	18,000	
Fluorene	22,000	
Phenanthrene	280,000	
Fluoranthene	156,000	
Pyrene	13,000	
Benzo[a]anthracene	4,800	
Benzo[a]pyrene	3,200	
Dibenzo[a,h]anthracene	1,800	
Indeno[1,2,3-cd]pyrene	1,400	
Chrysene	11,000	
PCBs	16,000	2.2/(5)
Heavy metals		23/(50)
Arsenic	23,000	
Chromium	530,000	
Lead	930,000	

[a] μg/kg = micrograms per kilogram.

physically inspected by a qualified professional and (2) confirmatory soil samples were not collected and analyzed from the excavation pit to verify the tank did not leak.

- *Surface staining.* Surface staining and stressed vegetation were observed near the former location of a back door near the location where solvents were used inside the manufacturing building, labeled "Main Building" on Figure 6.5.

The multiple Phase II subsurface investigations conducted at the facility identified four main sources of contamination resulting in the release of contaminants and requiring remediation. These sources of contamination included four areas with a track record of prior spills. Other sources or releases were identified during site evaluation but were not severe enough to require further action.

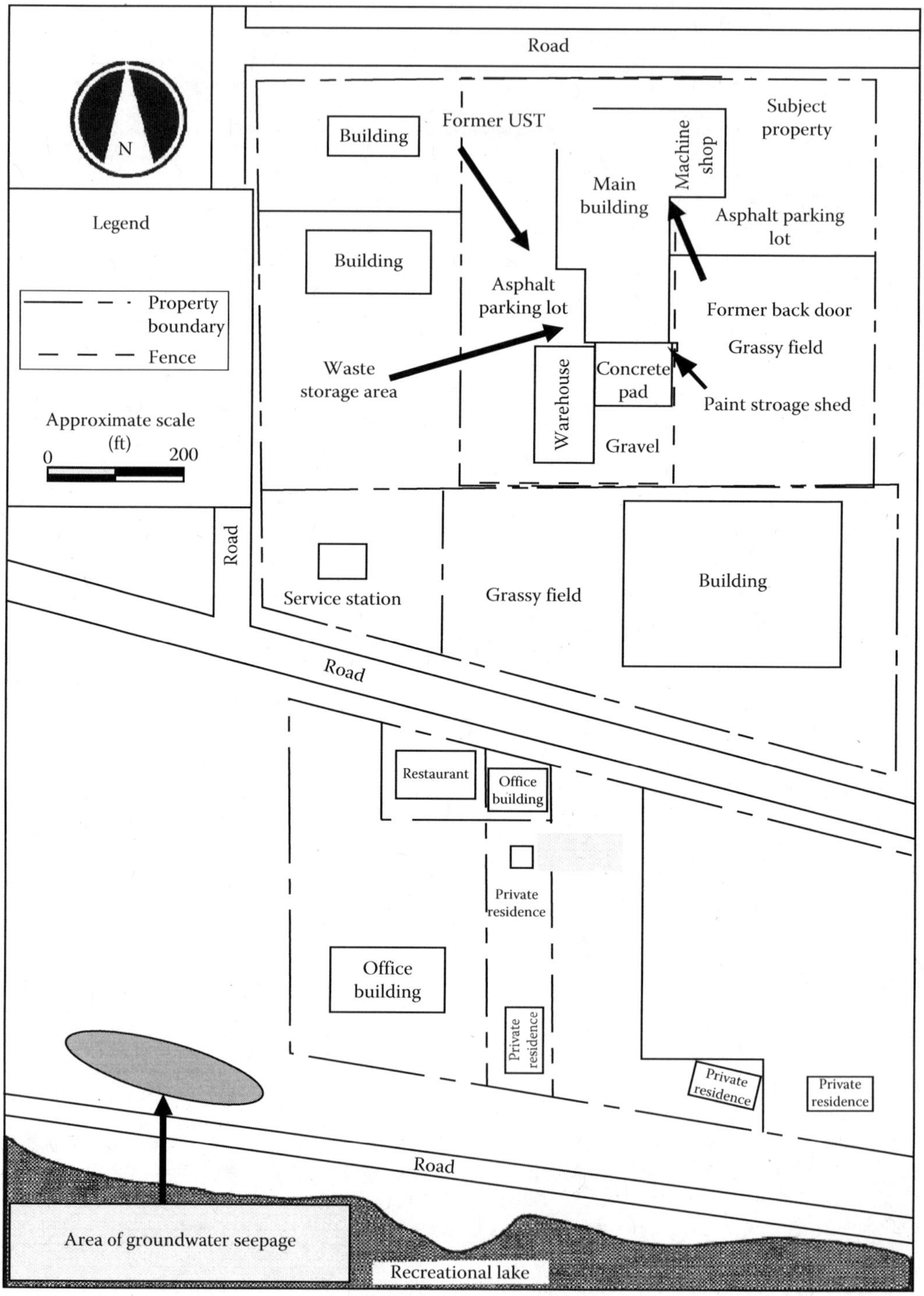

FIGURE 6.5
Map of Site 2 within an area of high vulnerability. (From Rogers, D.T., The importance of site observation and followup environmental site assessment—A case study, in: *Groundwater Management Book 12* National Groundwater Association, Dublin, OH, pp. 563–573, 1992.)

The initial subsurface investigation conducted involved drilling 12 soil borings in areas with the highest likelihood of detecting contamination. This operation was performed with similar objectives to those at Site 1: characterize the geology; identify the contaminants; and analyze the worst case samples. The results confirmed the presence of contamination at three of the four RECs identified in the Phase I environmental site assessment. Contaminant concentrations in near-surface soil were also detected at sufficient levels to require further investigation. The one location not pursued for further investigation was near the former underground storage tank. Four soil borings made at the location of this tank and the analysis of four soil samples taken from the soil beneath the tank did not confirm the presence of contamination above detectable concentrations.

Groundwater was encountered at a depth of 3.0–3.7 m (10–12 ft) beneath the surface of the ground during the initial investigation. Temporary monitoring wells were installed at select locations to evaluate the possible presence of groundwater impacts and to estimate the direction of groundwater flow. The analytical results suggested the presence of groundwater contamination likely originating from on-site sources. This finding was confirmed because there were levels of several contaminants exceeding applicable cleanup criteria, and the same contaminants were detected in near-surface unsaturated soil at the locations where the RECs were identified but were not detected in soil or groundwater at upgradient locations.

Six additional investigation phases were conducted to define the nature and extent of contamination. During these subsequent phases, a total of 132 additional soil borings were drilled, with many of the soil borings having multiple samples analyzed to help characterize the vertical extent of contamination. A total of 80 monitoring wells were installed to define the nature and extent of impacts to groundwater.

The general subsurface geology of the site immediately beneath the surface consisted of a sand deposit originating from a Pleistocene age glacial lake that occupied the region more than 12,000 years before the present. Specific geology beneath the site consisted of sand from the surface to a depth of 7.9–9.8 m (26–32 ft). Beneath this glacial lacustrine beach sand deposit was a ground moraine or lodgment till deposit extending to a depth of least 23 m (75 ft). Historical geological literature of the region indicated that the ground moraine or lodgment till deposit extended to depths of approximately 60 m (200 ft) beneath the ground in the area. Table 6.5 describes the geology between the ground surface and 23 m beneath the site.

During the investigation, multiple groundwater monitoring wells were installed at the same location but were screened at different depths within the saturated zone to evaluate the vertical distribution of contaminants within the aquifer. Several samples of the ground moraine deposit beneath the aquifer were also analyzed for the presence of contamination and for certain hydrologic parameters (such as hydraulic conductivity and grain size analysis) to evaluate whether the ground moraine deposit was an effective confining layer, preventing contaminant migration to deeper aquifers. In addition, three deep soil borings were drilled to a depth of 23 m in nonimpacted areas of the site to verify the horizontal distribution and thickness of the ground moraine deposit.

Other technical and geologically related information relevant to the site analysis included the following:

- Storm sewers in the immediate vicinity did not intersect any of the contamination
- Surface water drainage was controlled by storm sewers
- No buried utilities intersected contaminated areas

TABLE 6.5

Description of Geology for Site 2

Geologic Unit	Depth (m)	Color	Soil Class	Moisture Content	Description
Sand	0–7.9	Medium orange	SW	Dry to 3.0–3.6 m, then saturated	Medium- to coarse-grained sand with occasional pebbles. Evidence of bedding present. Thickness of bedding layers ranges between a few millimeters to 3 cm.
Clay with some silt	7.9–9.1	Light brown	SM	Saturated	Fine-grained sand. Immediately grades into blue clay. Sharp contact. Evidence of clay intraclasts in lower portion of sand indicating an erosional surface.
Clay	9.1–23	Light olive gray to blue gray	CH	Damp to dry	Blue to gray colored ground moraine clay. Upper portions very plastic. No visible signs of any silt or original depositional structures to indicate depositional layering or any sort. Very consistent in lithology and color with depth. No signs of larger grained materials such as pebbles.

- Potable water was supplied by the municipality and the source was more than 24 km (15 mi) away. However, some local residences used groundwater within the same aquifer for irrigation purposes
- Contaminated groundwater extended beyond the property boundary approximately 365 m (1200 ft)

A complex groundwater flow pattern existed at the site. The site was located on a groundwater divide where groundwater flowed to the east and slightly north in the northern portion of property, but flowed toward the southeast on the southern portion of the property. Figure 6.6 shows the on-site groundwater flow pattern.

The types of contaminants detected at the facility included

- VOCs including
  - DNAPLs commonly referred to as chlorinated solvents used to degrease and clean metal surfaces
  - LNAPLs used as solvents, paint thinners, and cleaning products, and are common constituents in fuels such as gasoline
- PCBs used in electrical equipment

The specific chemical compounds with their highest concentrations detected and estimated contaminant mass remediated are listed in Table 6.6.

The highly permeable soil and shallow groundwater depth (i.e., the vulnerable geology) allowed VOCs to rapidly infiltrate and migrate to groundwater. In addition, the high

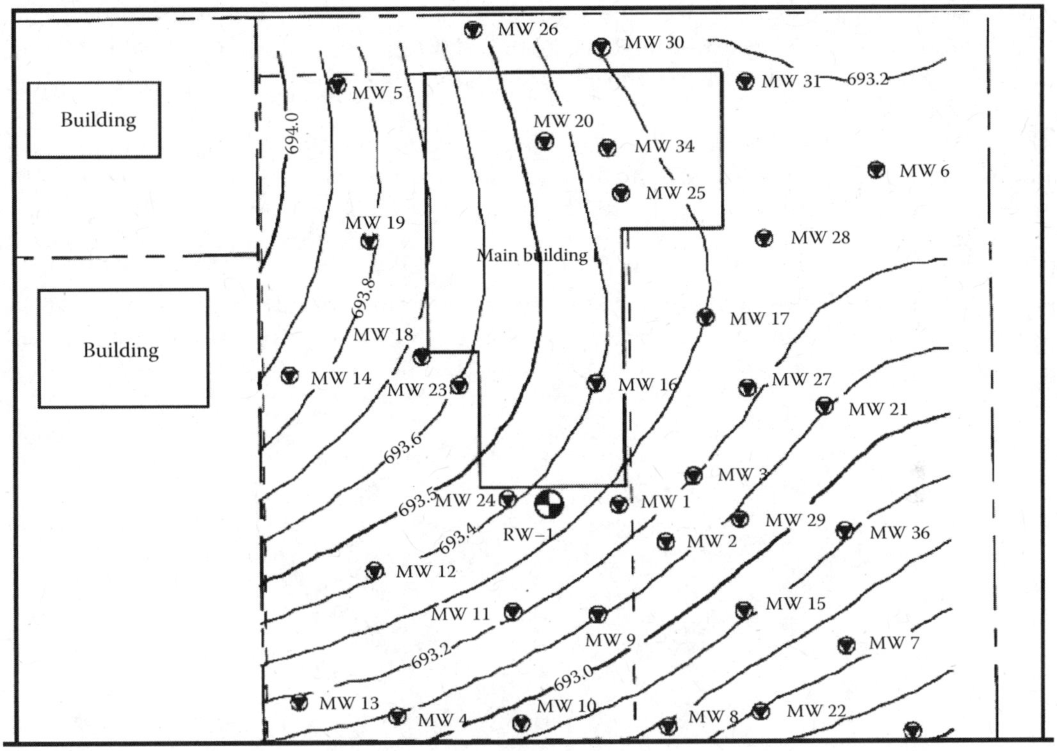

FIGURE 6.6
Groundwater flow pattern at Site 2.

TABLE 6.6
Contaminant Types, Concentration, Mass Remediated for Site 2 High Vulnerability Site

Contaminant	Maximum Concentration in Soil (μg/kg)	Maximum Concentration in Groundwater (μg/L)[a]	Estimated Contaminant Mass Remediated kg/(lb)
VOCs			45/(100)
DNAPL compounds			
Tetrachloroethene	22,000	3,250	
Trichloroethene	8,000	2,200	
Cis-1,2-dichoroethene	5,000	1,800	
Trans-1,2-dichloroethene	480	280	
1,1-Dichloroethene	640	220	
1,1,1-Trichloroethane	2,800	2,100	
Vinyl chloride	1,100		
LNAPL compounds			
Ethyl benzene	32,000	240	
Xylenes	28,000	130	
PCBs	7,000,000	Not detected	13.6/30

[a] μg/L = microgram per liter.

groundwater seepage velocities resulted in a VOC plume extending one-third of a mile to a downgradient spring, and ultimately discharged into a recreational lake and the Rouge River. The map in Figure 6.7 shows the extent of the VOCs in groundwater and the location of the downgradient spring.

PCBs were not detected in groundwater. Therefore, excavation and disposal of the PCB-contaminated soils at a licensed landfill was the remedial method of choice because the impacted soils were less than 1 m in depth, were not located beneath any buildings, could be conducted quickly, did not disturb ongoing facility operations, and was the lowest cost alternative.

Excavation was also the remedial action of choice for soils highly impacted with VOCs. This method was chosen because there was not a substantial volume of impacted soil with VOCs, as the sandy soils at the site had a low capacity for retaining VOCs. As a result, the

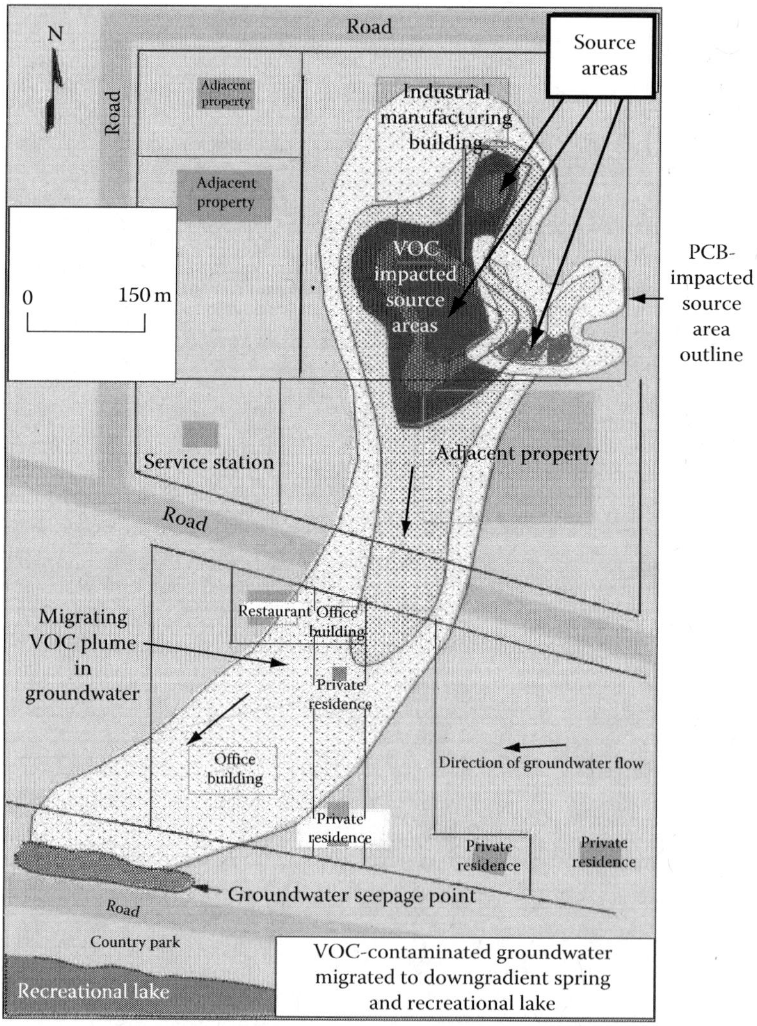

FIGURE 6.7
Contaminant migration at Site 2.

VOCs tended to migrate downward through the soil column and contaminate groundwater without adsorbing to soil grains.

The remedial method chosen for groundwater was air sparging and soil vapor extraction since the contaminants in groundwater were VOC compounds and did not extend past the mid-portion of the saturated zone. Air sparging involved the injection of air beneath the impacted groundwater and then letting the air rise naturally through the saturated zone. As the air migrated upward through the saturated zone, it volatilized the contaminants, and the vapors containing the VOCs were removed using a soil vapor extraction system in the vadose zone. The vapors were removed from the air by passing them through a granular activated carbon tank.

The VOCs contaminating groundwater at this site were DNAPL compounds having a specific gravity slightly greater than water. Therefore, when present at sufficient concentrations, DNAPL compounds may sink through the water column and contaminate lower portions of an aquifer. This sinking action did not occur at this site because its geology and hydrogeology—in effect a stratigraphic control—prevented the VOCs from migrating to the bottom of the aquifer. As listed and described in Table 6.5, the composition of the aquifer gradually became finer grained with depth, and the hydraulic conductivity decreased proportionately. This reduction of hydraulic conductivity, combined with low contaminant mass in groundwater, resulted in restricting contaminants to the upper portion of the aquifer. With the VOCs restricted to the upper portion of the aquifer, air sparging became the most practical remediation technique.

The time duration between the Phase I environmental site assessment and the receipt of the closure letter was approximately 14.5 years. The dollar costs for remediating this site stacked up this way: the PCB-contaminated soil, including investigation, was \$1.1 million; \$0.1 million for the VOC-contaminated soil; VOC-contaminated groundwater, including investigation, was \$6.6 million. The total cost for investigation and remediation of this site was \$7.8 million, which translates into a cost of \$27,180/kg (\$60,000/lb) of contaminant.

Remediation was verified by the regulatory authority through the collection and analysis of the soil samples from each area remediated. In addition, four iterations of groundwater samples over a period of 1 year were made until cleanup levels were achieved. Closure was then granted and a "No Further Action Required" letter was issued for the site. A risk assessment was also conducted for the site and was designed to determine if the residual contamination at the site would pose an ongoing and unacceptable risk to human health and the environment. Locations included in this evaluation were the on-site areas with persistent soil contamination, and the on-site and off-site areas with contaminated surface water and groundwater.

The results of the risk assessment indicated that a deed restriction was appropriate for the site property and banned the use of groundwater for any purpose. A cap consisting of asphalt pavement was also required for portions of the property where some soil contamination remained in place, and if any soil were to become disturbed or exposed, another evaluation must be conducted to evaluate the need for further remedial actions. Land use for the property was restricted to industrial use.

6.5.3 Site Comparison Analysis

A profound difference between these two examples is seen in the cost per kilogram to remediate the contaminants. Despite the smaller acreage, shorter time of industrial operations, and low contaminant mass released to the environment, the cost to remediate a kilogram of contaminant at the site of high vulnerability (Site 2) is 75 times greater than to remediate a kilogram of contaminant at the low vulnerability site (Site 1). This cost differential is solely due to the geology present at each site.

The high vulnerability site is located in a geologic area (a sand) that is environmentally vulnerable to contamination because (1) the highly permeable soil allows contaminants to infiltrate readily and migrate to groundwater and (2) contaminant plumes are transported at relatively high seepage velocities in the sand aquifer to potentially sensitive receptors that include potable water wells and surface water. The costs for the high vulnerability site would have been much higher had the contaminants migrated along the bottom portion of the aquifer; this is sometimes the case with the type of contaminants present (DNAPLs) since they are denser than water. Luckily, the decreased hydraulic conductivity within the lower portion of the aquifer prevented this from happening at the higher vulnerability site. In addition, the costs would have been significantly higher had there been a completed human pathway represented by the ingestion of contaminated groundwater. Table 6.7 summarizes the major differences between these two sites of environmental contamination. Please note that the vulnerability map ranking accurately predicted the relative costs of remediation.

TABLE 6.7

Site Comparison Table

Parameter	Site 1—Low Vulnerability	Site 2—High Vulnerability
Predominant geology	Clay	Sand
Presence of shallow groundwater	No	Yes
Size of site	6.5 ha (16 ac)	3.2 ha (8 ac)
Length of operation	70 years	30 years
Land use	Heavy industry	Heavy industry
Types of contaminants	VOCs, PNAs, PCBs, and heavy metals	VOCs and PCBs
Number of contaminants remediated	27	9
Contaminant mass remediated	227 kg	59 kg
Cleanup criteria	Same as Site 2 for overlapping compounds	Same as Site 1 for overlapping compounds
Cleanup cost[a]	\$400,000	\$7,800,000
Remedial methods (soil)	Excavation	Excavation
Remedial methods (groundwater)	Remediation not required	Air sparging and soil vapor Extraction
Other remedial control measures	None, unrestricted closure	Restricted closure included: • Deed restriction • Institutional controls • Industrial land use only
Timeframe	18 months	14.5 years
Cost per kilogram of contaminant	\$362	\$27,180
Vulnerability ranking using Table 6.2	13	65

[a] Cleanup costs include costs for investigation and remediation.

6.6 Summary and Conclusion

Through the construction of a vulnerability map, we have combined the knowledge gained from Chapters 2 through 5 to explain the relationship between the natural environment and human influence in urban areas. As stated in the introduction of Chapter 2, the arrangement, thickness, and composition of the sediment layers beneath our feet have a profound influence on where cities are located, how buildings are constructed, where roads are built—and perhaps most important to the development and redevelopment of our urban centers—how contaminants behave and how they affect the environment and people. The two case studies presented in this chapter have highlighted this relationship—but they are a just a small subset of the thousands of examples of this important concept and its multifaceted connections.

As demonstrated by comparing the two sites in this chapter and as we shall see in the next section of this book—once the environment has been contaminated at levels that pose a human or ecological risk—it is often very expensive to remediate, especially when groundwater is affected. Furthermore, it may be impossible to fully remediate some sites even with the most advanced technology. Therefore, minimizing wastes and preventing pollution have proven to be the most effective methods for reducing costs, and ultimately, preserving our environment. The two examples highlighted in this chapter are not uncommon. Tens of thousands of industrial and even commercial and residential sites in the United States have contaminated soil and groundwater to levels requiring one or more expensive remedial actions.

The realization that certain locations or areas within urban regions are especially vulnerable to contamination offers even greater promise for resolving future environmental issues. Geologic vulnerability analysis of urban regions produces essential information for evaluating the environmental and financial risks associated with development and redevelopment. By minimizing the impact of pollution once a release has occurred, certain geological features may play, if we so choose, a central role in the development and redevelopment of an urban area.

However, the story does not end here because geology alone is not responsible for dictating the environmental risks and the costs of remediating contamination. The physical chemistry of contaminants themselves also plays a central role. Until now the focus has been on learning the geology of urban areas and how the geology of a particular area or watershed influences environmental risk. The next section of this book introduces the next piece of the environmental risk puzzle—the contaminants themselves. And as will be demonstrated, the physical chemistry of specific contaminants shares top billing with the existing environmental geology when estimating the environmental risk.

References

Albinet, M. and J. Margat. 1970. Cartographie de la vulnerabilite a la pollution des nappes d"neau souterrians. *Bulletin BRGM (2nd Series)* 3:13–22.

Aller L., T. Bennett, J.H. Lehr et al. 1987. *DRASTIC: A Standardized System for Evaluating Ground Water Pollution Potential Using Hydrogeologic Settings*. United States Environmental Protection Agency USEPA-600/2-87-35. Ada, OK: USEPA.

Burn, S., M. Eiswirth, R. Correll et al. 2007. Urban infrastructure and its impact on groundwater contamination. In *Urban Groundwater—Meeting the Challenge*, ed. K.W.F. Howard, pp. 29–40. London: Taylor & Francis.

Eaton, T.T. and A. Zaporozec. 1997. Evaluation of groundwater vulnerability in an urbanizing area. In *Groundwater in the Urban Environment*, ed. J. Chilton et al., pp. 577–582. Rotterdam, the Netherlands: Balkema.

Farrand, W.R. 1982. Quaternary geology of southern (and northern) Michigan. Michigan Department of Natural Resources, Geological Survey Division, Lansing, MI.

Farrand, W.R. 1998. The glacial lakes around Michigan. Bulletin No. 4. Michigan Department of Natural Resources, Lansing, MI.

Focazio, M.J., T.E. Reilly, M.G. Rupert et al. 2001. Assessing ground-water vulnerability to contamination: Providing scientifically defensible information for decision makers. United States Geological Survey Circular 1224, Denver, CO.

Foster, S.S.D. and R. Hirata. 1988. *Groundwater Risk Assessment: A Methodology Using Available Data.* Lima, Peru: Pan American Center for Sanitary Engineering and Environmental Sciences, Ciencias del Ambiente (CEPIS).

Foster, S.S.D., B.L. Morris, and A.R. Lawrence. 1993. Effects of urbanization on groundwater recharge. In *Groundwater Problems in Urban Areas*, ed. W.B. Wilkinson, pp. 43–63. ICE Conference Proceedings, London, U.K.

Garcia-Fresca, B. 2007. Urban-enhanced groundwater recharge: Review and case study of Austin, Texas, USA. In *Urban Groundwater—Meeting the Challenge*, ed. K.W.F. Howard, pp. 3–18. London, U.K.: Taylor & Francis.

Hartig, J.H. and M.A. Zarull. 1991. Methods of restoring degraded areas in the Great Lakes. *Reviews of Environmental Contamination and Toxicology* 177:127–154.

Howard, K.W.F., S. Di Biase, J. Thompson et al. 2007. Stormwater infiltration technologies for augmenting groundwater recharge in urban areas. In *Urban Groundwater—Meeting the Challenge*, ed. K.W.F. Howard, pp. 175–188. London, U.K.: Taylor & Francis.

Kaufman, M.M., D.T. Rogers, and K.S. Murray. 2003. Surface and subsurface geologic risk factors to ground water affecting brownfield redevelopment potential. *Journal of Environmental Quality* 32:490–499.

Kaufman, M.M., D.T. Rogers, and K.S. Murray. 2005. An empirical model for estimating remediation costs at contaminated sites. *Journal of Water, Air and Soil Pollution* 167:365–386.

Keller, K.C., G. Van Der Kamp, and J.A. Cherry. 1989. A multiscale study of the permeability of a thick clay till. *Water Resources Research* 25:2299–2317.

Kibel, P.S. 1998. The urban nexus: Open space, brownfields, and justice. *Boston College Environmental Affairs Law Review* 25:589–618.

Loague, K., D.L. Corwin, and T.R. Ellsworth. 1998. The challenge of predicting nonpoint source pollution. *Environmental Science and Technology* 26:127–154.

Michigan Department of Environmental Quality. 2008. *Michigan Sites of Environmental Contamination.* Lansing, MI.

Michigan Department of Natural Resources. 1988. *List of Sites of Environmental Contamination.* Lansing, MI.

Mohrlok, U., C. Cata, and M. Bucker-Gittel. 2007. Impact on urban groundwater by wastewater infiltration into soils. In *Urban Groundwater—Meeting the Challenge*, ed. K.W.F. Howard, pp. 57–63. London, U.K.: Taylor & Francis.

Murray, J.E. and J.M. Bona. 1993. *Rouge River National Wet Weather Demonstration Project (RRNWWDP)*, Detroit: RRNWWDP.

Murray, K.S. 1996. Statistical comparison of heavy-metal concentrations in river sediments. *Journal of Environmental Geology* 27:54–58.

Murray, K.S. and D.T. Rogers. 1999a. Groundwater vulnerability, brownfield redevelopment and land use planning. *Journal of Environmental Planning and Management* 42:801–810.

Murray, K.S. and D.T. Rogers. 1999b. Evaluation of groundwater vulnerability in an urban watershed. *Proceedings of the 2nd International Congress on Water Resources and Environmental Research*, Brisbane, Australia, pp. 877–883.

Murray, K.S, A. Bazzi, C. Carter et al. 1997. Distribution and mobility of lead in soils at an outdoor shooting range. *Journal of Soil Contamination* 6:79–93.

Murray, K.S., M. Lybeer, D. Cauvet et al. 1999. Particle size and chemical control of heavy metals in bed sediment of the Rouge River, southeastern Michigan. *Journal of Environmental Science and Technology* 33:987–992.

Murray, K., D.T. Rogers, and M.M. Kaufman. 2006. Dissolved heavy metals in shallow ground water in a southeastern Michigan urban watershed. *Journal of the American Water Resources Association* 42(3):777–792.

Pierce, S.A., J.M. Sharp, and B. Garcia-Fresca. 2007. Evaluating groundwater allocation alternatives in an urban setting using a geographic information system data model and economic valuation technique. In *Urban Groundwater—Meeting the Challenge*, ed. K.W.F. Howard, pp. 197–211. London, U.K.: Taylor & Francis.

Robins, N., B. Adams, and S.S.D. Foster. 1994. Groundwater vulnerability mapping: The British perspective. *Hydrogeologie* 3:35–42.

Rogers, D.T. 1992. The importance of site observation and followup environmental site assessment—A case study. In *Groundwater Management Book 12*, pp. 563–573. Dublin, OH: National Groundwater Association.

Rogers, D.T. 1996. *Environmental Geology of Metropolitan Detroit*. Novi, MI: Clayton Environmental Consultants.

Rogers, D.T. 1997a. The influence of groundwater on surface water in Michigan's Rouge River watershed. *Proceedings of the American Water Resources Association, Conjunctive Use of Water Resources; Aquifer Storage and Recovery*, Long Beach, CA, pp. 173–180.

Rogers, D.T. 1997b. Surface geological map of the Rouge River watershed in southeastern Michigan. Wayne County, MI. 1:62,500. 2 Sheets.

Rogers, D.T. 1997c. Geologic sensitivity map of the Rouge River in southeastern Michigan. Wayne County, MI. 1:62,500. 1 Sheet.

Rogers, D.T. and K.S. Murray. 1997. Occurrence of groundwater in metropolitan Detroit. In *Groundwater in the Urban Environment*, ed. J. Chilton et al., pp. 155–161. Rotterdam, the Netherlands: Balkema.

Rogers, D.T. and K.S. Murray. 2002. The development and significance of a geologic sensitivity map of the Rouge River watershed in southeastern Michigan, USA. In *Geoenvironmental Mapping: Methods, Theory, and Practice*, ed. P.T. Bobrowsky, pp. 295–319. Rotterdam, the Netherlands: Balkema.

Rogers, D.T., K.S. Murray, and M.M. Kaufman. 2007. Assessment of groundwater contaminant vulnerability in an urban watershed in southeast Michigan, USA. In *Urban Groundwater—Meeting the Challenge*, ed. K.W.F. Howard, pp. 129–144. London, U.K.: Taylor & Francis.

Stiber, N.A., M.J. Small, and P.S. Fischbeck. 1995. The relationship between historic industrial Site use and environmental contamination. *Journal of the Air and Waste Management Association* 48:809–818.

Vuono, M. and R.P. Hallenbeck. 1995. Redeveloping contaminated properties. *Journal of Risk Management* 42:58–69.

Zaporozec, A. and T.T. Eaton. 1996. Ground-water resource inventory in urbanized areas. In *Hydrology and Hydrogeology of Urban and Urbanizing Areas*. American Institute of Hydrology (AIH) Annual Meeting Proceedings, St. Paul, MN.

Part II

Contamination

7

Common Contaminants in Urban Watersheds

7.1 Introduction

Pollution and *contamination* are now used synonymously to mean the introduction into the environment by humans of substances that are harmful or poisonous to human health and ecosystems (Van der Perk 2006). Over time the term pollution evolved to include not only substances, but also energy wastes such as heat, light, and noise. Because our concern in this book is primarily with chemical substances, we will use the term contamination throughout.

A chemical or substance becomes a *contaminant* when it is released into the environment either inadvertently or improperly—at the wrong place or in the wrong amounts. For example, milk becomes a contaminant when large quantities are released into a stream. In urban areas, contaminants are everywhere—in the air, soil, water, inside buildings, and in our homes. As stated in Chapter 4, most households contain chemicals that would be considered contaminants if they were released into the environment or disposed of improperly. These chemicals include the following (Figure 7.1):

- Cleaners
- Solvents
- Gasoline for lawn maintenance, snow removal, and other purposes
- Pesticides
- Herbicides
- Lawn fertilizers
- Some paints
- Oil
- Grease
- Dirt

In addition, the list of common household products shown in the following would become contaminants if they were improperly disposed:

- Computer equipment
- Televisions
- Some electrical equipment
- Most batteries
- Some building products
- Wood with certain applied preservatives or coatings

FIGURE 7.1
Household items that are potential contaminants. (From United States Geological Survey (USGS), *Volatile Organic Compounds in the Nation's Ground Water and Drinking-Water Supply Wells*, USGS Circular 1292, Reston, VA, 2006.)

Thousands of environmental contaminants exist, with the following categories of chemical and organic contaminants commonly present within urban areas:

- Volatile organic compounds (VOCs)
    - Dense nonaqueous phase liquids (DNAPLs)
    - Light nonaqueous phase liquids (LNAPLs)
- Polynuclear aromatic hydrocarbons (PNAs or PAHs)
- Semivolatile organic compounds (SVOCs)
- Polychlorinated biphenyls (PCBs)
- Pesticides and herbicides
- Heavy metals
- Others including the following:
    - Common fertilizers including nitrates, phosphorus, and potassium
    - Greenhouse gases
    - Carbon monoxide
    - Particulates (dust)
    - Ozone
    - Bacteria such as coliform bacteria
    - Viruses
    - Pharmaceuticals
    - Cyanide

- Asbestos
- Acids
- Bases
- Radioactive compounds
- Dioxins
- Emerging contaminants

We begin by briefly discussing the concept of contaminant toxicity. Understanding the general concepts of toxicity provides the basis for understanding the types, differences, and potential effects of exposure to contaminants in urban areas. The chapter concludes with a closer examination of each contaminant category listed above.

7.2 Contaminant Toxicity

Before discussing each contaminant group, a discussion of the meaning, complexity, and implications of toxicity posed by contaminants is appropriate. *Toxicity* or *potency* is the degree to which a chemical or substance is able to inflict damage to an exposed organism (USEPA 1989). Note that toxicity does not equal risk. The difference between toxicity and risk is based primarily upon the length of exposure to a chemical or substance, and whether the dosage received from this exposure is enough to cause harm. As we now know, toxic substances surround us and are present at most all locations on Earth. However, there is only risk if we are exposed to a substance or chemical long enough and at a high enough dose to cause harm. Toxicity does not assess risk—this determination is reserved for a risk assessment (Chapter 4).

There are three basic types of toxic categories:

1. Chemical or substance, including inorganic and organic substances such as acids and bases, flammable liquids, metals, etc.
2. Biological, including bacteria and viruses
3. Physical, including sound and vibration, heat and cold, light, radiation, etc.

For the purposes of this book, we are primarily concerned with chemical or substance toxicity.

Toxicity is measured by the effects on a whole organism, an individual organ, or the tissue of even a cell. Populations are most often used to measure toxicity since any one individual may have a different level of response to a toxin at a certain dose or concentration. The most common measure of chemical or substance toxicity is termed the LD_{50}; defined as the concentration or dose that is lethal to 50% of the population being tested (USEPA 1989). When direct data are not available, the LD_{50} is estimated by comparing the substance to other similar chemicals and organisms.

Another important measure of toxicity is the *lowest-observed-adverse-effect level* (LOAEL) or *threshold effect value*. The LOAEL is the lowest tested dose of a chemical or substance causing a harmful or adverse health effect (ATSDR 2009). Factors influencing the LOAEL include the following:

- The chemical's solubility in body fluids
- The particle size and state of the chemical
- Route of exposure
- Residence time of the chemical in the body
- Individual susceptibility

Dosage units are expressed as the mass of chemical per unit mass of body weight in mg/kg. These units are employed to evaluate the relative toxicities between animals of different species and size. To develop the most accurate human exposure limits, extensive animal studies are initially used to establish extrapolated human dosage limits for a specified chemical, and these estimates are refined by human health studies conducted on individuals known to have been exposed to the same chemical. The Occupational, Safety, and Health Administration (OSHA) has established a list of exposure limits for over 600 individual chemicals and refers to exposure limits as permissible exposure limits (PELs) (ACGIH 2009).

Most chemicals or substances have the potential to exhibit some adverse health effect on humans or other organisms, given a certain set of circumstances. *Adverse health effect* is defined as a change in body function or cell structure potentially leading to disease or health problems (ATSDR 2009).

The effect that a specific chemical or substance may exhibit on a living organism is dependent upon the following factors (USEPA 1989a; USEPA 2005):

- Nature of the chemical or substance
- Concentration
- Route of exposure
- Length of time of exposure
- Individual susceptibility

Chemicals and substances enter the human body through three *routes of exposure*—inhalation, ingestion, and dermal adsorption (USEPA 1989a). When exposure occurs, chemicals may produce one or more of these symptoms:

- Tissue irritation
- Eye irritation
- Rash
- Dizziness
- Bleeding
- Hair loss
- Loss or depressed vision
- Loss or depressed hearing

- Nausea
- Anxiety
- Narcosis
- Headache
- Vomiting
- Diarrhea
- Pain
- Fever
- Tremors
- Shortness or difficulty in breathing
- Psychotic behavior
- Euphoria
- Cancer
- Death
- And others

Human response to exposure of a chemical is described as acute or chronic. An *acute* response is generally characterized as a single high dose with rapid onset and disappearance of symptoms. *Chronic* response involves a stimulus lingering for a period of time (USEPA 2009a). Exposure to a contaminant may not trigger an immediate response. Instead, there is a *latency period*—a duration of time without observable effects. Certain chemicals or substances such as asbestos and forms of mercury have a latency period of up to 10–20 years (USEPA 2005).

Exposure to some chemicals or substances may result in the development of cancer, and any chemical, substance, radionuclide, or radiation contributing to the development or propagation of cancer is termed a carcinogen (United States Department of Health and Human Services 2005). *Cancer* exists when cells in the body become abnormal and grow or multiply out of control (ATSDR 2009). There is also often some selectivity involved with chemical exposure, as some chemicals may target certain organs, parts of the body, or reproduction. For example, *teratogens* may have an adverse affect on a developing fetus, *mutagens* may induce genetic changes that could affect future generations through reproduction, and *hepatotoxins* are chemicals posing a risk of liver damage.

Carcinogens are grouped into five general categories (ACGIH 2009):

- Group A1: Confirmed human carcinogen
- Group A2: Suspected human carcinogen
- Group A3: Confirmed animal carcinogen with unknown relevance to humans
- Group A4: Not classified as a human carcinogen
- Group A5: Not suspected as a human carcinogen

It is much more difficult to evaluate the toxicity of a mixture of contaminants compared to a single chemical compound or substance. This is because the interaction of a mixture of contaminants may produce enhanced or diminished effects. Common mixtures include gasoline, often containing more than 250 individual chemical compounds, industrial wastes, and a malfunctioning sewage treatment plant, typically discharging chemical and biological contaminants (USEPA 2008a).

The following sections briefly describe the major contaminant categories encountered in urban areas. For each contaminant category, specific contaminants, their common uses, physical attributes, and toxicities are presented.

7.3 Volatile Organic Compounds

VOCs are organic compounds that generally volatilize or evaporate readily under normal atmospheric pressure and temperatures (USGS 2006; USEPA 2008a). Table 7.1 is a list of the common VOCs and includes those sought out and analyzed when conducting a subsurface environmental investigation (USEPA 2008b). Included in Table 7.1 is the CAS registry number for each chemical compound. The CAS registry is maintained by the American Chemical Society and contains over 50 million disclosed substances, each with a unique numerical identifier and common use (American Chemical Society 2009).

VOCs have a high vapor pressure (evaporate quickly), low-to-medium solubility, and low molecular weight (USGS 2006). Most VOCs are considered toxic or harmful to humans and other organisms (USEPA 2009b). In addition, many VOCs are flammable and must be handled with extreme care.

The use of VOCs in the United States increased tenfold from 1945 to 1985 (USGS 2006). Some VOCs have had, and continue to have, very heavy usage. An example is gasoline, which contains numerous VOC compounds and its production and use continues to increase. Table 7.1 demonstrates many common uses of VOCs. As noted, many VOCs are flammable and are components of gasoline and other fuels such as diesel fuel, kerosene, and fuel oil. In addition, VOCs are commonly used as solvents, ingredients in paints, paint thinners, in the manufacturing process of pharmaceuticals, for caffeine extraction, nail polish removers, dry cleaning chemicals, mothballs, pesticides and fumigants, adhesives, refrigerants, and as a by-product of chlorination for water purification.

For hydrogeological purposes, VOCs are separated into two general categories—DNAPLs and LNAPLs (USGS 2009a,b). An additional category of VOCs are the trihalomethanes (THMs). Each is discussed separately in the following sections.

7.3.1 Light Nonaqueous Phase Liquids

LNAPLs are liquids lighter than water that do not mix or dissolve in water readily. Common LNAPL compounds include the VOCs: benzene, ethyl benzene, toluene, and xylenes. These compounds are also often symbolized by the acronym BTEX. BTEX compounds are common components of gasoline and are often used as indicator analytes when evaluating whether there has been a release of gasoline or other similar fuels to the environment (USEPA 1989b; USGS 2009b). LNAPLs also include other compounds lighter than water such as PAHs, and are discussed in the next section.

TABLE 7.1
Common Uses of VOCs

Compound (Listed Alphabetically)	CAS Registry Number	Common Uses
Acetone[a]	67-64-1	Nail polish remover, paint thinner, laboratory chemical
Benzene[b,c]	71-43-2	Gasoline component, solvent, pharmaceutical, dyes, and Plastics
Bromodichloromethane	75-27-4	Trihalomethane (by product of water purification)
Bromoform	75-25-2	Trihalomethane (by product of water purification)
Bromomethane	75-83-9	Fumigant
Carbon disulfide[d]	75-15-0	Cellulose manufacturing, soil fumigant
Carbon tetrachloride[e]	56-23-5	Solvent, dry cleaning, agriculture, formerly used in fire
Chlorobenzene	108-90-7	Extinguishers
Chloroethane	75-00-3	Solvent, insecticides
Chloroform	67-66-3	Solvent, pharmaceutical manufacturing
Chloromethane	74-87-3	Trihalomethane (by product of water purification)
Dibromochloromethane	124-48-1	Solvent, agricultural chemicals and cellulose
1,1-Dichloroethane	75-34-3	Trihalomethane (by-product of water purification)
1,2-Dichloroethane	107-06-2	Solvent, manufacture of plastic wrap and adhesives
1,1-Dichloroethene	75-35-4	Solvent, paint ingredient, fumigant, vinyl chloride production
cis-1.2-Dichloroethene	156-59-2	Manufacturing of plastics, adhesives, refrigerants
trans-1,2-Dichloroethene	156-60-5	Fat extraction from meat & fish, refrigerants, pharmaceuticals
Dichloromethane	79-09-2	Fat extraction from meat & fish, refrigerants, pharmaceuticals
1,2-Dichloropropane	78-87-5	Paint and stain remover, aerosol propellant, caffeine extraction
cis-1,3-Dichloropropene	10061-01-5	Solvent, stain remover, former soil fumigant
trans-1,3-Dichloropropene	10061-02-06	Solvent
Ethyl benzene	100-41-4	Solvent
2-Hexanone[f]	591-78-6	Gasoline component, solvent, styrene manufacturing
Methyl ethyl ketone (MEK)	78-93-3	Paint and paint thinner, used to dissolve oil and waxes
Methyl isobutyl ketone (MIBK)	108-10-1	Solvent
Methyl-tert-butyl ether (MTBE)	1634-04-4	Solvent, used in metal extraction
Naphthalene[g]	91-20-3	Gasoline additive, formerly used to dissolve gallstones
Styrene	100-42-5	Gasoline component, mothballs, insecticide
1,1,2,2-Tetrachloroethane	79-34-5	Gasoline component, coating, paint, rubber, adhesives
Tetrachloroethene (PCE)[e]	127-18-4	Solvent, ingredient in paints and pesticides
Toluene[h]	108-88-3	Solvent, dry cleaning, textile processing
1,1,1-Trichloroethane[e]	71-55-6	Gasoline component, solvent
1,1,2-Trichloroethane	79-00-5	Solvent, cosmetic ingredient, aerosol products, textile Processes
Trichloroethene (TCE)[e]	79-01-6	Solvent, manufacturing of plastic wraps

(*continued*)

TABLE 7.1 (continued)

Common Uses of VOCs

Compound (Listed Alphabetically)	CAS Registry Number	Common Uses
Vinyl acetate	108-05-4	Solvent, caffeine extraction, dry cleaning, paint and ink, rubber
Vinyl chloride	75-01-4	Polymer
Xylenes[i]	1330-20-7	Rubber, paper, and glass industries, PVC manufacturing
		Gasoline component, paint thinner ingredient, plasticizer

Source: United States Geological Survey (USGS), *Volatile Organic Compounds in the Nation's Ground Water and Drinking-Water Supply Wells*, USGS Circular 1292, Reston, VA, 2006; United States Environmental Protection Agency, *Latest Findings on National Air Quality: Status and Trends Through 2006*, EPA454/R-07-007, Research Triangle Park, NC, 2008c.

[a] Levy (2009).
[b] Lide (2008).
[c] ATSDR (2007a).
[d] Holleman and Wiberg (2001).
[e] Doherty (2000).
[f] ATSDR (1995a).
[g] ATSDR (2005a).
[h] ATSDR (2001).
[i] ATSDR (2007b).

BTEX compounds are aromatic hydrocarbons with a *benzene ring* forming the backbone of its molecular structure (Jensen 2009). As depicted within the compound benzene shown in Figure 7.2, the benzene ring consists of a hexagonal arrangement of six carbon atoms located at the vertices. Each atom is bonded to its adjacent atoms by a single covalent bond and by an unusual ring bond of electrons shared by all six carbon atoms. The chemical formulas and carcinogenicity for the BTEX compounds are listed in Table 7.2.

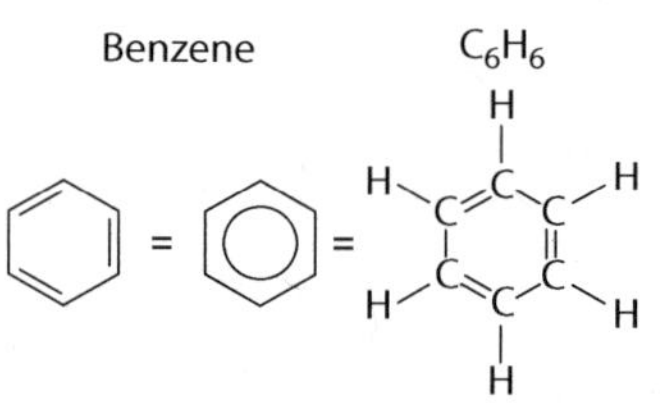

FIGURE 7.2
Benzene and the benzene ring.

TABLE 7.2

Chemical Formulas and Carcinogenicity for Each BTEX Compound

Compound	Carcinogenicity Rating	Chemical Formulas
Benzene	Group: A1	C_6H_6
Toluene	Group: A4	C_7H_8 or $C_6H_5CH_3$
Ethyl Benzene	Group: A1	C_8H_{10}
Xylenes[a]	Data not adequate to currently evaluate	C_8H_{10}, $C_6H_4(CH_3)_2$, or $C_6H_4C_2H_6$
MTBE	Data not adequate to currently evaluate	$C_5H_{12}O$

Source: United States Environmental Protection Agency, Integrated Risk Information System (IRIS), http://www.epa.gov/ncea/iris/intro.htm (accessed October 21, 2009), 2009b.

[a] Three isomers exist that include *o*-, *m*-, and *p*-xylene.

Health effects from exposure to benzene include headaches, dizziness, drowsiness, rapid heart rate, tremors, and unconsciousness. Exposure to high levels can result in death (ATSDR 2007a). Health effects from exposure to toluene include tiredness, confusion, loss of appetite, memory loss, loss of color vision, and nausea (ATSDR 2001); whereas exposure to ethyl benzene may cause irreversible damage to the inner ear and hearing loss, dizziness, and kidney damage (ATSDR 2007b). You are not safe around xylenes either, as exposure to them can result in headaches, lack of muscle coordination, dizziness, and confusion. Exposure to very high levels of xylenes can cause unconsciousness (ATSDR 2007c).

Methyl-tert-butyl ether (MTBE) is a LNAPL compound produced exclusively as a gasoline additive, and it belongs to a group of chemicals known as "oxygenates" because they raise the oxygen content of gasoline and thereby raise the octane level. MTBE is a colorless liquid at room temperature, and is very volatile and flammable (USEPA 2007). The purpose of adding MTBE to gasoline was to increase the efficiency of combustion in automobiles enabling them to run cleaner and emit fewer pollutants, thus improving air quality in urban areas. However, the use of MTBE has declined recently because (1) of health concerns and (2) MTBE has been detected in many groundwater aquifers used as drinking water sources in the United States (USEPA 2007). Many states now ban MTBE because of its propensity to contaminate groundwater and the high costs incurred to remove it from groundwater (United States Department of Energy 2009).

MTBE is also used to dissolve gallstones. Patients treated for gallstones using MTBE have MTBE delivered directly to the gall bladder through surgically inserted tubes. Health effects from exposure to MTBE may include nose and throat irritation, headaches, nausea, dizziness, and mental confusion. Currently, evidence suggesting that MTBE may cause cancer is lacking (ATSDR 1997; USEPA 2009b).

7.3.2 Dense Nonaqueous Phase Liquids

DNAPLs are liquids denser than water, that do not mix or dissolve readily in water (USGS 2006). DNAPL compounds include many common solvents and coal tar (Suthersan and Payne 2005). They are also commonly referred to as chlorinated solvents or halogenated VOCs because chlorine is in the atomic structure, and the most common uses of these compounds are for cleaning and degreasing (USGS 2006). Chlorinated solvents have been in use for nearly 100 years. They are still widely used by industry, and many household products contain them (Figure 7.3) (USGS 2006).

Halogenated VOCs are a group of organic compounds with a halogen atom as part of its molecular structure. Halogens include the elements Fluorine (F), Chlorine (Cl), Bromine (Br), or Iodine (I). Part of the uniqueness of halogenated VOCs is they tend to have a very weak tendency to form hydrogen bonds with water. This lack of affinity for water means the halogenated VOCs—especially those with fluorine or chlorine—tend to be hydrophobic and have low solubility (Suthersan and Payne 2005).

Common halogenated VOCs include the following:

- Tetrachloroethene (PCE)
- Trichloroethene (TCE)
- *cis*-1,2-Dichloroethene (DCE)

FIGURE 7.3
Household products containing chlorinated solvents. (From United States Geological Survey (USGS), *Volatile Organic Compounds in the Nation's Ground Water and Drinking-Water Supply Wells*, USGS Circular 1292, Reston, VA, 2006.)

- *trans*-1,2-Dichloroethene (DCE)
- Vinyl chloride
- 1,1,1-Trichloroethane (TCA)
- 1,1-Dichloroethene (DCE)
- Methylene chloride
- Carbon tetrachloride
- Chloroform
- Chlorobenzene
- 1,2-Dichlorobenzene

Many VOCs halogenated with chlorine have high electro negativities and form rather strong bonds with the carbon atoms in their structure. In addition, the substitution of a hydrogen atom by a chlorine atom enhances the inertness of the molecule. This inertness results in many halogenated VOCs being rather persistent when released into the environment. A degradation sequence, however, does exist for a group of the most commonly used chlorinated solvents, and is shown in Figure 7.4. This degradation sequence becomes very important when we discuss the fate and migration of these compounds in the next chapter.

Table 7.3 lists the chemical formulas and carcinogenicity ratings of selected chlorinated solvents (USEPA 2009b).

Health effects from overexposure to most chlorinated solvents include lung irritation, difficulty walking and speaking, poor coordination, dizziness, headache, nausea, sleepiness, unconsciousness, and even death (ATSDR 1997a, 2003a, 2006).

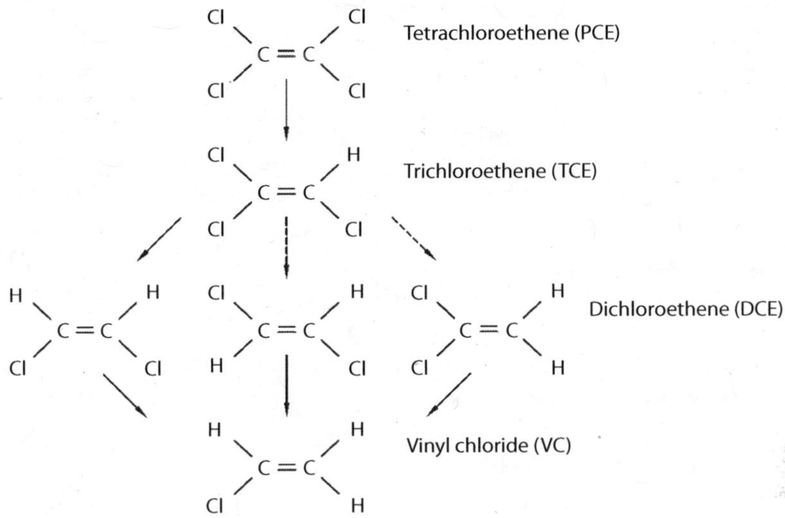

FIGURE 7.4
Degradation sequence of PCE.

TABLE 7.3
Chemical Formula and Carcinogenicity for Each BTEX Compound

Compound	Carcinogenicity Rating	Chemical Formula
Tetrachloroethene (PCE)	Currently under review	C_2Cl_4
Trichloroethene (TCE)	Currently under review	C_2HCl_3
cis-1,2-Dichloroethene (DCE)	Group: A4	$C_2H_2Cl_2$
trans-1,2-Dichloroethene (DCE)	Group: A4	$C_2H_2Cl_2$
Vinyl chloride	Group: A1	C_2H_3Cl
1,1,1-Trichloroethane (TCA)	Inadequate information	$C_2H_3Cl_3$ or CH_3CCl_3
1,1-Dichloroethene (DCE)	Group: A2	$C_2H_2Cl_2$
Methylene chloride	Inadequate information	CH_2Cl_2
Carbon tetrachloride	Group: A2	CCl_4
Chloroform	Group: A2	$CHCl_3$
Chlorobenzene	Inadequate information	C_6H_5Cl
1,2-Dichlorobenzene	Inadequate information	$C_6H_4Cl_2$

Source: United States Environmental Protection Agency, Integrated Risk Information System (IRIS), http://www.epa.gov/ncea/iris/intro.htm (accessed October 21, 2009), 2009b.

7.3.3 Trihalomethanes

THMs are a group of VOCs in which three of the four atoms of methane (CH_4) are replaced by halogen atoms. Many THMs are used as refrigerants, solvents, and are also by-products produced during water purification and chlorination. During water treatment, THMs can form as a by-product when chlorine and bromine are used to

TABLE 7.4

Chemical Formula and Carcinogenicity for Common THMs

Compound	CAS Registry Number	Carcinogenicity Rating	Chemical Formula
Chloroform	67-66-3	Group: A2	$CHCl_3$
Bromodichloromethane	75-27-4	Group: A2	$CHBrCl_2$
Dibromodichloromethane	124-48-1	Under Review	$CHBr_2Cl$
Bromoform	75-47-8	Under Review	$CHBr_3$

Source: United States Environmental Protection Agency, Integrated Risk Information System (IRIS), http://www.epa.gov/ncea/iris/intro.htm (accessed October 21, 2009), 2009b.

disinfect water for drinking or recreational use; they form from a reaction between chlorine and/or bromine with organic matter (EPA 2006). Chloroform, a common THM, has also been used as an anesthetic and is often used in swimming pools as a disinfectant (ATSDR 1997b). The most common THMs, CAS registry numbers, carcinogenicity rating, and chemical formula are listed in Table 7.4 (ATSDR 1989, 1997b). The structure of chloroform and bromodichloromethane are shown in Figures 7.5 and 7.6.

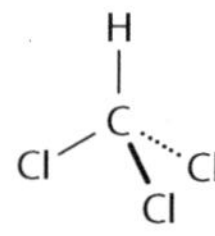

FIGURE 7.5 Molecular structure of chloroform.

According to USEPA (2009b), THMs may cause adverse health effects at high concentrations. Exposure to high concentrations of bromoform may interfere with normal brain function and cause sleepiness (ATSDR 2005). USEPA has established maximum allowable concentration of THMs in drinking water at 80 μg/L for the combined concentration of chloroform, bromoform, bromodichloromethane, and dibromochloromethane.

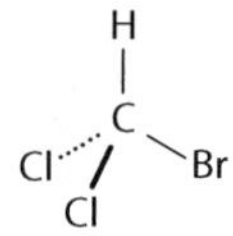

FIGURE 7.6 Molecular structure of bromodichloromethane.

7.4 Polynuclear Aromatic Hydrocarbons

PNAs or PAHs are a group of compounds formed synthetically or naturally during the incomplete combustion of coal, oil, tar, gas, wood, garbage, and may also be present in tobacco, medicines, dyes, plastics, pesticides, and charbroiled meat (United States Department of Health and Human Services [USDHH] 1995). Other sources of PAHs include heavy petroleum, including diesel fuels, kerosene, aviation fuels, heavy home-heating oils, oils, waste oil, and many lubricants (Missouri Department of Natural Resources 2006). Overall, there are more than 100 distinct PAH compounds.

Chemically, PAHs are composed of multiple benzene rings (Figure 7.7), are lighter than water, and are generally not very soluble in water. In addition, PAHs strongly attach themselves or sorb to soil grains, especially organic soils and clay (USEPA 1989). Because PAHs are not very soluble and are lighter than water, they are considered LNAPLs.

With two rings, naphthalene is the smallest of the PAHs. Naphthalene is a common PAH compound, and is characterized by its distinctive odor and common use in mothballs. Anthracene and phenanthrene are composed of three rings. Phenanthrene is an *isomer* of anthracene and they share many physical properties except boiling point and water solubility (Fetzer 2000). With its five rings, benzo(a)pyrene is the largest PAH molecule, and is the only known human carcinogen of the four compounds shown in Figure 7.7 (Group:A1).

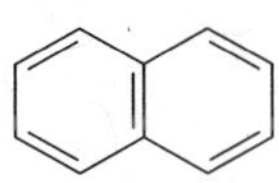

Naphthalene

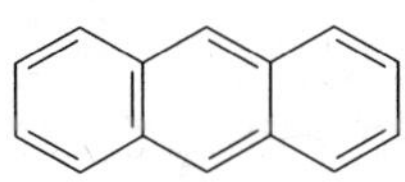

Anthracene

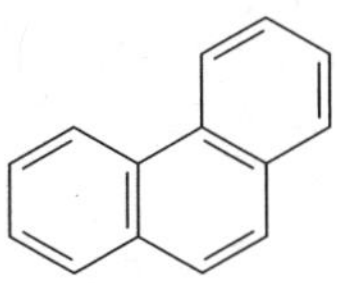

Phenanthrene

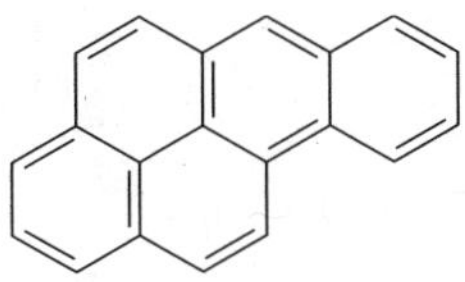

Benzo(a)pyrene

FIGURE 7.7
Molecular structure of the select PAH compounds; naphthalene, anthracene, phenanthrene, and benzo(a)pyrene. (From Fetzer, J.C., *The Chemistry and Analysis of Large Polycyclic Aromatic Hydrocarbons*, New York, John Wiley & Sons, 2000.)

TABLE 7.5
Common PAH Compounds

Compound (Alphabetically)	CAS Registry Number	Common Uses
Acenaphthene	83-32-9	Dyes, pesticides, pharmaceuticals, oil component
Acenaphthylene	208-96-8	Automobile exhaust, oil and lubricant component
Anthracene	120-12-7	Dyes, stains, wood preservatives, insecticides, oil
Benz[a]anthracene	56-55-3	Automobile exhaust, oil, and lubricant component
Benzo[b]fluoranthene	205-99-2	Automobile exhaust, oil, and lubricant component
Benzo[k]fluoranthene	92-24-0	Organic semiconductor, oil component
Benzo[ghi]perylene	198-55-0	Photoconductor, pitch, coal tar, tobacco smoke
Benzo[a]pyrene	50-32-8	Pitch, coal tar, automobile exhaust, tobacco smoke
Chrysene	218-01-9	Wood preservative, oil component, dyes
Dibenzo[a,h]anthracene	53-70-3	Wood preservative, insecticides, oil component
Fluoranthene	206-44-0	Automobile exhaust, oil component
Fluorene	86-73-7	Automobile exhaust, dyes, plastics, pesticides
Indeno[1,2,3-cd]pyrene	193-39-5	Pitch, coal tar, oil component
Naphthalene	91-20-3	Moth balls, oil and gasoline component
Phenanthrene	85-01-8	Cigarette smoke, oil, and lubricant component
Pyrene	129-00-0	Dyes, coal tar and pitch, oil component

Source: Agency for Toxic Substances and Disease Registry (ATSDR). *Polycyclic Aromatic Hydrocarbons*. General Contaminant Class. ATSDR ToxFAQs. Atlanta, GA, 1996; United States Environmental Protection Agency. Integrated Risk Information System (IRIS). http://www.epa.gov/ncea/iris/intro.htm, 2009b.

Common PAH compounds are listed in Table 7.5 along with their CAS registry number. Table 7.6 shows the carcinogenicity and chemical formula for each of the common PAHs. Animal studies have indicated PAHs can cause harmful effects on the skin, body fluids, and ability to fight disease (ATSDR 1996). Health effects from exposure to naphthalene may cause damage to red blood cells. Exposure to high levels of naphthalene may cause nausea, vomiting, diarrhea, dizziness, blood in the urine, and a yellow color to the skin (ATSDR 2005b).

7.5 Polychlorinated Biphenyls

PCBs are a group of synthetically produced compounds (USEPA 2009c). PCBs were produced in the United States from 1929 until 1979 when they were banned because of human

TABLE 7.6

Chemical Formula and Carcinogenicity for the Common PAH Compounds

Compound (Alphabetically)	Carcinogenicity Rating	Chemical Formula
Acenaphthene	Inadequate information	$C_{12}H_{10}$
Acenaphthylene	Inadequate information	$C_{12}H_8$
Anthracene	Group: A4	$C_{14}H_{10}$
Benz[a]anthracene	Group: A2	$C_{18}H_{12}$
Benzo[b]fluoranthene	Group: A2	$C_{20}H_{12}$
Benzo[k]fluoranthene	Inadequate information	$C_{18}H_{12}$
Benzo[ghi]perylene	Group: A4	$C_{20}H_{12}$
Benzo[a]pyrene	Group: A1	$C_{20}H_{12}$
Chrysene	Group: A2	$C_{18}H_{12}$
Dibenzo[a,h]anthracene	Group: A2	$C_{22}H_{14}$
Fluoranthene	Group: A1	$C_{16}H_{10}$
Fluorene	Group: A4	$C_{13}H_{10}$
Indeno[1,2,3-cd]pyrene	Group: A2	$C_{22}H_{10}$
Naphthalene	Group: A3	$C_{10}H_8$
Phenanthrene	Group: A2	$C_{14}H_{10}$
Pyrene	Group: A4	$C_{16}H_{10}$

Source: United States Department of Health and Human Services (USDHH), *Toxicological Profile for Polycyclic Aromatic Hydrocarbons*, Agency for Toxic Substances and Disease Registry, Atlanta, GA, 1995; United States Environmental Protection Agency, Integrated Risk Information System (IRIS), http://www.epa.gov/ncea/iris/intro.htm (accessed October 21, 2009), 2009b.

and environmental health effects (USEPA 2009d). Some of the physical properties of PCBs include the following (Phillips 1986; Barbalace 2009):

- Very low solubility in water
- High relative solubility in organic solvents, fats, and oil
- Low vapor pressures
- Strongly sorb to soil grains, especially organic-rich soils and clays
- Stable compounds
- Do not readily degrade
- Not flammable
- Resistant to oxidation, reduction, addition, elimination, and electrophilic substitution
- High boiling point

Due to the physical properties listed above, PCBs were widely used in many different industrial and commercial products, including the following (USEPA 2009c):

- Transformers and capacitors
- Other electrical equipment including voltage regulators, switches, reclosers, bushings, and electromagnets

- Oil used in motors and hydraulic systems
- Older electrical devices containing capacitors
- Fluorescent light ballasts
- Cable insulation
- Thermal insulation material including fiberglass, felt, foam, and cork
- Adhesives and tape
- Oil-based paint
- Caulking
- Plastics
- Carbonless copy paper
- Floor finish

PCBs are mixtures of chlorinated organic compounds called congeners (ATSDR 2001b). A *congener* is a related compound or compounds in a specific chemical family. There are a large number of congeners, with about 209 found in PCB mixtures (USEPA 2009d). PCBs were produced synthetically through electrophilic chlorination of a biphenyl molecule with chlorine gas. In the United States, PCBs are commonly known under the trade name aroclors. *Aroclors* are mixtures of PCBs distinguished by a four digit numbering system. The first two digits refer to the number of carbon atoms in a PCB molecule, and the second two digits indicate the percentage of chlorine by mass in the mixture (USEPA 2009d). For example, PCB aroclor 1254 means the mixture contains 12 carbon atoms and is 54% chlorine by weight (ATSDR 2001c).

As the degree of chlorination increases, the melting point increases and the vapor pressure and solubility decrease. The general chemical formula for PCBs is the following:

$$C_{12}H_{10-x}Cl_x$$

where x = 1–10.

Common PCB aroclors are listed in Table 7.7 (Phillips 1986; USEPA 2009c).

Figure 7.8 shows the basic structure of a PCB molecule.

TABLE 7.7

Common PCB Aroclors

PCB Aroclor	CAS Registry Number	Percent Chlorine	Average Number of Chlorine Atoms per Molecule	Average Molecular Weight
Aroclor 1016	12674-11-2	15.5–16.5	1.05	160
Aroclor 1221	11104-28-2	20.5–21.5	1.15	192
Aroclor 1232	11141-16-5	31.5–32.5	2.04	221
Aroclor 1242	53469-21-9	42	3.10	261
Aroclor 1248	12672-29-6	48	3.90	288
Aroclor 1254	11097-69-1	54	4.96	327
Aroclor 1260	11096-82-5	60	6.30	372
Aroclor 1262	37323-23-5	62	6.80	389
Aroclor 1268	11100-14-4	68	8.70	453

Currently, there is inadequate information to establish whether PCBs are a human carcinogen (USEPA 2009b), but the World Health Organization (2003) suspects they are (Group: A2). Notwithstanding cancer, they pose significant other dangers to human health. PCBs have been shown to cause cancer in laboratory animals, and have also been shown to cause a number of serious noncancer health effects in animals, including the immune system, reproductive system, nervous system, endocrine system, and others (USEPA 2009e).

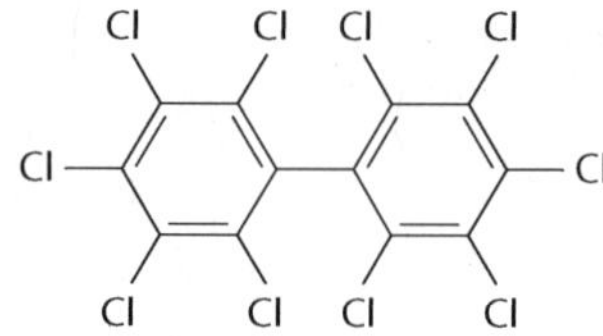

FIGURE 7.8
Basic PCB structure.

7.6 Semivolatile Organic Compounds

SVOCs are a group of organic-based compounds much less volatile than VOCs. A variety of SVOCs are used in clothing and building materials to provide flexibility, water resistance, or stain repellence, and to inhibit ignition or as flame retardants (ATSDR 2002a). Common groups of organic compounds associated with SVOCs are phthalates, phenols, amines, and esters.

7.6.1 Phthalates

The name phthalates is derived from phthalic acid—itself derived from naphthalene and previously discussed earlier in this chapter. Phthalates exhibit low water solubility, high oil solubility, and low volatility (Fetzer 2000). There are approximately 25 distinctive phthalate compounds, including the common ones shown in the following:

- Diethyl phthalate (DEP)
- Di-*n*-butyl phthalate (DBP)
- Bi or Bis(2-ethylhexyl) phthalate (DEHP)

The structural formula and CAS registration number are each of the three common phthalate compounds are listed in Table 7.8 (USEPA 2009b).

Figure 7.9 depicts the general structure of a phthalate molecule, and Figure 7.10 shows the structure of Di(2-exthylhexyl) phthalate (USEPA 2009b).

TABLE 7.8
Structural Formula and CAS Registration Numbers for Three Common Phthalates

Name	CAS Registration Number	Chemical Formula
Diethyl phthalate (DEP)	84-66-2	$C_6H_4(COOC_2H_5)_2$
Di-*n*-butyl phthalate (DBP)	84-74-2	$C_6H_4[COO(CH_2)_3CH_3]_2$
Di or Bis(2-ethylhexyl) phthalate (DEHP)	117-81-7	$C_6H_4[COOCH_2CH(C_2H_5)(CH_2)_3CH_3]_2$

Source: United States Environmental Protection Agency, Integrated Risk Information System (IRIS), http://www.epa.gov/ncea/iris/intro.htm (accessed October 21, 2009), 2009b.

where
O = oxygen
R and R′ = C_nH_{2n+1}
n = 4–15

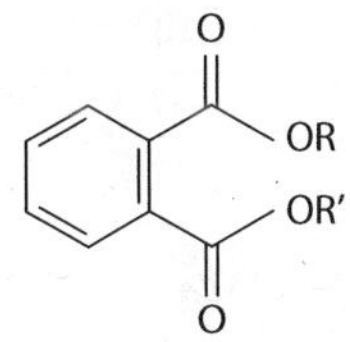

FIGURE 7.9
Basic phthalate molecule.

Phthalate compounds are primarily used as plasticizers. They are added to plastics to increase their flexibility, transparency, durability, and longevity, and are also used to soften polyvinyl chloride (PVC)—a common pipe material. More than a billion pounds of phthalates are used in a variety of products every year (ATSDR 2002a).

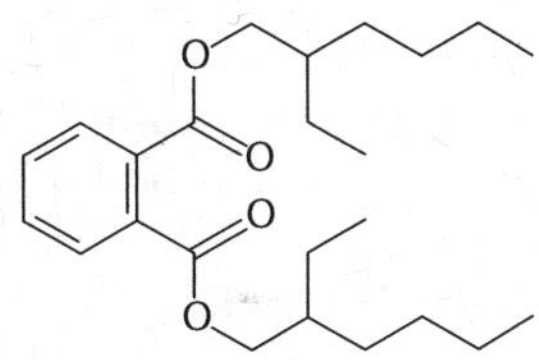

FIGURE 7.10
Molecular structure of Di(2-exthylhexyl) phthalate.

The variety of products using phthalates includes the following (ATSDR 2002a):

- Coatings
- Plastics
- Pharmaceuticals
- Gelling agents
- Stabilizers
- Dispersants
- Lubricants
- Binders
- Children's toys
- Modeling clay
- Eye shadow
- Moisturizer
- Nail polish
- Perfume
- Hair spray
- Detergents
- Packaging
- Waxes
- Paints
- Printing inks
- Fishing lures
- Caulk
- Shower curtains
- Vinyl upholstery
- Automobile interiors
- Food containers
- Many others

According to ATSDR (2002a) and USEPA (2009b), DEHP is a suspected human carcinogen (Group: A2), while DEP and DBP are not classified as human carcinogens (Group: A4).

According to ATSDR, DEHP is not toxic at the low levels usually present in the environment. In animals, DEHP has been found to damage the liver and kidneys and has affected the ability to reproduce in some animal species.

7.6.2 Phenol

Phenol is a group of organic compounds with a hydroxyl group (–OH) attached to a carbon atom in a benzene ring. Phenol compounds do occur naturally, and their presence in plant foliage discourages herbivores from consuming the plant material (Fetzer 2000; McMurry 2009). Figure 7.11 is a diagram showing the basic structure of a phenol molecule. The simplest phenol compound is carbolic acid (C_6H_5OH); also called phenol.

Phenol combined with formaldehyde forms a widely used polymer typically referred to as a phenolic resin. Phenolic resins have a wide range of industrial and commercial applications including the following:

- Billiard balls
- Countertops
- Plastic
- Electrical components
- Coatings
- Composites
- Abrasives
- Adhesives
- Felt bonding
- Foundry applications
- Friction products
- Refactory products
- Rubber additives used as reinforcing agents

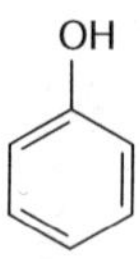

FIGURE 7.11 Basic structure of phenol.

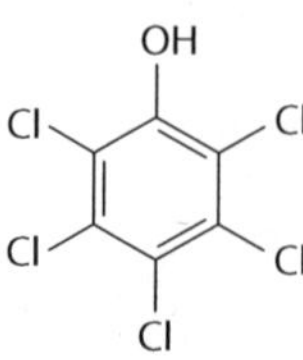

FIGURE 7.12 Basic molecular structure of pentachlorophenol.

Pentachlorophenol (PCP) is a phenol compound with the addition of chlorine into its molecular structure (Figure 7.12). PCP has been used as an herbicide, insecticide, fungicide, algaecide, and disinfectant. Other uses include leather, masonry products, and wood preservatives; and utility poles often employ PCP as a wood preservative (USEPA 2006a).

Bisphenol A or BPA is an organic phenolic compound with two phenol functioning groups. Figure 7.13 shows the basic structure of BPA. BPA is a very common compound and is used heavily in the production of plastic products including bottles for drinking water and numerous other consumer products (USDHH 2008). Some concern about health effects from exposure to BPA have been expressed by the USDHH (2008), as type 7 plastics (polycarbonates) may leach PBA into the liquids they hold. In 2009, Canada banned the use of PBA in polycarbonates in baby bottles (Carwile et al. 2009).

Table 7.9 lists the CAS registry number, structural formula, and carcinogenicity rating for phenol, PCP, and BPA.

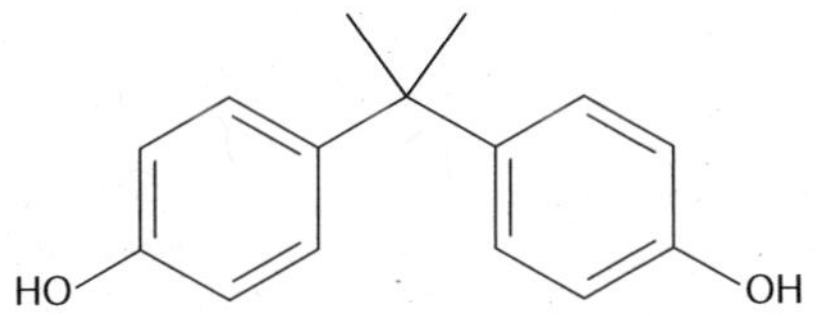

FIGURE 7.13
Basic structure of BPA.

TABLE 7.9
Data for Phenol, PCP, and BPA

Name	CAS Registration Number	Chemical Formula	Carcinogenicity Rating
Phenol	108-95-2	C_6H_5OH	Group: A4
Pentachlorophenol	87-86-5	$C_6H_4Cl_5O$	Group: A2
Bisphenol A (BPA)	80-05-7	$C_{15}H_{16}O_2$	Currently Under Evaluation

Source: United States Environmental Protection Agency, Integrated Risk Information System (IRIS), http://www.epa.gov/ncea/iris/intro.htm (accessed October 21, 2009), 2009b.

7.6.3 Amines

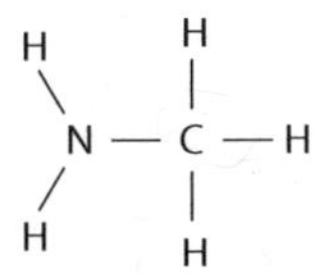

FIGURE 7.14
Basic molecular structure of methylamine.

Amines are organic compounds that contain nitrogen and are *basic.* Three widely used amine compounds include methylamine, dimethylamine, and trimethylamine, and they are prepared by the reaction of ammonia with methanol in the presence of a silicoaluminate catalyst (McMurry 2009). Amines are used in the manufacturing of dyes, plant growth regulators, resins, as a precursor in the manufacturing of tires, pharmaceuticals (ephedrine), pesticides, and surfactants (ATSDR 1999). Trimethylamine also forms naturally from decomposing plant and animal matter. Amines are often a gas at room temperature and pressure and are very soluble in water.

The basic structure of methylamine is shown in Figure 7.14. Table 7.10 presents the structural formula, CAS registration number, and carcinogenicity rating for methylamine, dimethylamine, and trimethylamine (USEPA 2009b).

7.6.4 Esters

Esters are a group of more than 50 compounds known collectively as acid derivatives. Ester compounds contain a modified carboxylic acid group (–COOH), in which the acidic

TABLE 7.10
Data for Methylamine, Dimethylamine, and Trimethylamine

Name	CAS Registration Number	Chemical Formula	Carcinogenicity Rating
Methylamine	74-89-5	CH_5N	Not Available
Dimethylamine	124-40-3	C_2H_7N	Not Available
Trimethylamine	75-50-3	C_3H_9N	Not Available

Source: United States Environmental Protection Agency, Integrated Risk Information System (IRIS), http://www.epa.gov/ncea/iris/intro.htm (accessed October 21, 2009), 2009b.

TABLE 7.11

Data for Benzyl Acetate and Ethyl Acetate

Name	CAS Registration Number	Chemical Formula	Carcinogenicity Rating
Benzyl acetate	140-11-4	$C_6H_5CH_2\ OCOOH_3$	Not Available
Ethyl acetate	141-78-6	$C_4H_8O_2$	Not Available

Source: United States Environmental Protection Agency, Integrated Risk Information System (IRIS), http://www.epa.gov/ncea/iris/intro.htm (accessed October 21, 2009), 2009b.

hydrogen atom has been replaced by a different organic functional group. Common uses of esters include the following (ATSDR 2002b, 2002c, 2002d):

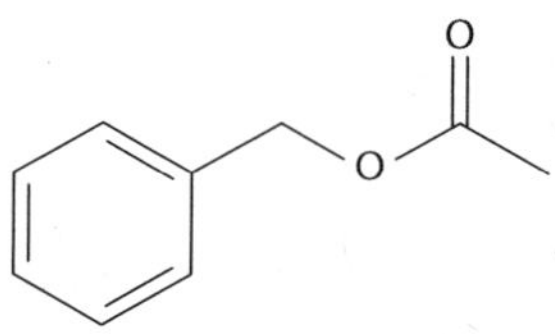

FIGURE 7.15
Molecular structure of benzyl acetate.

- Flame retardants
- Nail polish removers
- Glue
- Fragrances
- Clothing

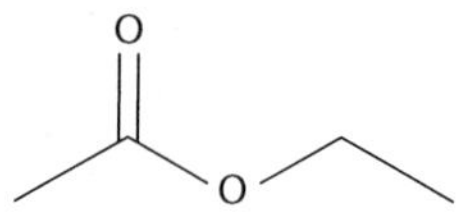

FIGURE 7.16
Molecular structure of ethyl acetate.

Polyester is the most common and most widely produced ester. They also occur naturally, and are found in flowers, in fats as triesters derived from glycerol and fatty acids, and in wine (ATSDR 2002b,c,d). The structures of two ester compounds, benzyl acetate and ethyl acetate are shown in Figures 7.15 and 7.16, respectively.

Table 7.11 presents the structural formula, CAS registration number, and carcinogenicity rating for benzyl acetate and ethyl acetate (USEPA 2009b).

At high concentrations, ester compounds can irritate the skin and mucous membranes and cause breathing difficulties (USEPA 2009b).

7.7 Heavy Metals

Heavy metals naturally exist in the environment at amounts termed their background levels. When human activities introduce additional heavy metals into the environmental and the background levels are exceeded, they then become contaminants.

In terms of their characteristics, heavy metals are generally not soluble in water, except for some varieties of arsenic and chromium VI at a neutral pH (Murray et al. 2004; SRC 2009). Heavy metals are not volatile but are often released into the atmosphere through the combustion of coal, automobile exhaust, metal production, and other methods (Murray et al. 2004).

The most common heavy metals occurring in urban areas include the following (Murray et al. 2004):

- Arsenic
- Barium
- Cadmium
- Chromium III
- Chromium VI
- Copper
- Lead
- Mercury
- Nickel
- Selenium
- Silver
- Zinc

Industrial watersheds are especially prone to heavy metal contamination. In an extensive study of the Rouge River watershed in southeastern Michigan, Murray et al. (2004) discovered elevated levels of barium, cadmium, chromium, copper, nickel, lead, and zinc in surface soil. The occurrence of heavy metals in soil and in groundwater in urban areas is discussed in greater detail in Chapter 9.

Table 7.12 lists the common uses for each of the heavy metals listed above. Table 7.13 lists the CAS registry number and carcinogenicity ratings for each of the heavy metals.

TABLE 7.12

Common Uses for Each of the Select Heavy Metals

Heavy Metal	Atomic Number	Uses
Arsenic	33	Wood preservative, poison, insecticide, pigments, chemical weapons
Barium	56	Superconductors, pigments, fireworks, lubricants, optics
Cadmium	48	Batteries, plastic stabilizer, pigments, metal plating, coatings, alloys
Chromium III	24	Pigments, inks, glass, steel additive, dyes, leather tanning, refractory, alloys
Chromium VI	24	Metal plating, corrosion resistance additive, wood preservative
Copper	29	Wire, building products, piping, jewelry, electromagnetics, brass and other alloys
Lead	82	Batteries, paint, ceramics, firearms, industrial coolant, electrodes, solder, construction materials, alloys, formerly a gasoline additive
Mercury	80	Electrical switches, thermometers, manometers, medical and dental applications, cosmetics, mercury-vapor lamps, formerly used in hat making
Nickel	28	Batteries, steel additive, magnets, coins, alloys
Selenium	34	Photocells, electronics, semiconductors, steel additive, alloys, copier and printing drums, glass manufacturing, pigments
Silver	47	Coins, electronics, circuit boards, alloys, mirrors, decorative items, jewelry, photographic films, batteries
Zinc	30	Metal plating, rust inhibitor (galvanization), brass alloy, batteries, cathodic protection, paint pigment, fire retardant, propellant, photocopying products, medical applications

Source: Pradyot, P., *Handbook of Inorganic Chemical Compounds*, McGraw Hill, New York, 2003; Krebs, R.E., *The History and Use of Earth's Chemical Elements: A Reference Guide*, Greenwood Publishing Group, Oxford, U.K., 2006.

TABLE 7.13

Data for Select Heavy Metals

Heavy Metal	CAS Registry Number	Carcinogenicity Rating
Arsenic	7440-38-2	Group: A1
Barium	7440-39-3	Group: A4
Cadmium	7440-43-9	Group: A2
Chromium III	16065-83-1	Group: A4
Chromium VI[a]	18540-29-9	Group: A1
Copper	7440-50-8	Not Available
Lead	7439-92-1	Group: A2
Mercury	7439-97-6	Group: A4
Nickel	7440-02-0	Not available
Selenium	7782-49-2	Group: A4
Silver	7440-22-4	Group: A4
Zinc	7440-66-6	Inadequate information

Source: United States Environmental Protection Agency, Integrated Risk Information System (IRIS), http://www.epa.gov/ncea/iris/intro.htm (accessed October 21, 2009), 2009b.

[a] Classified as Group: A1 inhalation route of exposure and Group: A4 by ingestion route of exposure.

As shown in Table 7.13, the toxicity and carcinogenicity of heavy metals varies widely. Additional but brief explanations of their general toxicities are given in the following:

- Arsenic is considered a human carcinogen (USEPA 2009b).
- Barium exposure can cause gastrointestinal disturbances and muscular weakness if it is ingested in a soluble form (ATSDR 2007e).
- Cadmium is considered a suspected or probable human carcinogen (USEPA 2009b; ATSDR 2008b).
- Chromium VI is considered a human carcinogen (ASTDR 2008a; USEPA 2009b). Chromium III is an essential nutrient that helps the body use sugar, protein, and fat.
- Copper is essential for good health in small amounts. High levels, however, can be harmful and cause nausea, vomiting, and diarrhea. Very high amounts can damage the liver and kidneys (ATSDR 2004a).
- Lead is considered a suspected or probable human carcinogen (USEPA 2009b). Lead can damage the nervous system and the brain (ATSDR 2007d).
- Mercury exposure may affect the brain and central nervous system (ATSDR 1999a).
- Nickel exposure most often results in an allergic reaction. Nickel may also affect the lungs (ATSDR 2005c).
- Selenium has beneficial and harmful effects. Low doses of selenium are necessary to maintain good health. High doses can cause harmful health effects such as, nausea, vomiting, and diarrhea. Prolonged exposure to selenium can cause a disease called selenosis (ATSDR 2003b).
- Silver exposure may result in a condition called arygia if exposure is prolonged. Arygia is a condition that causes a blue-gray discoloration of the skin and other body tissue (ATSDR 1999b).

- Zinc is essential for good health in small amounts. Too little zinc can cause hair and weight loss. Harmful effects do no usually occur until levels of ingestion exceed 10–15 times the recommended amount. Effects of overexposure to zinc include stomach cramps, nausea, and vomiting (ATSDR 2005d).

7.8 Other Contaminants

The remaining contaminants encountered in urban areas span a wide variety of organic and inorganic compounds and substances. These contaminants include the following:

- Pesticides and herbicides
- Fertilizers such as nitrates, phosphorus, and potassium
- Pharmaceuticals
- Bacteria such as coliform bacteria
- Viruses
- Cyanide
- Asbestos
- Acids
- Bases
- Radioactive compounds
- Dioxins
- Contaminants present only in the air such as
    - Greenhouse gases including carbon dioxide, methane, nitrous oxides, and others
    - Carbon monoxide
    - Particulates (dust)
    - Radon
    - Ozone
    - Many others
- Emerging contaminants
- And others

7.8.1 Pesticides and Herbicides

Many of the compounds previously discussed including several VOCs, SVOCs, and arsenic are/were used or are/were ingredients contained in pesticides and herbicides. Simply put, pesticides and herbicides are manufactured to kill things. USEPA (2009) defines a *pesticide* as preventing, destroying, repelling, or mitigating any pest. A *herbicide* is defined as a substance used to kill unwanted plants commonly referred to as weeds.

Pests are defined as living organisms occurring where they are not wanted or causing damage to crops or humans or other animals (USEPA 2009f). Examples include the following:

- Insects
- Mice or other animals
- Unwanted plants (weeds)
- Fungi
- Microorganisms, such as bacteria and viruses

Perhaps the most famous of all pesticides is the now banned substance called dichlorodiphenyltrichloroethane (DDT). It became famous as an environmental contaminant after the book *Silent Spring* written by Rachel Carson was published in 1962 (Carson 1962). The book catalogued the environmental impacts of the indiscriminant spraying of DDT in the United States and questioned the logic of releasing large amounts of chemicals into the environment without fully understanding their effects on ecology or human health. DDT (CAS registry number 50-29-3) is now considered a probable carcinogen (Group: A2) by USEPA (2009b) and is no longer used as a pesticide in the United States after its *use* was banned in 1972 (Note: DDT *production* was not stopped. The 2001 United Nations Environmental Program meeting held in Stockholm, Sweden (and put into effect in 2004) permitted its use for "vector control"—organisms that produce pathogens, such as mosquitoes). The structure of DDT is shown in Figure 7.17.

Common pesticides and herbicides, including DDT, are listed in Table 7.14 along with the respective CAS registry numbers and carcinogenicity ratings.

USEPA banned the use of the pesticide *chlordane* in 1983 because of potential environmental and human health concerns for all applications except termite control (ATSDR 1995b).

Malathion is an organophosphate insecticide widely used in agriculture, residential landscaping, and public recreation areas. Malathion is used widely to control mosquitoes, the West Nile

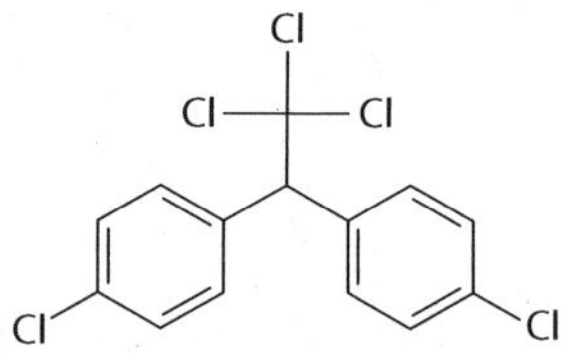

FIGURE 7.17
Basic molecular structure of DDT.

TABLE 7.14
Common Pesticides and Herbicides

Pesticide	CAS Registry Number	Carcinogenicity Rating	Chemical Formula
DDT	50-29-3	Group: A2	$C_{14}H_9Cl_5$
Chlordane	57-74-9	Group: A2	$C_{10}H_6Cl_8$
Malathion	121-75-5	Group: A4	$C_{10}H_{19}O_6\ PS_2$
Permethrin	52645-53-1	Group: A2	$C_{21}H_{20}Cl_2O_3$
Toxaphene	8001-35-2	Group: A2	$C_{10}H_{10}Cl_8$[a]
Herbicide			
Glyphosate	1071-83-6	Group: A4	$C_3H_8NO_5P$
2,4-D	94-75-7	Group: A4	$C_8H_6Cl_2O_3$
Pentachlorophenol	87-86-5	Group: A2	$C_6H_4Cl_5O$

[a] Represents a mean chemical formula since toxaphene is composed of a mixture of compounds.

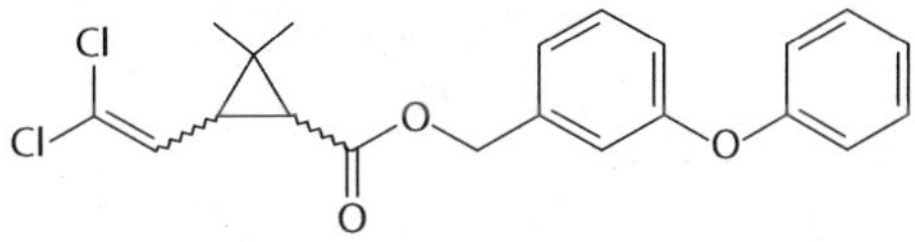

FIGURE 7.18
Basic structure of permethrin.

virus and was used in the 1980s in southern California to combat the Mediterranean Fruit Fly (ATSDR 2003c). Exposure to high amounts of malathion can cause difficulty breathing, chest tightness, vomiting, cramps, diarrhea, blurred vision, sweating, headaches, dizziness, loss of consciousness, and possibly death (ASTDR 2003c). If appropriate treatment is provided rapidly, there may be no long-term harmful effects.

Permethrin is a widely used synthetic insecticide and insect repellent used on cotton, wheat, maize, and alfalfa crops. It is also used to kill parasites on chickens and other poultry, and flea treatments for dogs. Permethrin is considered a neurotoxin and is highly toxic to both freshwater and estuarine aquatic organisms (ATSDR 2005e; USEPA 2006b). Figure 7.18 shows the basic molecular structure of permethrin.

Toxaphene is an insecticide composed of a mixture of over 670 chemicals (ATSDR 1997c). Toxaphene was one of the most widely used insecticides in the United States until 1982 when use dropped significantly and then was banned in 1990. It was primarily used in the southern states where it was applied to cotton to control pests, and also used elsewhere to control pests on livestock and to kill unwanted fish in lakes (ATSDR 1997c). Exposure to toxaphene may cause damage to the lungs, nervous system, and kidneys and can even cause death if exposure is extreme (ATSDR 1997c).

Glyphosate and *2,4D* are widely used herbicides in the United States. A common name or trade name for glyphosate is *Roundup*. USEPA (2009b) does not currently classify glyphosate and 2,4D as carcinogenic. However, these two compounds may affect the immune system, kidneys, and the liver (USEPA 2009b). *PCP* was discussed in Section 7.6.2.

A review of the chemical formulas of the compounds listed in Table 7.14 indicates many of these pesticides and herbicides are *organochlorines*—an organic compound containing at least one covalently bonded chlorine atom. Chlorine released into the environment presents special challenges because (1) it is highly toxic, (2) tends to be persistent due to the strength of the chlorine-carbon bonds, and (3) has an affinity for fatty tissues in vertebrates (fish and mammals), thus enabling bioaccumulation, a process where concentrations of a toxin increase at higher trophic levels in a food chain. Numerous studies have noted the presence of trace amounts of organochlorines in human breast milk (Calle et al. 2002).

7.8.2 Dioxins

Compounds generally referred to as *dioxins* represent a diverse set of halogenated substances and include other substances called furans. There are 75 dioxin isomers and 135 furan isomers. Dioxins are not intentionally produced and have no known use (ATSDR 2006c); they form unintentionally as a by-product of many industrial processes involving chlorine such as waste incineration and combustion. Dioxin compounds may also form as a by-product during the manufacture of chlorinated compounds and paper bleaching (ATSDR 2006c). A dioxin compound consists of two benzene molecules joined with two oxygen bridges. Figure 7.19 shows the basic structure of dioxin (National Research Council 2006).

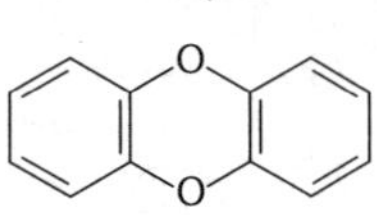

FIGURE 7.19
Basic structure of a dioxin molecule.

According to ATSDR (2006b), the most toxic dioxin compound is 2,3,7,8-tetrachlorodibenzo-*p*-dioxin or TCDD (CAS registry number 1746-01-6). Figure 7.20 shows the basic structure of TCDD.

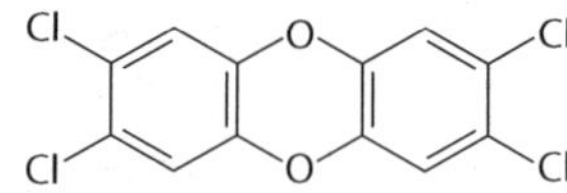

FIGURE 7.20
Basic molecular structure of TCDD.

Exposure to TCDD may lead to a condition known as chloroacne resulting in severe acne-like skin lesions occurring mainly on the face and upper body. USEPA (2009b) lists many of the dioxin compounds as either a known or suspected carcinogen. Dioxin compounds have been shown to bioaccumulate in humans and wildlife and they behave as teratogens and mutagens (National Research Council 2006).

7.8.3 Fertilizers

Fertilizers are chemical compounds designed to promote plant and fruit growth when applied (USEPA 2008a). The most common fertilizers are nitrogen (N), phosphorus (P), and potassium (K). Nitrogen is found primarily in an organic form in soils, but can also occur as nitrate. Because nitrate is very soluble and mobile, it is transported by surface water to rivers, lakes, and streams where it can promote algal growth. In many cases the algal growth is extensive. Nitrate can also contaminate drinking water. Phosphorus occurs in soil in organic and inorganic forms, but being more soluble than nitrate, can also be depleted in soil through surface water runoff. Phosphorus can also promote algal growth in rivers, lakes and streams, because it is a limiting nutrient in fresh water. Potassium (K) in fertilizers is commonly incorporated as potash—an oxide form of potassium that includes the compounds potassium chloride, potassium sulfate, potassium nitrate, and potassium carbonate. The term "potash" comes from the pioneer practice of extracting potassium carbonate (K_2CO_3) by leaching wood ashes and evaporating the solution in large iron pots. Fertilizers containing potassium generally do not promote algal growth (USEPA 2008a).

The United States Department of Agriculture (USDA) has tracked fertilizer use in the United States since 1960. According to the USDA, approximately 140 lb of fertilizer containing N-P-K are applied each year per acre of land farmed, and this amount has increased more than 200% since 1960 (USDA 2007).

In urban areas, fertilizers containing N–P–K are common and routinely applied to residential lawns and golf courses to help maintain green healthy grass and gardens (USEPA 2008a). According to USEPA (2009b), nitrogen, phosphorus, and potassium fertilizers are not currently known to cause cancer. Exposure to high concentrations may cause nausea and vomiting.

7.8.4 Cyanide

Cyanide is any chemical compound containing the cyano group—a carbon atom triple-bonded to a nitrogen atom. This bond is depicted in Figure 7.21. Common cyanide compounds include hydrogen cyanide, potassium cyanide, and sodium cyanide. Certain bacteria, fungi, and algae can produce cyanide, and cyanide is present in a number of foods and plants (ATSDR 2006c).

Cyanide compounds occur as gases, liquids, and solids. Inorganic cyanides are commonly salts of the cyano anion CN^-. Organic compounds containing the cyano group are called nitriles. Those compounds that are able to release the cyano group CN^- ion are highly toxic to humans and animals (ATSDR 2006d).

$$[:C\equiv N:]^-$$

FIGURE 7.21
Structure of a cyano group.

Hydrogen cyanide is a colorless gas with a faint, bitter, and almond-like odor. Sodium cyanide and potassium cyanide are both white solids with a bitter, almond-like odor when volatilization occurs. Cyanide and hydrogen cyanide have industrial applications in metal plating, metallurgy, some mining applications for extraction of precious metals, and in the manufacturing of plastics (ATSDR 2006d).

According to USEPA (2009b), cyanide and related compounds are not classified as carcinogens (Group: A4), but they are highly toxic. According to USEPA (2009b) and ATSDR (2006d), exposure to cyanide can cause rapid breathing, low blood pressure, headaches, coma, and death.

7.8.5 Asbestos

Asbestos (CAS Registry Number 1332-21-4) is the name given to a group of six different fibrous minerals that include the following (ATSDR 2001d):

- Amosite
- Chrysotile
- Crocidolite
- And the fibrous forms of the following:
    - Tremolite
    - Actinolite
    - Anthophyllite

Asbestos is controversial because of its many tiny fibers. If these fibers shed out from asbestos-containing materials they create an airborne dust, which when inhaled can penetrate lung tissue and cause asbestosis, mesothelioma, and lung cancer. The U.S. Occupational Safety and Health Administration (OSHA) and Environmental Protection Agency (EPA) make no distinction between the two kinds of asbestos—chrysotile and amphibole—though geologists and some public health experts say that chrysotile's soft and nonbarbed fibers are less damaging to the lungs than the splintery fibers of amphibole minerals.

The term asbestos describes a variety of fibrous, nonflammable minerals with flexibility and high tensile strength. Their unique properties were used mostly between the 1940s and 1970s in fireproof insulation, vinyl flooring, pipe insulation, ceiling tiles, brake linings, and roof coatings. Chrysotile, a serpentine mineral, is also known as "white asbestos" and makes up about 95% of asbestos found in buildings in the United States. The other asbestos minerals—crocidolite, amosite, anthophyllite, tremolite, and actinolite—are amphiboles not commonly used in commercial products.

OSHA began regulating workplace asbestos in 1970, around the time when miners and construction workers began reporting serious lung disease. At that time, the United States used about 800,000 Metric Tons of asbestos per year. OSHA published the Asbestos Standard for the Construction Industry, which outlined four categories of asbestos contamination, and specified precautions and disposal techniques for each class. For instance, if asbestos insulation is removed (Class 1), contractors and supervisors trained in asbestos removal must be onsite wearing respirators and protective clothing.

In the late 1970s and early 1980s, USEPA followed suit by banning any new uses for asbestos, designating it a class A human carcinogen (on par with secondhand cigarette smoke) and drafting the Asbestos Hazard Emergency Response Act (AHERA). In 1986, President Reagan signed the Act, which also put into effect the Toxic Substances Control Act, requiring all public and private schools to survey their structures for asbestos and implement an "appropriate response action."

Ironically, ridding a building of asbestos can make the air inside more hazardous than before. When asbestos materials are torn out from walls, ceilings and between pipes, asbestos dust contaminates the air. Had the asbestos materials not been touched in the first place, the fibers would remain harmlessly contained in the ceiling tiles of the insulation.

Responding to EPA and OSHA regulations, worried schools districts and homeowners spent billions ripping asbestos products out of their buildings. In 1999, removing asbestos materials in the United States cost about $3 billion. Throughout the 1990s, New York City schools alone spent more than $100 million on asbestos removal, without performing an analysis of the mineralogy of the asbestos.

Asbestos legislation continues to evolve, as does research on the suite of asbestos minerals. USEPA's response to asbestos has changed markedly over the years. In 1983 the agency's asbestos handbook stated that removing the material is always appropriate, while their 1990 handbook acknowledged that asbestos removal may cause more contamination than leaving it in place. Now, according to its Web site, "EPA's advice on asbestos is neither to rip it all out in a panic nor to ignore the problem under a false presumption that asbestos is risk free ... asbestos material in buildings should be located [and] it should be appropriately managed (Geotimes 2001).

Asbestos minerals have strong, flexible, and separable fibers that can be spun and woven. The main commercial and industrial value of asbestos lies in its ability to be heat resistant. For this reason asbestos was widely applied in the United Stated in manufactured goods and building construction as a heat insulator for these products:

- Roofing, ceiling, and floor tiles
- Window caulking
- Pipe wrap and pipe insulation
- Paper products
- Friction products, such as automobile brakes, clutches, and transmissions
- Heat-resistant fabrics
- Packaging
- Gaskets
- Coatings
- Some vermiculite and talc containing products
- Fire resistant doors

Exposure to asbestos causing an adverse health effect is almost always through the inhalation route. Asbestos affects the lungs and may lead to a condition called asbestosis. USEPA classifies asbestos as a human carcinogen (Group: A1), and it has been linked to a form of cancer called mesothelioma (USEPA 2009b). According to ATSDR (2001d), we are all exposed to some asbestos in the air we breathe, especially in urban areas. Fortunately, the concentrations in air are generally very low, being on the order of 0.00001–0.0001 fibers of asbestos per milliliter of air.

7.8.6 Acids and Bases

An *acid* increases the concentration of the hydrogen ion H^+ when dissolved in water and lowers the pH (potential hydrogen) of the solution. The bare hydrogen ion, H^+, is short for the hydronium ion, H_3O^+, since a bare H^+ does not exist in a solution. Conversely, a *base* increases the concentration of the hydroxide ion OH^- when dissolved in water, and raises the pH of the solution (Meyers 2003). Common acids and bases are listed in Table 7.15 (Meyers 2003).

We recognize acids and bases by their simple properties, such as taste, and conclude the sour taste of a lemon indicates it must be acidic. Bases tend to taste bitter. On the pH scale, any substance with a pH less than 7 (the neutral point) is acidic and any substance having a pH greater than 7 is basic. Acids and bases are widely used in industry, and are present in many widely consumed foods and drinks. Stronger acids and bases are used in household cleaners and detergents, especially those used on glassware and in ovens (USEPA 2008a).

Acids or bases are only toxic if they are strong, meaning they are of relatively low or high pH. Exposure to strong acids and strong bases causes respiratory irritation and burning, and causes skin burns. Significant exposure may cause severe burns and even death (USEPA 2009b). Currently, adequate information is not available to evaluate the potential carcinogenic effects of common acids and bases (USEPA 2009b).

Ammonia is a common basic chemical widely used as a household cleaning agent and in many industrial applications (ATSDR 2004). Ammonia is present naturally throughout the environment in air, soil, and water. Exposure to high levels of ammonia may cause lung, skin, and throat irritation. Some people with asthma may react more negatively to the inhalation of ammonia (ATSDR 2004b).

Hydrochloric acid (also referred to as hydrogen chloride) is a common acid widely used in industry as a cleaning agent, in the manufacturing of PVC, in making steel, and making leather. Hydrochloric acid is also present in humans and other organisms as a gastric acid

TABLE 7.15

Common Acids and Bases

Common Acids		Common Bases	
Acid	**Chemical Formula**	**Base**	**Chemical Formula**
Acetic acid	$HC_2H_3O_2$	Ammonia	NH_3
Benzoic acid	$HC_7H_5O_2$	Aniline	$C_6H_5NH_2$
Boric acid	H_3BO_3	Dimethylamine	$(CH_3)_3NH$
Carbonic acid	H_2CO_3	Ethylamine	$C_2H_5NH_2$
Cyanic acid	HCNO	Hydrazine	H_2H_4
Formic acid	$HCNO_2$	Hydroxylamine	NH_2OH
Hydrocyanic acid	HCN	Methylamine	CH_3NH_2
Hydrofluoric acid	HF	Pyridine	C_5H_5N
Hydrogen sulfide	H_2S	Urea	NH_2CONH_2
Hydrochloric acid	HCl	Potassium hydroxide	KOH
Nitric acid	HNO_3	Sodium bicarbonate	$NaHCO_3$
Phosphoric acid	H_3PO_4	Sodium hydroxide	NaOH
Pyruvic acid	$HC_3H_3O_3$	Calcium hydroxide	$Ca(OH)_2$
Sulfuric acid	H_2SO_4		

(ATSDR 2004c), and sometimes exists as an acid mist. This mist may cause skin and lung irritation, and skin burns can occur if you are exposed to a highly concentrated mist for a prolonged period (ATSDR 2004c).

7.8.7 Radioactive Compounds

Radioactive decay occurs when an unstable atomic nucleus spontaneously loses energy by emitting ionizing particles and radiation. This decay, or loss of energy, results in an atom of one type (parent nuclide) transforming into an atom of a different type (daughter nuclide). All elements with an atomic number greater than 80 possess radioactive isotopes, and all isotopes of elements with an atomic number greater than 83 are radioactive.

Some radioactive compounds deserving special attention include the following (Kathren 1991):

- Beryllium
- Calcium
- Carbon
- Potassium
- Cadmium
- Cesium
- Iodine
- Strontium
- Palladium
- Tin
- Radon
- Radium
- Thorium
- Uranium
- Plutonium

Radon (CAS registry number 10043-92-2) is the most common radioactive compound present in urban areas, and it has the potential for adverse human health effects (ATSDR 2000). USEPA (2009b) classifies radon as a human carcinogen, and exposure to radon for a long period of time at elevated concentrations may cause cancer.

Radon is a decay product of uranium, found naturally in the Earth's crust. It is one of the heaviest substances existing as a gas under normal conditions of pressure and temperature. The highest average radon concentrations in the United States are found in Iowa, southeastern Pennsylvania, and Appalachian Mountain areas. During the decay process, alpha, beta, and gamma radiation is released. Alpha particles can travel only short distances and cannot penetrate your skin. Beta particles can penetrate through your skin but not your whole body. Gamma particles can penetrate your whole body. Radon is normally present at very low levels in outdoor air but may be present at higher levels in indoor air, especially in basements and buildings with poor ventilation and in well water (ATSDR 2008c).

7.8.8 Greenhouse Gases

Greenhouse gases are gases in the atmosphere capable of absorbing and emitting radiation within the thermal infrared range (Karl and Trenberth 2003; USEPA 2009g). Greenhouse gases cause the Earth to warm up, and since the start of the industrial revolution in the early eighteenth century, levels of greenhouse gases in the Earth's atmosphere have increased (United States Department of State 2006). Figure 7.22 shows the greenhouse gas effect. The primary effects of greenhouse gases are climate change related and not directly related to toxicity.

Greenhouse gases are not generally investigated at specific sites of environmental contamination. It is important, however, to discuss greenhouse gases because of their potential impacts on climate change. Figure 7.23 shows the increase of carbon dioxide in the atmosphere since 1960 (NOAA 2009). The yearly variation shown in Figure 7.23 is seasonal and is attributed to extraction of carbon dioxide from plant matter during photosynthesis.

The greenhouse gases include the following compounds (USEPA 2009g):

- Carbon dioxide (CO_2)
- Methane (CH_4)
- Nitrous oxide (N_2O)
- Fluorinated gases including the following:
    - Hydrofluorocarbons (HFCs)
    - Perfluorocarbons (PFCs)
    - Sulfur hexafluoride (SF_6)—an industrially produced gas with a global warming potential (GWP) almost 24,000 times that of CO_2. The GWP measures the relative contribution of a gas to global warming based on its ability to absorb infrared radiation, its residence time in the atmosphere, and the specific wavelengths of energy absorbed. For comparison purposes, CO_2 has a GWP = 1.

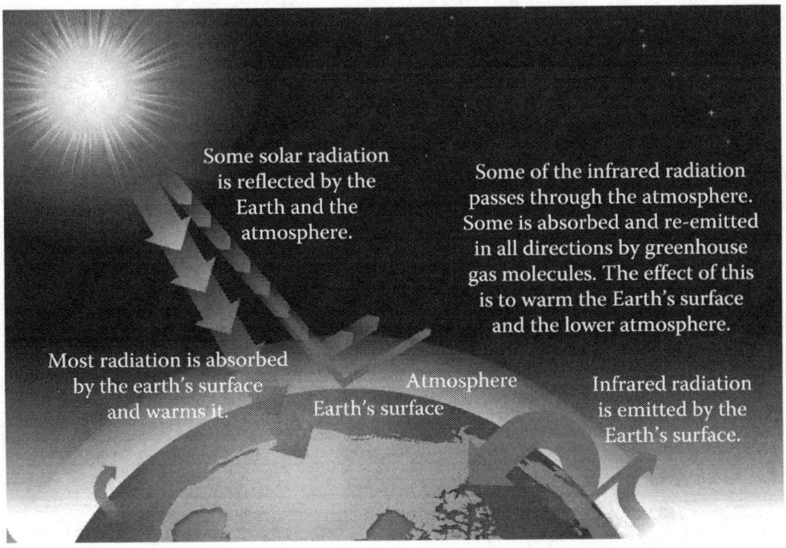

FIGURE 7.22
Greenhouse gas effect. (From United States Environmental Protection Agency, Greenhouse Gases, http://www.epa.gov/climatechange/indicators/pdfs/CI-greenhouse-gases.pdf (accessed June 28, 2010), 2010a.)

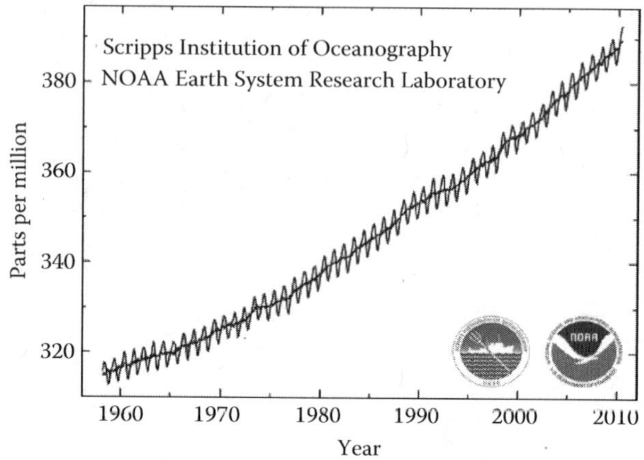

FIGURE 7.23
Carbon dioxide concentrations from 1960 to 2008 measured at Mauna Loa, Hawaii. (From National Oceanic and Atmospheric Administration (NOAA), Atmospheric CO_2 at Mauna Loa observatory, http://www.esrl.noaa.gov/gmd/ccgg/trends (accessed June 28, 2010), 2010.)

Greenhouse gases originate from a variety of sources, such as (USEPA 2009g)

- Fossil fuel combustion
    - Coal
    - Gasoline and diesel fuel in automobiles and trucks
    - Aviation fuels
    - Home heating fuels such as home heating oil and kerosene
- Industrial processes–refineries and cement making. Cement making is an often overlooked source; it accounts for 5%–7% of anthropogenic CO_2 (Worrell and Galitsky 2004).
- Waste disposal facilities
- Electrical generation
- Mining
- Residential and commercial sources
- Agriculture

Figure 7.24 shows a breakdown of yearly greenhouse gas emissions by country of origin.

Increasing carbon dioxide levels in the atmosphere have been linked to the burning of fossil fuels (USEPA 2009g), and the evidence is clear. Global fossil fuel emissions have increased to 8000 million Metric Tons of carbon in 2004 compared to 2500 million Metric Tons in 1950. Figure 7.25 shows the percent contribution of each type of greenhouse gas on a global scale (USEPA 2010b).

7.8.9 Carbon Monoxide

Carbon monoxide (CO) is a colorless, odorless, nonirritating gas that is very toxic to humans and animal life (USDHH 2009). Carbon monoxide is formed naturally and

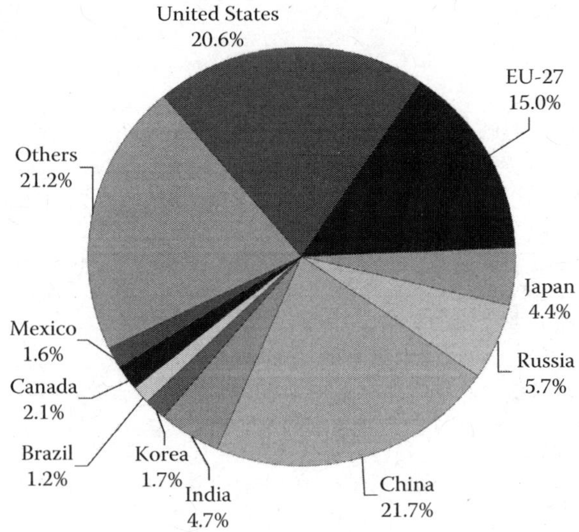

FIGURE 7.24
2007 Greenhouse gas contribution by country. (From Europeon Commission Directorate—General for Energy and Transport, Chapter 4.2, Environment, http://ec.europa.eu/energy/publications/doc/statistics/part_4_energy_pocket_book_2010.pdf (accessed August 8, 2010), 2010.)

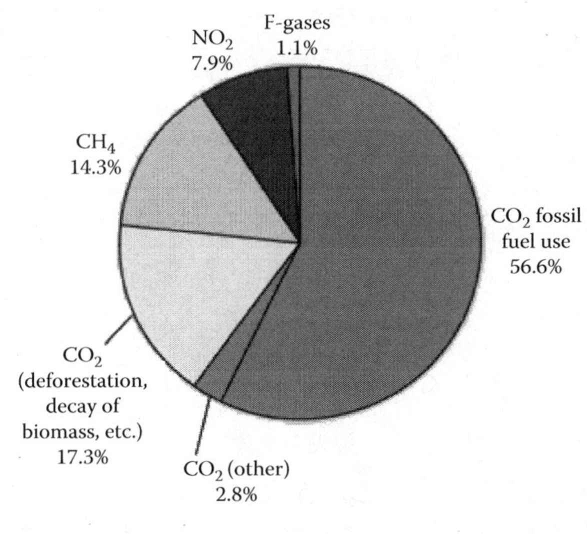

FIGURE 7.25
Greenhouse gas contributions. (From United States Environmental Protection Agency, Global Greenhouse Gas Data, http://www.epa.gov/climatechange/emissions/globalghg.html, (accessed July 2, 2010), 2010b.)

anthropogenically. The human body produces a small amount of CO when red blood cells convert protoporphyrin into bilirubin. Synthetically, most carbon monoxide is created by the incomplete combustion of fossil fuels, with the highest percentage coming from automobile exhaust. Inside homes, significant sources of carbon dioxide emissions can be natural gas and home heating oil furnaces, hot water heaters, appliances, wood

FIGURE 7.26
Structure of carbon monoxide.

burning stoves, and fireplaces. Carbon monoxide is also created as a by-product of several industrial processes, including metal melting and chemical synthesis (USDHH 2009). Figure 7.26 shows the structure of carbon monoxide represented in three different ways.

Carbon monoxide is one of the most common causes of fatal air poisoning (USDHH 2009). When inhaled, carbon monoxide combines with hemoglobin to produce carboxyhemoglobin. A condition known as anoxemia arises because carboxyhemoglobin cannot deliver oxygen to body tissues efficiently. Exposure to high levels of carbon monoxide may cause headache, nausea, vomiting, dizziness, lethargy, seizures, coma, and even death (USDHH 2009). Concentrations of less than 700 ppm of CO may cause up to 50% of the body's hemoglobin to convert to carboxyhemoglobin. In the United States, OSHA has established a long-term exposure standard for carbon monoxide at 50 ppm.

7.8.10 Ozone

Ozone (O_3) is a gas present in Earth's upper atmosphere and at ground level. The ozone occurring at ground level is considered an air pollutant by USEPA (2009h); whereas the ozone occurring in the upper atmosphere is not considered a pollutant. In the stratosphere, ozone is beneficial to life on Earth because it absorbs the sun's ultraviolet (UV) radiation (USEPA 2009h). Nearer to the surface, however, the story is different. Ozone from the ground level to a height of approximately 10 km (6 mi) in the troposphere is the main constituent of urban smog (USEPA 2009h). Figure 7.27 shows the two structural forms of ozone (USEPA 2009h).

Ozone in the stratosphere is gradually being destroyed by anthropogenically produced chemicals referred to as ozone-depleting substances (ODS), including the following:

- Chlorofluorocarbons (CFCs)
- Hydrochlorofluorocarbons (HCFCs)
- Halons
- Methyl bromide
- Carbon tetrachloride
- Methyl chloroform

Figure 7.28 shows the two locations in the atmosphere where ozone occurs.

These substances were, and still are used in coolants, foaming agents, fire extinguishers, solvents, pesticides, and aerosol propellants. Once released into the environment, they tend to degrade very slowly (USEPA 2009h). It should also be noted that HFCs have been substituted widely for CFCs. While this substitution has slowed the harmful effects

FIGURE 7.27
Basic molecular structure of ozone.

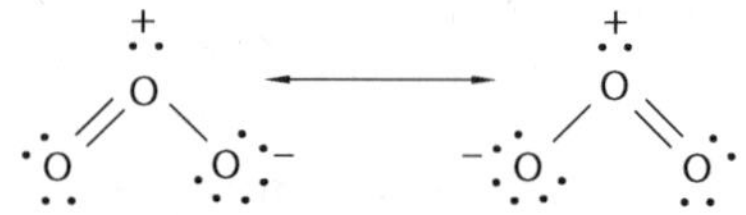

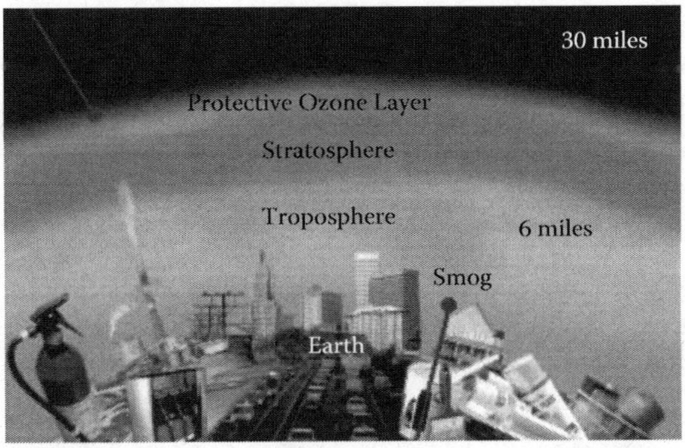

FIGURE 7.28
Ozone in the atmosphere: Good up high, bad nearby. (From United States Environmental Protection Agency, Ozone–Good up high bad nearby. Office of Air and Radiation, EPA-451/K-03-01, Washington, DC, http://www.epa.gov/oagps001/gooduphigh/ozone.pdf (accessed July 2, 2010), 2003.)

brought on by free chlorine in the upper atmosphere, HFCs are potent greenhouse gases and their use should be carefully monitored, and ultimately reduced.

Ground-level ozone is not emitted directly into the air, but is created by chemical reactions between oxides of nitrogen (NO_x) and VOCs in the presence of sunlight. Emissions from automobile exhaust, gasoline vapors, chemical solvents, electrical generating facilities, and certain factories are some of the major sources that emit compounds leading to the generation of ozone (USEPA 2009h). Table 7.16 shows the percent contributions of these major sources.

Ground-level ozone is a concern in urban regions of the United States. During the summer, strong sunlight and hot weather produces the conditions necessary for producing harmful levels of ozone. Overexposure to ground-level ozone can result in difficulty breathing and other respiratory affects (USEPA 2009h); especially for the elderly, very young, and those with existing respiratory ailments.

TABLE 7.16
Sources and Contribution of NO_x and VOCs

Compound	Source	Percent Contribution
NO_x	Motor vehicles	56
	Utilities	22
	Industrial, commercial, and residential fuel consumption	17
	Other	5
	Total	**100**
VOCs	Industrial and commercial processes	50
	Motor vehicles	45
	Consumer solvents	5
	Total	**100**

7.8.11 Sulfur Dioxide

Sulfur dioxide (SO_2) is released naturally through volcanic eruptions. Anthropogenically, significant sources of sulfur dioxide include automobile exhaust, the burning of coal, and some industrial processes (USEPA 2008c). Sulfur dioxide is a component of smog. Other components of smog include ozone, carbon monoxide, particulate matter (PM), VOCs, and nitrous oxides (USEPA 2008c). The basic structure of sulfur dioxide varies slightly (Pradyot 2003) (Figure 7.29).

Figure 7.30 shows an example of smog. When released into the atmosphere, sulfur dioxide often reacts with water vapor and eventually forms sulfuric acid (H_2SO_4). When precipitation occurs with a pH lower than that of natural rain (5.6) it is considered *acid rain* (USEPA 2008c). Figure 7.31 is a diagram showing acid rain development.

According to USEPA (2008c), acid rain and smog are serious environmental problems affecting large parts of the United States, especially heavily urbanized areas. A total of 21 out of the 27 U.S. urban areas listed in Table 2.4 experience acid rain. Figure 7.32 shows the areas of the United States having levels of smog capable of causing an adverse health affect in humans (USEPA 2008c).

7.8.12 Particulate Matter

PM is also known as particle pollution. PM is a complex mixture of very small particles and liquid droplets. Particle pollution consists of a number of components, including the following (USEPA 2009i):

- Acids
- Nitrates

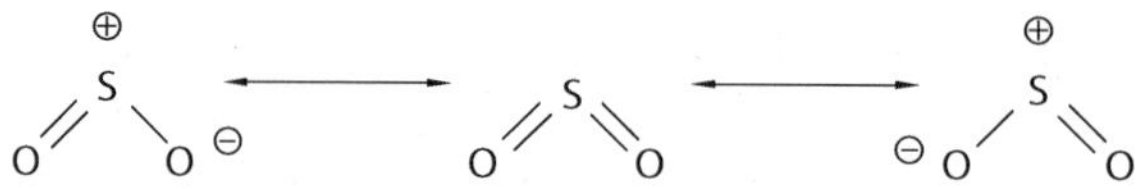

FIGURE 7.29
Basic molecular structure of sulfur dioxide.

FIGURE 7.30
Smog in an urban area. (Photo by Daniel T. Rogers.)

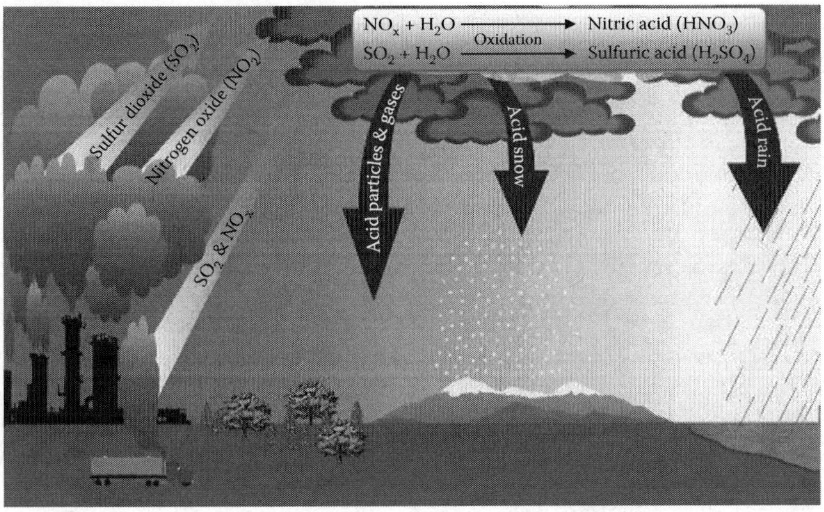

FIGURE 7.31
Development of acid rain. (From New York State Department of Environmental Conservation, Acid deposition large graphic, http://www.dec.ny.gov/chemical/41293.html (accessed June 27, 2010), 2010.)

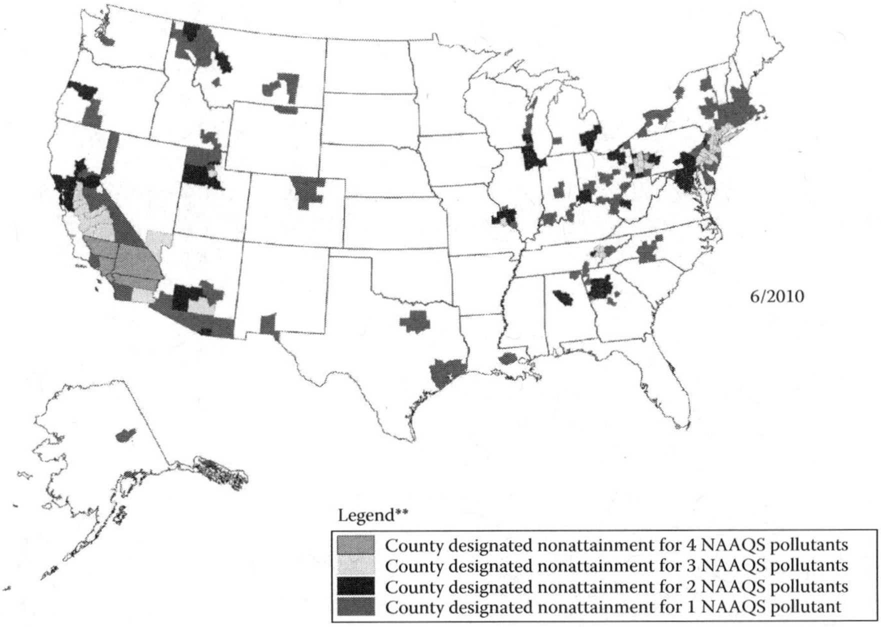

FIGURE 7.32
Nonattainment areas of the United States for smog-forming air pollutants in 2006. (From United States Environmental Protection Agency, Counties designated "nonattainment" for Clean Air Act's national ambient air quality standards (NAAQS) http://www.epa.gov/oaps001/greenbk/mapnpoll.pdf (accessed June 28, 2010), 2010c.)

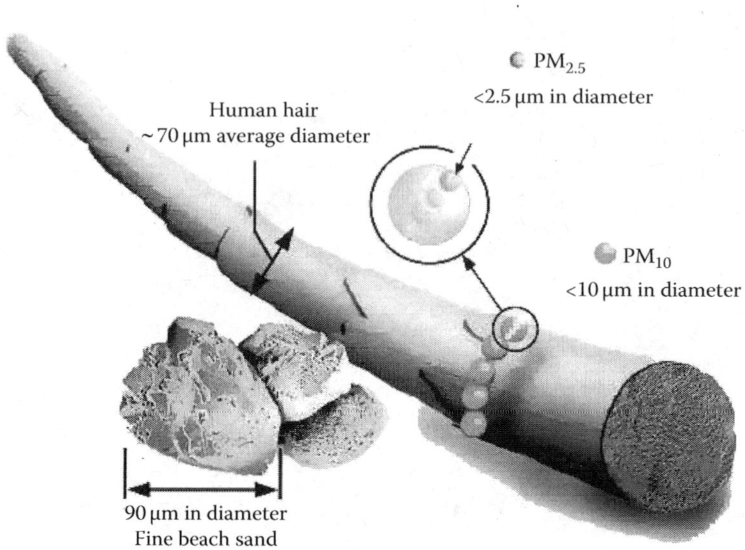

FIGURE 7.33
Particle size representation. (From United States Environmental Protection Agency, *Particulate Matter: Basic Information*, http://www.epa.gov/particlepollution/basic.html (accessed June 28, 2010), 2010d.)

- Sulfates
- Organic chemicals
- Metals
- Soil or dust particles

The size of the particles is directly related to their potential for causing adverse health effects. Particles less than 10 micrometers (μm) or less are small enough to pass through the nose and throat and enter the lungs while breathing (USEPA 2009i). Once inhaled, these particles can affect the heart and lungs and cause substantial adverse health effects. Figure 7.33 is a diagram of particle size relative to a human hair (USEPA 2009i).

USEPA groups PM into two categories (USEPA 2009i):

- Particles ranging in size from 2.5 to 10 μm are considered inhalable coarse particles; most likely to be present near roadways and dust-producing industries.
- Particles less than 2.5 μm are considered fine particles and are present in smoke and haze. These particles can be emitted from forest fires or power plants, certain industries, cigarette smoke, and automobiles.

Adverse health effects caused by inhalation of PM include the following (USEPA 2009i):

- Increased respiratory irritation, coughing, and difficulty breathing
- Decreased lung function
- Aggravated asthma
- Development of chronic bronchitis

- Irregular heartbeat
- Nonfatal heart attacks
- Premature death in people with heart of lung disease

An additional concern is the potential exposure to other contaminants sorbed onto PM, which are then ingested or inhaled. This topic will be covered in greater detail in Chapter 10.

7.8.13 Bacteria, Parasites, and Viruses

Although we do not always think of bacteria, parasites, and viruses as everyday environmental contaminants, they are common in urban areas and can cause many diseases and adverse health effects. They can contaminate water supplies and are also present in the air we breathe, in soil, on and within food, and building surfaces. Therefore, because they are so widespread and can produce adverse health effects, we will cover the basics.

Bacteria are organisms made up of just one cell. They are capable of multiplying themselves through a process called "binary fission," whereby a single bacterium grows to approximately twice its normal size and then splits into two daughter cells that are exact copies of the original bacterium. Bacteria live everywhere, even inside most organisms. Most bacteria are harmless and some are beneficial by destroying other harmful bacteria within our bodies (Madigan et al. 2008). However, some may cause disease such as tuberculosis. One of the more common harmful bacteria is a group called *E. coli*, which is short for *Escherichia coli*. Most *E. coli* are harmless but a strain called serotype O157:H7 can cause food poisoning in humans. Three basic shapes of bacteria exist—rounded, rod shaped, and spirals. *E. coli* bacteria are rod-shaped.

The ability for *E. coli* to survive for a brief period outside the body creates the potential for the bacteria to spread and infect other people. The spread of *E. coli* usually occurs when there is poor sanitation or when untreated sewerage is discharged from municipal wastewater treatment plants, which occasionally occurs during flood events (USEPA 2009k). Adverse health effects from exposure to *E. coli* bacteria typically include gastroenteritis, urinary tract infections, skin rashes, and neonatal meningitis (USEPA 2009k).

A *virus* is a subcelluar infectious agent capable of replicating itself inside the cells of another organism (Madigan et al. 2008). They are typically 100 times smaller than a bacterium. Viruses consist of two parts: (1) DNA or RNA molecules carrying genetic information and (2) a protein coat protecting the genes. Some viruses may also have an outside layer of fat surrounding them while they are outside a cell. Viruses spread in many ways, including aerosol routes (coughing and sneezing), through infected or contaminated water, and exchange of body fluids. Viruses cause diseases such as the common cold, influenza, chickenpox, mumps, ebola, and HIV.

A *parasite* is an organism living on or within a different organism (the host) at the expense of the host organism. Common examples of parasites causing adverse health effects in humans are Cryptosporidium and Giardia; both may cause severe intestinal disorder (Center for Disease Control 2009). Exposure to these two parasites occurs by consuming affected water (Center for Disease Control 2009). In humans, Giardia creates uncomfortable, but curable gastrointestinal symptoms, whereas Cryptosporidium can create life-long gastrointestinal symptoms in persons with weak immune systems and may result in death.

7.8.14 Emerging Contaminants

Through research, we are now learning about the presence in the environment of many chemicals and microbes that historically were not considered contaminants (USGS 2009c). Emerging contaminants originate from urban and agricultural sources and impact soil and groundwater at many urban locations. The potential health risks posed by emerging contaminants are not fully known. Many emerging contaminants enter the environment from residential waste products, and this fact has prompted a shift in traditional thinking that held most releases of contaminants were from industrial sources (Barnes et al. 2008).

Emerging contaminants include a wide variety of compounds consisting of the following (Barnes et al. 2008; USGS 2009c):

- Pharmaceuticals and drugs including
  - Antibiotics
  - Steroids
  - Antibacterial chemicals
  - Hormones
  - Narcotics
  - Many other legal and illegal drugs
- Insect repellants
- Solvents
- Detergents
- Plasticizers
- Fire retardants
- Veterinary antibiotics
- Others

Common pharmaceuticals and drugs with the capability of becoming contaminants if not properly disposed include the following:

- Hormones, such as testosterone
- Antibiotics, such as penicillin
- Sildenafil citrate, commonly known as Viagra
- Benzoylmethylecgonine, commonly known as cocaine

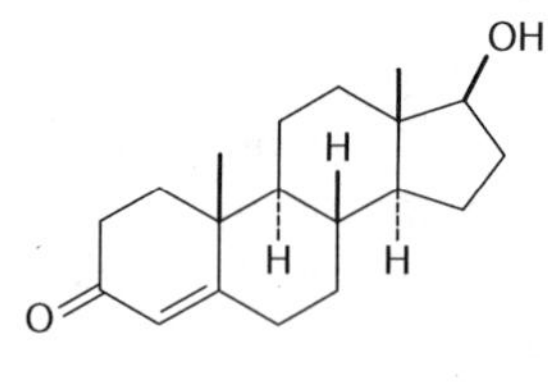

FIGURE 7.34
Structure of testosterone.

Benzoylmethylecgonine or cocaine is a stimulant affecting the central nervous system, and also acts as an appetite suppressant. Antibiotics inhibit the growth of bacteria. Sildenafil citrate is an arterial stimulant (Barnes et al. 2008). Testosterone is a male sex hormone, an anabolic steroid, and affects the growth of muscle mass. Figures 7.34 through 7.37 show the structures of testosterone, penicillin, sildenafil nitrate, and benzoylmehylecgonine, respectively.

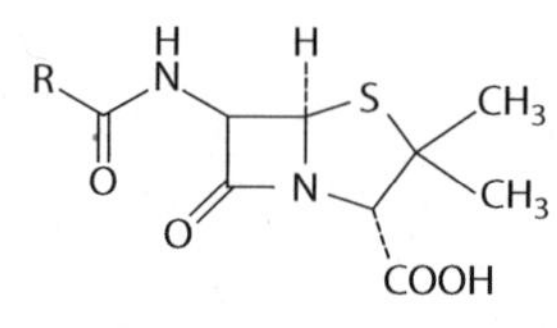

FIGURE 7.35
Structure of the antibiotic penicillin.

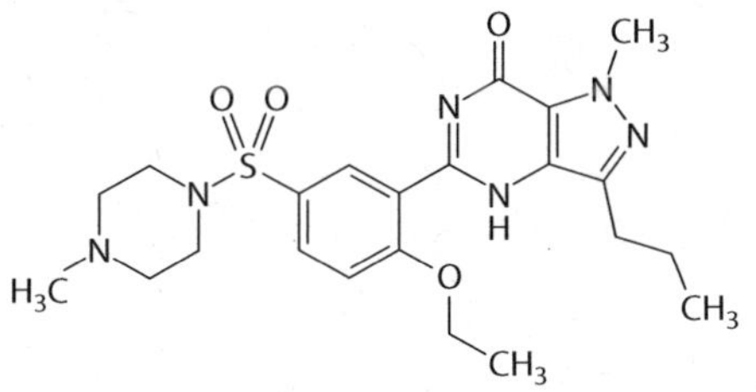

FIGURE 7.36
Structure of sildenafil citrate.

Table 7.17 lists the CAS number and molecular formula for benzoylmethylecgonine, penicillin, sildenafil citrate, and testosterone.

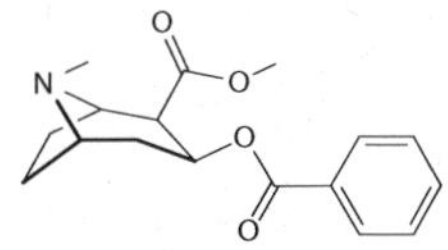

FIGURE 7.37
Structure of benzoylmethylecgonine.

Two other emerging contaminants of note include a group of compounds called perchlorates and the compound 1,4-Dioxane. *Perchlorates* are colorless and odorless salts. They are a group of compounds including the following:

- Magnesium perchlorate ($MgClO_4$)
- Potassium perchlorate ($KClO_4$)
- Ammonium perchlorate (NH_4ClO_4)
- Sodium perchlorate ($NaClO_4$)
- Lithium perchlorate ($LiClO_4$)

The basic structure of the perchlorate ion is shown in Figure 7.38. Perchlorates are very reactive and are commonly used in explosives, fireworks, road flares, and rocket motors (ATSDR 2008c). Perchlorate may also be present in bleach as an impurity. Adverse health affects of exposure to perchlorates include the ability of the thyroid gland to uptake iodine. Iodine is needed to produce hormones that regulate many body functions. USEPA does not currently list any of the perchlorate compounds as human carcinogens (USEPA 2009b).

1,4-Dioxane ($C_4H_8O_2$) is a clear liquid that easily dissolves in water. It is one of three isomer varieties of dioxanes, and is primarily used as an industrial solvent, with less widespread use in cosmetics, shampoos, and detergents (ATSDR 2007f). The ability of 1,4-Dioxane to dissolve so easily in water and its penchant for not being biodegradable results in 1,4-Dioxane easily contaminating surface water and groundwater. Inadequate information is available for classifying the carcinogenicity of 1,4-Dioxane (USEPA 2009b). Figure 7.39 shows the basic structure of the dioxane isomers.

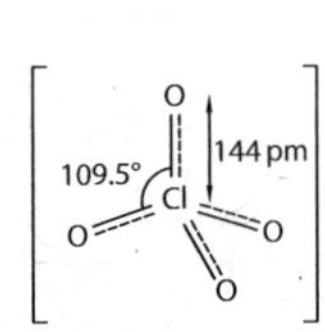

FIGURE 7.38
Structure of the perchlorate ion.

TABLE 7.17

Date for Select Emerging Pharmaceuticals and Drugs

Compound	CAS Registration Number	Molecular Formula
Benzoylmethylecgonine	50-36-2	$C_{17}H_{21}NO_4$
Sildenafil citrate	139755-83-2	$C_{22}H_3N_8O_4S$
Testosterone	58-22-0	$C_{19}H_{28}O_2$
Penicillin	Not listed	$R–C_9H_{11}N_2O_4S$[a]

[a] R indicates a variable group attached to molecule.

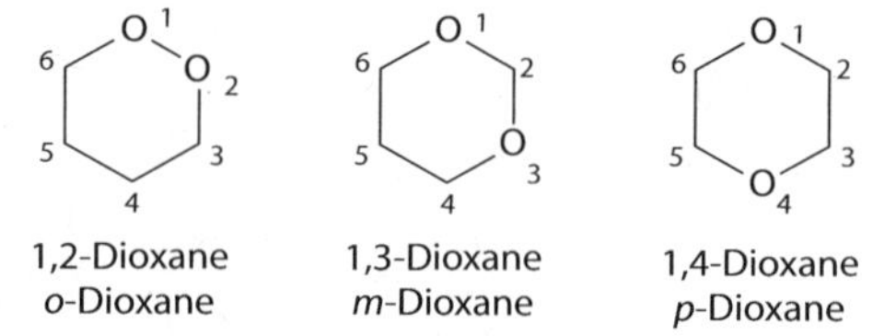

FIGURE 7.39
Structure of dioxane isomers.

7.9 Summary and Conclusion

There are thousands of contaminants existing everywhere, and many of the more unsavory ones are concentrated within urban areas. They are in the air we breathe, the water we drink, the food we eat, and in the dirt we play in. They are organic and inorganic. Many naturally occur—many do not. Some we know about and some we do not. Some are more toxic than others. Some may cause cancer and some may not. All of them have the ability to cause some adverse health effect in humans, to other organisms, or negatively impair or impact the environment if the exposure and dose are just right. Otherwise, they would not be considered contaminants.

This now leads us to the next set of questions we will explore. How do contaminants behave once they are released into the environment? Do they degrade? Where would we go to find them? How long do they last?

Evaluating the behavior of contaminants in the environment is commonly referred to as fate and transport assessment. Analysis of the fate and transport of contaminants once released into the environment is crucial for accurately assessing the risk posed by a specific compound. Just because a contaminant exists does not mean there will be a risk to human health or the environment. There must be a completed pathway—the contaminant must be transported from its point of release to a place where exposure can occur. Chapter 8 discusses this central concept, and describes the fate and transport of many contaminants introduced in this chapter.

References

American Chemical Society. 2009. CAS Registry Numbers. http://www.cas.org/expertise/cascontent/registry/regsys/html (accessed October 20, 2009).

American Conference of Governmental Industrial Hygienists (ACGIH). 2009. *Guide to Occupational Exposure Values*. ACGIH, Cincinnati, OH.

Agency for Toxic Substances and Disease Registry (ATSDR). 1989. *Bromodichloromethane*. CAS Registry Number 75-27-4. ATSDR ToxFAQs. Atlanta, GA.

Agency for Toxic Substances and Disease Registry (ATSDR). 1995a. *2-Hexanone*. CAS Registry Number 591-78-6. ATSDR ToxFAQs. Atlanta, GA.

Agency for Toxic Substances and Disease Registry (ATSDR). 1995b. *Chlordane*. CAS Registry Number 57-74-9. ATSDR ToxFAQs. Atlanta, GA.

Agency for Toxic Substances and Disease Registry (ATSDR). 1996. *Polycyclic Aromatic Hydrocarbons*. General Contaminant Class. ATSDR ToxFAQs. Atlanta, GA.

Agency for Toxic Substances and Disease Registry (ATSDR). 1997a. *Tetrachloroethene*. CAS Registry Number 127-18-4. ATSDR ToxFAQs. Atlanta, GA.

Agency for Toxic Substances and Disease Registry (ATSDR). 1997b. *Chloroform*. CAS Registry Number 127-18-4. ATSDR ToxFAQs. Atlanta, GA.

Agency for Toxic Substances and Disease Registry (ATSDR). 1997c. *Toxaphene*. CAS Registry Number 8001-35-2. ATSDR ToxFAQs. Atlanta, GA.

Agency for Toxic Substances and Disease Registry (ATSDR). 1999a. *Mercury*. CAS Registry Number 7439-97-6. ATSDR ToxFAQs. Atlanta, GA.

Agency for Toxic Substances and Disease Registry (ATSDR). 1999b. *Silver*. CAS Registry Number 7440-22-4. ATSDR ToxFAQs. Atlanta, GA.

Agency for Toxic Substances and Disease Registry (ATSDR). 2000. *Radon Toxicity: Who Is at Risk?* ATSDR, Atlanta, GA.

Agency for Toxic Substances and Disease Registry (ATSDR). 2001a. *Toluene*. CAS Registry Number 108-88-3. ATSDR ToxFAQs. Atlanta, GA.

Agency for Toxic Substances and Disease Registry (ATSDR). 2001b. *Polychlorinated Biphenyls*. ATSDR ToxFAQs. Atlanta, GA.

Agency for Toxic Substances and Disease Registry (ATSDR). 2001c. *PCB Aroclor 1254*. CAS Registry Number 11097-69-1. ATSDR ToxFAQs. Atlanta, GA.

Agency for Toxic Substances and Disease Registry (ATSDR). 2001d. *Asbestos*. CAS Registry Number 1332-21-4. ATSDR ToxFAQs. Atlanta, GA.

Agency for Toxic Substances and Disease Registry (ATSDR). 2002a. *Di(2-ethylhexyl) Phthalate*. CAS Registry Number 117-81-7. ATSDR ToxFAQs. Atlanta, GA.

Agency for Toxic Substances and Disease Registry (ATSDR). 2002b. *Toxicological Profile for Flame Retardant Ester Compounds*. ATSDR, Atlanta, GA.

Agency for Toxic Substances and Disease Registry (ATSDR). 2002c. *Benzyl Acetate*. CAS Registry Number 140-11-4. ATSDR ToxFAQs. Atlanta, GA.

Agency for Toxic Substances and Disease Registry (ATSDR). 2002d. *Ethyl Acetate*. CAS Registry Number 141-78-6. ATSDR ToxFAQs. Atlanta, GA.

Agency for Toxic Substances and Disease Registry (ATSDR). 2003a. *Trichloroethene*. CAS Registry Number 79-01-6. ATSDR ToxFAQs. Atlanta, GA.

Agency for Toxic Substances and Disease Registry (ATSDR). 2003b. *Selenium*. CAS Registry Number 7782-49-2. ATSDR ToxFAQs. Atlanta, GA.

Agency for Toxic Substances and Disease Registry (ATSDR). 2003c. *Malathion*. CAS Registry Number 121-75-5. ATSDR ToxFAQs. Atlanta, GA.

Agency for Toxic Substances and Disease Registry (ATSDR). 2004a. *Copper*. CAS Registry Number 7440-50-8. ATSDR ToxFAQs. Atlanta, GA.

Agency for Toxic Substances and Disease Registry (ATSDR). 2004b. *Ammonia*. CAS Registry Number 7664-41-7. ATSDR ToxFAQs. Atlanta, GA.

Agency for Toxic Substances and Disease Registry (ATSDR). 2004c. *Hydrochloric Acid*. CAS Registry Number 7647-01-0. ATSDR ToxFAQs. Atlanta, GA.

Agency for Toxic Substances and Disease Registry (ATSDR). 2005a. *Bromoform*. CAS Registry Number 75-25-2. ATSDR ToxFAQs. Atlanta, GA.

Agency for Toxic Substances and Disease Registry (ATSDR). 2005b. *Naphthalene*. CAS Registry Number 91-20-3. ATSDR ToxFAQs. Atlanta, GA.

Agency for Toxic Substances and Disease Registry (ATSDR). 2005c. *Nickel*. CAS Registry Number 7440-02-0. ATSDR ToxFAQs. Atlanta, GA.

Agency for Toxic Substances and Disease Registry (ATSDR). 2005d. *Zinc*. CAS Registry Number 7440-66-6. ATSDR ToxFAQs. Atlanta, GA.

Agency for Toxic Substances and Disease Registry (ATSDR). 2005e. *Permethrin: Toxicologic Information about Pesticides*. CAS Registry Number 52645-53-1. ATSDR, Atlanta, GA.

Agency for Toxic Substances and Disease Registry (ATSDR). 2006a. *Methyl-tert-butyl Ether*. CAS Registry Number 1634-04-4. ATSDR ToxFAQs. Atlanta, GA.

Agency for Toxic Substances and Disease Registry (ATSDR). 2006b. *Vinyl Chloride*. CAS Registry Number 75-01-4. ATSDR. Atlanta, GA.

Agency for Toxic Substances and Disease Registry (ATSDR). 2006c. *Dioxins: Chemical Agent Briefing Sheet*. ATSDR, Atlanta, GA.

Agency for Toxic Substances and Disease Registry (ATSDR). 2006d. *Cyanide*: CAS Registry Numbers 74-90-8; 143-33-9; 151-50-8; 592-01-8; 544-92-3; 506-61-6; 460-19-5; and 506-77-4. ATSDR, Atlanta, GA.

Agency for Toxic Substances and Disease Registry (ATSDR). 2007a. *Benzene*. CAS Registry Number 71-43-2. ATSDR ToxFAQs. Atlanta, GA.

Agency for Toxic Substances and Disease Registry (ATSDR). 2007b. *Ethybenzene*. CAS Registry Number 100-41-4. ATSDR ToxFAQs. Atlanta, GA.

Agency for Toxic Substances and Disease Registry (ATSDR). 2007c. *Xylene*. CAS Registry Number 1330-20-7. ATSDR ToxFAQs. Atlanta, GA.

Agency for Toxic Substances and Disease Registry (ATSDR). 2007d. *Lead*. CAS Registry Number 7439-92-1. ATSDR ToxFAQs. Atlanta, GA.

Agency for Toxic Substances and Disease Registry (ATSDR). 2007e. *Barium*. CAS Registry Number 9440-39-3. ATSDR ToxFAQs. Atlanta, GA.

Agency for Toxic Substances and Disease Registry (ATSDR). 2007f. *1,4-Dioxane*. CAS Registry Number 123-91-1. ATSDR ToxFAQs. Atlanta, GA.

Agency for Toxic Substances and Disease Registry (ATSDR). 2008a. *Chromium*. CAS Registry Number 7440-47-3. ATSDR ToxFAQs. Atlanta, GA.

Agency for Toxic Substances and Disease Registry (ATSDR). 2008b. *Cadmium*. CAS Registry Number 7440-43-9. ATSDR ToxFAQs. Atlanta, GA.

Agency for Toxic Substances and Disease Registry (ATSDR). 2008c. *Radon*. CAS Registry Number 714859-67-7. ATSDR ToxFAQs. Atlanta, GA.

Agency for Toxic Substances and Disease Registry (ATSDR). 2008d. *Perchlorates*. CAS Registry Numbers 10034-81-8, 7778-74-7, 7790-98-9, 7601-89-0, and 7791-03-9. ATSDR ToxFAQs. Atlanta, GA.

Agency for Toxic Substances and Disease Registry (ATSDR). 2009. ATSDR Glossary of Terms. Center for Disease Control. http://atsdr.cdc.gov/glossary.html (accessed October 22, 2009).

Barbalace, R.C. 2009. The Chemistry of Polychlorinated Biphenyls. EnvironmentalChemistry.com. September 2003. http://www.environmentalchemistry.com/ypgi/chemistry/pcb.html (accessed October 26, 2009).

Barnes, K.K., D.W. Kolpin, E.T. Furlong et al. 2008. A national reconnaissance of pharmaceuticals and other organic wastewater contaminants in the United States. *Journal of Science in the Total Environment* 42:192–200.

Calle, E.E, H. Frumkin, S.J. Henley et al. 2002. Organochlorines and breast cancer risk. *CA: A Cancer Journal for Clinicians* 52:301–309.

Carson, R. 1962. *Silent Spring*. Boston, MA: Houghton Mifflin.

Carwile, J.L, H.T. Luu, L.S. Bassett et al. 2009. Use of polycarbonate bottles and urinary bisphenol a concentrations. *Environmental Health Perspectives* 117:1368–1372.

Center for Disease Control (CDC). 2009. *Cryptosporidium Fact Sheets*. Atlanta, GA: Department of Health and Human Services, CDC.

Doherty, R.E. 2000. A history of the production and use of carbon tetrachloride, tetrachloroethylene, trichloroethylene, and 1,1,1-trichloroethane in the United States: Part 1—historical background; carbon tetrachloride and tetrachloroethylene. *Journal of Environmental Forensics* 1:69–81.

Europeon Commission Directorate—General for Energy and Transport. Chapter 4.2. Environment. http://ec.europa.eu/energy/publications/doc/statistics/part_4_energy_pocket_book_2010.pdf (accessed August 8, 2010).

Fetzer, J.C. 2000. *The Chemistry and Analysis of Large Polycyclic Aromatic Hydrocarbons*. New York: John Wiley & Sons.

Geotimes, 2001. Earth materials and public health. American Geological Institute. http://www.agiweb.org/geotimes/nov01/feature_earthsubst.html (accessed July 24, 2010).

Holleman, A.F. and E. Wiberg. 2001. *Inorganic Chemistry*. San Diego, CA: Academic Press.

Jensen, W.B. 2009. The origin of the circle symbol for aromaticity. *Journal of Chemical Education* 86:423–425.

Karl, T.R. and K.E. Trenberth. 2003. Modern global climate change. *Science* 302:1719–1723.

Kathren, R. 1991. *Radioactivity and the Environment*. Leiden, the Netherlands: Taylor & Francis.

Kaufman, M.M., D.T.Rogers, and K.S. Murray. 2005. An empirical model for estimating remediation costs at contaminated sites. *Water, Air, and Soil Pollution* 167:365–386.

Krebs, R.E. 2006. *The History and Use of Earth's Chemical Elements: A Reference Guide*. Oxford, U.K.: Greenwood Publishing Group.

Levy, A.B. 2009. Acetone. In *Ullmann's Encyclopedia of Industrial Chemistry*, 7th edn. New York: John Wiley & Sons.

Lide, D.R. 2008. Physical constants of organic compounds. In *Handbook of Physics and Chemistry*, 89th edn. Boca Raton, FL: CRC Press.

Madigan, M.T., J.M. Martinko, P.V. Dunlap et al. 2008. *Brock Biology of Microorganisms*, 12th edn. New York: Prentice-Hall.

McMurry, J.E. 2009. *Fundamentals of Organic Chemistry*. New York: Brook Cole.

Meyers, R. 2003. *The Basics of Chemistry*. Westport, CT: Greenwood Press.

Missouri Department of Natural Resources. 2006. *Missouri Risk-Based Corrective Action for Petroleum Storage Tank Sites Sampling for Polynuclear Aromatic Hydrocarbons*. Hazardous Waste Fact Sheet. Jefferson City, MO.

Murray, K. S., D.T. Rogers, and M.M. Kaufman. 2004. Heavy metals in an urban watershed in southeastern Michigan. *Journal of Environmental Quality* 33:163–172.

National Oceanic and Atmospheric Administration (NOAA). 2009. Atmospheric CO_2 at Mauna Loa observatory. http://www.esrl.noaa.gov/gmd/ccgg/trends (accessed June 28, 2010).

National Research Council. 2006. Health risks from dioxin and related compounds. Washington, DC: National Academies Press.

New York State Department of Environmental Conservation. 2010. Acid Deposition Large Graphic. http://www.dec.ny/chemical/41293.html (accessed June 27, 2010).

Phillips, D.J. 1986. *PCBs and the Environment*, Vol. 2. Boca Raton, FL: CRC Press.

Pradyot, P. 2003. *Handbook of Inorganic Chemical Compounds*. New York: McGraw Hill.

SRC. 2009. Environmental fate data base (EFDB). CHEMFATE Chemical Search. http://srcinc.com/what-we-do/efbd.aspx (accessed November 10, 2009).

Suthersan, S.S. and F.C. Payne. 2005. *In Situ Remediation Engineering*. Boca Raton, FL: CRC Press.

United States Department of Agriculture (USDA). 2007. *United States Fertilizer Use and Cost*. Washington, DC.

United States Department of Energy. 2009. Status of Impact of State MTBE Bans. USDOE. http://www.eia.doe.gov/oiaf/servicerpt/mtbeban/table1.html (accessed November 4, 2009).

United States Department of State. 2006. *United States Climate Action Report*. Fourth Climate Action Report to the United Nations Framework on Climate Change. Washington, DC.

United States Department of Health and Human Services (USDHH). 1995. *Toxicological Profile for Polycyclic Aromatic Hydrocarbons*. Atlanta, GA: Agency for Toxic Substances and Disease Registry.

United States Department of Health and Human Services (USDHH). 2005. *Report on Carcinogens*, 11th edn. Washington, DC: Public Health Service. National Toxicology Program.

United States Department of Health and Human Services (USDHH). 2008. *NTP—CERHR Monograph on the Potential Human Reproductive and Developmental Effects of Bisphenol A*. National Institute of Health Publication Number. 08-594. Washington, DC.

United States Department of Health and Human Services (USDHH). 2009. *Draft Toxicological Profile for Carbon Monoxide*. ATSDR. Washington, DC.

United States Environmental Protection Agency. 1989a. *Transport and Fate of Contaminants in the Subsurface*. USEPA Center for Environmental Research Information. EPA/625/4-89/019. Cincinnati, OH.

United States Environmental Protection Agency. 1989b. *Risk Assessment Guidance for Superfund. Volume I. Human Health Evaluation Manual (Part A)*. EPA/540/1-89/002. Office of Emergency and Remedial Response. Washington, DC.

United States Environmental Protection Agency. 2003. Ozone—Good up high bad nearby. Office of Air and Radiation. EPA-451/K-03-01. Washington, DC. http://www.epa.gov/oagps001/gooduphigh/ozone.pdf (accessed July 2, 2010).

United States Environmental Protection Agency. 2005. *Guidelines for Carcinogenic Risk Assessment*. EPA/630/P-03/001F. Washington, DC.

United States Environmental Protection Agency. 2006a. *Pentachlorophenol Consumer Fact Sheet*. Washington, DC.

United States Environmental Protection Agency. 2006b. *Permethrin Consumer Fact Sheet*. EPA 738-F-06-012. Washington, DC.

United States Environmental Protection Agency. 2007. *Overview of Methyl-tert-butyl ether*. Washington, DC.

United States Environmental Protection Agency. 2008a. *EPA's Report on the Environment*. EPA/600/R-07/045F. Washington, DC.

United States Environmental Protection Agency. 2008b. *Test Methods for Evaluating Solid waste, Physical/Chemical Methods, Final Update IV*. SW-846. Washington, DC.

United States Environmental Protection Agency. 2008c. *Latest Findings on National Air Quality: Status and Trends Through 2006*. EPA454/R-07-007. Research Triangle Park, NC.

United States Environmental Protection Agency. 2009a. Waste and Cleanup Assessment Glossary. http://www.epa.gov/oswer/riskassessment/glossary.htm (accessed October 22, 2009).

United States Environmental Protection Agency. 2009b. Integrated Risk Information System (IRIS). http://www.epa.gov/ncea/iris/intro.htm (accessed October 21, 2009).

United States Environmental Protection Agency. 2009c. Polychlorinated Biphenyls Fact Sheet: Basic i\ Information. http://www.epa.gov/epawaste/hazard/tsd/pcbs/about.htm (accessed October 26, 2009).

United States Environmental Protection Agency. 2009d. Polychlorinated Biphenyls Fact Sheet: Laws and Regulations. http://www.epa.gov/epawaste/hazard/tsd/pcbs/pubs/laws.htm (accessed October 26, 2009).

United States Environmental Protection Agency. 2009e. Health Effects of PCBs. http://www.epa.gov/epawaste/hazard/tsd/pcbs/pubs/effects.htm (accessed October 26, 2009).

United States Environmental Protection Agency. 2009f. What is a Pesticide? http://www.epa.gov/pesticides/about.htm (accessed October 28, 2009).

United States Environmental Protection Agency. 2009g. *Climate Change-Greenhouse Gas Overview*. Washington, DC.

United States Environmental Protection Agency. 2009h. Ozone—Good up high bad nearby. Air Quality Planning and Standards. Washington, DC.

United States Environmental Protection Agency. 2009i. *Particulate matter*. Air Quality Planning and Standards. Washington, DC.

United States Environmental Protection Agency. 2009k. *E. coli Bacteria*. Total coliform rule-basic information. Washington, DC.

United States Environmental Protection Agency. 2010a. Greenhouse Gases. http://www.epa.gov/climatechange/indicators/pdfs/CI-greenhouse-gases.pdf (accessed June 28, 2010).

United States Environmental Protection Agency. 2010b. Global Greenhouse Gas Data. http://www.epa.gov/climatechange/emissions/globalghg.html (accessed July 2, 2010).

United States Environmental Protection Agency. 2010c. Counties Designated "Nonattainment" for Clean Air Act's National Ambient Air Quality Standards (NAAQS) http://www.epa.gov/oaps001/greenbk/mapnpoll.pdf (accessed June 28, 2010).

United States Environmental Protection Agency. 2010d. *Particulate Matter: Basic Information*. http://www.epa.gov/particlepollution/basic.html (accessed June 28, 2010).

United States Geological Survey (USGS). 2006. *Volatile Organic Compounds in the Nation's Ground Water and Drinking-Water Supply Wells*. USGS Circular 1292. Reston, VA.

United States Geological Survey (USGS). 2009a. Toxic Substance Hydrology Program. DNAPL. http://www.usgs.gov/definitions/dnapl_def.html (accessed October 21, 2009).

United States Geological Survey (USGS). 2009b. Toxic Substance Hydrology Program. BTEX. http://www.usgs.gov/definitions/btex.html (accessed October 21, 2009).

United States Geological Survey (USGS). 2009c. *Emerging Contaminants—Fact Sheet*. USGS. Washington, DC.

Van der Perk, M. 2006. *Soil and Water Contamination: From Molecular to Catchment Scale*. Balkema: Proceedings and Monographs in Engineering, Water and Earth Sciences. London, U.K.: Taylor & Francis.

World Health Organization (WHO). 2003. *Polychlorinated Biphenyls: Human Health Aspects*. Geneva, Switzerland: WHO.

Worrell, E. and C. Galitsky. 2004. *Energy Efficiency Improvement Opportunities for Cement Making: An Energy Star Guide for Energy and Plant Managers*. Environmental Technologies Division. Lawrence Berkeley National Laboratory. LBNL-54036.

8

Contaminant Fate and Transport

8.1 Introduction

We now know what contaminants are, where they come from, and what they can do to our bodies if we are exposed to them. We know that a contaminant only presents a risk to human health and environment if there is a completed exposure pathway. So, where do contaminants end up when they are released into the environment, and how do humans and the environment become exposed to contaminants?

The answer to these questions is obtained through an understanding of the process called **contaminant fate and transport**, defined as the sequence of anthropogenic and natural events involving a contaminant source, its mobilization or transport, and its ultimate fate or resting place, termed a **sink** (Rogers et al. 2007). Sources of contamination include many human activities performed primarily at the surface resulting in the release of toxic substances into the environment. These toxic substances may be transported over time or remain relatively close to their source before they are degraded, transformed, or destroyed because (1) contaminants released into the environment are often mixtures, (2) each contaminant is unique chemically, and (3) the geologic environment in which the contaminants are released is also unique. It is, therefore, necessary to understand the physical chemistry of each contaminant, the microbiologic environment, and the geologic environment into which they are released when characterizing their fate and transport (Rogers et al. 2007).

During their transport and before reaching their final sink, certain contaminants may reside at multiple intermediate sinks for different periods of time. Intermediate sinks include surface water, groundwater, and the atmosphere, and the contaminants held within these water-containing sinks will flow and ultimately reach the oceans. Aquifers with a very low hydraulic conductivity are for practical purposes considered final sinks, as are inland bogs and some wetlands. Sediment and soil can function as intermediate or as final sinks, since erosion may move both of these unconsolidated materials. The oceans are almost always a final sink of contamination, although wave action and ocean currents may occasionally bring contamination onshore.

To a large extent, the level of human health and/or environmental risk is a function of two fundamental concepts introduced in this chapter: *mobility* and *persistence*. **Mobility** is a measure of a substance's potential to migrate. **Persistence** is a measure of a substance's ability to remain in the environment before being degraded, transformed, or destroyed (Rogers et al. 2007). In Chapter 7, it was noted that substances with higher toxicity pose greater potential for adverse health risks, but the risk to humans and the environment grows exponentially if the chemical is both mobile and persistent. For example, a highly toxic but immobile chemical may affect a few people in a warehouse through inhalation, whereas a mobile and persistent chemical of moderate toxicity can contaminate a public water supply or migrate to a different sink. In these locations, the potential for widespread human exposures and ecosystem damages is much higher and may persist for decades under certain conditions.

The following sections discuss how and where contaminant releases occur; how their migration through the soil, groundwater, and atmosphere proceeds; and how they end up in sinks. The chapter concludes with a brief description of the fate and transport for each contaminant group.

8.2 Contaminant Releases into the Environment

The fate and transport of contaminants begins with their release into the environment. Contaminant releases originate from numerous sources and under different circumstances; they vary on degree, concentration, duration, mass, volume, and whether a single contaminant is released or if a mixture of contaminants are released. Each of these factors influences their fate and transport. For instance, some releases may be very small and avoid detection. On the other hand, some releases are so large and sudden (e.g., a tanker spill) that they become the leading story of the next newscast. Sudden and large releases increase the probability of severe environmental impairment or destruction, especially if they occur in or near a sensitive ecological area.

Historically, there was no regulation of the disposal of wastes containing contaminants. Until the mid-twentieth century, the most convenient and least costly method of waste disposal was "up the stack or down the river" (Haynes 1954). We now know that the perception of a contaminant leaving your immediate vicinity and being gone forever is not true.

Today, many releases of contaminants are permitted but carefully monitored, such as wastewater discharges from industrial and municipal sources and those to the atmosphere from industrial and commercial sources and power plants. The legislation governing such releases (the Clean Water Act of 1972 and the Clean Air Act of 1990) permits a point source (such as a wastewater plant or a smokestack) to release specific compounds at low concentrations to ensure there are no adverse health and environmental effects detectable on the media they are released into (water or air). A majority of the other contaminant releases into the environment are unintentional or accidental and occur from numerous sources under a multitude of circumstances, and these include (Fetter 1993; Rogers 1996; USGS 2006a)

- Permitted releases of contaminants in wastewater and air emissions
- Spills and leaks from several types of containers or operations, including
  - Drums of various sizes and shapes
  - Pipelines
  - Above ground storage tanks
  - Underground storage tanks
  - Tanker trucks and other transport vehicles
  - Tanker ships
  - Railroad tanker cars
  - Aircraft

- Accidents or collisions involving automobiles where gasoline and oil may be released, railroad derailments, etc.
- Septic systems
- Sewer leaks
- Landfills and dumps
- Automobile exhaust
- Sumps and dry wells
- Former disposal lagoons
- Animal feed lots
- Injection wells
- Fertilizer application
- Application of pesticides and herbicides
- Application of deicing compounds (i.e., road salts)

Exploring and evaluating fate and transport of contaminants in the environment must consider the complex interplay between the environment and the contaminant. And, as Figure 8.1 shows, fate and transport is also influenced by the method and location of contaminant release and the volume, mass, and duration of the release. We will break down and describe these factors influencing the migration of contaminants, degradation of contaminants, and then describe how the fate and transport of each group of substances is influenced by the group's specific chemistry.

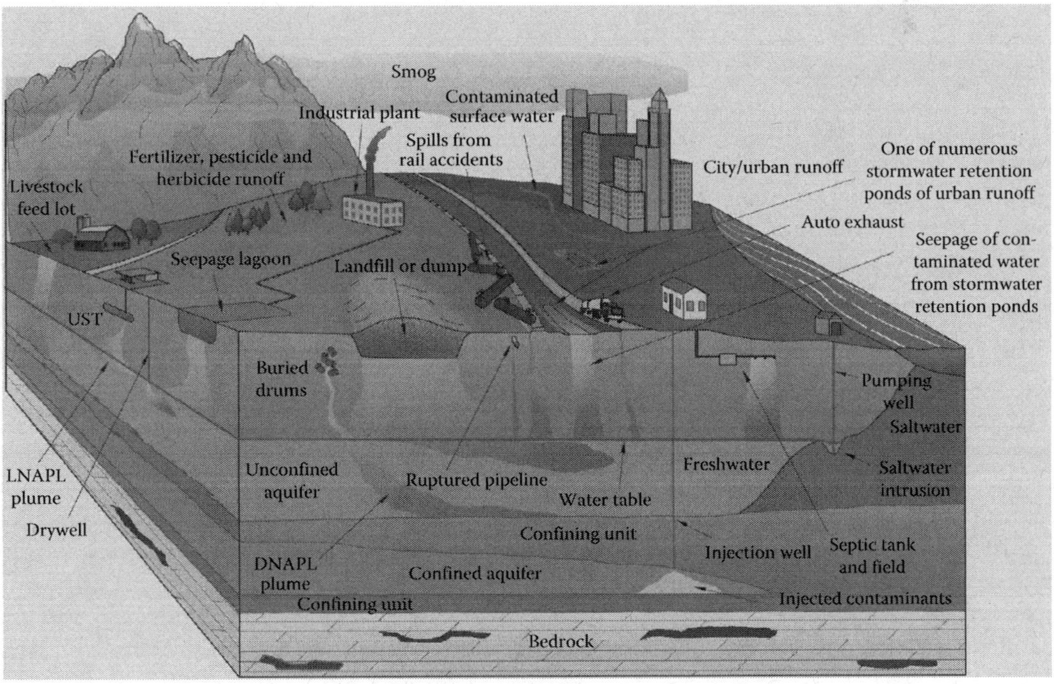

FIGURE 8.1
(See color insert.) Sources and locations of contaminant release.

8.3 Principles of Contaminant Fate and Transport

Interaction with the environment begins immediately after a contaminant has been released. Once released, a contaminant can do three things (Manahan 1993; Hemond and Fechner-Levy 2000):

- Stay put
- Migrate in soil, water, air, or a combination of media
- Degrade, transform, or get destroyed

The factors controlling these three outcomes depend upon the physical chemistry of the contaminant and the characteristics of the receiving environment (USGS 2006a; Rogers et al. 2007). For example, some contaminants may change form, as when many of the chemical compounds in gasoline immediately evaporate after their exposure to the atmosphere. Other contaminants may degrade in a matter of minutes after being released, while some may last for thousands of years or sometimes longer, as with certain radioactive compounds. Therefore, computing a mass balance should be the first action when assessing any particular release, as this will ensure that the mass or volume of contaminant released is accurately measured. Conducting a mass balance also serves to validate or refute our current understanding of the behavior of the different environmental elements influencing the contaminant once it has been released.

A simple mass balance is expressed as

$$\text{Amount released} = \text{amount recovered} + \text{amount lost to the environment (air + water + soil)} \tag{8.1}$$

Once in the environment, contaminants can and often move between soil, surface water, groundwater, and the atmosphere, and they can also degrade. Many contaminants degrade quickly if conditions are favorable, yet others persist and last for years or decades depending upon the conditions present (USGS 2006a). Factors degrading contaminants fall into two broad categories: biotic degradation and abiotic degradation. **Biotic** degradation involves microorganisms or fungi and occurs when an organism, such as a bacterium, uses a contaminant as a source of food and either degrades or transforms the contaminant (USGS 2006a). **Abiotic** degradation involves other processes not including microorganisms. Examples of abiotic degradation include photolysis—the degradation as a result of exposure to sunlight (USGS 2006a)—and hydrolysis. As described in Chapter 3, hydrolysis involves cleaving a molecule into two parts by the addition of a molecule of water.

When examining contaminant degradation in the environment, it is important not to confuse dilution with degradation. If given enough time, contaminants may become diluted, and this process results in a decrease of the contaminant concentrations per unit volume of the media being measured. However, dilution is not degradation, since it does not involve a chemical transformation of the contaminant. We will discuss degradation in more detail in Section 8.3.2.

8.3.1 Basic Contaminant Transport Concepts

Contaminant transport in the environment is dominated by three physical transport mechanisms (USEPA 1996a; USGS 2006a): advection/convection, molecular diffusion, and dispersion. **Advection** is the horizontal transport of any property by the atmosphere and water. Common examples are the transfer of heat by wind and sediment transport within a flowing stream. **Convection** is the vertical advection of air, water, or other fluid as a result of thermal differences. We introduced the concept of convection in Chapter 2 as the driving force behind plate tectonics. **Molecular diffusion** is the movement of a chemical from an area of higher concentration to an area of low concentration due to the random motion of the chemical molecules. **Dispersion** (also referred to as hydrodynamic dispersion) is the tendency for contaminants to spread out from the path of the expected advective flow (USGS 2006a). Occasionally, the effects of diffusion and dispersion are treated together, but for the purposes of this book we treat them separately.

The rate of advective transport of a contaminant is often expressed in terms of flux density. **Flux density** is the mass of a chemical transported across an imaginary surface of a unit area per unit of time. Equation 8.2 shows this relationship (Hemond and Fechner-Levy 2000), which is independent of the media involved (soil, surface water, groundwater, or the atmosphere):

$$J = CV \tag{8.2}$$

where

J is the flux density = (mass/[length × width] × time) or $[M/L^2 \times T]$
C is the concentration of the chemical per cubic liter or meter of media $[M/L^3]$
V is the velocity [length/time] or [L/T]

An example of molecular diffusion is shown in Figure 8.2 (Payne et al. 2008). From the release time to infinity, a contaminant released into a fluid such as air or water will diffuse throughout the fluid at random locations.

Fick's First Law of Diffusion (Equation 8.3) can be used to predict the diffusive flux of a contaminant (solute) across an imaginary plane as a function of the rate of change in concentration with distance (Hemond and Fechner-Levy 2000).

$$J = -D\left(\frac{dC}{dx}\right) \text{(one dimension)} \tag{8.3}$$

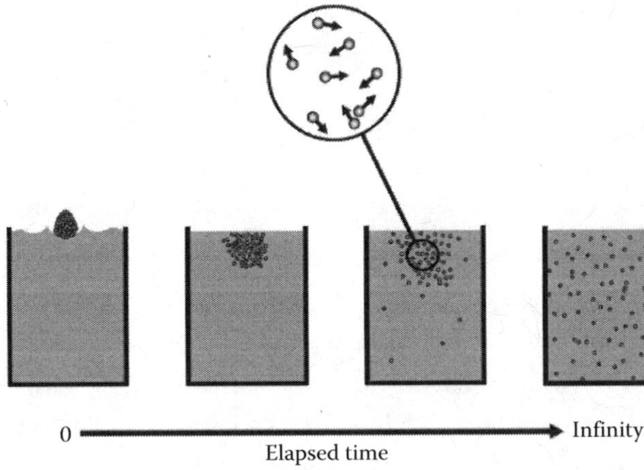

FIGURE 8.2
Molecular diffusion. (Adapted from Payne, F.C. et al., *Remediation Hydraulics*, CRC Press, Boca Raton, FL, 2008. With permission.)

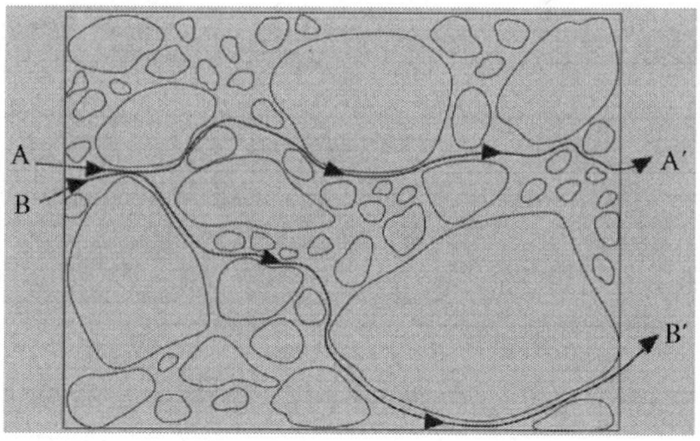

FIGURE 8.3
Effects of dispersion on subsurface migration in a porous medium.

where
J is the flux density $[M/L^2 \times T]$
D is the Fickian mass transport coefficient $[L^2/T]$
C is chemical concentration $[M/L^3]$
x is the distance over which a concentration change is being considered [L]

Note: In simple calculations, the minus sign is often omitted if the direction of Fickian transport is clear.

As a contaminated fluid flows through a porous medium, it will mix with non-contaminated water. The result will be a dilution of the contaminant by a process known as dispersion (Fetter 1993). Longitudinal dispersion occurs along streamlines as the contaminated water moves downgradient. Lateral or transverse dispersion occurs at right angles to longitudinal dispersion. Figure 8.3 shows the basic causes of longitudinal and lateral dispersion. As a fluid moves through pores in an aquifer, it follows a tortuous path, moving faster through the center of the pore and slower along the edges of the soil particles. As a result, some of the fluid will travel a longer path.

General factors that influence or control the rate of migration of a contaminant include (Hornsby 1990):

- Physical properties of the contaminants themselves
- The geological environment where the release occurs
- Climatological factors
- Vegetation factors

Specific physical properties affecting migration of contaminants in soil include (USGS 2006a; Rogers et al. 2007; Payne et al. 2008):

- Solubility in water. The more soluble a contaminant is in water, the more mobile it will be in the subsurface environment.
- Vapor pressure. As vapor pressure increases, affinity to volatilize increases and the more likely a contaminant will be present in the gas phase.
- Molecular weight. The higher the molecular weight, the greater the energy requirement to transport the contaminant in the horizontal direction. Increased molecular weight may induce the contaminant to migrate downward in areas of steep slopes.

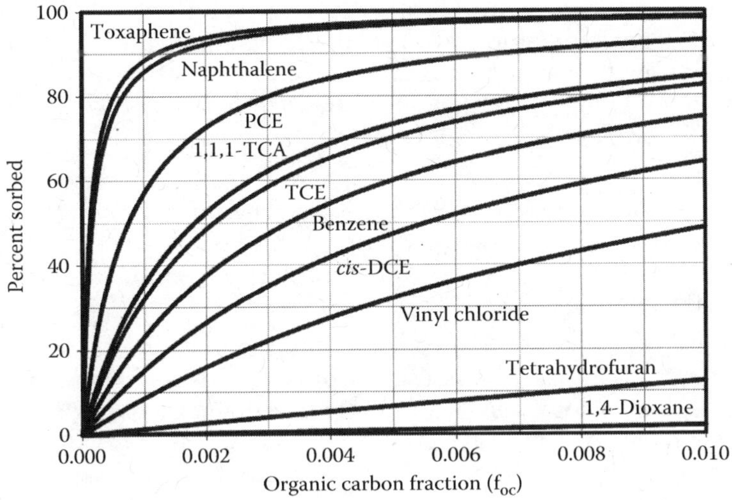

FIGURE 8.4
Sorbed percent of select VOCs and its relation to increasing amounts of organic carbon within geologic media. (Adapted from Payne, F.C. et al., *Remediation Hydraulics*, CRC Press, Boca Raton, FL, 2008. With permission.)

- Chemical stability and persistence. Stable compounds have more time to migrate and may migrate further if conditions are favorable before they are degraded, transformed, or destroyed.
- Sorptive properties. Sorptive processes include adsorption and absorption. The lower the sorptive properties, the higher the migration potential. This relationship is shown in Figure 8.4, which graphs the percent sorbed (nonaqueous) of select volatile organic compounds (VOCs) in relation to organic carbon content within the subsurface. For this relationship to exist, the number of sorption sites within the geological materials has not been exceeded (Payne et al. 2008).

Specific geologic factors affecting migration of contaminants in soil include (Rogers 1996; USGS 2006a; Rogers et al. 2007):

- Composition. As noted in Chapters 2 and 3, soil type has a significant influence on the migration of contaminants. In general, coarse-grained soils, such as sand and gravel, do not impede the migration of contaminants nearly as much as soils composed of clay.
- Porosity and permeability. Direct evidence of migration potential is hydraulic conductivity. As described in Chapter 3, soils composed of sand and gravel have a hydraulic conductivity generally 100 to 1000s of times higher than soils composed of clay. Although soils or sediments composed of clay may have a high relative porosity, they are usually not as permeable because the porosity is generally not interconnected, or the pore spaces are simply too small to allow the flow of water. In some cases, however, as demonstrated in Chapter 3, secondary porosity such as root fragments and vertical fractures can make a seemingly impervious clay deposit much more permeable than expected.
- Organic carbon content. As just discussed, an increased total organic content in soil may impede the migration of certain types of contaminants as long as the

contaminant mass does not exceed the holding capacity of the soil. Soils or sediments composed of sand and gravel generally have lower organic carbon content compared to soils composed of clay. An example of high relative organic carbon content is shown in Figure 2.36 where a sand deposit is interbedded with layers of ash primarily composed of organic carbon.

- Soil chemistry. The pH, redox potential, and other soil chemistry factors influence the contaminant migration of many different types of compounds. Many metals, for instance, are particularly sensitive to pH differences in soil, and these differences—along with the characteristics of each metal—influence their migration patterns in the environment.
- Stratigraphy. This is where heterogeneity and the anisotropic nature of the geologic sediments play a significant role at both a micro- and macroscale. As shown in Chapter 2, the geology beneath the surface can and does change dramatically in just a few meters in any direction. The result is differing sediment types and chemical composition, including pH and redox potential, of subsurface layers acting to impede or enhance contaminant migration.
- Unconformities. From Chapter 3, the presence of unconformities influences the migration of groundwater as well as the migration of contaminants. Hydrogeologically, the presence of an unconformity indicates there is a surface or plane in the subsurface geologic environment. This space often produces a significant difference in the hydraulic conductivities within the soils or sediments above and below the unconformity, especially if these units are fine-grained sediments such as silts or clays. As a result, contaminants released in this type of location use the unconformity as a sink and migrate much further than expected.

Specific climatological factors affecting migration of contaminants in soil include (USGS 2006a)

- Freeze-thaw cycles. Freeze-thaw cycles (Chapter 3) can lead to the development of vertical fractures in the soil to depths approaching 10 m. These vertical fractures are a type of secondary porosity and, if present, can greatly increase the migration potential of contaminants vertically through the soil column. In addition, the freezing of near surface soils may trap contaminants at the surface and lead to increased contaminant loading during warmer periods when the ice melts.
- Rainfall. Because water is the universal solvent, geographic locations receiving abundant rainfall play a significant role in enhancing the migration of contaminants, especially if they are soluble. Rainfall also enhances the migration of contamination through the physical transport of particles with sorbed contamination on their surfaces.
- Snowfall. Airborne deposition of contaminants may become temporarily trapped in seasonal snowpack. Increased contaminant loading to the environment may occur during warmer periods when the snowpack melts (Wania et al. 1998).
- Wind. Many locations within the United States contain significant amounts of wind-blown deposits, especially in the southwest (Chapter 2). Contaminants with a high sorption potential may become attached to fine wind-blown sediment grains and transported over long distances (Section 7.8.12). In addition, volatile contaminants released as a gas are routinely transported by wind.

- Water vapor. Humidity plays a significant role in the water cycle by affecting air movement and by reacting with contaminants in the gas phase and contaminants released into the atmosphere attached or sorbed onto particulate matter. Contamination goes along for the ride when precipitation formed around contaminated condensation nuclei is transported from the atmosphere to the lithosphere and then entrained by surface runoff.
- Fog. Fog can be an effective agent for the transfer of acid rain by transferring it to vegetation or other surface materials through direct contact.
- Flood events and hurricanes. Due to their catastrophic nature and magnitudes, floods and hurricanes may not only increase contaminant migration but can also cause significant releases. During the 1993 floods of the Mississippi River, numerous barrels containing hazardous waste were swept away and deposited in the Gulf of Mexico.
- Solar energy. Sunlight breaks down some contaminants through a process called photolysis. In toxic microorganisms, ultraviolet light passes easily through cell walls, cytoplasm, and nuclear membranes and prevents DNA replication.

Specific vegetative factors that affect migration of contaminants in soil include (USGS 2006a)

- Roots. These pathways within shallow subsurface geological materials can enhance the migration of contaminants (Chapter 3). Certain plants can uptake contaminants. Many types of plants and trees have the capability with their root systems to assist in the removal of contaminants from shallow subsurface soil. Contaminants may be stored in plant tissues or are transformed through biologic processes.
- Microorganisms. Microorganisms in the soil often biodegrade many different types of contaminants by using the contaminants themselves as a source of food.

8.3.2 Basic Contaminant Degradation Concepts

The degradation of specific compounds in the environment is expressed in terms their half-life. **Half-life** is the average amount of time required to degrade half or 50% of a specific contaminant population (USEPA 1996a). Contaminants degrade through biotic or abiotic processes, and the processes controlling their rate of decay depend upon the following factors (USEPA 1996a; USGS 2006a):

- The nature of the release. This group of factors includes
  - Media receiving the release. Was it into the atmosphere, surface or subsurface soil, surface water, ocean, directly to groundwater, or a combination of media?
  - Amount (volume or mass) of the release
  - Number of contaminants released. Was it a single contaminant—or a mixture of several contaminants?
  - Physical state of the release (i.e., liquid, solid, or gas)
  - Time duration of the release

- Geologic environment. The most significant geologic factors include
  - Soil composition and other physical characteristics such as permeability, porosity, moisture content, composition, extent and distribution, thickness, total organic carbon content, pH, redox potential, dissolved oxygen (if saturated), and other parameters
  - Depth to bedrock, type, composition, distribution, fractures, permeability, porosity of bedrock, and other parameters
  - Terrain and topography
  - Potential surface and subsurface migration pathways
- Climatic factors
  - Release location. Different climates—deserts, mountains, humid areas, or temperate regions can influence the type and rate of degradation
- Surface water features
  - Distance to surface water bodies and their type. Immature streams, mature streams, rivers, lakes, wetlands, and bogs can differ in pH due to the rock composition of their channels and bottoms and the amount of organic matter they receive from outside, inputs termed allochthonous
- Weather conditions at the time of the release. Weather conditions are often important and sometimes overlooked as potentially significant. Those conditions affecting degradation and migration include
  - Temperature
  - Humidity
  - Precipitation
  - Wind speed and direction
- Biologic factors
  - The type, distribution, and amount of microorganisms will influence the rate and can determine if degradation even occurs
- Anthropogenic factors. Anthropogenic factors are often overlooked and frequently significant. These include
  - Physical landscape alteration (i.e., buildings, roads, parking lots, etc.)
  - Surface water drainage modifications including stormwater control and wetland destruction
  - Alteration of native vegetation
  - Introduction of invasive vegetation
  - Regional contaminant loading, including sources, duration, type, release points, and physical state of contaminants (i.e., solid, liquid, or gas)
  - Developmental history of the area and region

8.3.2.1 Biotic Degradation

Microbes have the ability to oxidize a variety of organic contaminants including many VOCs, polynuclear aromatic hydrocarbons (PAHs), and other compounds. This capability arises from their enormous variety, populations, rapid growth, and diversity of environmental niches. Soluble organic compounds with low molecular weights such as alcohols and organic acids are metabolized and degraded rapidly by microbes, perhaps because these compounds also occur naturally and microbes have evolved to degrade them more efficiently (Hemond and Fechner-Levy 2000). Halogenated synthetic or anthropogenic compounds, however, are not easily degraded by microbes (USEPA 2006), and some contaminants that escape biotic degradation bioaccumulate in the bodies of organisms.

The rate of biodegradation by microorganisms generally slows if the organic contaminants possess the following (Hemond and Fechner-Levy 2000):

- High molecular weight
- Low water solubility
- Presence of benzene or aromatic rings
- A large amount of branching within the molecular structure
- Presence of halogen atoms in the structure (chlorine, fluorine, bromine, or iodine)

8.3.2.2 Abiotic Degradation

Abiotic degradation refers to degradation processes accomplished without microorganisms. Common abiotic degradation processes include

- Photolysis
- Hydrolysis
- Reduction–oxidation
- Radioactive decay

Photolysis (sometimes referred to as photodegradation or photochemical degradation) occurs in the presence of sunlight. Common examples of this process include the fading of colored and dyed objects and the transformation of plastic object textures from pliable to brittle. Photolysis is most common in the atmosphere, surface water, and at Earth's surface. The degree of degradation caused by sunlight depends on the wavelength spectrum of the light, intensity of light exposure, and duration (Hemond and Fechner-Levy 2000). If the energy per photon is sufficient to break a specific chemical bond, photolysis can be initiated. Once begun, increased light intensity will result in a faster rate of degradation. Ultraviolet light is especially effective at degrading many organic contaminants (USEPA 1996a).

Degradation by photolysis is often observed in organic compounds with double bonds between their carbon atoms. Many VOCs, PAHs, and semivolatile organic compounds (SVOCs) characterized by a benzene ring fit this pattern. Under favorable conditions, the degradation of many organic compounds by photolysis may occur in a short period of time—from a few hours to a few days (Lyman et al. 1990).

The process of **hydrolysis** occurs when a water molecule breaks. Contaminant degradation by hydrolysis also destroys a molecule of contaminant. Two types of chemical compounds are susceptible to degradation by hydrolysis (Schwartzenbach et al. 1993):

- Alkyl halides, straight-chained or branch-chained hydrocarbons, where one or more hydrogen atoms have been replaced by a chlorine, fluorine, bromine, or iodine atom. Using an "X" to represent a halogen atom and an "R" to represent the hydrocarbon group, the basic hydrolysis reaction is shown in Equation 8.4:

$$H_2O + R - X \rightarrow R - OH + H^+ + X^- \tag{8.4}$$

- Esters, compounds containing a modified carboxylic acid group (–COOH), where the acid hydrogen atom has been replaced by a different organic functional group. The process of hydrolysis within this group converts the ester compound into the "parent" organic acid and an alcohol (Equation 8.5):

$$H_2O + CH_3COOC_2H_5 \rightarrow CH_3COOH + C_2H_5OH \tag{8.5}$$

In reduction–oxidation degradation reactions (redox), electrons are transferred from one atom to another. **Chemical reduction** is defined as the addition of electrons and **chemical oxidation** is defined as the loss of electrons (Hemond and Fechner-Levy 2000). In a reaction involving atoms A and B, if atom A gains an electron, it is reduced, and atom B, having donated an electron, is the **reductant**. Because atom B loses an electron, B is oxidized and atom A is the **oxidant**. Each reaction involving the loss or gain of an electron is termed a **half reaction**.

The oxidation of contaminants can occur very rapidly through combustion or incineration. Here, fire transforms the contaminants through oxidation at greatly elevated temperatures and is represented by applications in cooking, heating, and transportation (Hemond and Fechner-Levy 2000).

8.3.3 Fate and Transport of Contaminants in Soil

Folklore holds that the presence of soil protects groundwater quality by filtering contaminants before they reach and impact groundwater (Hornsby 1990). Soil does have a limited ability to filter contamination; however, it does not do a perfect job of holding, filtering, degrading, transforming, or destroying contaminants. These capabilities also depend upon a number of factors related to the chemistry of the contaminant and the geological environment where the contaminant is released.

Soil is defined as the unconsolidated mineral matter on the immediate surface of Earth (Soil Science Society of America 1987). Basic to an understanding of soil are the factors affecting its development and ultimate physical structure. The composition, texture, and thickness of soil are influenced by its source material, plant growth, micro- and macroorganisms, climate, topography, process of formation (e.g., alluvial, fluvial, and glacial), and physical and chemical weathering since original formation (Brady and Well 1999). Structurally, soil is composed of three phases: soil gases, soil water, and organic and inorganic solids. The gas and water phases may comprise 25%–50% of the total volume of a surface soil, especially at shallow depths (USEPA 1999).

Contaminants released into the soil can migrate within all three phases. Once a contaminant is resident in soil, these factors determine its migration rate (Schnoor 1996; USEPA 1999; Kaufman et al. 2009):

- Contaminant mass released
- Duration of the release
- Physical chemistry of the contaminant
- Physical chemistry of the soil (e.g., pH, redox potential and mineralogy)
- Amount of water present
- Permeability of the soil
- Retention capacity of the soil

- Distribution of plant matter
- Biological interaction between the contaminant and indigenous microorganisms

All these factors must be well understood before an accurate assessment of the fate and transport of a contaminant in soil can be made.

Contaminants migrate through the soil by two basic processes: diffusion and mass flow. The rates of diffusion and mass flow greatly depend upon the local geology and the physical chemistry of the contaminant. Diffusion of substances through soil and aquifer materials occurs in response to differences in energy from one point to another. These energy gradients may be caused by differences in temperature or chemical concentrations within the contaminated area. In most cases, however, the principal process moving a contaminant through soil is mass flow or advection because contaminants generally want to move downward through the soil under the force of gravity (USGS 2006a).

Contaminant-specific physical and chemical attributes affecting the migration of contaminants in soil include (USEPA 1999, 1996b,c; Wiedemeier et al. 1999)

- Solubility
- Vapor pressure
- Density
- Chemical stability
- Persistence
- Adsorption potential

In soil, solid phase contaminants migrate much more slowly than liquid phase contaminants and tend to remain relatively close to their point of release or deposition (USEPA 1999). Before they can migrate a significant distance, solid phase contaminants must change phase or undergo a transformation process. For example, heavy metals—a solid phase contaminant—typically remain at their point of release or deposition. If they undergo oxidation, however, their solubility and other properties enabling migration may increase (Lindsay 1979). And, once a contaminant begins to dissolve in water, it may also be subject to further transformation reactions induced by indigenous bacteria present in the surface soil (Sutherson and Payne 2005).

If a source continues to emit contaminant in solid or liquid form that dissolves in water, the underlying soil will eventually become saturated. The leading edge of contamination will migrate either horizontally or vertically or both as long as the retention capacity of the soil is exceeded (USEPA 1999). When the contaminant release stops, the migration of the liquid contaminant will significantly decrease as the soil regains its retention capacity (Kaufman et al. 2009).

Since soil is also composed of gas, contaminant migration through the vapor phase is often observed with contaminants having higher relative vapor pressures. VOCs are frequent participants in this type of migration. Capillary forces can also induce the migration of liquid phase contaminants.

8.3.4 Fate and Transport of Contaminants in Surface Water

The transport of contaminants in surface water is dominated by turbulent advective flow. Because the rate of flow in a river or stream varies significantly by location and over time, estimating the contaminant flow involves averaging the streamflow variations and contaminant concentrations over a specified time interval. Conducting measurements at multiple locations also provides a more accurate measure of the rates of streamflow and contaminant transport.

Molecular diffusion also influences contaminant migration in surface water. Turbulent flow is characterized by water moving in constantly changing and unpredictable patterns. The swirls resulting from turbulent flow are called **eddies**, and they appear in many sizes, volumes, and velocities. Random mixing of the water within eddies creates turbulent diffusion and also influences mass transport. Wave action can create similar eddying effects in lakes and other nonflowing water bodies.

Analytically, the transport rates for chemicals in surface water are expressed in terms of flux density. Flux density is the mass of a chemical transported across an imaginary surface of a unit area per unit of time (Equation 8.2; Hemond and Fechner-Levy 2000). Fick's First Law (Equation 8.3) is also used to describe the flux density of mass transport by turbulent dispersion (Hemond and Fechner-Levy 2000).

Figure 8.5 depicts a municipal or industrial waste water plant discharging to surface water and many of the ensuing contaminant fate and transport processes, including (USGS 1995a):

- Transport of discharged wastewater solute downstream
- Mixing due to turbulent advection and turbulent diffusion
- Photolysis

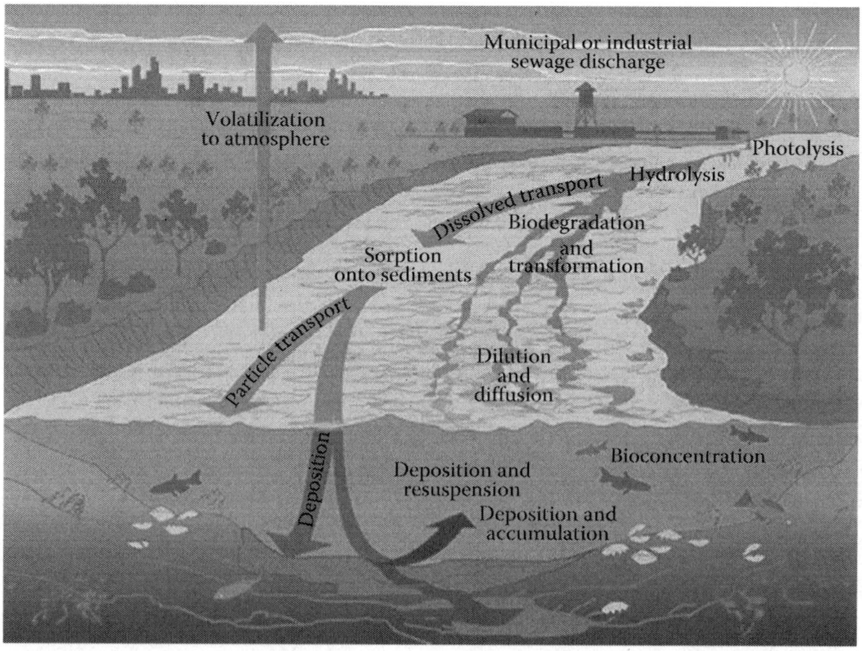

FIGURE 8.5
(See color insert.) Fate and transport effects in surface water. (From United States Geological Survey (USGS), Contaminants in the Mississippi River, USGS Circular 1133, Washington, DC, 1995a.)

- Hydrolysis
- Biodegradation
- Transformation
- Volatilization to the atmosphere
- Sorption of contaminants in sediment
- Bioaccumulation of contaminants by animal and plant life
- Dilution

Molecular diffusion and groundwater discharge and recharge are not shown in the figure.

Since most urban areas of the United States obtain their potable water from surface sources, wastewater discharges are a concern. Treatment costs rise when the source of supply is contaminated, and the risks of biological contamination also increase. Moreover, as we noted in Chapters 2 and 3, a majority of urban areas are located along rivers and streams and other surface water bodies such as the Great Lakes, and many of these water bodies have already been degraded—some significantly (USGS 1995b).

8.3.4.1 Contaminant Persistence and Bioaccumulation

When certain contaminants are released to surface waters through overland flow, stormwater runoff, or wastewater discharge, they may accumulate in sediments. Compounds with a higher likelihood of accumulating have the following physical characteristics:

- Low solubility
- High molecular weight
- Low potential to degrade
- High sorption potential

Contaminants having these physical chemistry attributes typically do not sustain themselves in surface water unless the rate of flow is substantial. Even then, they may be carried along the bottom of the stream or river until the carrying capacity of surface water is insufficient and the contaminants settle to the bottom. As shown in Figure 8.5, certain locations in the stream bottom become a sink for these contaminants as they accumulate (USGS 1996a). If the source of contamination persists, greater amounts of the contaminant will be deposited. The accumulation of contaminants in sediments increases the exposure risk to aquatic and terrestrial plant and animal life. If any of the contaminants exhibit bioaccumulation properties, contamination may proceed up the food chain from bottomdwelling macroinvertebrates to small fish and eventually to larger fish, predatory birds, and other organisms. Humans are situated at the top of food chain, and the potential risks to human health must be considered when evaluating the fate and transport of contaminants in surface water (USEPA 2009a, 2009b).

Contaminants considered to be bioaccumulative include (USEPA 2009b)

1. Mercury
2. Polychlorinated biphenyls (PCBs)
3. Chlordane
4. Dioxins
5. Dichlorodiphenyltrichloroethane (DDT)

Surface water bodies under advisory because of one or more contaminants include (USEPA 2008a)

- 43% of the nation's total lake acres (excluding the Great Lakes), representing approximately 7.3 million ha (18 million ac) of surface water
- 39% of the nation's total river miles, or approximately 2,250,000 km (1.4 million miles)
- 42% of the nation's contiguous coastal waters
- 100% of the Great Lakes and their connecting waters

Figure 8.6 shows the number of advisories in each state and the states issuing statewide advisories in 2008 (USEPA 2009b). Every state had at least one advisory in 2008. Figure 8.7 shows the number of advisories by contaminant. Please note the sharp increase in the area of surface water covered by PCB and mercury advisories, while there has been relative areal stability for DDT, dioxins, chlordane, and others (USEPA 2009b).

8.3.5 Fate and Transport of Contaminants in Groundwater

Transport of contaminants in groundwater is dominated by three factors: advection, dispersion, and molecular diffusion. When applied to groundwater, **advection** is the movement of contaminants by the bulk motion of groundwater flow, dispersion is the tendency for contaminants to spread out from the path of the expected advective flow,

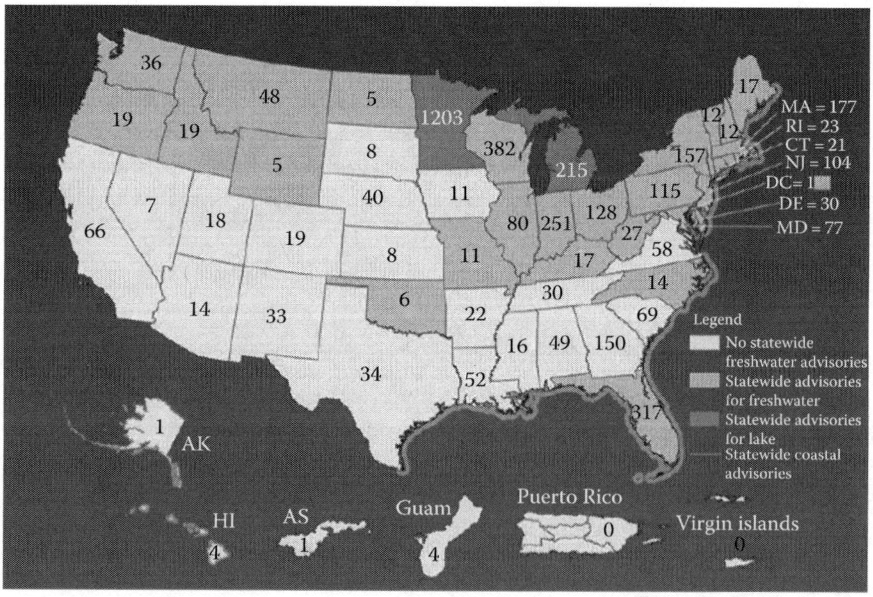

FIGURE 8.6
Health advisories for bioaccumulative contaminants in surface water by state. (From United States Environmental Protection Agency (USEPA), *Biennial National Listing of Fish Advisories for 2008*, EPA-823-F-09-007, Washington, DC, 2009b.)

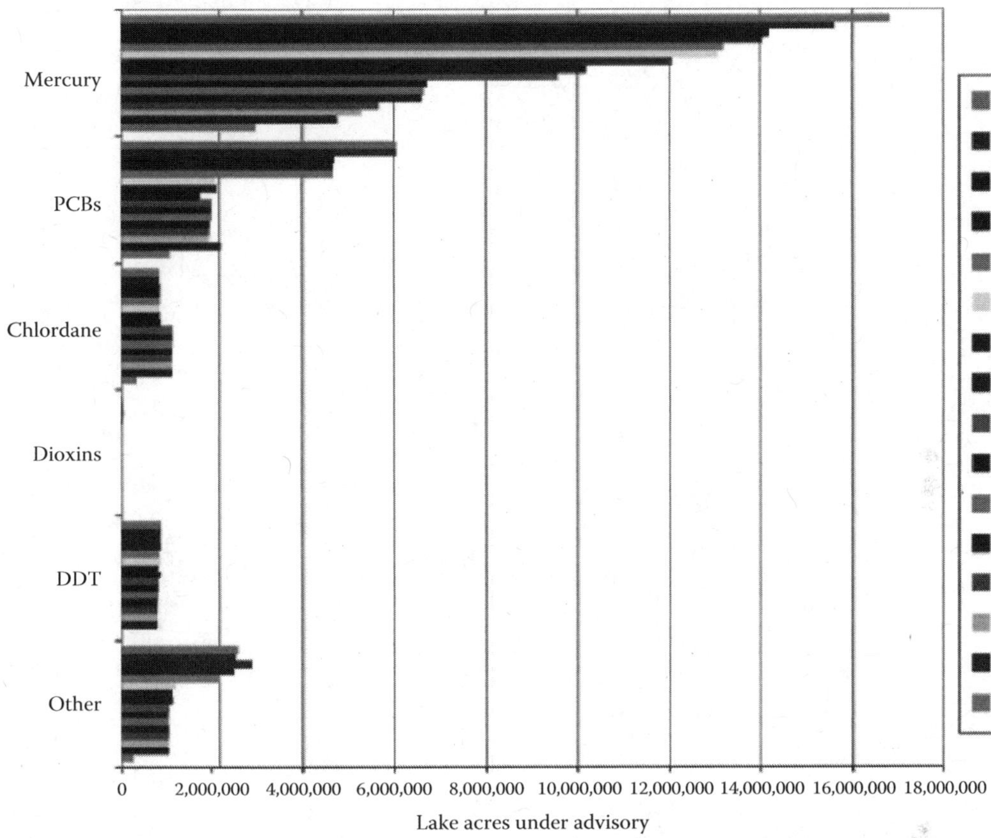

FIGURE 8.7
Advisories by contaminant from 1993 to 2008. (From United States Environmental Protection Agency (USEPA), *Biennial National Listing of Fish Advisories for 2008*, EPA-823-F-09-007, Washington, DC, 2009b.)

and diffusion is the action of spreading of molecules from areas of high concentration to areas of low concentration at the molecular level. Groundwater flow lacks turbulent diffusion because velocities are typically much slower. In some instances, however, groundwater does display turbulent dispersion, especially in karst topography (Chapters 2 and 3), where water flowing beneath the surface flows and behaves much like a stream at the surface.

Figure 8.8 shows a spill from an underground storage tank (USGS 2006a). Here, advective transport of contaminants in groundwater is occurring at the water table boundary. Diffusion, biodegradation, volatilization, and recharge from surface precipitation affecting the contaminant migration are also shown. The effects of dispersion and diffusion are represented by the spreading of the contaminant plume as it migrates from a hole or ruptures at the bottom of the tank (USGS 1998, 2006a).

The representation of dispersion in Figure 8.8 is overly simplistic, because the geology of unconsolidated sediments is very complex and typically displays a high degree of heterogeneity and anisotropic distribution patterns. As a result, contaminants migrating in unconsolidated deposits do not migrate uniformly but migrate within the physical parameters of advection dictated by the particular subsurface geology.

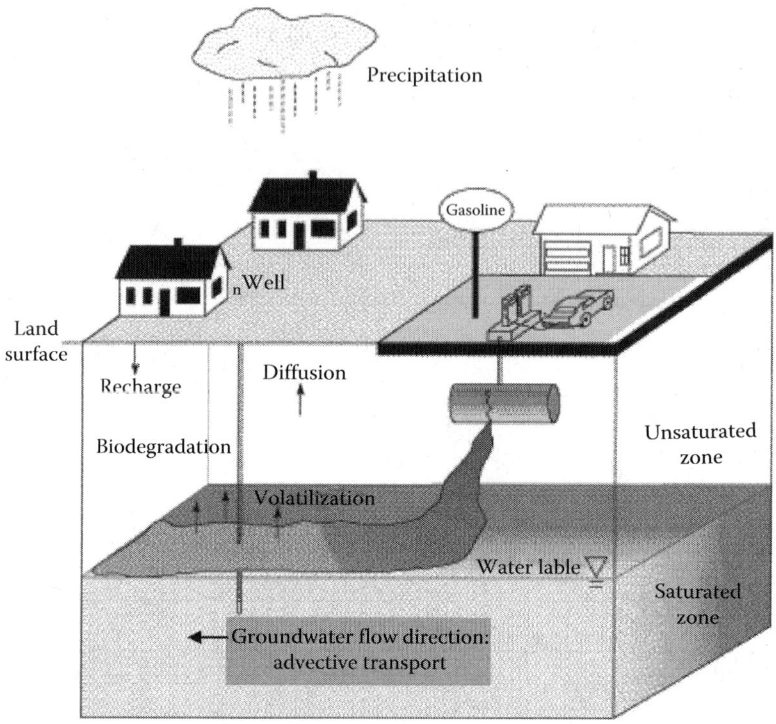

FIGURE 8.8
Advective transport and other processes effecting the migration of contaminants in groundwater. (From United States Geological Survey (USGS), Simulating transport of volatile organic compounds in the unsaturated zone using the computer model R-UNSAT, USGS Fact Sheet 019-98, Washington, DC, 1998.)

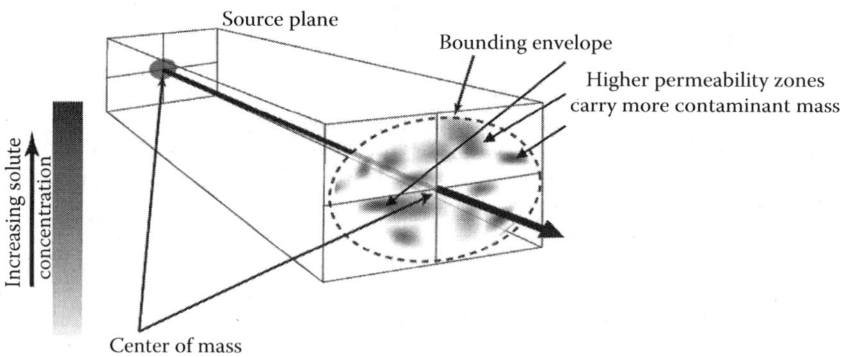

FIGURE 8.9
Contaminant migration in heterogeneous and anisotropic geologic media. (Adapted from Payne, F.C. et al., *Remediation Hydraulics*, CRC Press, Boca Raton, FL, 2008. With permission.)

This concept is represented in Figure 8.9, where a contaminant (solute) is shown migrating in the more permeable layers. More highly permeable layers have a higher hydraulic conductivity and behave as preferred groundwater and contaminant migration pathways. These layers are essentially super highways for groundwater and contaminant transport, and in some instances the hydraulic conductivity is from 100 to sometimes 1000 times greater over distances of just a few centimeters.

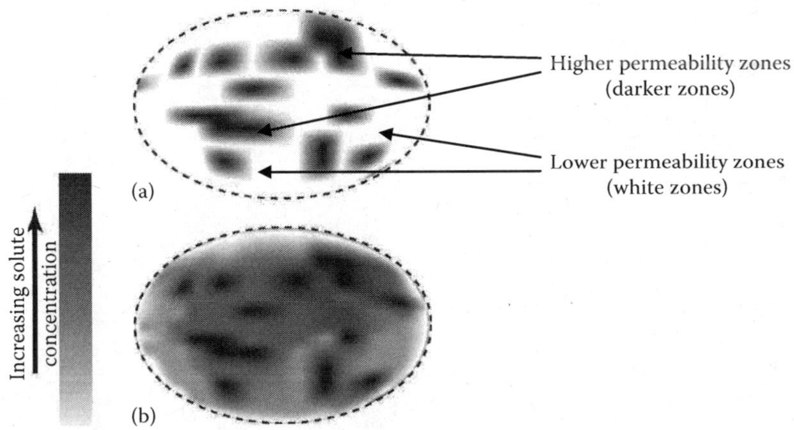

FIGURE 8.10
Effects of diffusion over time within a contaminant plume. (a) depicts an immature contaminant plume and (b) depicts a mature contaminant plume and the effects of diffusion. (Adapted from Payne, F.C. et al., *Remediation Hydraulics*, CRC Press, Boca Raton, FL, 2008. With permission.)

Several examples of this affinity for contaminants to migrate within more permeable layers are shown in this book (Figures 2.24, 2.28, and 2.36). Zones with higher permeability move water more quickly and have a higher flux density. If more water moves through these higher permeability zones, then a potentially greater contaminant mass also moves through. Figure 8.9 shows this relationship at the right of the diagram, downgradient from the source.

As contamination continues to migrate along these flow paths of higher permeability dictated by the subsurface geology, diffusion of contaminants into less permeable zones occurs (Figure 8.10). The top portion of the figure represents the flow paths of contaminants in the early stages of migration, and the later stages of migration are shown in the figure's bottom portion. Over time, the contaminant (solute) has diffused into the less permeable, lower hydraulic conductivity geologic materials (shown as in Figure 8.10b; Payne et al. 2008).

The transport of contaminants in groundwater is also influenced by many of the same factors affecting the migration of contaminants in unsaturated soil or the vadose zone:

- Physical chemistry of the contaminants
    - Solubility
    - Molecular weight
    - Vapor pressure
    - Stability and persistence
- Sorption potential
- Type, distribution, and amount of microorganisms
- Tendency to biodegrade
- Dissolved oxygen content of groundwater
- Geological factors
    - Stratigraphy (including thickness and distribution of geological units down to microstratigraphic scales at the centimeter or even millimeter)

    - Presence of unconformities
    - Soil or sediment chemistry
    - Organic carbon content
    - Porosity and permeability
    - Composition
- Climate factors
    - Freeze and thaw cycles
    - Recharge from surface precipitation
    - Flood events
    - Seasonal climatic variations
- Vegetative factors including types and distribution of surface vegetation, root networks, and water requirements

Sorption potential has a significant effect on the migration of contaminants in groundwater because it slows the migration of contaminants even as the flow rate of the transporting groundwater remains constant. This effect is termed **retardation** (USGS 2006a), and the degree of retardation present depends upon the specific contaminants sorptive affinity and the amount of total organic carbon in the aquifer matrix (Bedient et al. 1994) (Figure 8.4).

Many contaminants are captured by pumping wells or migrate to surface water if the travel times and/or distances are short enough before they degrade. Figure 8.11 shows an

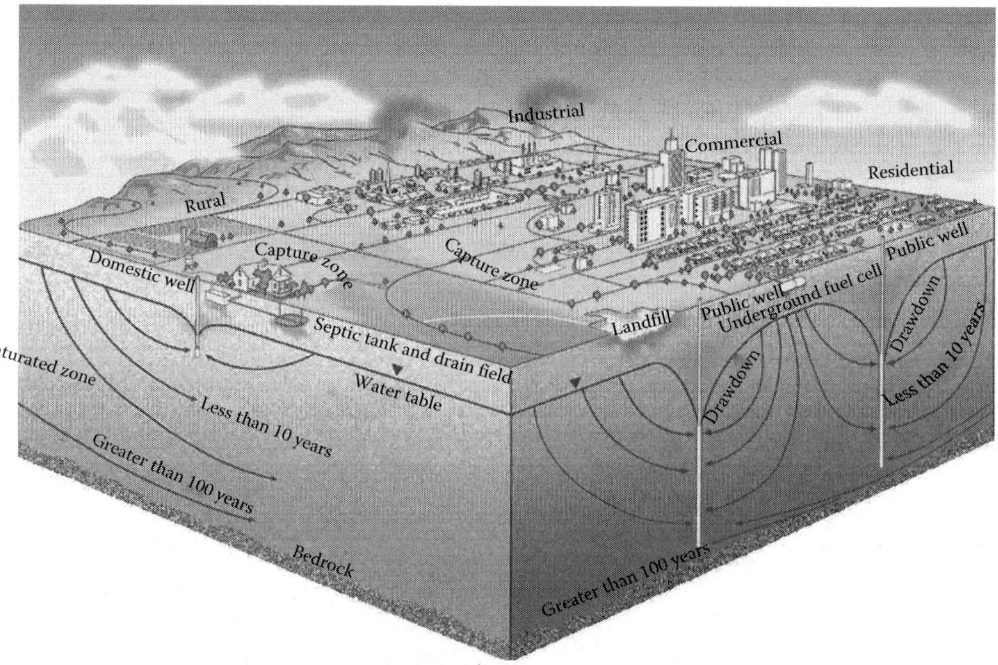

FIGURE 8.11
Potential groundwater travel time and capture zone beneath an urban area. (From United States Geological Survey (USGS), Volatile organic compounds in nation's ground water and drinking-water supply wells, National Water-Quality Assessment Program, USGS Circular 1292, Washington, DC, 2006a.)

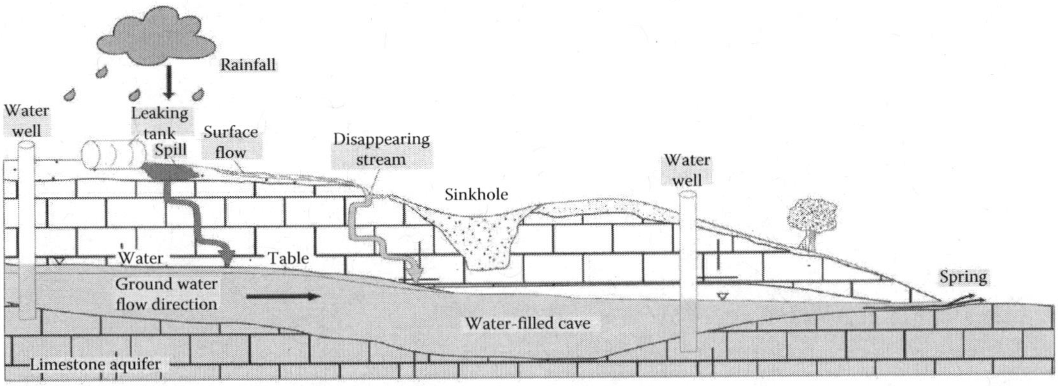

FIGURE 8.12
Contaminant migration in a karst aquifer. (From United States Geological Survey (USGS), Ground-water quality protection, Open-File Report 95-376, Nashville, Tennessee, http://www.pubs.usgs.gov/of/1995/ofr-95376 (accessed December 2009), 1995b.)

example of travel times and capture zone in groundwater beneath urban areas. Any contaminant reaching groundwater within the area marked *capture zone* has the potential to enter the public water supply if the contaminant does not degrade before reaching a public water supply well. Several sources of contamination listed in Section 8.1 are also shown: septic tanks, underground storage tanks (USTs), landfills, industrial facilities, and power plants. These contaminant sources are typical for any urban area within the United States and pose distinct threats to contaminate a public or private water supply.

8.3.5.1 Karst Topography

In karst topographical settings, the transport of groundwater contamination may behave similarly to surface water (Ford and Williams 2007). Some karst formations may exhibit turbulent advective flow because they have flow rates approaching the velocities observed in surface water flow (Heath 1983). Figure 8.12 shows an example of contaminant flow in a karst aquifer (USGS 1996a).

8.3.6 Fate and Transport of Contaminants in the Atmosphere

Different contaminants affect different portions of the atmosphere (Chapter 7). For instance, chlorofluorocarbons (CFCs) affect the protective ozone layer. Figure 8.13 shows the layers of the atmosphere and the location of the ozone layer within the stratosphere.

Contaminant behavior in the atmosphere is very similar to the behavior of contaminants observed in surface water. Advective transport, turbulent diffusion, and molecular diffusion also influence contaminant migration in the atmosphere (Hemond and Fechner-Levy 2000). Figure 8.14 shows smoke from a fire billowing up (turbulent diffusion) into the atmosphere and the horizontal movement of the smoke by advective transport. We also see a type of advective transport called convection; in this process, air rises due to thermal differences in the atmosphere. Turbulent advective mixing by wind and convection of the atmosphere is most significant within the troposphere (Schlatter 2009). These forces are very effective at transporting contaminants in the gas phase and also do a good job of moving solid particulate matter in the atmosphere (USEPA 2008a).

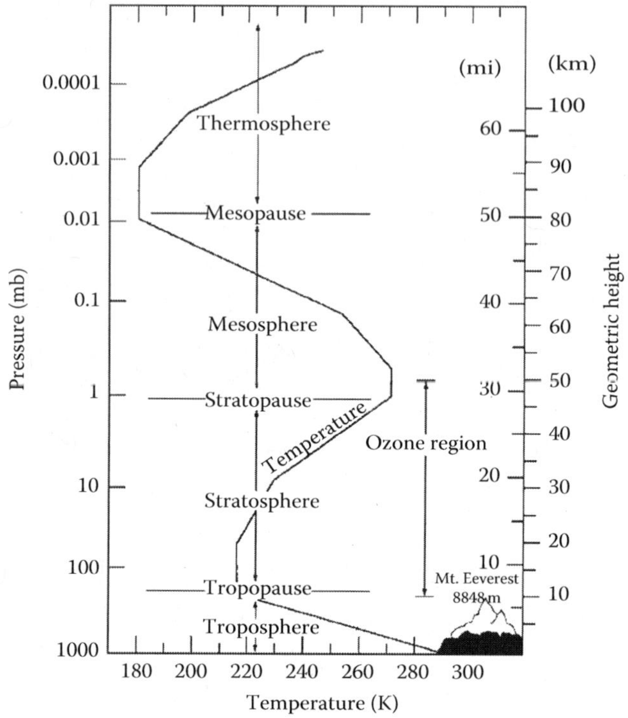

FIGURE 8.13
Layers of the atmosphere. (From United States Standard Atmosphere, *The Standard Atmosphere of the United States*, National Oceanic and Atmospheric Administration, National Aeronautics and Space Administration, and the United States Air Force, Washington, DC: United States Government Printing Office, 1976.)

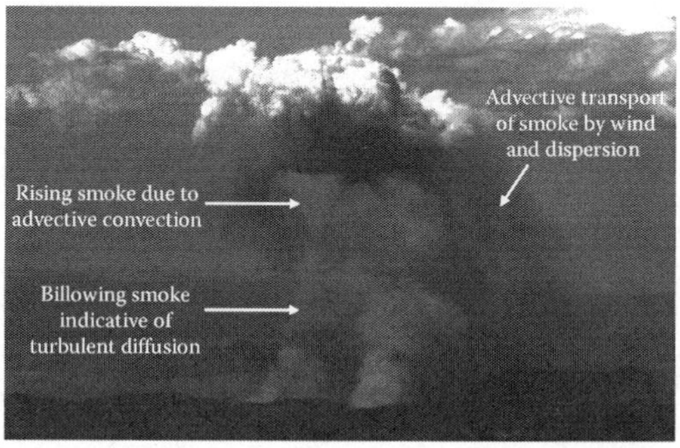

FIGURE 8.14
Smoke rising into the atmosphere from a fire. (Photo by Daniel T. Rogers.)

Anthropogenic sources of atmospheric contaminants are significant and present themselves as an array of different contaminants released in high volumes annually (USEPA 1998a). Most contaminants are released into the atmosphere from anthropogenic sources at or near the surface, with most of the impacts occurring in the troposphere, and to a lesser degree, the stratosphere (USEPA 1991). Contaminants released near the ground surface can mix throughout the troposphere in a few weeks, but it can take years or decades for them to reach the stratosphere (Hemond and Fechner-Levy 2000). Records of atmospheric contaminants and their effects can be traced to the thirteenth century when King Edward I banned the burning of kiln coal in London due to its impacts on air quality (Wilson 1996).

Temperature and *pressure* are two important factors affecting the migration of contaminants in the atmosphere. The reason temperature and pressure play a much more significant role in atmospheric fate and transport than in soil and water is because of the ideal gas laws—a combination of Boyle's and Charles' gas laws describing the relationships between temperature, density, and pressure. Temperature and pressure typically decrease with increasing altitude in the troposphere. The temperature in the lower portion of the stratosphere is relatively constant and helped give rise to its name meaning "stratified" (Figure 8.13). Table 8.1 lists the standard temperatures and atmospheric pressure within the atmosphere (United States Standard Atmosphere 1976).

The composition of the atmosphere is detailed in Table 8.2 (United States Standard Atmosphere 1976).

Oxygen is a recent addition to the atmosphere in geological terms. The origin of oxygen began with algae production approximately 2.45 billion years ago (Farquhar et al. 2000; Raub and Kirschvink 2008). The presence of oxygen in the atmosphere plays a significant role and affects contamination in the environment through oxidation reactions and rates of combustion (USEPA 2008a).

Contaminants initially released into the atmosphere often do not remain in the air; they settle out and contaminant the soil or surface water. Some contaminants, however, remain in the atmosphere for long periods of time, and other contaminants initially released into soil or water may volatilize and contaminate the air. In some cases, air contaminants may settle out of the atmosphere and adsorb onto a soil grain on the land surface only

TABLE 8.1

Standard Atmospheric Temperature and Pressure with Increasing Altitude

Altitude			Temperature	
ft	m	Pressure (atm)	°F	°C
0	0	1.000	59.0	15.0
2,000	610	0.943	51.9	11.0
4,000	1,219	0.888	44.7	7.0
6,000	1,826	0.836	37.6	3.1
8,000	2,438	0.786	30.5	−0.8
10,000	3,048	0.738	23.3	−5.0
15,000	4,572	0.564	5.5	−14.7
20,000	6,096	0.459	−12	−24.4
30,000	9,144	0.297	−48	−44.4
40,000	13,123	0.185	−67	−55
60,000	18,288	7.1×10^{-2}	−67	−55
80,000	24,384	2.7×10^{-2}	−67	−55
100,000	30,480	1.0×10^{-2}	−67	−55
140,000	42,672	2.0×10^{-3}	74	23.3
180,000	54,864	5.7×10^{-4}	170	76.7
220,000	67,056	1.7×10^{-4}	92	33.3
300,000	91,440	1.5×10^{-5}	27	−2.8
380,000	115,824	7.7×10^{-7}	188	86.7

TABLE 8.2

Composition of the Atmosphere

Gas	Chemical Symbol	Mean Molecular Weight (m/mol)	Concentration Parts per Million by Volume
Nitrogen	N_2	28.013	780,840
Oxygen	O_2	31.999	209,460
Argon	Ar	39.948	9,340
Carbon dioxide	CO_2	44.010	384
Neon	Ne	20.180	18.18
Helium	He	4.003	5.24
Methane	CH_4	16.043	1.774
Krypton	Kr	83.798	1.14
Hydrogen	H_2	2.106	0.56
Nitrous oxide	N_2O	44.012	0.32
Xenon	Xe	131.293	0.09
Ozone	O_3	47.998	0.01–0.10

to be picked up later by the wind and sent airborne again. Factors controlling whether a contaminant remains in the atmosphere include

- Physical and chemical factors of the contaminants including vapor pressure, molecular weight, solubility, and reactivity
- Geography and local topography
- Climate and weather conditions

The average person inhales approximately 20,000L of air per day (USEPA 2008a). Each year, the World Health Organization (2009) estimates that 2.4 million people die from causes directly attributable to air pollution, with the elderly and young children at the most risk. Specific diseases caused from prolonged exposure to air contaminants are chronic and often do not immediately appear after exposure. These diseases include heart disease, lung cancer, and bronchitis. The burning of fossil fuels in power plants and automobile, truck, and bus exhaust account for 90% of all air pollution in the United States (USEPA 2008a). Figure 8.15 shows some of the significant sources, methods of transport, and removal of air pollutants in the atmosphere (USEPA 2008a).

Deposition of contaminants onto the land from the atmosphere occurs in two different ways:

1. Dry deposition. Dry deposition is typically dust or particulate matter settling out of the air. The amount of dry deposition depends upon the amount of suspended particles, wind speed and duration, and particle size. Figure 8.16 shows a dust storm potentially depositing a significant amount of dry material. Contaminants are often present within dry deposition events, especially in urban areas where they may be sorbed onto the surfaces of particulate matter in the air (USEPA 2008a).
2. Wet deposition. Wet deposition occurs when snow, fog, or a rain droplet forms and then dissolves or carries a contaminant to the surface. Acid rain is a good example of wet deposition. Figure 8.17 shows the wet deposition of contaminants.

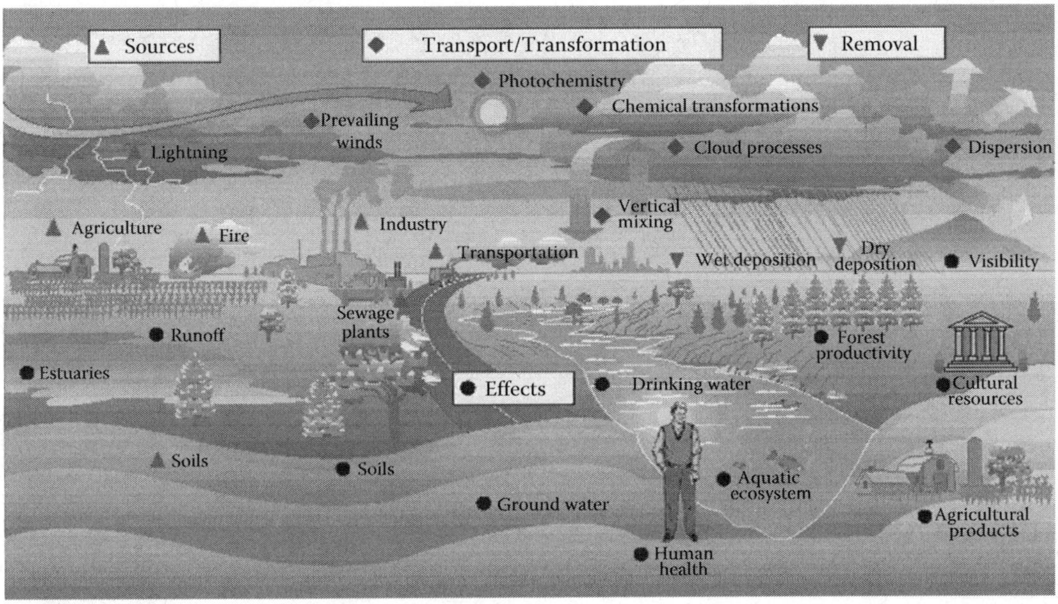

FIGURE 8.15
(See color insert.) Sources, transport methods, and removal of air pollution. (From United States Environmental Protection Agency (USEPA), *Latest Findings on National Air Quality: Status and Trends through 2006*, EPA454/R-07-007, Research Triangle Park, NC, 2008a.)

FIGURE 8.16
Dust storm that demonstrates transport and dry deposition of particulates. (From National Oceanic and Atmospheric Administration (NOAA), The May 29th dust storm, http://www.crh.noaa.gov/ddc/?n=dust (accessed June 26, 2010), 2010.)

8.4 Fate and Transport of Contaminants

A brief discussion of the fate and transport behavior of each contaminant group presented in Chapter 7 follows.

8.4.1 VOCs

VOCs are organic compounds that generally volatilize or evaporate readily under normal atmospheric pressure and temperatures. They usually have a high vapor pressure,

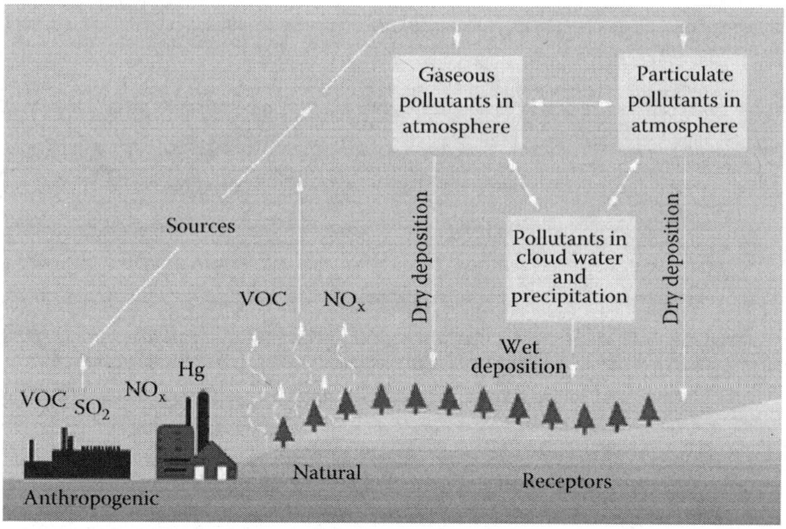

FIGURE 8.17
Process of wet and dry deposition of air contaminants. (From United States Environmental Protection Agency (USEPA), What is acid rain? http://www.epa.gov/acidrain/what/index.html (accessed June 29, 2010), 2010.)

low-to-medium solubility, and low molecular weight. As a result of these chemical characteristics, VOCs are common air, soil, and water contaminants (USGS 2006a; USEPA 2008a). Automobile exhaust contains VOCs. When combined with other common air pollutants and sunlight, urban smog will form if atmospheric conditions are favorable—that is, there is an ample supply of the combined sources of VOCs and other smog-forming contaminants. This type of smog formation produces **photochemical smog**, and is shown in Figure 8.18.

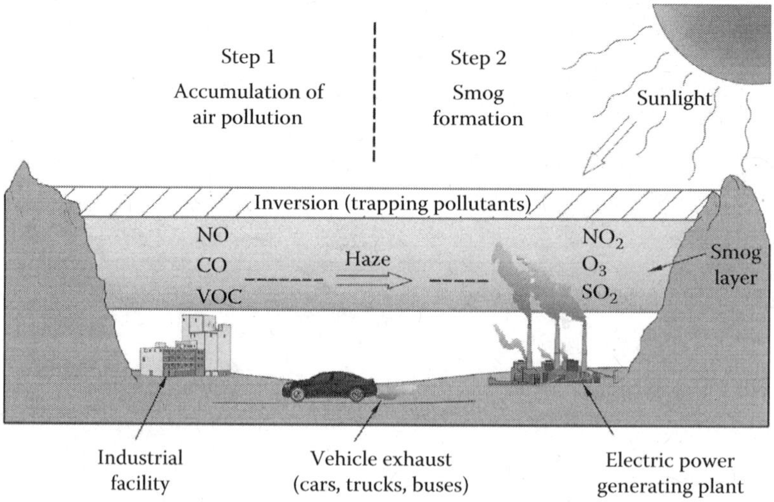

FIGURE 8.18
Formation of photochemical smog.

VOCs are also released directly onto the ground surface through leaks or spills at, or near the surface. Sources of these leaks include underground storage tanks, service stations, refineries, and pipelines. Because of these surface and shallow subsurface releases, VOCs are common groundwater contaminants and have been detected in the groundwater of numerous aquifers in the United States (USGS 2006a). A study of groundwater in the United States detected VOCs at a concentration of 0.02 μg/L in >50% of approximately 3500 samples collected from 100 different groundwater aquifers across the country (USGS 2006a). The VOCs detected most often included bromoform, bromodichloromethane, chloroform, chloromethane, 1,1-dichloroethane, dichlorodifluoromethane, methylene chloride, dibromodichloromethane, methyl-tert-butyl ether (MTBE), trichloroethene (TCE), tetrachloroethene (PCE), 1,1,1-trichloroethane (TCA), *trans*-1,2-dichloroethene (DCE), toluene, and trichlorofluoromethane.

This same study indicates the vulnerable nature of many aquifers of the United States, and their location corresponds with many urban areas. In fact, of the 28 major urban areas listed in Table 2.4, 27 have detectable concentrations of VOCs (only Kansas City is missing). This extent represents a population of 125.4 million—over 41% of the entire U.S. population that are potentially exposed to VOCs in their water supply. Figure 8.19 shows the locations where VOCs were detected, and Figure 8.20 shows the detection frequency of the most commonly detected VOCs from the USGS study (USGS 2006a).

FIGURE 8.19
Occurrence of VOCs in groundwater aquifers of the United States. (From United States Geological Survey (USGS), Volatile organic compounds in nation's ground water and drinking-water supply wells, National Water-Quality Assessment Program, USGS Circular 1292, Washington, DC, 2006a; United States Geological Survey (USGS), Pesticides in the nation's streams and groundwater, 1992–2001—A summary, USGS Fact Sheet 2006-3028, Washington, DC, 2006a.)

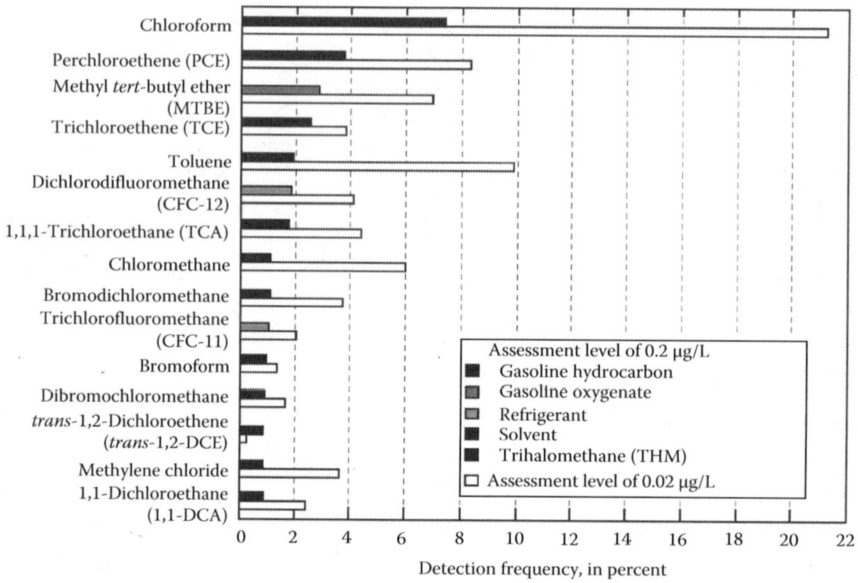

FIGURE 8.20
Frequency of detection: most common VOCs. (From United States Geological Survey (USGS), Volatile organic compounds in nation's ground water and drinking-water supply wells, National Water-Quality Assessment Program, USGS Circular 1292, Washington, DC, 2006a.)

VOCs exist as light nonaqueous phase liquids (LNAPLs) and dense nonaqueous phase liquids (DNAPLs). Because LNAPLs are lighter than water, they tend to float on top of groundwater as depicted in Figure 8.8, whereas the heavier than water DNAPLs tend to sink through the water column in an aquifer if conditions are favorable (Figure 8.21; USGS 2006a).

Contaminant degradation rates vary widely and depend on many factors including (1) the nature of the release, (2) physical chemistry of the contaminants themselves, (3) the

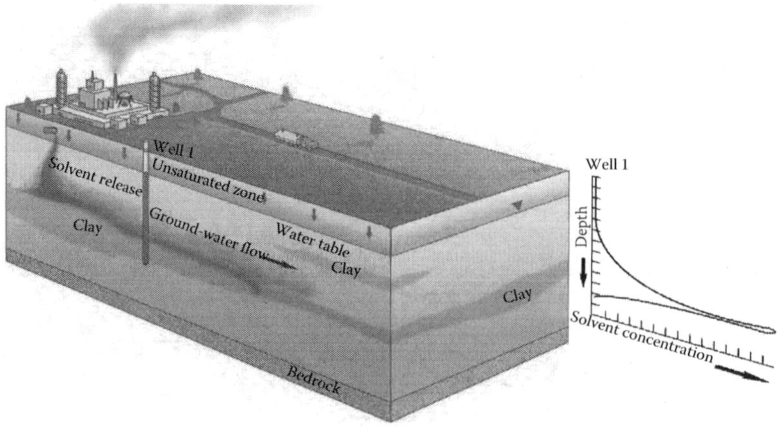

FIGURE 8.21
Migration of DNAPL compounds. (From United States Geological Survey (USGS), Volatile organic compounds in nation's ground water and drinking-water supply wells, National Water-Quality Assessment Program, USGS Circular 1292, Washington, DC, 2006a.)

geological environment where the contaminants are released, and (4) the presence, type, and distribution of microorganisms (Howard et al. 1997; USEPA 1998b; McKone and Enoch 2002; USGS 2006a).

Degradation rates also vary by media. In general, organic compounds in the atmosphere, including VOCs, degrade more quickly than the same organic compounds released and migrating to subsurface soil and groundwater (USEPA 1998a,b).

VOCs not degrading very easily include many of the chlorinated solvents or DNAPL compounds (USEPA 1996a). Many chlorinated VOCs including PCE, TCE, 1,1,1-TCA, DCE, and vinyl chloride are considered very persistent in the environment once released and can remain present for decades (USEPA 1996a). Other VOCs, such as benzene, toluene, ethyl benzene, and xylenes, are not typically as persistent in the environment and have been known to biodegrade in a few months to years if conditions are favorable (Rogers 1995; USEPA 1996a).

The VOC MTBE is persistent in the environment. Its relatively high solubility in water and low sorptive properties compared to other common VOC contaminants has resulted in significant MTBE-contaminated groundwater supplies at many urban locations throughout the United States (USGS 2006a; USEPA 2009a).

The trihalomethane VOCs include chloroform, bromoform, bromodichloromethane, and dibromochloromethane. These compounds have been detected in the groundwater of many U.S. aquifers (USGS 2006a). Trihalomethanes have high relative vapor pressures and commonly evaporate quickly when in contact with the atmosphere. Therefore, exposure to trihalomethanes is of special concern during showering and washing. They degrade by photolysis when exposed to direct sunlight and can also be degraded by microorganisms (ATSDR 1997a, 2005a).

8.4.2 PAHs

PAHs are LNAPL compounds, and being lighter than water they float on surface water and groundwater. They do not readily dissolve in water and have low vapor pressures compared to most VOCs. PAHs are common constituents of automobile exhaust, especially from diesel fuel (USGS 2006a), and are common air contaminants in urban areas. In addition, PAHs have a high sorptive affinity and can therefore attach to particulate matter (ATSDR 1996).

Many PAH compounds biodegrade under favorable conditions. The half-life of PAHs is shortest in the atmosphere due to photochemical degradation and lasts just a few days or weeks (ATSDR 1996). The half-life of PAH compounds in soil and groundwater is longer and may last for several years or decades. When released to soil, PAHs tend to migrate more slowly than VOCs because of their higher molecular weight and sorption to soils with high organic content.

8.4.3 PCBs

Once in the environment, PCBs do not readily degrade—they remain in the environment for long periods of time and often cycle between air water and soil. PCBs can be carried long distances attached to particulate matter and have been detected in snow and sea water far away from any known point of release (USEPA 2009b). This transport capability is confirmed by their worldwide detection. The lighter the PCB compound (fewer number of chlorine atoms in its structure; Section 7.5), the farther it can be transported from its release point.

PCBs are not very soluble, have high sorptive potential, and low vapor pressures. This combination of physical properties largely determines their environmental distribution, as they are only present at high concentrations in water or air when sorbed to particulate matter. Sinks where PCBs are frequently detected include soil near release points and sediments in rivers, streams, and lakes (USEPA 2009b). PCBs bioaccumulate in the leaves and above-ground parts of plants and food crops, and in aquatic organisms and fish where PCBs are present in sediments (Section 8.3.4). As a result, humans and other organisms ingesting impacted plant material, food crops, or organisms containing PCBs may bioaccumulate PCBs in their body tissues (ATSDR 2001a).

8.4.4 SVOCs

SVOCs are much less volatile than VOCs, but notable exceptions here are the amine compounds that exist as a gas at room temperature and standard pressure (ATSDR 2002a). In terms of solubility, phthalates and phenols do not readily dissolve in water, whereas amines and esters may dissolve, become mobile, and reach groundwater (ATDSR 1999c, 2002a,b,c,d).

When released into the environment, SVOCs are commonly detected in soil because they have high sorptive potentials. Many SVOCs, including pentachlorophenol, are degraded by microorganisms under favorable conditions and are also susceptible to photolysis and hydrolysis (ATSDR 2001, 2002a). Amines and esters degrade in minutes when exposed to direct sunlight (ATSDR 1999).

8.4.5 Heavy Metals

Heavy metals are released directly to air, water, and soil. In most cases, these contaminants do not remain in the atmosphere for long periods of time because they have high specific gravity and become deposited onto the land surface shortly after being emitted. Mercury, however, has been detected as far away as 50 mi from its source after being released into the atmosphere (USEPA 1997). Lead is also considered a common air pollutant (USEPA 2008a).

Major sources of lead include metal melting facilities, battery manufacturing, and leaded gasoline and fuels. The good news is there has been a 92% decrease in atmospheric lead concentrations over the period from 1980 through 2008. Contributing to the observed decrease has been the removal of lead from gasoline and fuels, increased efficiency in air pollution control equipment, and the regulation of lead emissions sources (USEPA 2008a).

The solubility of heavy metals in water is very low, except for some varieties of arsenic and chromium VI at a neutral pH. Due to their inability to form a solution, the preferred sinks for heavy metals are soil and sediments. Many metals undergo some transformation such as oxidation after being released into the environment but are not destroyed and remain in the environment (ATSDR 1999a,b, 2003, 2004a, 2005b,c, 2007a,b,c, 2008a,b). Due to their low solubility and high specific gravities, they tend to remain near release points. This is why increased concentrations of heavy metals are present in the near surface soil of urban areas as a result of releases from anthropogenic sources (Murray et al. 2004).

Some heavy metals such as mercury accumulate in sediment of lakes, rivers, and streams. Mercury may undergo a transformation through a process known as methylation after being released into the environment, typically when it reaches a surface water body. The process of methylation transforms elemental mercury to methyl mercury (CH_3Hg). Methyl mercury is the most toxic form of mercury and has the potential to bioaccumulate in aquatic organisms, including fish (USGS 2000).

On land and in sediments, other heavy metals such as barium, cadmium, chromium, lead, copper, and zinc can accumulate in plant matter if they are present at sufficient concentrations in soil within the root zone of plants (Nedunuri et al. 1998). Removal of contaminants from near-surface soil is not uncommon and sometimes the preferred alternative. Suitable locations for removal include closed landfill sites requiring a vegetative cap to minimize erosion potential, or sites where contamination is shallow, relatively static, and has become a chronic problem (Singer et al. 2003). **Phytoremediation**—the removal of contamination using plants—has been applied to other contaminants including organic compounds, but with limited success.

8.4.6 Other Contaminants

The remaining contaminants encountered in urban areas span a wide variety of organic and inorganic compounds. Their fate and transport is briefly discussed here.

8.4.6.1 Pesticides and Herbicides

Pesticides and herbicides are released into the environment for specific purposes and are most commonly detected in near-surface soils and surface water rather than groundwater (USGS 2006b). The highest concentrations of pesticides exist in the nation's streams and sediments within urban areas where they have been detected in 83% of streams and 70% of sediments (USGS 2006b). Figure 8.22 breaks down the pesticides detected in streams and bed sediments by land use. While the total tonnage of pesticide and herbicide use in

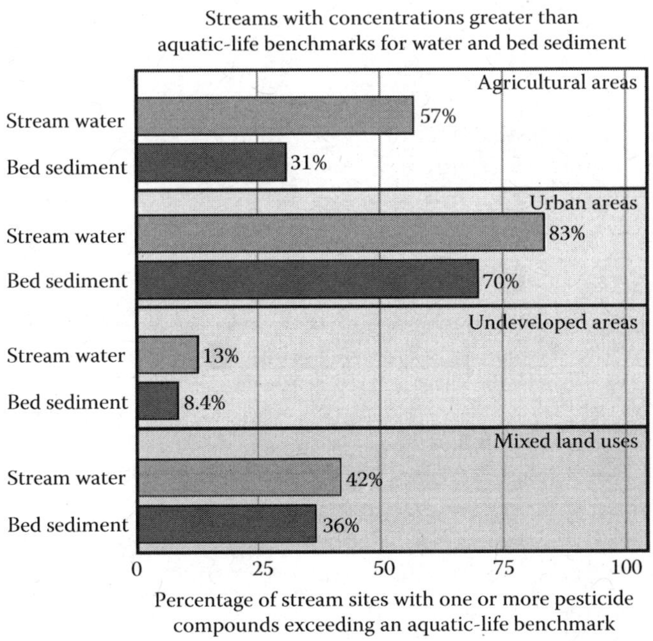

FIGURE 8.22
Pesticide detections in streams and sediments in the United States. (From United States Geological Survey (USGS), Pesticides in the nation's streams and groundwater, 1992–2001—A summary, USGS Fact Sheet 2006-3028, Washington, DC, 2006b.)

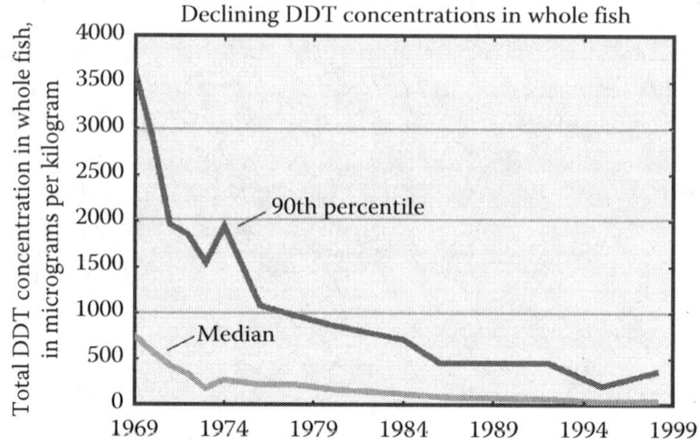

FIGURE 8.23
DDT concentrations in fish from 1969 to 1999. (From United States Geological Survey (USGS), Pesticides in the nation's streams and groundwater, 1992–2001—A summary, USGS Fact Sheet 2006-3028, Washington, DC, 2006b.)

the United States remained constant during the 1980s and early 1990s, Figure 8.23 shows that the concentration of the pesticide DDT has steadily declined in fish tissue during this same period (USGS 2006b). Since DDT is bioaccumulative, this event marks some progress.

Permethrin and toxaphene are two other widely used compounds in insecticides and pesticides, respectively. The fate and transport concern with permethrin centers around its use as an insect repellent, application to crops, and flea treatment for pets. All these activities involve human exposure (ATSDR 2005c,d), and with a half-life of approximately 28 days, there is often adequate time for human contact.

The now-banned pesticide toxaphene strongly sorbs to soil particles (Figure 8.4) and is not very soluble in water (SRC 2009). Common sinks of toxaphene include sediments and soil, where it bioaccumulates in fish and mammals. Toxaphene is also in the atmosphere since it evaporates when in a solid form or dissolved (ATSDR 1997b). It is estimated that the half-life of toxaphene is more than 10 years in soil (SRC 2009), so there is a good chance it is still present at appreciable concentrations in the environment.

8.4.6.2 Dioxins

Since one method of dioxin formation is through incineration and combustion, dioxin compounds are present in the atmosphere and have been detected around the globe (ATSDR 2006a). When dioxins are released at the surface, they sorb to soil particles and are often detected in sediments in lakes, rivers, and streams acting as sinks (ATSDR 2006a). Dioxin compounds are considered bioaccumulative contaminants (Section 8.3.4), with the potential to build up in the food chain and yield detectable concentrations in the tissues of many animals (ATSDR 2006a; USEPA 2009d). Due to the presence of chlorine in their atomic structure, dioxins do not readily degrade once they are formed and released into the environment (USEPA 2006).

8.4.6.3 Fertilizers

The most common fertilizers include nitrogen (N), phosphorus (P), and potassium (K). A major sink for fertilizers is surface water because they are applied to the soil surface and are considered soluble in water and mobile—especially in the case of nitrates and phosphorus. Once in surface water, nitrate and phosphorus can promote excessive algal growth. Significant algal growth can deplete the dissolved oxygen in surface water and cause suffocation and death to aquatic organisms. The solubility of some fertilizers, combined with the hydraulic linkage between surface water and groundwater, may lead to groundwater contamination.

The natural process of enrichment of surface waters with plant nutrients is termed **eutrophication**. When anthropogenic activities such as fertilization or sewage discharges accelerate this natural process, **cultural eutrophication** occurs (McGucken 2000).

8.4.6.4 Cyanide

In the atmosphere, cyanide is most often present as hydrogen cyanide. When present in surface water, cyanide compounds will form hydrogen cyanide and then enter the atmosphere through evaporation. When released to soil, cyanide compounds are considered fairly mobile when the retention capacity of the soil is exceeded and may migrate and contaminate groundwater. Cyanide compounds are degraded by microorganisms when present at low concentrations. When concentrations of cyanide compounds are elevated, they tend to be toxic to microorganisms and resist degradation (ATSDR 2006b).

The half-life of cyanide in the atmosphere ranges between 1 and 3 years (ATSDR 2006b). In soil and water, the half-life of cyanide compounds is much more difficult to estimate because the concentration, distribution, and presence of microorganisms available to degrade the cyanide compounds vary.

8.4.6.5 Asbestos

Asbestos fibers do not degrade, evaporate, or dissolve in water and remain virtually unchanged in the environment (ATSDR 2001c). Asbestos originates from naturally occurring minerals and is therefore present in the environment. Average background concentrations of asbestos in air range from 0.00001 to 0.0001 fibers per milliliter of air and are highest in urban areas (ATSDR 2001c).

Small diameter asbestos fibers can remain suspended in the atmosphere for a long period of time compared to larger fibers (those larger than 10 microns; ATSDR 2001c). Since asbestos was widely used in building materials, it is most common in urban areas and where natural deposits are present. Asbestos can become airborne through the disturbance of asbestos-containing materials during demolition or remodeling activities. The asbestos threat is tied to its mineralogy (Chapter 7).

8.4.6.6 Acids and Bases

When released into soil, acids and bases neutralize rapidly. They are diluted when they come into contact with water if a difference in pH levels exists. Therefore, if environmental impairment occurs, it must be realized rapidly before the acid or base becomes neutralized. This impairment occurs with the majority of sudden and accidental releases but does not hold true for acid rain, which generates effects with slower onsets.

Acids and bases may migrate a significant distance—sometimes more than 1.6 km (1 mi)—when released in the atmosphere and may cause significant impairment to living organisms exposed to their vapors (ATSDR 2004b,c). The characteristics of acid rain are covered during the discussion of sulfur dioxide later in this chapter.

8.4.6.7 Radioactive Compounds

Radioactive compounds occur naturally, with the most common being radon. Radon is produced from the decay of uranium (ATSDR 2008c) and is present in air, water, and soil. Radon may build up in basements, especially if cracks exist, or other subsurface structures located above natural deposits having higher relative uranium levels. The half-life of radon is approximately 4 days (ATSDR 2008c). Most of the human exposure attributed to other radioactive compounds results from medical devices, diagnostic treatments, testing equipment such as x-ray machines, and cancer therapy (Kathren 1991; ATSDR 2000).

8.4.6.8 Greenhouse Gases

Greenhouse gases decay very slowly and are primarily atmospheric contaminants. Some quantities of these gases are naturally removed from the atmosphere, such as the removal of carbon dioxide during photosynthesis. However, the anthropogenic addition of carbon dioxide and other greenhouse gases into the atmosphere has greatly exceeded the capacity of the natural environment to remove them (USEPA 2009c). As a result, greenhouse gas concentrations have been increasing (NOAA 2009). Figure 8.24 shows the process of photosynthesis.

8.4.6.9 Carbon Monoxide

Carbon monoxide is created when fuel is not burned completely. Carbon monoxide is formed naturally and anthropogenically. The most significant source of carbon monoxide is from automobile exhaust (USEPA 2008a). Inside homes, significant sources of carbon dioxide emissions are natural gas and oil furnaces, hot water heaters, appliances, wood burning stoves and fireplaces. Carbon monoxide is also created as a by-product of several industrial processes including metal melting and chemical synthesis (USDHH 2009). From 1980 to 2008, there was a 79% decrease in carbon dioxide in the United States. This decrease was attributed to improved air pollution control equipment for stationary and mobile sources of air pollution (USEPA 2008a).

8.4.6.10 Ozone

Ozone is a gas occurring in Earth's upper atmosphere and at ground level. As noted in Chapter 7, ozone in the upper atmosphere is greatly beneficial to life on Earth because it filters ultraviolet radiation, but ozone occurring at ground level is considered an air pollutant (USEPA 2009e). Ground-level ozone is not emitted directly into the air—it is created by chemical reactions between oxides of nitrogen and VOCs in the presence of sunlight. Emissions from automobile exhaust, gasoline vapors, chemical solvents, electrical

$$6H_2O + 6CO_2 \xrightarrow{\text{Sunlight}} C_6H_{12}O_6 + 6O_2$$

$$\text{Water} + \text{Carbon dioxide} \longrightarrow \text{Glucose (a carbohydrate)} + \text{Oxygen}$$

FIGURE 8.24
Photosynthesis and the removal of carbon dioxide.

generating facilities, and some factories trigger the production of ground-level ozone (Figure 7.29; USEPA 2009e). This variety of ozone is a concern in urban regions of the United States during the summer, because strong sunlight and hot weather can generate higher levels (USEPA 2009e).

Since the Clean Air Act of 1990, atmospheric ozone concentration in the United States has declined 25% (USEPA 2008a). Better control of stationary and mobile sources of air pollution such as automobile exhaust is behind this improvement (USEPA 2008a).

8.4.6.11 Sulfur Dioxide

Sulfur dioxide is a component of smog, and it also combines with nitrous oxide compounds to eventually form sulfuric acid—commonly referred to as acid rain (USEPA 2008a). It is removed from the atmosphere during precipitation and is neutralized quickly in soil if the pH of the soil is greater than 7. Some areas of the northeastern United States have soils lacking the ability to effectively neutralize the effects of acid rain, and there have been adverse effects on aquatic life and vegetation in the region. Efforts to reduce sulfur dioxide emissions from stationary and mobile sources have resulted in a decrease of 71% since 1980 in the United States (USEPA 2008a). Nevertheless, the pH of rain in the eastern United States remains acidic (Figure 8.25; NADP 2009).

8.4.6.12 Particulate Matter

Urban areas have the highest concentrations of particulate matter, which is a significant distributor of contaminants in the atmosphere. Contaminants such as SVOCs, some VOCs,

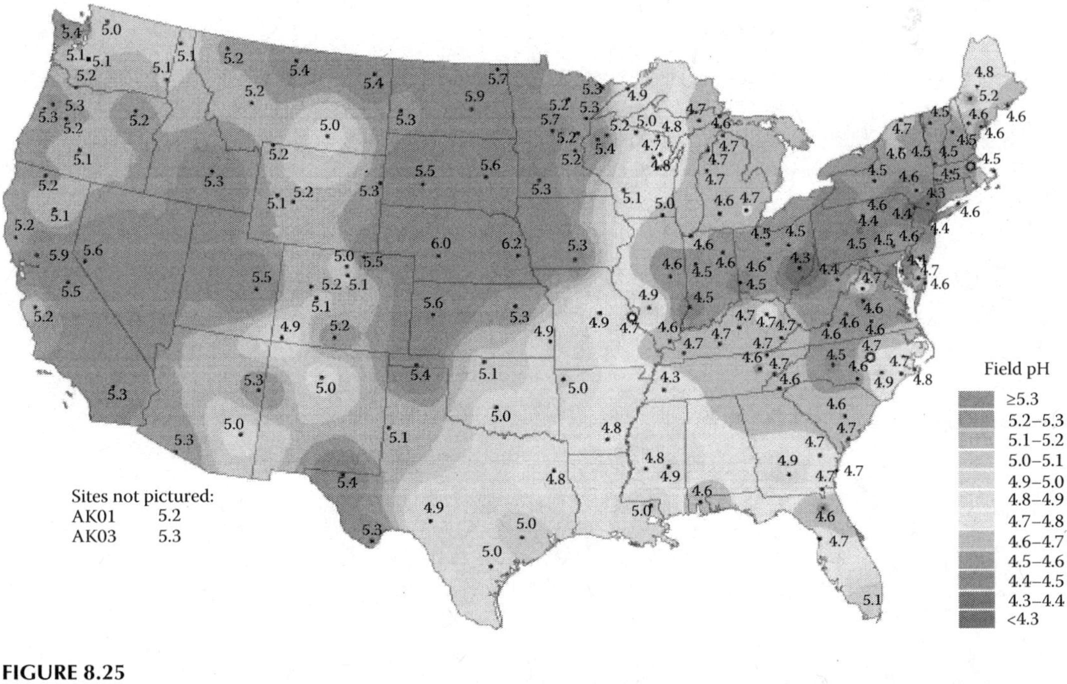

FIGURE 8.25
(See color insert.) pH of precipitation in the United States in 2004. (From National Oceanic and Atmospheric Administration (NOAA), Atmospheric CO2 at Mauna Loa observatory, http://www.esrl.noaa.gov/gmd/ccgg/trends/co2_data_mlo.html (accessed November 11, 2009), 2009.)

PCBs, and many pesticides and herbicides may sorb to a soil particle and travel a significant distance through wind action (USEPA 1998a). The size of particulate matter is significant because the largest sizes tend to settle to the ground surface first. Smaller particles can travel around the globe and remain suspended for years if favorable conditions exist (USEPA 2008a). Here is some good news: there was a 10%–20% decline in atmospheric particulate matter within the urban areas of the United States between 2001 and 2007 (USEPA 2008a).

8.4.6.13 Bacteria, Parasites, and Viruses

Bacteria, parasites, and viruses are present in large numbers everywhere in the environment. They are in and on the food we eat, in and on our bodies, in the air we breathe and the water we drink, in soil, and at depths within Earth. Many are beneficial, but some have the potential to adversely affect our health and well-being (Madigan et al. 2008). Ominously, these pathogens have evolved mechanisms to persist in the environment. Bacteria such as *Escherichia coli* can persist in secondary, nonhost habitats for prolonged periods (Ksoll et al. 2007). Giardia and Cryptosporidium form cysts and oocysts, respectively, and these structures sustain these parasites outside their hosts until they are ingested through drinking water (Wallis et al. 1996). Viruses, such as influenza, are more likely to persist in the environment at colder temperatures. The dry, cold conditions pull moisture out of droplets released by coughs and sneezes, which allows the virus to linger in the air. At winter temperatures, the virus's outer covering, or envelope, hardens to a rubbery gel that could shield the virus as it passes from person to person, the researchers have found. At warmer temperatures, however, the protective gel melts to a liquid phase (Shaman and Kohn 2009).

8.4.6.14 Emerging Contaminants

Emerging contaminants including many pharmaceuticals, 1,4-dioxane, and perchlorates are resistant to degradation and can remain in the environment for long periods of time (ATSDR 2007d, 2008d). Many pharmaceuticals and drugs, 1,4-dioxane, and perchlorates have been detected in groundwater where they can migrate long distances due to their relatively high solubility and resistance to degradation.

Emerging contaminants are difficult to study because many have entered the environment from nontraditional sources such as residential septic systems, as opposed to industrial sources (ATSDR 2008d; Barnes et al. 2008).

8.5 Summary and Conclusion

There are thousands of contaminants existing everywhere. After they are released, many of them migrate in air, soil, and water. Some are persistent, while others resist degradation. The geology and hydrogeology of urban areas play a significant role in affecting the fate and transport of contaminants and determine their final disposition.

Automobile exhaust is responsible for a significant amount of the pollution released every year in the United States and throughout the world. Contaminants associated with automobile use affects the air we breathe, the water we drink, and may also affect our global climate.

Halogenated contaminants (those containing chlorine, fluorine, bromine, or iodine within their structure) have the ability to remain in the environment for long periods of time (e.g., chlorinated solvents). Several of these compounds have the ability to accumulate in the bodies of living organisms (e.g., PCBs) and through bioaccumulation may expose humans after they work their way up the food chain. Some contaminants can change from a gas to a liquid or a liquid to a gas and cycle between the soil, air, and water if they last long enough (e.g., some VOCs). Other contaminants not changing form can be found in sinks or areas where they accumulate (e.g., river and lake sediment).

Heavy metals pose special challenges in urbanized watersheds because they are persistent, highly toxic, and some, like hexavalent chromium, exhibit high mobility in groundwater. In Chapter 9, case studies of heavy metals in the soil and groundwater are presented for the highly urbanized Rouge River watershed of southeastern Michigan.

References

Agency for Toxic Substance and Disease Registry (ATSDR). 1996. *Polycyclic Aromatic Hydrocarbons.* General Contaminant Class. ATSDR ToxFAQs. Atlanta, GA.

Agency for Toxic Substances and Disease Registry (ATSDR). 1997a. *Chloroform.* CAS Registry Number 127-18-4. ATSDR ToxFAQs. Atlanta, GA.

Agency for Toxic Substances and Disease Registry (ATSDR). 1997b. *Toxaphene.* CAS Registry Number 8001-35-2. ATSDR ToxFAQs. Atlanta, GA.

Agency for Toxic Substances and Disease Registry (ATSDR). 1999a. *Mercury.* CAS Registry Number 7439-97-6. ATSDR ToxFAQs. Atlanta, GA.

Agency for Toxic Substances and Disease Registry (ATSDR). 1999b. *Silver.* CAS Registry Number 7440-22-4. ATSDR ToxFAQs. Atlanta, GA.

Agency for Toxic Substances and Disease Registry (ATSDR). 1999c. *Dimethylamine.* CAS Registry Number 124-40-3. ATSDR ToxFAQs. Atlanta, GA.

Agency for Toxic Substances and Disease Registry (ATSDR). 2000. *Radon Toxicity: Who Is at Risk?* ATSDR. Atlanta, GA.

Agency for Toxic Substances and Disease Registry (ATSDR). 2001a. *Polychlorinated Biphenyls.* ATSDR ToxFAQs. Atlanta, GA.

Agency for Toxic Substances and Disease Registry (ATSDR). 2001b. *Pentachlorophenol.* CAS Registry Number 87-86-5. ATSDR ToxFAQs. Atlanta, GA.

Agency for Toxic Substances and Disease Registry (ATSDR). 2001c. *Asbestos.* CAS Registry Number 1332-21-4. ATSDR ToxFAQs. Atlanta, GA.

Agency for Toxic Substances and Disease Registry (ATSDR). 2002a. Di(2-ethylhexyl) phthalate. CAS Registry Number 117-81-7. ATSDR ToxFAQs. Atlanta, GA.

Agency for Toxic Substances and Disease Registry (ATSDR). 2002b. *Toxicological Profile for Flame Retardant Ester Compounds.* ATSDR. Atlanta, GA.

Agency for Toxic Substances and Disease Registry (ATSDR). 2002c. *Benzyl Acetate.* CAS Registry Number 140-11-4. ATSDR ToxFAQs. Atlanta, GA.

Agency for Toxic Substances and Disease Registry (ATSDR). 2002d. *Ethyl Acetate.* CAS Registry Number 141-78-6. ATSDR ToxFAQs. Atlanta, GA.

Agency for Toxic Substances and Disease Registry (ATSDR). 2003. *Selenium.* CAS Registry Number 7782-49-2. ATSDR ToxFAQs. Atlanta, GA.

Agency for Toxic Substances and Disease Registry (ATSDR). 2004a. *Copper.* CAS Registry Number 7440-50-8. ATSDR ToxFAQs. Atlanta, GA.

Agency for Toxic Substances and Disease Registry (ATSDR). 2004b. *Ammonia.* CAS Registry Number 7664-41-7. ATSDR ToxFAQs. Atlanta, GA.

Agency for Toxic Substances and Disease Registry (ATSDR). 2004c. *Hydrochloric Acid*. CAS Registry Number 7647-01-0. ATSDR ToxFAQs. Atlanta, GA.

Agency for Toxic Substances and Disease Registry (ATSDR). 2005a. *Bromoform*. CAS Registry Number 75-25-2. ATSDR ToxFAQs. Atlanta, GA.

Agency for Toxic Substances and Disease Registry (ATSDR). 2005b. *Nickel*. CAS Registry Number 7440-02-0. ATSDR ToxFAQs. Atlanta, GA.

Agency for Toxic Substances and Disease Registry (ATSDR). 2005c. *Zinc*. CAS Registry Number 7440-66-6. ATSDR ToxFAQs. Atlanta, GA.

Agency for Toxic Substances and Disease Registry (ATSDR). 2005d. *Permethrin: Toxicologic Information about Pesticides*. CAS Registry Number 5264553-1. ATSDR. Atlanta, GA.

Agency for Toxic Substances and Disease Registry (ATSDR). 2006a. *Dioxins: Chemical Agent Briefing Sheet*. ATSDR. Atlanta, GA.

Agency for Toxic Substances and Disease Registry (ATSDR). 2006b. *Cyanide*. CAS Registry Number 74-90-8, 143-33-9, 151-50-8, 592-01-8, 544-92-3, 506-61-6, 460-19-5, 506-77-4. ATSDR ToxFAQs. Atlanta, GA.

Agency for Toxic Substances and Disease Registry (ATSDR). 2007a. *Lead*. CAS Registry Number 7439-92-1. ATSDR ToxFAQs. Atlanta, GA.

Agency for Toxic Substances and Disease Registry (ATSDR). 2007b. *Barium*. CAS Registry Number 9440-39-3. ATSDR ToxFAQs. Atlanta, GA.

Agency for Toxic Substances and Disease Registry (ATSDR). 2007c. *Arsenic*. CAS Registry Number 7440-38-2. ATSDR ToxFAQs. Atlanta, GA.

Agency for Toxic Substances and Disease Registry (ATSDR). 2007d. *1,4-Dioxane*. CAS Registry Number 123-91-1. ATSDR ToxFAQs. Atlanta, GA.

Agency for Toxic Substances and Disease Registry (ATSDR). 2008a. *Cadmium*. CAS Registry Number 7440-43-9. ATSDR ToxFAQs. Atlanta, GA.

Agency for Toxic Substances and Disease Registry (ATSDR). 2008b. *Chromium*. CAS Registry Number 7440-47-3. ATSDR ToxFAQs. Atlanta, GA.

Agency for Toxic Substances and Disease Registry (ATSDR). 2008c. *Radon*. CAS Registry Number 14859-67-7. ATSDR ToxFAQs. Atlanta, GA.

Agency for Toxic Substances and Disease Registry (ATSDR). 2008d. *Perchlorates*. CAS Registry Number 10034-81-8, 7778-74-7, 7790-98-9, 7601-89-0, 7791-03-9. ATSDR ToxFAQs. Atlanta, GA.

Barnes, K.K., D.W. Kolpin, E.T. Furlong et al. 2008. A national reconnaissance of pharmaceuticals and other organic wastewater contaminants in the United States. *Journal of Science in the Total Environment* 42:192–200.

Bedient, P. B. H.S. Rifai, and C.J. Newell. 1994. *Ground Water Contamination: Transport and Remediation*. Englewood Cliffs, NJ: Prentice Hall.

Brady, N.C. and R.R.Well. 1999. *The Nature and Properties of Soils*, 12th edn. Upper Saddle River, NJ: Prentice Hall.

Farquhar, J., B. Huiming, and M. Thiemens. 2000. Atmospheric influence of earth's earliest sulfur cycle. *Science* 289:756–758.

Fetter, C.W. 1993. *Contaminant Hydrogeology*, 2nd edn. Upper Saddle River, NJ: Prentice Hall.

Ford, D.C. and P. Williams. 2007. *Karst Hydrogeology and Geomorphology*, 2nd edn. New York: John Wiley & Sons.

Haynes, W. 1954. *American Chemical Industry—A History*, Vols. I–IV. New York: Van Nostrand.

Heath, R.C. 1983. *Basic Ground-Water Hydrology*. United States Geological Survey, Water Supply Paper 2220. Alexandria, VA: United States Government Printing Office.

Hemond, H.F. and E.J. Fechner-Levy. 2000. *Chemical Fate and Transport in the Environment*. London, U.K.: Academic Press.

Hornsby, A.G. 1990. *How Contaminants Reach Groundwater*. Gainsville, FL: University of Florida Institute of Food and Agriculture.

Howard, P.H., E.M. Michalenko, D.K. Basu et al. 1997, *Handbook of Environmental Fate and Exposure Data for Organic Chemicals, Vol. 5, Solvents 3*. Chelsea, MI: Lewis Publishers.

Kathren, R. 1991. *Radioactivity and the Environment*. Leiden, the Netherlands: Taylor & Francis.

Kaufman, M.M., D.T. Rogers, and K.S. Murray. 2009. Using soil risk factors for assessing the potential for groundwater contamination to the lower Great Lakes, USA. *Environmental Geology* 56:1009–1021.

Ksoll, W.B., S. Ishii, M.J Sadowsky et al. 2007. Presence and sources of fecal coliform bacteria in epilithic periphyton communities of Lake Superior. *Applied and Environmental Microbiology* 73:3771–3778.

Lindsay, W.L. 1979. *Chemical Equilibria in Soils*, NY: John Wiley & Sons.

Lyman, W.J., W.F. Reehl, and D.M. Rosenblatt. 1990. *Handbook of Chemical Property Estimation Methods*. Washington, DC: American Chemical Society.

Madigan, M.T., J.M. Martinko, P.V. Dunlap et al. 2008. *Brock Biology of Microorganisms*, 12th edn. New York: Prentice-Hall.

Manahan, S.E. 1993. *Fundamentals of Environmental Chemistry*. Chelsea, MI: Lewis Publishing.

McGucken, W. 2000. *Lake Erie Rehabilitated: Controlling Cultural Eutrophication, 1960s–1990s*. Akron, OH: University of Akron Press.

McKone, T.E. and K.G. Enoch, 2002. *CalTox™, A Multimedia Total Exposure Model Spreadsheet User's Guide*. Lawrence Berkeley National Laboratory. LBNL 47399. University of California, Berkeley, CA.

Murray, K.S., D.T. Rogers, and M.M. Kaufman. 2004. Heavy metals in an urban watershed in southeastern Michigan. *Journal of Environmental Quality* 33:163–172.

National Atmospheric Deposition Program (NAPD). 2009. Hydrogen ion concentrations as pH from measurements made at field laboratories, 2004. NAPD. http://napd.sws.uiuc.edu/lib/brochures/FieldChem.pdf (accessed December 17, 2009).

National Oceanic and Atmospheric Administration (NOAA). 2009. Atmospheric CO2 at Mauna Loa observatory. http://www.esrl.noaa.gov/gmd/ccgg/trends/co2_data_mlo.html (accessed November 11, 2009).

National Oceanic and Atmospheric Administration (NOAA). 2010. The May 29th Dust Storm. http://www.crh.noaa.gov/ddc/?n=dust (accessed June 26, 2010).

Nedunuri, K.V., L.E. Erickson, and R.S. Govindaraju. 1998. Modeling the role of active biomass on the fate and transport of heavy metals in the presence of root exudates. *Journal of Hazardous Waste Research* 1:1–25.

Payne, F.C., J.A. Quinnan, and S.T. Potter. 2008. *Remediation Hydraulics*. Boca Raton, FL: CRC Press.

Raub, T.D. and J.L. Kirschvink. 2008. A pan-precambrian link between deglaciation and environmental oxidation. In *Antarctica: A Keystone in a Changing World*. Proceedings of the 10th International Symposium on Antarctic Earth Sciences, eds. A.K. Cooper, P.J. Barrett, B. Stagg et al., pp. 83–90. Washington, DC: National Academies Press.

Rogers, D.T. 1995. *Intrinsic Bioremedation of Gasoline-Contaminated Groundwater—A Case Study*. Air and Waste Management Association. Annual Meeting, San Antonio, TX.

Rogers, D.T. 1996. *Environmental Geology of Metropolitan Detroit*. Novi, MI: Clayton Environmental Consultants.

Rogers, D.T., K.S. Murray, and M.M. Kaufman. 2007. Assessment of groundwater contaminant vulnerability in an urban watershed in southeast Michigan, USA. In *Urban Groundwater—Meeting the Challenge*, ed. K.W.F. Howard, pp. 129–144. London, U.K.: Taylor & Francis.

Schlatter, T.W. 2009. *Atmospheric Composition and Vertical Structure*. Boulder, CO: National Oceanic and Atmospheric Administration (NOAA).

Schnoor, J.L. 1996. *Environmental Modeling: Fate and Transport of Pollutants in Water and Soil*. New York: John Wiley & Sons.

Schwarzenbach, R.P., P.M. Gschwend, and D.M. Imboden. 1993. *Environmental Organic Chemistry*. New York: John Wiley & Sons.

Shaman, J. and M. Kohn. 2009. Absolute humidity modulates influenza survival, transmission, and seasonality. *Proceedings of the National Academy of Sciences USA* 106:3243–3248.

Singer, A.C., D.E. Crowley, and I.P. Thompson. 2003. Secondary plant metabolites in phytoremediation and biotransformation. *Trends in Biotechnology* 21:123–130.

Soil Science Society of America. 1987. *Glossary of Soil Science Terms*. Madison, WI.

SRC. 2009. Environmental fate data base (EFDB). CHEMFATE Chemical Search. http://srcinc.com/what-we-do/efbd.aspx (accessed November 10, 2009).

Sutherson, S.S. and F.C. Payne. 2005. *In Situ Remediation Engineering*, Boca Raton, FL: CRC Prees.

United States Department of Health and Human Services (USDHH). 2009. *Draft Profile for Carbon Monoxide*. ATSDR. Washington, DC.

United States Environmental Protection Agency (USEPA). 1991. *Air Pollution and Health Risk*. EPA/450/3-90-022. Washington, DC.

United States Environmental Protection Agency (USEPA), 1996a. *Transport and Fate of Contaminants in the Subsurface*. EPA/625/4-89/019. Washington, DC.

United States Environmental Protection Agency (USEPA). 1996b. *Bioscreen Natural Attenuation Decision Support System*, Version 1.4. Office of Research and Development. Washington, DC.

United States Environmental Protection Agency (USEPA). 1996c. *Soil Screening Guidance: Technical Background Document*. Office of Solid Waste and Emergency Response. Washington, DC.

United States Environmental Protection Agency (USEPA). 1997. *Mercury Study Report to Congress: Volume III: Fate and Transport of Mercury in the Environment*. EPA-454/R-97-005. Washington, DC.

United States Environmental Protection Agency (USEPA). 1998a. *Taking Toxics Out of the Air*. EPA451/K-98-001. Washington, DC.

United States Environmental Protection Agency (USEPA). 1998b. *Chemical Fate Half-Lives for Toxics Release Inventory (TRI) Chemicals*. Washington, DC.

United States Environmental Protection Agency (USEPA). 1999. *Fundamentals of Soil Science as Applicable to Management of Hazardous Wastes*. Office of Research and Development. EPA/540/S-98/500. Washington, DC.

United States Environmental Protection Agency (USEPA). 2006. *An Inventory of Sources and Environmental Releases of Dioxin-Like Compounds in the United States for the Years 1987, 1995, and 2000*. EPA/600/P-03/002f. Washington, DC.

United States Environmental Protection Agency (USEPA). 2008a. *Latest Findings on National Air Quality: Status and Trends through 2006*. EPA454/R-07-007. Research Triangle Park, NC.

United States Environmental Protection Agency (USEPA). 2008b. *EPA's Report on the Environment*. EPA/600/R-07/045F. Washington, DC.

United States Environmental Protection Agency (USEPA). 2009a. Overview of methyl tertiary butyl ether (MTBE). http://www.epa.gov/mtbe/faq/htm (accessed December 10, 2009).

United States Environmental Protection Agency (USEPA). 2009b. *Biennial National Listing of Fish Advisories for 2008*. EPA-823-F-09-007. Washington, DC.

United States Environmental Protection Agency (USEPA). 2009c. *Climate Change-Greenhouse Gas Overview*. Washington, DC.

United States Environmental Protection Agency (USEPA). 2009d. Integrated Risk Information System (IRIS). http://www.epa.gov/ncea/iris/intro.htm (accessed October 21, 2009).

United States Environmental Protection Agency (USEPA). 2009e. Ozone—Good up high bad nearby. Air Quality Planning and Standards. Washington, DC.

United States Environmental Protection Agency (USEPA). 2010. What is acid rain? http://www.epa.gov/acidrain/what/index.html (accessed June 29, 2010).

United States Geological Survey (USGS). 1995a. Contaminants in the Mississippi River. USGS Circular 1133. Washington, DC.

United States Geological Survey (USGS). 1995b. Ground-water quality protection. Open-File Report 95-376. Nashville, Tennessee. http://www.pubs.usgs.gov/of/1995/ofr-95376 (accessed December 2009).

United States Geological Survey (USGS). 1998. Simulating transport of volatile organic compounds in the unsaturated zone using the computer model R-UNSAT. USGS Fact Sheet 019-98. Washington, DC.

United States Geological Survey (USGS). 2000. Mercury in the environment. Fact Sheet 146-00. USGS. Washington, DC.

United States Geological Survey (USGS). 2006a. Volatile organic compounds in nation's ground water and drinking-water supply wells. National Water-Quality Assessment Program. USGS Circular 1292. Washington, DC.

United States Geological Survey (USGS). 2006b. Pesticides in the nation's streams and groundwater, 1992–2001—A summary. USGS Fact Sheet 2006-3028. Washington, DC.

United States Standard Atmosphere. 1976. *The Standard Atmosphere of the United States*. National Oceanic and Atmospheric Administration, National Aeronautics and Space Administration, and the United States Air Force. Washington, DC: United States Government Printing Office.

Wallis, P.M., S.L. Erlandsen, J.L. Isaac-Renton et al. 1996. Prevalence of *Giardia* cysts and *Cryptosporidium* oocysts and characterization of *Giardia* spp. isolated from drinking water in Canada. *Applied Environmental Microbiology* 62:2789–2797.

Wania, F., J.T. Hoff, C.Q. Jia et al. 1998. The effects of snow and ice on the environmental behavior of hydrophobic organic chemicals. *Journal of Environmental Pollution* 102:79–95.

Wiedemeier, T. H., H.S. Rifai, C.J. Newell et al. 1999. *Natural Attenuation of Fuels and Chlorinated Solvents in the Subsurface*. New York: John Wiley & Sons.

Wilson, R.1996. Introduction, In *Particles in Our Air: Concentration and Health Effects*, ed. R. Wilson and J. Spengler, pp. 1–14. Cambridge, MA: Harvard University Press.

World Health Organization. 2009. Health statistics and health information systems. Mortality Database Tables. http://www.who.int/healthinfo/morttables/en/index/html (accessed December 9, 2009).

9

Metal Contamination in Urban Watersheds

9.1 Introduction

Metals naturally occur and are present in the soils, air, and water of all urban watersheds. Although metal contamination is briefly mentioned in Chapter 7, we emphasize metals in this chapter for a variety of reasons. First, metals are common and important contaminant constituents in urban streams and stream sediment. Second, as noted in Chapter 7, metals are commonly produced as by-products of many types of industrial, commercial, and even residential developments. Third, metals are released into the environment. This is particularly true in southeast Michigan, which developed as the focal point for the automotive industry in the country and where metals are used as pigments in paints, as rust inhibitors, metal plating, and in raw materials. Fourth, metals are interesting to study because they are subject to a number of reactions in soil and sediment including sorption and precipitation and are greatly influenced by the redox conditions in these environments. Fifth, metals can be used to inexpensively screen for contamination at old industrial sites (brownfields), without knowing the site history. For example, the USEPA lists 14 metals (Sb, As, Ba, Be, Cd, Cr, Cu, Pb, Hg, Ni, Se, Ag, Th, and Zn) as part of their 129 most common pollutants (USEPA 2003). These metals can all be analyzed simultaneously and inexpensively using ICP-MS (inductively coupled plasma mass spectrometry), as opposed to the higher cost and longer time required to randomly test for the presence of hundreds of organic compounds at an undocumented brownfield site.

Historically, soil scientists and geologists have studied near-surface soil for agricultural or farming purposes to better understand natural ecosystems (Thornton 1991) or to focus on the health effects in urban areas associated with one or two metals (Mielke et al. 1983). As a result, there has been little information available on the background level of metals in the near-surface soil in urban areas.

In this chapter, we focus on anthropogenic activities related to 100 years of industrialization in the metropolitan Detroit area and explore how this history has contributed to the contamination of the soil, stream sediment, and groundwater by many of the 14 metals listed above. Two case studies are presented: the first addresses heavy metal contamination in soil and evaluates the background and anthropogenic levels of metals and the second focuses on dissolved metals in groundwater.

9.2 Heavy Metals in Soil

Metals naturally occur in soil in one or more of seven different ways: (1) dissolved in soil solution, (2) occupying exchange sites on inorganic constituents, (3) adsorbed in inorganic constituents, (4) associated with insoluble soil organic matter, (5) precipitated as pure mixed solids, (6) as secondary minerals, and (7) in the structure of primary minerals. The metals that have been introduced into the environment through human (anthropogenic) activities are associated with the first five (Shuman 1991).

Migration of metals in soil is influenced by physical and chemical characteristics of each specific metal and by several environmental factors. The most significant environmental factors appear to be (1) soil type, (2) total organic content, (3) redox potential, and (4) pH (McLean and Bledsoe 1992; Jaagumagi 1993; Murray et al. 1999). Metals in soil solution also migrate through mass transfer by leaching to groundwater, plant uptake, and volatization, which are important migration mechanisms when considering the mobility of arsenic, mercury, and selenium (Mattigod et al. 1981).

With respect to soil type and organic content, clay-rich soils generally have a higher retention capacity than soils with little or no clay (e.g., soils composed primarily of sand). Soils with a high organic content also have a higher retention capacity than soils with a lower relative organic content (Stevenson 1991). Oxidizing conditions generally increase the retention capacity of metals in soil, while reducing conditions will generally reduce the retention capacity of metals (McLean and Bledsoe 1992). In terms of pH, cationic metals, which include lead, copper, nickel, and zinc, have a higher retention capacity in soil with a pH greater than 7 compared to soils with a pH less than 7 (Lindsay 1979; Harter and Lehmann 1983). However, oxyanion metals, which include arsenic, selenium, and hexavalent chromium, have a higher retention capacity with a pH of less than 7 compared to soil with a pH greater than 7 (Lindsay 1979; Neal et al. 1987). Through the interaction of these factors, lead, copper, and silver demonstrate the highest capacity for retention in soil. Conversely, arsenic, chromium, and mercury are mobile if concentrations are high enough and favorable soil conditions are present (McLean and Bledsoe 1992).

9.3 Case Study: Metals in Soil, Rouge River Watershed

The contamination of urban soils can pose a significant threat to human health if certain contaminants are present at sufficient concentrations or at locations where human or ecological exposure can occur. Since the majority of the U.S. population lives in these areas, this contamination is a significant concern. Yet, the characterization of these soils has always posed a difficult challenge. Urban soils have the greatest potential to be disturbed by human activity, complicating site investigations; more importantly, background information is frequently limited, and, as pointed out by Alkhatib and O'Connor (1998), the sheer cost of analyzing soil samples discourages investigators from collecting additional background samples.

9.3.1 Study Objectives, Study Area, and Methods

Because of its industrial base, the Rouge River watershed in southeastern Michigan contains thousands of sites of environmental concern. Environmental site investigations have been conducted at locations with known or suspected impacts to evaluate the presence of heavy metals and other compounds (MDEQ 1998). In 2004, Murray et al. adopted a watershed-level approach to characterize the background levels of metals in both the near-surface and subsurface soils in southeast Michigan. The watershed approach is particularly important in these types of studies because many of the processes that contribute to the occurrence and distribution of metals in the soils of an urban environment (movement of water; location of industry) are closely tied to drainage patterns. Consequently, the watershed concept has become a key factor in modern urban planning (RRRAP 1994; Murray and Rogers 1999).

Specific objectives of this study were to determine if there are significant differences between the metal concentrations: (1) in surface and near-surface soils (<0.5 m in depth), the shallow subsurface soils (0.5–10.0 m in depth) and in soils at depths greater than 10.0 m; (2) among the various soil types related to the glacial history of southern Michigan; and (3) among major land use designations, such as residential, commercial, and industrial properties. Another important goal of the study was to simply evaluate the background concentration of metals in the soil of an urban environment. Sites selected for this study included parks, elementary and high schools, community colleges, churches, banks, new residential developments as well as older residences, courthouses, town hall, law firms, malls, golf courses, vacant property, and industrial properties. As shown in Table 2.4, the geology of the Rouge River watershed is sufficiently similar to many other urban areas to be considered a model. Consequently, the results of this study have implications for many cities, especially Cleveland, Chicago, Saint Louis, Salt Lake City, Seattle, and Los Angeles, which are sufficiently similar geologically. Internationally, cities such as London, Paris, and Mexico City are also in this category.

The data used in this study were derived from several sources including project files compiled by the Michigan Department of Environmental Quality (MDEQ), now the Department of Natural Resources and Environment (DNRE), which oversees the investigation and cleanup of hazardous waste sites in Michigan. Files from more than 3000 known or suspected sites of environmental contamination were reviewed at the MDEQ southeast Michigan district headquarters. Although soil samples were analyzed for contaminants at the vast majority of the 3000 sites, only a few hundred sites met the rigorous standards established for inclusion in this study. The methodology used in this study to characterize urban metal concentrations in the soil relied on samples collected over a period of 10 years. This feature of the study helped average the variability of metal concentrations caused by the constantly changing near-surface urban soils versus the more stable subsurface soils.

Each site was carefully evaluated and screened to eliminate obvious data bias. Specific sites excluded from the study included those with restricted access (a copper fabricating facility) and industrial properties with extremely high concentrations of a particular metal, for example, a lead smelter, a chrome-plating operation, and a gun range. Three sites were eliminated because the near-surface soil was considered fill material of an unknown age and origin. Performing this type of careful evaluation and screening was covered in detail in Chapter 4.

This screening resulted in a final dataset of 3786 soil samples analyzed for heavy metals at 171 sites. Each of these sites was then classified with respect to land use and designated as residential, commercial or industrial. The metals evaluated for this study included antimony, arsenic, barium, beryllium, cadmium, chromium, copper, lead, mercury, nickel, selenium, silver, thallium, and zinc. Figure 9.1 shows the location of the 171 sites in the study watershed where samples were collected from the three land-use categories.

Table 9.1 shows the total number of sites classified as residential, commercial, or industrial and the number of samples collected from each of the soil units. The total number of sites located on the various soil units exceeds 171 because some sites contained more than one soil unit, and they were collected from soil borings that may have come from deeper soils, which were texturally distinct from the surface soil. Table 9.1 also indicates the percentage of sites in each land use and soil unit category.

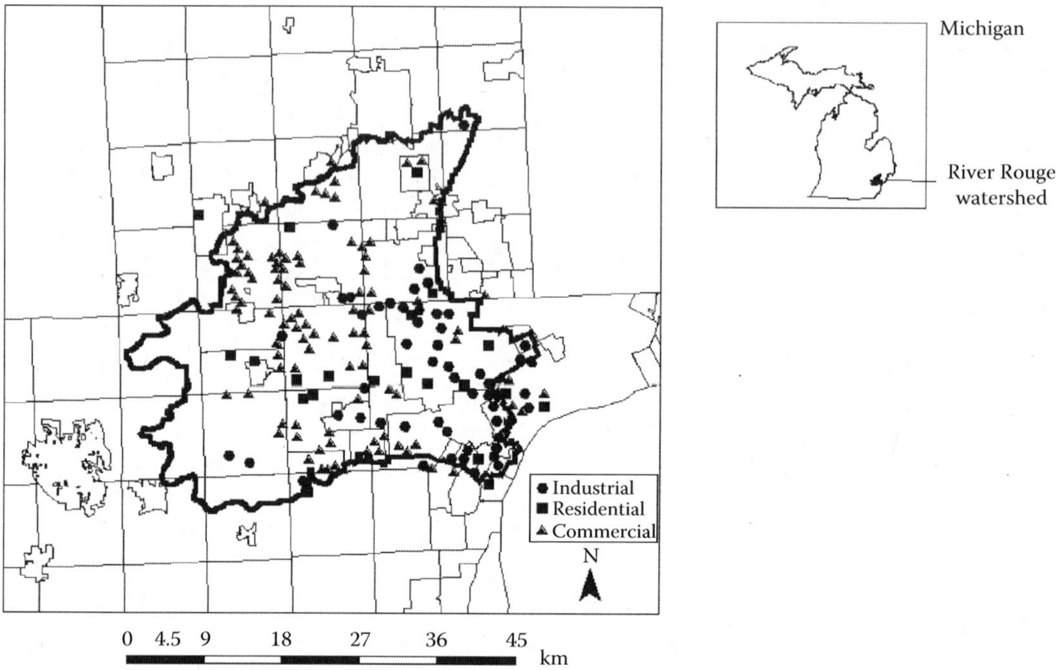

FIGURE 9.1
Land use sampling sites within the Rouge River watershed. (From Murray, K. et al., *J. Environ. Qual.*, 33, 163, 2004. With permission, American Society of Agronomy.)

TABLE 9.1
Site and Sample Location by Land Use Category and Geologic Unit

	Number of Sites	Percentage of Sites	Number of Samples	Percentage of Samples
Land use category				
Residential	28	16.4	847	22.3
Commercial	95	55.5	1634	43.2
Industrial	48	28.1	1305	34.5
Total	**171**	**100.0**	**3786**	**100.0**
Soil units				
Moraine	9	3.8	74	3.9
Outwash	6	2.6	66	3.5
Sandy silty clay	64	27.5	620	32.0
Sand	56	24.0	387	19.8
Silty clay	18	7.7	135	7.0
Upper clay	21	9.0	230	11.8
Lower clay	59	25.4	428	22.1
Total	**233**	**100.0**	**1940**	**100.0**

TABLE 9.2
Metal Analytical Methods

Metal	EPA Method
Arsenic	7061
Barium	6010
Cadmium	6010
Chromium	6010
Copper	6010
Lead	6010
Mercury	7470
Nickel	6010
Selenium	7740
Silver	6010
Zinc	6010

To be consistent with previous work, all results reported by Murray et al. (2004) were total recoverable levels of metals. Each metal analysis selected for inclusion in this study followed identical laboratory quality control procedures established by the MDNRE and mandated by the State of Michigan under Public Act 451, Part 201. The near-surface samples were collected from the upper 0.5 m of the surficial soil in the vicinity of the site being investigated. Soil collection standards typically require the collection of a soil sample at the base of the soil's A horizon using a stainless-steel hand trowel, or a manual or mechanically driven sampler. Subsurface samples were collected at depths ranging from 0.5 to 20 m typically during the installation of groundwater monitoring wells or soil borings used to determine the areal and vertical extent of contamination during a site investigation. Soil samples were generally collected using a 0.6 or 1.2 m-long steel sampler that was hydraulically pushed into the ground using a Geoprobe®, or by a steel split-spoon sampler, which was pounded into the ground at various depths using a truck-mounted drill rig equipped with hollow-stem augers. Subsurface soil samples would have limited exposure to automobile emissions and road runoff and should therefore have less of an anthropogenic signature. The samples were analyzed using USEPA 6000 or 7000 series methods (USEPA 1983) and followed all USEPA protocol (SW 846 Test Methods). Specific analytical methodologies for each metal are presented in Table 9.2.

The surficial soils and topographic relief within the study area result from several glacial advances and retreats during the recent geologic past. The resulting glacial drift has produced moraines, outwash deposits from braided streams, lake bed plains, and adjacent beach deposits. Each of these glacially derived deposits produces a texturally characteristic soil type, for example, moraines composed of glacial till are prominent in the northwestern part of the watershed and are mixtures of clay, sand, and gravel deposited by ice during glacial periods. Due to the unsorted nature of these deposits, they may have locally low permeability and moderate porosity where the clay content is high. Physical features associated with glacial terrains also include outwash plains, eskers, kames, irregular drainage patterns, wetlands, and lakes. Soils associated with many of these features consist of well-drained loams and sandy loams, with some areas of poorly drained sandy soils. Erosion potential is the highest in this area because of the steep slopes created by the terminal moraines.

Glacial outwash deposits are present in the northeastern portion of the watershed and are present between the linear moraine deposits (Farrand 1982, 1988; Rogers 1997a). Outwash consists of deposits from flowing meltwater at the margins of glaciers. They are generally well-sorted and contain large amounts of sand and gravel, with minor silt and clay. Soils are medium textured and moderately well drained and have a moderate slope. Beach and fluvial deposits formed along the western perimeter of a former glacial lake during the retreat of the Lake Erie lobe. These deposits are found in a northeast-southwest trending belt in the middle of the watershed. They tend to form very uniform, well-drained sandy soils.

Lake beds are the prominent feature in the remainder (southeast part) of the watershed (Farrand 1982; Rogers 1997a). These lake beds, characterized by low, gentle slopes consist of clay units of lacustrine origin. They are distinguished within the Rouge River watershed by the percentage of silt and sand present and are classified as a sandy and silty clay, a sandy clay and an upper clay. The associated soils are loams or clay loams, which are poorly drained. Beneath the lake bed plains lies a diamicton or lower clay layer also historically described as a ground moraine. For this study, it was assumed that samples collected from this lower clay unit, which is not exposed at the surface, represents the natural background conditions within the watershed. Consequently, the occurrence of metals within the lower clay unit is due solely to nonanthropogenic causes. This is a reasonable assumption since the unit is typically 3 m below the surface and isolated from surface activities by at least 1.5 m of low-permeability clay (Rogers 1997b; Murray and Rogers 1999). Moreover, because groundwater is not known to occur in this lower clay, there is negligible potential of metals migrating from anthropogenic sources via groundwater pathways (Rogers 1996; Rogers and Murray 1997). The location and distribution of the soil units just described are shown in Figure 5. 20, and Figure 5.11 has a geologic cross section showing the distribution of the lower clay unit. A series of diagrams (Figure 5.15a through h) depicts the evolution of deposition of the glacial deposits within the watershed during the last 14,500 years.

The land use at each sampling location was well documented. Significant changes in land use with time, however, may complicate the classification if, for example, redevelopment changed a former industrial site into one zoned for commercial. Thus, in urban areas such as Detroit, understanding the history of land use is important, and the changing land use patterns present significant challenges in evaluating the occurrence and distribution of metals in near-surface soil relative to land use. Although each soil sample was categorized by its current land use classification (residential, commercial, and industrial), every attempt was made through the use of aerial photography (black and white, both digital and stereo pairs at a scale of 1:24,000 for years 1990, 1995, and 2000) obtained from the Southeast Michigan Council of Governments (SEMCOG) to ensure that land use had not changed substantially over the 10 year period of this study. In addition, a Phase I environmental site assessment (Chapter 4) was conducted at each selected site. The Phase I included evaluating and confirming historical land use at least 60 years prior to the date of the investigation, and in many locations addressed a period over 100 years. Obviously, significant changes in land use would tend to skew results. The inherent variability within urban soils resulting from moving, backfilling, covering, and mixing was addressed by collecting a spatially dispersed sample set (distributed across the watershed) over the period from 1992 to 2002, and using aerial photography to evaluate changes in land use at questionable sites.

Due to the differing nature of the various MDEQ investigations, not all samples were analyzed for all parameters. Within the sample set, three metals, antimony, beryllium, and thallium, were analyzed in less than 10% of the samples collected. Consequently, these metals are reported separately in Table 9.3 along with their respective range of detection levels. They have also been eliminated from the statistical results presented with the other metals as their sample detection limits would have disproportionately skewed the results. It is important to note that the range of detection for each of the low-occurrence metals

TABLE 9.3

Range of Concentration for Low-Occurrence Metals

Metal	Number of Samples Analyzed	Range Detected (mg/kg)
Antimony	7	3.7–6.1
Beryllium	6	0.5–1.5
Thallium	4	0.45–1.23

was consistent with the range of detection reported by Shackette and Boergnen (1984). A review of the occurrence data suggests that all of these metals are likely the result of anthropogenic sources rather than from natural occurrence.

9.3.2 Results

The analytical results for the eleven remaining metals are presented in Tables 9.4 and 9.5. Table 9.4 indicates each metal's horizontal (west-to-east distribution across the six soil units within the watershed) and vertical (surface, subsurface, and lower clay) distribution across the watershed. Table 9.5 reports each metal's mean concentration in both surface and subsurface soils across each of the three land use categories and is split into separate tables for surface soil (Table 9.5a) and subsurface soil (Table 9.5b).

A few trends are apparent from these results. As shown in Table 9.4, surface concentrations are generally greater than subsurface and concentrations generally increase in a west-to-east direction across the watershed. This latter trend is commensurate with a west-to-east increase in urbanization and industrialization. The moraine unit shows consistently lower levels of metals than any of the other soil units, and has mean concentrations of metals statistically similar to that found in the lower clay layer, which is assumed to contain naturally occurring metal concentrations. Although the moraine has locally high permeability, it generally contains substantially more clay than either the outwash or the sand units. The low concentration of metals present in this unit is thus attributed to fewer anthropogenic sources in the western, more rural part of the watershed. More interesting, however, is the relatively high concentration of metals found in the sand unit. The sand unit varies in thickness from less than a meter to more than 10 m and is highly permeable with a hydraulic conductivity ranging from 10^{-4} to 10^{-1} cm/s (Rogers and Murray 1997). Yet, the sand unit contains statistically higher concentrations of metals than either the moraine or outwash units, which are both exposed at the surface westward of the sand.

The explanation for this apparent paradox is the location of the sand within the watershed. The sand is located in the center of the watershed along the urban fringe. This is an area that has undergone rapid urbanization and industrialization over the past 20 years and represents the transition between the rural west and the more heavily industrialized eastern part of the watershed. Contamination derived from spills, leaking underground or above ground tanks can quickly pass through the vadose zone within the sand to reach the water table, typically at a depth of no more than 3 m below the ground surface in southeast Michigan. Consequently, this part of the watershed has the highest incidence of groundwater contamination (Murray and Rogers 1999; Kaufman et al. 2003), and the sand unit, despite its lack of clay and organic material contains a higher concentration of the metals most often associated with industry: arsenic, barium, cadmium, chromium, copper, nickel, and selenium than even the sandy silty clay unit that is located immediately to the west. The pivotal position and characteristics of the sand unit is underscored by metal concentrations that are typically 50% less than the metal concentrations found in the two easternmost clay-rich units.

The silty clay unit, which is exposed at the surface immediately to the east of the sand unit contains the highest arsenic concentrations of all the soil units with its highest concentrations present in both surface and subsurface soils in industrial areas. Although arsenic concentrations were expected to be uniformly high across the entire watershed, as a function of either the weathering of natural arsenic-bearing minerals associated with the underlying Marshall Sandstone or from the atmospheric deposition associated with

TABLE 9.4

Metal Distribution in the Geologic Units (in mg/kg)[a]

	Westernmost Geologic Units			Easternmost Geologic Units		
Layer	**Moraine**	**Outwash**	**Sandy and Silty Clay**	**Sand**	**Silty Clay**	**Clay**
Arsenic						
Surface	1.9 (1.8), 19	3.6 (3.4), 6	4.6 (1.3), 6	5.5 (6.1), 93	7.9 (6.2), 138	6.5 (7.6), 221
Subsurface	ND[b]	ND	12 (2.8), 15	3.1 (3), 65	5.3 (4.6), 142	7.2 (4.2), 59
Lower clay	ND	4.1 (none), 1	ND	2.4 (1.92), 6	9.2 (3.2), 30	5.8 (2.1), 63
Barium						
Surface	31 (59), 19	36 (53), 6	205 (290), 6	80 (61), 93	122 (226), 90	121 (226), 126
Subsurface	ND	ND	58 (26), 12	20 (19), 61	66 (3.7), 249	61 (20), 56
Lower clay	ND	64 (none), 1	ND	22.5 (8.9), 10	73.8 (5.8), 31	51 (14.8), 63
Cadmium						
Surface	0.38 (0.59), 25	0.14 (0.12), 6	0.94 (1.36), 10	0.9 (9.65), 127	3.5 (34.2), 184	3.9 (0.59), 25
Subsurface	ND	ND	0.5 (0.5), 71	0.4 (0.53), 98	1.0 (3.7), 249	0.94 (1.32), 76
Lower clay	ND	0.18 (none), 1	0.57 (0.11), 49	0.2 (0.14), 12	0.28 (18.5),124	0.50 (0.9), 58
Chromium						
Surface	6.5 (6.8) [25]	12 (8.6), 6	17 (17.0), 33	30.5 (70), 109	63 (298), 147	52.5 (252), 226
Subsurface	ND	ND	9.8 (6.2), 72	11 (14.6), 107	21 (26), 252	49 (47.6), 61
Lower clay	ND	13 (none), 1	6.6 (1.1), 23	9.7 (6.1), 12	14 (0.24), 93	14.5 (6.1), 63
Copper						
Surface	7.5 (9.6), 19	16 (4.0), 4	31 (13), 6	61 (98), 103	85 (234), 143	70 (236), 389
Subsurface	ND	ND	18 (3.8), 12	10 (16), 73	24 (41), 155	37 (26), 78
Lower clay	ND	18 (none), 1	6.6 (1.1), 23	9.8 (5.96), 6	20 (3.9), 28	19.5 (4.74), 63
Lead						
Surface	8.8 (13), 37	8.0 (4.7), 16	90 (15), 13	65 (78), 129	108 (335), 337	162 (405),1343
Subsurface	ND	ND	22 (104), 145	6.3 (11.5), 446	30 (68), 518	58 (142), 202
Lower clay	4.1 (none) [1]	7.9 (2.1), 3	11.3 (10.1), 75	4.1 (none), 1	7.9 (2.1), 3	11.3 (10.1), 75
Mercury						
Surface	0.07 (0.01), 11	0.06 (0.04), 6	0.2 (0.05), 5	0.1 (0.25), 92	0.17 (0.27),109	0.27(0.41),122
Subsurface	ND	ND	0.06 (0.03), 10	0.06 (0.11), 54	0.98 (0.08),102	0.07 (0.09), 60
Lower clay	ND	0.1 (none), 1	ND	0.015 (0.002),4	0.1 (0.04), 26	0.09 (0.04), 63
Nickel						
Surface	0.09 (0.59), 25	0.14 (0.12), 6	0.94 (1.36), 10	17.5 (260), 54	28 (83), 100	51 (150), 138
Subsurface	ND	ND	11.5 (7.85), 26	8.4 (16), 49	22.6 (10), 19	20 (2.5), 3
Lower clay	ND	ND	ND	ND	ND	0.09 (2.2), 8
Selenium						
Surface	0.37 (0.125), 11	0.06 (0.22), 8	0.33 (0.45), 5	0.4 (0.35), 92	0.96 (1.46), 99	1.85 (2.9), 119
Subsurface	ND	ND	0.12 (0.05), 8	0.5 (0.5), 56	0.75 (0.69), 98	1.1 (0.8), 59
Lower clay	ND	0.05 (none), 1	ND	0.6 (0.2), 4	0.23 (0.19), 24	0.93 (0.93), 58
Silver						
Surface	0.37 (0.125), 11	0.06 (0.22), 8	0.33 (0.45), 5	0.4 (0.35), 92	0.96 (1.46), 99	1.85 (2.9), 119
Subsurface	ND	ND	0.12 (0.05), 8	0.5 (0.5), 56	0.75 (0.69), 98	1.1 (0.8), 59
Lower clay	ND	0.05 (none), 1	ND	0.6 (0.2), 4	0.23 (0.19), 24	0.93 (0.93), 58

TABLE 9.4 (continued)

Metal Distribution in the Geologic Units (in mg/kg)[a]

	Westernmost Geologic Units			Easternmost Geologic Units		
Layer	**Moraine**	**Outwash**	**Sandy and Silty Clay**	**Sand**	**Silty Clay**	**Clay**
Zinc						
Surface	49 (76), 18	49 (16), 4	138 (112), 6	160 (96), 102	221 (416), 141	251 (439), 251
Subsurface	ND	ND	42 (79), 15	26 (40), 77	77 (115), 147	84.5 (85), 78
Lower clay	ND	42 (none), 1	31.7 (4.52) [2]	31 (22), 8	56 (14.5), 32	36.4 (6.4), 62

[a] Concentrations are derived from Michigan Department of Environmental Quality (MDEQ) files. The first number listed in each cell is the measured concentration in mg/kg; values in parentheses are standard deviations; those following the comma are the sample counts.

[b] ND = no data.

TABLE 9.5

Mean Concentration of Metals in Surface and Subsurface Soils Relative to Land Use

	Commercial			Residential			Industrial		
Metal	**N**	**X**	**Φ**	**N**	**X**	**Φ**	**N**	**X**	**Φ**
(a) Surface soil									
As	205	5.1	5.5	77	6.3	4	201	7	9
Ba	151	69	66	71	128	135	118	148	222
Cd	234	2.2	1.5	80	1.1	1.1	151	4.5	13
Cr	282	27	83	67	31	52	197	55	159
Cu	205	32	60	82	30	39	377	113	269
Pb	418	93	300	535	160	250	893	150	380
Hg	167	0.2	0.24	58	0.08	0.07	120	0.2	0.3
Ni	151	16	9.8	35	24	29	132	58	150
Se	164	0.6	0.3	57	0.8	0.8	113	1.9	2.2
Ag	152	0.5	0.2	54	0.8	0.5	114	2.3	2.2
Zn	202	130	310	81	120	124	239	257	534
(b) Subsurface soil									
As	95	5.6	4.0	78	2.3	4.2	108	7.1	4.2
Ba	90	55	25	70	61	37	82	76	24
Cd	306	0.5	1.5	86	0.4	0.5	102	1.9	2.5
Cr	292	11.6	14	96	25.4	38	104	27	35
Cu	98	20	14	94	26	57	126	14	17
Pb	906	20	60	220	34	77	185	39	72
Hg	88	0.05	0.01	53	0.1	0.8	85	0.08	0.05
Ni	3	36	3	18	11	7	38	14.5	24
Se	90	0.8	0.6	53	0.35	0.2	82	0.9	0.9
Ag	67	0.6	0.4	48	0.5	1.1	72	2.2	4
Zn	94	67	137	98	66	80	15	60	60

Note: N = number of samples; X = Mean (mg/kg); Φ = Standard Deviation.

coal-fired power plants, this was not the case. In fact, relatively modest levels of arsenic were present in both surface and subsurface soils across all three land use categories and all soils with the exception of the silty clay. The distribution of arsenic within the watershed may thus result from localized industrial sources of arsenic and conditions that serve to increase its mobility. These conditions include an adequately high concentration, a pH greater than 7, and the oxidation state of the arsenic. For example, As [V] is the dominant form of arsenic under aerobic conditions found in the near-surface soils. In this oxidized state arsenic will strongly bind with soil and sediment, particularly in clay-rich soils. In an anaerobic environment, conditions associated with the subsurface clay-rich soils, arsenic generally forms insoluble and nonmobile sulfides. However, in soils with a pH greater than 7, arsenic tends to bind less strongly to soil particles and becomes more mobile. As an example, residential soils in the Delray community of southwest Detroit have arsenic concentrations 10 times higher than other locations in southeast Michigan. One likely cause of this high arsenic concentration is the neighborhood's proximity to Zug Island, an industrial complex with a long history of steel manufacturing fueled by the burning of coal (Figure 9.2) (Peterman and Murray 2010).

The silty clay unit also contains substantially higher levels of barium, cadmium, lead, and zinc than any of the other soils units. Because the silty clay is located in the more industrialized part of the watershed, the higher concentration of these metals suggests anthropogenic sources. This premise is supported by the data presented in Table 9.5, which indicates almost uniformly higher metal concentrations in industrial land uses in both surface and subsurface soils. Two notable exceptions were the high lead and mercury levels present in residential areas in the surface and shallow subsurface soils, respectively. Lead in the surface soils in residential areas was present at levels 16 times that found in the lower clay layer. The high lead levels are most likely due to deposition of lead dust from sources such as lead-based paints in older residences common in urban areas and the former use of leaded gasoline (Mielke et al. 1983; Mielke et al. 1984; Mielke 1999). Subsurface lead concentrations in residential areas are also relatively high, with mean values at more than twice the background levels found in the lower clay layer. Mean values of surface lead in industrial areas are nearly as high as in residential areas (15.5 times background) with

FIGURE 9.2
Coal piles ready for combustion near a residential neighborhood in southwest Detroit, MI. (Photo by Kent S. Murray.)

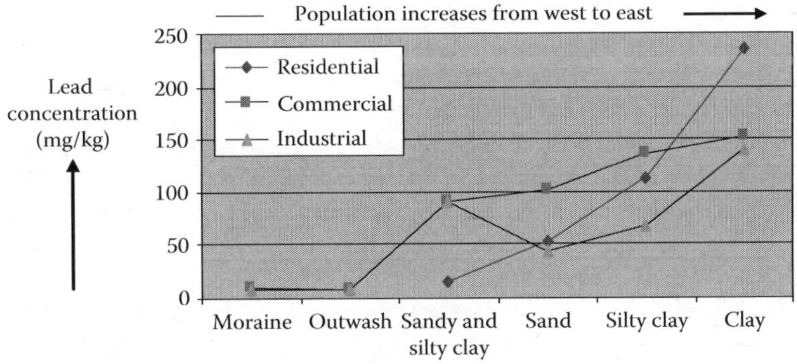

FIGURE 9.3
West-to-east concentration (mg/Kg) of lead in surface soil by land use. (From Murray, K. et al., *J. Environ. Qual.*, 33, 163, 2004. With permission, American Society of Agronomy.)

several industrial sites (and surrounding neighborhoods) exhibiting lead concentrations hundreds of times the background concentration. In addition, subsurface lead in industrial areas was nearly five times higher than the background level (Figure 9.3), underscoring the impact of industry's role in contributing to soil lead contamination.

Mercury was present in the near-surface soils in commercial areas at levels four times the background level of 0.09 mg/kg present in the lower clay unit. Throughout most of the subsurface, mercury concentrations are equivalent to background levels, while mercury concentrations at commercial and industrial sites are more than twice the background levels. The high incidence of mercury at industrial sites is probably related to the production of chlorine, caustic soda, and hydrogen. It may also be related to former automotive paint industries or the production of electrical equipment (Jaagumagi 1993). The high incidence of mercury at commercial sites is more problematic and may reflect changing land use in an urban area with an industrial base that is more than 100 years old.

Concentrations of copper and zinc at the surface were significantly lower at residential and commercial properties compared to industrial properties. This suggests the primary sources of copper and zinc were industry related. In addition, concentrations of many other heavy metals, such as cadmium, chromium, nickel, selenium, and silver were also detected in surface soil at industrial properties at higher concentrations than commercial and residential properties. This again suggests the elevated concentrations of these metals may have an anthropogenic source, presumably, industrial sites. On the other hand, concentrations of barium, mercury, and arsenic did not vary significantly between the land use categories; implying that these metals do not have a significant anthropogenic source.

Concentrations of heavy metals in the Lower Clay Unit differ only slightly from the heavy metal concentrations observed in previous Michigan studies. In fact, most metals (arsenic, barium, cadmium, chromium, lead, mercury, nickel, silver, and zinc) were present at mean concentrations slightly less than the previous Michigan studies. Only copper and selenium were present at mean concentrations greater than those of previous studies; however, these elevated concentrations would not be considered statistically significant. These results support our contention that metal concentrations in the Lower Clay Unit within the Rouge River watershed can be reasonably interpreted to represent naturally occurring concentrations of heavy metals and confirm their use as background heavy metal concentrations for this study.

Comparing the results of the land use data to background concentrations, it was expected that the metals would be the highest within industrial and the lowest for residential areas. In surface soils, industrial land use was indeed the highest for each metal, with the modest exception of lead as discussed above, and arsenic in the residential soils of Delray. However, instead of residential having the lowest metal concentrations, the commercial category had the lowest concentrations for 7 of the 11 metals included in this analysis. The exceptions were cadmium, copper, mercury, and zinc. In subsurface soils of commercial properties, these four metals plus nickel continued to be present at higher concentrations than subsurface soils from residential areas.

A closer look at lead in surface and subsurface soil confirms these general trends. For example, in surface soil, mean Pb concentrations differ significantly across the six different soil groups and gradually increase in concentration in a west-to-east direction.

This increase is commensurate with an increase in anthropogenic sources of lead, including a higher incidence of older homes containing lead-based paint, a higher occurrence of residual lead in the soil derived from the former use of leaded gasoline, and a higher incidence of industrial sources. The only deviation from this trend was a lower level of lead in the sand unit than in the sandy and silty clay unit located to the west of the sand. This discrepancy, however, can be explained by the lack of adsorptive characteristics relative to the sandy and silty clay unit. As discussed above, the sand unit contains virtually no clay yet still has mean surface concentrations of lead that are nearly eight times that found in the moraine and outwash units that contain on average 20% or more clay. Mean lead concentrations at the surface were also significantly different across all land uses, with mean lead levels in surface soil derived from residential areas approximately 16 times that of the mean lead concentrations found in the lower clay unit. Additionally, mean lead concentrations at the surface were significantly different within any one land use category, for example, commercial land use, when evaluated across all soil units, suggesting that the west-to-east trend is independent of land use and is thus more likely a function of an overall increase in urbanization and industrialization.

In the shallow subsurface soils, mean lead concentrations were also significantly different across all soil units, with mean lead concentrations in the upper clay, which is exposed at the surface in the easternmost part of the watershed, including the City of Detroit, nearly twice the level of the mean lead concentration found in the adjacent silty clay unit and more than five times the background concentration of lead found in the lower clay unit (Table 9.4). The mean lead concentrations at industrial sites in shallow subsurface soils were substantially different from the mean lead values in subsurface soil at either commercial or residential sites and were still three times the mean background concentration of lead present in the lower clay unit.

9.4 Case Study: Dissolved Metals in Shallow Groundwater, Rouge River Watershed

In general, metals have limited mobility in soil and groundwater due to cation exchange or sorption onto mineral grains (Fetter 1993; Bedient et al. 1994). However, the concentrations of naturally occurring as well as anthropogenically introduced metals in near-surface soils, which can then migrate to shallow groundwater, can vary significantly due to physical and chemical processes operating across geographic regions.

Within groundwater, the concentrations of heavy metals are most significantly influenced by (1) presence of heavy metals in soil that has come into contact and reacts with groundwater and (2) the chemical nature of the groundwater itself (Fetter 1993). Metals have increased mobility in groundwater if the groundwater chemistry exhibits a low cation-exchange capacity or if they are attached to a mobile colloid (Bedient et al. 1994). Conditions that promote metal mobility would, therefore, include a hydrogeologic environment that consists of a subsurface saturated environment composed of an acidic, sandy soil with low-organic and clay content (Dowdy and Volk 1983).

9.4.1 Study Objectives, Study Area, and Methods

This study (Murray et al. 2006) measured dissolved heavy metal concentrations in shallow groundwater within the same region—the Rouge River watershed. The primary objectives of this study were (1) to characterize the distribution of heavy metals in groundwater in different geologic and land use environments within the watershed, (2) identify the most mobile metals and the locations associated with their mobility, (3) determine the amount of contamination represented by the measured mean concentrations of heavy metals in the shallow groundwater, and (4) assess the potential public health and ecosystem impacts from any dissolved metal contamination found. The evaluation of heavy metal mobility was enabled through the direct comparison of the mean concentrations of heavy metals measured in the soil by the Murray et al. (2004) study and groundwater.

The discharge of near-surface shallow groundwater to the surface waters of the Rouge River is significantly influenced by the presence of the lower clay unit, which has effectively limited the vertical migration of groundwater to lower aquifers because (1) it is thick (80 m to greater than 300 m thick), (2) vertical groundwater flow in the clay is insignificant, (3) the unit completely underlies the watershed, and (4) has a relatively uniform hydraulic conductivity, which has been measured to be less than 1×10^{-7} cm/s (Rogers 1997b). Glacial deposits similar to the lower clay unit have also been described as occurring with a similar hydraulic conductivity at other locations of northern North America (Freeze and Cherry 1979; Keller at al. 1989, 1991). Because of the lower clay unit, near-surface shallow groundwater within the watershed has no effective migratory outlet other than surface water. Therefore, any dissolved contaminants in shallow groundwater will eventually be discharged to the surface waters of the Rouge River and subsequently flow into the Detroit River and eventually into the lower Great Lakes.

The most significant hydrogeologic units (other than the lower clay unit) within the watershed are the Moraine Unit, Outwash Unit, and the Sand Unit (Figure 5.20). These three units significantly influence surface water because they (1) contain a large volume of groundwater, (2) account for the majority of the baseflow in the Rouge River, (3) contain the majority of tributaries of the river, and (4) are the source of groundwater recharge from precipitation and surface water infiltration (Rogers 1997b; Rogers and Murray 1997; Murray and Rogers 1999). In addition, these geologic units are also significant because shallow groundwater within these geologic units is also used as a source of potable water (Mozola 1954, 1969; Rogers 1996), creating a potential pathway for human ingestion of shallow groundwater. Inspection of the groundwater flow direction at each site revealed a good correspondence with the direction of surface water flow (Figure 6.2).

As with the soil metal study discussed earlier, the data used in the study of dissolved metals in the groundwater were also derived primarily from project files compiled by the MDEQ. Also, as discussed in Section 9.3, each site selected was carefully evaluated and screened to eliminate obvious data bias, including the same sites excluded during the soils metal investigation, i.e., the Cu-fabricating facility, several industrial properties with extremely high concentrations of a particular metal, the Pb smelter, several chrome-plating facilities, and a gun range. In addition, the three sites eliminated previously because the near-surface soil was considered fill material of an unknown age and origin were also eliminated here. They were not included because the shallow groundwater occurring beneath these sites was not believed to be in a naturally occurring geologic formation.

This screening resulted in a final dataset of 1140 groundwater samples collected from 126 sites, with the high sample-to-site ratio helping to ensure the replication of results. The selected sites included parks, elementary and high schools, community colleges, churches, banks, new residential developments as well as older residences, courthouses, a town hall, law firms, malls, golf courses, vacant property, and industrial properties. Samples were collected over a period of 10 years from 1990 to 2000, which helped average the variability of metal concentrations caused by the constantly changing urban region.

Each of these sites was then classified with respect to land use and designated as residential, commercial, or industrial. The metals evaluated for this study included As, Ba, Cd, Cr, Cu, Pb, Hg, Ni, Se Ag, and Zn. Table 9.6 shows the number and percent of sites within the land use categories as well as the distribution of the samples obtained within each geologic unit.

Shallow groundwater within the study area flows toward and discharges into the Rouge River (Rogers and Murray 1997; Rogers 1997b; Murray and Rogers 1999). Inspection of the groundwater flow direction at each site revealed a good correspondence with the direction of surface water flow.

Monitoring wells were typically installed using a drilling rig equipped with an 11 cm inner-diameter hollow stem auger. Monitoring wells were constructed and sampled according to procedures outlined by USEPA (1986, 1990). In the laboratory, analysis of the groundwater samples for dissolved metals was accomplished using USEPA 6000 or 7000 series methods (USEPA 1983) and following all USEPA protocols (SW 846 test methods).

TABLE 9.6

Site and Sample Distribution

	Number of Sites	Percent of Sites	Number of Samples	Percent of Samples
Land use				
Commercial	70	56	580	51
Industrial	32	25	438	38
Residential	24	19	122	11
Total	**126**	**100**	**1140**	**100**
Geologic unit				
Outwash	4	3	26	2
Moraine	6	5	17	2
Sandy and Silty Clay	16	13	189	17
Sand	50	39	364	32
Silty Clay	30	24	195	17
Clay	20	16	349	30
Total	**126**	**100**	**1140**	**100**

To evaluate the mobility of heavy metals, the ratio of the mean concentrations of each metal in groundwater to surface soil was computed. The metal concentrations within surface soil were chosen since this is where the highest concentrations occur, and this location also accounts for the longest vertical distance metals could migrate before reaching groundwater. Although the measurement units of the mean concentrations within soil (mg/kg) and groundwater (mg/L) are different, the ratio of mean groundwater concentration to mean surface soil concentration provides a consistent and unbiased way to compare the amounts of different metals present along this vertical profile. These ratios are then ranked to identify the most mobile metals and their associated geologic units.

To demonstrate the magnitude of contamination, another ratio is calculated between the mean concentration of each metal and the metal's **maximum contaminant level** (MCL). Because numerous drinking water supply wells exist within the watershed, the presence of a specific metal within groundwater is considered significant if its mean concentration within a geologic unit or land use category exceeds the current drinking water standard—the MCL in mg L^{-1} set by the USEPA (1992, 2003).

9.4.2 Results

Table 9.7 contains the mean concentrations of heavy metals within the vertical soil/groundwater column for each of the six geologic units of the watershed. The mean concentrations of the sites within the three categories of land use are also presented. Significant concentrations of metals in groundwater exceeding the MCL are denoted by an asterisk.

As shown in Table 9.7, there are 66 total possible combinations of metals and geologic units (11 metals * 6 geologic units) within the study area. Subtracting the seven instances of missing data leaves 59 combinations. Significant metal concentrations occur in 19 of these 59 cases (32%), which suggest groundwater contamination by heavy metals is widespread throughout the watershed. The metals showing significant concentrations within the most geologic units are Pb (5), Cd (4), and Cr (4). Mean concentrations for each of these metals in the sand exceeds the MCL, with Cr exhibiting the highest mean concentration. Copper, Ba, and Zn do not exceed their MCLs within groundwater across any of the geologic units; Ni has no established standard—the 0.1 mg/L standard was remanded in 1995 (USEPA 2003).

All of the metals tested—with the exception of Se—exhibited the highest mean concentrations at industrial sites. In the Murray et al. (2004) soil study, the mean concentration of metals generally increased across the watershed in a west-to-east direction. This same pattern of increase is also seen with metals dissolved in groundwater (Figure 9.4a and b), and with the vertical scale of contamination in Figure 9.4b indicating these metals occur at generally higher mean concentrations than those metals graphed in Figure 9.4a. The MCLs were exceeded by five metals (As, Cd, Cr, Pb, Hg) at industrial properties by As Cd, Pb, and Hg at commercial sites and Cd and Pb at residential locations. Thus, the MCLs were exceeded by cadmium and lead within all land use categories.

All of the metals exceeding the MCL within the industrial land use category (As, Cd, Cr, Pb, Hg) also exceeded their MCLs within the clay unit. This characteristic illustrates the historical urban development pattern of the region, which saw the heaviest concentrations of industry located within the older, more urbanized eastern portion of the watershed.

TABLE 9.7

Mean Concentrations of Heavy Metals by Geologic Unit and Land Use

	Surface mg/kg	Subsurface mg/kg	GW mg/L	GW/Surface Ratio	Surface mg/kg	Subsurface mg/kg	GW mg/L	GW/Surface Ratio
	Arsenic (MCL = 0.05)				**Barium (MCL = 2.0)**			
Geologic unit								
Moraine	1.9	Na	0.04[a]	0.021	31.0	Na	Na	Na
Outwash	3.6	Na	0.01[a]	0.003	36.0	Na	0.5	0.014
Ssc	4.6	12	0.003	0.001	205.0	58.0	0.18	0.001
Sand	5.5	3.1	0.075[a]	0.014	80.0	20.0	0.29	0.004
Sc	7.9	5.3	0.23[a]	0.029	122.0	66.0	0.36	0.003
Clay	6.5	7.2	0.35[a]	0.054	121.0	61.0	1.76	0.015
Land use								
Commercial	5.1	5.6	0.14[a]	0.027	69.0	55.0	0.61	0.009
Industrial	7.0	7.1	0.27[a]	0.039	148.0	76.0	1.21	0.008
Residential	6.3	2.3	0.008	0.001	128.0	61.0	0.25	0.002
	Cadmium (MCL = 0.005)				**Chromium (MCL = 0.1)**			
Geologic unit								
Moraine	0.38	Na	0.015[a]	0.039	6.5	Na	0.008	0.001
Outwash	0.14	Na	0.0004	0.003	12.0	Na	0.039	0.003
Ssc	0.94	0.5	0.0045	0.005	17.3	9.8	0.05	0.003
Sand	0.9	0.4	0.06[a]	0.067	30.5	10.9	2.17[a]	0.071
Sc	3.5	1.0	0.008[a]	0.002	63.6	21.0	0.89[a]	0.014
Clay	3.9	0.94	0.038[a]	0.009	52.5	48.2	3.1[a]	0.059
Land use								
Commercial	2.2	0.5	0.005[a]	0.002	27.0	11.6	0.02	0.001
Industrial	4.5	1.9	0.076[a]	0.017	55.0	27.0	4.87[a]	0.089
Residential	1.1	0.4	0.02[a]	0.018	31.0	25.4	0.37	0.001

		Copper (MCL = 1.3)				Lead (MCL = 0.015)		
Geologic unit								
Moraine	7.5	Na	Na	Na	8.8	Na	0.014	0.002
Outwash	16.0	Na	0.014	0.001	8.0	Na	0.019[a]	0.014
Ssc	31.0	18.0	0.01	0.000	90.4	21.5	0.035[a]	0.000
Sand	61.0	10.0	0.04	0.001	65.2	6.3	0.21[a]	0.003
Sc	85.0	24.0	0.19	0.002	107.8	29.9	0.12[a]	0.001
Clay	70.0	37.0	0.06	0.001	162.4	58.3	0.08[a]	0.000
Land use								
Commercial	32.0	20.0	0.05	0.002	93.0	20.0	0.045[a]	0.000
Industrial	113.0	14.0	0.04	0.000	150.0	39.0	0.053[a]	0.000
Residential	30.0	26.0	0.03	0.001	160.0	34.0	0.042[a]	0.000
		Mercury (MCL = 0.002)				**Nickel (MCL = none established)**		
Geologic unit								
Moraine	0.07	Na	Na	Na	0.09	Na	0.1	1.11
Outwash	0.06	Na	0.0003	0.005	0.14	Na	Na	Na
Ssc	0.2	0.06	0.0002	0.001	0.94	11.5	Na	Na
Sand	0.1	0.06	0.0007	0.007	17.5	8.4	0.043	0.002
Sc	0.17	0.98	0.009[a]	0.053	28.0	22.6	0.77	0.028
Clay	0.27	0.07	0.034[a]	0.126	51.0	20.0	0.08	0.002
Land use								
Commercial	0.2	0.05	0.003[a]	0.015	16.0	36.0	0.12	0.008
Industrial	0.2	0.08	0.02[a]	0.1	58.0	14.5	0.27	0.005
Residential	0.08	1.1	0.0004	0.005	24.0	11.0	Na	Na
		Selenium (MCL = 0.05)				**Silver (MCL = 0.1)**		
Geologic unit								
Moraine	0.37	Na	Na	Na	0.37	Na	Na	Na
Outwash	0.06	Na	0.005	0.083	0.4	Na	0.0015	0.004
Ssc	0.33	0.12	0.004	0.012	2.3	0.3	0.0005	0.000

(continued)

TABLE 9.7 (continued)

Mean Concentrations of Heavy Metals by Geologic Unit and Land Use

	Surface mg/kg	Subsurface mg/kg	GW mg/L	GW/Surface Ratio	Surface mg/kg	Subsurface mg/kg	GW mg/L	GW/Surface Ratio
Sand	0.4	0.5	0.006	0.015	0.5	1.0	0.0068	0.014
Sc	0.96	0.75	0.073[a]	0.076	1.23	2.7	0.013	0.011
Clay	1.85	1.1	0.04	0.022	1.72	0.41	0.033	0.019
Land use								
Commercial	0.6	0.8	0.047	0.078	0.5	0.6	0.008	0.016
Industrial	1.9	0.9	0.029	0.015	2.3	2.2	0.02	0.009
Residential	0.8	0.35	0.005	0.006	0.8	0.5	0.009	0.011
		Zinc (MCL = 5.0)						
Geologic unit								
Moraine	49.0	Na	0.23	0.005				
Outwash	49.0	Na	0.11	0.002				
Ssc	138.0	42.0	0 03	0.000				
Sand	160.0	26.0	0.08	0.001				
Sc	221.0	77.0	0.34	0.002				
Clay	251.0	84.5	0.45	0.002				
Land use								
Commercial	130.0	67.0	0.31	0.002				
Industrial	257.0	60.0	0.34	0.001				
Residential	120.0	66.0	0.1	0.001				

Note: MCL in mg^{-1}.

[a] Concentration > MCL.

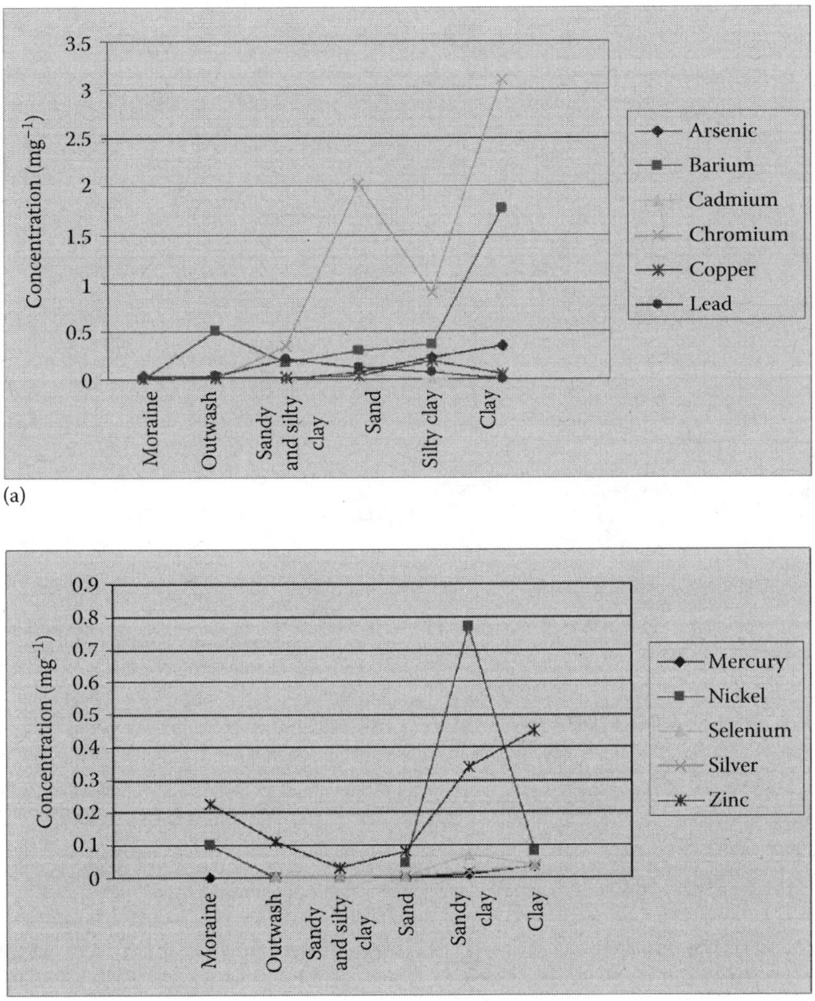

FIGURE 9.4
West-to-east distribution of heavy metals in groundwater for (a) arsenic, barium, chromium, copper, and lead; (b) mercury, nickel, selenium, silver, and zinc (west is to the left on each graph). (From Murray, K. et al.: Dissolved heavy metals in shallow ground water in a southeastern Michigan urban watershed. *J. Am. Water Resour. Assoc.*, 2006. 42. 777–792. Copyright Wiley-VCH Verlag GmbH & Co. KGaA. Reproduced with permission.)

Metal mobility is assessed by computing the ratio between the mean concentrations of metals in groundwater and at the soil surface. Columns 1–3 of Table 9.8 show the metals with the 10 highest ratios and the geologic unit where this ratio occurs; column 4 will be referenced during the discussion on contamination.

High ratios are observed for several metals within the clay and silty clay units (Hg, Se, Cr, and As), indicating some vertical mobility does occur for these metals within clay-rich soils and subsoils. Two of these same metals (Se and Cr) are also mobile within units consisting of larger-grained soil particles, which have higher hydraulic conductivities (sand, moraine, and outwash). Cadmium also exhibits high ratios within the sand and

TABLE 9.8

Metal Mobility and Contamination Ranking (n = 59)

Metal	Ratio (gw-Surface)	Geologic Unit	Measured Conc./ MCL (Rank) (mg/L)
Hg	0.126	Clay	17.0 (3)
Se	0.083	Outwash	(36)
Se	0.076	Silty clay	(18)
Cd	0.067	Sand	(5)[a]
Cr	0.071	Sand	(2)
Cr	0.059	Clay	(1)
As	0.054	Clay	(9)
Hg	0.053	Silty clay	(12)
Cd	0.039	Moraine	(14)
As	0.029	Silty clay	4.6 (11)

[a] Missing top 10 places: 4th = Pb in sand @ 14.0; 6th = Cr in silty clay @ 8.9; 7th = Pb in silty clay @ 8.0; 8th = Cd in clay @ 7.6; 10th = Pb in clay @ 5.3.

moraine geologic units. There is generally lower mobility observed for Ag, Ba, Cu, Ni, Pb, and Zn.

Referring again to Table 9.8, groundwater contamination is represented by the ratio between the mean concentration of the metal and its MCL in column 4. The rank of each metal appears in parentheses after the ratio. The two highest contamination ratios and ranks belong to Cr in clay and sand, respectively. Other significant rankings exist for Cd and Pb in the sand unit, which were ranked fourth and fifth, respectively. Thus, three metals had measured mean concentrations which were over 10 times the MCL within a geologic unit with high hydraulic conductivity.

These results demonstrate that the groundwater resource within the Rouge River watershed is significantly polluted with heavy metals. Most significant are the occurrences of elevated concentrations of heavy metals within the sand unit, with the mean concentrations of Cr, Cd, and Pb exceeding the MCL by 22, 12, and 14 times, respectively. The main commercial uses of Cr include chrome alloy in chromium metal products and in chrome plating, and to a lesser extent as compounds in paints, dyes, explosives, ceramics, glass, photography, and paper. The principal use of Cd is as an alloy in electroplating, in Ni-Cd batteries, solders, electronic equipment, photographic supplies, glass, ceramics, and plastics (Jaagumagi 1993). The high concentration of Cr and Cd in groundwater confirms the principle of increased metal mobility in acidic, sandy soil.

The presence of elevated levels of Pb in groundwater may be due to the industrial mix within the watershed. Metal industries use Pb in smelting and electroplating, but the single largest use of Pb is in the production of lead acid batteries. Other uses include paints, glassware, electronic equipment, plastics, solder, and construction materials. Notwithstanding, this result is somewhat surprising, as Pb has a soil retention and adsorption capacity higher than any other metal and is generally considered immobile in subsurface soil (McLean and Bledsoe 1992). However, the mobility of Pb within southeast Michigan soils was suggested by Murray et al. (1997) in their study of Pb at an outdoor shooting range. Because of its low MCL, even small amounts of Pb which manage to migrate to groundwater can generate a high-measured mean concentration to MCL ratio.

The detection of Cr at such high relative concentrations in groundwater compared to other metals in this study indicates that the species of Cr present is likely Cr^{+6}, hexavalent chromium. Moreover, Murray et al. (2008) found high levels of Cr^{+6} in the shallow groundwater. Hexavalent chromium has far greater mobility in soil and solubility in water than other forms of chromium (Fetter 1993). It is also considered the most toxic form of chromium to humans and ecosystems at the higher and lower levels of the food chain (Krishan 1997; Nussey et al. 2000; Proctor et al. 2002).

Compounding the contamination problem within the sand unit are the physical characteristics of its aquifer and surface drainage. The relatively high hydraulic conductivity of the sand unit, coupled with the high density of streams there (see Figure 5.20) creates a short residence time for contamination. In the humid microthermal climate of this region, the stream system is effluent, meaning groundwater recharges streams through baseflow. Thus, any contaminant entering groundwater within the sand unit will flow in the general direction of the surface drainage, and would enter a surface stream within a relatively short time period due to the high drainage density. For example, using a conservative estimate of the hydraulic conductivity for the sand unit from Murray and Rogers (1999) of $1 * 10^{-2}$ cm/s, the groundwater would flow 315 m in 1 year: (0.01 cm/s * 86,400 s/day * 365 days/year/1000 cm/m = 315 m/year). Given the surface drainage density, there is a good likelihood that contamination within the groundwater would reach a stream within a relatively short period of time. Once in the surface water, the flow will carry the contamination within hours to the Detroit River and to Lake Erie.

Another significant finding of this study is the presence of elevated concentrations of metals (Hg, Se, Cr, and As) within the groundwater of the clay and silty clay geologic units, as mobility of metals in clay is generally believed to be very low. Several factors could account for this result, including very high concentrations of these metals within the clay soil units, cracks within the clay unit which allow for some downward migration of contamination (Murray et al. 1997), and the long duration and continuous deposition of these metals within these areas due to the development history of the watershed.

From public health and environmental perspectives, the contamination of the regional groundwater along with potential damage to the hydraulically connected Great Lakes ecosystem poses significant challenges for watershed management efforts. Significant numbers of residential wells (over 5000) exist within the watershed, and many homeowners do not test their water since it is not a legal requirement. Moreover, the testing services provided by Michigan's county public health departments only test for bacteria and anions, such as chloride and nitrate. If homeowners want to test their wells for metals, they must pay $10.00 per metal, and it is doubtful if most homeowners know about this option. There are also a large number of public supply (Class II) wells that provide water for over 25 people. Wayne County, which comprises over 40% of the Rouge River watershed land area does not test for chromium in its Class II wells. Consequently, there may be a need to reevaluate the wellhead protection guidelines communities follow, given the already existing contamination, and the ability for metals to reach groundwater through all types of soil.

Surface water may also be directly impacted by metals transported by stormwater runoff. The presence of some metals at the surface—many of which are sorbed onto small soil particles—means that surface water bodies will receive these contaminants when they are entrained by overland flow. Wetland vegetation may be capable of removing some percentage of the heavy metals; however, most lakes in southeastern Michigan no longer have this capability due to development. Since copper and chromium VI are highly toxic to fish, this is another concern. The process of stormwater runoff within urban watersheds is presented in Chapters 12 and 13.

9.5 Conclusions

The Murray et al. (2004) study of metals in soil represents an initial effort to characterize the metal concentrations in surface and near-surface soils in an urban environment in southeast Michigan. The results of this study have (1) demonstrated that metal concentrations in an urban environment are the greatest at the surface, (2) characterized a directional increase of metal concentrations in a west-to-east trend across the watershed commensurate with a general west-to-east increase in urbanization and industrial activity, and (3) identified lead as the heavy metal with the highest surface concentration with mean levels present at more than 16 times background at residential sites and 15.5 times greater than background at industrial sites. More importantly, lead concentrations have been detected in surface soil at specific industrial sites and adjacent residential neighborhoods at concentrations hundreds of times the level that may occur naturally in the soil. Similarly, arsenic concentrations in the Rouge River watershed for certain residential areas are 10 times greater than the background concentrations, due most likely to their proximity to Zug Island.

Heavy metal concentrations beneath the surface (at depths between 0.5 and 10 m), although elevated, were in general agreement with those of previous studies conducted in Michigan and the United States. Heavy metal concentrations in soils at depths greater than 10 m are in complete agreement with previous studies of heavy metal concentrations in rural areas of Michigan and the United States. This suggests that soils at depths greater than 10 m are generally only affected to a minor extent from anthropogenic heavy metal sources if certain conditions are satisfied.

The results of the Murray et al. (2004) soil metals study have important implications for land use planning and for future site investigations in any urban area when heavy metal contamination is suspected. First, the evaluation of human and ecological risks can be achieved by concentrating investigation efforts on heavy metals in the surface soils. The results of the study have demonstrated that heavy metal concentrations are the highest at the surface and quickly decrease with increasing depth. Therefore, concentrating investigative and remedial efforts at the surface will efficiently identify the most elevated heavy metal concentrations derived from anthropogenic sources. Second, innovative regional strategies for the cleanup of sites with metal contamination need to be developed. Third, the identification of potential hydrologic and ecological impacts of metals in the near-surface zone is essential for the protection of groundwater. Since this study has identified that the occurrence of heavy metals in an urban environment is the highest at the surface, the results can be integrated into evaluations of interflow, baseflow, increased leaching from acid rain, infiltration/inflow to public water supplies, runoff to surface water systems, and the potential revegetation capacity at redeveloped sites. Fourth, the mapping of the surface geology in urban areas can pinpoint soils particularly sensitive and prone to contamination. Finally, future studies related to the occurrence of heavy metals in other urban areas should consider the effects of historic land use activities and the potential human health issues and risks associated with heavy metals in the surface soils.

With respect to dissolved metals in groundwater, results from the 2006 Murray et al. study indicate urbanization and its accompanying industrial activities have contributed to heavy metal impacts across all land use categories (industrial, commercial, and residential) and throughout the varied surface geology of the Rouge River watershed. Chromium is prevalent at industrial sites, and the high levels of chromium in groundwater strongly suggest the chromium is in a hexavalent form—which is much more mobile and soluble than its trivalent form—and has greater potential to damage the environment and impact human health. This is especially significant since hexavalent chromium has a high contaminant risk factor, which will be examined in great detail in Chapter 10. Significantly high contamination ratios were demonstrated by Cr, Cd, and Pb within the watershed's sand unit, but there were also several metals detected at concentrations exceeding their MCLs in clay-related soil units, indicating the existence of some mobility within a soil medium previously considered virtually immobile for metals.

These findings create considerable environmental concern because of the hydraulic connection between the watershed's surface waters and the Great Lakes. Moreover, many private and public wells use the groundwater within the region for potable water, thus creating a potential public health risk.

In the next chapter, a model is developed and applied within an urbanized watershed to help predict the risks posed by specific contaminants to soils, groundwater, and air.

References

Alkhatib, E. and T. O'Connor, 1998. Background levels of priority pollutant metals in soil. *American Environmental Laboratory* 10:6–9.

Bedient, P.B., R.S. Hanadi, and C.J. Newell. 1994. *Ground Water Contamination*. Englewood Cliffs, NJ: Prentice Hall.

Dowdy, R.H. and V.V. Volk. 1983. Movement of heavy metals in soil. In *Chemical Mobility and Reactivity in Soil Systems*, Journal of the Soil Science Society of America, pp. 229–239. Madison, WI.

Farrand, W.R. 1982. *Quaternary Geology of Southern (& Northern) Michigan*. Michigan Department of Natural Resources, Geological Survey Division. Lansing, MI. 1:500,000, 2 Sheets.

Farrand, W.R. 1988. *The Glacial Lakes around Michigan*. Michigan Department of Natural Resources Bulletin. 4. Lansing, MI.

Fetter, C.W. 1993. *Contaminant Hydrogeology*, 2nd edn. Upper Saddle River, NJ: Prentice Hall.

Freeze, R.A. and J.A. Cherry. 1979. *Groundwater*. Englewood Cliffs, NJ: Prentice-Hall.

Harter, R.D. and R.G. Lehmann. 1983. Use of kinetics for the study of exchange reactions in soils. *Journal of the Soil Science Society of America* 47:666–669.

Jaagumagi, R. 1993. *Development of the Ontario Provincial Sediment Quality Guidelines for Arsenic, Cadmium, Chromium, Copper, Iron, Lead, Manganese, Mercury, Nickel and Zinc*. Toronto, ON: Ontario Ministry of the Environment.

Kaufman, M.M., K.S. Murray, and D.T. Rogers. 2003. Surface and subsurface geologic risk factors to groundwater affecting brownfield redevelopment potential. *Journal of Environmental Quality* 32:490–499.

Keller, K.C., G. Van Der Camp, and J.A. Cherry. 1989. A multiscale study of the permeability of a thick clay till. *Water Resources Research* 25:2299–2317.

Keller, K.C., G. Van Der Camp, and J.A. Cherry. 1991. Hydrogeochemistry of a clayey till: Spatial variability. *Water Resources Research* 27:2543–2554.

Krishan, P. 1997. *Chromium Toxicity Effects on Microorganisms and Treatment Processes*. NTIS: DE00761432. College Station, TX: Texas A&M University.

Lindsay, W.L. 1979. *Chemical Equilibria in Soils*. New York: John Wiley and Sons.

Mattigod, S.V., G. Sposito, and A.L. Page. 1981. Factors affecting the solubilities of trace metals in soils. In *Chemistry in the Soil Environment*, ed. D.E. Baker, pp. 203–221. ASA Special Publication Number. 40. American Society of Agronomy. Madison, WI.

McLean, J.E. and B.E. Bledsoe, 1992. *Ground Water Issue Behavior of Metals in Soils*. Technology Innovation Office. Office of Solid Waste and Emergency Response. Washington, DC: USEPA.

Michigan Department of Environmental Quality (MDEQ) 1998. Natural Resources Environmental Protection Act 451 (NREPA), as amended. Lansing, MI.

Mielke, H.W., J.C. Anderson, K.J. Berry et al. 1983. Lead concentrations in inner city soils as a factor in the child lead problem. *American Journal of Public Health* 73:1366–1369.

Mielke, J.H., B. Blake, S. Burroughs et al. 1984. Urban lead levels in Minneapolis: The case of the Hmong children. *Environmental Research* 34:64–76.

Mielke, J.H. 1999. Lead in the inner cities. *American Scientist* 87:62–73.

Mozola, A.J. 1954. *A Survey of Groundwater Resources in Oakland County, Michigan*. Michigan Geological Survey. Publication Number 48. Part II. Lansing, MI.

Mozola, A.J. 1969. *Geology for Land and Groundwater Development in Wayne County, Michigan*. Michigan Geological Survey. Lansing, MI.

Murray, K.S., A. Bazzi, C. Carter et al. 1997. Distribution and mobility of lead in soils at an outdoor shooting range. *Journal of Soil Contamination* 6:79–93.

Murray, K.S., D. Cauvet, M. Lybeer et al. 1999. Particle size and chemical control of heavy metals in bed sediment from the Rouge River, southeast Michigan. *Environmental Science and Technology* 33:987–992.

Murray, K.S. and D.T. Rogers, 1999. Groundwater vulnerability, brownfield redevelopment and land use planning. *Journal of Environmental Planning and Management* 42:801–810.

Murray, K.S., D.T. Rogers, and M.M. Kaufman. 2004. Heavy metals in an urban watershed in southeastern Michigan. *Journal of Environmental Quality* 33:163–172.

Murray, K.S., D.T. Rogers, and M.M. Kaufman. 2006. Dissolved heavy metals in shallow groundwater in an urban watershed in southeastern Michigan. *Journal of the American Water Resources Association* 42:777–792.

Murray, K.S., M.M. Kaufman, and D.T. Rogers. 2008. Heavy metal contamination in an urban watershed in southeastern Michigan, USA. In *Proceedings of the Sixth International Conference on Remediation of Chlorinated and Recalcitrant Compounds* (Monterey, CA), Paper H-007. Columbus, OH: Battelle.

Peterman, C. and K.S. Murray, 2010. *The Delray Initiative*. Geological Society of America Annual Meeting. Denver, CO.

Neal, R.H., G. Sposito, K.M. Holtzclaw et al. 1987. Selenite absorption on alluvial soils: Soil composition and pH effects. *Journal of the Soil Science Society of America* 51:1165–1169.

Nussey, G., J.H.J van Vuren, and H.H. du Preez. 2000. Bioaccumulation of chromium, manganese, nickel and lead in the tissues of the Moggel, Labeo umbratus (Cyprinidae), from Witbank Dam, Mpumalanga. *Water SA* 26:269–284.

Proctor, D.M., J.M. Otani, B.L. Finley et al. 2002. Is hexavalent chromium carcinogenic via ingestion? A weight-of-evidence review. *Journal of Toxicology and Environmental Health, Part A*. 65:701–746.

Rogers, D.T. 1996. *Environmental Geology of Metropolitan Detroit*. Novi, MI: Clayton Environmental Consultants.

Rogers, D.T. 1997a. *Surficial Geologic Map and Cross Section of the Rouge River Watershed*, Michigan. River Rouge National Wet Weather Demonstration Project (RRNWWDP) Wayne County, MI. 1:62,500. 2 Sheets.

Rogers, D.T. 1997b. The Influence of groundwater and surface water in Michigan's Rouge River watershed. *Proceedings of the American Water Resources Association Conference. Conjunctive Use of Water Resources: Aquifer Storage and Recovery*. Long Beach, CA.

Rogers, D.T. and K.S. Murray, 1997. Occurrence of groundwater in metropolitan Detroit, U.S.A. In *Groundwater in the Environment, Vol. 1*, ed. J. Chilton et al. Rotterdam, the Netherlands: Balkema.

Rouge River Remedial Action Plan Update (RRRAP) 1994. Detroit, MI: Wayne County Department of Environment.

Shacklette H.T. and J.G. Boerngen. 1984. *Element Concentrations in Soil and other Materials of the Conterminous United States*. United States Geological Survey Professional Paper 1270. Washington, DC: United States Government Printing Office.

Shuman. L.M. 1991. Chemical forms of micronutrients in soils. In *Micronutrients in Agriculture*, ed. J.J. Mortvedt, Soil Society of America Book Series #4, pp. 113–114. Madison, WI: Soil Society of America.

Stevenson, F.J. 1991. Organic matter-micronutrient reactions in soil. In *Micronutrients in Agriculture*, ed. J.J. Mortvedt, Soil Society of America Book Series #4. pp. 145–186. Madison, WI: Soil Society of America.

Thornton, I. 1991. Metal contamination of soil in urban areas. In *Soils in the Urban Environment*, ed. P.G. Bullock et.al. Oxford: Blackwell Scientific Publications.

USEPA. 1983. *Test Methods for Evaluating Solid Waste* (SW-846). Washington, DC: USEPA.

USEPA. 1986. *Resource Conservation and Recovery Act Ground Water Monitoring Technical Enforcement Guidance Document*. OSWER-9950-1, Office of Solid Waste and Emergency Response. Washington, DC.

USEPA. 1990. *Ground Water. Volume 1. Ground Water and Contamination*. EPA-823-R-97-006. USEPA. Washington, DC.

USEPA. 1992. *National Secondary Drinking Water Regulations*. EPA/816-F-03-016, USEPA Office of Water. Washington, DC.

USEPA. 2003. *National Primary Drinking Water Regulations*. EPA/816-F-03-016, Office of Water, Washington, DC.

10

Contaminant Risk Factors

10.1 Introduction

What are the risks posed by contaminants to humans once they are released? The first step to answering this question involves combining the following three factors related to the physical chemistry of the contaminant: (1) toxicity or potency, (2) mobility, and (3) persistence. Next, the composite physical chemistry attributes are considered within the context of a region's geological vulnerability. This framework provides a powerful tool for assessing the environmental risk of an urban region, but it fails to consider the known risks of release related to specific surface activities. In this chapter, we introduce the concept of surface risk, one component of the environmental risk picture within any urban region.

The risks posed by contaminants are not equal. Although it sounds like a paradox, an extremely toxic contaminant may not pose as much risk as a moderately toxic contaminant. This outcome can occur if the extremely toxic contaminant does not migrate and degrades quickly, and the moderately toxic contaminant exhibits higher mobility and lasts for decades before degrading. The release location is also a factor in determining risk. If the extremely toxic contaminant is released at a certain location and under certain conditions, it may inflict significant harm before it degrades. On the other hand, the moderately toxic contaminant may have much more opportunity to inflict harm because it lasts longer and is mobile.

Contaminants also behave differently in soil, water, and air. It is logical, therefore, to assess contaminant risk as a function of each environmental media. We begin the evaluation process by assessing the probability that a release will occur given certain land use criteria. This is a critical step in the evaluation process, because not only must a contaminant be present for there to be risk, it must also be released for there to be potential exposure. After examining the potential risks of a release, we will then examine how the media releases target and develop contaminant risk factors for groundwater, soil, and air. When these factors are combined with surface risk and geological vulnerability, the total environmental risk can be estimated.

This is where the predictive power lies, knowledge of the physical attributes of the chemicals used combined with the geology of a region and the probability of a release occurring. Together, these pieces form the basis of a scientifically grounded environmental assessment process that can lead to the successful development or redevelopment of any urban region.

10.2 Surface Risk Factor

Surface risk, as employed here, is the probability of any given site contaminating the environment given the best available data from public sources. Surface risk evaluations have traditionally used spatially generalized categories of land use to represent various levels of risk, such as industrial, commercial, residential, and recreational (Baringer et al. 1990; Eckhardt and Stackelberg 1995; Secunda et al. 1998).

The use of general land use categories for vulnerability assessments is problematic because of their inadequate **spatial resolution**, defined as the smallest identifiable element in a sequence (Tobler 1988). In urban and urbanizing areas, mixed land uses within small areas such as city blocks are common, so the variable risks may be obscured by generalizations when the capture zones for water supply wells, termed **wellhead protection zones**, are delineated. For example, the 10 year capture zone is the subsurface and surface areas from where water (and any contamination it carries) will reach the well over a time period of 10 years. Figure 10.1 shows an example capture zone of a water supply well in an urban area (Wisconsin Department of Natural Resources 1999). Figure 10.2 shows an example of a recharge area in cross section (USGS 1998).

To demonstrate this problem of inadequate spatial resolution, we can consider a water well in an area designated as for residential land use as depicted in Figure 10.3 (Kaufman et al. 2003). Within this zone of low risk is a single and small industrial establishment engaged in metal plating. Metal plating activities exhibit a high incidence rate of soil and groundwater contamination, but this specific risk is masked by the generalization of the area within the capture zone as a lower-risk residential category. This is depicted in Figure 10.3a.

Figure 10.3b illustrates how two adjacent, but different zones create edge effects. At the edge, the residential zone becomes exposed to the higher risks associated with industrial land, but the sources and amounts of the actual risk from the industrial land near the edge

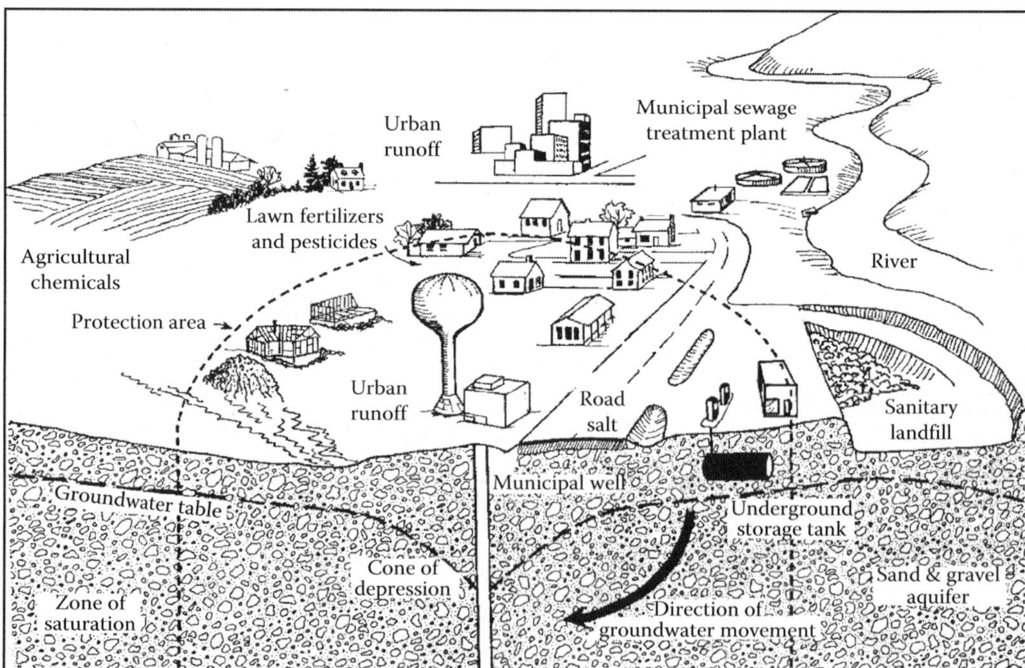

FIGURE 10.1
Example capture zone. (From Wisconsin Department of Natural Resources (WDNR), *Wellhead Protection*, WDNR Publication PUB-DG-039 99REV, Madison, WI, 1999.)

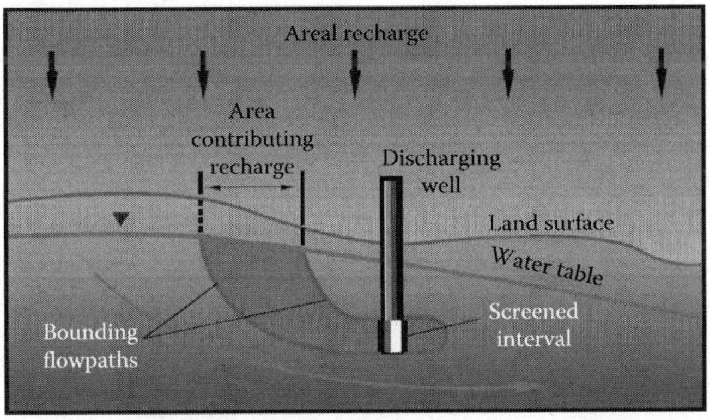

FIGURE 10.2
Example recharge area. (From United States Geological Survey (USGS), *Estimating Areas Contributing to Recharge of Wells*, USGS Circular 1174, Denver, CO, 1998.)

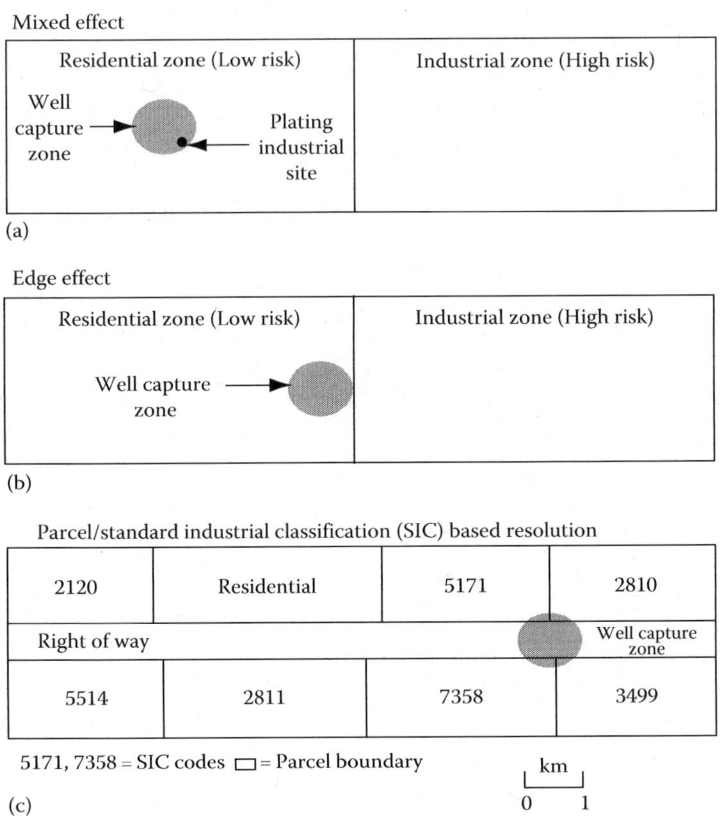

FIGURE 10.3
Effects of different spatial resolutions across multiple land use types on surface risk.

are unknown. Figure 10.3c shows that the use of the Standard Industrial Classification (SIC) code or the equivalent for each establishment within a general land use category permits a greatly improved parcel-level spatial resolution of the relative risks of contamination.

The SIC code is a four-digit code defined as follows (United States Office of Management and Budget 1987)*:

- The first two digits identify a major group, such as agriculture, retail trade, and manufacturing.
- The third digit denotes industry groups within each major category, such as agricultural crop production.
- The fourth digit identifies a specific industry code, such as metal plating.

A normalized measure of risk between different establishment types is achieved through the use of contamination incident rates (Kaufman 1997). Incidence rates are obtained by

1. Assigning an SIC code to each source of contamination appearing on a known list of contaminated sites. Lists of known contaminated sites are available through private companies or are publicly available through either local municipalities or state or federal environmental agencies, such as the Michigan Department of Environmental Quality (MDEQ 2008).
2. Obtaining the total number of establishments for each SIC code within the study area (United States Bureau of the Census 1997).
3. Dividing the number of contaminated sites with a specific SIC code by the total number of establishments with the same SIC code in the study area.

To scale the scores equivalently to the other risk factors discussed in the following sections, these rates are multiplied by 10 and converted to scores between 0 and 10. These scores are then summed for a circular area encompassing each water well within the study area.

Figure 10.4 shows an example surface risk calculation for a brownfield site situated above an area where the subsurface geology is composed of sand. The computed risk includes only those lighter-shaded establishments contained within the geological unit composed of sand, because the bounding geological units (sandy clay, sandy, and silty clay) have much lower hydraulic conductivities yielding much lower contaminant migration potentials.

The legend box on Figure 10.4 shows the SIC and risk scores for three of the many establishments within the radius (Kaufman et al. 2003). For example, SIC 2822 represents an establishment producing synthetic rubber with a risk score of 4.00. This particular risk score was calculated by dividing the two SIC 2822 establishments known to be sites of contamination by the five SIC 2822 establishments in the region and then multiplying the result by 10. The risk scores for the other establishment types shown in Figure 10.4 are computed similarly.

---

* The four-digit SIC codes were replaced by the six-digit North American Industry Classification System (NAICS) in 1997. The NAICS was developed to allow for a higher level of comparability in business statistics among Mexico, Canada, and the United States. Some governmental agencies still use the SIC code, and the authors prefer them because the descriptions are more complete than those provided by the NAICS.

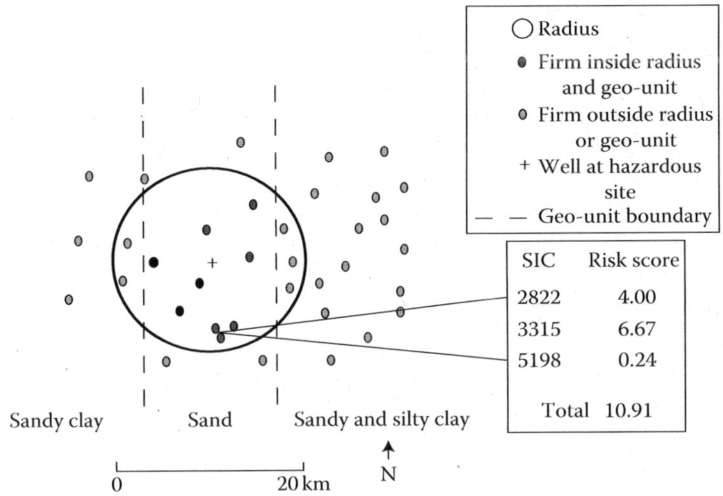

FIGURE 10.4
Calculation of surface risk using the risk values computed for specific establishment types within the capture radius around a contaminated brownfield site.

The circle shown in Figure 10.4 represents the 10 year capture zone of a pumping well located in the center of the circle. Circular areas may be effective in modeling wellhead protection areas under small regional hydraulic gradients and low groundwater velocities (Baringer at al 1990; Camp and Outlaw 1998). In this example, the capture zone was calculated by using existing data from hydrogeologic investigations conducted near the study site, under the assumptions of a pumping rate that induces a hydraulic gradient of 0.1 m/m in all directions toward the pumped well, a relatively level water table, and predominately horizontal groundwater flow. It is also possible to derive the capture zone if the necessary information on gradient, hydraulic conductivity, and effective porosity is obtained at the site.

After an appropriate radius is established, an automated procedure for the surface risk can be performed. This can be achieved by (1) geocoding a file containing the street addresses of public, commercial, and industrial establishments (labeled "Firms" in Figure 10.4) each coded by its SIC designation and (2) using the capabilities of a geographic information system (GIS) to sum the risk scores for each firm within the specified radius on a digital map image (Kaufman 2000).

Table 10.1 lists the 10 year capture zones and mean surface risk values for four of the most common soil types found in the Rouge River Watershed in southeastern Michigan.

TABLE 10.1
Surface Risk Values for the Rouge Watershed

Geologic Unit	10 Year Capture Zone (km)	Mean Surface Risk Value
Clay	0.03	0.04
Silty clay	0.06	33.70
Sandy silty clay	1.07	0.42
Sand	597.00	343.64

This watershed is a good example because it is heavily urbanized and includes a broad spectrum of geological units including clay, silty clay, sandy silty clay, and sand (Kaufman et al. 2003).

The mean surface risk values are highest in the sand for three reasons: (1) there is a high density of sites with a propensity for releases of contaminants to occur, (2) the larger capture zone for sand translates into potentially more sites of contamination than the other geological units, and (3) the subsurface geology (i.e., sand) does not significantly impede the migration of contaminants.

Industrial location patterns also help to explain other mean surface risk values in this watershed. For example, the silty clay unit has a mean surface risk value of 33.7 and the sandy silty clay unit has a surface risk value of 0.42, yet the silty clay unit has a 10 year capture zone smaller than the sandy silty clay unit. This discrepancy is due to the significantly greater number of potential contaminant sources located within the silty clay unit.

The next section examines individual contaminants and their migration potential within different subsurface geological environments.

10.3 Groundwater Contaminant Risk Factor

The adverse risk posed to groundwater by the contaminants themselves is often overlooked or underrepresented (Kaufman et al. 2005; Rogers et al. 2007a). Considering specific types of contamination using vulnerability models is important because each contaminant has unique physical chemical properties significantly influencing its behavior when released into the environment (Chapters 7 and 8).

Contaminant fate and transport evaluations require interdisciplinary analyses involving chemical, geological, hydrological, and biological factors (USEPA 1989, 1992). Correspondingly, the development of groundwater vulnerability models for specific contaminants requires an interdisciplinary process (Rogers et al. 2007a). The critical physical/chemical attributes influencing contaminant risk are associated with mobility and persistence and include the following factors (USEPA 1989; USEPA 1996a; Wiedemeier et al. 1999; USGS 2006; Rogers et al. 2007a):

1. Solubility
2. Vapor pressure
3. Density
4. Chemical stability
5. Persistence
6. Adsorption potential

As noted previously, toxicity is an important factor when examining risk. Mobility and persistence are also critically important because these two factors dictate a chemical's ability to migrate from its point of release in the environment to a distant point where human exposure may occur, such as a drinking water supply or a surface water body. The environmental risk posed by specific contaminants to contaminant groundwater, termed **Contaminant Risk Factor for Groundwater** (CRF_{GW}) can be developed as a function of these three factors (Kaufman et al. 2005; Rogers et al. 2007a).

Contaminants released into the environment only pose a risk to humans if there is a completed exposure pathway. In general terms, the CRF_{GW} is expressed in Equation 10.1 (Kaufman et al. 2005; Rogers et al. 2007a).

$$CRF_{GW} = \text{toxicity} \times \text{mobility} \times \text{persistence} \tag{10.1}$$

Toxicity values are obtained from the USEPA Integrated Risk Information System (IRIS) (2009a). This database is updated weekly, and often more frequently. The toxicity values selected should be the most conservative for each exposure pathway—ingestion, dermal adsorption, and inhalation. Using the most conservative value is appropriate since exposure to contaminated groundwater can occur in each of these pathways. For instance, dermal adsorption can occur during washing, ingestion can occur through drinking, and inhalation can occur during showering.

Mobility is derived from Henry's law constant and the retardation factor shown in Equation 10.2 (Kaufman et al. 2005; Rogers et al. 2007a).

$$M = (H)(R) \tag{10.2}$$

where
M is the mobility
H is the Henry's law constant
R is the retardation factor

Henry's law constant (H) (atm/mol/m^3) is a measure of the tendency for substances to volatilize, and is very useful in assessing the mobility of specific contaminants because solubility affects the volatilization of contaminants into the atmosphere (Sander 1999). It is related to vapor pressure (VP) (atm.), molecular weight (MW) (g/mol), and solubility in water (W_s) (g/L).

Henry's law constants can be obtained from several sources including USEPA (1996a); Sander (1999); Wiedemeier (1999); Suthersan and Payne (2005); Payne et al. (2008).

The retardation factor is represented by Equation 10.3.

$$R = 1 + \frac{(\rho b)(Kd)}{\eta} \tag{10.3}$$

where
R is the retardation factor
ρb is the bulk density of aquifer matrix (g/cm^3)
Kd is the distribution coefficient (mL/g)
η is the effective porosity (calculated as a percent value)

The distribution coefficient is calculated using Equation 10.4.

$$Kd = (Foc)(Koc) \tag{10.4}$$

where
Kd is the distribution coefficient
Foc is the organic carbon partition coefficient (kg/kg)
Koc is the fraction of total organic carbon in soil (L/kg)

Values for the organic carbon partition coefficient can be obtained from numerous sources including USEPA (1996a), Wiedemeier (1999), USEPA (2002a), Suthersan and Payne (2005). To obtain the best representation, values for the fraction of organic carbon should be collected in the field. If field collection is not possible, standard values and ranges can be obtained from USEPA (1996a), Wiedemeier (1999); USEPA (2002a); Suthersan and Payne (2005).

The retardation factor represents the ratio between the rate of groundwater movement and the rate of contaminant movement. When the retardation value equals 1, the rate of groundwater movement equals the rate of contaminant movement and no retardation is expected. A retardation value > 1 indicates groundwater movement is greater than contaminant movement, so increasing values indicate greater contaminant retardation (USEPA 1989, 2009b).

Persistence values are obtained from the literature and expressed as first-order decay rates in years (Howard et al. 1991; USEPA 1996a,b, 2000). The first-order decay rates selected for each compound represent the most conservative values of the spectrum of data available.

Finally, the CRF_{GW} is calculated in Equation 10.5 by multiplying the inverse of the chemical compound's toxicity (T), by the inverse of its mobility (M) and its persistence (P).

$$CRF_{GW} = \frac{1}{(T)} \times \frac{1}{(M)} \times (P) \tag{10.5}$$

where

CRF_{GW} is the contaminant risk ractor for groundwater
T is the toxicity
M is the mobility
P is the persistence

The inverse of the toxicity value must be used because the integer values assigned for toxicity decrease with increasing carcinogenicity (USEPA 2009a). The inverse of the mobility values must also be used because the calculated values of retardation decrease with increasing mobility.

Here is an example calculation of the CRF_{GW} for a chemical XYZ in a geological unit composed of sand:

$$CRF_{GW} \text{ for chemical XYZ} = \frac{1}{(T)} \times \frac{1}{(M)} \times (P)$$

Step 1: Obtain toxicity value:

The toxicity of XYZ chemical was obtained from the literature and has a value of 0.04.

Step 2: Determine the mobility value:

We need the Henry's law constant and the retardation factor. Let us calculate the retardation factor first. We start with the distribution coefficient (Kd) using Equation 10.4:

$$Kd = (Foc)(Koc)$$

Foc was obtained through analysis of several soil samples in the study area and was found to be 0.0003 kg/kg. Koc for chemical XYZ was obtained from the literature and has a value of 58.9 L/kg (USEPA 2002a). We can now calculate the distribution coefficient

$$\text{Kd} = (\text{Foc})(\text{Koc}) = (0.0003)(58.9) = 0.017$$

The remaining values necessary to calculate the retardation value are the bulk density of the aquifer material and the effective porosity. The bulk density of the aquifer material was obtained through the collection and analysis of soil samples and was 1.7 g/cm^3. The effective porosity was estimated from literature values to be 25% or 0.25.

All the necessary information has been obtained, and the retardation value of 1.11 is calculated using Equation 10.3:

$$\text{XYZ chemical retardation} = R = 1 + \frac{(\rho b)(\text{Kd})}{\eta}$$

$$\text{XYZ chemical retardation} = R = 1 + \frac{(1.7)(0.017)}{0.25} = 1.11$$

Now we can complete the mobility calculation by plugging in the Henry's law constant (H) obtained from the literature into Equation 10.2. The Henry's law constant for XYZ chemical is 0.228.

$$M = (H)\ (R)$$

$$M = 0.228 \times 1.11 = 0.253$$

Step 3: Determine the persistence value. The persistence value was obtained from the literature and is 0.2 years.

Step 4: With toxicity, mobility, and persistence values obtained, the CRF_{GW} is calculated using Equation 10.5.

$$CRF_{GW} \text{ for chemical XYZ} = \frac{1}{(T)} \times \frac{1}{(M)} \times (P)$$

$$CRF_{GW} \text{ for chemical XYZ} = \frac{1}{(0.04)} \times \frac{1}{(0.253)} \times (0.2) = 19.76$$

$$CRF_{GW} \text{ for chemical XYZ} = 19.76$$

Tables 10.2 through 10.4 display the groundwater contaminant risk factors (CRF_{GW}) for common VOC LNAPL, VOC DNAPL, and PAH compounds, respectively, in the Rouge River watershed (Kaufman et al. 2005; Rogers et al. 2007a). Table 10.5 summarizes those values and includes other selected compounds. The values span the four most common types of soils found in urban areas of the United States, including clay, silty clay, sandy silty clay, and sand. As such, their relative magnitudes can be used as a starting point for risk assessments in other urbanized watersheds with similar geological units.

In Table 10.2, the CRF_{GW} values for benzene are greater than other volatile organic compounds (VOCs) within the LNAPL group. This is because benzene exhibits the highest combined values of toxicity, mobility, and persistence in groundwater compared to the other LNAPL compounds examined.

TABLE 10.2

CRF_{GW} for Common VOC LNAPLs

LNAPL Compound	Soil Type	CRF_{GW}
Benzene	Clay	3.64
	Silty clay	14.50
	Sand silty clay	15.40
	Sand	19.40
Toluene	Clay	0.63
	Silty clay	4.10
	Sand silty clay	4.53
	Sand	7.53
Ethyl benzene	Clay	0.27
	Silty clay	2.10
	Sand silty clay	2.40
	Sand	4.80
Xylenes	Clay	1.10
	Silty clay	7.80
	Sand silty clay	8.10
	Sand	13.90
Mean LNAPL CRF_{GW}	Clay	1.41
	Silty clay	7.12
	Sand silty clay	7.60
	Sand	11.40

Table 10.3 lists the CRF_{GW} for common VOCs from the DNAPL group (Kaufman et al. 2005; Rogers et al. 2007a).

Examination of the CRF_{GW} for DNAPL compounds indicates they are much greater than those of the LNAPL compounds listed in Table 10.2. The comparatively higher CRF_{GW} of DNAPLs stems not from their relative toxicity, but from their higher persistence and mobility in groundwater compared to the LNAPL compounds (Rogers et al. 2007a).

Table 10.4 lists the CRF_{GW} for common PAHs.

The CRF_{GW} for PAH compounds are significantly lower than the CRF_{GW} for DNAPL and LNAPL VOCs. This difference occurs because PAHs strongly sorb to soil particles and are much less soluble in water compared to the VOCs examined (Rogers et al. 2007a).

Table 10.5 contains a summary of CRF_{GW} for the LNAPL and DNAPL VOCs, PAHs, and includes the CRF_{GW} for total PCBs, the pesticide chlordane, and the heavy metals chromium VI, lead, mercury, and arsenic.

The contaminant with the highest CRF_{GW} is chromium VI, followed by DNAPL VOCs, LNAPL VOCs, mercury, and lead, with their CRF_{GW} ranging from 100 to 10,000 times less than chromium VI. Contaminants with the lowest CRF_{GW} include chlordane, PAHs, and the PCBs having a CRF_{GW} more than a billion times lower than chromium VI.

The distributions of CRF_{GW} are shown in Figure 10.5.

Examination of the CRF_{GW} values in Figure 10.5 reveals a grouping of contaminants into three distinct levels. Chlordane, PCBs, and PAHs form the group with low CRF_{GW}; lead, mercury, and LNAPLs constitute the moderate range; and Chromium VI and DNAPL compounds appear on the top of the figure with the highest CRF_{GW}. Chlordane, PCBs, and PAHs are very toxic and persistent contaminants in the environment, but they strongly

TABLE 10.3

CRF_{GW} for Common VOC DNAPLs

LNAPL Compound	Soil Type	CRF_{GW}
Tetrachloroethene (PCE)	Clay	98.0
	Silty clay	601.0
	Sand silty clay	657.0
	Sand	1048.0
Trichloroethene (TCE)	Clay	148.0
	Silty clay	933.0
	Sand silty clay	1018.0
	Sand	1647.0
cis-1,2-Dichloroethene	Clay	228.0
	Silty clay	702.0
	Sand silty clay	730.0
	Sand	851.0
trans-1,2-Dichloroethene	Clay	131.0
	Silty clay	495.0
	Sand silty clay	520.0
	Sand	647.0
Vinyl chloride	Clay	860.0
	Silty clay	1911.0
	Sand silty clay	1962.0
	Sand	2132.0
1,1,1-Trichloroethane (1,1,1-TCA)	Clay	2.1
	Silty clay	11.0
	Sand silty clay	12.0
	Sand	17.3
Mean DNAPL CRF_{GW} for degradation sequence from PCE to vinyl chloride[a] (Figure 7.4)	Clay	333.0
	Silty clay	773.0
	Sand silty clay	1091.0
	Sand	1274.0

[a] Represents cumulative risk.

sorb to soil and are not very soluble in water. As a result, they have very low CRF_{GW} compared to the other contaminants listed.

The middle grouping has moderate CRF_{GW} resulting from a combination of factors unique to each contaminant. LNAPLs have moderate mobility and the ability to degrade in the environment. Mercury is not very mobile but may become transformed in the environment to methyl mercury, a change allowing it to be adsorbed by organisms that increases its environmental risk. Lead has low solubility in water but is very persistent. Arsenic has a much higher CRF_{GW} compared to lead because it is more soluble and toxic.

The group consisting of chromium VI and DNAPL compounds has the highest CRF_{GW} values. Some of these values may be more than a million times greater than the contaminants in the lowest group. This magnitude of difference occurs because chromium VI and DNAPL compounds have relatively high toxicity, mobility, and persistence in the environment.

TABLE 10.4

CRF_{GW} for Select PAHs

PAH Compound	Soil Type	GWCRF
Naphthalene	Clay	0.0004
	Silty clay	0.003
	Sand silty clay	0.004
	Sand	0.01
Chrysene	Clay	0.000002
	Silty clay	0.00002
	Sand silty clay	0.00002
	Sand	0.0009
Benzo[b]fluoranthrene	Clay	<0.00001
	Silty clay	<0.00001
	Sand silty clay	<0.00001
	Sand	<0.00001
Benzo[k]fluoranthrene	Clay	<0.00001
	Silty clay	<0.00001
	Sand silty clay	<0.00001
	Sand	<0.00001
Phenanthrene	Clay	0.00002
	Silty clay	0.0002
	Sand silty clay	0.0002
	Sand	0.0007
Benzo[g,h,i] perylene	Clay	<0.00001
	Silty clay	<0.00001
	Sand silty clay	<0.00001
	Sand	<0.00001
Benzo(a)pyrene	Clay	<0.00001
	Silty clay	<0.00001
	Sand silty clay	<0.00001
	Sand	<0.00001
Mean PAH CRF_{GW}	Clay	0.0001
	Silty clay	0.0008
	Sand silty clay	0.001
	Sand	0.002

Source: Kaufman, M.M. et al., *Water, Air, and Soil Pollut.*, 167, 365, 2005; Rogers, D.T. et al., Assessment of groundwater contaminant vulnerability in an urban watershed in southeast Michigan, USA, in *Urban Groundwater—Meeting the Challenge*, Howard, K.W.F. (ed), Taylor & Francis, London, U.K., 2007a, 129–144.

Another characteristic gleaned from these data is the correspondence between the increase in the risk factor for each contaminant and the increase in mean grain size from clay to sand. Larger mean grain sizes increase permeability and raise the mobility factor, especially if a specific contaminant's physical chemistry has a high relative solubility and low sorptive potential. If this relationship is correct, then contaminants with a high

TABLE 10.5

Summary of CRF_{GW} for LNAPL and DNAPL VOCs, PAHs, Other Select Compounds

Compound	Soil Type	CRF_{GW}
Mean LNAPL CRF_{GW}	Clay	1.41
	Silty clay	7.12
	Sand silty clay	7.60
	Sand	11.40
Mean DNAPL CRF_{GW} for degradation sequence from PCE to vinyl chloride	Clay	333.00
	Silty clay	773.00
	Sand silty clay	1091.00
	Sand	1274.00
Mean PAH CRF_{GW}	Clay	0.0001
	Silty clay	0.0008
	Sand silty clay	0.001
	Sand	0.002
PCBs	Clay	0.00002
	Silty clay	0.00026
	Sand silty clay	0.0003
	Sand	0.0009
Pesticide chlordane	Clay	0.0035
	Silty clay	0.035
	Sand silty clay	0.041
	Sand	0.046
Chromium VI	Clay	948.00
	Silty clay	2080.00
	Sand silty clay	2116.00
	Sand	2300.00
Lead	Clay	0.03
	Silty clay	0.30
	Sand silty clay	0.34
	Sand	1.10
Mercury	Clay	1.57
	Silty clay	2.52
	Sand silty clay	2.77
	Sand	3.10
Arsenic	Clay	9.88
	Silty clay	11.26
	Sand silty clay	21.12
	Sand	50.25

Source: Kaufman, M.M. et al., *Water, Air, and Soil Pollut.*, 167, 365, 2005; Rogers, D.T. et al., Assessment of groundwater contaminant vulnerability in an urban watershed in southeast Michigan, USA, in *Urban Groundwater—Meeting the Challenge*, Howard, K.W.F. (ed), Taylor & Francis, London, U.K., 2007a, 129–144.

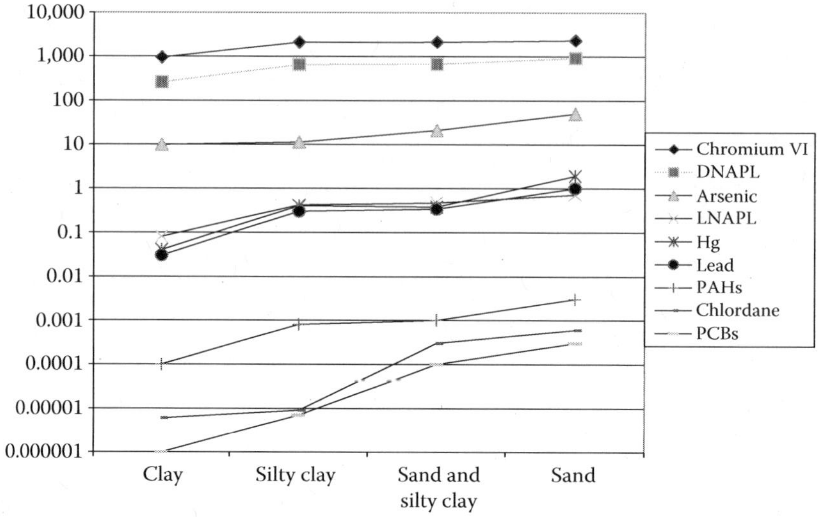

FIGURE 10.5
Distribution of CRF_{GW}.

CRF_{GW} should be detected in groundwater at greater distances from their source than contaminants with low CRF_{GW}. Specifically, sites contaminated with chromium VI and VOC DNAPLs should exhibit greater contamination extents than sites contaminated by the other chemical compounds evaluated.

To test this hypothesis, 83 contaminated sites within the Rouge River watershed and 127 additional sites located outside the Rouge River watershed were evaluated. The Rouge sites were located on different types of geological units, and the external sites were located in areas of varied subsurface geology and contaminant type (Kaufman et al. 2005). Of the 127 sites located outside the Rouge River watershed, 117 were distributed among 32 different states, with the remaining 10 sites located in other countries including Canada, Italy, England, France, Belgium, South Africa, and Australia. All 210 sites were located in urban areas with varying geology composed of unconsolidated sediments originating from lacustrine, fluvial, or glacial processes. The critical information gathered from each of the 210 sites is described in Table 10.6 (Kaufman et al. 2005; Rogers et al. 2007a,b). A summary of the data is listed in Table 10.7 (Kaufman et al. 2005; Rogers et al. 2007a,b).

A scan down the rightmost column (average extent from source) of Table 10.7 strongly validates the hypothesis: sites contaminated with chromium VI and VOC DNAPLs do exhibit greater contamination extents than sites contaminated by the other chemical compounds. The top 4 average extents are 2200 m (eight world sites, chromium in sand), 1012 m (eight Rouge watershed sites, DNAPL in sand), 975 m (15 world sites, DNAPL in sand), and 625 m (four Rouge watershed sites, DNAPL in moraine). The next ranked average extent is almost twice as small (328 m) found in eight LNAPL sand sites within the Rouge watershed. Strong empirical evidence indicates a positive association between a chemical's CRF_{GW} and its likelihood to migrate.

Since contamination affects other media such as soil and air, we must also develop Contaminant Risk Factors for those media.

TABLE 10.6

Critical Information Obtained from Sites of Environmental Contamination

Category	Description
Contaminant	General chemical category (e.g., DNAPL)
Type of facility	Primary activity at the site (dry cleaning, foundry, etc.)
Geology	Composition, stratigraphy, and other information on subsurface units
Remedial technology for groundwater	Contaminant abatement method (e.g., air sparging, pump and treat, in situ chemical or biological treatment, natural attenuation)
Remedial technology for soil	Contaminant abatement method (e.g., excavation, capping, soil vapor extraction, institutional controls)
Mass (kg)	Total mass of contamination at site
Cost	Total cost of investigation and remediation from start to finish
Media remediated	Soil, water, or other (e.g., building decontamination, demolition)
Extent (m)	Measured extent of contamination in each media affected
Cost/kg	Cost of investigation and remediation per kilogram of contaminant
Years of operation	Number of years the facility had been in operation
Geologic vulnerability	Geologic vulnerability rating (see Table 6.1)
Surface risk factor	Average surface risk (Section 10.2 and Table 10.1)
CRF_{GW}	CRF_{GW} calculated for each contaminant (see Section 10.3, Equation 10.5)
Soil cost	Amount of the total remediation cost attributed to soil
Soil cost/kg	Cost per kilogram to remediate soil
Groundwater cost	Amount of the total remediation cost attributed to groundwater
Groundwater cost/kg	Cost per kilogram to remediate groundwater
Groundwater mass (kg)	Mass of groundwater remediated at the site

10.4 Soil Contaminant Risk Factor

In most cases, contaminants released to the ground surface migrate downward through the upper soil layers. Over time, they may or may not contaminate groundwater. Therefore, evaluating the potential for a chemical to contaminate the soil should also be conducted whenever there is a potential for it to contaminate groundwater. Development of the Soil Contaminant Risk Factors (CRF_{SOIL}) should be a high priority in urban watersheds because it provides an additional piece for characterizing the total risks posed by contaminants in the environment.

The migration potential of a contaminant in soil is dependent upon the same physical and chemical attributes as those found in groundwater: solubility, vapor pressure, density, chemical stability, persistence, and adsorption potential. There is also a biological interaction between the contaminant and the soil environment to which the chemical is released (Schnoor 1996) and, as with the CRF_{GW}, is accounted for within the persistence factor.

Given the similarities between the two processes, developing the CRF_{SOIL} requires rather simple modifications to the CRF_{GW} equation. The CRF_{SOIL} is calculated by multiplying the inverse of a chemical's toxicity (T), by its mobility (M) and persistence (P) (Equation 10.6). The change in the equation is reflected in the mobility factor, where it

TABLE 10.7
Data Summary

Contaminant of Concern	Number of Sites	Soil Type	Number of Sites	Geologic Setting	Average Cost ($/Kg)	Average Extent from Source (m)
Rouge River watershed data						
DNAPL	23	Moraine	4	Glacial	145,000	625
		SSC[a]	3	GL[c]	3,260	42
		SC[b]	2	GL	1,366	50
		Sand	8	GL	116,400	1,012
		Clay	6	GL	721	30
LNAPL	27	Moraine	5	Glacial	6,411	270
		SSC	4	GL	669	43.5
		SC	3	GL	518	45.6
		Sand	8	GL	2,627	328
		Clay	7	GL	319	38
PAHs	22	Moraine	0	Glacial	—	—
		SSC	2	GL	203	32
		SC	7	GL	841	27.4
		Sand	5	GL	444	30
		Clay	8	GL	964	16.3
Lead	11	Moraine	0	Glacial	—	—
		SSC	2	GL	538	42
		SC	3	GL	230	25
		Sand	2	GL	442	27
		Clay	4	GL	68.7	52.5
Worldwide data						
DNAPL	27	Clay	9	Fl[d], GL, L[e]	474	40
		SC	3	Fl, GL	224	95
		Sand	15	Fl, GL, L	98,269	975
LNAPL	27	Clay	10	Fl, GL, L	197.8	35
		SC	4	Fl, GL	416	23
		SSC	3	GL	700	25
		Sand	10	Fl, GL, L	1,255	175
PAHs	26	Clay	10	Fl, GL, L	366	30.5
		SC	2	GL	125	22
		SSC	3	Fl	340	20
		Sand	11	Fl, GL	213	25
Lead	13	Clay	5	Fl, GL, L	190	125
		SC	4	Fl, GL, L	930	81
		Sand	4	Fl	500	100
Chromium	19	Clay	11	Fl, GL, L	474	40
		Sand	8	Fl, GL, L	81,713	2,200
Mercury	3	Clay	1	GL	1,000	10
		SC	2	Fl	3,000	75
Chlordane	2	Clay	1	GL	600	10
		Sand	1	GL	830	15

TABLE 10.7 (continued)

Data Summary

Contaminant of Concern	Number of Sites	Soil Type	Number of Sites	Geologic Setting	Average Cost ($/Kg)	Average Extent from Source (m)
PCBs	8	SC	3	Fl	1,200	5
		Sand	5	Fl, GL, L	2,053.9	13.75
Arsenic	2	Clay	1	Fl	780	30
		Sand	1	GL	960	40

[a] SSC, sandy and silty clay.
[b] SC, sandy clay.
[c] GL, glacial lacustrine.
[d] Fl, fluvial.
[e] L, lacustrine.

is not necessary to multiply by the inverse because the calculated values increase with increasing retardation for soil.

$$\text{CRF}_{\text{SOIL}} \text{ for chemical XYZ} = \frac{1}{(\text{T})} \times (\text{M}) \times (\text{P}) \tag{10.6}$$

where

CRF_{SOIL} is the soil contaminant risk factor
T is the toxicity
M is the mobility
P is the persistence

Here is an example calculation of a CRF_{SOIL} for a chemical XYZ in a geological unit composed of sand:

Step 1: Obtain toxicity value:
The toxicity of XYZ chemical was obtained from the literature and has a value of 0.04.
Step 2: Determine the mobility value:
Calculate the distribution coefficient (Kd) using Equation 10.4:

$$\text{Kd} = (\text{Foc})(\text{Koc})$$

Foc was determined through analysis of several soil samples in the study area and was found to be 0.0003 kg/kg. Koc for chemical XYZ was obtained from the literature and has a value of 58.9 L/kg (USEPA 2002a). Therefore, we now have enough information to calculate the distribution coefficient as follows:

$$\text{Kd} = (\text{Foc})(\text{Koc}) = (0.0003)(58.9) = 0.017$$

The remaining values necessary to calculate the retardation value are the bulk density of the aquifer material and the effective porosity. The bulk density of the aquifer material was obtained through the collection and analysis of soil samples and was 1.7 g/cm^3. The effective porosity was obtained from literature values and was 25% or 0.25.

All the necessary information has been obtained, and the retardation value of 1.11 is calculated using Equation 10.3:

$$\text{XYZ chemical retardation} = R = 1 + \frac{(\rho b)(Kd)}{\eta}$$

$$\text{XYZ chemical retardation} = R = 1 + \frac{(1.7)(0.017)}{0.25} = 1.11$$

Plug in Henry's law constant (H) obtained from the literature into Equation 10.2; the Henry's law constant for XYZ chemical is 0.228.

$$M = (H)(R)$$

$$M = 0.228 \times 1.11 = 0.253$$

Step 3: Determine the persistence value. The persistence value was obtained from the literature and is 0.2 years.
Step 4: With toxicity, mobility, and persistence values obtained, the CRF_{SOIL} is calculated using Equation 10.6:

$$CRF_{SOIL} \text{ for chemical XYZ} = \frac{1}{(T)} \times (M) \times (P)$$

$$CRF_{SOIL} \text{ for chemical XYZ} = \frac{1}{(0.04)} \times (0.253) \times (0.2) = 1.265$$

$$CRF_{SOIL} \text{ for chemical XYZ} = 1.265$$

Contaminant Risk Factors in soil for common LNAPL VOCs in the Rouge River watershed are listed in Table 10.8 (Rogers et al. 2007b; Kaufman et al. 2009). As with the $CRF_{GW,}$ the values span the four most common types of soils found in urban areas of the United States, so their relative magnitudes can be used as a starting point for risk assessments in other urbanized watersheds with similar geological units.

As shown in Table 10.8, the values for LNAPL compounds are greatest for soil composed of clay, and reflect the tendency of LNAPLs to sorb more strongly to finer grained soils.

Table 10.9 lists the CRF_{SOIL} for common DNAPL VOCs (Kaufman et al. 2009).

The CRF_{SOIL} for DNAPLs are much greater than those of LNAPL compounds listed in Table 10.8. Not only are the DNAPL compounds more toxic, but they are also much more persistent in the environment than the LNAPL compounds (Kaufman et al. 2009). For instance, the LNAPL compound benzene—a very toxic chemical—has a half-life of 0.2 years, whereas the DNAPL compound tetrachloroethene—also very toxic—has a half-life of 18 years because of its sequential degradation to vinyl chloride. If released into the environment at the same time, the DNAPL tetrachloroethene would last approximately 90 times longer than the LNAPL benzene.

Table 10.10 lists the CRF_{SOIL} for common PAHs (Kaufman et al. 2005; Rogers et al. 2007b).

TABLE 10.8

CRF_{SOIL} Values for Common LNAPL VOCs

LNAPL Compound	Soil Type	CRF_{SOIL}
Benzene	Clay	10.00
	Silty clay	2.50
	Sand silty clay	2.30
	Sand	1.90
Toluene	Clay	12.76
	Silty clay	1.97
	Sand silty clay	1.79
	Sand	1.07
Ethyl benzene	Clay	74.80
	Silty clay	9.60
	Sand silty clay	8.40
	Sand	4.20
Xylenes	Clay	23.13
	Silty clay	3.43
	Sand silty clay	3.10
	Sand	1.80
Mean LNAPL CRF_{SOIL}	Clay	30.17
	Silty clay	4.37
	Sand silty clay	3.89
	Sand	2.24

The CRF_{SOIL} values for PAH compounds are significantly higher than the CRF_{SOIL} for LNAPL VOCs, because PAHs strongly sorb to soil particles and are much less soluble in water (Kaufman et al. 2009).

Table 10.11 summarizes the CRF_{SOIL} for the LNAPL and DNAPL VOCs, PAHs, and lists the CRF_{SOIL} for total PCBs, the pesticide chlordane, and the heavy metals chromium VI, lead, mercury, and arsenic (Rogers et al. 2007b; Kaufman et al. 2009).

Contaminants with the highest CRF_{SOIL} are mercury, chlordane, PCBs, arsenic, and PAHs, respectively. These contaminants strongly sorb to soil particles and tend not to be very soluble in water. DNAPL compounds also have rather elevated CRF_{SOIL}, especially in soil composed of clay because they also strongly sorb to fine-grained soils. LNAPL VOCs and chromium VI had the lowest CRF_{SOIL}. LNAPL VOCs have rather short half-lives in soil and chromium VI does not strongly sorb to soil grains and is rather soluble.

When the data in Table 10.11 are graphed, it becomes clear that the highest CRF_{SOIL} in each group occurs in soils composed of clay. Clay soils have the lowest permeability and have more surface area available for contaminants with high sorptive potential. Therefore, contaminants with low water solubility and high relative sorptive potentials will have high CRF_{SOIL} values in clay-rich soils (Figure 10.6).

Figure 10.6 indicates mercury has the highest CRF_{SOIL} and chromium VI has the lowest. Chromium VI's low CRF_{SOIL} results from its high relative water solubility and low sorptive potential. On the other hand, mercury has a high sorptive potential and low water solubility, and checks in with a high CRF_{SOIL}. It should also be noted that for every contaminant the CRF_{SOIL} value is the highest in clay, a consequence of this soil texture acting to impede the migration of contaminants.

TABLE 10.9

CRF_{SOIL} Values for Common DNAPL VOCs

LNAPL Compound	Soil Type	CRF_{SOIL}
Tetrachloroethene (PCE)	Clay	3,955.0
	Silty clay	647.0
	Sand silty clay	591.0
	Sand	371.0
Trichloroethene (TCE)	Clay	255.0
	Silty clay	40.0
	Sand silty clay	37.0
	Sand	23.0
cis-1,2-Dichloroethene	Clay	3,698.0
	Silty clay	1,205.0
	Sand silty clay	1,159.0
	Sand	993.0
trans-1,2-Dichloroethene	Clay	491.0
	Silty clay	111.0
	Sand silty clay	99.0
	Sand	87.0
Vinyl chloride	Clay	5,727.0
	Silty clay	2,575.0
	Sand silty clay	2,508.0
	Sand	2,308.0
1,1,1-Trichloroethane (1,1,1-TCA)	Clay	13.7
	Silty clay	2.5
	Sand silty clay	2.3
	Sand	1.6
Mean DNAPL CRF_{SOIL} for degradation sequence from PCE to vinyl chloride[a] (Figure 7.4)	Clay	2,825.0
	Silty clay	915.6
	Sand silty clay	878.8
	Sand	756.4

[a] Represents cumulative risk.

Table 10.12 compares the CRF_{GW} with the CRF_{SOIL}.

From Table 10.12, it is clear that the contaminants exhibiting high soil risk factors (e.g., chlordane, mercury, and PCBs) have a low groundwater contaminant risk factor. This behavior is primarily attributed to their low mobility. Computing the ratio of the data in columns 3 and 4 yields the likelihood of a particular contaminant type contaminating soil and/or groundwater. This computation is shown in column 5, and the values range from 1.0×10^{-9} to 766—a difference in magnitude of more than a billion.

A ratio value of 1.0 in column 5 indicates a particular chemical compound has an equal probability of contaminating soil and groundwater; values less than 1.0 indicate a higher likelihood of contaminating soil only, and values greater than 1.0 mean the compound is more likely to contaminate groundwater. The high mean (arithmetic average) CRF_{GW}/CRF_{SOIL} ratios of chromium VI and DNAPL clearly suggest these compounds have the

TABLE 10.10

CRF_{SOIL} Values for Select PAHs

PAH Compound	Soil Type	CRF_{SOIL}
Naphthalene	Clay	11.8
	Silty clay	1.24
	Sand silty clay	1.06
	Sand	0.37
Chrysene	Clay	2,407.00
	Silty clay	240.00
	Sand silty clay	204.00
	Sand	61.00
Benzo[b]fluoranthrene	Clay	146.00
	Silty clay	14.60
	Sand silty clay	12.40
	Sand	3.70
Benzo[k]fluoranthrene	Clay	126.00
	Silty clay	12.60
	Sand silty clay	10.70
	Sand	3.21
Phenanthrene	Clay	70.00
	Silty clay	7.00
	Sand silty clay	5.90
	Sand	1.80
Benzo[g,h,i] perylene	Clay	1,080.00
	Silty clay	108.00
	Sand silty clay	91.00
	Sand	27.50
Benzo(a)pyrene	Clay	797.30
	Silty clay	79.70
	Sand silty clay	67.50
	Sand	20.20
Mean PAH CRF_{SOIL}	Clay	662.58
	Silty clay	66.16
	Sand silty clay	55.23
	Sand	16.82

Source: Kaufman, M.M. et al., *Environ. Geol.*, 56, 1009, 2009.

highest likelihood of contaminating groundwater, whereas the low mean ratios of PCBs, mercury, and chlordane position them as posing the greatest contamination threat to soil. This conclusion is supported by the average extents of contamination shown in column 6, which were measured using analytical data from hundreds of sites (Kaufman et al. 2005, Rogers et al. 2007a; Kaufman et al. 2009). The measured extents of contamination listed in the rightmost column represent contaminants at concentrations exceeding the applicable regulatory action level from an identified anthropogenic source. As predicted by the computed CRF_{GW}/CRF_{SOIL} ratios, the average extents of contamination were the largest for chromium VI and DNAPL compounds.

TABLE 10.11

Summary of CRF_{SOIL} Values for the LNAPL and DNAPL VOCs, PAHs, and CRF_{SOIL} for Other Selected Compounds

Compound	Soil Type	CRF_{SOIL}
Mean LNAPL CRF_{SOIL}	Clay	30.17
	Silty clay	4.37
	Sand silty clay	3.89
	Sand	2.24
Mean DNAPL CRF_{SOIL} for degradation sequence from PCE to vinyl chloride	Clay	3,408.00
	Silty clay	1,116.00
	Sand silty clay	1,073.00
	Sand	965.00
Mean PAH CRF_{SOIL}	Clay	662.58
	Silty clay	66.16
	Sand silty clay	55.23
	Sand	16.82
PCBs	Clay	17,992.00
	Silty clay	2,800.00
	Sand silty clay	2,524.00
	Sand	1,457.60
Pesticide chlordane	Clay	28,850.00
	Silty clay	2,663.00
	Sand silty clay	2,215.00
	Sand	1,876.00
Chromium VI	Clay	10.20
	Silty clay	4.00
	Sand silty clay	3.50
	Sand	3.00
Lead	Clay	234.00
	Silty clay	11.70
	Sand silty clay	9.94
	Sand	3.03
Mercury	Clay	41,800.00
	Silty clay	19,140.00
	Sand silty clay	17,600.00
	Sand	15,600.00
Arsenic	Clay	7,800.00
	Silty clay	2,950.00
	Sand silty clay	2,850.00
	Sand	2,400.00

Source: Kaufman, M.M. et al., *Environ. Geol.*, 56, 1009, 2009.

CRF_{SOIL} values were significantly higher for each contaminant group where the soil was composed of clay, but lowest for each contaminant group in sandy soils. This observation is the result of several physical/chemical factors, including

- Density
- Grain size

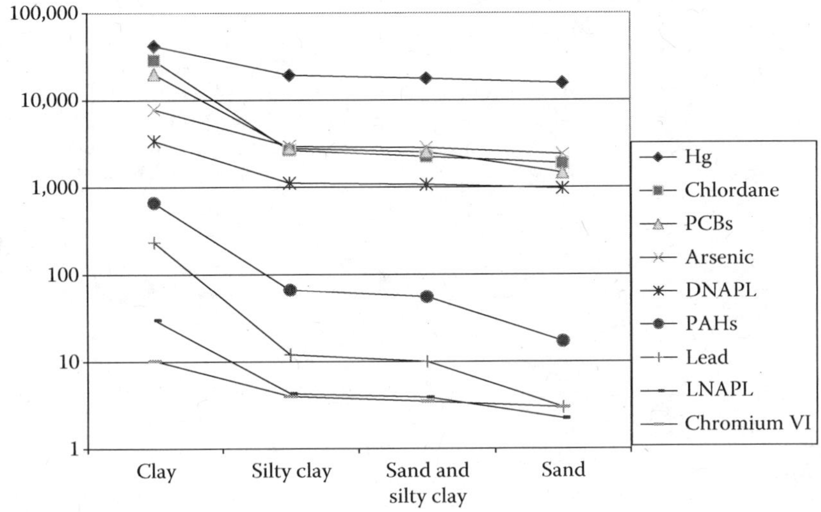

FIGURE 10.6
CRF_{SOIL} distribution by contaminants type.

- Composition
- Permeability
- pH
- Redox

These factors work to increase the retention capacity of soils composed of clay and decrease the capacity in sandy soils. For example, clay soils generally have a higher contaminant retention capacity and a much lower permeability than sandy soils. Thus, the CRF_{SOIL} reflects a decrease in risk as the mean grain size of soil increases.

10.5 Air Contaminant Risk Factor

Atmospheric contaminants vary dramatically and originate from anthropogenic and natural sources. Anthropogenic sources are most prevalent in urban areas and include stationary emitters such as manufacturing facilities, power generating plants, dry cleaners, and mobile emitters such as automobiles, trucks, buses and airplanes (USEPA 2008). Natural sources include particles from volcanic eruptions, biological decay, forest fires, and pollen. VOCs are also emitted by plants and trees. As an example of atmospheric contaminant variety, consider the VOC isoprene emitted by some tree species (e.g., oaks) (Sharkey et al. 2008). Isoprene is carcinogenic to humans if inhaled as a concentrated vapor within a closed space (USEPA 2002b), but concentrations in the atmosphere do not reach levels of concern.

Contaminants are released into the atmosphere as a gas, particulate matter, or as part of a water droplet through sorption or solution (USEPA 2008). Primary contaminants include heavy metals, volatile and semivolatile compounds and particulate matter. Secondary

TABLE 10.12

Comparison of CRF_{SOIL} to CRF_{GW}

Contaminant Type	Geological Unit	Groundwater Contaminant Risk Factor[a] (CRF_{GW})	Soil Contaminant Soil Risk Factor (CRF_{SOIL})	Groundwater/ Soil Contaminant Risk Factor Ratio CRF_{GW}/CRF_{SOIL}	Average Extent of Contamination from Source (m)
DNAPL[b]	Clay	333	3,408	0.09	30
	SC[c]	773	1,116	0.69	42
	SSC[d]	1,091	1,073	1.01	50
	Sand	1,419	965	1.47	1,012
	Moraine	1,274	982	1.30	625
LNAPL[b]	Clay	0.08	30.17	0.002	38
	SC	0.42	4.37	0.09	43.5
	SSC	0.47	3.89	0.12	45.6
	Sand	0.72	2.24	0.32	328
	Moraine	0.65	2.52	0.25	270
PAHs[b]	Clay	0.0001	576.44	1.7×10^{-7}	25
	SC	0.0008	57.55	3.7×10^{-6}	25
	SSC	0.001	48.79	2.0×10^{-5}	30
	Sand	0.002	12.49	1.6×10^{-4}	40
	Moraine	0.002	17.47	1.1×10^{-4}	40
Lead	Clay	0.03	234	1.2×10^{-4}	25
	SC	0.3	11.7	0.02	25
	SSC	0.34	9.94	0.03	25
	Sand	1.1	3.03	0.36	30
	Moraine	1.0	3.56	0.28	30
Mercury	Clay	1.57	41,800	3.70×10^{-5}	30
	SC	2.52	19,140	1.31×10^{-4}	20
	SSC	2.77	17,600	1.57×10^{-4}	20
	Sand	3.10	15,600	1.98×10^{-4}	No data
	Moraine	2.85	16,328	1.74×10^{-4}	No data
Chromium VI	Clay	948	10.2	93	40
	SC	2,080	4	520	No data
	SSC	2,116	3.5	604	No data
	Sand	2,300	3.0	766	2,200
	Moraine	2,190	3.2	684	No data
Arsenic	Clay	9.88	7,800	1.26×10^{-3}	30
	SC	11.26	2,950	3.81×10^{-3}	No data
	SSC	21.12	2,850	7.41×10^{-3}	No data
	Sand	50.25	2,400	2.09×10^{-2}	40
	Moraine	43.65	2,615	1.66×10^{-3}	No data
Chlordane	Clay	0.0035	28,850	1.21×10^{-7}	25
	SC	0.035	2,663	1.31×10^{-5}	No data
	SSC	0.041	2,215	1.85×10^{-5}	No data
	Sand	0.046	1,876	2.45×10^{-5}	No data
	Moraine	0.043	1,950	2.21×10^{-5}	25
Total PCBs	Clay	0.000026	17,992	1.0×10^{-9}	25
	SC	0.00026	2,800	9.3×10^{-8}	25

TABLE 10.12 (continued)

Comparison of CRF_{SOIL} to CRF_{GW}

Contaminant Type	Geological Unit	Groundwater Contaminant Risk Factor[a] (CRF_{GW})	Soil Contaminant Soil Risk Factor (CRF_{SOIL})	Groundwater/ Soil Contaminant Risk Factor Ratio CRF_{GW}/CRF_{SOIL}	Average Extent of Contamination from Source (m)
	SSC	0.0003	2,524	1.2×10^{-7}	25
	Sand	0.0009	1,457	6.2×10^{-7}	25
	Moraine	0.0008	1,548	5.2×10^{-7}	25

[a] Data from Kaufman, M.M. et al. *Water, Air, and Soil Pollut.*, 167, 365–386, 2005.
[b] Contaminant risk factors listed represent an average value of specific compounds listed in Tables 10.5 and 10.11.
[c] SC, sandy clay
[d] SSC, sandy and silty clay

contaminants are formed in the atmosphere through photochemical or chemical reactions (USEPA 2008). Many of these contaminants travel between the air, soil, and surface water and change states. For example, contaminants initially released into the soil or water may volatilize and become airborne. Or in some cases, contaminants airborne in the vapor phase may become deposited on the soil and sorb onto a soil particle only to become airborne again by wind action.

Given the composition and behavior of air pollutants, developing a model for evaluating the contaminant risks to air must address contaminants in the vapor phase and contaminants attached to particulate matter. Once airborne, a combination of these seven factors control whether there is an adverse effect from an air contaminant (McKone and Enoch 2002; USEPA 2008):

- Toxicity
- Mobility
- Persistence
- Volume released
- Time period of the release
- Distance to a specific receptor being evaluated
- Wind speed and direction

The last four factors in this list are environmental factors dependent upon site-specific criteria, so they are not included in the development of the air contaminant risk factor (CRF_{AIR}). These environmental factors are applicable when evaluating the impacts of an actual release, as done with the definition of the nature and extent of a contaminant plume within soil and groundwater.

Three familiar factors remain (toxicity, mobility, and persistence). In addition, there are similarities between the physical and chemical attributes controlling the migration potential of a contaminant in air, groundwater, and soil including solubility, vapor pressure, density, chemical stability, persistence, and adsorption potential (Kaufman et al. 2005; USGS 2006; Rogers et al. 2007a). On the basis of these similarities, the environmental risk posed by specific contaminants to air, termed the Air Contaminant Risk Factor (CRF_{AIR}), can be computed by modifying the equations of the CRF_{GW} and CRF_{SOIL} (Kaufman et al. 2005; Rogers et al. 2007b).

The CRF_{AIR} is expressed in general terms as Equation 10.7 (Kaufman et al. 2005; Rogers et al. 2007a).

$$CRF_{AIR} = Toxicity \times [(Mobility_{gas} \times Persistence_{gas}) + (Mobility_{particulate} \times Persistence_{particulate})] \tag{10.7}$$

Toxicity values are obtained from the USEPA IRIS (2009a). The values selected should be the most conservative for each exposure pathway—ingestion, dermal adsorption, and inhalation. Using the most conservative value is appropriate since exposure to contaminated air can occur in each of these pathways. For instance, dermal adsorption of soil with sorbed contaminants can occur during any outdoor activity, ingestion can occur through hand to mouth contact and from swallowing, and inhalation can occur through inhalation of contaminants in the gas or particulate matter, especially if the particulate matter is very fine.

Mobility for air is represented separately for contaminants in the gas phase and for solid particulate matter, and is combined with persistence. This separation is necessary because the half-life of many contaminants is significantly different depending on whether the contaminant is in the gas phase or solid particulate matter. For instance, xylene has a half-life of 6 days in the vapor phase, but its half-life is 1 year when present in the particulate phase.

To evaluate mobility in the gas phase, Henry's law constant alone is sufficient because a retardation factor is not necessary. The mobility of atmospheric contaminants in the gas phase can be expressed as in Equation 10.8.

$$M = (H) \tag{10.8}$$

where

M is the mobility
H is the Henry's law constant

As with the CRF_{GW} and CRF_{SOIL}, Henry's law constants are derived from Equation 10.2, or from the literature including USEPA (1996a); Sander (1999); Wiedemeier (1999); Suthersan and Payne (2005); Payne et al. (2008).

The mobility of a contaminant sorbed to particulate matter is represented by Equation 10.9.

$$M_{(particulate)} = \frac{1}{SpG} \times Koc \tag{10.9}$$

where

$M_{(particulate)}$ is the mobility of a contaminant sorbed to particulate matter
SpG is the specific gravity
Koc is the partitioning coefficient

Specific gravity is an important determinant of mobility because as the specific gravity of a contaminant increases, its buoyancy in the atmosphere decreases. This relationship explains why the inverse of the specific gravity is used. The partitioning coefficient is a measure of a contaminant's tendency to sorb to particulate matter. This factor is important because the higher the partitioning coefficient, the higher the likelihood a contaminant will attach to particulate matter. For example, a contaminant with a low specific gravity and high sorptive potential has the potential to be very mobile in the atmosphere compared to one with a high specific gravity and low sorptive potential.

Values for the partitioning coefficient can be obtained from numerous sources, including USEPA (1996a), Wiedemeier (1999), USEPA (2002a), and Suthersan and Payne (2005).

Persistence of a contaminant in the air is generally expressed in the gas phase and particulate phase (McKone and Enoch 2002; USEPA 2008). Persistence times for most organic compounds are much less in the gas phase than when they are attached to particulate matter.

Finally, the CRF_{AIR} is calculated by multiplying the inverse of the chemical compound's toxicity (T), by mobility (M) and persistence (P) in the gas and particulate phases (Equation 10.10). Multiplying by the inverse of the mobility value is not necessary.

$$CRF_{AIR} = \frac{1}{(T)} \times [(M_{gas} \times P_{gas}) + (M_{particulate} \times P_{particulate})] \tag{10.10}$$

where

CRF_{AIR} is the soil contaminant risk factor
T is the toxicity
M_{gas} is the mobility of a contaminant in the gas phase
P_{gas} is the persistence of a contaminant in the gas phase
$M_{particulate}$ is the mobility of a contaminant sorbed to particulate matter
$P_{particulate}$ is the persistence of a contaminant sorbed to particulate matter

The inverse of the toxicity value must be used because the integers assigned for toxicity values decrease with increasing carcinogenicity (USEPA 2009). The presence of mobility and persistence factors for gas and particulates accounts for the significant differences in the half-lives depending on whether the contaminant is present as a gas, or if it is sorbed onto particulate matter (USEPA 2009).

Using Equation 10.10, an example calculation of the CRF_{AIR} for a Chemical XYZ is written as

$$CRF_{AIR} \text{ for chemical XYZ} = \frac{1}{(T)} \times [(M_{gas} \times P_{gas}) + (M_{particulate} \times P_{particulate})]$$

Step 1: Obtain toxicity value:

The toxicity of XYZ chemical was obtained from the literature and has a value of 0.04.

Step 2: Determine the mobility and persistence values for the gas phase:

Mobility in the gas phase (M_{gas}) = Henry's law constant (H) = 0.228
Persistence in the gas phase (P_{gas}) = literature value of 4 days

$$M_{gas}(0.228) \times P_{gas}(4) = 0.228 \times 4 = 0.912$$

Step 3: Determine the mobility and persistence values for the particulate phase using Equation 10.9:

$$\text{Mobility in the particulate phase}\,(M_{\text{particulate}}) = \frac{1}{\text{SpG}} \times \text{Koc}$$

SpG for Chemical XYZ = 0.88
Koc for Chemical XYZ = 1.92
Therefore

$$\text{Mobility in the particulate phase}\,(M_{\text{particulate}}) = \frac{1}{0.88} \times 1.92 = 2.18$$

Persistence in the particulate phase (P_{gas}) = literature value of 1.0 years

$$M_{\text{particulate}}(2.18) \times P_{\text{particulate}}(0.2) = 2.18 \times 1.0 = 2.18$$

Step 4: Determine the CRF_{air} using Equation 10.10:

$$CRF_{AIR}\text{ for chemical XYZ} = \frac{1}{(T)} \times [(M_{gas} \times P_{gas}) + (M_{\text{particulate}} \times P_{\text{particulate}})]$$

$$CRF_{AIR}\text{ for chemical XYZ} = \frac{1}{(0.04)} \times [(0.228 \times 4) + (2.18 \times 1.0)]$$

$$CRF_{AIR}\text{ for chemical XYZ} = \frac{1}{(0.04)} \times [(0.91) + (2.18)]$$

$$CRF_{AIR}\text{ for chemical XYZ} = 77.25$$

This sample calculation for CRF_{AIR} demonstrates how the risk posed by the chemical in the particulate phase is much greater than the risk posed by the chemical in the gas phase (2.18 vs. 0.912). This occurs in most cases because the contaminant is much more persistent as a particulate than in the gas phase, and as a result, there is an increased probability of exposure with a particulate form of a chemical. This trend is common with most contaminants, especially organics.

Table 10.13 displays the CRF_{AIR} values for LNAPLs, DNAPLs, PAHs, selected metals, PCBs, and the pesticide chlordane.

As shown in column 2, lead had the lowest contaminant risk factor for air compared to any of the contaminants evaluated. The reason is not because lead is less toxic, but because lead is not volatile and has the second highest specific gravity of any compound evaluated. The specific gravity of lead is 11.34; mercury is the only compound evaluated with a higher specific gravity (13.6). Lead is routinely released into the environment from air emission sources (USEPA 2008), but most airborne lead is rapidly deposited on the ground surface.

The particulate to gas ratios shown in column 5 are significant because the contaminants shown here are commonly found in U.S. urban areas, where particulate matter tends to concentrate. With the exception of benzene and mercury, the risk of exposure to contaminants in the atmosphere is highest from those contaminants sorbed to particulate matter.

TABLE 10.13

Summary of CRF_{air} Values and Exposure Risk Potential

Chemical Compound	Contaminant Risk Factors for Air (CRF_{air})	Percent of Potential Exposure Risk from Gas Phase	Percent of Potential Exposure Risk from Particulate Phase	Approximate Ratio of Exposure Risk Particulate:Gas
LNAPL compounds				
Benzene	36.05	55.69	44.31	4:5
Ethyl benzene	18.18	11.00	89.00	9:1
Toluene	7.75	13.22	86.88	8.5:1.5
Xylenes	18.95	2.07	97.93	97:3
Mean LNAPL	20.23	81.98	318.02	
		20.49%	79.51%	4:1
DNAPL compounds				
PCE	91.09	1.63	98.37	98:2
TCE	9.37	1.05	98.94	99:1
111-TCA	7.71	0.08	99.92	99:1
cis-12-DCE	34.10	0.18	99.82	99:1
Trans -12-DCE	18.90	0.10	99.90	99:1
Vinyl chloride	243.97	8.30	91.70	92:8
Mean DNAPL	67.52	11.34	588.66	98:2
		1.89%	98.11%	
PAHs				
Naphthalene	30.44	0.003	99.997	1,000:1
Chrysene	162.09	0.0003	99.9997	1,000:1
Benzo (b) fluoranthene	167.19	0.0004	99.9996	1,000:1
Benzo (k) fluoranthene	97.74	0.01	99.99	1,000:1
Phenanthrene	40.40	0.0009	99.9991	1,000:1
Benzo (g,h,i) perylene	184.13	0.0002	99.9998	1,000:1
Benzo (a) pyrene	94.59	0.000008	99.999992	100,000:1
Mean PAHs	124.82	0.0148	699.985	
		0.003%	99.997%	1,000:1
Metals				
Arsenic	44.50	3.6	96.4	24:1
Chromium VI	210.25	44.59	55.41	5:4
Mercury	166.33	83.36	16.64	1:5.6
Lead	3.45	12.68	87.32	7:1
Mean metals	106.13	144.23	255.77	
		36.06%	63.94%	16:9
Mean PCBs	37.02	0.009	99.99	1,000:1
Chlordane	173.02	0.017	99.983	500:1

This is dramatically demonstrated by PAHs and PCBs—compounds having a mean particulate to gas ratio of 1000–1. Also displaying impressively large particulate to gas ratios are chlordane (500:1) and DNAPL compounds (100:1). If an equal weighting were used between the gas phase and particulates, the differential would be much greater. This is because the values of persistence in the gas phase used to calculate the CRF_{AIR} values are in days and the values of persistence in the particulate phase are in years.

The LNAPL VOCs provide a good example of how the presence of particulate matter can transform a short-lived group of chemicals into a significant atmospheric risk. This group of compounds includes benzene, toluene, ethyl benzene, and xylenes and is commonly referred to as BTEX. They are common components of gasoline and highly volatile, toxic, and flammable, but degrade rapidly in the gaseous phase due to photolysis. Their persistence increases significantly when sorbed onto a particle, and this is reflected by their moderate values for CRF_{AIR} in column 2. Without this sorption, these values would be very low.

10.6 Discussion and Implications

Table 10.14 compares the contaminant risk factors for groundwater, soil, and air. DNAPLs and chromium VI have the highest potential to contaminate groundwater, mercury, PCBs, and chlordane have the highest potential to contaminate and remain in soil or sediment, and several compounds including LNAPLs and DNAPLs, PCBs, PAHs, chlordane, arsenic, chromium VI, and mercury have the potential to contaminate the air, especially when attached to particulate matter.

An examination of the DNAPL compounds listed in Tables 10.3, 10.9, and 10.13, reveals that the compound vinyl chloride exhibits the highest groundwater, soil, and air contaminant risk factors compared to the other DNAPL compounds evaluated. This dubious distinction occurs because vinyl chloride has the following physical/chemical attributes: (1) it is the most toxic DNAPL compound evaluated (USEPA 2009a) and is also the most mobile, (2) it is the most soluble DNAPL compound in water compared to the other DNAPL compounds evaluated, and (3) it has the highest vapor pressure so it evaporates readily (ATSDR 2006b).

Chromium VI is also a compound of special note. It has the highest groundwater and air risk factors because of its high toxicity and relatively high solubility in water. Because

TABLE 10.14

Contaminant Risk Factors for Groundwater, Soil, and Air for Each Contaminant Group

Contaminant Type	Groundwater CRF (CRF_{GW})	Soil CRF (CRF_{SOIL})	Air CRF (CRF_{AIR})
DNAPL	978.00	1,508.00	67.52
LNAPL	0.47	10.16	20.23
PAH	0.001	142.00	124.82
Total PCBs	0.0004	5,264.00	37.00
Chlordane	0.0001	7,501.00	173.00
Arsenic	27.24	3,723.00	44.00
Chromium VI	1,926.00	5.17	210.00
Lead	0.55	52.44	3.45
Mercury	2.56	22,013.00	166.00

of its high solubility, we would expect chromium VI to have the lowest soil risk factor; and it does as a result of its low sorptive potential. Examining the contaminant risk factors for many compounds helps explain their fate and transport, especially where contaminants are found after being released into the environment (their sinks) and why they have time to migrate or degrade after being released. For instance, mercury and PCBs tend to accumulate in river and lake sediments and soil, but are rarely detected in water. When they are detected in water, it is at typically low concentrations. This fate and transport is predicted by the contaminant risk factors scores for mercury and PCBs—they score very high for soil, and very low for groundwater (Figure 10.7).

Two more examples of the predictive power of the risk scores are seen with the DNAPL compounds and chromium VI in groundwater. As shown in Table 10.7, groundwater contaminant plumes for DNAPL compounds and chromium VI are far longer than any of the other compounds evaluated. When the CRF_{GW} and the geological vulnerability assessment described in Chapter 6 are combined, it results in a synergistic effect. This outcome is likely when chromium VI or DNAPLs are released (often continuously over a long duration) into a vulnerable geological environment (e.g., soils composed of sand with shallow groundwater used as a source of potable water). In these cases, significant adverse environmental and human health effects may occur (Rogers et al. 2006).

Another view of the data is provided by the pie charts shown in Figure 10.8. For each type of contaminant risk factor, certain contaminants favor certain environmental media.

Table 10.15 provides an interpretation of the data presented by the pie charts above, and can be used as a summary reference for CRFs by major contaminant groups and compounds of selected interest.

Additional support for the validity of the risk scores would be their ability to explain a high amount of the variation in remediation costs, and to also explain a high amount of the variation in remedial costs across a wide range of geological environments. When the air, groundwater, soil, and surface risk factors are combined, a total risk characterization of a chemical's environmental risks is achieved. Does more risk result in higher remediation cost?

To answer this question, a stepwise regression was used to search for the variables that could explain the most variance in remediation costs at 79 sites in an urbanized

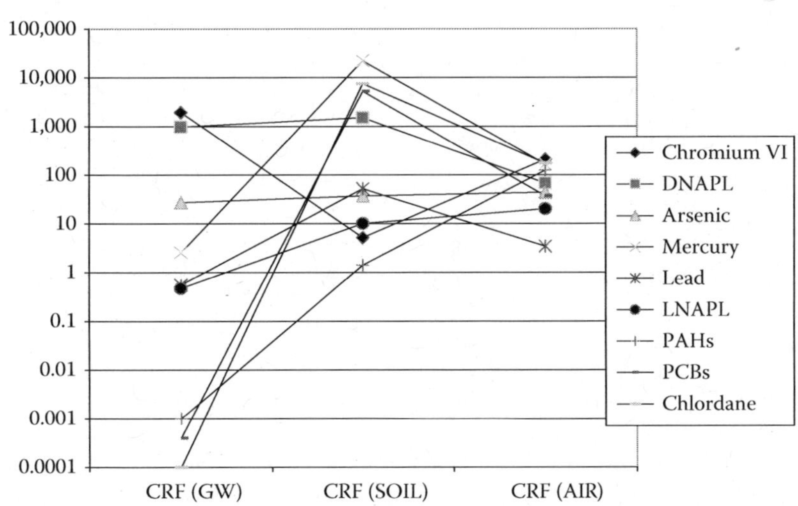

FIGURE 10.7
Groundwater, soil, and air CRF distribution.

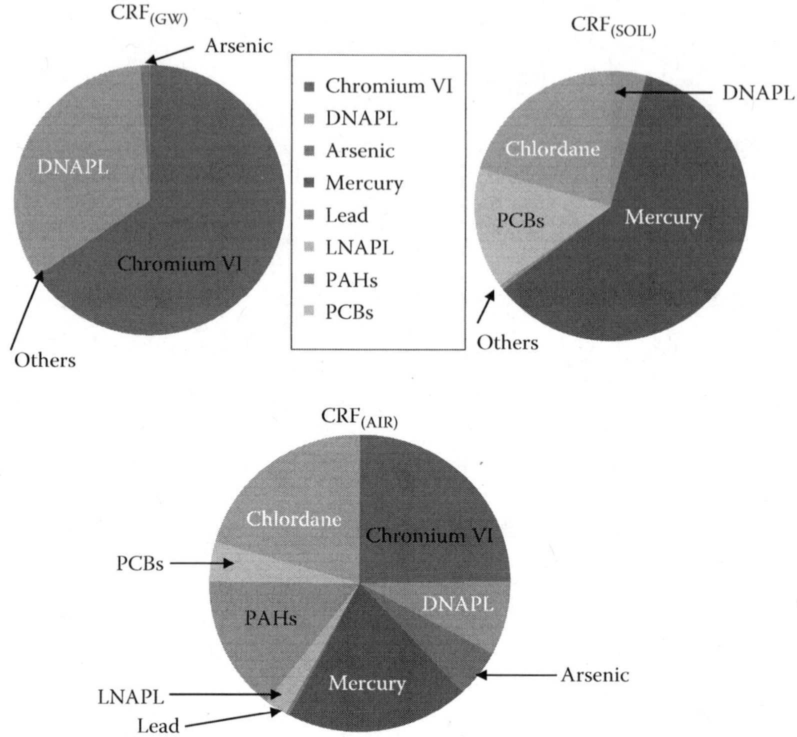

FIGURE 10.8
Pie charts of groundwater, soil, and air CRFs.

TABLE 10.15
Relative CRF Risk Rankings of Contaminants by Media

Compound	CRF $_{GW}$	CRF $_{SOIL}$	CRF $_{AIR}$
Chromium VI	Extremely high	Very low	Very high
DNAPL	High	High	Moderate
Arsenic	Low	Very low	Moderate
LNAPL	Very low	Very low	Moderate
Mercury	Very low	Extremely high	Very high
Lead	Very low	Very low	Low
PAHs	Extremely low	Very low	Very high
Chlordane	Extremely low	Very high	Very high
PCBs	Extremely low	Very high	Moderate

watershed within a variety of geological environments. Using as input the variables listed in Table 10.6, early iterations of the stepwise procedure on the Rouge River watershed data set retained extent of contamination and the total risk scores to explain remediation cost. Examination of the distribution of the dependent and independent variables indicated the data were positively skewed. After a logarithmic transformation, and the multiplication of the total risk scores by plume length, the River Rouge data set indicates a strong correlation between the remediation cost and the total risk given by Equation 10.11 and represented by Figure 10.9 (Kaufman et al. 2005).

$$\text{Lg cost} = 5.107 + 0.4949 \text{ lg risk weight} \tag{10.11}$$

When Equation 10.11 was tested against 127 sites located worldwide in various geological environments, there was a very strong correlation between the estimated values derived from Equation 10.11 for remediation cost and the actual cost values ($r = 90$; $F = 334.1$, $p < .0001$). The results are shown in Figure 10.10. It is reasonable to conclude that total risk scores do help explain much of the variation in remediation costs.

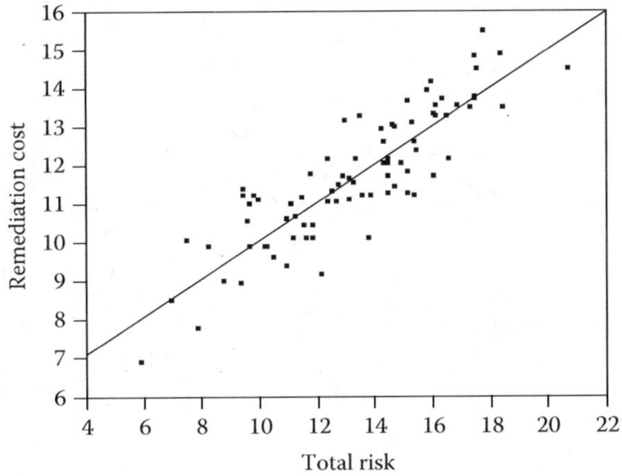

FIGURE 10.9
Regression analysis of remedial cost and CRF_{GW}. (With kind permission from Springer Science+Business Media: Kaufman, M.M. et al., *Water, Air, and Soil Pollut.*, 167, 381, 2005.)

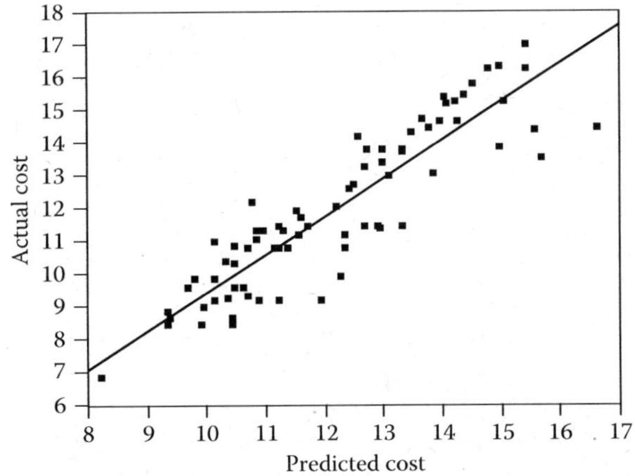

FIGURE 10.10
Actual vs. predicted remedial costs using the CRF_{GW}. (With kind permission from Springer Science+Business Media: Kaufman, M.M. et al., *Water, Air, and Soil Pollut.*, 167, 382, 2005.)

For all the sites tested, those contaminated with chromium VI and DNAPL compounds located within a geological setting of sand exhibited the highest extent of contamination and cost of remediation per kilogram of contaminant recovered.

There still is much work needed on risk scores. Contaminant risk factors of other contaminants such as ozone, sulfur dioxide, greenhouse gases, viruses, bacteria, acids and bases, and fertilizers have not yet been calculated. The main reasons for this are (1) there is a lack of data required to calculate a CRF, (2) some uncertainties exist in assessing risk for certain contaminants such as bacteria and carbon dioxide, and (3) the appropriateness and scientific justification for the use of weighting factors in calculating CRFs for these compounds are still unknown.

10.7 Summary and Conclusion

CRFs for air, groundwater, and soil provide a powerful tool for assessing risks posed by contaminants anywhere in the environment. They are derived by combining the three physical/chemical attributes of toxicity, mobility, and persistence. CRF analysis has demonstrated why some contaminants prefer certain locations or sinks. For instance, CRF analysis has shown why PCBs and mercury prefer soil or sediments instead of being present as dissolved constituents in groundwater.

Combining geological vulnerability analysis with CRF analysis can reveal and explain potential synergistic effects. Examples of synergy occur when DNAPLs and chromium VI are released in a vulnerable geologic setting, such as one with a sandy soil, shallow groundwater, and current use as a potable source of water. Under these conditions, the potential for adverse human health and ecological impact greatly increases, and is unfortunately often realized. With CRF and geological vulnerability analysis now in hand, these situations can be avoided in the future with proper urban planning.

CRF analysis employing the risks posed by surface releases, added to the risk factors for groundwater, soil, and air also has predictive power for evaluating and assessing future cleanup costs for sites of contamination. This ability has major implications for the redevelopment of urban areas, and we explore this in later chapters.

References

Agency for Toxic Substances and Disease Registry (ATSDR). 2006b. *Vinyl Chloride*: CAS/Registry Number 75-01-4. ATSDR. Atlanta, GA.

Baringer, T.H., D. Dunn, W.A. Battaglin et al. 1990. Problems and methods involved in relating land use to groundwater quality. *Water Resources Bulletin* 26:1–9.

Camp, C.V. and J.E. Outlaw. 1998. Stochastic approach to delineating wellhead protection areas. *Journal of Water Resources, Planning and Management* 124:199–208.

Eckhardt, D.A.V. and P.E. Stackelberg. 1995. Relation of groundwater quality to land use on Long Island, New York. *Ground Water* 33:1019–1033.

Howard, H.H., R.S. Boethling, W.F. Jarvis et al. 1991. *Handbook of Environmental Degradation Rates.* Boca Raton, FL: CRC Press.

Kaufman, M.M. 1997. Spatial assessment of risk to groundwater from surface activities. *Proceedings of the Applied Geography Conferences* 20:135–142.

Kaufman, M.M. 2000. Automated procedure for groundwater risk assessment from surface sources. *Proceedings of the Applied Geography Conferences* 23:277–284.

Kaufman, M.M., K.S. Murray, and D.T. Rogers. 2003. Surface and subsurface geologic risk factors to ground water affecting brownfield redevelopment. *Journal of Environmental Quality* 32:490–499.

Kaufman, M.M., D.T. Rogers, and K.S. Murray. 2005. An empirical model for estimating remediation costs at contaminated sites. *Water, Air, and Soil Pollution* 167:365–386.

Kaufman, M.M., D.T. Rogers, and K.S. Murray. 2009. Using soil and contaminant properties to assess the potential for groundwater contamination to the lower Great Lakes, USA. *Environmental Geology* 56:1009–1021.

McKone, T.E. and K.G. Enoch, 2002. CalTox™, *A Multimedia Total Exposure Model Spreadsheet Users Guide*. Lawrence Berkeley National Laboratory. LBNL 47399. University of California, Berkeley, CA.

Michigan Department of Environmental Quality (MDEQ). 2008. *State List of Contaminated Sites*. MDEQ. Lansing, MI.

Payne, F.C., J.A. Quinnan, and S.T. Potter. 2008. *Remediation Hydraulics*. Boca Raton, FL: CRC Press.

Rogers, D.T., M.M. Kaufman, and K.S. Murray. 2006. Improving environmental risk management through historical impact assessments. *Journal of the Air and Waste Management Association* 56:816–823.

Rogers, D.T., K.S. Murray, and M.M. Kaufman. 2007a. Assessment of groundwater contaminant vulnerability in an urban watershed in southeast Michigan, USA. In *Urban Groundwater—Meeting the Challenge*, ed. K.W.F. Howard, pp. 129–144. London, U.K.: Taylor & Francis.

Rogers, D.T., K.S. Murray, and M.M. Kaufman. 2007b. *Improving the Assessment of the Potential for Groundwater Contamination Using an Analytical Risk Factor Model*. International Union of Geodesy and Geophysics World Congress. Perugia, Italy.

Sander, R. 1999. Compilation of Henry's law constants for inorganic and organic species of potential importance in environmental chemistry (Version 3). http://www.henrys-law.org (accessed December 20, 2009).

Schnoor, J.L. 1996. *Environmental Modeling: Fate and Transport of Pollutants in Water and Soil*. New York: John Wiley & Sons.

Secunda, S., M.L. Collin, M.L. Melloul et al. 1998. Groundwater vulnerability assessment using a composite model combining DRASTIC with extensive agricultural land use in Israel's Sharon region. *Journal of Environmental Management* 54:39–57.

Sharkey, T.D., A.E. Wiberley, and A.R. Donohue. 2008. Isoprene emission from plants: Why and how. *Annals of Botany* 101:5–18.

Suthersan, S.S. and F.C. Payne. 2005. *In Situ Remediation Engineering*. Boca Raton, FL: CRC Press.

Tobler, W. 1988. Resolution, resampling and all that. In *Building Database for Global Science*, ed. H. Mounsey and R. Tomlinson, pp. 129–137. London, U.K.: Taylor & Francis.

United States Environmental Protection Agency (USEPA). 1989. *Transport and Fate of Contaminants in the Subsurface*. EPA/625/4-89/019. Washington, DC.

United States Environmental Protection Agency (USEPA). 1992. *Dense Nonaqueous Phase Liquids*. Office of Research and Development. Washington, DC.

United States Environmental Protection Agency (USEPA). 1996a. *Bioscreen. Natural Attenuation Decision Support System*, Version 1.4. Office of Research and Development. Washington, DC.

United States Environmental Protection Agency (USEPA). 1996b. *Soil Screening Guidance: Technical Background Document*. Office of Solid Waste and Emergency Response. Washington, DC.

United States Environmental Protection Agency (USEPA). 2000. *Biochlor: Natural Attenuation Decision Support System*. Version 1.0 Office of Solid Waste and Emergency Response. Washington, DC.

United States Environmental Protection Agency (USEPA). 2002a. *Supplemental Guidance for Developing Soil Screening Levels for Superfund Sites*. Office of Solid Waste and Emergency Response. OSWER 9355.4-24. Washington, DC.

United States Environmental Protection Agency (USEPA). 2002b. *Health Assessment of 1,3-Butadiene*. National Center for Environmental Assessment, USEPA/600/P-98/001F. Washington, DC.

United States Environmental Protection Agency (USEPA). 2008. *Latest Findings on National Air Quality: Status and Trends through 2006*. EPA454/R-07-007. Research Triangle Park, NC.

United States Environmental Protection Agency (USEPA). 2009a. Integrated Risk Information System (IRIS). http://www.epa.gov/ncea/iris/intro.htm (accessed October 21, 2009).

United States Environmental Protection Agency (USEPA). 2009b. EPA online tools for site assessment calculation: Retardation factor calculator. Ecosystems Research Division. USEPA. http://www.epa.gov/athens/learn2model/part two/onsite/retard.html (accessed December 28, 2009).

United States Geological Survey (USGS). 1998. *Estimating Areas Contributing to Recharge of Wells*. USGS Circular 1174. Denver, CO.

United States Geological Survey (USGS). 2006. *Volatile Organic Compounds in Nation's Ground Water and Drinking-Water Supply Wells*. National Water-Quality Assessment Program. USGS Circular 1292. Washington, DC.

United States Geological Survey (USGS). 1987. United States Office of Management and Budget. *Standard Industrial Classification Manual*. Washington, DC: United States Government Printing Office.

Wiedemeier, T.H., H.S. Rifai, C.J. Newell et al. 1999. *Natural Attenuation of Fuels and Chlorinated Solvents in the Subsurface*. New York: John Wiley & Sons.

Wisconsin Department of Natural Resources (WDNR). 1999. *Wellhead Protection*. WDNR Publication PUB-DG-039 99REV. Madison, WI.

11

Remediation: Techniques and Cost

11.1 Introduction

Contaminants released into the environment often require cleanup to reduce or eliminate the risks posed by their presence. This cleanup, termed **remediation**, meaning remedy, varies widely and depends on many different factors.

More than $100 billion dollars is spent annually investigating and remediating contaminated sites in the United States (USEPA 2004). In 2008, the USEPA estimated that there were well over 350,000 sites of environmental contamination in the United States requiring remediation. Excluded from this count were those sites where remediation is in progress or has been completed. In addition, there are an estimated 500,000–1 million abandoned industrial facilities or brownfield sites also excluded from USEPA's list, because a cost estimate to investigate and remediate brownfield sites has not been conducted. Most of these abandoned industrial sites are located in urban areas of the United States and concentrated within the 29 urban areas listed in Table 2.4. Since these sites possess some degree of contamination and will likely require some form of remediation (Rogers et al. 2006), it is likely that the environmental costs to investigate and remediate sites of environmental contamination in United States will exceed the latest available USEPA estimate of (U.S.) $250 billion.

The following sections describe some of the more common remedial technologies for soil, groundwater, surface water, sediment, and air. As with any remedial strategy, controlling the source of contamination is critical for success, because failure to control the source of contamination will result in recontamination. The chapter concludes with a cost analysis of remediating common contaminants in the environment, including many described in Chapter 7.

11.2 Remediation Overview

The objective of any remediation project is to prevent, remove, treat, change, destroy, or transform the potentially harmful contaminants from the medium or media of concern so that any risks posed by the contaminants have been effectively eliminated or reduced to an acceptable level (USEPA 2005a). As noted in Chapter 4, there are several different studies conducted at a site of environmental contamination that enable investigators to acquire relevant information and navigate through the required field procedures for achieving the desired goal or objective. Even with a given set of checks and balances, however, many remedial projects are unsuccessful and must be revised or abandoned in favor of a different technological approach.

The most common cause of remediation failure is the improper characterization of the site. Not identifying a source of contamination, underestimating contaminant mass, not identifying a complete list of contaminants of concern, executing an incomplete sampling plan, or improper sampling techniques and methods can lead to remediation being unsuccessful and result in increased cost, time, and effort to achieve the remedial objective. Other causes for remediation failure include technology misapplication and design, environmental changes such as a change in groundwater flow direction, sudden change in local climatic conditions such as a drought or flood, changes in geochemistry, and additional releases.

Some of the factors controlling or influencing the extent of the remediation required include

- Contaminant or contaminants released
- Concentration
- Nature and extent
- Media impacted (air, surface water, groundwater, soil, sediment)
- Geology
- Hydrogeology
- Hydrology
- Location
- Potential receptors
- Land use
- Future land use
- Climate
- Time necessary to achieve objective(s)
- Results of risk assessment and feasibility study
- Cost

Driven by the need for more effective and less costly cleanup costs, new technologies are constantly being developed to remediate contaminated soil and groundwater. Over the past several decades, approximately 2000 new remedial technologies for environmental contamination have emerged (USEPA 1995, 2007). The impetus for these new technologies includes (1) ineffectiveness of early remedial methods, (2) excessive cost of early methods, and (3) regulatory goals requiring remediation to standard-based levels that in many instances were to pristine pre-contaminant or background conditions.

In the mid-1990s, there was a transition from the standard-based remediation cleanup goals toward risk-based cleanup goals (ASTM 1995). Remediating a site to risk-based criteria involves calculating site-specific cleanup levels based on the risks posed by the presence of specific contaminants. This process incorporates (1) the toxicity and nature of each contaminant, (2) the site's geological and hydrogeological setting, (3) fate and transport mechanisms, (4) future land use, and (5) analysis of potential receptors. In most cases, applying risk-based cleanup goals translated into lower amounts of contaminant remediation to achieve closure for a specific contaminant at a particular site, and this guideline lowered the cost and time required to remediate a site (USEPA 2002).

The overview of the most commonly available remedial technologies begins with a description of the methods used to remediate contaminated soil.

11.3 Common Soil Remediation Technologies

Common soil remediation technologies include (USEPA 2002, 2007; Rogers et al. 2008, 2009):

- Excavation
- Active and passive bioremediation
- Monitored natural attenuation
- Mechanical soil aeration
- Capping
- Land use restrictions
- Soil vapor extraction (SVE)
- Phytoremediation
- Soil washing
- In situ thermal desorption
- Ex situ thermal desorption or incineration
- Electrokinetics
- Solidification/Stabilization
- Fracturing
- Vitrification
- Chemical oxidation
- Chemical dehalogenation

A brief description of each common soil remediation technology follows.

11.3.1 Soil Excavation

This technique involves the removal of affected soil through excavation and disposal of the soil at a licensed landfill (USEPA 2002). It is commonly the first method employed at a contaminated site.

Soil excavation is not always cost effective, and in some cases cannot be conducted. These situations listed below may result in the selection of a different soil remediation strategy:

- Impacted soil extends to depths too deep to effectively excavate.
- Excavation cannot be conducted because a damaged structure(s) is present, and moving it is either too expensive, or may cause unsafe conditions if excavation of soil were conducted.
- The volume of soil to be excavated is too large and costly.
- The excavation itself may cause unsafe conditions (i.e., excavating flammable soils).
- Soil contaminants present are of a certain mixture and concentration and require special handling, treatment, restrictions, or may be banned from disposal in certain landfills.

11.3.2 Active and Passive Bioremediation

Active bioremediation involves the introduction of microorganisms, along with additional nutrients into contaminated soil, to allow natural processes to degrade the contaminants into harmless substances (USEPA 2001a). Active bioremediation is typically used on soil where the existing soil bacteria may have been destroyed or overwhelmed by a contaminant spill or where time constraints require a more speedy recovery. Active bioremediation may also be used to target a specific high boiling point hydrocarbon such as fuel oil, or heavy diesel fuel. Passive bioremediation generally relies on the existing soil bacteria. This method has often been referred to as landfarming (Figure 11.1), because the contaminated soil is spread out to a thickness of no more than a half a meter, and then periodically tilled to bring the soil in contact with air (see Section 11.3.4). A summary of a form of active bioremediation using treatment cells is found in Murray and Clark (1993) (Figures 11.2 and 11.3). An unconventional form of passive bioremediation of hydrocarbon contaminated soil using treatment galleries was successfully demonstrated by Murray et al. (1997). In either active or passive bioremediation, before the microbes can effectively degrade and eventually destroy the contaminants, these criteria must be satisfied (USEPA 2001a):

FIGURE 11.1
Example of landfarming in the central valley of California. (Photo by Kent S. Murray.)

FIGURE 11.2
Constructing a treatment cell in San Diego. (Photo by Kent S. Murray.)

FIGURE 11.3
Electronic valves control fresh air circulation in a treatment cell. (Photo by Kent S. Murray.)

- Contaminants must be biodegradable. Many light nonaqueous phase liquid (LNAPL) compounds such polynuclear aromatic hydrocarbons (PAHs) can only be degraded through the use of active bioremediation techniques.
- A suitable temperature range exists for the natural processes to begin and sustain themselves.
- Nutrients (fertilizers) are present to support microbial growth.
- Enough oxygen is present to assist with the decomposition process.

When these conditions are not satisfied or are not optimal, the microbes do not grow, grow slowly, or die. One way to improve unfavorable conditions is to pump air, nutrients, or other substances (such as a carbohydrate like molasses) into the impacted soil where the microbes have been placed to enhance degradation (Murray and Clark 1993; USEPA 2001a).

11.3.3 Monitored Natural Attenuation

Monitored natural attenuation relies on natural processes to *attenuate* (render less harmless) contaminants in soil. Natural attenuation occurs at most contaminated sites; however, favorable conditions must exist to achieve successful remediation in a reasonable amount of time (USEPA 2001b). Impacted soil is tested periodically to ensure that the attenuation of contaminants is occurring at a satisfactory rate. Natural attenuation of LNAPL compounds occurs much more quickly and efficiently than dense nonaqueous phase liquid (DNAPL) compounds, which require much more evaluation (USEPA 1998). Figure 11.4 shows an example of natural attenuation; a contaminant present in the soil evaporates from the soil into the atmosphere where it is destroyed by sunlight through photolysis.

This technique is often used when the source and the majority of the contamination have been removed, and it is not practical to use more costly remediation methods to remove the remaining residual concentrations (USEPA 2001b).

11.3.4 Mechanical Soil Aeration

Mechanical soil aeration utilizes photolysis to remediate contaminants in near surface soil. This technique is most effective when the contaminants of concern have a high relative vapor pressure and are degradable by photolysis. The process is essentially similar to the

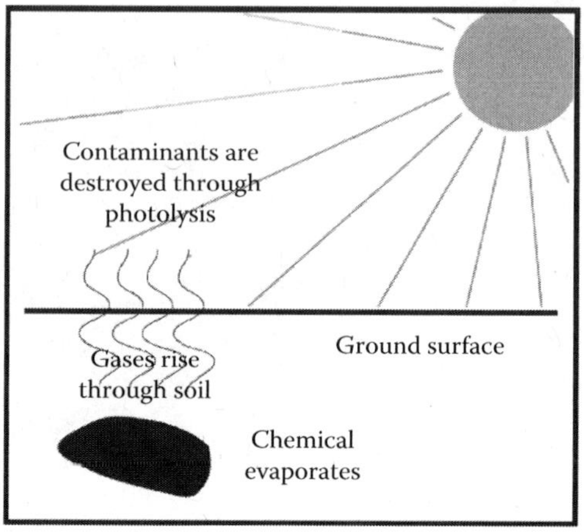

FIGURE 11.4
Example of natural attenuation. (From United States Environmental Protection Agency (USEPA), *A Citizen's Guide to Monitored Natural Attenuation*, EPA 542-F-01-004, Office of Solid Waste and Emergency Response, Washington, DC, 2001b.)

landfarming techniques associated with passive bioremediation and it involves mechanical turning or tilling of the impacted soil so it can more readily evaporate, or induce more efficient photolysis through direct exposure to sunlight (USEPA 2007).

11.3.5 Capping

Capping is routinely used at sites where the highest concentrations of contamination have been removed or remediated, but an additional level of protection is still required. Sites eligible for capping include those contaminated beyond the level adequately addressed by monitored natural attenuation, or where natural attenuation is not expected to occur, as with many heavy metals. Capping can take several different forms including asphalt, concrete, geosynthetic liner, and many others, but must be maintained regardless of the material selected. Some caps are constructed using many different layers, which is typical for many closed landfills (USEPA 2002, 2003). An example of an engineered multilayer barrier cap is shown in Figure 11.5.

11.3.6 Land Use Restrictions

Land use restrictions are common at many remediated sites of environmental contamination. They are employed to minimize the potential for exposure to low-levels of

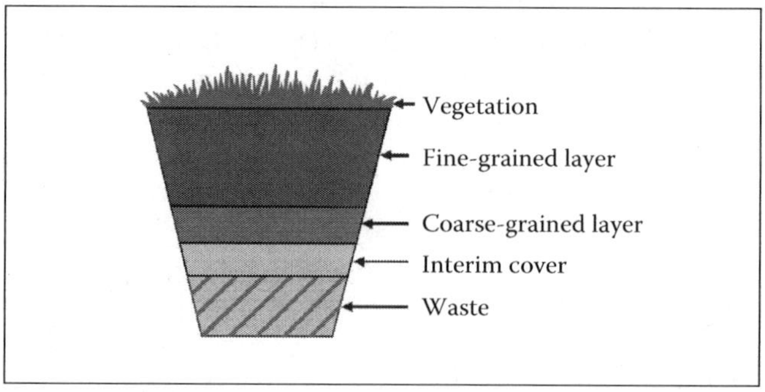

FIGURE 11.5
Example multilayered cap. (From United States Environmental Protection Agency (USEPA), *Evapotranspiration Landfill Cover Systems Fact Sheet*, EPA 542-F-03-015, Office of Solid Waste and Emergency Response, Washington, DC, 2003.)

contamination or to eliminate a potential exposure pathway. As described in Chapter 4, they are usually accompanied with some sort of risk evaluation or risk assessment to support the conclusion that a land use restriction is sufficient to lower the risk posed by the presence of contamination to an acceptable level. Common examples of this practice include maintaining a site for industrial use, prohibiting the use of groundwater, or restricting access to certain areas of a site. For example, the City of Livonia, Michigan, has an ordinance restricting the use of groundwater throughout the city limits, primarily due to widespread groundwater contamination.

11.3.7 Soil Vapor Extraction

SVE removes soil contaminants as vapors from above the water table. This technology is limited to contaminants with the ability to evaporate rather readily, such as volatile organic compounds (VOCs). In addition, the geology must be composed of material permitting the movement of air through the zone where the contaminants exist, such as sand or gravel. The captured contaminant vapors are removed before the air is discharged into the atmosphere (Figure 11.6).

11.3.8 Phytoremediation

Phytoremediation uses plants to remediate contaminated soil. Certain plants can degrade, destroy, or transform many types of contaminants (USEPA 1999a):

- Some heavy metals, such as copper
- Pesticides
- Perchlorates
- Certain PAHs

An additional benefit of plant use is realized at some sites where vegetation acts as a ground cover and prevents or minimizes the effects of rain erosion, and helps prevent the

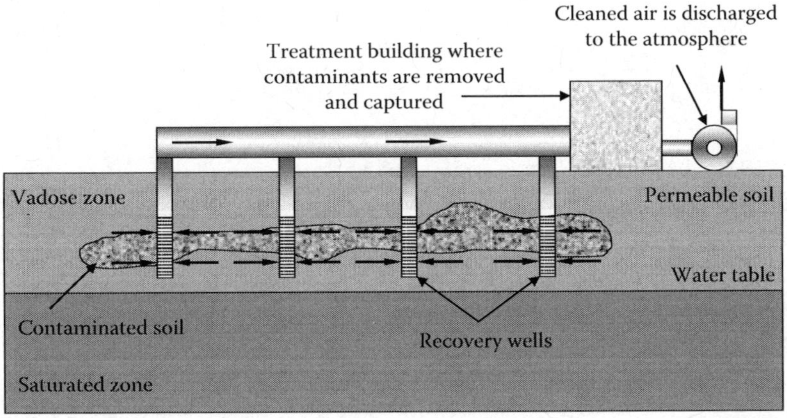

FIGURE 11.6
Example of an SVE system. (From Federal Remediation Technologies Roundtable, Soil vapor extraction, http://www.frtr.gov/matrix2/section1/list-of-fig.html (accessed June 24, 2010), 2010.)

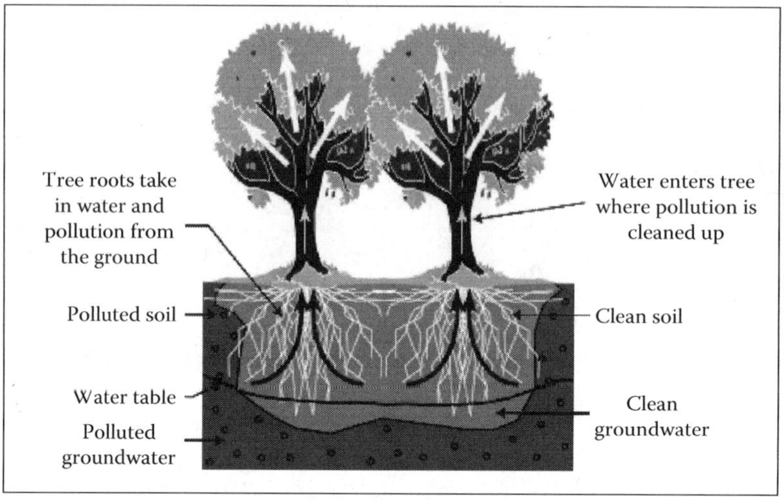

FIGURE 11.7
Example of phytoremediation. (From United States Environmental Protection Agency (USEPA), *Phytoremediation Resource Guide*, EPA 542-B-99-003, Office of Solid Waste and Emergency Response, Washington, DC, 1999a.)

mechanical migration of contaminants. Vegetation can also reduce the amount of water migrating through the contaminant zone in the soil, thus decreasing the leaching or dissolving of contaminants into deeper soil and eventually to groundwater (Figure 11.7). The use of phytoremediation techniques has been limited by the uncertainty regarding the collection and disposal of the resulting contaminated vegetative material at the end of each growing season.

11.3.9 Soil Washing

Soil washing essentially "scrubs" soil to remove and separate the contaminant from soil particles using detergents or other chemicals depending on the type of contaminant, concentration, and soil type. Contaminants sorbed to soil particles can be removed or separated by a detergent or chemical used to lower the sorptive potential of the contaminants. Some heavy metals, fuels, and pesticides can be remediated using this technique (Figure 11.8).

11.3.10 In Situ Thermal Treatment

In situ thermal treatment involves injecting a form of heat into subsurface soils to mobilize the contaminants for easier recovery. Common heat sources include

- Steam
- Hot air
- Hot water
- Electrical resistance
- Radio frequency
- Thermal conduction

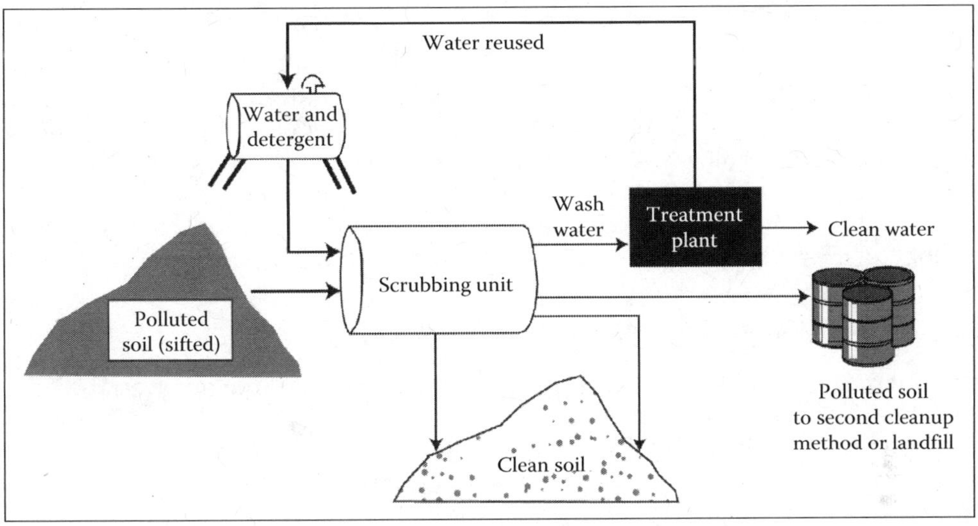

FIGURE 11.8
Example of soil washing. (From United States Environmental Protection Agency (USEPA), *A Citizen's Guide to Soil Washing,* EPA 542-F-01-008, Office of Solid Waste and Emergency Response, Washington, DC, 2001c.)

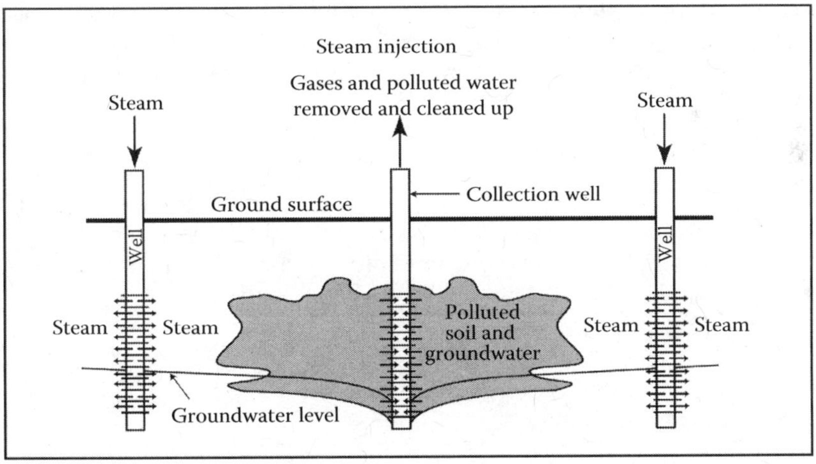

FIGURE 11.9
Example of in situ thermal treatment. (From United States Environmental Protection Agency (USEPA), *A Citizen's Guide to In Situ Thermal Treatment,* EPA 542-F-01-0012, Office of Solid Waste and Emergency Response, Washington, DC, 2001d.)

An added benefit of thermal technologies occurs when the heat destroys certain contaminants. In most cases, the added heat energy acts to evaporate the contaminants, and makes collecting the contaminants using SVE much more efficient (Figure 11.9).

11.3.11 Ex Situ Thermal Treatment

Ex situ thermal treatment is similar to in situ thermal treatment with the key distinctions being the contaminated soil is excavated and treated thermally at the surface. After treatment, the soil is either returned into the ground or transported to a landfill.

11.3.12 Electrokinetics

Electrokinetics is a remedial technology used to separate and extract heavy metals and other contaminants from saturated and unsaturated soil, sludges, and sediments. It works by inducing the migration of contaminants in subsurface soil through an imposed electrical field via electroosmosis. A low voltage current is applied to initiate the migration of contaminants. The configuration involves the application of an electrical potential between electrode pairs implanted in the ground on each side of a contaminated soil mass (Cauwenberghe 1997).

11.3.13 Solidification/Stabilization

Solidification/Stabilization is conducted in situ and ex situ and involves the addition or injection of a material, such as concrete or bentonite clay to permanently immobilize or entomb the contaminants. This technique does not remove contaminant mass—it is designed to eliminate or minimize the potential of the contaminants from migrating and potentially causing harm to human health or the environment.

11.3.14 Fracturing

Fracturing involves cracking dense contaminated soils or rock so other remedial technologies can work more efficiently. The cracks—called *fractures*—create pathways through which contaminants can be more quickly and efficiently removed from the ground. The fractures are created by injecting either high pressure water or air into the contaminated zone (Figure 11.10).

11.3.15 Vitrification

Vitrification is a form of solidification/stabilization, and uses electrical power to heat and transform contaminated subsurface soil into a glass-like substance. This process is

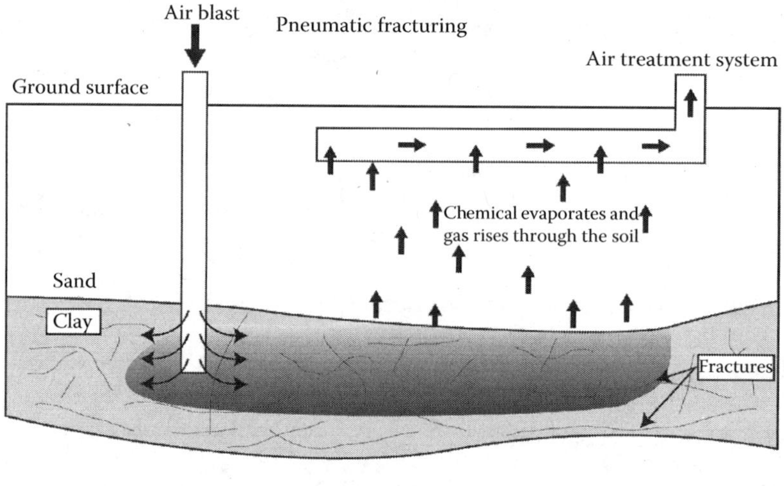

FIGURE 11.10
Example of pneumatic fracturing. (From United States Environmental Protection Agency (USEPA), *A Citizen's Guide to Fracturing*, EPA 542-F-01-0015, Office of Solid Waste and Emergency Response, Washington, DC, 2001e.)

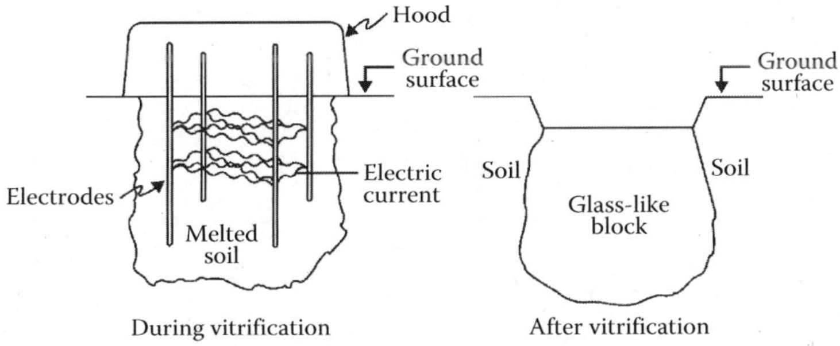

FIGURE 11.11
Example of vitrification. (From United States Environmental Protection Agency (USEPA), *A Citizen's Guide to Vitrification*, EPA 542-F-01-0015, Office of Solid Waste and Emergency Response, Washington, DC, 2001f.)

accomplished by four rods called electrodes. The electrodes are placed into the contaminated soil and an electrical current passed between them results in the melting of intervening soil.

When the electrodes are turned off, the soil cools and vitrifies into a glass-like substance that entombs the contaminants. Vitrified soil often has significantly less volume because the original grain to grain contacts have been fused together and any porosity of the vitrified material is eliminated (USEPA 2001f).

An example of vitrification is shown in Figure 11.11.

11.3.16 Chemical Oxidation

In chemical oxidation, the addition of an oxidizing agent to contaminated soil results in two possible outcomes: (1) the complete destruction of the contaminants or (2) it produces a source of oxygen capable of inducing more rapid rates of bioremediation as long as the correct dosage is used. This technology can be very effective for remediating many organic compounds, including some DNAPL compounds, provided the correct dosages are applied and the chemicals can be delivered to the location where the contaminants reside in the subsurface. Common oxidizing agents include (USEPA 2007):

- Ozone
- Hydrogen peroxide
- Sodium percarbonate
- Sodium permanganate
- Potassium permanganate
- Sodium persulfate

11.3.17 Chemical Dehalogenation

Chemical dehalogenation is a remedial method that removes *halogens* from contaminants. As noted in Chapter 7, halogens are contaminants containing halogen atoms within their atomic structure. Halogens include fluorine, chlorine, bromine, and iodine. Common halogenated contaminants include several DNAPL VOCs, PCBs and dioxins.

The process of chemical dehalogenation involves the excavation of contaminated soil. The soil is sifted and crushed to remove larger objects and enable better remedial treatment.

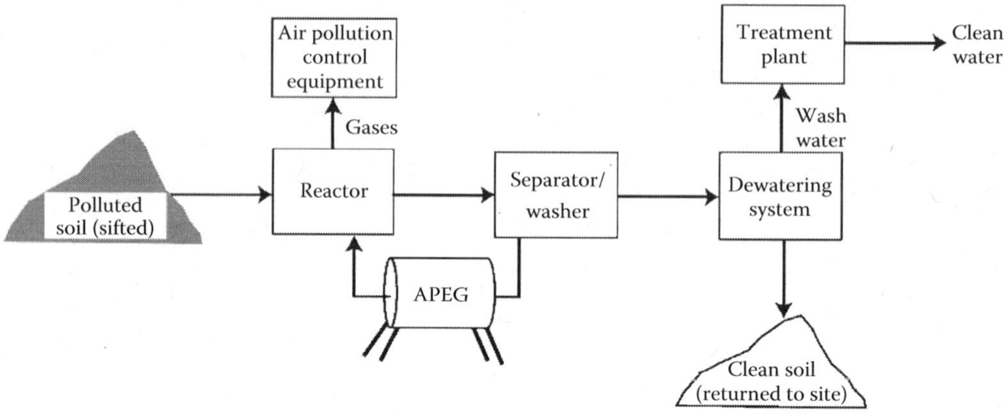

FIGURE 11.12
Example of the glycolate dehalogenation process. (From United States Environmental Protection Agency (USEPA), *A Citizen's Guide to Chemical Dehalogenation*, EPA 542-F-01-0010, Office of Solid Waste and Emergency Response, Washington, DC, 2001g.)

This sifted soil is then mixed with chemical agents and heated in a reactor. During this process, a chemical reaction occurs that removes the halogen atom from the molecular structure of the contaminant, and depending on the completeness of the reaction, destroys or renders the contaminant less harmful.

There are two common types of chemical dehalogenation: glycolate dehalogenation and base-catalyzed dehalogenation (USEPA 2001g). Figure 11.12 is an example of glycolate dehalogenation.

11.4 Summary of Soil Remediation Technologies

Each soil remediation technology described in Section 11.3 has advantages and disadvantages depending upon the different conditions existing at any given site. Frequently, a combination of technologies is used to achieve the objective. At sites where there are widespread impacts with different types of contaminants, rarely if ever, is one technology used (USEPA 2007).

Figure 11.13 summarizes the different soil remediation technologies applied at 977 CERCLA Superfund sites in the United States (USEPA 2007). A total of 53% of the technologies used ex situ technologies and 47% used in situ technologies. Soil excavation was excluded because it was the technique most commonly employed, and it remains the most common technology used for remediation of soil. This prevalence exists in part because soil excavation is the least costly method when relatively small volumes (less than 5000 ton) of contaminant are present and accessible. However, in a recent study of metal-contaminated soil in southeast Michigan, Murray et al. (2004) discovered that the most extensive contamination is restricted to the upper 0.5 m of soil, thus making soil excavation not only cost effective, but quite effective at quickly and efficiently removing the soil posing the greatest risk to human health. In addition, soil excavation may be the only economical technology available, especially when different types of contaminants are present (USEPA 2007; Rogers et al. 2009).

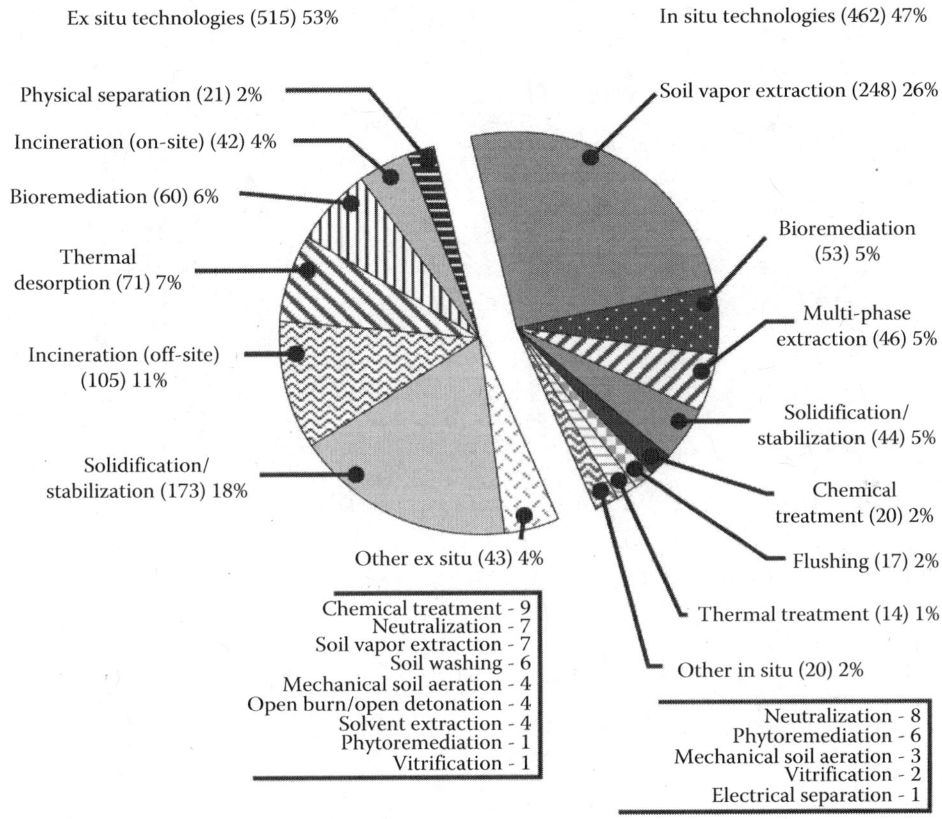

FIGURE 11.13
Soil remediation techniques at 978 CERCLA Superfund sites. (From USEPA, *Treatment Technologies for Site Cleanup: Annual Status Report*, 12th edn. EPA-542-R-07-012. Office of Solid Waste and Emergency Response. Washington, DC, 2007.)

The three most important controlling factors governing the selection of the most efficient soil remediation technologies employed are (1) cost, (2) the geological environment where the contaminants reside, and (3) the physical chemistry of the contaminants. Another important factor is whether a mixture of different types of contaminants and geology exists. If the geology, contaminant type, and distribution are straightforward (e.g., they lack synergy), selecting an appropriate soil remedial technology is a simpler task. It is also important to evaluate whether the contaminants are classified from a regulatory perspective as hazardous or nonhazardous because this distinction routinely plays an important role in the selection of a remedial technology. For instance, if a contaminated mass of soil is classified as hazardous, the disposal cost at a landfill licensed to accept the soil is considerably more than for soil classified as nonhazardous. In this situation, other technologies would be considered due to the increased cost of transporting and disposing a hazardous waste.

Figure 11.14 shows the relationship between uniform geology, soil remediation technology, and hazardous and nonhazardous classification for VOCs (Rogers et al. 2009). As indicated by the horizontal dotted line, if the soil volume is less than 5000 tons, soil excavation is the preferred remedial option. When soil volumes are greater than 5000 tons, the

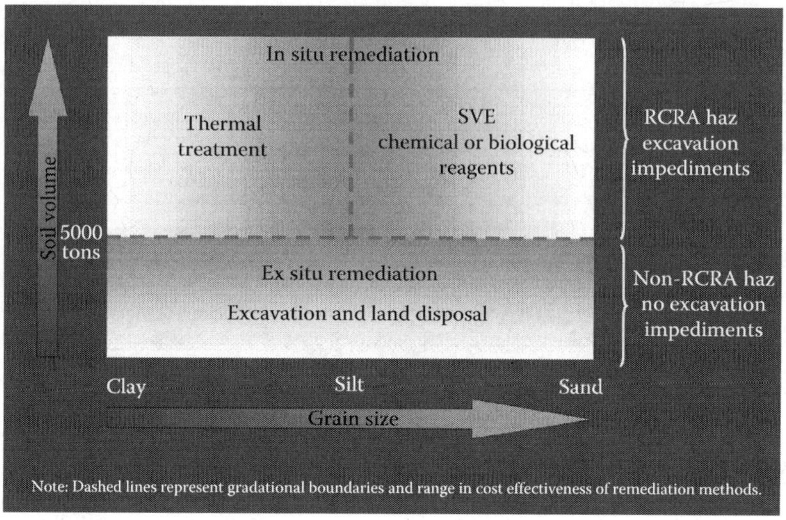

FIGURE 11.14
Selection of VOC soil remediation methods based on geology and volume.

probability of other technologies being selected increases primarily due to cost, time, or access considerations. As soil grain size increases from left to right, the preferred technologies change depending on the dominant geology of the site.

11.5 Common Groundwater Remediation Technologies

The first method used to remediate impacted groundwater was called **pump and treat**. Contaminated groundwater was pumped to the surface where it was treated to remove a large percentage, or all of the contamination. The treated water was then reinjected into the aquifer or discharged to the sanitary sewer system or surface water body. Early treatment technologies were adopted from existing methods used for industrial wastewater and publicly owned treatment plants. These techniques included solids removal, flocculation, bioreactors, activated carbon filtration, disinfection, and reverse osmosis. By the middle to late 1980s, it became apparent that pump and treat technologies applied to groundwater were very expensive.

Not only was pump and treat expensive, it was not broadly effective. Groundwater pump and treat was an effective means to ensure a contaminant plume did not spread if it were designed and operated correctly, but it was not cost effective in restoring an aquifer's groundwater to pre-contaminated levels. Since this discovery, several new methods termed *in situ technologies* have been developed. Instead of bringing the contaminated water to the surface to be treated, the treatment is delivered where the contamination resides—in the saturated zone. The goal of the in situ technologies was to destroy, chemically reduce, or transform the contaminants into harmless compounds within the plume of contaminants (USEPA 2007). These new technologies included

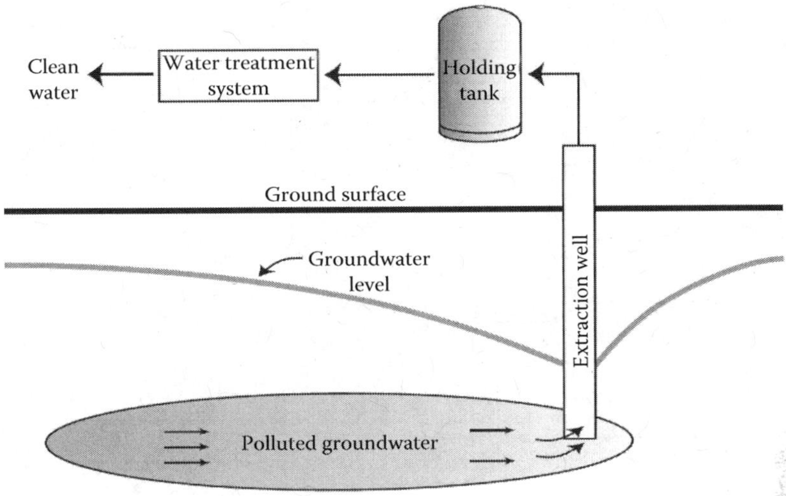

FIGURE 11.15
Groundwater pump and treat system. (From United States Environmental Protection Agency (USEPA), *A Citizen's Guide to Pump and Treat*, EPA 542-F-01-0025, Office of Solid Waste and Emergency Response, Washington, DC, 2001h.)

air sparging, installation of permeable reactive barriers (PRBs), injection of biological agents, and injection of chemical reagents.

Please do not bury pump and treat systems yet. Pumping and treating groundwater can be effective at removing significant quantities of contaminant mass under certain favorable conditions. One potential application of pump and treat is with young and immature groundwater plumes. These types of plumes may display only minor amounts of diffusion and their contaminants may not have reached the lower permeable zones (Figure 8.9). Here, a focused groundwater pump and treat system could be effective (Payne et al. 2008).

With pump and treat technologies, the method used for contaminant removal from groundwater depends on the type of contaminant. For instance, if the contaminants are VOCs, the contaminants may be removed by stripping the VOCs from groundwater using air in a vessel called an *air stripper.* Contaminants may also be removed by circulating the contaminated groundwater through containers filled with activated carbon. The VOCs in groundwater become attached or sorb to the activated carbon and are removed from the groundwater (Figure 11.15).

Other common groundwater remediation technologies include (USEPA 2002, 2007):

- In situ soil flushing
- Air sparging
- PRBs
- Injection of biological agents
- Injection of chemical agents
- Monitored natural attenuation
- Institutional controls
- Multiphase extraction

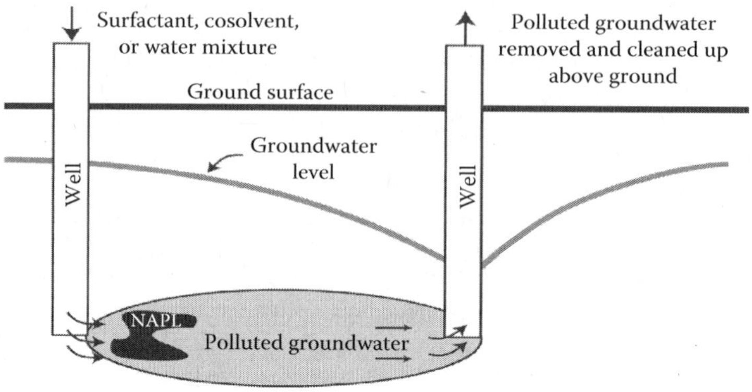

FIGURE 11.16
Example of in situ soil flushing. (From United States Environmental Protection Agency (USEPA), *A Citizen's Guide to Soil Flushing*, EPA 542-F-01-0011, Office of Solid Waste and Emergency Response, Washington, DC, 2001i.)

11.5.1 In Situ Soil Flushing

In situ soil flushing represents a modification to the traditional pump and treat system and is employed to increase contaminant removal efficiency. With this method, water or chemicals are pumped or percolated into saturated soil with the objective of flushing or driving the contaminants to a location where they can be removed, usually by a pumping well (Figure 11.16).

11.5.2 Air Sparging

Air sparging is an in situ groundwater remediation technology that uses injected air to volatize contaminants in groundwater. As the injected air rises through the saturated zone and reaches the unsaturated zone, the vapors containing the contaminants are removed from the ground using a SVE system (Figure 11.17).

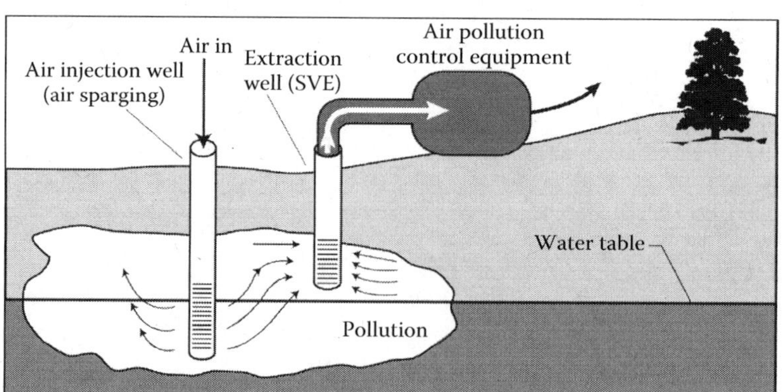

FIGURE 11.17
Example air sparging system. (From United States Department of Energy, *Remediation Technologies Screening Matrix and Reference Guide*, Version 4.0, Washington, DC, 2002.)

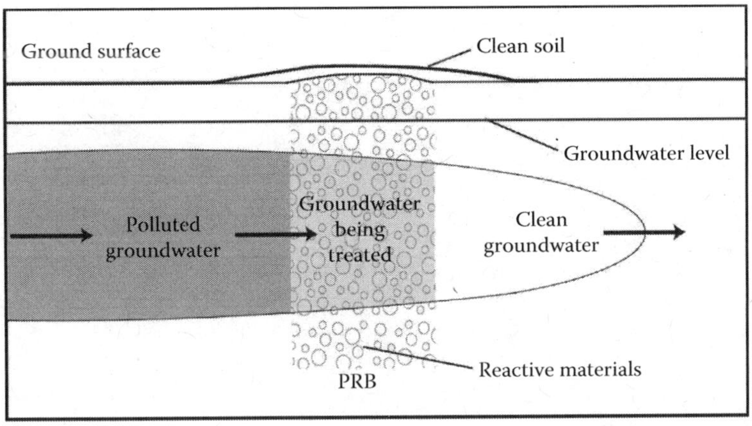

FIGURE 11.18
Example of a PRB. (From United States Environmental Protection Agency (USEPA), *A Citizen's Guide to Soil Vapor Extraction and Air Sparging*, EPA 542-F-01-006, Office of Solid Waste and Emergency Response, Washington, DC, 2001k.)

11.5.3 Permeable Reactive Barriers

A PRB is a wall or fence-like structure constructed beneath the surface of the ground within the saturated zone downgradient of groundwater contamination. The wall is composed of chemicals that degrade or destroy the targeted contaminants. PRBs are considered passive remedial systems, and generally take a long period of time to achieve remedial objectives because they rely on the natural flow of groundwater to pass through the reactive barrier. As contaminated groundwater migrates through the barrier, the contaminants are degraded or destroyed (USEPA 2001k) (Figure 11.18).

11.5.4 Injection of Biological Agents

Injection of biological agents employs technology similar to the remediation of contaminated soil with microorganisms. The difference lies in the delivery system. Here, instead of microorganisms, the injection consists of chemicals used to promote microbial growth. The chemicals used include oxygen or other food sources such as molasses or other carbohydrates (USEPA 2007).

11.5.5 Injection of Chemical Agents

Injection of chemical agents is similar to the injection of biological agents, but the objective is not to promote microbial growth. The intention is to create or promote a chemical reaction designed to transform or destroy the contaminants. This outcome can be accomplished by injecting compounds such as ferrous sulfate, calcium polysulfide, or zero-valent iron into groundwater contaminated with chromium VI. When the chromium VI comes into contact with any of the three agents it is reduced by a chemical reaction into chromium III. Chromium III is much less soluble in groundwater than chromium VI, resulting in the removal of chromium VI from groundwater and a lowering of its dissolved concentration (USEPA 2007).

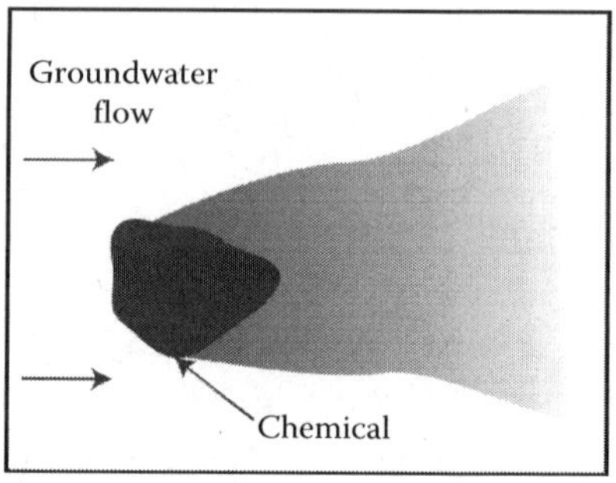

FIGURE 11.19
Example of natural attenuation of groundwater. (From United States Environmental Protection Agency (USEPA), *A Citizen's Guide to Monitored Natural Attenuation*, EPA 542-F-01-004, Office of Solid Waste and Emergency Response, Washington, DC, 2001b.)

11.5.6 Monitored Natural Attenuation

Monitored natural attenuation of contaminants in groundwater is similar to their natural attenuation in soil. Natural attenuation in groundwater, however, relies upon dilution to assist in lowering contaminant concentrations and can be effective with certain types of contaminants provided no continuing source of groundwater contamination exists. This method also relies on natural physical degradation and natural biodegradation to reduce contaminant concentrations (Figure 11.19).

11.5.7 Institutional Controls

Institutional controls are used to remove potential receptor pathways, and include the restriction or prohibition of groundwater use in impacted areas. These measures apply at locations where groundwater impacts are extensive, or when remediating groundwater is not practical or cost prohibitive. Institutional controls often require state and local municipality approval.

11.5.8 Multiphase Extraction

Some sites have high concentrations of contaminants present in the subsurface as pure product. In these cases, there is likely more than one contaminant phase present, including the vapor phase, dissolved phase, pure product or free phase. The initial effort is focused on removing the most contaminant mass as quickly as possible—the free product. Other phases in situ can be removed by the same extraction well. This combination provides increased efficiency and lowers the cost and time to remove large amounts of contaminant mass. The technique involves placing multiple types of extraction pumps into the same borehole to extract contaminants present in each phase (Figure 11.20).

11.6 Summary of Groundwater Remediation Technologies

Selecting the appropriate groundwater remediation technology follows a similar process as selecting the most appropriate soil remediation technology. The three most important factors are (1) cost, (2) the geology and hydrogeology where the contaminants reside, and (3) the physical chemistry of the contaminants targeted for remediation. As with soil remediation, if a mixture of different types of contaminants is present in groundwater,

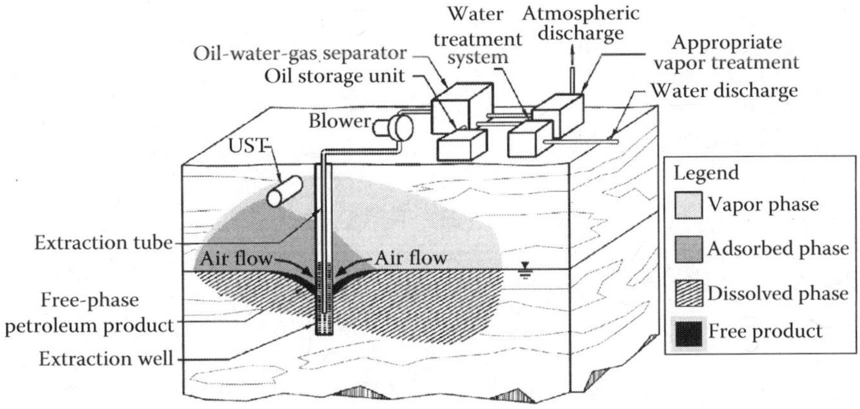

FIGURE 11.20
Example of a multiphase extraction well. (From United States Environmental Protection Agency (USEPA), *Cost and Performance Report for LNAPL Recovery*, EPA 542-R-05-016, Office of Solid Waste and Emergency Response, Washington, DC, 2005b.)

selecting more than one technology may be required to achieve the remedial objectives. With specific contaminants, certain technologies provide more efficient remediation. For most urban areas of the United States, Figure 11.21 shows the spectrum of geological conditions and contaminant types where certain remedial technologies are effective (Rogers et al. 2009). For instance, air sparging may be an appropriate technology for remediating LNAPL VOCs in groundwater if the soil is permeable enough (e.g., sands and gravels); however, even the presence of small lenses of fine sand or silt can severely limit the effectiveness of this technique (Peterson et al. 2001; Peterson and Murray 2003). Additionally, air sparging is not an appropriate technology for remediating chromium VI because chromium and most other metals do not evaporate.

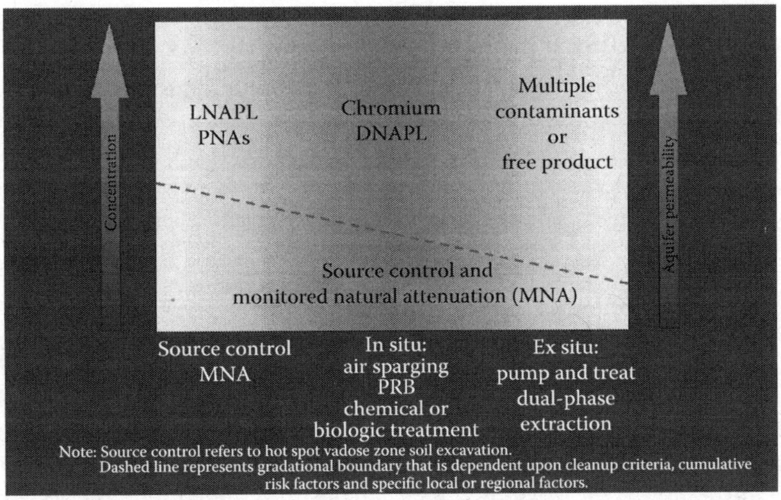

FIGURE 11.21
Remedial matrix of geology, contaminant type, and remedial technology.

Another factor to consider when examining the remediation of groundwater is the time required to complete the remediation. Remediating groundwater may require more than 20–30 years or more to complete. Factors influencing the time required and also likely to impact the cost to remediate groundwater include (USEPA 1999b; Rogers et al. 2006):

- Geology and hydrogeology: Various types of geology influence the technology selected and also dictate the rate of contaminant recovery. For example, low permeability aquifers may require longer remediation efforts.
- Contaminant concentration and distribution: Extensive and mature groundwater plumes are likely to have contaminants diffused into the lower permeable zones within the affected aquifer. In these situations, remediation time may be extended.
- Cleanup goals: Achieving lower residual contaminant concentrations with cleanup can require more time.
- Technology selected: PRBs often take longer because they are more passive remedial systems relying heavily on natural groundwater flow to deliver contaminants to the remediation area.
- Remedial-system design and maintenance: The rate of cleanup is one important factor here, but other factors including structural impediments such as buildings and roads must also be considered. In addition, the spacing of extraction or injection wells may not be optimized due to engineering issues and other uncontrollable circumstances, such as underground infrastructure.
- Whether there are multiple types of contaminants present: Contaminant plumes with multiple phases and types of contaminant (e.g., dissolved phase and free phase and DNAPS VOCs and metals), may have to be remediated in steps or phases and require additional time.
- Location: Urban areas often dictate or limit the technologies available for shortening the time required for groundwater remediation. For instance, buildings, roads, and other structures may lengthen the time required to investigate and conduct source control measures. As a result, groundwater remediation is initiated later and takes longer to complete.
- Investigative errors and delays: Failure to fully characterize the nature and extent of contamination can be disastrous and will result in an ineffective remediation attempt.
- Source control: If all the sources of contamination have not been identified and abated, remediation of groundwater may be ineffective and fail.

Selected groundwater remedies at Superfund sites are presented in Figure 11.22.

Groundwater pump and treat is the most common remedial method selected for the treatment of groundwater at Superfund sites (70% frequency). Other in situ methods, however, such as air sparging, PRBs, and chemical treatment are now being selected more frequently (USEPA 2007). This trend is due to an increase in the efficiency of more innovative technologies and their acceptance by remediation professionals. In addition, many groundwater pump and treat systems are expensive and have not achieved cleanup goals (USEPA 2007). Figure 11.23 shows a breakdown of other groundwater remediation technologies that are now being either selected or tested.

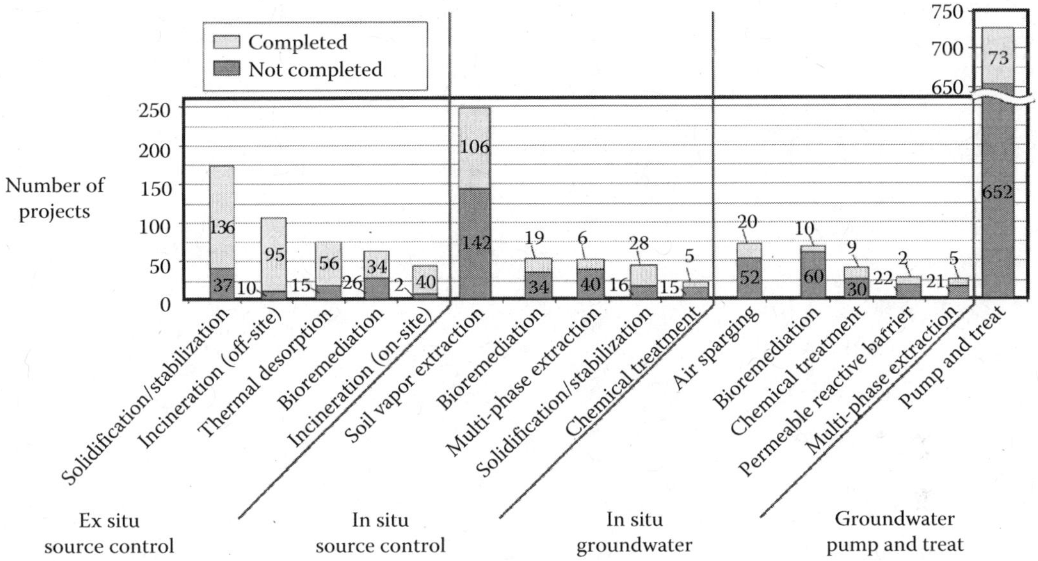

FIGURE 11.22
Selected remedies at Superfund sites. (From United States Environmental Protection Agency (USEPA), *Treatment Technologies for Site Cleanup: Annual Status Report*, 12th edn., EPA-542-R-07-012, Office of Solid Waste and Emergency Response, Washington, DC, 2007.)

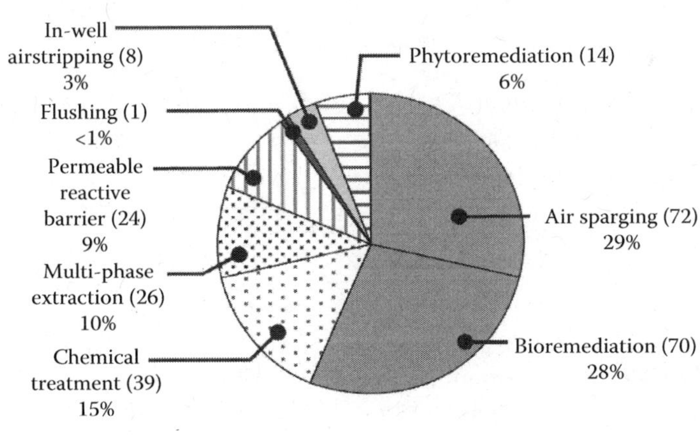

FIGURE 11.23
Breakdown of Superfund sites using in situ technologies. (From United States Environmental Protection Agency (USEPA), *Treatment Technologies for Site Cleanup: Annual Status Report*, 12th edn., EPA-542-R-07-012, Office of Solid Waste and Emergency Response, Washington, DC, 2007.)

11.7 Sediment Remediation

Remediating contaminated sediment presents difficulties not encountered when remediating soil. These difficulties include the location of the contamination beneath the surface of water bodies and within sensitive ecosystems. Remediating contaminated sediment therefore requires careful study and planning.

Common remedial methods for remediating contaminated sediment include (USEPA 2005b; Sediments Focus Group 2007):

- Source control
- Excavation
- Dredging
- Bioremediation
- Capping
- Natural attenuation
- Institutional control

11.7.1 Source Control

Source control may be more difficult to achieve with contaminated sediments, because the contamination may have a distant origin and/or emanate from numerous locations. To ensure recontamination does not occur, careful evaluation and study are necessary before initiating remedial activities (USEPA 2005c).

11.7.2 Excavation

Excavation of contaminated sediments is usually conducted within floodplains, stream and river banks, and beach areas when water does not cover the contaminated materials. Access to these areas is relatively easy when they are contaminated and require remediation, and excavation is usually preferred over other technologies if the volume of contamination is manageable. In these cases, the impacted sediments can be removed quickly before they become covered with water; otherwise excavation becomes potentially impossible and much more costly (USEPA 2005c).

11.7.3 Dredging

Dredging involves excavating submerged contaminated sediments. Multiple techniques and a variety of dredging equipment have been developed to address the array of situations encountered. Figure 11.24 shows a common dredging technique with an excavator on a floating barge (USEPA 2005c).

Dredging submerged sediment often causes contaminated sediment particles to become suspended in the water column. This may cause contamination to spread in the downgradient direction of water flow.

To avoid this situation, extreme measures are sometimes conducted (USEPA 2005c; Sediments Focus Group 2007):

- Temporarily draining the surface water requiring remediation
- Rerouting the surface water around the affected area in small streams or rivers requiring remediation
- Installing sheet pile walls around the affected areas and pumping the water from inside the sheet pile wall to provide access to the affected area

FIGURE 11.24
Example of dredging. (From United States Environmental Protection Agency (USEPA), Hudson River PCBs Superfund site, http://hunsondredgingdata.com/Monitoring/Water?currentweek=08-16-2002 (accessed June 26, 2010), 2010a.)

11.7.4 Bioremediation

Bioremediation of contaminated sediments involves the same processes as those used for bioremediating soil and groundwater. However, many contaminants often present in sediments such as PCBs and mercury are not readily remediated using microbes (USEPA 2005c).

11.7.5 Capping

Sometimes, no active remedial method is available to effectively remove all contaminated sediment without potentially casing more harm, as occurs with the suspension of sediments in the water column. In these situations, capping may be preferred. Capping contaminated and submerged sediments must be conducted slowly so any disturbance is kept to a minimum (USEPA 2005c).

11.7.6 Natural Attenuation

Natural attenuation may be an affective remedial option in the following situations: (1) the source(s) have been eliminated, (2) the contaminated sediments are not heavily impacted, and (3) other methods will not substantially reduce the risk posed by the presence of the contaminated sediments (USEPA 2005c).

11.8 Surface Water Remediation

The preferred approach to remediating surface water has been through the enactment of pollution prevention regulations. Since the National Pollutant Elimination System (NPDES) was enacted in 1972, significant improvements have been made in improving the surface waters of the United States (USEPA 2008a, 2010b). The NPDES process requires municipalities and industry to meet specified standards in order to discharge wastewater or stormwater to the surface waters of the United States (USEPA 2010b).

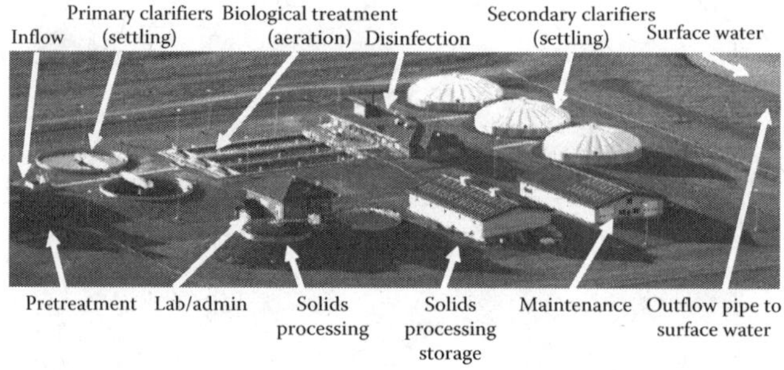

FIGURE 11.25
Wastewater treatment plant. (From Portage Lake Water and Sewer Authority, Houghton, Michigan, http://www.plwsa.org/index/php (accessed June 25, 2010), 2010.)

A significant component of the flows within many surface streams is the discharge from wastewater treatment plants (WWTP). An aerial view of a municipal WWTP is shown in Figure 11.25.

The level of treatment applied before wastewater is discharged to surface water depends on the concentration and types of contaminants present. The spectrum of wastewater treatment spans simple methods, such as constructing a settling pond to remove solids and suspended particles before treatment, to a very complex multistep process removing several contaminants before discharge. Figure 11.26 shows the multistep process used by municipal wastewater treatment systems involving the removal of solids and biological treatment before the water is discharged.

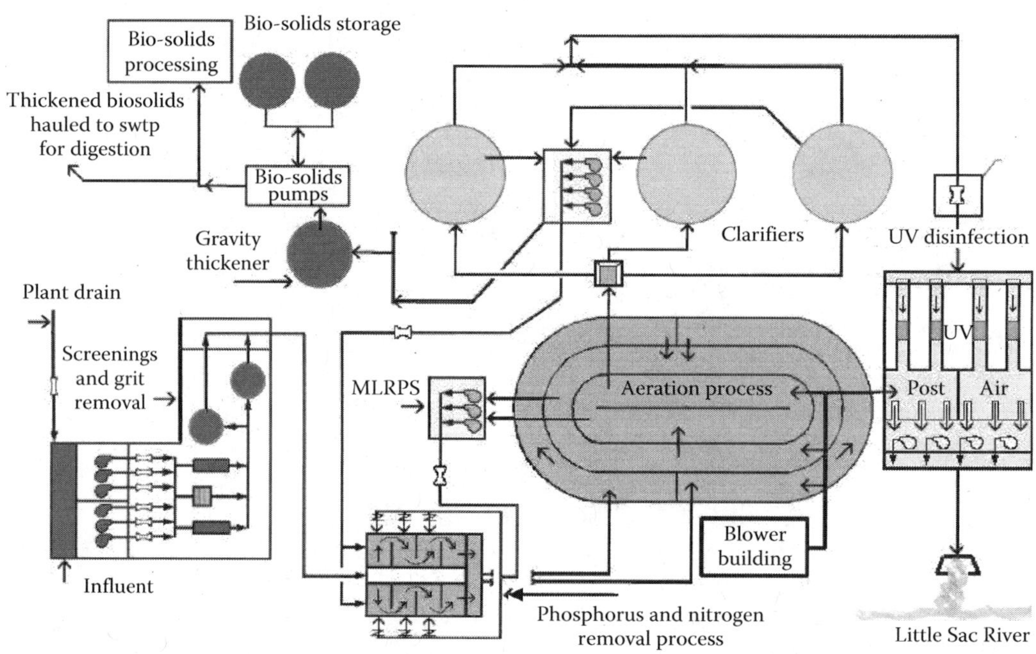

FIGURE 11.26
Schematic of a municipal wastewater treatment system. (From City of Springfield Missouri, Northwest Wastewater Treatment Plant, http://www.springfieldmo.gov/sanitary/nw_treatment.html (accessed November 27, 2010), 2010.)

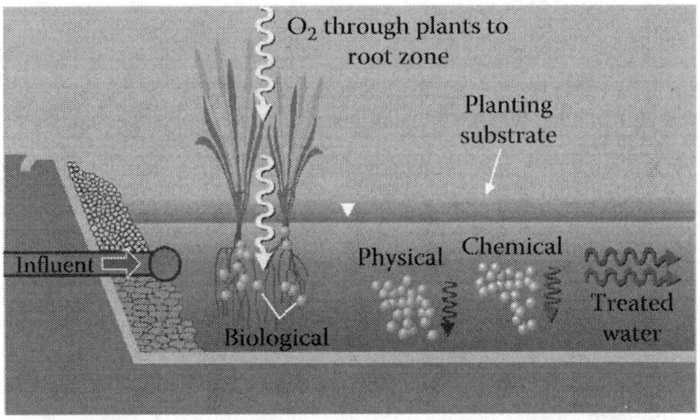

FIGURE 11.27
Simplified ENS for remediation of surface water. (From Eifert, W.H., *Simple, Reliable and Cost-Effective Solutions for Water Management and Treatment: A Technology Overview with Case Studies*, Roux Associates, New York, 2010b, p. 15. With permission.)

Engineered wetlands or Engineered Natural Systems (ENS) for water treatment is another effective method for the remediation of surface water under certain conditions. This process involves a combination of natural biological, chemical, and physical processes to remediate common surface water contaminants. Constructing an effective ENS requires matching the available degradation methods with the contaminant properties at the site, with a typical ENS utilizing several different natural methods to destroy contaminants including photolysis, bioremediation, phytoremediation, chemical degradation, settling, and many others. Figure 11.27 shows a simplified schematic of an ENS.

11.8.1 Spills

Remediating impacted surface water from spills is most effective when a quick response confines the release to as small an area as possible. Releases to surface water are most often related to petroleum products such as fuels, refined oil, and unrefined or crude oil. Common techniques for responding and addressing a surface water fuel or oil spill include (NOAA 2010):

- Doing nothing: If there is no possibility the spill will reach shore or sensitive areas and natural attenuation will likely remediate the spill through these natural processes: photolysis, biodegradation, physical dispersion through wave action and surface currents, and dilution.
- Containing the spill with booms and collecting the material with skimmer pumps. As noted in Chapter 7, fuels and oil are LNAPL compounds and therefore float. Using booms and skimmers can remove significant contaminant mass.
- Using dispersants to break up, spread, and enhance natural degradation. Dispersants lower the surface tension of LNAPL compounds and can help reduce the environmental impacts.
- Introducing biological agents: Adding biological agents such as carbohydrates may hasten degradation of the affected area.

Figures 11.28 and 11.29 show the remedial response activities to a small oil spill on a river. Here, containment and cleanup of the spill is accomplished with booms and adsorbent pads.

FIGURE 11.28
Containment booms and adsorbent pads. (Photo by Daniel T. Rogers.)

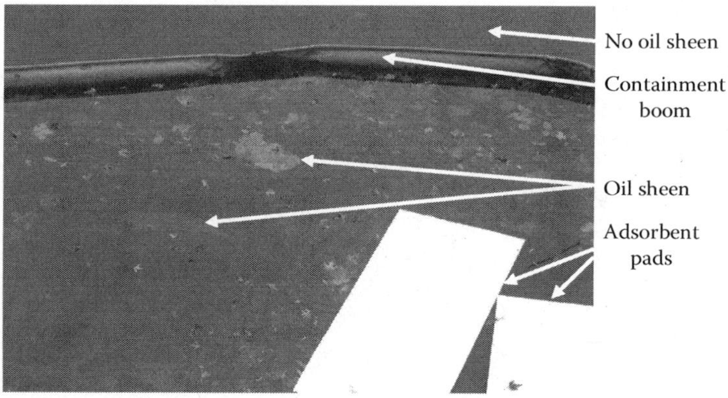

FIGURE 11.29
Oil sheen containment booms and adsorbent pads. (Photo by Daniel T. Rogers.)

As noted in Chapter 3, groundwater often discharges to surface water and accounts for much of the baseflow in surface streams. Surface water can therefore become contaminated from impacted groundwater discharging to surface water. This event, shown in Figure 11.30, is often poorly understood and underestimated as a pollutant source to surface water (Murray et al. 1997; Rogers 1997; Rogers and Murray 1997; USEPA 2000).

When pollution originates from dispersed locations as in Figure 11.30, it is referred to as nonpoint source pollution (Novotny 2003). Remediation of nonpoint source pollution is challenging and can only be effective if

- A monitoring network is established.
- Continuous monitoring is conducted using mass balance techniques to establish what fraction of the total contaminant load originated from nonpoint sources.
- The geology and hydrogeology of a region or watershed is well understood. Knowing the characteristics of baseflow is critical (e.g., recharge and discharge areas).
- Potential sources of pollution are identified across three dimensions, including soil, groundwater, surface water, the atmosphere, and any human activities.

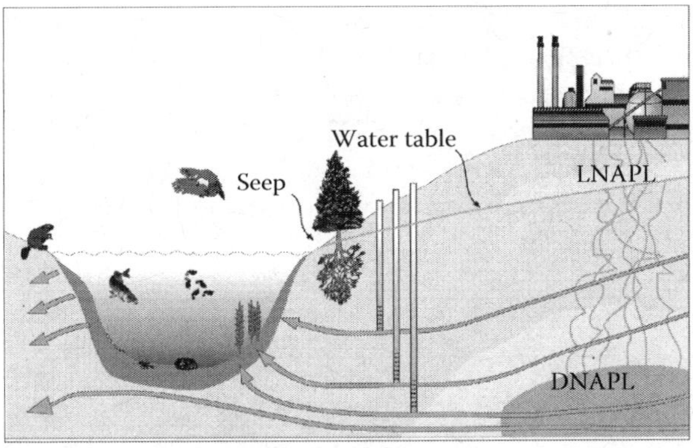

FIGURE 11.30
Contaminated groundwater migrating toward surface water. (From United States Environmental Protection Agency (USEPA), *Proceedings of the Ground-Water Surface-Water Interactions Workshop*, EPA/542/R-00-007, Office of Solid Waste and Emergency Response, Washington, DC, 2000.)

- There is complete identification of the contaminants. Identifying potential contaminants will assist with where and how to monitor potential contaminant pathways (e.g., shallow groundwater, deep groundwater, air, or sediment).
- Urban factors, including stormwater discharge and modifications to the natural geological setting are assessed and measured.
- Historical modifications to the urban environment such as redevelopment and landfilling are accounted for.
- Pollution prevention programs and initiatives are created to minimize or eliminate future impacts.
- There is adequate public awareness, and the changes required are convenient to implement. A relevant example is the growth of recycling programs due to the increased convenience of curbside recycling.

11.9 Air Remediation

As with surface water, the preferred approach to cleaning up polluted air is pollution prevention. When accidental releases occur, they are addressed through emergency response actions involving evacuation in the downwind direction, air monitoring, and source control (e.g., stopping the release).

Releases of air contaminants are regulated under the Clean Air Act administered by the USEPA and enforced by state and local agencies. Many states have also enacted their own stricter regulations. Air pollution is regulated two ways: (1) through a process requiring numerous operations to obtain a permit or (2) through inspection, such as motor vehicle emission inspections and testing, which confirm emission standards are not exceeded. Emitters of air pollution may be regulated by both methods. Industrial operations emitting atmospheric pollutants must obtain a permit, and may be required to install air pollution control devices to capture pollutants before they are released to the atmosphere (USEPA 2008b).

There are two categories of air pollutant sources: (1) mobile sources, such as automobiles, trucks, buses, farm machinery, and airplanes and (2) stationary sources, such as industrial facilities, chemical production facilities, pharmaceutical companies, and refineries (USEPA 2008b). Motor vehicle exhaust accounts for 90% of air contamination (USEPA 2008b); however, there have been significant improvements in air quality over the last few decades. Programs and incentives used to reduce the amount and type of contaminants from motor vehicle exhaust include

- Increasing fuel efficiency
- Decreasing the amount of emissions
- Installing air pollution control devices on vehicles (such as catalytic converters)
- Increasing public awareness (e.g., driving less and not refueling during critical periods)
- Reformulating fuels (e.g., eliminating the use of lead as an additive)
- Developing more efficient engines
- Requiring inspections and routine maintenance, if necessary (e.g., vehicle emissions inspections)

Stationary sources are divided into two categories: gaseous contaminants and particulate matter. Removal of contaminants from the gaseous phase is accomplished through various technologies, including (USEPA 2008b):

- Contact condenser
- Surface condenser
- Thermal incinerator
- Catalytic incinerator

Removal of particulates is usually accomplished with the installation of a dust collector (also referred to as a baghouse), or by a similar device called a wet scrubber. The process of collecting dust (particulates) using a baghouse involves capturing the particulate matter at or very near the emission source by placing the source under a vacuum. The particulates are captured using a filter before the air stream is exhausted to the atmosphere. Capture efficiencies are very high and routinely exceed 95% (USEPA 2008b) Figure 11.31.

11.10 Cost of Remediation

Remediation is costly, and not only in the financial sense. Remediation takes time, patience, and scientific understanding. Some of the factors controlling whether remediation is conducted, how it is conducted, and the resulting costs are (Rogers et al. 2006, 2009):

- Geology: Different types of geology can influence the extent of contamination and methods of remediation.
- Hydrogeology: The presence of groundwater often plays a significant role in assessing risk, migration pathways, and the selection of appropriate technology.

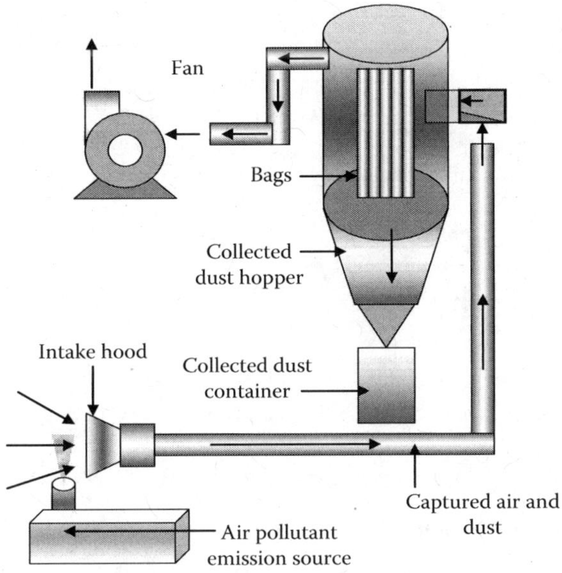

FIGURE 11.31
Generalized schematic of a dust collector.

- Contaminant type: The type of contaminants influences migration pathways, mobility, risk, and cleanup targets.
- Nature and extent: Nature and extent of contamination impacts receptor analysis and potential routes of exposure and cleanup targets.
- Contaminant mass and concentration: The mass and concentration of contaminants and the potential presence of free phase influences remedial technology selection.
- Medias to be remediated (e.g., soil, groundwater, surface water, or sediment): Some media are more difficult and costly to remediate than others.
- Land use: Residential land typically requires the most stringent cleanup targets.
- Cleanup targets: Contaminant type, location, and potential receptors and exposure pathways exert the most influence over establishing cleanup targets.
- Technology selected: Some technologies may be more costly than others (i.e., pumping and treating groundwater vs. natural attenuation).
- Time: Time may influence cost, especially if the release involves actual human exposure or migration to a sensitive ecological receptor that requires immediate remedial action.
- Legal action: Litigation can delay cleanup efforts and lead to more costly remediation due to delays and potential disputes over technology types.

To evaluate whether certain contaminants cost more to remediate than others and the effects of geology on remediation, 280 remedial actions conducted at sites of environmental contamination were investigated (Rogers et al. 2006, 2009). A few characteristics of each site included

- Of the 280 sites, 246 were located in the United States and 34 were located in other countries.
- Sites were located in 34 different states: AL, AK, AR, CA, CO, DE, FL, GA, IL, IN, IA, KS, KY, MA, MD, MI, MN, MO, MS, NC, NH, NJ, NV, NY, OH, OK, OR, RI, TN, TX, UT, VA, WA, WI.
- International sites were located in 14 countries: Australia, Belgium, Brazil, Canada, China, England, France, Germany, Italy, Malaysia, Mexico, Russia, South Africa, Ukraine.
- Property size ranged from 1.8 ac to more than 300 ac.
- All were located in urban areas.
- Many of the facilities operated for more than 100 years.
- Operations at most facilities included heavy manufacturing.

The data collected from each site included (Rogers et al. 2006, 2009)

- Contaminant discovery facts
- Estimated date of release
- Release circumstances
- Release operation location
- Source of the release
- Contaminant(s) of concern
- Media(s) impacted
- Media(s) remediated
- Surface and subsurface geology
- Topography
- Surface and subsurface hydrological setting
- Fate and transport analysis
- Analysis of potential receptors
- Remedial technology selected
- Contaminant mass remediated
- Contaminant mass remaining in place
- Timeframe from discovery to remediation
- Cost of investigation
- Cost of remediation
- Projected future cost to closure if not yet achieved
- Project future timeframe to closure if not yet achieved
- Regulatory involvement

Contaminants remediated included Arsenic, Chromium VI, Lead, Mercury, Perchlorate, Chlordane, LNAPL VOCs, DNAPL VOCs, PAHs, and PCBs. Several other contaminants were detected, but were not at sufficiently high enough concentrations to be the focus or target of a remedial action. Some of these contaminants also occur naturally. Concentrations detected, however, exceeded background concentrations and were therefore considered anthropogenic.

These contaminants included antimony, barium, beryllium, cadmium, copper, manganese, nickel, selenium, silver, zinc, phenols, phthalates, other SVOCs, and asbestos.

Of the 280 sites evaluated

- 193% or 69% had soil contamination only.
- 87% or 31% had soil and groundwater contamination.
- 21% or 7% had significant free product.

All the sites were located on unconsolidated sediments consisting of glacial, fluvial, lacustrine, alluvial, or marine sediments composed of gravel, sand, silt, clay, or mixtures of these materials.

11.10.1 Results

For the contaminant groups, Table 11.1 summarizes their frequencies of detection and required remediation, and percent of the total cost. Table 11.2 contains the number of sites where each contaminant group was detected, whether groundwater was affected, if free product was present, as well as the cost of the remediations. The data in Table 11.1 indicate that PAHs were the group of compounds most often detected but were remediated just one-third of the time, and accounted for only 1% of the total cost. Although PAHs were often detected, remediation was not always required, and, when remediation was required, the costs were lower than anticipated (Table 11.2). As shown in Tables 11.1 and 11.2, LNAPL VOCs had a frequency of remediation equivalent to PAHs, but at a much higher cost. Over 90% of the cost to remediate LNAPL VOCs was associated with free product.

Tables 11.3 and 11.4 show the number of sites and selected frequency of the remedial technologies where soil and groundwater were remediated, respectively. As depicted in Table 11.3, excavation of soil was the preferred remedial option, especially when the volume of soil was less than 5000 tons and there was no impediment to excavation. When there was a large volume of impacted soil or excavation impediments existed, SVE was the preferred remedial option for VOCs in course-grained soils.

A combination of alternatives including, risk assessment, institutional controls, capping, and thermal treatment were the preferred alternatives for large volumes of fine-grained

TABLE 11.1

Summary of Frequency of Detection, Remediation, and Cost

Contaminant	Frequency of Detection (%)	Frequency of Remediation (%)	Percent of Total Cost
PAHs	83	26.6	1.1
LNAPL VOCs	60	25.9	44.4
DNAPL VOCs	33.4	24.1	37.7
Lead	20	9.0	1.3
PCBs	8.3	5.0	2.5
Chromium VI	10	4.5	10.0
Mercury	4	2.5	1.8
Perchlorate	0.7	0.8	0.3
Arsenic	5.4	0.8	0.8
Chlordane	1.5	0.8	0.1
Total		100	100

TABLE 11.2

Site Breakdown by Contaminant Type and Cost

Contaminant	Total Number of Sites	Number of Groundwater Sites	Number of Free Product Sites	Cost (Million Dollars)
PAHs	76	8	6	6.4
LNAPL VOCs	72	33	15	253.9
DNAPL VOCs	67	37	0	215.9
Lead	25	0	0	7.4
PCBs	14	0	0	14.6
Chromium VI	12	6	0	57.2
Mercury	7	0	0	10.8
Perchlorate	3	2	0	1.1
Arsenic	2	1	0	4.8
Chlordane	2	0	0	0.2
Total	280	87	21	572.3

TABLE 11.3

Breakdown of Soil Remediation

Soil Remediation Technology	Number of Sites	Frequency Selected (%)
Excavation	259	93.5
Risk assessment	85	32.8
Institutional controls	70	25.1
SVE	49	17.6
Capping	48	17.2
Thermal treatment	5	1.8
In situ chemical treatment	5	1.8

TABLE 11.4

Breakdown of Groundwater Remediation

Soil Remediation Technology	Number of Sites	Frequency Selected (%)
Risk assessment	73	84.1
Institutional controls	63	72.4
In situ chemical treatment	41	47.1
Pump and treat	35	40.2
MNA or NA	31	35.6
Air sparging	8	9.2
Thermal treatment	2	2.4

soil or contaminated soil with excavation impediments. The multiple use of remedial technologies was also selected when groundwater was impacted, especially with Chromium VI or DNAPL. As shown in Table 11.4, groundwater pump and treat is not currently a preferred remedial method. The exception here occurs when there is free product or an incompatible overlapping multicompound contaminant plume (e.g., heavy metals and DNAPL). In situ chemical treatment is currently the preferred remedial method for chromium VI and DNAPL contaminated groundwater (e.g., calcium polysulfide, ferrous sulfate).

TABLE 11.5

Total Cost and Mass Remediated by Media

Remediated Media or Phase	Percentage of Total Cost	Percentage of Total Mass Recovered
Soil	24.3	17.6
Groundwater	41.6	2.0
Free product	34.1	80.4
Total	100	100

Table 11.5 displays the percentages of total cost and mass recovered by the media remediated, and also highlights the cost inefficiency of remediating groundwater. Although 41% ($238 million) of the total cost was devoted to remediating groundwater, only 2% of the total contaminant mass was recovered. This cost disparity highlights the difficulty of cost effective remediation once groundwater has become impacted.

Table 11.6 shows the cost to remediate a kilogram of contaminant by soil type. Data on arsenic, chlordane, mercury, and perchlorate are not provided due to the lack of an adequately sized data set. Figure 11.32 displays these data graphically.

TABLE 11.6

Cost of Remediating a Kilogram of Contaminant by Soil Type

Contaminant	Geology	Remedial Cost ($/kg)
Chromium VI	Clay	1,000
	Silty clay	1,500
	Sandy and silty clay	1,500
	Sand	77,633
DNAPL	Clay	721
	Silty clay	1,366
	Sandy and silty clay	3,263
	Sand	66,024
LNAPL	Clay	319
	Silty clay	518
	Sandy and silty clay	778
	Sand	527
PCBs	Clay	4,700
	Silty clay	4,800
	Sandy and silty clay	5,500
	Sand	5,700
PAHs	Clay	922
	Silty clay	843
	Sandy and silty clay	303
	Sand	442
Lead	Clay	68
	Silty clay	230
	Sandy and silty clay	538
	Sand	445

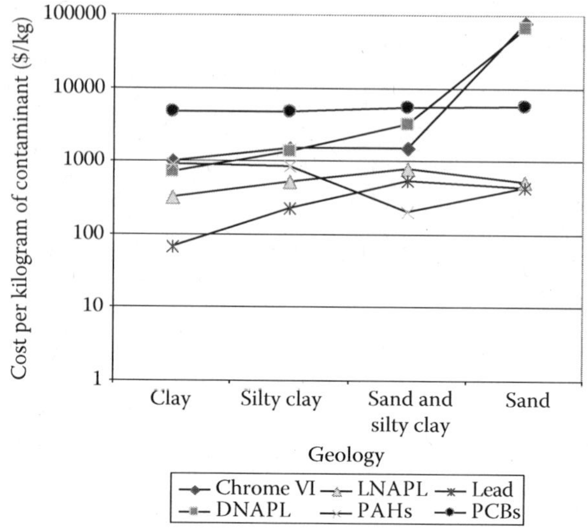

FIGURE 11.32
Remediation cost per kilogram of contaminant by soil type.

Figure 11.32 shows that remediation costs are highest for DNAPL and Chromium VI. The figure also illustrates how contaminants with a high CRF_{GW} released into a vulnerable geologic environment produce a synergistic effect. The cost of remediating chromium VI and DNAPL VOCs in geologically vulnerable areas results in a cost of remediating a kilogram of contaminant that is greater than $77,000 and $66,000, respectively. The cost of remediating chromium VI and DNAPL VOCs in the soils with lower geological vulnerability is over 50 times less.

The cost to remediate a kilogram of PCBs is the third highest for the contaminants evaluated. This elevated cost results from two factors: (1) PCB remediation is heavily regulated and requires intensive investigation and review and (2) PCBs are occasionally detected in sediments requiring a remedial action.

LNAPL VOCs, lead, and PAHs are not influenced by geology to the extent exhibited by chromium VI and DNAPL VOCs. These contaminants do not exhibit a sufficiently high mobility or persistence, which results in a lower CRF_{GW}. As a result, these contaminants are more cost effective to remediate on a per kilogram basis.

Overall, cost effectiveness in remediating groundwater has improved, but still remains high. On average, groundwater is six times more expensive to remediate than soil, and this cost differential increases to a factor of 10 when remediating contaminated groundwater with Chromium VI or DNAPL. The time required for remediation is a key factor contributing to the cost differential between soil and groundwater. When only soil was contaminated, the average time required to investigate and remediate a site was 2 years; if groundwater was contaminated, the same operations took 13 years.

11.11 Summary and Conclusion

This chapter has highlighted the different remedial approaches for cleaning up our air, water, and soil. The approaches taken for remediating the air and surface water are heavily weighted toward pollution prevention and permitting, and they have been effective at significantly reducing the amount of contaminants entering the air and surface waters of the United States.

An expanding variety of technologies now exist for remediating soil, water, sediments, and air. The most important factor, however, for ensuring a successful remediation effort is the proper characterization of the site. Investigators must identify all sources of contamination, produce accurate estimates of contaminant mass, and follow sound sampling procedures. Technology alone cannot rescue a poor game plan.

Sites where groundwater is contaminated are six times more expensive to remediate than those where soil alone is affected. Analysis of a large sample of contaminated sites revealed there are very high costs associated with remediating sites where the contaminant of concern was chromium VI and DNAPL VOCs. A synergistic effect is realized when these contaminants are released in a geologically vulnerable area. The remedial cost patterns at sites of environmental contamination also demonstrate the importance and predictive power of using Contaminant Risk Factors (CRFs) to assist in future sustainable development and redevelopment of urban areas across the United States.

Pollution prevention techniques implemented on a watershed-wide and site-specific basis are a logical first step toward sustainable development. Consideration of the geology and contaminant-specific physical chemistry can help determine what chemicals should be used, where they should be used, and how they should be used. In the next chapter, we present guidelines for achieving sustainable urban watersheds by building on the scientific foundation of geology and contaminant management.

References

American Society for Testing Materials (ASTM) 1995. *Standard Guide for Risk-Based Corrective Action*. ASTM Publication E1739-95. West Philadelphia, PA.

Cauwenberghe, L.V. 1997. *Electrokinetics*. Ground-Water Remediation Technologies Analysis Center. Pittsburgh. PA.

City of Springfield, Missouri. 2010. Northwest Wastewater Treatment Plant. http://www.springfieldmo.gov/sanitary/nw_treatment.html (accessed November 27, 2010).

Eifert, W. 2010a. *Engineered Natural Systems: A Technology Overview with Case Studies*, p. 15. New York: Roux Associates.

Eifert, W.H. 2010b. *Simple, Reliable and Cost-Effective Solutions for Water Management and Treatment: A Technology Overview with Case Studies*. New York: Roux Associates.

Federal Remediation Technologies Roundtable. 2010. Soil vapor extraction. http://www.frtr.gov/matrix2/section1/list-of-fig.html (accessed June 24, 2010).

Murray, K.S. and D.C. Clark. 1993. Treatment cell concept for the bioremediation of hydrocarbon contaminated soils. In *Principles and Practices for Diesel Contaminated Soils*, Vol. II, ed. P.T. Kostecki, E.J. Calabrese, and C. Barkan, pp. 83–95. Amherst, MA: Association for the Environmental Health of Soils.

Murray, K. S., J.B. McNeal, and S.B. McNeal. 1997. Air treatment galleries and soil venting: a solution to soil and groundwater contamination at remote sites. *Environmental Protection* 8:24–28.

Murray, K.S., A. Farkas, J. Heiden et al. 1997. Surface water quality analysis. Rouge River, southeast Michigan. *Michigan Academician* 29:159–171.

Murray, K.S., D.T. Rogers, and M.M. Kaufman. 2004. Heavy metals in an urban watershed in southeastern Michigan. *Journal of Environmental Quality* 33:163–172.

National Oceanic and Atmospheric Administration (NOAA). 2010. Spill response and restoration. Office of Response and Restoration. NOAA's National Ocean Service. http://response.retoration.noaa.gov (accessed January 1, 2010).

Novotny, V. 2003. *Water Quality: Diffuse Pollution and Watershed Management*. NY: John Wiley & Sons.

Payne, F.C., J.A. Quinnan, and S.T. Potter. 2008. *Remediation Hydraulics*. Boca Raton, FL: CRC Press.

Peterson, J.W. and K.S. Murray. 2003. Grain-size heterogeneity and subsurface stratification in air sparging of dissolved-phase contamination: Laboratory experiments-field implications. *Environmental and Engineering Geoscience* 9:71–82.

Peterson, J.W., K.S. Murray, Y.I. Tulu et al. 2001. Air-flow geometry in air-sparging of fined-grained sands. *Hydrogeology Journal* 9:168–176.

Portage Lake Water and Sewer Authority. 2010. Houghton, Michigan. http://www.plwsa.org/index/php (accessed June 25, 2010).

Rogers, D.T. 1997. The influence of groundwater on surface water in Michigan's Rouge River watershed. *Proceedings of the American Water Resources Association, Conjunctive Use of Water Resources; Aquifer Storage and Recovery*, pp. 173–180. Long Beach, CA.

Rogers, D.T. and K.S. Murray 1997. Occurrence of groundwater in Metropolitan Detroit. In *Groundwater in the Urban Environment*, ed. J. Chilton et al., pp. 155–161. Rotterdam, FL: Balkema.

Rogers, D.T., M.M. Kaufman, and K.S. Murray. 2006. Improving environmental risk management through historical impact assessments. *Journal of the Air and Waste Management Association* 56:816–823.

Rogers, D.T., M.M. Kaufman, and K.S. Murray. 2008. *Empirical Analysis of Contaminant Risk at Brownfield Sites*. United States Environmental Protection Agency National Brownfields Conference Proceedings. Detroit, MI.

Rogers, D.T., M.M. Kaufman, and K.S. Murray 2009. *An Analysis of Remedial Technology Effectiveness at Brownfield Sites*, abstract, p. 525. Association of American Geographers Annual Meeting. Las Vegas, NV.

Sediments Focus Group. 2007. *Guide to the Assessment and Remediation of State-Managed Sediment Sites*. Association of State and Territorial Solid Waste Management Officials. Washington, DC: USEPA.

United States Department of Energy. 2002. *Remediation Technologies Screening Matrix and Reference Guide*. Version 4.0. Washington, DC.

United States Environmental Protection Agency (USEPA). 1995. *Remediation Technologies for LNAPL Contaminated Groundwater*. Office of Underground Storage Tanks. Washington, DC.

United States Environmental Protection Agency (USEPA). 1998. *Technical Protocol for Evaluating Natural Attenuation of Chlorinated Solvents*. EPA/600/R-98/128. Office of Research and Development. Washington, DC.

United States Environmental Protection Agency (USEPA). 1999a. *Phytoremediation Resource Guide*. EPA 542-B-99-003. Office of Solid Waste and Emergency Response. Washington, DC.

United States Environmental Protection Agency (USEPA). 1999b. *Groundwater Cleanup: Overview of Operating Experience at 28 Sites*. EPA 542-R-99-006. Office of Solid Waste and Emergency Response. Washington, DC.

United States Environmental Protection Agency (USEPA). 2000. *Proceedings of the Ground-Water Surface-Water Interactions Workshop*. EPA/542/R-00-007. Office of Solid Waste and Emergency Response. Washington, DC.

United States Environmental Protection Agency (USEPA). 2001a. *A Citizen's Guide to Bioremediation*. EPA 542-F-01-001. Office of Solid Waste and Emergency Response. Washington, DC.

United States Environmental Protection Agency (USEPA). 2001b. *A Citizen's Guide to Monitored Natural Attenuation*. EPA 542-F-01-004. Office of Solid Waste and Emergency Response. Washington, DC.

United States Environmental Protection Agency (USEPA). 2001c. *A Citizen's Guide to Soil Washing*. EPA 542-F-01-008. Office of Solid Waste and Emergency Response. Washington, DC.

United States Environmental Protection Agency (USEPA). 2001d. *A Citizen's Guide to In Situ Thermal Treatment*. EPA 542-F-01-0012. Office of Solid Waste and Emergency Response. Washington, DC.

United States Environmental Protection Agency (USEPA). 2001e. *A Citizen's Guide to Fracturing*. EPA 542-F-01-0015. Office of Solid Waste and Emergency Response. Washington, DC.

United States Environmental Protection Agency (USEPA). 2001f. *A Citizen's Guide to Vitrification*. EPA 542-F-01-0015. Office of Solid Waste and Emergency Response. Washington, DC.

United States Environmental Protection Agency (USEPA). 2001g. *A Citizen's Guide to Chemical Dehalogenation*. EPA 542-F-01-0010. Office of Solid Waste and Emergency Response. Washington, DC.

United States Environmental Protection Agency (USEPA). 2001h. *A Citizen's Guide to Pump and Treat*. EPA 542-F-01-0025. Office of Solid Waste and Emergency Response. Washington, DC.

United States Environmental Protection Agency (USEPA). 2001i. *A Citizen's Guide to Soil Flushing*. EPA 542-F-01-0011. Office of Solid Waste and Emergency Response. Washington, DC.

United States Environmental Protection Agency (USEPA). 2001j. *A Citizen's Guide to Soil Vapor Extraction and Air Sparging*. EPA 542-F-01-006. Office of Solid Waste and Emergency Response. Washington, DC.

United States Environmental Protection Agency (USEPA). 2001k. *A Citizen's Guide to Permeable Reactive Barriers*. EPA 542-F-01-005. Office of Solid Waste and Emergency Response. Washington, DC.

United States Environmental Protection Agency (USEPA). 2002. *A Citizen's Guide to Cleanup Methods*. EPA 542-F-01-007. Washington, DC.

United States Environmental Protection Agency (USEPA). 2003. *Evapotranspiration Landfill Cover Systems Fact Sheet*. EPA 542-F-03-015. Office of Solid Waste and Emergency Response. Washington, DC.

United States Environmental Protection Agency (USEPA). 2004. *Cleaning Up the Nation's Waste Sites: Markets and Technology Trends*. 2004 edn. USEPA. EPA-542/R-04/101. Washington, DC.

United States Environmental Protection Agency (USEPA). 2005a. *Road Map to Understanding Innovative Technology Options for Brownfield Investigation and Cleanup*, 4th edn. EPA 542-B-05-001. Office of Solid Waste and Emergency Response. Washington, DC.

United States Environmental Protection Agency (USEPA). 2005b. *Cost and Performance Report for LNAPL Recovery*. EPA 542-R-05-016. Office of Solid Waste and Emergency Response. Washington, DC.

United States Environmental Protection Agency (USEPA). 2005c. *Contaminated Sediment Remediation Guidance for Hazardous Waste Sites*. EPA-540-R-05-012. Office of Solid Waste and Emergency Response. Washington, DC.

United States Environmental Protection Agency (USEPA). 2007. *Treatment Technologies for Site Cleanup: Annual Status Report*, 12th edn. EPA-542-R-07-012. Office of Solid Waste and Emergency Response. Washington, DC.

United States Environmental Protection Agency (USEPA). 2008a. *EPA's Report on the Environment*. EPA/600/R-07/045F. Washington, DC.

United States Environmental Protection Agency (USEPA). 2008b. *Latest Findings on National Air Quality: Status and Trends through 2006*. EPA454/R-07-007. Research Triangle Park, NC.

United States Environmental Protection Agency (USEPA). 2010a. Hudson River PCBs Superfund site. http://hunsondredgingdata.com/Monitoring/Water?currentweek=08-16-2002 (accessed June 26, 2010).

United States Environmental Protection Agency (USEPA). 2010b. National Pollutant Discharge Elimination System (NPDES). http://www.epa.gov/npdes/index.cfm (accessed January 5, 2010).

Part III

Sustainable Development

12

Urbanization and the Disruption of Matter and Energy Flows within Watersheds

12.1 Introduction

The preceding chapters have emphasized that any effort to attain sustainable urban watersheds must include a thorough investigation and understanding of the interactions between geology and contamination. Since nature does not isolate geology and contamination from other physical processes such as the hydrologic cycle, it is important to investigate the linkages between geology and contamination to other processes occurring within urban watersheds. This will help ensure when planning measures are implemented they are informed by the broadest foundation of the applicable science.

Watersheds are physical systems, consisting of interconnected components performing work over a defined geographic scale. Much of the work watershed systems accomplish involves the transport of soil and nutrients by wind and water. The energy required for these tasks comes from the sun and goes through numerous conversions. Before human activities impacted watersheds, their balances of matter and energy were in dynamic equilibrium. Occasionally, fluctuations occurred during floods and other disturbances, but, over time, the streams within the watersheds returned to a condition where they flowed along a gradient reflecting the balance between their erosion, transport, and deposition. Over many years, an overhead view of the streams in this balanced state would reveal local changes in response to natural fluctuations in sediment load and runoff—but overall the channel would maintain its average morphology. A balance between the recharges from precipitation and discharges to surface water also kept the aquifer systems in a long-term balance.

With the advent of large-scale agriculture and urbanization, the flows of matter and energy into and within watersheds changed dramatically. Sediment loads into urban streams increased significantly as soil was dislodged from construction sites (Wolman and Schick 1967), and floods occurred more frequently as urban development created more impervious surface and greater volumes of runoff (Klein 1979). Water quality also declined from the discharges of factories, urban and agricultural runoff, automobile exhaust, contributions from contaminated groundwater, and atmospheric deposition. The severity of pollution in North American streams was symbolized by the Cuyahoga River in Ohio catching fire in 1969.

In the United States, the response to these problems of excess flows of matter and energy and to the threats caused by pollutants has been significant. Erosion and sedimentation controls have been implemented at the state and local levels. Developing property within sensitive floodplain zones has become more difficult, and the Clean Water Act passed in 1972 created an effective mechanism for controlling pollution discharges from point sources. There is, however, a long way to go before our streams and aquifers achieve anything close to the type of balance found in predevelopment landscapes.

To broaden the scientific foundation necessary for sustainable planning efforts, this chapter builds on the basic concepts of watersheds introduced in Chapter 3 by beginning with a description of the structure and function of watersheds. An overview of the general impacts of urbanization follows, and the chapter concludes by investigating the specific disruptions of matter and energy flows in urban watersheds incurred by these human-generated and urban-related processes: (1) water infrastructure and its impacts on ecosystems, (2) increased stormwater runoff, (3) accelerated erosion and sedimentation, and (4) the creation of excess sensible heat. The information garnered will help steer the pollution prevention measures necessary to begin the return to dynamic equilibrium in our streams, aquifers, and watersheds.

12.2 Watershed Structure

To identify the components of watershed structure, it is instructive to return to the words used by John Wesley Powell when he defined watersheds: "… that area of land, a bounded hydrologic system, within which all living things are inextricably linked by their common water course…"

Reduced to their basic parts, watersheds are composed of land, water, air, and living things. They are, however, not simple things—but complex three-dimensional systems (Figure 12.1).

12.2.1 The Atmosphere

Starting at the top of Figure 12.1, the atmosphere initiates the input of water to watersheds through precipitation. Once discharged from clouds, precipitation has several travel options: (1) continue its direction downward into soil as infiltration, (2) follow the nearest downward gradient on the surface and become runoff, or (3) just stop and park on the surface of a leaf where it may drip to the ground and follow paths 1 and 2 or evaporate. Several factors influence the routing:

- Rainfall intensity (measured in mm/h)
- Rainfall type (drizzle or large drops)

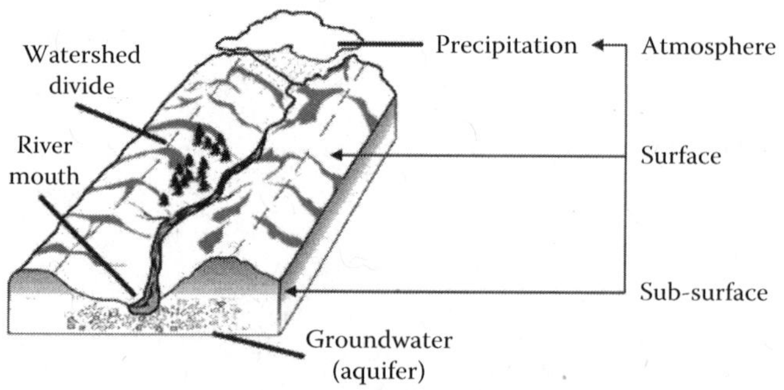

FIGURE 12.1
Three dimensions of a watershed. (From Kentucky, State of Kentucky, Division of Water website. http://www.water.ky.gov/watersheds (accessed June 3, 2010), 2010.)

- Presence of antecedent soil moisture
- Amount and type of vegetation present
- Soil type and texture
- Amount of impervious surface

High intensities of rainfall may fall on soils not fully capable of infiltrating the water. If the soils are dry and permeable (e.g., larger-grained, such as sandy loam), then infiltration rates will be higher. Some clay soils will initially infiltrate water at a high rate, but the rate declines rapidly as the clay particles expand and reduce the porosity and permeability. Urban soils tend to be compacted (Gregory et al. 2006) and have a generally low-organic content that varies by their age. Soils in older urban residential areas tend to have higher organic contents than those within newer developments (Scharenbroch et al. 2005). These characteristics reduce infiltration and increase runoff. Tree canopies intercept and fragment drops of precipitation and reduce its kinetic energy, lowering their erosive capacities. The differences in volume (and mass) between a droplet of drizzle and a large drop from a thunderstorm are quite significant:

$$\text{Volume of a sphere} = 4/3\pi r^3 \text{ (the 4/3 is a constant and can be dropped for this comparison).}$$

For a droplet of drizzle 0.02 mm in diameter:

$$\text{Radius, } r = 0.01\,\text{mm, volume} = \pi r^3 = (3.14 * 0.01)^3 = 0.000031\,\text{mm.}$$

For a larger drop 2.0 mm in diameter:

$$\text{Radius} = 1.0\,\text{mm, volume} = \pi r^3 = (3.14 * 1.0)^3 = 30.959144 \text{ or } 30.1\,\text{mm.}$$

The volume difference is

$$30.1/0.000031 = 1{,}000{,}000.$$

With a volume 1 million times greater, the 2.0 mm drop has more mass and more potential to dislodge soil.

Certain types of trees are very effective at intercepting precipitation, and vegetation can reduce runoff significantly. In urban areas, species such as the Brisbane Box tree exhibit interception rates as high as 66.5% (Xiao and McPherson 2002). A study in Dayton, Ohio, found that a modest increase in the existing tree canopy cover could reduce surface runoff by 5% (Sanders 1986).

In urban areas, there is more impervious surface created by roads and rooftops. These constructions steer precipitation into a runoff path, rather than one of infiltration, and replace the vertical movement of large quantities of water with horizontal movement.

The atmosphere is also a source of contaminants into watersheds. Significant atmospheric deposition of lead and other metals such as arsenic, chromium, and mercury occurs over many watersheds and is higher in urban watersheds (Murray et al. 1997; Sabin et al. 2006). Acidic precipitation also acts as a contaminant to the surface waters and groundwater within watersheds. As a result of the chemical reaction between water vapor and CO_2 in the atmosphere, the pH of natural rainfall is slightly acidic at 5.6. Events of acidic precipitation with a pH below 4.5 are common (USGS 2010), and pH values of 3.5 have been recorded in the eastern U.S. and Europe (Menz and Seip 2004). Due to the logarithmic measurement scale, these values are 10–100 times more acidic than natural precipitation.

The impacts of acidic precipitation on the landscape are significant. Acidic precipitation interferes with the ability of vegetation to take up calcium and other soil base cations, has led to widespread forest damage (Park et al. 2008), and damages to stream and lake ecosystems (Keller et al. 2007; Deyton et al. 2009). Groundwater can also be impacted by acidic precipitation, since the lower pH precipitation mobilizes many heavy metals in soil and increases the risks of their transport into deeper groundwater (Kjoller et al. 2004).

12.2.2 The Surface and Subsurface

The collection and organization of precipitation performed by watersheds is enabled by their topography and influenced by the regional climate and geology. In Figure 12.1, the higher elevations shown as the watershed divide define the boundary of a surface watershed. It is within the "basin" created by the bounding high elevations where all precipitation is collected. Any precipitation falling to the left of the watershed divide shown is captured by the adjacent watershed.

Climate and geology interact to produce the drainage density and drainage pattern of a watershed. Drainage density refers to the length of all channels within a specified area—typically a square kilometer; drainage pattern is the arrangement of the stream channels. Rivers such as the Nile do not have many tributaries once they flow out of the Ethiopian Highlands through the Sahara Desert (Figure 12.2a). In a region with higher precipitation, the Amazon River requires several tributaries with more annual discharge than the Mississippi River to collect the water (Figure 12.2b).

Regional geology largely determines the routes taken by flowing surface water. Over time, a stream system develops a particular drainage pattern in response to the local topography and subsurface geology. Three of the most common drainage patterns (dendritic, parallel, and rectangular) illustrate the relationship between geology and drainage pattern. In the sedimentary geological environments where many urban watersheds are located, the dendritic (treelike) drainage pattern is the most common form (Figure 12.3a).

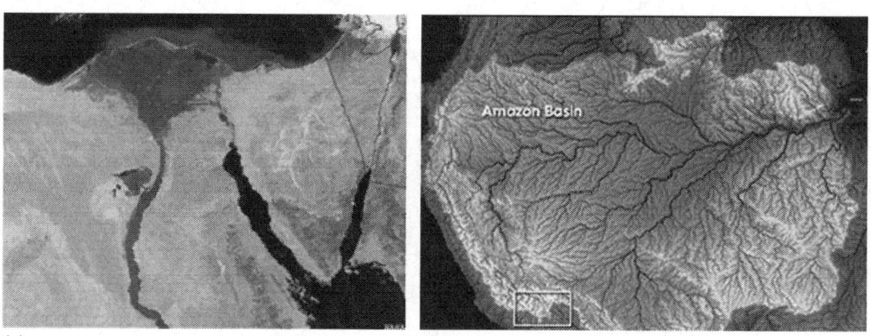

(a) (b)

FIGURE 12.2
Comparative drainage densities resulting from climate. (a) Nile. (From NASA/GSFC/LaRC/JPL, MISR Team.) (b) Amazon. (From University of Maryland's Global Land Cover Facility and NASA http://earthobservatory.nasa.gov/IOTD/view.php?id=7823)

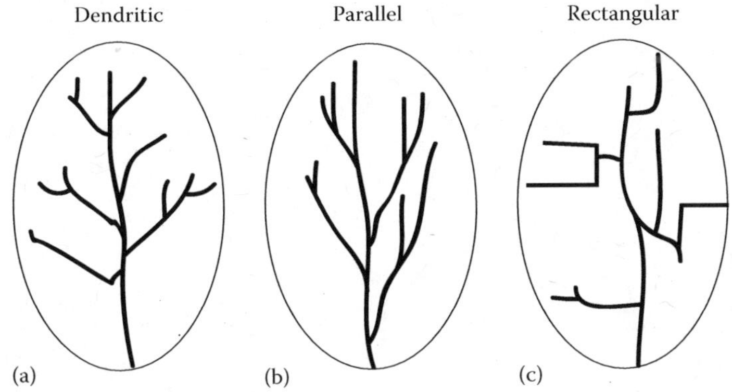

FIGURE 12.3
Drainage patterns. (a) Dendritic, (b) parallel, (c) rectangular.

Here, the subsurface geology has a similar resistance to weathering, and so there is no apparent control over the direction the tributaries take. Parallel drainage patterns (Figure 12.3b) form where there is a pronounced slope to the surface. A parallel pattern also develops in regions of parallel, elongate landforms where there are outcropping resistant rock bands. Tributary streams tend to stretch out in a parallel-like fashion following the slope of the surface. A parallel pattern may also indicate the presence of a major fault cutting across an area of steeply folded bedrock. The rectangular drainage pattern (Figure 12.3c) is found in regions that have undergone faulting. Streams follow the path of least resistance and become concentrated in places where exposed rock is the weakest. Movement of the surface due to faulting offsets the direction of the stream, causing the tributary streams to make shape bends and enter the main stream at high angles (Ritter et al. 2002).

In the subsurface zone of watersheds, aquifers lend form and structure to the land surface above. Different bodies of water such as lakes and wetlands are the representation of exposed groundwater in humid environments. In terms of structure, if too much water is pumped from an aquifer, the land above it will subside. Land subsidence has occurred in urban areas of Texas, California, Colorado, Delaware, and Louisiana (USGS 2000).

12.2.3 Zonation of Surface Watersheds

Watersheds are organized as a nested hierarchy, having smaller basins set inside the next larger basin. The Missouri, Ohio, and Arkansas basins, for example, are nested within the Mississippi watershed. Stream discharge increases as each basin adds its runoff to the next. As a result, the average discharge of most watersheds increases with total drainage area (Figure 12.4).

There is also a hierarchy to the stream channels within a drainage network. Streams occupying a drainage basin form a hierarchical network of channels that, in humid areas, hold increasingly larger volumes of water as one moves toward the mouth of the basin. A stream's order is its rank, or relative position, within the network (Strahler 1952). A first-order stream is a channel with no tributaries, and channels of this type occur primarily at the higher elevations (headwaters) of the watershed. A second-order stream is a channel fed by at least two first-order tributaries. The joining of two-second order streams forms a third-order stream. Ranking continues until the highest-ordered channel is reached (Figure 12.5). First and second-order streams are located in the headwaters and typically convey small volumes of water.

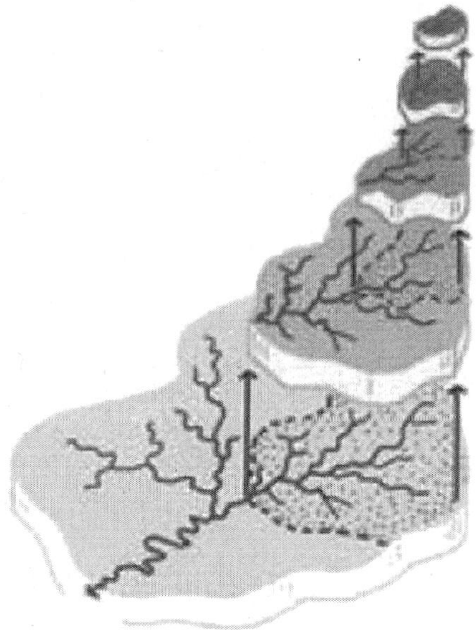

FIGURE 12.4
Nested watersheds. (From Marsh, W.M., *Landscape Planning: Environmental Applications*, 5th edn., New York, John Wiley & Sons, 2010.)

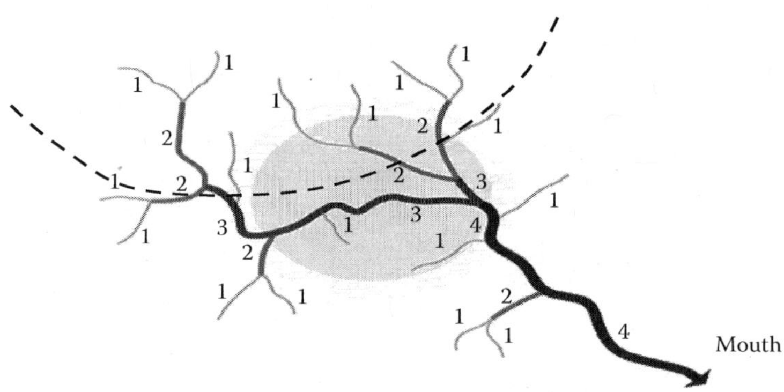

FIGURE 12.5
Stream orders (the dashed line demarcates the headwater region). (From Federal Interagency Stream Restoration Working Group (FISRWG), Stream corridor restoration—principles, processes and practices, adopted as Part 653 of the National Engineering Handbook, USDA Natural Resources Conservation Service, http://www.nrcs.usda.gov/technical/stream.restoration (accessed May 29, 2010), 2001.)

Stream orders are important for watershed sustainability, because they indicate locations where there is a potential mismatch between the naturally evolved stream capacity and the higher runoff volumes created by urban development. Lower-order streams are also more vulnerable, because they are unable to dilute contaminants.

12.3 Watershed Function

Watersheds exist at the crossroads of Earth's great environmental spheres: the atmosphere, biosphere, hydrosphere, and lithosphere. Many of the interactions between the physical systems operating within and between these spheres are present in watersheds. For example, when soil erosion occurs after a rainstorm, the hydrologic cycle operating within the hydrosphere interacts with the geomorphic system of the lithosphere. A geomorphic system is characterized by the removal, transport, and deposition of material by wind or water.

The activity of individual systems and their interactions with other systems produces work—the application of force over distance. In watersheds, most of the work is accomplished by the movement of materials through the motion of fluids (wind and flowing water). On sloped terrain, gravitational force also performs some of the work. The work produced at the watershed scale can also be thought of as their function. Erosion, therefore, is a function of watersheds, as is providing pathways and sinks for nutrients.

Performing work also requires the presence of energy necessary to create a force. Over most of the nearly 4.6 billion years of Earth's history, nature has crafted its energy allocations in a sustainable way. Humans, however, have introduced extra energy into natural systems. This is not surprising, for, unlike animals, humans must produce their means of subsistence. The development of large-scale agriculture, manufacturing, and residences represents a tremendous amount of energy. It is this extra energy input by humans into the watershed system that causes malfunctions (Table 12.1).

Some of the key functions of watersheds are identified at the general level of systems interaction shown in Table 12.1. Human inputs and the resulting malfunctions are shown in the third column. The list of functions performed by watersheds expands significantly if subsystems of the major systems listed in the table are considered. For example, when the subsystems of precipitation and runoff interact within the hydrologic system, one of the outcomes is infiltration, which has the function of providing soil moisture recharge.

Table 12.1 also demonstrates another general property of systems—feedback. **Feedback** is an outcome of a process within a system that affects the overall function of the system.

Negative feedback results when an output from a process in a system slows down or dampens the overall operation of the system. For example, if a surplus of organic matter is input into a stream by a storm, more organisms with the purpose of decomposing organic matter will come on the scene and work until the original balance is achieved. Barring

TABLE 12.1

Watershed Functions and Associated Malfunctions from Human Inputs

System	Interaction: Outcome/Human Input	Function/Malfunction
Climatic	Ecologic: formation of biomes	Provision of species habitat
	Global warming—excess energy in the atmosphere	Species extinction
Ecologic	Hydrologic: Inputs of nutrients into streams	Nutrient transport
	Excess nutrients input into streams by wastewater discharges	Cultural eutrophication
Hydrologic	Tectonic/geomorphic: Formation of drainage networks	Movement of water from headwaters to mouth
	Modification of stream channels (straightening, deepening, and widening)	Flooding, unnatural channel adjustments leading to more artificial channel modifications
Tectonic/Geomorphic	Climatic: uplift from magma plumes creates more precipitation due to a higher elevation of the land mass (this occurred with the East African rift)	Erosion, transport, and deposition of surface materials (rock and debris)
	Building on unstable soil	Accelerated erosion rates

additional surplus inputs during this period, the organic surplus is reduced and then followed by a decline in the numbers of organic-consuming organisms. The system returns to equilibrium. **Positive feedback** occurs when the outcome of a process within a system speeds up or magnifies the system's activity or work output. Anthropogenic measures designed to modify a stream channel often initiate positive feedback. When a dam for sediment control is placed in a stream, the downstream segment of the stream becomes hungry for sediment. Since a stream's equilibrium depends largely on the amount of water and sediment it transports, the stream hunts for more sediment. Downstream of the dam the stream initiates the erosion of its own banks. This outcome sometimes prompts engineers to cover the natural streambanks with concrete. Besides the ecological insult to the bankside habitat, this measure reduces the stream's width, and lower volumes of runoff become sufficient to trigger a flood. It can get nastier. Positive feedback may form a loop, with each successive iteration producing a more unstable condition. For these reasons, identification of the existing and potentially destructive positive feedbacks imparted by humans on watersheds is a necessary step for achieving sustainable watersheds. In urban watersheds, this is especially challenging because of their complexity and the large quantities of energy and matter humans contribute.

12.4 Urbanization and Watersheds: Disruption of Matter and Energy Flows

Urbanization encompasses the processes contributing to urban growth, especially an increased number of people coming to the cities to live and the subsequent city expansion through the annexation of surrounding land and adjacent communities (Rugg 1972). Globally, the trend toward urbanization is increasing. By 2030, nearly two-thirds of the world's population will live in urban areas (Figure 12.6). In North America, the degree of

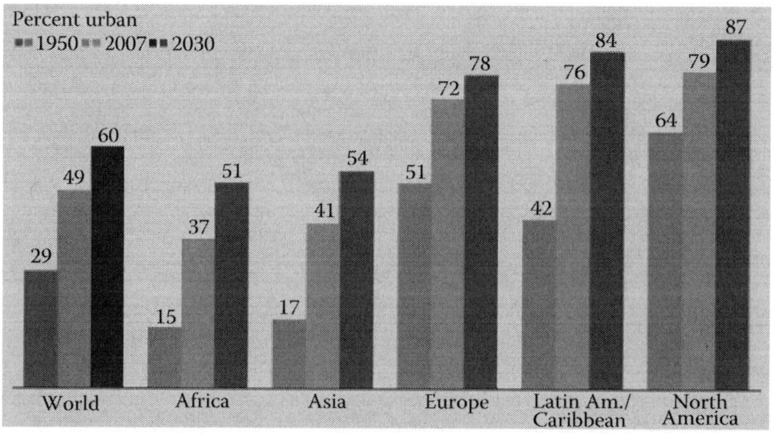

FIGURE 12.6
World urbanization trends 1950–2030. (From United Nations, *World Urbanization Prospects: The 2005 Revision*, New York, 2006; Haub, C., *World Population Data Sheet*, Population Reference Bureau, Washington, DC, 2007.)

population concentration is characterized by urban areas occupying only 2.6% of the total land use, but containing 79% of the total population (Lubowski et al. 2006).

The urban concentration of people, industry, commerce, and transport comes with environmental costs. Cities are warmer, the soil is poorer, surface waters and aquifers are more polluted, and ecosystems are under more stress. Although urbanized land is just one component of the mixture of land uses present in urbanized watersheds, the effects or urbanization are often felt throughout the entire watershed. If sustainable urbanized watersheds are the goal, it is therefore essential to understand the disruption of their matter and energy flows induced by urbanization at the watershed scale. To achieve this understanding, examples of significant human impacts within the four environmental spheres (biosphere, hydrosphere, lithosphere, and atmosphere) are investigated.

12.4.1 Water Infrastructure and Ecosystems

Ecosystems enable organisms to exchange energy and matter between themselves and the outside environment. Their basic structure consists of organisms arranged vertically by trophic level into food chains and food webs (Figure 12.7).

These structures organize the intake of external solar energy, produce food, and return energy in the form of respiration to the environment. At the bottom of the food chain are the producers, organisms capable of photosynthesis. Matter is transferred upward through the trophic levels of the food chain or food web by the addition of body mass. The arrows between some of the organisms in the food web signify the interrelationships between multiple organisms in an ecosystem. For example, without predators, the population of certain species becomes uncontrolled, as was the case when rabbits were introduced in Australia.

Cities take in energy from the outside environment, and mediate flows of matter and energy within themselves and the surrounding areas. Not surprisingly, there is a growing trend in the literature and professional planning practice to treat cities as ecosystems (Newman and Jennings 2008). In urban areas with intensive land use, ecosystems are affected by human impacts and the results are often observable. There is some logic then to classifying the impacts humans make on ecosystems as a useful first step toward understanding the depth and breadth of the specific problems confronting urban watersheds.

Human impacts on ecosystems can be grouped into six major classes (Odum 1971):

- Reduction
- Fragmentation
- Substitution

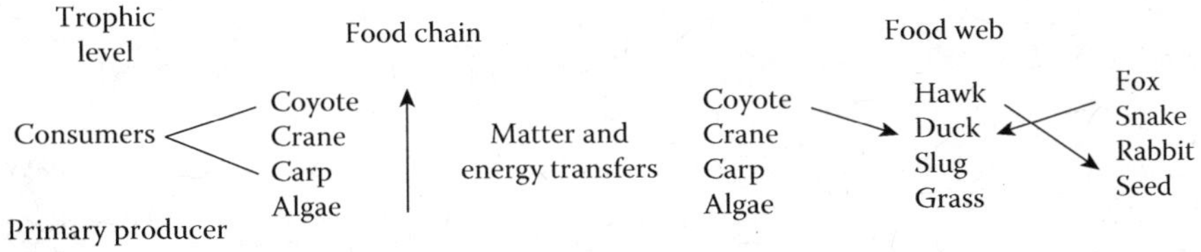

FIGURE 12.7
Food chain and food web.

- Simplification
- Contamination
- Overgrowth

Reduction is the loss of areal coverage of an ecosystem or community. **Fragmentation** consists of breaking ecosystems into spatially separated units or fragments. **Substitution** occurs when one or more organisms/species are replaced by others. **Simplification** is the reduction in the number of species in an ecosystem or community. **Contamination** is the introduction of harmful substances into an ecosystem, typically anthropocentric-manufactured chemicals. **Overgrowth** refers to the importation of excess nutrients into an ecosystem.

The necessity to harness water for use by large concentrations of population and to process the wastewater produced has input massive quantities of energy and materials into urban watersheds. Construction of water infrastructure has changed the course of rivers (the Mississippi) and created a false sense of security about flood risk (Hurricane Katrina). In addition, the discharge of millions of gallons of effluent by urban wastewater plants into streams often constitutes a significant percentage of their total flow.

Due to their scope and permanence, many of the impacts resulting from these activities can be readily identified. Here, a matrix is employed to accomplish this task (Table 12.2).

For the purposes of this analysis, water infrastructure is divided into five human-engineered systems: water supply, sanitary sewers, stormwater, overland flow, and navigation/flood control. The cells of the matrix contain one impact associated with each human-engineered system. Impacts without broad significance are assigned "minimal."

Water supply. This system includes water treatment facilities and the pipes comprising the water distribution network. Also included are the land and infrastructure required to accomplish groundwater recharge, such as the conjunctive use system practiced in the Los Angeles basin (Faunt 2009). Dams, reservoirs, pipes, and pumps required for interbasin water transfers to urban areas are not included. Conventional landfills are included, because one of the outputs from the water supply system is a fair quantity of nontoxic lime sludge.

Result: Overall, the impacts from this system are minimal, but there must be land set aside for treatment facilities. Proper siting of these structures avoids the interruption of habitat corridors and the displacement of sensitive ecosystems, such as wetlands. If withdrawals for the water supply are made from a local stream, care must be taken not to go beyond a 20% reduction below bankfull stage. Otherwise, temperatures can rise in the stream and disrupt its ecosystem (Poole and Berman 2001). Since warmer water holds less dissolved oxygen, many fish adapted to colder water (e.g., trout) and requiring a certain level of oxygen may be threatened if water temperature rises.

Sanitary sewers. Includes wastewater treatment plants and the sanitary sewer pipe network delivering sewage to them. Landfills for handling the sludge produced are also included, since it is toxic and is disposed in sanitary landfills, incinerated, or processed into biosolids for agricultural applications.

Result: The main threat here is from infiltration/inflow (I & I), which refers to clear water from rain and snowmelt improperly draining into the sanitary sewer system. Much of the inflow results from cracked pipes, with older urban areas being more prone to leaks. During heavy rainstorms, I & I may cause the sanitary sewers to quickly fill with clear

TABLE 12.2

Ecosystem Impacts across Major Engineered Systems in Urban Watersheds

System Impact	Water Supply	Sanitary Sewers	Stormwater	Overland Flow	Navigation/ Flood Control
Reduction	Loss of land	Loss of land	Loss of riparian habitat	Loss of land	Loss of riparian habitat
Fragmentation	Minimal	Minimal	Detention ponds interrupt habitat corridors for terrestrial species	Roads create habitat patches too small to support some species	Streams are split by dams; sedimentation occurs
Substitution	Minimal, unless withdrawals cause stream temperature to rise	Minimal	"Rough fish" replace fish with higher oxygen requirements	Loss of dissolved oxygen in streams	Creates access problems for waterfowl; other species may move in
Simplification	As above with substitution	Minimal	Decline in species diversity in the receiving waters	Decline in species diversity in the receiving waters	Dams can block fish migrations
Contamination	Leaking landfills	Infiltration/ Inflow	Delivery of heavy metals to water bodies	Pathogens delivered to water bodies	Removal of bankside vegetation can allow more pollutants to enter streams
Overgrowth	Minimal	Sewer overflows induce the growth of macrophytes	Cultural eutrophication from increased input of phosphorus	Excess nitrogen and phosphorus are carried to streams and lakes	As above with contamination

rain water that should have been directed to the storm sewer. When too much of this clear water mixes with untreated sewage, the volume flowing through the sewer pipes may exceed the capacity of the sanitary sewer system pipes. This condition can produce sewer overflows into basements or streets.

Stormwater. Consists of storm drains of different types, for example, culverts, field tiles, and buried storm sewer pipes. Constructed wetlands and ponds for detention, retention, or infiltration are included, along with the weirs accompanying some of these implementations.

Result: The open components of stormwater control systems (e.g., detention ponds) and the untreated discharge from pipes create significant ecosystem impacts. Detention ponds consume valuable property and require ongoing maintenance to function properly. Poorly timed discharges from detention pond networks have exacerbated local floods (Debo and Ruby 1982). Contaminants in stormwater can affect riparian habitat and lead to species substitution. Rough fish, such as carp, may replace trout and other species requiring a higher water quality. Stormwater may also contain high levels of coliform bacteria, oil, and grease and particles with adsorbed heavy metals (USEPA 2004).

Overland flow. This item encompasses the ephemeral drainage network created during wet-weather events. Roads, rooftops, parking lots, and other surfaces acting to intensify the drainage network are included.

Result: This system has similar impacts as stormwater. The density and pattern of roads in urban areas often creates small patches of habitat incapable of supporting a wide diversity of birds (Ortega-Álvarez and MacGregor-Fors 2009).

Navigation/Flood control. Any measure used for **channelization** applies: locks, straightening, deepening, or widening as well as dams, reservoirs, dikes, levees, or rip-rap.
Result: Channelization often involves the removal of bankside vegetation, causing a litany of environmental and ecosystem impacts to ensue (Brooker 1985). Dams can increase erosion and have significant impacts on fish migrations by creating environments more conducive to nonnative and exotic plant, fish, snail, insect, and animal species (WCD 2000).

This analysis demonstrates how a finer delineation of watershed impacts is obtained when major systems are decomposed into subsystems. For example, contamination emerges as a critical concern, since it spawns several other impacts including substitution, simplification, reduction, and overgrowth. Navigation and flood control measures that straight-jacket streams with concrete also have many negative consequences. There are also additional on-site (within the watershed) and external impacts. On-site, the process of building infrastructure has ecosystem consequences, because there is accelerated erosion when building roads, laying pipe, or constructing buildings. Outside of the watershed, the need to supply the electricity for the pumps used for water transfers also creates impacts. For instance, the California State Water Project is the largest single user of energy in California. In the process of delivering water from the San Francisco Bay-Delta to Southern California, the project uses 2%–3% of all electricity consumed in the state (Cohen et al. 2004). How much coal is burned to support this activity?

12.4.2 Stormwater

Stormwater is nonpoint pollution occurring in urban areas. During wet weather events, particles containing contaminants from the soil and road surface are carried overland to the nearest water body or released by storm drains downstream. The problem is exacerbated by urbanization, which increases the amount of impervious surface (Walesh and Videkovich 1978).

Stormwater is best characterized as an interrelated problem of water quantity and water quality. From a quantitative perspective, the extra volumes of runoff have two major impacts: (1) the creation of impervious surfaces such as concrete and asphalt lowers the surfaces' resistance to flow and increases the competence (ability to carry particles) of the flowing stormwater. Surface resistance is incorporated as the roughness coefficient in Manning's equation used to compute the average open-channel flow velocity (Equation 12.1). As the equation demonstrates, when smooth pavement replaces rough ground, the value of the roughness coefficient (n) decreases, and average velocity increases.

$$V = \frac{100}{n} * R^{2/3} * S^{1/2} \tag{12.1}$$

where

V is the average velocity (m/s)
R is the hydraulic radius (m), defined as A/P, where A is the cross-sectional area of the flow and P is the length of the wetted portion of flow boundary (the wetted perimeter)
S is the slope of the water surface (m/m)
n is the roughness coefficient (available from published tables, dimensionless)

(2) the extra volumes of runoff generated can cause local flooding, especially in first-order streams. Many early studies on stormwater problems reported these events (Novotny 2003).

From a qualitative perspective, stormwater is polluted water, and the type of land use influences the quality of stormwater (Table 12.3).

Industrial land uses exhibit the highest levels of total suspended solids. Solids are responsible for transporting heavy metals in stormwater, and the higher levels of metals present at industrial sites confirm this relationship. Fecal coliform levels are highest at residential sites, and this is likely the result of animal wastes.

After stormwater traverses a given land use, there is a direct relationship between stormwater quantity and the quality of the receiving water body. Figure 12.8 illustrates this relationship.

When impervious area increases, runoff volume increases and this leads to a deterioration of stormwater quality. A decline in the quality of the receiving water body ensues. As indicated by the small boxes at the bottom of columns 1 and 2, there is also a relationship between stormwater and groundwater. As shown in column 1, when receiving water quality decreases, there is a decline in groundwater quality. The box at the bottom of column 2 identifies a relationship between increases in impervious surface and a decline in groundwater quality. Problems to ecosystems (denoted by "C" on the diagram) result from all of the quantitative and qualitative characteristics of stormwater. Reducing stormwater flow

TABLE 12.3

Land Use and Stormwater Quality

Parameter[a]	Residential	Commercial	Industrial
% Impervious	37.0	83.0	75.0
Temperature (°C)	16.4	16.0	17.9
Total Suspended Solids (mg/L)	49.0	42.0	78.0
Fecal Coliform (mpn/100 ml)	8345.0	4300.0	2500.0
Total Nitrogen Kjeldahl (mg/L)[b]	1.4	1.6	1.4
Total Phosphorus (mg/L)	0.3	0.22	0.26
Cadmium (μg/L)	0.5	0.89	2.0
Chromium (μg/L)	4.6	6.0	14.0
Lead (μg/L)	12.0	18.0	25.0
Zinc (μg/L)	31.5	59.0	112.0

Source: Pitt, R. et al., *The National Stormwater Quality Database* (NSQD, version 1.1), Department of Civil and Environmental Engineering, University of Alabama, Tuscaloosa, AL, 2005.

[a] Sample size varied slightly for some of the parameters. Overall, the average sample size for residential was 1050 sites, commercial had 500 sites, and industrial 525.

[b] Kjeldahl nitrogen is the sum of ammonia-nitrogen and organic nitrogen.

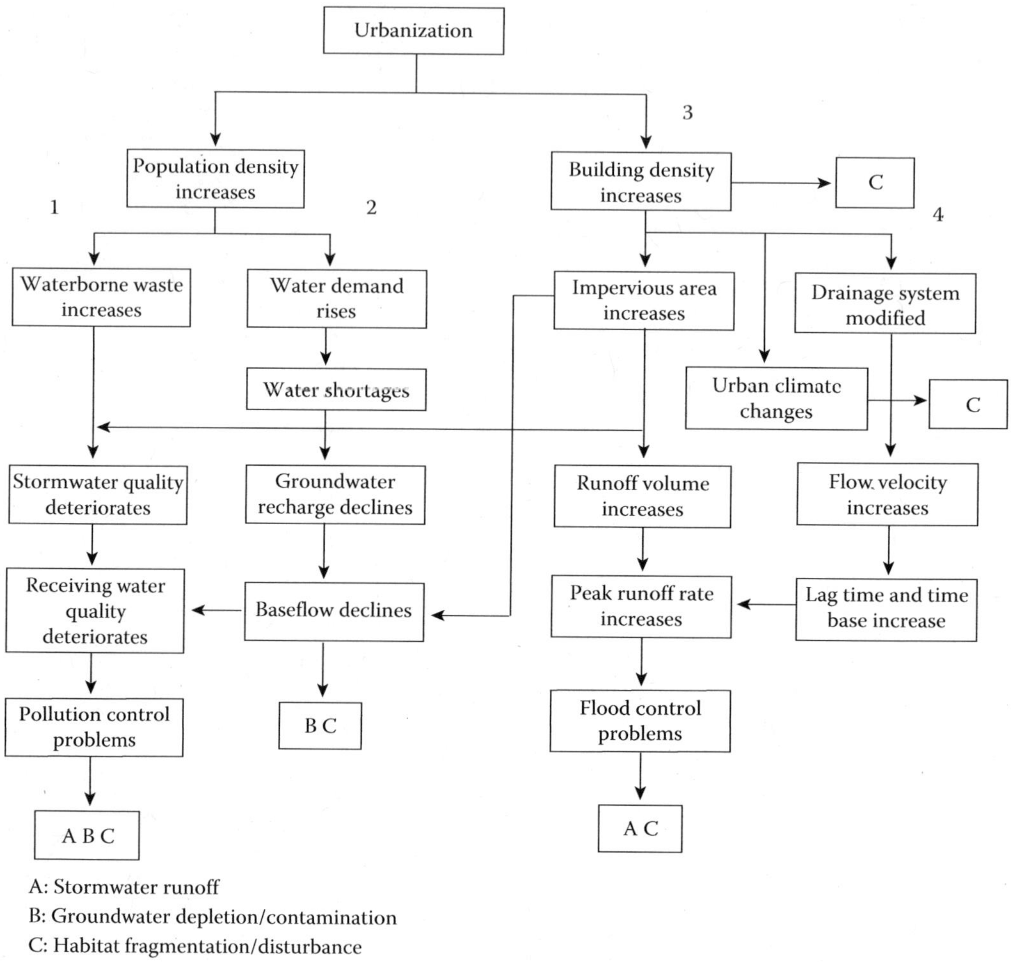

FIGURE 12.8
Water problems associated with urbanization. The numbers are used to assist with column references in the discussion. (Reprinted from *Landsc. Urban Plann.*, 36, Kaufman, M. and Marsh, W., Hydro-ecological implications of edge cities, 277–290, 1997, with permission from Elsevier.)

volumes would therefore result in less contamination delivered into streams, wetlands, lakes, and seas.

12.4.2.1 Case Study

The experience of one community in the Rouge watershed in southeast Michigan illustrates the environmental disruption caused by stormwater and by the first attempts to control its impacts.

Novi, Michigan, lies in the northwest section of the Rouge watershed (Figure 12.9).

The northwest section of the watershed is dominated by moraine soils with good drainage, but starting in the early 1970s Novi began to have drainage problems when it rained. The city had imported clay excavated from the construction of Detroit's freeways, which unlike most freeways in the Unites States were built below the ground. This clay from the old lakebeds in the southeastern part of the watershed had poor infiltration capacity

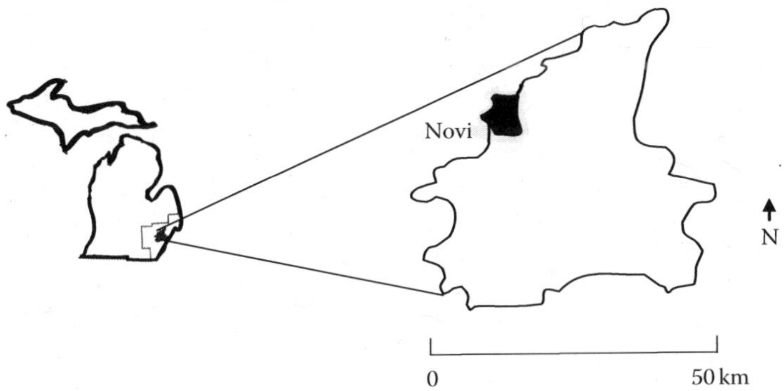

FIGURE 12.9
Novi, Michigan, within the Rouge watershed.

(Rogers 1997). During Novi's rapid urbanization, when this soil type was combined with the creation of impervious surface, the conditions for generating high amounts of runoff were created.

The first response in Novi, as in many communities, was to require new developments to build on-site detention basins to hold the water and attenuate the peak flows (Figure 12.10). The goals were to maintain predevelopment streamflows and improve water quality through the settling of some sediments during the brief detention period.

As development in Novi accelerated during the 1970s, the number of on-site detention basins grew. By 1980, there were almost 50 detention storage facilities. Despite this rise in detention basin use, stormwater problems continued. Major arterial roads were often immersed by stormwater runoff, and the large number of detention basins was becoming a maintenance problem. Local associations with no expertise in on-site detention management took over the upkeep of the basins after subdivisions were completed. On-site detention also created an additional expense to developers, a health hazard, and a confiscation of developable land (Kaufman 1994).

After a bond issue in 1980 designed to raise money for additional drainage failed, the City hired an engineering firm. The firm developed a plan to use 28 regional detention basins for holding the excess runoff. At the proposed rate of construction, there would have been over 600 on-site detention ponds when the City was fully developed. In an urbanized area of 80 km^2, this translated into an average of 7.5 ponds/km^2. Faced with this prospect, residents in 1982 approved the regional detention plan and the flooding stopped.

The timing, placement, and character of human interventions are critical to minimizing environmental impacts. The use of on-site detention as the first response fragmented

FIGURE 12.10
Detention basin. (Photo by Daniel T. Rogers.)

the landscape, and the ponds did not have the capacity to hold the extra runoff generated by urbanization. Regional detention as the second response provided more storage and stopped the floods, but at some cost to the sensitive wetland ecosystems included as part of the plan. Neither method captured stormwater before it left its source—the rooftops and lawns on individual parcels.

12.4.3 Erosion and Sedimentation

The construction of new roads, buildings, and other infrastructure accompanies urbanization. When land is disturbed by construction activity, soil erosion increases dramatically. Erosion rates as high as 40,000 times the preconstruction rate have been observed. The typical magnitude is between 20 and 40, which equates to a loss of 27–64 tons (30–70 tons) of soil per acre per year (Goldman et al. 1986).

Formal laws mandating erosion and sedimentation control exist in many countries. **Sedimentation** is the deposition of eroded material in a sink. In urban streams, where erosion rates at nearby construction sites have accelerated erosion, more sedimentation occurs (Wolman and Schick 1967). At the federal level in the United States, sediment is considered the principal contaminant in the water systems, a point explicitly recognized by the Clean Water Act of 1972. Many states have also adopted erosion and sedimentation control laws and ordinances. Unfortunately, enforcement of their provisions is not always carried out.

In 2000, a study of 30 construction sites in the urbanizing region of east-central Michigan found that procedures used to stabilize slopes, stabilize soil, and manage water were not sufficient to control soil from leaving the site. Only four of the sites successfully implemented the Best Management Practices recommended by the state and local erosion and sedimentation control ordinances (Kaufman 2000). None of the sites studied were inspected after rain events, which would have provided the best indication of the efficacy of the erosion control measures used. As seen with stormwater management, the timing of human interventions and the selection and placement of the proper controls are critical to avoiding more damage.

Soils take a beating in urban areas. In part due to the removal of topsoil during construction activity, the top layer of the urban soil horizon may have low organic content (Scharenbroch et al. 2005). This lack of organic material inhibits their ability to contain certain contaminants and prevent their migration. As noted earlier, urban soils are also compacted and as a result exhibit lower infiltration capacities. In addition, they are often overfertilized and contribute significant levels of nitrogen and phosphorus to stormwater (Stow et al. 2001; Collins et al. 2010).

In some countries, erosion is a catastrophe. Tanzania has reported losses of one-third of its Gross Domestic Product from soil erosion (Beyadi 2010). In one major study of soil erosion, it was found that, during the last 40 years, nearly one-third of the world's arable land has been lost by erosion and continues to be lost at a rate of more than 10 million ha/year (Pimental et al. 1995). In many countries, soil erosion is similar to high blood pressure—a silent killer. The geographic scale of erosion and its gradual nature make it difficult to observe on a daily basis. Quite often the damage is done before anybody notices.

12.4.4 The Urban Heat Island

An **urban heat island** (UHI) results from the excess energy input into the atmosphere from anthropogenic activities. The primary source of the additional energy comes from

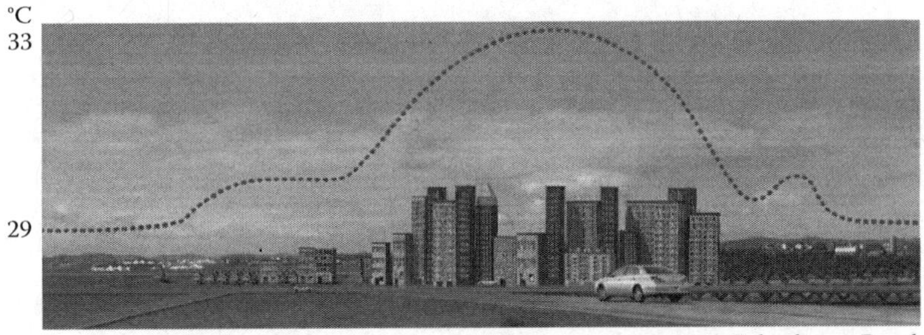

FIGURE 12.11
Temperature profile by land use. (From http://eospso.gsfc.nasa.gov/eos_homepage/for_scientists/data_products/OurChangingPlanet/PDF/Page_299_new.pdf)

the replacement of vegetation with structures and pavement, along with some input from traffic (Yalcin and Yeteman 2009). Ambient air temperatures rise and create the profile across the different types of land uses as shown in Figure 12.11.

One well-documented effect of the UHI is an increase in the magnitude of thunderstorms over major urban areas (Chagnon 1978). These more intense precipitation events exacerbate the "first flush" effect—the rapid removal of the most recently deposited contaminants from road surfaces (Deletic 1998). Air quality also suffers. The higher levels of particulate matter present in urban areas means contaminants will be airborne longer (Chapter 10).

Additional research has revealed other significant impacts, including increased groundwater temperature (Yalcin and Yeteman 2009) and increased temperatures in surface streams as a result of the flow of stormwater over warm urban pavement (Leblanc et al. 1997).

The extent of the UHI is impressive. The lighter areas in Figure 12.12 show the elevated air temperatures detected in the heavily urbanized sections of New York City.

Consider this scenario: it is a hot summer day in an industrial city. Within this city are hundreds, perhaps thousands of contaminated brownfield sites. Late in the day, an intense afternoon thunderstorm occurs and delivers 20 min of hard downpours. The drop sizes are large, and they dislodge significant amounts of soil particles onto which are adsorbed heavy metals from the brownfield sites. Another cruel injustice occurs: the intense thunderstorms affect the area of the city with the highest percentage of impervious surface, and the volume of runoff generated is massive. Tonnes of contaminated soil particles are entrained by the stormwater and delivered into several tributaries within the urbanized watershed.

Unfortunately, this is not fiction. At uncapped contaminated sites where there is exposed soil, stormwater acts as a catalyst for spreading contamination (Murray et al. 2004). Eventually, some of the contaminated soil particles may percolate into groundwater, especially if acidic precipitation mobilizes them.

12.5 Summary and Conclusion

By examining some of the major environmental impacts specific to urbanized areas and their watersheds, we obtain a broader scientific foundation required for the planning efforts

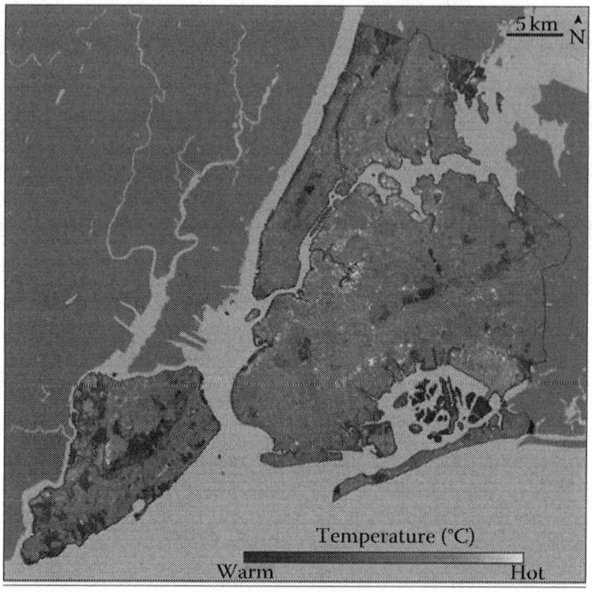

FIGURE 12.12
Satellite image of the urban heat island, New York City, USA. (From NASA http://earthobservatory.nasa.gov/IOTD/view.php?id=6800)

required to create sustainable watersheds. The disruptions of matter and energy created by the processes examined in this chapter are interrelated with each other and linked to geology and contamination through the flows of water and air and the transport of soil.

Many human engineering inputs are necessary. We cannot have urban areas without roads, but we can find ways to design them better and reduce their negative impacts. For instance, the use of porous pavement in areas not requiring a high load bearing capacity can reduce runoff and increase infiltration. As the examples in this chapter have shown, the prerequisites for achieving sustainability are to identify the reducible impacts and then select the proper time and location for injecting the appropriate feedback into the system. Opting not to produce a contaminant in a production process is one example of constructive negative feedback. Locating a production facility that must use contaminants in a less vulnerable geologic environment is another.

The next chapter explores the options available for preventing pollution through a variety of source control methods for point and nonpoint sources.

References

Beyadi, A. 2010. Environmental cost of soil erosion amounts to one-third of GDP. *Tanzania Daily News*, May 24.

Brooker, M.P. 1985. The ecological effects of channelization. *The Geographical Journal* 151:63–69.

Chagnon, S. 1978. Urban effects on severe local storms at St. Louis. *Journal of Applied Meteorology* 17:578–586.

Cohen, R., B. Nelson, and G. Wolff. 2004. *Energy Down the Drain: The Hidden Costs of California's Water Supply*. New York: Natural Resources Defense Council.
Collins, K.A., T.J. Lawrence, E. K. Stander et al. 2010. Opportunities and challenges for managing nitrogen in urban stormwater: A review and synthesis. *Ecological Engineering*, in press doi:10.1016/j.ecoleng.2010.03.
Debo, T.N. and H. Ruby. 1982. Detention basins—An urban experience. *Public Works* 113:42–43.
Deletic, A. 1998. The first flush load of urban surface runoff. *Water Research* 32:2462–2470.
Deyton, E.B., J.S. Schwartz, R.B. Robinson et al. 2009. Characterizing episodic stream acidity during stormflows in the Great Smoky Mountains National Park. *Water, Air, and Soil Pollution* 196:3–18.
Faunt, C.C., ed. 2009. *Groundwater Availability of the Central Valley Aquifer, California*. U.S. Geological Survey Professional Paper 1766.
Federal Interagency Stream Restoration Working Group (FISRWG). 2001. Stream corridor restoration—Principles, processes and practices, adopted as Part 653 of the National Engineering Handbook, USDA Natural Resources Conservation Service. http://www.nrcs.usda.gov/technical/stream.restoration (accessed May 29, 2010).
Goldman, S.J., K. Jackson, and T.A. Bursztynsky. 1986. *Erosion and Sediment Control Handbook*. New York: McGraw-Hill.
Gregory, J.H., M.D. Dukes, P.H. Jones et al. 2006. Effect of urban soil compaction on infiltration rate. *Journal of Soil and Water Conservation* 61:117–124.
Haub, C. 2007. *World Population Data Sheet*. Washington, DC: Population Reference Bureau.
Kaufman, M.M. 1994. *Determinants of Community Response to Stormwater Pollution in an Urbanized Watershed*, PhD dissertation. University of Michigan, Ann Arbor.
Kaufman, M.M. 2000. Erosion control at construction sites: The science-policy gap. *Environmental Management* 26:89–97.
Kaufman, M.M. and W.M. Marsh. 1997. Hydro-ecological implications of edge cities. *Landscape and Urban Planning* 36:277–290.
Keller W, N.D. Yan, J.M. Gunn et al. 2007. Recovery of acidified lakes: Lessons from Sudbury, Ontario, Canada. *Water, Air, and Soil Pollution* 7:317–322.
Kentucky, 2010. State of Kentucky, Division of Water website. http://www.water.ky.gov/watersheds (accessed June 3, 2010).
Kjoller, C., D. Postma, and F. Larsen. 2004. Groundwater acidification and the mobilization of trace metals in a sandy aquifer. *Environmental Science and Technology* 38:2829–2835.
Klein, R.D. 1979. Urbanization and stream quality impairment. *Water Resources Bulletin* 15:948–963.
LeBlanc, R.T., R.D. Brown, and J.E. FitzGibbon. 1997. Modeling the effects of land use change on the water temperature in unregulated urban streams. *Journal of Environmental Management* 49:445–469.
Lubowski, R.N., M.Vesterby, S. Bucholtz et al. 2006. *Major Uses of Land in the United States, 2002 USDA Economic Information Bulletin 14*. Washington, DC.: USDA.
Marsh, W.M. 2010. *Landscape Planning: Environmental Applications*, 5th edn. New York: John Wiley & Sons.
Menz, F.C. and H.M. Seip. 2004. Acid rain in Europe and the United States: An update. *Environmental Science & Policy* 7:253–265.
Murray, K.S., J. Heiden, A. Farkas et al. 1997. Heavy metal concentrations and distribution in river sediments, Rouge River, southeast Michigan. *Michigan Academician* 29:537–555.
Murray, K.S., D.T. Rogers, and M.M. Kaufman. 2004. Heavy metals in an urban watershed in southeastern Michigan. *Journal of Environmental Quality* 33:163–172.
Newman, P. and I. Jennings. 2008. *Cities as Sustainable Ecosystems: Principles and Practices*. Washington DC: Island Press.
Novotny, V. 2003. *Water Quality: Diffuse Pollution and Watershed Management*. New York: John Wiley & Sons.
Odum, E.P. 1971. *Fundamentals of Ecology*. Philadelphia, PA: Saunders.
Ortega-Alvarez, R. and I. MacGregor-Fors. 2009. Living in the big city: Effects of urban land-use on bird community structure, diversity, and composition. *Landscape and Urban Planning* 30:189–195.

Pimental, D.C., P. Harvey, P.K. Resosudarmo et al. 1995. Erosion and economic costs of soil erosion and conservation benefits. *Science* 267:1117–1123.

Park, B.B., R.D. Yanai, T.J. Fahey et al. 2008. Fine root dynamics and forest production across a calcium gradient in northern hardwood and conifer ecosystems. *Ecosystems* 11: 325–341.

Pitt, R., A. Maestre, and R. Morquecho. 2005. *The National Stormwater Quality Database* (NSQD, version 1.1). Dept. of Civil and Environmental Engineering, University of Alabama, Tuscaloosa.

Poole, G.C. and C.H. Berman. 2001. An ecological perspective on in-stream temperature natural heat dynamics and mechanisms. *Environmental Management* 27:787–802.

Ritter, D.F., R.C. Kochel, and J.R. Miller. 2002. *Process Geomorphology*, 4th edn. Long Grove, IL: Waveland Press.

Rogers, D.T. 1977. *Surficial Geologic Map and Cross Section of the Rouge River Watershed*, Michigan. 1:62,500. River Rouge National Wet Weather Demonstration Project (RRNWWDP), Wayne County, MI. 2 Sheets.

Rugg, D.S. 1972. *Spatial Foundations of Urbanism*. Dubuque, IA: Wm. C. Brown.

Sabin, L.D., J.H. Lim, K.D. Stolzenbach et al. 2006. Atmospheric dry deposition of trace metals in the coastal region of Los Angeles, California, USA. *Environmental Toxicology and Chemistry* 25:2334–2341.

Sanders, R.A. 1986. Urban vegetation impacts on the hydrology of Dayton, Ohio. *Urban Ecology* 9:361–376.

Scharenbroch, B.C., J.E. Lloyd, and J.L. Johnson-Maynard 2005. Distinguishing urban soils with physical, chemical, and biological properties. *Pedobiologia* 49:283–296.

Stow, C.A., M.E. Borsuk, and D.W. Stanley. 2001. Long-term changes in watershed nutrient inputs and riverine exports in the Neuse River, North Carolina. *Water Research* 35:1489–1499.

Strahler, A.N. 1952. Dynamic basis of geomorphology. *Geological Society of America Bulletin* 63:923–938.

United Nations, 2006. *World Urbanization Prospects: The 2005 Revision*. New York.

United States Environmental Protection Agency (USEPA), 2004. National water quality inventory report http://www.epa.gov/owow/305b/2004report (accessed June 1, 2010).

United States Geological Survey (USGS), 2000. Land subsidence in the United States, USGS fact sheet-165-00. http://water.usgs.gov/ogw/pubs/fs00165/ (accessed June, 2, 2010).

United States Geological Survey (USGS), 2010. Water science for schools. http://ga.water.usgs.gov/edu/acidrain.html (accessed November 28, 2010).

Walesh, S.G. and R.M. Videkovich. 1978. Urbanization: Hydrologic-hydraulic damage effects. *Journal of Hydraulic Engineering* 104:644–659.

Wolman, M.G. and A.P. Schick. 1967. Effects of construction on fluvial sediment, urban and suburban areas of Maryland. *Water Resources Research* 3:451–464.

WCD (World Commission on Dams). 2000. *Dams and Development: A New Framework for Decision-Making*, London, U.K.: Earthscan.

Xiao, Q. and E.G. McPherson. 2002. Rainfall interception by Santa Monica's municipal urban forest. *Urban Ecosystems* 6:291–302.

Yalcin, T. and O Yetemen. 2009. Local warming of groundwaters caused by the urban heat island effect in Istanbul, Turkey. *Hydrogeology Journal* 17:1247–1255.

13

Pollution Prevention

13.1 Introduction

It is difficult and costly to remediate a contaminant after it has been released into the environment. The contaminant may spread into soil, water, and air and often causes harm to the environment before cleanup can occur. These reasons underscore why preventing the release of contaminants is a prerequisite for creating a sustainable environment.

Preventing contaminants from entering the environment is called **pollution prevention**. To be successful, this process includes reducing or eliminating waste at the source by modifying production processes, promoting the use of nontoxic or less toxic substances, implementing conservation techniques, and reusing materials instead of putting them into the waste stream (USEPA 2010a).

The legal framework for this pollution prevention effort is embodied in the Pollution Prevention Act of 1990 (42 U.S.C. §13101 et seq.), which focused the industry on these measures (USEPA 1990):

- Pollution should be prevented or stopped at the source whenever feasible.
- Pollution that cannot be prevented should be recycled.
- Pollution that cannot be prevented or recycled should be treated in an environmentally safe manner.
- Releases of pollution into the environment should be conducted only as a last resort and should be conducted safely.

Sources of pollution can be designated as point sources or nonpoint sources (USEPA 2003). **Point source pollution** originates from identifiable sources, such as smokestacks or sewage outfall pipes. **Nonpoint source pollution** emanates from diffuse or unknown sources and is the leading cause of water pollution in the United States (USEPA 2002). Examples of nonpoint source pollution include

- Contaminated groundwater from an unknown source
- Contaminated stormwater runoff originating from parking lots, roads, and lawns
- Air deposition of contaminants and particulates
- Erosion
- Runoff from agricultural areas

The road to sustainability must incorporate effective pollution prevention that yields observable results. In the United States, the observation of major pollution events has been a catalyst for significant levels of response (Kaufman and Marsh 1995). As a corollary to this, observable progress in preventing pollution, we postulate, will provide additional incentives to continue those efforts. At the onset and throughout, science must guide the planning process, and the results should be published and open for critical review by scientists, professionals, and the public. Other components of this process include (1) maximizing resource efficiency, (2) implementing existing and developing technical innovations, (3) minimizing the use of toxic chemicals, and (4) education (USEPA 2009).

This chapter begins with a quick assessment of the recent pollution prevention efforts in the United States. Next, a framework developed by the authors for preventing pollution at industrial point sources is presented. Since point and nonpoint sources are characterized by common transport media and transport processes, portions of this framework are then applied to the source reduction efforts for the nonpoint pollution variants of stormwater and erosion.

13.2 Pollution Prevention in the United States

The institutional framework for pollution prevention in the United States is embodied in the Pollution Prevention Results Task Force (PPRTF) formed in 2003. Membership in the Task Force is comprised of representatives from state and local programs, EPA Headquarters and Regional Offices, pollution prevention Resource Exchange Centers, and the National Pollution Prevention Roundtable. The Task Force was created with the objective of developing a National Pollution Prevention Results Data System designed to gather and aggregate data from the state and local programs (USEPA 2009).

The third and most recent report issued by the PPRTF in 2009 indicates that there were substantial reductions of material waste, airborne pollution, and water pollution between 2004 and 2006. A total of 3,492,661,249 kg (7.7 billion lb) of pollution was prevented from entering the environment. Of this total, 2,267,961,850 kg (5 billion lb) came from reducing materials and waste, 90,718,474 kg (2 billion lb) from reducing air emissions, and 27,215,542 kg (60 million lb) from reducing water pollution. The total cost savings of these pollution prevention efforts are estimated to be $6.4 billion (USEPA 2009). Although idiosyncrasies of the data prevent the ability to ascertain trends, the quantities represented indicate there is significant effort underway in pollution prevention. Perhaps the best indicator of this effort is the steady rise in the number of facilities participating in pollution prevention planning: 1100 in 2004; 1900 in 2005; and 2600 in 2006.

These numbers alone, however, do not tell the complete story of pollution prevention—there are additional quantitative and qualitative considerations. From a quantitative perspective, the quantities of pollution reduction must be weighed against the quantities of pollution produced. If the annual rates of reduction are consistently surpassing the annual rates of pollution released into the environment, then improvements may be indicated. Progress, however, depends not only on consistent quantitative reductions but also on the *qualitative* nature of the pollution released, including the types, toxicity, persistence, and mobility of contaminants released, and an accounting of the impacts on the geologic, ecologic, hydrologic, and atmospheric environments where these contaminants enter. A full characterization of progress, therefore, would need to document the changing status of ecosystems, receiving water quality (surface water and groundwater), soil conditions, and air pollution levels realized through pollution prevention efforts. Moreover, due to the human pathways present with many contaminant releases, there would need to be evidence of fewer emergency hospital visits and a trend toward less chronic occupational-related diseases, such as the skin and lung problems associated with the use of chromium.

It is beyond the scope of this book to carry out this type of analysis. What can be managed here is the presentation of a procedure for pollution prevention that has had success with industrial point source applications, and an extension of this procedure to nonpoint sources. These experiences can then be applied to help achieve the broader quantitative and qualitative goals of pollution prevention outlined above.

13.3 Implementing Pollution Prevention Techniques: Industrial Point Sources

Successful implementation of pollution prevention involves careful planning. Within the broad array of planning venues and forms (e.g., urban, environmental, strategic, and business), there are common threads to the planning process:

- Identification of what you want to do (goals/objectives)
- Collection of data
- Specification of methods for achieving the goals/objectives
- Implementation using the selected methods
- Assessment of the results

The planning process shown here is cyclical. Assessment may lead back to more data collection, or if implementation fails with the methods selected, new methods can be developed and implemented until success is achieved. Sometimes, the outcome changes the entire goal of the project, especially in cases where you bit off more than you could chew.

Science should be infused into the planning process wherever appropriate. Accurate measurement is a foundation of good science, so to ensure scientific standards for data collection are met, the procedures should include a statistically sound specification of the sample size. As noted in earlier chapters, the samples should also be collected, transported, and analyzed according to existing professional standards and methods.

Another area where science must be incorporated into the execution of a plan is the **experimental design**. Science is fundamentally about identifying and explaining variation, and the experimental design—which is the assignment of subjects to experimental groups—provides the roadmap. Although it may sound obvious, at contaminated sites there are two types of locations: contaminated and uncontaminated. Assigning these locations into two groups allows investigators to study the similarities and differences between them. This separation is how we learned that contamination tends to occur more frequently near low points in buildings, which is described in greater detail later in this chapter. We then use this knowledge to help design the most effective measures for pollution prevention—the "where" of intervention.

Successful source control also requires an understanding of the process producing the pollution. Processes occur over time, so the specification of where to intervene should be accompanied by the proper timing of our pollution prevention efforts—the "when" of intervention. If loading docks are areas in a facility more prone to a contaminant release, then busy times at these locations require special diligence.

The initiation of pollution prevention efforts occurs within organizations and represents a form of change, or innovation. To succeed it is important to get the science right and the innovation must (1) be testable and implementable at small scales, (2) represent an observable improvement over existing conditions, (3) be culturally acceptable (in the corporate/organizational and social senses), (4) be economically feasible, and (5) be convenient to implement. These five conditions characterize successful innovations (Rogers 1995).

What follows is a planning process based on the experiences of implementing a successful pollution prevention effort at a major manufacturing company (Rogers et al. 2006). This process contains the basic elements of plans, incorporates scientific aspects of the geologic environment and contaminant properties, and recognizes the social context for implementation.

13.3.1 Step 1: Establishing Objectives and Gathering Background Data

Establishing objectives or goals for pollution prevention provides a baseline for measuring success. Making decisions with better information can avoid the specification of arbitrary objectives and goals, and this information can be obtained with targeted data collection and evaluation. This workflow will help with the attainment of goals and focus on limited resources where they will produce maximum benefit. It is recommended these data are collected:

- Mass and volumes of each type of solid wastes generated, including
  - Solid wastes such as wood and paper products, plastic, metals, glass, etc.
  - Regulated solid wastes
  - Regulated hazardous wastes
- Mass and volumes of liquid wastes generated, including
  - Stormwater volumes and content
  - Sanitary discharge volumes
  - Industrial wastewater volumes and contents
  - Regulated liquid wastes in containers
  - Regulated liquid hazardous wastes in containers
- Mass and volumes of air emissions generated, including
  - VOC emissions
  - Particulate emissions
  - Heavy metal emissions
  - All other identifiable air emissions
- Energy
  - Sources, types, consumption, and rates through time
  - Energy loss
- Purchasing habits
  - Bulk containers compared to smaller containers and amounts
  - Types of containers (e.g., gallon containers vs. spray cans)
  - Packaging (plastic vs. cardboard or biodegradable material)

- Production efficiencies
- Product packaging
- Shipping (rail vs. truck)

Where possible, a mass balance should be calculated to ensure the accuracy of the data. Once the data have been collected and evaluated, steps 2 through 4 should be completed before firmly establishing goals and objectives for any pollution prevention program.

13.3.2 Step 2: Inventory of Hazardous Substances

The next step performed to develop a pollution prevention plan at any location—whether a manufacturing facility or a household—is to inventory and assess hazardous substance use. Chapter 7 covered many common contaminants present in urban areas, including households. As they are often very close to industrial sites, households can assist with any inventory since the average American household stores 3–10 gal of hazardous materials (Smolinske and Kaufman 2007). To inventory chemicals, CAS Registration Numbers should be used since many products display either trade names or synonyms. If available, material safety data sheets (MSDS) often provide valuable information.

Inspecting the label on chemical containers is required, especially when the container is a mixture of chemical products. In these cases, the name of the product is often a trade name and is not very helpful in identifying the specific chemicals contained, so inspection of the label is the only effective method. An example of a label on a chemical product is shown in Figure 13.1.

Chemicals should be inventoried by chemical group, such as VOCs, PAHs, SVOCs, PCBs, metals, acid, bases, and whether they are present in gas, liquid, or solid form. Transformers containing PCBs should also have appropriate labels as shown in Figure 13.2. After the

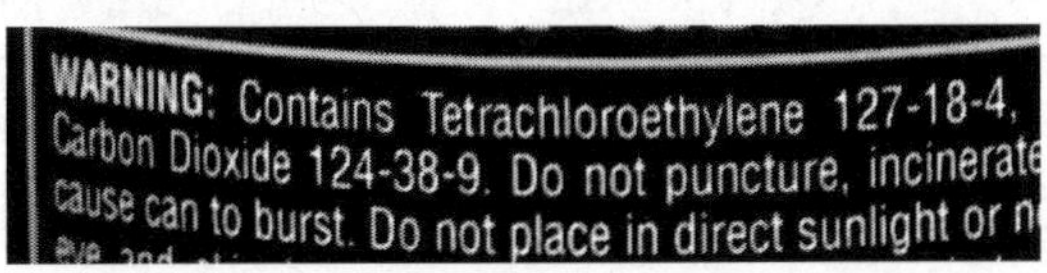

FIGURE 13.1
Chemical labels. (Photo by Daniel T. Rogers.)

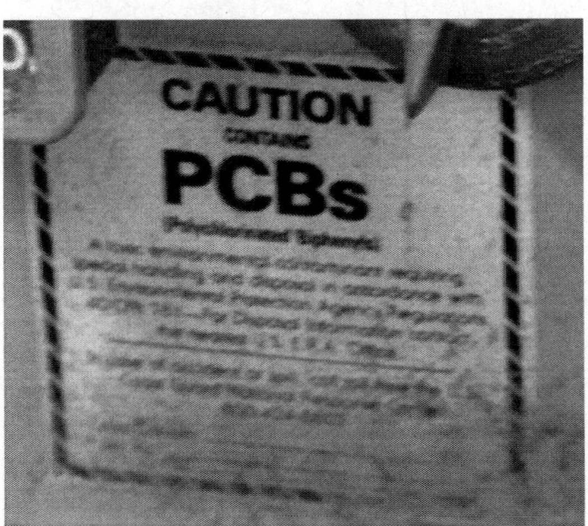

FIGURE 13.2
Chemical labels. (Photo by Daniel T. Rogers.)

hazardous substance inventory has been completed, a map should be created showing the following: location of where each hazardous substance enters the property, where the substances are stored before use, where they are used, and where they are stored after being used and before any residuals are discarded. Since many contaminant release locations occur from low points in buildings, highlighting pits, sumps, and floor drains on the map can also help provide valuable information for pollution prevention efforts (Rogers et al. 2006).

13.3.3 Step 3: Assessing CRFs of Substances Present

The hazardous substances having the highest contaminant risk factors (CRFs) should now be evaluated. This evaluation helps to prioritize the pollution prevention efforts. Many facilities and households may have more than 100 different hazardous substances. Prioritizing is therefore critical to achieve maximum benefit.

As noted in Chapter 10, chromium VI and DNAPL VOCs have the highest CRFs for groundwater, with PCBs, mercury, chlordane, and PAHs having the highest CRFs for soil. Concentrating pollution prevention efforts on those contaminants with the potential to significantly impact groundwater such as chromium VI and DNAPL VOCs will have the most long-term benefit; Chapter 14 presents these scenarios.

13.3.4 Step 4: Preliminary Assessment of Geologic Vulnerability

Most urban areas have not been geologically mapped, so the details necessary to accurately evaluate the vulnerability at any location may not be available. The suggestions offered here can help determine if any given area presents enough potential risk to warrant a more detailed examination, and whether an aggressive pollution prevention initiative should be pursued:

- Source, location, and type of potable water. Contact the local municipality to evaluate the source of potable water for the area in question. In addition, inquire whether there are any groundwater extraction wells of any type within at least a 1.6 km (1 mi) radius of the location being evaluated. If any wells exist, request a copy of the installation records for further examination.
- Nearest surface water body. Determine where the nearest surface water body is located with respect to the location being evaluated. Topographic maps (7.5 min) can help with this procedure.
- Stormwater collection and discharge. The local municipality should have information about stormwater collection, its treatment, and the discharge locations for the area being evaluated.
- General geological conditions. Examining well records could provide valuable information on soil type, stratigraphy, and depth to groundwater if they can be obtained in the vicinity of the location being evaluated.

If the source of potable water in the area being evaluated is obtained from groundwater and extraction wells are nearby, environmental risk should be considered high. In this case, the geologic vulnerability rating will likely exceed a score of 50, and pollution prevention efforts should become a top priority (Chapter 6). The other factors listed above will likely require examination and evaluation by a qualified professional before an appropriate geologic vulnerability rating can be determined.

13.3.5 Step 5: Preventing Pollution through ESPM Methods

Pollution prevention can be implemented with a stepwise evaluation process that proceeds from the most preventative measure to the least preventative measure. This process is referred to as ESPM and consists of the following steps:

- Elimination: Not using potentially harmful chemicals
- Substitution: Using a less potentially harmful chemical instead of a harmful one
- Prevention: Using engineering controls and other measures to minimize the potential for a release; employed if eliminating or substituting a potentially harmful chemical is not possible.
- Minimization: Reducing usage of harmful chemicals or a reduction in generated wastes through process changes, recycling, or other methods.

After steps 1 through 4 have been completed, a focused and achievable plan for pollution prevention can be developed and implemented. Facilities where a synergistic effect could be realized between the environment and the contaminant properties should receive the highest initial effort. This effort is characterized by aggressive pollution planning initiatives for reducing the potential risks (Rogers et al. 2006).

13.3.5.1 Elimination

The most aggressive form of pollution prevention is *elimination* of hazardous chemical use. Where possible, elimination of hazardous substance use is the preferred pollution prevention method because it is the easiest to manage and has the greatest benefit to the environment. As noted in Chapters 7 and 8: if there is no hazardous chemical use, there is not risk. In most cases, total elimination of hazardous chemical use is very rare, so the elimination of chemicals should focus on contaminants with high CRFs. For example, chromium VI and DNAPL VOCs have high CRFs for groundwater, while PAHs, chlordane, PCBs, and mercury have high CRFs in soil.

A crucial step in the process of eliminating hazardous substances is to develop a chemical ordering procedure. This procedure protects against unauthorized hazardous substances making their way into operations at the facility without prior knowledge. All proposed chemicals or substances should undergo a review process to evaluate whether they are acceptable for use.

13.3.5.2 Substitution

The next most aggressive form of pollution prevention is *substitution*. Substitution involves using alternative chemicals with the goal of greatly reducing risk. For instance, if a facility uses DNAPL VOCs for cleaning, an effective substitute may be citrus-based cleaners. The *Solvent Alternatives Guide* provides options and guidance for evaluating available chemical substitutes for common solvents (USEPA 2010b).

Other alternatives exist for chemical substitution. For instance, mercury-containing devices such as switches, thermometers, and monometers can be substituted by digital devices. Liquid transformers containing PCBs can be substituted by dry transformers or with transformers not containing any detectable concentration of PCBs.

Other examples of substitution include

- Using paints without VOCs and certain heavy metals
- Using biodegradable oils
- Using paraffin as a lubricant instead of oil

13.3.5.3 Prevention

Prevention of contamination entering the environment can be accomplished through engineering controls. Engineering control methods include release prevention, release detection, release containment, and release cleanup immediately after a spill.

The following examples (denoted by bullets) highlight some ways activities and operations can be modified to minimize the potential for the release of hazardous substances to the environment:

- Evaluating and eliminating potential points of release. Sumps, pits, trenches, floor drains, and chemical storage and usage areas are common points for hazardous substance release. Conducting an inventory of these locations and locations of chemicals present at a facility will assistance in identifying areas where releases may occur.

Figure 13.3 provides an example of a potential release area and illustrates several practices that should be avoided, including

- Storing liquid wastes in improperly labeled containers. The center drum on the left side has a nonhazardous label, yet it is also labeled as containing a waste solvent.
- Storing liquid wastes in upside down containers.
- Storing liquid wastes on bare ground.
- Storing liquid wastes outside without a roof to prevent contact with stormwater and corrosion.
- Storing liquid wastes in drums where the structural integrity has been compromised.

FIGURE 13.3
Improper storage of drums containing liquid waste. (Photo by Daniel T. Rogers.)

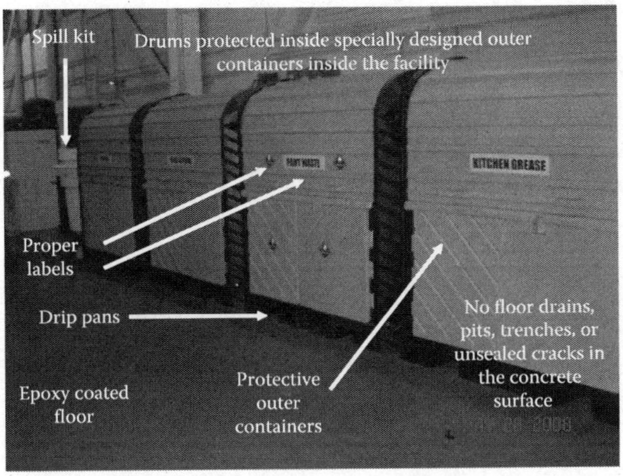

FIGURE 13.4
Proper storage of drums containing liquid waste. (Photo by Daniel T. Rogers.)

- Modifying liquid waste storage areas. Figure 13.4 demonstrates one way to properly store liquid chemicals and substances prior to disposal. This storage method for liquid wastes has multiple and redundant engineered systems to prevent a release and contain a release if one does occur. These engineered systems include
  - Storing liquid wastes inside and under a roof.
  - Coating the floor with epoxy.
  - Providing secondary containment if a release were to occur. In this example, drip pans are located beneath each outer container.
  - Locating the liquid waste storage area at a location without floor drains, sumps, trenches, and pits.
  - Using redundant storage containment. Sealed drums containing liquid wastes are located inside the outer containers pictured in this example.
  - Properly labeling the contents and potential hazards.

Figure 13.5 shows an example of an outdoor liquid waste storage area. A roof has been constructed and treated with an impervious surface so in the event of an accidental spill, the liquid will not migrate beyond the containment structure or through the concrete surface. This surface is shown in Figure 13.6.

- Sealing sumps, pits, trenches, floor drains, and liquid storage areas with an impervious surface. In most instances, leaks will occur from structures composed of concrete storing or conveying liquid wastes. Leaks can occur from seams, cracks, or in some cases from the direct migration of liquid through the concrete surface itself. These surfaces should be sealed to prevent a release to the subsurface. Figure 13.7 is an example of a sealed floor drain; Figures 13.4 and 13.6 also represent sealed liquid storage areas.
- Providing secondary containment. Engineering redundant systems can effectively prevent an uncontrolled release to the environment. Figure 13.8a is an example of an aboveground storage tank (AST) containing diesel fuel with no secondary

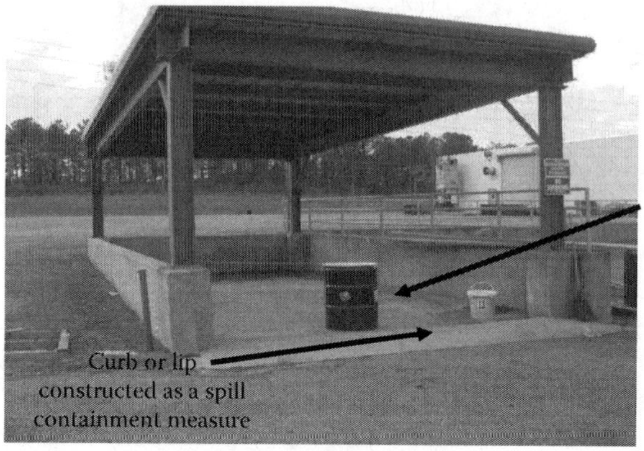

FIGURE 13.5
Example of an outdoor drum storage area. (Photo by Daniel T. Rogers.)

FIGURE 13.6
Impervious surface on an outdoor liquid waste storage area. (Photo by Daniel T. Rogers.)

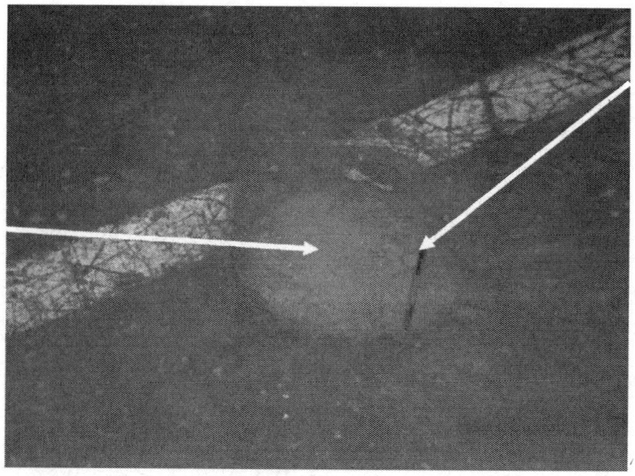

FIGURE 13.7
Example of a sealed floor drain. (Photo by Daniel T. Rogers.)

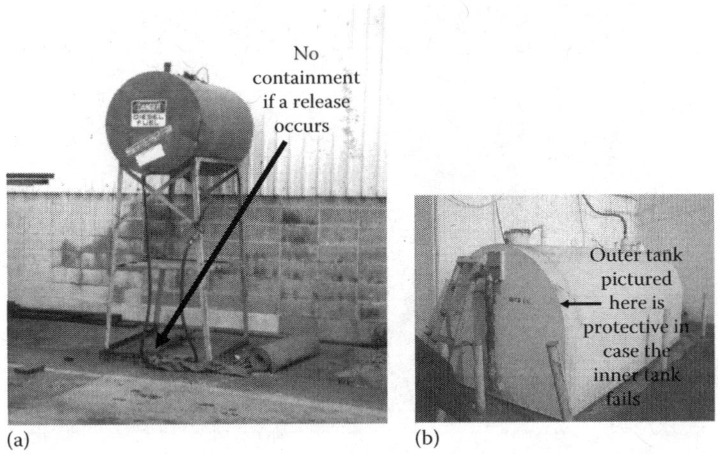

FIGURE 13.8
(a) AST with no containment. (b) AST with containment. (Photo by Daniel T. Rogers.)

containment in case the tank ruptures. Figure 13.8b is an example where an outer tank provides adequate protection if the inner tank fails.

- Spill containment and cleanup. Despite the existence of engineering controls, accidental releases do occur. Therefore, proper response is necessary to prevent the uncontrolled release of a hazardous substance and to protect human health and the environment. Spill stations outfitted with an assortment of tools, containers, personal protective gear, and instructions can ensure that small spills of liquids not presenting an immediate threat to human health are addressed quickly and safely. These stations should be located near hazardous substances. Figure 13.9

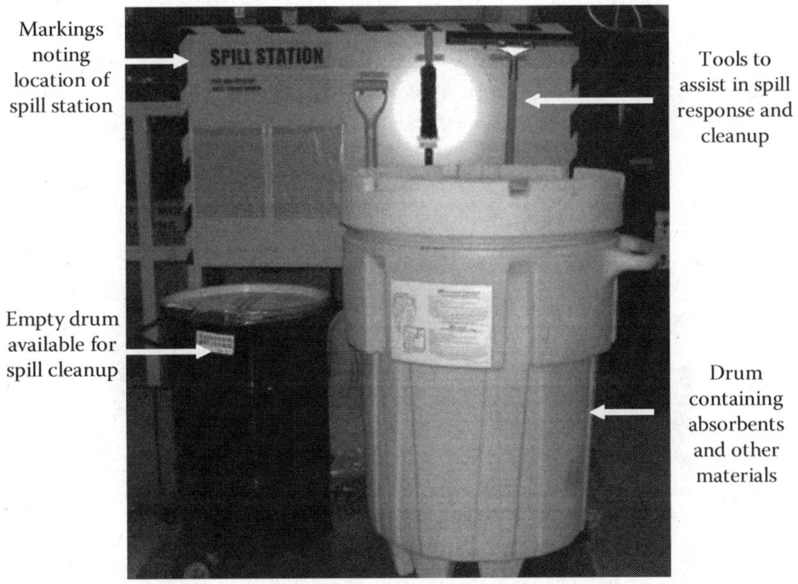

FIGURE 13.9
Example of a spill station. (Photo by Daniel T. Rogers.)

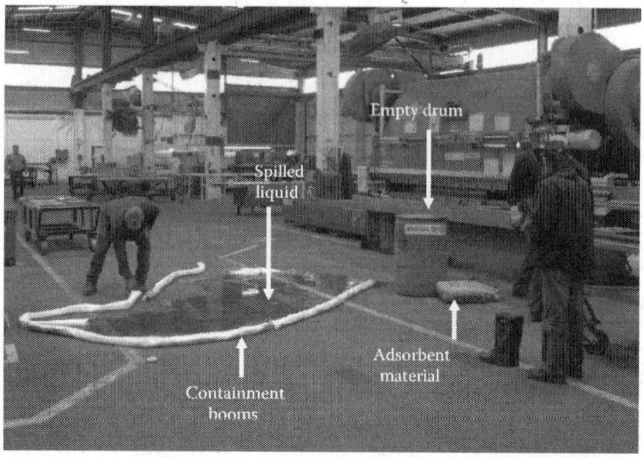

FIGURE 13.10
Spill response and training exercise. (Photo by Daniel T. Rogers.)

shows a typical spill station; refer to Figure 13.4 for an example of a spill kit located near a liquid waste storage area.

- Education and training. Education and training are critical for the prevention, response, and cleanup of spills. These efforts can also prevent the response to a spill or accidental release of a hazardous substance when evacuation and immediate notification are the necessary courses of action, and qualified emergency personnel are required on-site. Figure 13.10 shows a spill response and training exercise.

13.3.5.4 Minimization

Waste minimization involves using less hazardous substances through conservation efforts or process changes. The result is a reduction in the amount of wastes requiring disposal. In some instances, minimization also includes recycling. For instance, reducing the discharge or generation of liquids through process changes, recycling of water, or the evaporation/recycling of liquid or solid waste can greatly lower the volumes and mass of waste streams. An example of waste minimization is shown in Figure 13.11. Here, the

FIGURE 13.11
Reducing heavy metals in wastewater. (Photo by Daniel T. Rogers.)

additional treatment of wastewater using reverse osmosis reduces the amount of heavy metals in the discharged wastewater.

Recycling of metals, plastic, glass, wood products, and many other materials can greatly minimize the amount of solid waste generated and disposed of in a landfill. Often, these activities result in significant cost savings. The future in this arena is promising, as the discovery of new options for the beneficial reuse of waste materials parallels the appearance of new waste sources (USEPA 2009).

Energy reduction strategies evaluate the energy types consumed and their usage rates by location over time. In addition, energy loss and potential recovery/recycling may also come into play.

Purchasing habits can greatly assist minimization efforts. When supplies are purchased in bulk containers, the number of containers requiring disposal is reduced. It is also possible to realize reductions in energy usage and air emissions through production efficiencies. For instance, switching to more efficient gas turbines in certain production processes may save energy, while conducting energy-intensive activities at night when energy demand is low can lower air emissions. At off-peak times, power companies are more likely to substitute a nonfossil fuel such as hydroelectric for coal.

Product packaging and transportation methods are also important areas where many improvements can be made from an environmental perspective. For instance, using packaging made from biodegradable material instead of plastic can have a significant positive impact on the environment. In transportation, the use of rail or barges to transport products consumes much less fossil fuel and reduces air emissions.

13.3.6 Step 6: Assessing Results

Practicing pollution prevention has qualitative and quantitative benefits that can result in significant financial gains and cost avoidance. The evidence of quantitative pollution prevention benefits can be found by tracking reductions, especially if data have been collected before a pollution prevention program was initiated. Quantitative evaluation is reflected by the reductions of USEPA (2009)

- Waste volumes
- Raw materials
- Energy usage

On-site qualitative benefits are realized by USEPA (2009)

- Release avoidance
- Additional protection of human health and the environment
- Lowering regulatory reporting requirements
- Lowering environmental liability

Across an urbanized watershed, qualitative improvements will be seen in ecosystem health, receiving water quality (surface water and groundwater), soil conditions, and air pollution levels.

Continuous improvement, evaluation, and inspection should be developed as part of program assessment, because the lack of an effective pollution prevention program can result in significant liability and cost.

One final note: pollution prevention represents an innovation within an organizational context. With this in mind, the key "in-house" actions for achieving environmental project objectives, realizing cost savings, and reducing environmental risk are (1) establish easily identifiable objectives with the involvement of senior management, (2) perform an effective accounting of cost savings and tracking of other related benefits, and (3) communicate the results across all levels of the organization. In addition, it is beneficial to have compliance personnel trained in environmental science, with these personnel positioned at a high level within the organization (Rogers et al. 2006).

13.3.7 Financial Incentives for Industrial Pollution Prevention

The major financial incentives for implementing pollution prevention at most industrial facilities are the direct and indirect cost savings realized from a successful program. Direct costs savings accrue from reducing the amounts of solid waste generated and water consumed. Indirect cost savings occur when the long-term liabilities associated with proper waste disposal at a licensed facility are lowered because the amount of waste decreases. Other long-term liability reductions such as savings on litigation costs and reduced contaminant releases can also significantly reduce the costs associated with cleanup. Finally, waste reduction often reduces the amount of regulatory reporting requirements, and this outcome can significantly cut costs.

Implementation of an effective pollution prevention plan at any manufacturing facility requires the commitment, input, and cooperation of every employee. Representatives from purchasing, production, maintenance, human resources, and environmental departments must work together in close cooperation to identify, implement, and measure every aspect of a pollution prevention program. Tracking progress is necessary to sustain a program and create more involvement with its execution and outcomes. Many pollution prevention programs fail because they do not fully quantify their benefits.

13.4 Implementing Pollution Prevention Techniques for Nonpoint Sources in Urbanized Watersheds

Stormwater and erosion are related, as both involve the transport of materials by fluids. Stormwater originates from precipitation and is the polluted overland flow of water in urban areas. As precipitation flows over roads, rooftops, parking lots, construction sites, and lawns, it becomes contaminated with oil and grease, pesticides, litter, and pollutants from vehicles. The EPA estimates that over 10 trillion gallons of untreated stormwater make their way into U.S. surface waters each year (USEPA 2004). Dense urbanization exacerbates the problem, since the amount of pollution present in stormwater runoff is correlated with the amount of impervious cover (Schueler 1994). As noted in Chapter 12, erosion in urban watersheds is accelerated by disturbing the soil, mainly through construction activities. The dislodged soil particles are transported by water (stormwater) into storm drains, where they make their way to a receiving water body.

The Clean Water Act (CWA) of 1972 gave the USEPA the authority to regulate point source discharges through the National Pollution Discharge Elimination System (NPDES) program. In 1987, a survey of the nation's waters indicated point source control alone was not sufficient to achieve the "fishable and swimmable" goal of the CWA because nonpoint sources, especially agricultural and urban runoff, were contributing substantial amounts of pollution (Humenik et al. 1987). Also in 1987, the USEPA initiated the NPDES Stormwater program, requiring municipalities with separate storm sewer systems located in incorporated areas with populations of 100,000 or more to obtain NPDES permits for stormwater discharges (USEPA 1996). In 1999, Phase II of this program was extended to smaller municipalities and required permit holders to implement post-construction stormwater management programs using best management practices (BMPs) in *new development and redevelopment (emphasis by the authors).* Examples of BMPs for stormwater management include education, road salt management, street cleaning, and erosion control measures, such as silt fences and covering exposed soil.

Since the overland flow of water in urban areas (stormwater) forms the transport component of erosion, stormwater becomes the focal point of pollution prevention methods. The goal here is not to develop a list of stormwater BMPs; the EPA has a "menu of BMPs" designed to help communities with their implementation of the Phase II stormwater rules (USEPA 2010c). Instead, our focus is that given the limitations of the Phase II Stormwater controls (they only apply to new development or redevelopment), how can we achieve effective pollution prevention methods for stormwater in older urban areas?

To help answer this question, we now apply the pollution prevention framework used for point source control to nonpoint sources.

13.4.1 Applying the Source Control Framework to Nonpoint Sources

13.4.1.1 Step 1: Establishing Objectives and Gathering Background Data

Objectives for nonpoint pollution control will vary according to specific watershed conditions and needs. To place this effort on a firm foundation and to obtain the best information, the following general objectives are recommended:

- *Integrate nonpoint pollution control fully with watershed management/watershed restoration efforts.* There is a growing movement to implement nonstructural source control measures for stormwater (e.g., bioretention, swales, porous pavement) — collectively referred to as "low-impact development". Unfortunately, the adoption of these procedures at larger scales across watersheds is being hampered by (1) uncertainties in performance and cost, (2) insufficient engineering standards and guidelines, (3) fragmented responsibilities, (4) lack of institutional capacity, (5) lack of legislative mandate, (6) lack of funding and effective market incentives, and (7) resistance to change (Roy et al. 2008). Integration of these efforts with watershed-wide initiatives can help overcome some of these obstacles and increase their extent of implementation.
- *Manage the land to achieve better water quality.* Most watershed surface areas are 95% land, so proper land management is necessary to improve water quality. For example, a study of 27 water suppliers conducted by the Trust for Public Land and the American Water Works Association in 2002 found that the more forest cover in a watershed, the lower the treatment costs. According to the study, for every 10% increase in forest cover in the source area, treatment and chemical costs decreased approximately 20%, up to about 60% forest cover. Approximately 50%–55% of the variation in treatment costs was explained by the percent of forest cover in the source area (Ernst et al. 2004).

- *Integrate quantity and quality.* Reducing stormwater quantity will improve the quality of receiving waters.
- *Recognize the linkages between and within point and nonpoint sources, and optimize the control measures based on these linkages.* Point and nonpoint sources are linked through surface runoff, groundwater flow, and particulate transport. For example, using phytoremediation at brownfield sites provides vegetative cover and minimizes erosion and surface runoff.
- *Opt for nonstructural (low-cost and nonpermanent) controls first.* Many urban areas built expensive and permanent combined sewer overflow (CSO) control basins (e.g., the Rouge watershed) before adequately characterizing the potential of runoff reduction from nonstructural controls within the watershed (Kaufman and Marsh 1995).
- *Try to maximize the vertical movement of nonpolluted or minimally degraded water.* Many stormwater strategies advocate moving stormwater vertically, but this action can degrade groundwater and ultimately surface waters if the water is highly contaminated. Locations for infiltration should be chosen and prepared carefully so they have adequate organic content to assist with biodegradation and the adsorption of organics and heavy metals.
- *Pay close attention to erosion control in the headwaters region of the watershed.* The highest energy gradients are in the first-order stream segments found in the headwaters, so lowering erosion here lowers the kinetic energy of the stream and minimizes channel erosion. Reducing the erosion capacity of the streams will reduce the deposition of materials downstream, and also in the lakes and impoundments functioning as settling basins.

To support the general objectives above, the following data can be collected and generated:

- *Current measures of water quality at several locations throughout the watershed.* Universities and watershed organizations often conduct regular stream sampling and benthic monitoring, so check with them first. Benthic monitoring collects samples of materials dislodged from stream beds and assesses the species diversity. The presence of certain insects, for instance stoneflies, may indicate high water quality. Remember to also collect samples from wells within the watershed. This sampling is often free, with collection vials provided by the County Public Health Department. At a minimum, the parameters tested for in streams and groundwater should include total suspended solids, pH, dissolved oxygen, temperature, phosphorus, nitrate, and fecal coliform.
- *A stream order map for the watershed.* As noted earlier, lower-order streams are more vulnerable to contamination, so identifying their locations can assist with prioritizing source control efforts. Some universities have done this already.
- *Hydrographs for all gauged streams within the watershed.* A hydrograph is a plot of discharge (Q) over a specified time interval (Figure 13.12). These are available in different time increments (day, month, and year) from the U.S. Geological Survey. Hydrographs will help characterize the flood patterns within the watershed. Flood waters often carry massive amounts of contamination, so knowing where floods occur more frequently can help direct efforts to minimize stormwater runoff, which contributes large volumes of water to streams during wet-weather events.

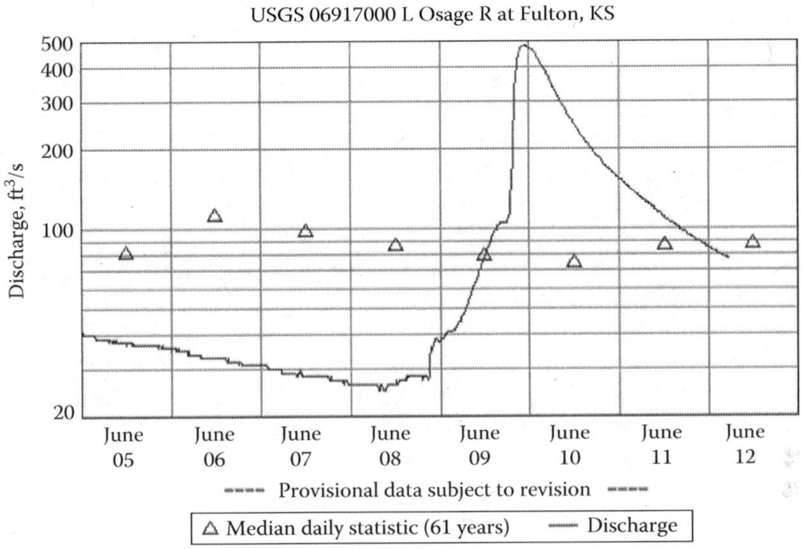

FIGURE 13.12
Daily hydrograph for a stream in Kansas, United States. (http://waterdata.usgs.gov/ks/nwis/uv?)

- *Aerial photographs of the watershed taken during "leaf on" conditions.* Try to get the most recent images available. The Natural Resource Conservation Service has aerial imagery for many urban areas. Land cover and tree patterns can assist with strategies to provide more interception storage and more infiltration capacity. Knowing the locations of brownfields and their proximity to surface water and groundwater recharge zones is also important for minimizing the impacts of contaminated runoff. In addition, aerial photos can identify locations along urban streams where public access is limited. In heavily urbanized watersheds, rail lines and industrial facilities often block access to the water. If the properties are abandoned, these areas can be restored to improve access.
- *A map of current water drainage infrastructure.* Many communities have digital maps of their sanitary and storm sewer lines, surface drains, and the locations of detention basins and access structures. These maps can help you identify where runoff volumes are large and natural drainage patterns have been altered.
- *A map showing the locations of brownfields in the watershed, and the results of the soil and groundwater testing from these sites.* The brownfield map can help identify areas where there are potential storage areas for stormwater runoff and for increasing the local density of vegetation. Soil and groundwater data can be used to characterize the extent and pattern of contamination, and the erosion potential at these abandoned sites.

13.4.1.2 Step 2: Inventory of Hazardous Substances

With point sources, reading product labels was the important activity. For nonpoint sources, determining the locations within the watershed where hazardous substances may be released and transported by erosion and stormwater is the essential action. During wet weather, lawns release nitrogen, phosphorus, and pesticide residues; feedlots contribute bacteria; parking lot runoff has oil, metals, and antifreeze; construction sites release soil; and industrial facilities may have on hand barrels of hazardous material capable of being swept away in a flood. It is important to know the locations of the hazardous substances and their relationship to the ephemeral drainage network capable of transporting their releases. For example, some erosion may reach a catch basin 30 m (100 ft.) away, yet in other cases the eroded material will never make it to a storm drain because it reaches a natural depression first. Aerial photos and topographic maps can help identify potential routes of overland flow; field checking, however, is required for verification. In addition, dye testing is often used to help identify infiltration/inflow of precipitation into sewers and to reveal the paths of stormwater and wastewater. During this process, nontoxic dyed water is introduced into roof drain leaders, driveway drains, or area drains. In some instances, dyed water is injected into the ground around foundations to check for the illegal connection of foundation drains. After introducing the dyed water, the downstream sanitary sewer access structure is checked for its presence to ascertain the path.

13.4.1.3 Step 3: Assessing CRFs of Substances Present

The locations where the highest CRFs are present should now be evaluated to help with the prioritization of the pollution prevention efforts. The results of the water sampling should guide the selection of locations. For example, if aquatic ecosystems have already been damaged and the groundwater samples produce high levels of chromium, this contaminant needs to be addressed as a priority.

13.4.1.4 Step 4: Preliminary Assessment of Geologic Vulnerability

The vulnerability of soils to the discharge of stormwater also needs to be considered. Some suggestions for assessing the soil vulnerability in an urbanized watershed include

- Do not rely upon the county soil surveys—use them for general orientation to the soils in the watershed. Urban soils are highly disturbed, and frequently the soil surveys do not capture their variation because the survey may be 10–20 years old.
- Instead, conduct a soil sampling program that enables a correlation between soil type, infiltration rates, and routes of overland flow during wet-weather events. The objective is to reveal those locations in the watershed where larger-grained vulnerable soils and less-vulnerable fine-grained soils exist. Use the soil data available from environmental site investigations to help determine the locations for additional sampling (Chapter 4). Once the sites for sampling have been selected, compare their locations to the major pathways of overland flow in the watershed, such as steep unvegetated slopes, large parking lots, and major streets. Then sample those sites within or along the drainage paths. At the sampled sites, use an infiltrometer to measure the infiltration capacity of the soil and have the soil samples analyzed for type, texture, and organic content by the local agricultural extensions of the nearest state university.
- Delineate floodplains and other areas where groundwater levels are likely to be closer to the surface. Try to avoid the placement of infiltration activities in these zones and concentrate on areas where the depth to groundwater is higher, thus giving more time for organisms and structures in the soil to reduce the contamination.

- If groundwater is used as a source of potable water in the area being considered for enhanced infiltration, and extraction wells are nearby, environmental risk should be considered high.

13.4.1.5 Step 5: Preventing Pollution through ESPM Methods

13.4.1.5.1 Elimination

Elimination of stormwater and erosion are impossible since they are triggered by natural processes. From the perspective of creating a sustainable watershed, elimination translates into reducing the amounts of stormwater volume and eroded material transported, as well as lowering the use of contaminants entrained by stormwater. These actions must obtain results that work toward returning streams to a state of dynamic equilibrium, reducing the sedimentation within lakes and reservoirs, and preserving soil for future generations.

Stormwater management in older urban areas presents special problems. House lots and vegetated spaces are typically smaller, downspout discharges are often routed directly onto paved surfaces (e.g., driveways), and sidewalks add to the amount of impervious surface. Compounding the problem is the presence of CSOs. A CSO occurs when the capacity of a pipe carrying sanitary and storm flows (a combined sewer) is exceeded and a special sewer, called an interceptor, is used to transport the untreated overflow to the nearest water body. Over 45 million people in 746 U.S. communities within 32 states are affected by CSOs. EPA estimates that about 850 billion gallons of untreated wastewater and storm water are released as CSOs each year in the United States (USEPA 2004).

In urban areas, overland flow originates from impervious surfaces, such as rooftops and roads, and sloped residential lots. The size of these areas is frequently small—often less than 0.10 ha (1/4 ac) —with flows greatly influenced by the micro-topography. For instance, at various locations, stormwater and sediment sinks exist with the micro-topography of a watershed; certain soils may infiltrate the entire storm volume, surface depressions store and infiltrate runoff, and locally dense vegetation canopies can intercept as much as 30% of the incoming precipitation (Cape et al. 1991).

The incorporation of micro-topography in the urban landscape has immense practical significance for preventing stormwater pollution. If the capabilities of the local micro-topography are exploited to store and/or redirect stormwater, large volumes of stormwater in urban areas can be kept out of the human-built drainage network. Since 95% of the total flow volume within combined sewers is stormwater, source reduction applied to stormwater management can help developed urban areas avoid substantial investment costs in structural stormwater management infrastructure (e.g., regional detention basins to contain CSOs, or sewer separation) (Kaufman 1999). In addition, reducing the volume of water treated results in substantial savings for communities and their residents.

Table 13.1 shows the water balance of an urban parcel (e.g., house, business, or industrial site) during a precipitation event. Inflows, outflows, and storage components of the water balance are routed through the four spheres of the geosphere (atmosphere, biosphere, hydrosphere, lithosphere) at a small geographic scale (0.1–0.3 ha). The micro-topographic elements of the landscape affecting the pathways of water are identified in the third column. For instance, precipitation moves along two paths: the first path is from the atmosphere to the biosphere, where vegetation affects the flows, and the second path is from the atmosphere to the lithosphere, where soil and pavement influence the flows.

Strategies to reduce stormwater runoff volumes must include the modification of those landscape features capable of rerouting stormwater back to its natural downward path. Structural (engineering-based) approaches to this problem result in building more pipe

TABLE 13.1

Urban Parcel Water Balance during Precipitation[a]

Component	Paths	Micro-Topography
Inflows		
Precipitation	Atmosphere → biosphere	Vegetation soil, pavement (driveways, sidewalks, and road surfaces)
	Atmosphere → lithosphere	
Runoff	Lithosphere → lithosphere	Soil, grass surfaces, pavement, roofs, gutters
Outflows		
Runoff	Lithosphere → lithosphere → hydrosphere	Soil, grass surfaces, pavement, storm sewers
Evaporation	Lithosphere → atmosphere	Ground surface
Evapotranspiration	Biosphere → atmosphere	Vegetation
Storage		
Interception	Atmosphere → biosphere	Vegetation
Infiltration	Atmosphere → lithosphere → hydrosphere[b]	Soil, vadose zone water
Surface depression storage	Atmosphere → lithosphere	Ground surface

[a] Including the road frontage.
[b] Interflow.

capacity, or providing basins for storage. An earth science approach looks at the scale of the processes generating the runoff, the source location and direction of the flows, and the landscape components affecting the transmission system at the relevant scales.

In Table 13.2, opportunities for stormwater storage or diversion at the micro-topographic scale are shown for each landscape feature influencing stormwater flows. The objective is to maximize storage at the site closest to the origin of runoff. At the micro-topographic scale of the house lot or single parcel, this translates into keeping stormwater on the lot and out of the storm sewers (Kaufman 1999). This is accomplished by directing water downward, instead of horizontally.

As Table 13.2 shows, increasing interception and infiltration and minimizing overland flow from impervious surfaces form the core components of a stormwater retrofitting strategy in developed urban areas. Increasing the organic content of soils (by topfilling with humus and cutting with a mulching mower), using available storage in surface depressions, and the use of swales can provide large amounts of infiltration if the lot layout permits their construction.

Flat roofs (such as those on parking lots) converted into vegetated green roof systems can reduce stormwater volumes and delay its release, and also help reduce the urban heat island effect. One study showed that green roofs not only reduced the amount of stormwater runoff, but they also extended its duration over a period of time beyond the actual rain event (van Woert et al. 2005). In another study, rain and roof runoff data collected from seven rains during October and November 2002 showed that the green roofs delayed the start of runoff by an average of 5.7h. The green roofs retained an average of 45% of the rain from the seven storms evaluated and delayed the peak runoff by 2h. Roof temperature data collected between April 2002 and February 2003 showed that the green roof

TABLE 13.2

Opportunities for Stormwater Storage and Diversion at the Micro-Topographic Scale

Microscale Landscape Feature	Opportunities for Storage/Diversion
Branches	Certain tree canopy shapes (such as beech) act to funnel water downward to zones containing more permeable soil near the main root zone (Mosley 1982)
Leaves	Conifers have generally higher interception capacity (Cape et al. 1991)
Soil	Increasing the organic content of soil increases its water retention capacity (Hudson 1994)
Median strips	Concave or flat median strips can increase stormwater storage
Surface depressions	Downspouts can be routed to slight depressions to increase infiltration, instead of discharging to paved surfaces. These areas can also delay stormwater from entering the pipe network
Root buttresses	Roots from larger trees and shrubs create friable soil more capable of infiltrating water (Bartens et al. 2008)
Flat rooftops	Can become storage areas with the addition of vegetation or rooftop cisterns
Swales	Store overland flow and prevent entry into street drains
Downspouts	Should be disconnected from footing drains and routed to cisterns or natural depression storage areas for infiltration or use as graywater; "in-line" storage bins can store up to 100 gal per downspout
Driveways	Use porous pavement to reduce runoff
Sidewalks	Use porous pavement

maximum surface temperatures averaged 6°C higher in the winter and more than 19°C lower in the summer (DeNardo et al. 2005).

Downspout diversion provides excellent potential for stormwater flow reduction in highly urbanized communities lacking available space for large-scale detention facilities. In the Beecher Water District, a highly urbanized area near Flint, Michigan, a downspout diversion program was implemented from 1996 to 1997. Here, downspouts were connected to footing drains and into sanitary sewer pipes, resulting in frequent sanitary sewer pipe overflows during wet weather. Downspout diversion contributed to a reduction of over 35% in the mean flow volumes within the sanitary sewer collection network and reduced overtime costs associated with overflow maintenance (Kaufman and Wurtz 1997). A similar reduction of 35% was also achieved in Dearborn, Michigan (Montrief, pers. com. 2010). In Portland, Oregon, the Downspout Disconnection Program has disconnected 50,000 downspouts, and these disconnections are removing more than 1 billion gallons of stormwater annually from the combined sewer system (City of Portland 2010). Where downspouts are already disconnected, many can be rerouted to prevent direct discharge onto paved surfaces.

The construction of large-scale CSO basins in several areas (e.g., Boston, Atlanta, Detroit) represents a major investment (Figure 13.13). Using source control to keep stormwater out of the sewer pipes before constructing these basins—as Portland has done—may help reduce infrastructure costs.

It may not be practical or affordable to convert driveways and sidewalks to porous pavement. There are, however, opportunities for some conversions whenever homeowners and commercial establishments repave their driveways and parking lots.

13.4.1.5.1.1 Elimination of Pollutants As with point sources, the elimination of chemicals should focus on contaminants with high CRFs. Many lake associations ban the use of

FIGURE 13.13
Construction of a deep-shaft combined sewer basin. (Photo by Kent S. Murray.)

fertilizers and pesticides due to concerns over cultural eutrophication. It would be good practice if every homeowner considering the use of a timed fertilization program offered by "lawn care experts" would conduct an inexpensive soil sample to assess the level of nutrients already present in the soil. Most lawns do not need four to six fertilization/herbicide applications per season.

Heavy metals are a major concern in the near-surface soils and groundwater of urban areas (Murray et al. 2006). Eliminating the use of lead in paints and chromium in some industrial applications will help lower their presence in stormwater runoff. Some elimination of metals from stormwater can also be achieved by protecting the soil from erosion in areas where metal contamination is high.

13.4.1.5.2 Substitution

Contaminants released into stormwater by lawns can be significantly reduced through better lawn management. Most state universities have agricultural extension services, and these organizations offer excellent advice on how to reduce the amount of fertilizer used and identify the least environmentally offensive types.

Road salt is a significant contaminant of surface waters and groundwater. A survey of 23 springs in the Greater Toronto Area of southern Ontario recorded chloride contamination levels, resulting from the winter application of road deicing salt, ranging from <2 to $>1200\,\mathrm{mg\ L^{-1}}$ (Williams et al. 2000). In another study, runoff from the application of road salts as a deicer led to the widespread contamination of groundwater and effectively changed the chemistry of the Rouge River due to groundwater discharge to this surface water body (Xun and Murray 2002). Many communities, especially those with lakes present, substitute sand for road salt.

13.4.1.5.3 Prevention

Soil stabilization—especially at construction sites—can help prevent erosion and the transport of contaminated soil by stormwater. One activity not used enough for preventing soil erosion is the inspection of erosion control measures after rainstorms. Another frequent problem is the failure of communities to require the collection of soil samples for textural analysis before development occurs at a site (Kaufman 2000). Basic information about soil type and texture can help direct specific erosion control measures. For example, the performance of certain geotextiles varies due to the difference in the characteristics of suspended solids associated with different soils (Barrett et al. 1998).

Special consideration should also be given to road placement in developing areas and wellhead protection. Roads should be designed so as to not join partial areas of runoff, and stormwater runoff should be considered when creating protection zones around wells.

13.4.1.5.4 Minimization

Since homes store an average of 3–10 gal of hazardous waste, all of the substitution recommendations made in the point source section apply here as well. Recycling paper, plastic, and metal cans is also an effective way to reduce inputs of undesired substances into streams. In addition, individual households can advance pollution minimization through the purchase of environmentally friendly products and packaging, increasing their efficiencies of energy and water use, and selecting less polluting transportation forms such as carpooling and mass transit.

13.4.1.6 Step 6: Assessing Results

Efforts to reduce erosion and stormwater runoff can be evaluated on several levels. Visually, improvements across the watershed will be seen in ecosystem health, receiving water quality (surface water and groundwater), and soil conditions. Hydrologically, there would be a lower frequency of floods and a less "flashy" stream response, that is, a slowing down of the relatively short time required to reach peak runoff typical of urban watersheds. Testing of the stormwater runoff would reveal consistently lower amounts of suspended solids, nitrogen, phosphorus, heavy metals, bacteria, and other contaminants responsible for degrading water quality. Groundwater quality would improve based on systematic well testing, and baseflow rates would increase as more water was infiltrated. The increased baseflow would also help to keep stream channels at higher levels throughout the year and make aquatic ecosystems less vulnerable to the effects of higher water temperatures.

The results of any sampling/monitoring efforts undertaken to reduce erosion and stormwater within watersheds should be published. This helps educate the public about these efforts and may increase public involvement. It also makes the organizations conducting the work more accountable.

13.5 Summary and Conclusion

Pollution prevention benefits everyone, especially the environment. By evaluating the processes generating waste, the types of wastes generated, identifying specific chemicals used, calculating CRFs, and evaluating geologic vulnerability, an effective and efficient pollution prevention program can be developed. This program maximizes the protection of human health and the environment and can result in significant financial benefits, future cost avoidance, and liability reduction.

Point sources and nonpoint sources share similar characteristics, specifically their transport media and transport processes. This common ground allows for the application of a successful point source prevention framework to nonpoint sources.

The next chapter presents six case studies that demonstrate the costs resulting from not having the correct information to develop an effective pollution prevention program. The outcome in some cases is damaging for the environment and costly for all involved.

References

Barrett, M.E., J.F. Malina Jr., and R.J. Charbeneau. 1998. An evaluation of geotextiles for temporary sediment control. *Water Environment Research* 70:283–290.

Bartens, J., S.D. Day, J.R. Harris et al. 2008. Can urban tree roots improve infiltration through compacted subsoils for stormwater management? *Journal of Environmental Quality* 37:2048–2057.

Cape, J.N., A.H.F. Brown, S.M.C. Robertson et al. 1991. Interspecies comparisons of throughfall and stemflow at three sites in northern Britain. *Forest Ecology and Management* 46:165–177.

City of Portland, Bureau of Environmental Services, 2010. Downspout Disconnection Program. http://www.portlandonline.com/bes/index.cfm?c=43081&a=177712 (accessed June 12, 2010).

Denardo, J.C., A.R. Jarrett, H.B. Manbeck et al. 2005. Stormwater mitigation and surface temperature reduction by green roofs. *Transactions of the ASABE* 48:1491–1496.

Ernst, C., R. Gullick, and K. Nixon. 2004. Protecting the source: Conserving forests to protect water. *Opflow* 30(5):1, 4–7.

Hudson, B.D. 1994. Soil organic matter and water quality. *Journal of Soil and Water Conservation* 49:189–193.

Humenik, F., M. Smolen, and S. Dressing. 1987. Pollution from nonpoint sources. *Environmental Science and Technology* 23:737–742.

Kaufman, M.M. 1999. Micro-topographic opportunities for stormwater management in urban landscapes. *Papers and Proceedings of the Applied Geography Conferences* 22:86–95.

Kaufman, M.M. 2000. Erosion control at construction sites: The science-policy gap. *Environmental Management* 26:89–97.

Kaufman, M.M. and W.M. Marsh. 1995. CSO control is no longer mere engineering. *American City and County* 110(2):10.

Kaufman, M.M. and M. Wurtz. 1997. Hydraulic and economic benefits of downspout diversion. *Journal of the American Water Resources Association* 33:491–497.

Montrief, Dean. Superintendent of Water and Sewerage, City of Dearborn, MI. Workshop Presentation at the University of Michigan-Dearborn, July 20, 2010.

Mosley, M.P. 1982. The effect of a New Zealand beech forest canopy on the kinetic energy of water drops and on surface erosion. *Earth Surface Processes and Landforms* 7:103–107.

Murray, K.S., D.T. Rogers, and M.M. Kaufman. 2006. Dissolved heavy metals in shallow groundwater in an urban watershed in southeastern Michigan. *Journal of the American Water Resources Association* 42:777–792.

Rogers, E.M. 1995. *Diffusion of Innovations*, 4th edn. New York: The Free Press.

Rogers, D.T., M.M. Kaufman, and K.S. Murray. 2006. Improving environmental risk management through historical impact assessments. *Journal of the Air and Waste Management Association* 56:816–823.

Roy, A.H., S.J. Wenger, T.D. Fletcher et al. 2008. Impediments and solutions to sustainable, watershed-scale urban stormwater management: Lessons from Australia and the United States. *Environmental Management* 42:344–359.

Schueler T.R. 1994. The importance of imperviousness. *Watershed Protection Techniques* 1:1–11.

Smolinske, S. and M.M. Kaufman. 2007. Consumer perception of household hazardous materials. *Clinical Toxicology* 45:1–4.

United States Environmental Protection Agency (USEPA). 1990. *Pollution Prevention Act of 1990*. Code of Federal Regulation 42 CFR, Chapter 133. Washington, DC.

United States Environmental Protection Agency (USEPA). 1996. *Overview of the Storm Water Program.* EPA 833-R-96-008. Office of Water, Washington, DC.

United States Environmental Protection Agency (USEPA). 2002. *National Water Quality Inventory: Report to Congress: 2002 Recycling Report*. EPA-841-R-07-001. Washington, DC.

United States Environmental Protection Agency (USEPA). 2003. *National Management Measures to Control Nonpoint Source Pollution from Agriculture.* EPA 841-B-03-004. Washington, DC.

United States Environmental Protection Agency (USEPA). 2004. *Report to Congress: Impacts and Control of CSOs and SSOs*, April 26, 2004, EPA 833-R-04-001, Office of Water, Washington DC.

United States Environmental Protection Agency (USEPA). 2009. *Road to Sustainability.* USEPA National Pollution Prevention Roundtable. Washington, DC.

United States Environmental Protection Agency (USEPA). 2010a. Pollution prevention. http://www.epa.gov/p2 (accessed January 10, 2010).

United States Environmental Protection Agency (USEPA). 2010b. *Solvent Alternatives Guide (SAGE).* USEPA Air Pollution Prevention and Control Division. Research Triangle Institute. http://www.p2pays.org/ref/19/1816/index.cfm.htm (accessed January 22, 2010).

United States Environmental Protection Agency (USEPA). 2010c. *National Menu of Stormwater Best Management Practices.* http://www.epa.gov/npdes/stormwater/menuofbmps (accessed June 10, 2010).

Van Woert, N.D., D.B. Rowe, J.A. Andresen et al. 2005. Green roof stormwater retention. *Journal of Environmental Quality* 34:1036–1044.

Williams, D.D., N.E. Williams, and Y. Cao. 2000. Road salt contamination of groundwater in a major metropolitan area and development of a biological index to monitor its impact. *Water Research* 34:127–138.

Xun, Z. and K.S. Murray. 2002. Change in groundwater chemistry along the Rouge River, southeastern Michigan, USA. *Groundwater* 23:131–133.

14

Case Studies: Successful and Unsuccessful Redevelopment of Contaminated Sites

14.1 Introduction

There are several reasons for the selection of the six case studies presented in this chapter. The primary reason is to provide students, professionals, organizations, policy makers, and concerned individuals with real-world examples of the interactions between geology and contamination at the site scale. Reviewing these cases illustrates why the effective application of sound science in the form of geologic theory and practice and the knowledge of contaminant characteristics are critical to successful site planning and future pollution prevention efforts. Moreover, the lessons outlined here and throughout this book can be successfully applied to future urban planning projects within the framework of a watershed approach at broader geographic scales. Since land represents 90%–95% of the surface area of urban watersheds, achieving sustainability requires effective land use planning at the sites within the watershed coupled with measures designed to prevent pollution from those sites.

Another important reason behind the selection of these case studies is their representativeness—they capture significant amounts of variation in terms of urban geology and contamination. What they have in common is their location within heavily urbanized areas of the United States, and all were, or are, brownfield sites occupied by light or heavy manufacturing activities. In addition, all are located near large surface water bodies or rivers. Their variability is expressed by their size, length of operation, geological environment, and contaminants of concern (COC). In terms of size, they range from less than 1 ha (2 ac) to more than 114 ha (250 ac). Operational histories range from less than 35 years to nearly 250 years. The geology beneath each site consists of unconsolidated sedimentary deposits encompassing thicknesses from less than 15 m (50 ft) to more than 305 m (1000 ft), with origins from glacial, alluvial, fluvial, and lacustrine geological processes. Their COC represent a broad array including Light Nonaqueous Phase Liquids (LNAPLs), Dense Nonaqueous Phase Liquids (DNAPLs), Volatile Organic Compounds (VOCs), Polycyclic or Polynuclear Aromatic Hydrocarbons (PAHs), polychlorinated biphenyls (PCBs), SVOCs, and several heavy metals (arsenic, cadmium, copper, chromium VI, lead, manganese, nickel, and mercury).

Each study has been rigorously performed and is a matter of public record (Rogers et al. 2006). The lessons learned are summarized at the end of each case study.

14.2 Case Study 1: Michigan

Case Study 1 is located in an urban area in Michigan (MDNRE 2010). It is an idle brownfield site. According to available records, it was first developed as a furniture manufacturing

TABLE 14.1
Site Characteristics of Case Study 1

Characteristic	Site Information
Location	Michigan
Years of operation	More than 100 years
Size of property	5 ac
Geologic setting	Glacial and lacustrine sediments
Soil type and stratigraphy	Interbedded medium to fine sand from the surface to a depth of approximately 27 m (90 ft) where a clay deposit is encountered
Contaminants detected	DNAPL and LNAPL VOCs, PAHs, Pb, Mn, PCBs
COC in groundwater	DNAPL VOCs, Mn, and PAHs
Depth to groundwater	4.6 m (15 ft)
Highest CRF_{GW} ranking	1274 DNAPL VOCs
Highest CRF_{SOIL} ranking	1457 PCBs
Geovulnerability ranking	78

facility in the late 1800s and evolved into a foundry manufacturing parts for the automobile industry in the early 1900s.

The history of chemical use was significant throughout the operational history at the facility, specifically

- LNAPL and DNAPL VOCs for cleaning, painting, staining, and varnishing
- PAHs in paint, lubricating, and machining
- Heavy metals from paints
- PCBs in electrical equipment and capacitors
- Heavy metals from foundry operations

Table 14.1 lists some basic information concerning the site including its size, years of operation, COC, dominant subsurface geology, depth to groundwater, and whether groundwater is present in an aquifer. Figure 14.1 shows the urban setting of the site.

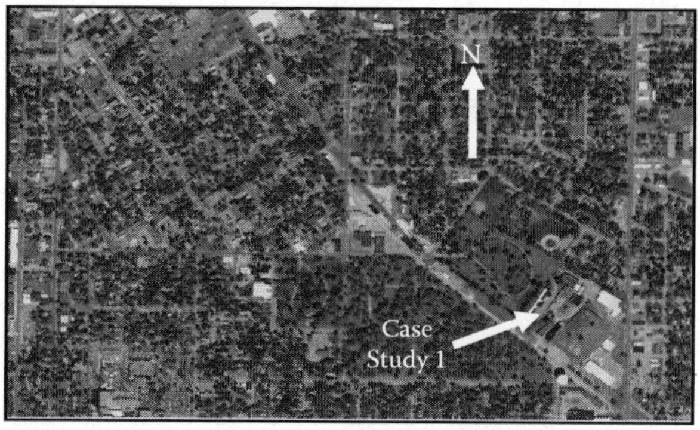

FIGURE 14.1
Aerial image of Michigan site showing urban setting.

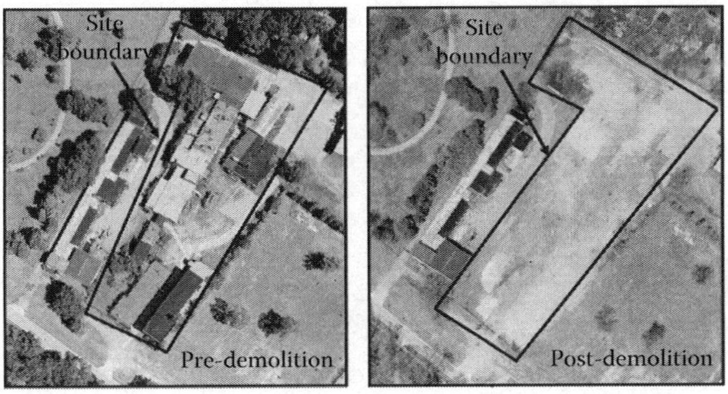

FIGURE 14.2
Pre- and post-demolition site maps (=site boundary).

Operations at the facility ceased in 2001 and the buildings were demolished in 2007. Figure 14.2 shows pre- and post-demolition aerials of the site.

Buildings were demolished to

- Lower carrying cost.
- Minimize trespass liabilities.
- Assist in evaluating potential sources of contamination. There had been more than 10 expansions conducted during facility history. Historical engineering drawings indicated numerous potential sources and many were not available for inspection. These sources included former pits, sumps, floor drains, tanks, and chemical storage areas. The historical map shown in Figure 14.3 was one of the many older maps consulted that aided with the assessment of the site by identifying potential sources of leaks and spills.
- Lower investigation cost by providing unobstructed access to all potential areas of concern (AOCs).
- Minimize the potential of not properly investigating all potential contaminant sources.

A Phase I environmental site assessment (Phase I ESA) was conducted and identified more than 20 recognized environmental conditions (RECs). Figure 14.4 depicts the identified RECs, and when this map is compared with the historical map of Figure 14.3 it demonstrates a concentration of RECs along the building fringes—locations frequently used for sumps, pits, and trenches.

A Phase II subsurface investigation confirmed the presence of contamination. COC included manganese in soil and groundwater, PAHs in soil and PAH-free product floating on top of groundwater, and DNAPL VOCs in soil and groundwater. Subsequent Phase II investigations confirmed the presence of a significant DNAPL VOC groundwater plume. However, the results of the Phase II also indicated the presence of multiple DNAPL VOC plumes, with at least one of the groundwater plumes originating from an upgradient source. This finding indicated a broader regional assessment of groundwater and potential sources must be conducted—a task made easier by an accurate geologic map. Figure 14.5 shows the available geologic map for the area near the site.

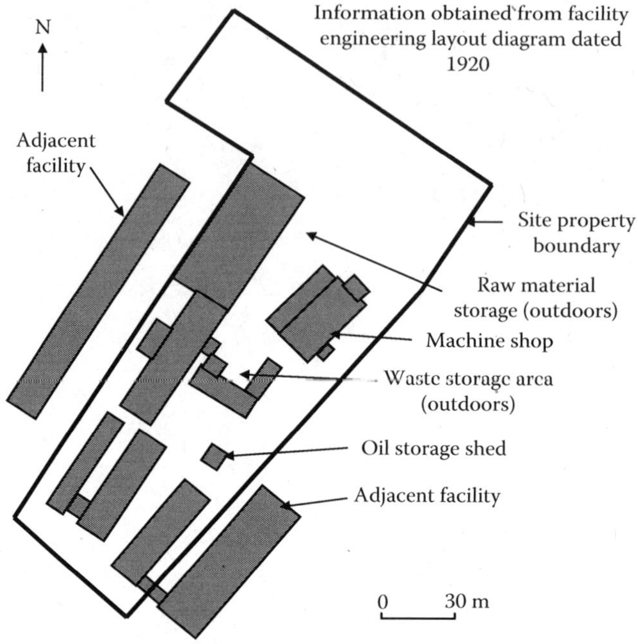

FIGURE 14.3
Historical map of site from 1920.

The areas under and immediately adjacent to the site have not been mapped at all, and the areas mapped do not provide information on the stratigraphy or identify appropriate geological units. This lack of a basic geologic map increases the difficulty of environmental investigations. As we noted in Chapter 2, urban geology is often not considered as appealing as the national parks, but is complex and fundamental to understanding the physical context for contamination events. Without this foundation, a proper assessment of vulnerability cannot be made, leading to poor decisions resulting in the location of facilities posing risks to the subsurface and surface environments in highly vulnerable areas—as is the case here.

For decades, this site has been located in a geologically vulnerable area with significant historical use of chemical compounds and a high groundwater contaminant risk factor. This combination resulted in a large contaminant plume affecting a potable aquifer, rendering its water off limits to not only the affected area—but to the general region. Unfortunately, this event is not an isolated occurrence at this location. Several contaminant plumes were identified from multiple sources during investigative activities, and as depicted in Figure 14.5, they are migrating toward a surface water body. As we know from Chapter 3, this phenomenon commonly occurs and, if not addressed, could eventually contribute to the deterioration of surface water quality. Figure 14.6 shows a planimetric (overhead) view of the DNAPL VOC contaminant plumes.

Aquifers are 3D in nature and represent a stack of related structural units. Understanding their vertical structure and the relationships between the units are critical to managing the resource. New mapping techniques have been developed at much higher resolutions with the ability to manipulate horizontal and vertical exaggeration. These capabilities enable

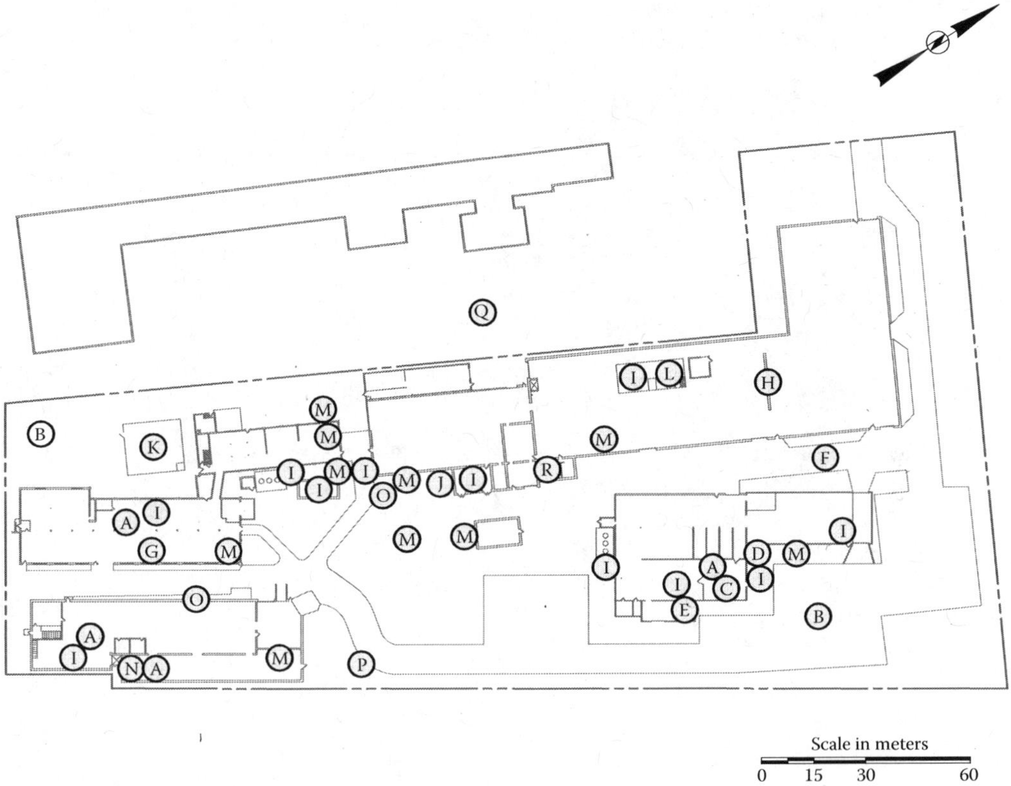

FIGURE 14.4
Identified RECs (each REC appears as a letter inside a circle).

vertical relationships to be emphasized without any negative impact on the generally understood shape of the surface in question. Figure 14.7 shows **high-resolution aquifer profiling**—a 3D cross section of the site and adjacent areas depicting at least three DNAPL contaminant plumes.

The 3D map and graph combination shown in Figure 14.7 provides additional information relevant to the remediation effort, including the relative subsurface distributions of the pollutants and their respective concentrations.

After reviewing the data and maps, it was determined that the cost of investigation and remediation would be extensive and could not be accurately measured. Timely restoration of the affected aquifer is unlikely because

- The contaminants are migrating along the bottom of the aquifer. DNAPL compounds are slightly denser than water and when present at a sufficient concentration may sink through the water column as they migrate by advection in the downgradient direction of groundwater flow. The location of contaminants at the bottom of the aquifer renders remedial methods more difficult and costly.
- The COC are DNAPL VOCs, and these substances do not degrade quickly. According to historical chemical usage records, DNAPL VOCs have not been used in more than 25 years. Since the highest concentrations detected were more

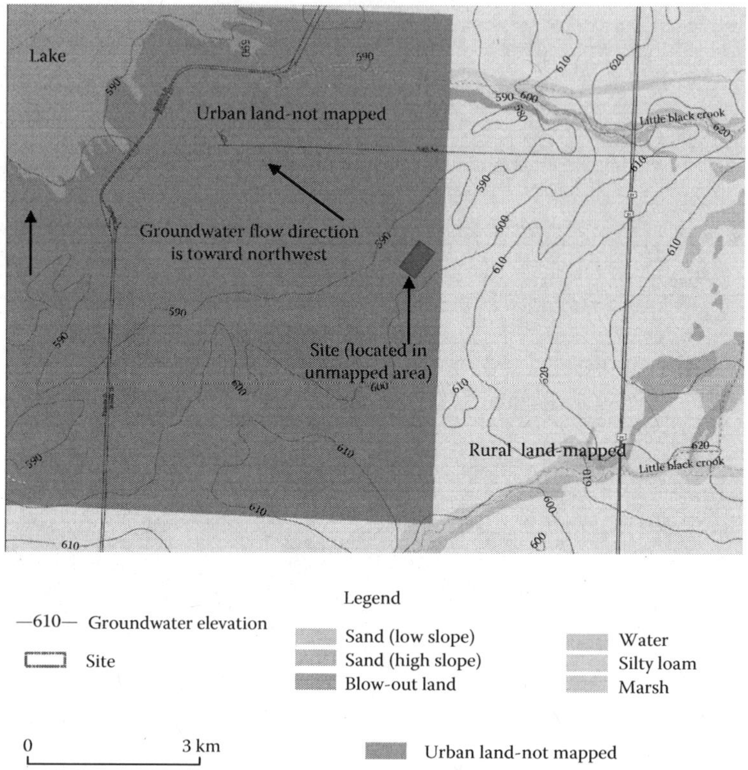

FIGURE 14.5
(See color insert.) Available geologic map showing lack of available data in urban area shaded gray.

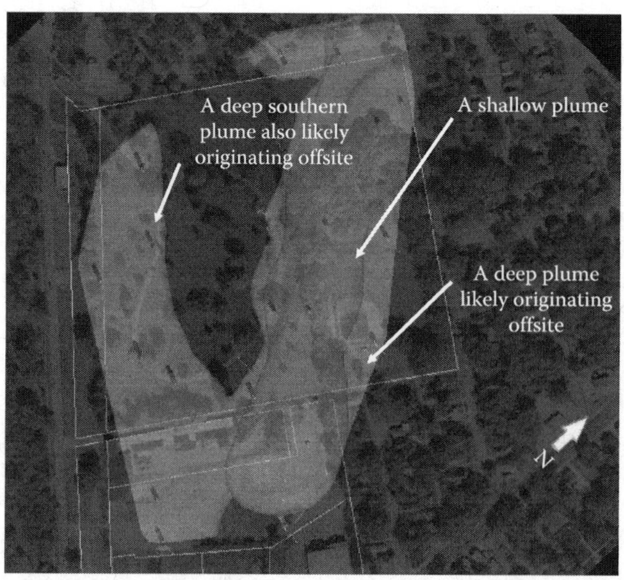

FIGURE 14.6
(See color insert.) Surface map showing contaminant plumes.

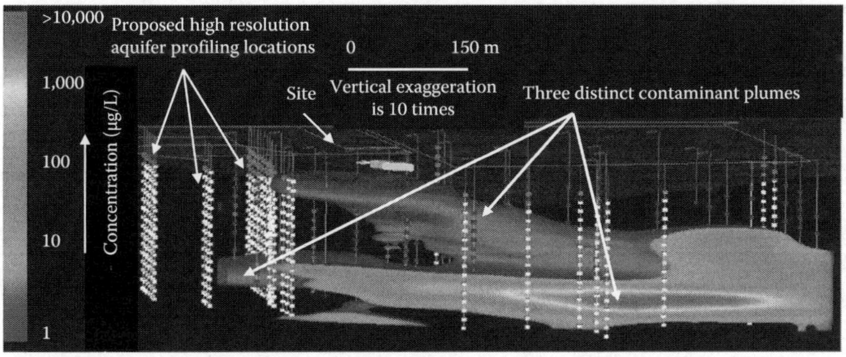

FIGURE 14.7
(See color insert.) Three-dimensional view of DNAPL VOC contaminant plumes.

than 305 m (1000 ft) downgradient from any suspected source, it confirmed the degradation rate was extremely slow. Therefore, natural attenuation alone will likely not be effective.

- Not all contaminant sources have been identified, removed, and abated. Several potential upgradient sources have been identified, but, until they have been fully investigated and removed, groundwater remediation cannot be effectively implemented.
- Remedial technologies will be effective at reducing contaminant mass but will not achieve timely and cost-effective restoration. One goal would be to reduce contaminant mass sufficiently so there is no detectable deterioration when the contaminant plumes reach surface water.

The lessons learned from this case study include

- Geologic maps are essential to all remediation efforts. In areas where there are synergistic effects between vulnerable geology and chemicals with high contaminant risk factors, geologic maps can help with the determination of contaminant extents.
- Low points within and near buildings such as concrete sumps, pits, trenches, and floor drains are potential sources of contamination.
- The fate and migration of DNAPL VOCs in a vulnerable geologic environment can present difficult challenges for remediation efforts.
- Because of surface and groundwater interaction, different geographic scales of analysis—e.g., the site scale and watershed scale—must be considered to achieve an accurate depiction of the contamination risks.
- Geologic forensic techniques and 3D mapping such as high-resolution aquifer profiling can help provide a complete and comprehensive evaluation of the site history, potential source evaluation, and knowledge of contaminant fate and transport.
- Three-dimensional maps depict the extent of contamination in an easy-to-understand and communicable format.

14.3 Case Study 2: Illinois

Case Study 2 is located in an urban area in Illinois (IEPA 2010). It is an idle brownfield site. According to available records, it was first developed as a chrome plating facility in the late 1960s and operated until 2003. Table 14.2 lists some basic information about the site.

The chemical primarily used at the facility was the heavy metal chromium VI. Secondary chemicals used included mercury and lead. For cleaning, small amounts of solvents consisting of an LNAPL mixture were used; however most of the cleaning conducted at this facility consisted of a caustic dip tank and bath.

The facility was demolished after operations ceased. Demolition of the structure provided access to the subsurface for sampling, lower carrying cost, and reduced the potential trespass liabilities. A Phase I ESA was conducted and identified five RECs, all associated with the use and storage of chromium VI and LNAPL VOCs. Subsequent Phase II investigations revealed concentrations of chromium that greatly exceeded the applicable regulatory criteria. In subsurface soil, there were very high concentrations (>10,000 ppm), and in perched groundwater the levels were greater than 440 ppm.

To estimate the nature and extent of the contaminant mass, more than 75 soil borings were drilled, over 20 shallow monitoring wells were installed, and nearly 500 samples of soil and groundwater were collected and analyzed. The estimated mass of chromium VI exceeded 13,000 lb. During the course of this extensive investigation, other contaminants were detected including LNAPL VOCs, lead, and mercury, but only chromium VI required a remedial action. The majority of the releases originated from a concrete sump and trench located inside the facility.

Figures 14.8 and 14.9 show the location of the site before and after it was demolished, respectively.

Figure 14.10 shows a 3D cross section of the site and adjacent areas depicting the extent of chromium VI impacts exceeding the applicable regulatory limit of 420 ppm. Figure 14.11 offers a different perspective of the extent of chromium VI and includes the locations of the contaminant sources.

Inspection of the 3D maps shown in Figures 14.10 and 14.11 offers some insights into the behavior of contaminants when they lack synergy with the geologic environment. Although the site exhibited significant levels of chromium VI in soil, the vertical extent of

TABLE 14.2

Site Characteristics of Case Study 2

Characteristic	Site Information
Location	Illinois
Years of operation	Approximately 35 years
Size of property	Less than 2 ac
Geologic setting	Lacustrine sediments
Soil type and stratigraphy	Clay interbedded with some fine silt layers
Contaminants detected	LNAPL VOCs, mercury, lead, and CrVI
COC	CrVI
Depth to groundwater	Groundwater present along building footings
Highest CRF_{GW} ranking	CrVI 2080
Highest CRF_{SOIL} ranking	CrVI 10.2
Geovulnerability ranking	13

FIGURE 14.8
Illinois site before demolition.

FIGURE 14.9
Illinois site after demolition.

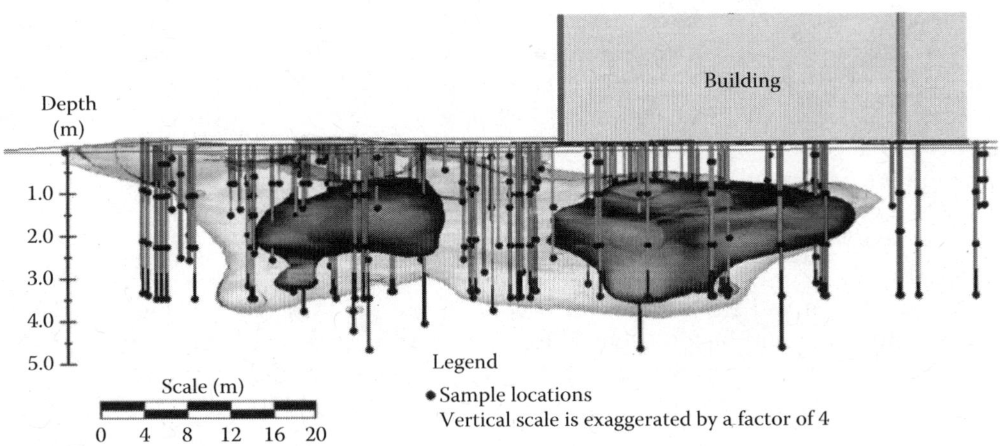

FIGURE 14.10
(See color insert.) Cross section showing soil boring locations and extent and magnitude of CrVI impacts (darker color indicates higher concentration).

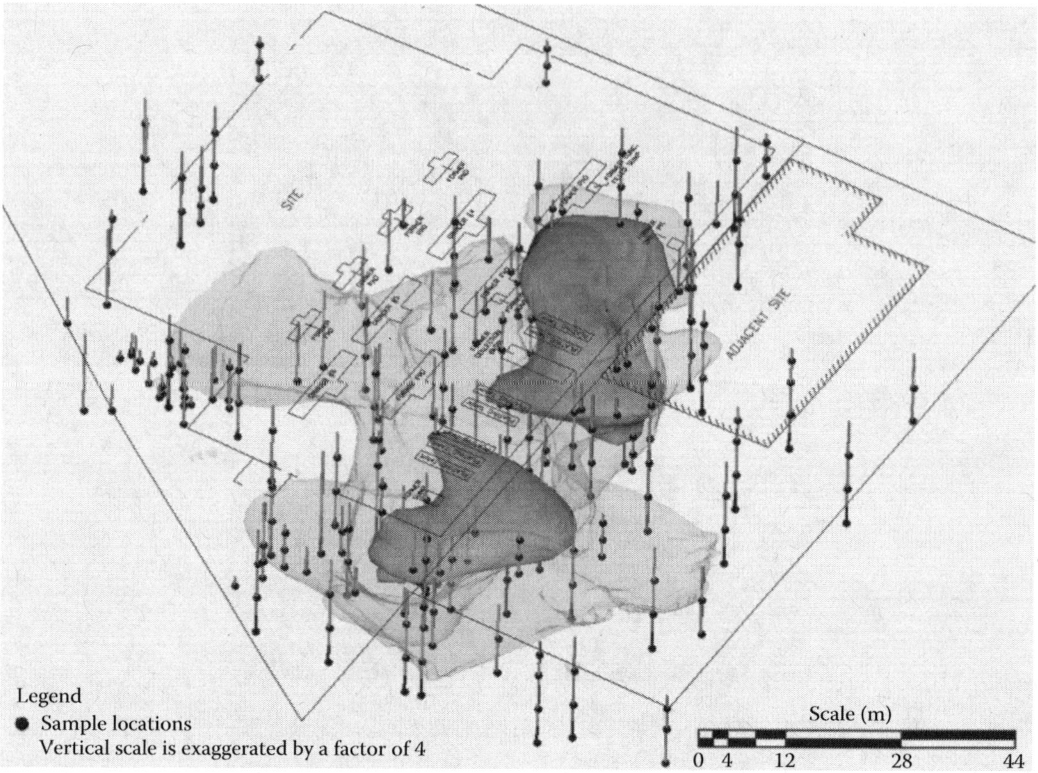

FIGURE 14.11
(See color insert.) Three-dimensional map view showing soil boring locations, source areas, extent and magnitude of on-site and off-site CrVI impacts (darker color indicates higher concentration).

the chromium plumes shown in Figure 14.10 is about 3 m (10 ft), indicating the chromium VI did not migrate significantly downward. This lack of migration occurred because the site is not within a geologically vulnerable area; it is located in an area of thick lacustrine deposits composed of clay. Since there is little or no groundwater present in this subsurface zone, a synergistic effect was not realized.

During the development of the site, a large amount of fill material was placed beneath the building, paved areas, and along building footings. The fill material consisted of sand and gravel that subsequently became water saturated and aided in spreading the contamination once released. Contamination did not extend to depths greater than 5 m (15 ft) beneath the surface of the ground (Figure 14.10) and traveled approximately 15 m (50 ft) beyond the release points shown in Figure 14.11.

Although the groundwater present at the site was anthropogenic in origin, it took a network of monitoring wells to arrive at this conclusion. As with Case Study 1, no detailed urban geologic map had been prepared for the urban region containing the site. By providing information concerning the distribution, stratigraphy, and hydrology of the shallow unconsolidated sediments, the urban geologic map would have reduced monitoring well-drilling costs.

The cost of remediation exceeded $1 million, with the lack of any completed contaminant pathways and a geologic environment consisting of clay helping to keep the costs down. Because the contamination was shallow and the buildings were demolished, excavation was very economical and became the primary remedial technology. Soils with lower contaminant concentrations were treated with an application of ferrous sulfate—a chemical reactant—to immobilize and change the valance state of the chromium VI to chromium III. Chromium III is much less soluble in water, is not nearly as mobile as chromium VI, is not a recognized carcinogen, and is much less damaging to plants and animals.

The lessons learned from this case study include

- Geologic maps are essential to all remediation efforts. In areas where there are no synergistic effects between the geologic environment and chemicals with high contaminant risk factors, geologic maps can help reduce monitoring well-drilling costs.
- Low points in and around buildings can act as conduits of contamination.
- Anthropogenic disturbance (i.e., backfilling building foundations and footings with porous materials) can enhance contaminant migration.
- Three-dimensional maps depict the extent of contamination in an easy-to-understand and communicable format.
- This site is underlain by thick lacustrine clay. The low vertical extent of large amounts of the highly mobile chromium VI suggests there is limited migration potential in this geologic environment.
- There can be successful and cost-effective remediation when contamination occurs in a favorable geological setting.

14.4 Case Study 3: Indiana

Case Study 3 is located in an urban area in Indiana (IDEM 2005). It became a brownfield site when it was shut down after more than 100 years of operation. According to available records, it was first developed as a foundry in the late 1800s. Table 14.3 lists some basic information concerning the site.

TABLE 14.3

Site Characteristics of Case Study 3 Located in Indiana

Characteristic	Site Information
Location	Indiana
Years of operation	More than 100 years
Size of property	40 ac
Geologic setting	Lacustrine and fluvial sediments
Soil type and stratigraphy	Clay interbedded with some fine silt and sand layers
Contaminants detected	LNAPL VOCs, lead, PCBs, and PAHs
COC	Lead and PCBs
Depth to groundwater	Less than 1.5 m (5 ft) beneath the surface of the ground
Highest CRF_{GW} ranking	Benzene 19.4
Highest CRF_{SOIL} ranking	PCBs 2524
Geovulnerability ranking	71

Figure 14.12 shows the location of the site within a heavily urbanized area of mixed land use. Heavy industry dominates the areas to the north and west of the site and residential land use dominates the areas east and south. A shipping canal flowing north to a larger surface water body borders the site along the western property boundary. The geology of the site is dominated by lacustrine deposits of interbedded clay, silt, sand, and dune deposits. Depths of unconsolidated deposits are believed to be greater than 15 m (50 ft). Detailed geologic mapping in three dimensions has not been conducted for the region. The direction of groundwater flow is west northwest toward the ship canal.

Figure 14.13 is a lower-altitude aerial photograph showing the facilities on-site. Please note the presence of the transformer area at the right-center of the photograph. Until their ban in 1979, PCBs were widely used as a fire retardant and insulator in the

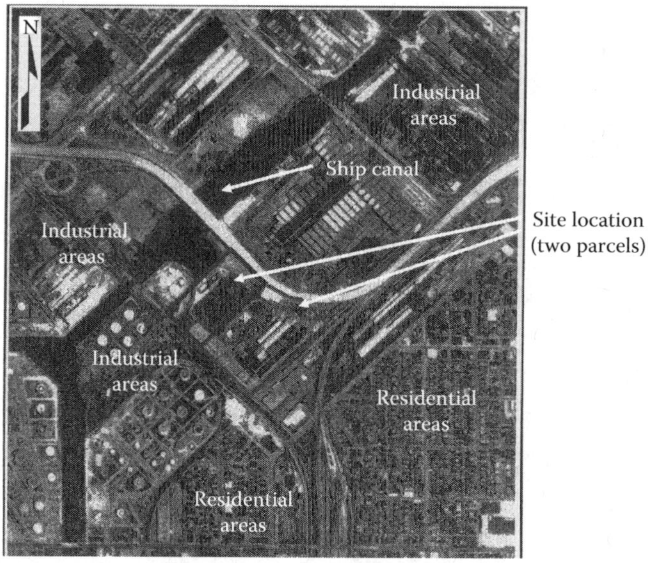

FIGURE 14.12
Aerial photograph showing the location and surrounding region of the Indiana site.

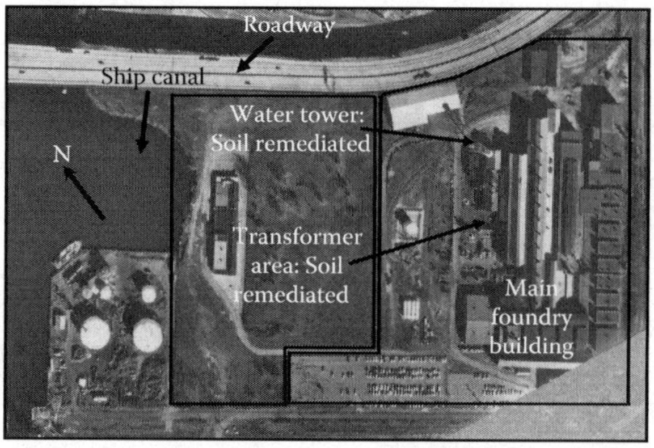

FIGURE 14.13
Facilities on-site.

manufacture of transformers and capacitors due to their ability to withstand exceptionally high temperatures.

Investigation of this site began with a Phase I ESA that identified 25 RECs and 15 AOCs. Three Phase II subsurface investigations were conducted to find and define the nature and extent of contamination at the site. Investigative actions included drilling more than 80 soil borings, installing temporary groundwater monitoring wells, and collecting and analyzing more than 100 samples for selected compounds including DNAPL and LNAPL VOCs, PAHs, SVOCs, PCBs, and heavy metals (As, Ba, Cd, Cr, Cu, Pb, Hg, Mg, Mn, Se, Ag, and Zn). Figure 14.14 shows the locations of the RECs and AOCs.

This site is located in a geologically vulnerable area of shallow groundwater hydraulically connected and very close to a large surface water body (vulnerability rating = 71). Luckily, a synergistic effect between a high-risk chemical released into a geologically vulnerable area did not occur because this site did not use significant amounts of chemicals. As a result, the extent of contamination did not extend more than 6 m (20 ft) beyond release points and did not extend to depths greater than 1 m beneath the ground surface. A geologic map would have helped predict this outcome, improved the soil and groundwater sampling protocols, and would have also saved significant time and financial resources.

Of the several contaminants detected in soil, the only areas requiring remediation were one impacted with PCBs and another impacted with lead. The area impacted with PCBs was located near the transformer area (Figure 14.13). The area impacted with lead was located beneath the facility's water tower painted with a lead-based paint removed by weathering (Figure 14.13). Contaminants were detected in the shallow groundwater, but not at levels requiring further action. The remedial option selected was excavation of impacted soil. Excavation was selected since both impacted locations were accessible and the contamination was shallow and restricted to the upper 1 m of soil. Remediation was completed within 1 week. The total cost of investigation and remediation of this site was $400,000 and was completed in 1 year from initiation of the Phase I ESA to closure by the regulatory agency.

The lessons learned from this case study include

- Geologic maps are essential to all remediation efforts. In areas of vulnerable geology and chemicals with low groundwater risk factors, geologic maps provide valuable information that help guide investigative and remedial efforts.
- Applying urban investigative and geologic forensic techniques such as comprehensive site history and knowledge of contaminant fate and transport help to increase the effectiveness of the remediation effort.
- A characterization of the fate and transport of lead and PCBs. Lead and PCBs had been released at this facility and did not impact groundwater even though groundwater was shallow (within 1 m of the surface). These substances also did not migrate a significant distance laterally because of the nature of the compounds (high soil contaminant risk factor) and the absence of anthropogenic pathways or disturbance.
- There is a lack of synergy at sites between chemicals and the geologic environment where chemicals with a high risk to groundwater were not used significantly during the operational history of the facilities.
- Successful and cost-effective remediation is possible when no high-risk chemicals are present. Modest environmental impacts and timely remediation can also be achieved.

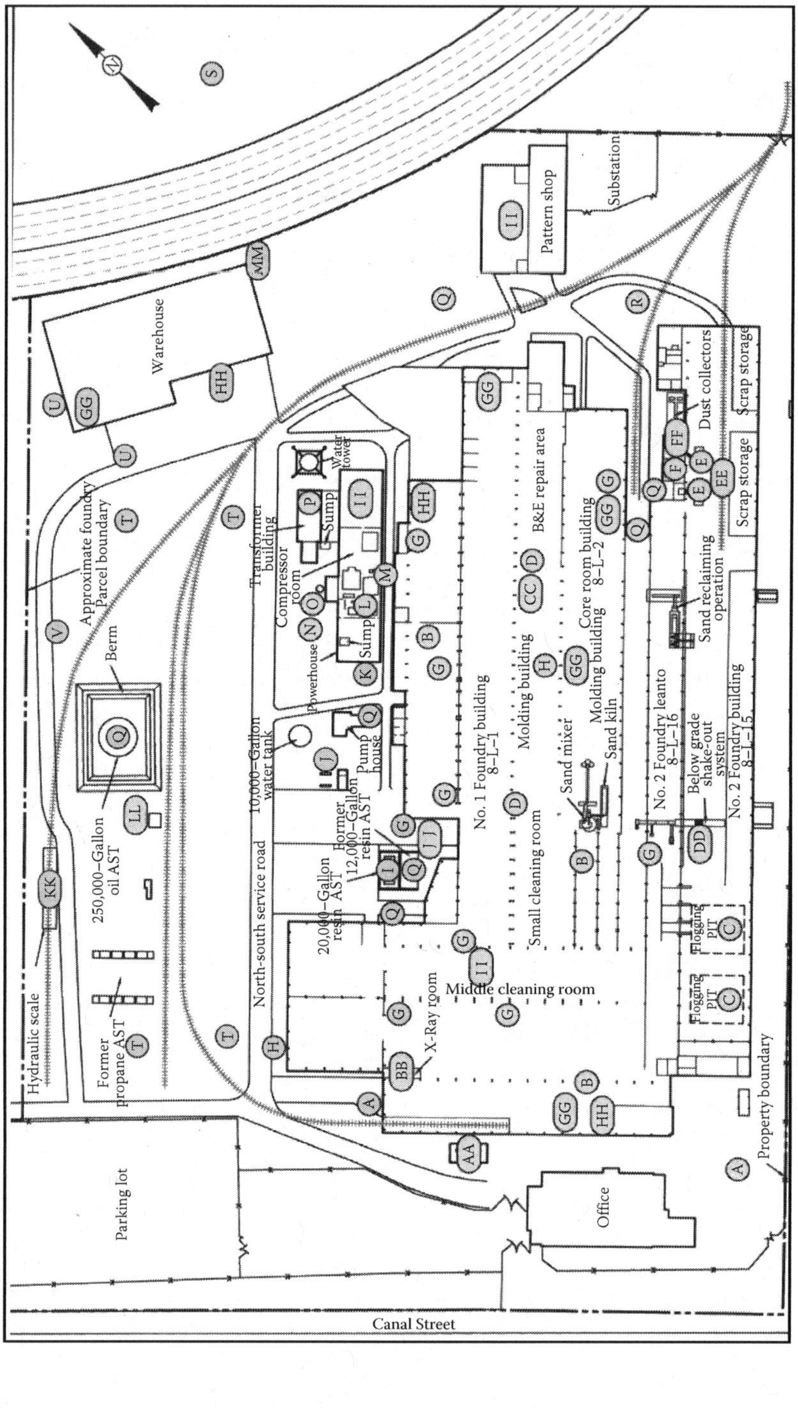

FIGURE 14.14
Location of RECs for Case Study 3.

14.5 Case Study 4: Wisconsin

Case Study 4 is located in an urban area in Wisconsin (WDNR 2003). It became a brownfield site when the plant was shut down after more than 80 years of operation. According to available records, it was first developed as a heavy manufacturing facility in the 1920s. Table 14.4 lists some basic information concerning the site, and Figure 14.15 shows the site.

This site is located in a geologically vulnerable area of moderate magnitude (vulnerability rating = 55). Shallow groundwater is very near and hydraulically connected to a large surface water body via anthropogenic means (storms sewers). No detailed urban geologic map has been prepared for the region.

Investigation of this site began with a Phase I ESA that identified 25 recognized environmental concerns and 15 locations of less significant concern. Three Phase II subsurface investigations were conducted to find and define the nature and extent of contamination at the site. Operations included drilling more than 300 soil borings, installing more than 100 groundwater monitoring wells, and collecting and analyzing more than 1000 samples for

TABLE 14.4

Site Characteristics of Case Study 1 Located in Wisconsin

Characteristic	Site Information
Location	Wisconsin
Years of operation	More than 100 years
Size of property	40 ac
Geologic setting	Lacustrine and fluvial sediments
Soil type and stratigraphy	Clay interbedded with some fine silt and sand layers
Contaminants detected	LNAPL and DNAPL VOCs, PCBs, PAHs, arsenic, lead, cadmium, and copper
COC for soil	LNAPL and DNAPL VOCs, PAHs, SVOCs, PCBs, and lead
COC for groundwater	DNAPL VOCs
Depth to groundwater	Less than 1.5 m beneath the surface of the ground
Highest CRF_{GW} ranking	DNAPL VOCs 1274
Highest CRF_{SOIL} ranking	PCBs 1457
Geovulnerability ranking	55

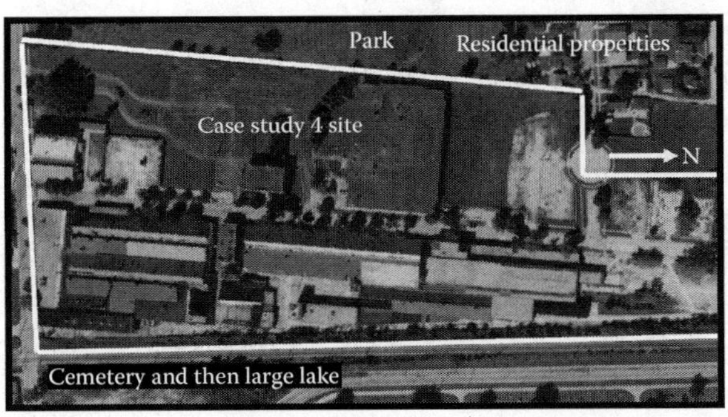

FIGURE 14.15
Site map of Case Study 4.

selected compounds that included DNAPL and LNAPL VOCs, PAHs, SVOCs, PCBs, and heavy metals (As, Ba, Cd, Cr, Cu, Pb, Hg, Se, Ag, and Zn).

Several contaminants were detected in soil, including LNAPL and DNAPL VOCs; PAHs; SVOCs; PCBs; and the heavy metals arsenic, cadmium, lead, and copper. Contaminants detected in groundwater included DNAPL VOCs and low levels of LNAPL VOCs. Contaminants requiring remediation included LNAPL and DNAPL VOCs, PAHs, SVOCs, PCBs, lead and cadmium in soil and DNAPL VOCs in groundwater. Figure 14.16 shows the location of soil borings and monitoring wells and the extent of contaminant impacts requiring remediation.

The geology of the site was dominated by a thin sand layer that became thinner toward the southern portion of the site. The thickness of the sand was less than 3 m in the northern portion and only 1 m thick in the southern portion. Beneath the sand was a thick sequence of clay more than 18 m (60 ft) thick and believed to be a glacial lodgment till deposit. The

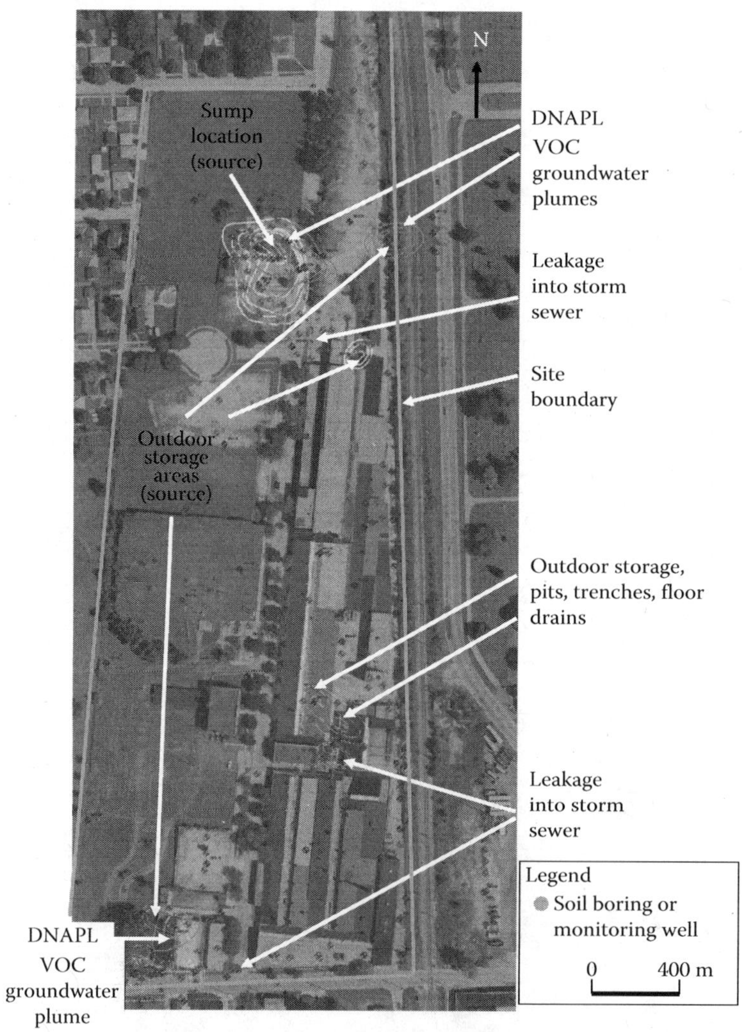

FIGURE 14.16
(See color insert.) Extent of impacts at Case Study 4.

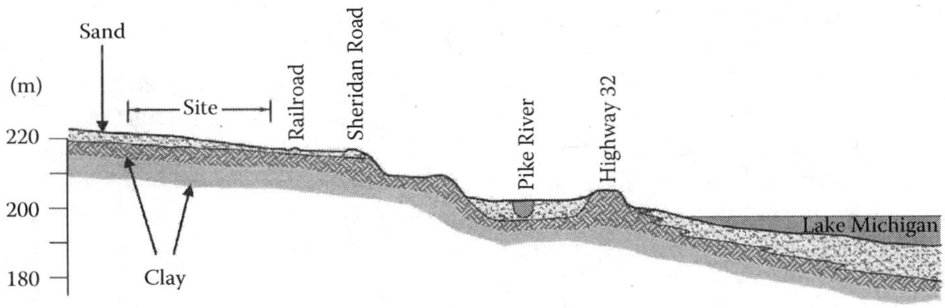

FIGURE 14.17
Geologic cross section of Case Study 4 site.

sand deposit originated from beach and reworked beach deposits from fluvial processes. Groundwater was encountered at depths of less than 1.5 m (5 ft) across the site. Figure 14.17 is a west to east geologic cross section of the site showing the location and relative thickness of the geologic layers and the proximity of the site to the large surface water body.

Storm sewers and building foundations significantly influenced groundwater flow across the site. The leakage of groundwater into the storm sewers at several locations provided an anthropogenic pathway for contaminants in groundwater and resulted in the discharge of contaminants into a large surface water body located less than 1 km (0.6 mile) away. On a more positive note, the building foundations effectively prevented the development of large groundwater contaminant plumes because they extended only to the base of the sand, where further downward migration was prevented by the basal clay.

Sources of contamination included sumps, pits, trenches, floor drains, and outdoor storage of waste chemicals; many of these are identified in Figure 14.16. The remedial method selected for soil was excavation since the contamination was accessible after demolition and did not extend to depths greater than 3 m. Since the impacts were shallow due to the anthropogenic influences just mentioned, excavation and pumping were used for the remediation of groundwater.

The ultimate disposition of the site was influenced by its location within a residential urban area and its proximity to a large surface water body. Continued use as an industrial site was not feasible because of the site's residential location and lack of appropriate highway access, so the surface structures were demolished and a portion of the site was converted into a city park. Demolition had the effects of

- Speeding redevelopment.
- Lowering carrying cost.
- Minimizing trespass liabilities.
- Assisting in identifying and evaluating potential sources of contamination. There had been more than 25 expansions conducted during facility history. Historical engineering drawings indicated numerous potential historical sources of contamination, and many were not available for inspection. These included former pits, sumps, floor drains, tanks, and chemical storage areas.
- Lowering remediation cost by providing unobstructed access to all potential AOCs.
- Minimizing the potential for missing—or not properly investigating—all potential contaminant sources.

Figures 14.18 and 14.19, respectively, show aerial and ground-level views of the site after demolition of the buildings.

This site did experience some synergistic effect between the moderate vulnerability of the geologic environment and the use of high-risk chemicals. Damage was limited by the leakage of groundwater into the storm sewers, the lack of significant groundwater volumes, and building foundations inhibiting groundwater flow. These anthropogenic influences prevented a larger environmental impact commensurate with the risks present. Despite this good fortune, investigation and remedial costs were expensive—the project lasted 12 years after conducting the Phase I ESA, with expenditures totaling $9.3 million,

The lessons learned from this case study include

- Geologic maps are essential to all remediation efforts. In areas where there is high anthropogenic disturbance, geologic maps can help investigators distinguish between natural and artificial drainage patterns.

FIGURE 14.18
Photograph after site demolition.

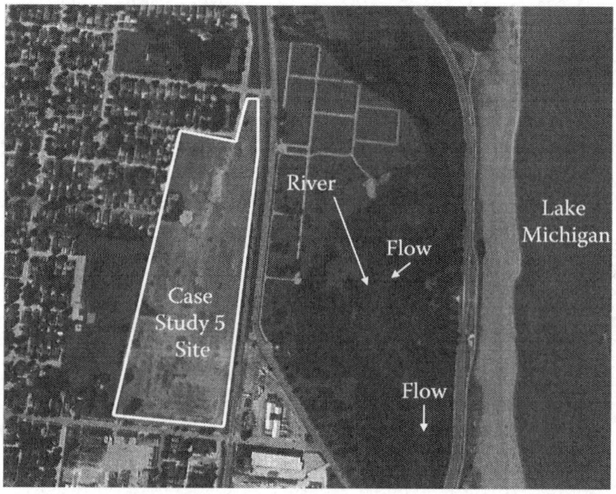

FIGURE 14.19
Aerial photograph after site demolition and remediation.

- The application of effective urban investigative and geologic forensic techniques such as conducting a comprehensive site history and the knowledge of contaminant fate and transport can identify potential anthropogenic effects at a site.
- Anthropogenic influences may reduce the synergistic effects resulting from certain combinations of geology and high-risk chemicals. This effect is apparent where groundwater is shallow, in short supply, and the constructed infrastructure works to prevent the downward movement of the contamination (e.g., storm sewers and building foundations).
- Anthropogenic infrastructure can influence migration pathways, complicating environmental investigative and remedial efforts.
- The fate and migration of lead, PCBs, PAHs, and SVOCs released into a geologic environment of shallow groundwater indicates these compounds did not migrate a significant distance laterally. Their relative immobility arises from the nature of the compounds—they have a high soil contaminant risk factor.
- Outdoor storage areas offer potential sites for contaminant release.
- Low points in and around buildings can act as effective conduits of contamination; sources of contamination were sumps, trenches, and floor drains.
- Successful redevelopment of a damaged site is possible. Here, a city park and conversion to other land uses is in progress.

14.6 Case Study 5: California

Case Study 5 is located in an urban area in California (DTSC 2006). It became a brownfield site when shut down after 36 years of operation. According to available records, it was first developed as a chrome plating facility in 1966 and operated until 1999. Table 14.5 lists some basic information concerning the site.

TABLE 14.5

Site Characteristics of Case Study 1 Located in California

Characteristic	Site Information
Location	California
Years of operation	36 years
Size of property	2 ac
Geologic setting	Alluvial and fluvial sediments
Soil type and stratigraphy	Sand interbedded with some fine silt and clay layers.
Contaminants detected	Chromium VI, lead, cadmium, nickel, and mercury
COC for soil	Lead, mercury, cadmium, and chromium VI
COC for groundwater	Chromium VI
Depth to groundwater	Approximately 21 m (70 ft) beneath the surface of the ground
Highest CRF_{GW} ranking	2300 chromium VI
Highest CRF_{SOIL} ranking	Mercury 15,600
Geovulnerability ranking	63

Investigation at this site began with an assessment of a subsurface sump used to temporarily store waste plating fluids. Results of soil samples collected near the sump indicated the presence of chromium VI at elevated levels, and this finding triggered further investigation. Subsequent Phase II investigations revealed groundwater had been impacted at a depth of 21 m (70 ft) beneath the ground surface, and an additional chromium VI source was located upgradient and adjacent to the facility. This establishment was also performing chrome plating.

Several phases of investigation had been conducted to evaluate other potential sources of contamination and whether other contaminants were present. Contaminants detected in soil included cadmium, chromium VI, lead, mercury, and nickel. Contaminants detected in groundwater from site operations were restricted to chromium VI. The chromium VI contamination in groundwater extends more than 3.2 km (2 mi) downgradient from the source. Figure 14.20 shows the location of soil borings and monitoring wells and the preliminary extent of contaminant impacts requiring remediation.

The geology of the site consists of an interbedding of sand, silt, and small clay lenses of groundwater encountered at a depth of approximately 21 m beneath the surface. Groundwater is present in a complex network of aquifers that are used as a potable source of water. The aquifers are also used to store billions of gallons of water for potential future use. Groundwater beneath the site is flowing west toward a river located less than 4 km (2.5 mi) from the site.

With its vulnerability rating of 65, this site is prone to groundwater contamination. As demonstrated here, even though groundwater exists at a relatively deeper depth than the other sites, contaminants can easily impact deep groundwater under favorable conditions and reach surface water discharge locations far downgradient. Figure 14.22 shows the interpolated chromium VI groundwater plume from this site reaching the riverbed over 3.2 km (2 mi) away. If a geologic map for this urban area existed, it would have simplified the interpolation of the contaminant plume.

The technologies selected to remediate soil were excavation of the near-surface soil and in situ chemical treatment of subsurface soil. Excavation of surface soil was selected because in situ chemical treatment and other technologies were not feasible due to the presence of multiple heavy metals requiring remediation. In situ treatment was selected for subsurface soil at depths between 6.7 m (22 ft) and 21 m (70 ft) because chromium VI was the only contaminant requiring remediation. The in situ treatment of the deeper soil consisted of injecting a chemical reactant (ferrous sulfate) that reduced the chromium VI to chromium III. This chemical reaction immobilized the chromium so it could no longer migrate and impact groundwater. Figure 14.21 shows a geologic cross section of this site and the areas selected for soil excavation and in situ chemical treatment.

Since the remedial alternative selected for the shallow soil was soil excavation, demolition of the building was necessary to gain unobstructed access to impacted areas. Demolition of the building and removal of the footings had the following added benefits:

- Enhanced redevelopment potential
- Lowered carrying cost
- Minimized trespass liabilities
- Assisted in confirmation of contaminant sources
- Lowered overall remediation cost by providing unobstructed access to all potential AOCs
- Minimized the potential for missing—or not properly investigating—all potential contaminant sources

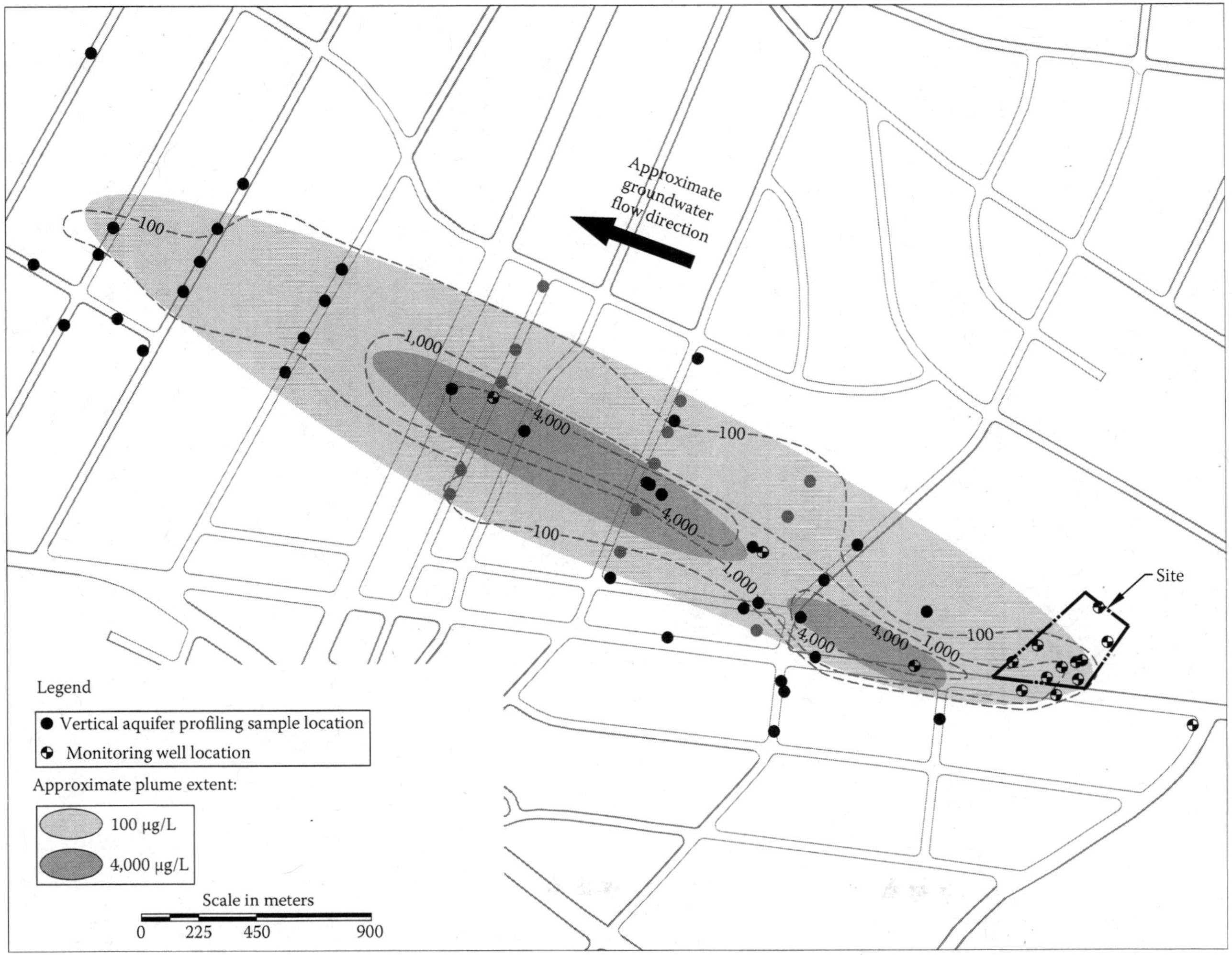

FIGURE 14.20
Location of groundwater sample collection points and preliminary area requiring groundwater remediation.

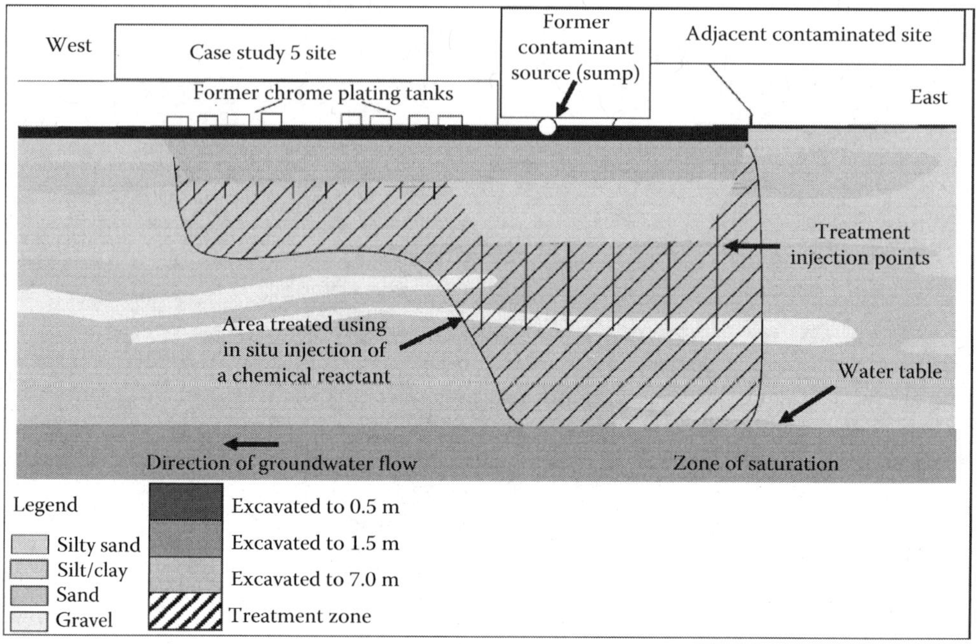

FIGURE 14.21
Geologic cross section and areas selected for soil excavation and in situ treatment.

The selected technology for groundwater remediation was to pump and treat groundwater within seven discrete zones using calcium polysulfide. The treated groundwater was reinjected to create a circulation pattern within each of the seven treatment zones. Each treatment zone was located along the longitudinal axis of the contaminant plume at approximate intervals ranging from 500 to 1000 m. The spacing was contingent upon contaminant concentrations (i.e., higher concentrations earned a tighter spacing) and access to the subsurface. Care was taken to ensure that the location of aboveground treatment equipment was placed in areas that created minimal disturbance and maintained security.

This site exhibited a synergistic effect between a geologically vulnerable area and a chemical with a high groundwater contaminant risk factor. When groundwater is affected, costs soar; this is borne out by the more than $25 million it took to remediate this site's soil and groundwater.

The lessons learned from this case study include

- Geologic maps are essential to all remediation efforts. In areas where there is high geologic vulnerability and highly mobile contaminants, the synergistic effects produced by their interaction can be better predicted by knowing the aquifer structure and subsurface flow patterns.
- Application of effective urban investigative techniques such as collecting groundwater samples using a cone penetrometer (CPT) method in easements and aquifer profiling assists with the investigation and remedial efforts.
- Contamination occurring at sites where there is vulnerable geology and high-risk contaminants is likely to exhibit synergistic effects, and be costly to remediate.

- The fate and transport of chromium VI at this site indicates it is more mobile than other heavy metals in the environment. Lead, nickel, cadmium, and mercury were also released at this site, but only chromium VI was mobile enough to contaminate groundwater.
- A low building location—in this case a sump—was the source of the contamination.
- Successful remediation is possible using a simple and established method (excavation) and an innovative method (in situ chemical treatment for deeper soil and groundwater).

14.7 Case Study 6: New Jersey

Case Study 6 is located in an urban area in New Jersey (NJDEP 2010). It became a brownfield site when shut down after a long operational history that began in 1760. Approximately 23 ha (50 ac) of the 114 ha (250 ac) site were used for industrial operations and were foundry related. The remaining 91 ha (200 ac) were used for other purposes such as agriculture, trucking, landfilling, and a shooting range or were forested. Table 14.6 lists some basic information concerning the site, and Figure 14.22 shows an aerial photograph of the site taken in 2006.

Investigation of this site began with a Phase I ESA that identified more than 200 RECs. Several Phase II subsurface investigations were conducted to find and define the nature and extent of contamination at the site. These activities included drilling more than 300 soil borings, installing groundwater monitoring wells, and collecting and analyzing more than 1000 samples of soil, groundwater, and sediment for selected compounds that included DNAPL and LNAPL VOCs, PAHs, SVOCs, PCBs, and heavy metals (As, Ba, Cd, Cr, Cu, Pb, Hg, Se, Ag, and Zn).

Several contaminants were detected in soil, including LNAPL and DNAPL VOCs, PAHs, SVOCs, PCBs, and heavy metals (arsenic, cadmium, lead, and mercury). Contaminants

TABLE 14.6

New Jersey Site Characteristics

Characteristic	Site Information
Location	New Jersey
Years of operation	Nearly 250 years
Size of property	Over 250 ac
Geologic setting	Fluvial sediments
Soil type and stratigraphy	Silt and fine-grained sand interbedded with clay
Contaminants detected	LNAPL and DNAPL VOCs, PCBs, PAHs, arsenic, cadmium, chromium III, lead, and mercury
COC	PCBs, PAHs, lead, and mercury
Depth to groundwater	Less than 1.5 m beneath the surface of the ground
Highest CRF_{GW} ranking	Mercury 3.1
Highest CRF_{SOIL} ranking	Mercury 15,600
Geovulnerability ranking	67

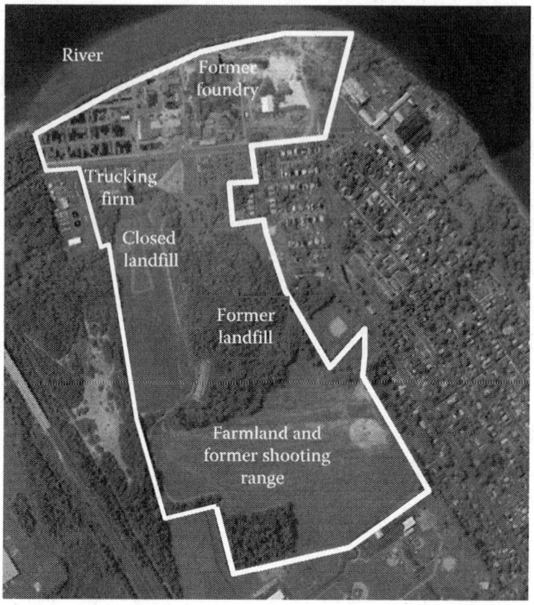

FIGURE 14.22
Urban setting of New Jersey site.

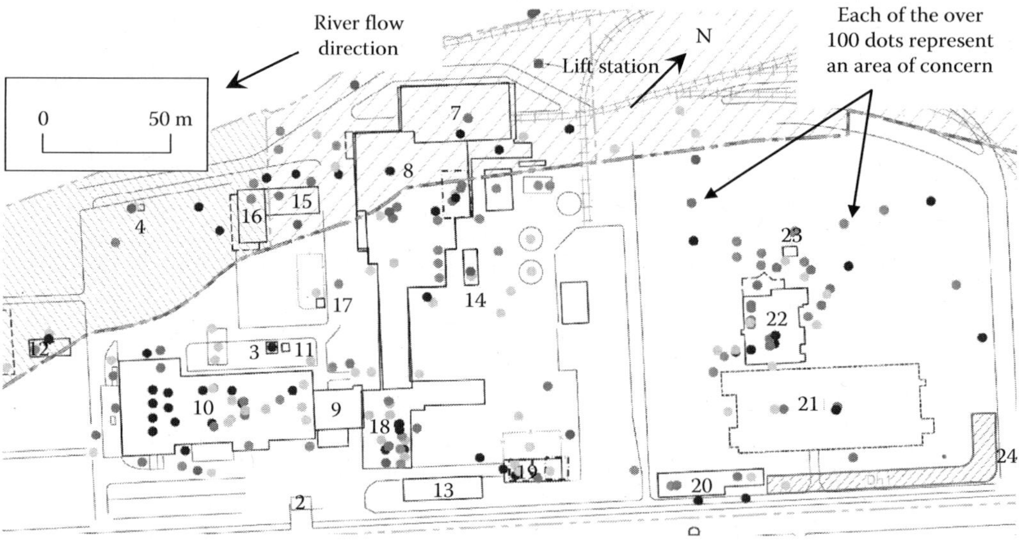

FIGURE 14.23
Location of RECs for the foundry portion of the site.

detected in groundwater included low levels of LNAPL VOCs, PAHs, and SVOCs, with none requiring remediation except for several areas with PAH-free product. Contaminants requiring remediation in soil included PAHs, PCBs, lead, and mercury.

Figure 14.23 shows most of the RECs are associated with the foundry located along the river on the west-northwest portion of the site. Within and near the foundry, the sources of contamination included sumps, pits, trenches, floor drains, and outdoor storage of waste materials where accidental spills had occurred.

The geology of the site was dominated by fluvial sediments composed of interbedded sands, silts, and clays to depths greater than 15 m (50 ft) below the ground surface. Groundwater is encountered within 1.5 m (5 ft) of the surface at most locations near the river and less than 6 m (20 ft) at locations furthest from the river. The direction of groundwater flow is generally west toward the river. The absence of a detailed geological map of unconsolidated sediments for this urban region increased the difficulty of characterizing the subsurface flow patterns.

This case study demonstrates the behavior of contaminants with low groundwater contamination risk factors. Despite its location within a geologically vulnerable area (vulnerability rating = 67), a synergistic effect between the geologic environment and contamination did not occur. Even though shallow groundwater was present and the contaminants did not have far to migrate to impact a surface water body, the relative immobility of the contaminants released prevented their migration into the groundwater.

The remedial method selected for soil was excavation since the contamination was accessible after demolition and did not extend to depths greater than a meter. The remedial methods selected for groundwater were removal of PAH-free product and monitored natural attenuation (MNA).

For nonfoundry areas of the site, cleanup of near-surface soil was required in the area located along the east side where a former shooting range was located. Here, shallow soils were impacted with lead. Other areas required a deed restriction and institutional controls including nondisturbance of certain areas (areas previously used as landfill areas) and maintaining a direct contact barrier such as asphalt pavement.

This case study was costly, because of the size of the site and its nearly 250 year duration of industrial operations. The precise total cost of investigation and remediation have not been determined because environmental work at the site is incomplete.

The lessons learned from this case study include

- Geologic maps are essential to all remediation efforts. In areas where there is high geologic vulnerability and relatively immobile contaminants, the potential for synergistic effects produced by their interaction can be better predicted by knowing the subsurface flow patterns.
- There are benefits to "front loading" the investigation by applying effective urban investigative and geologic forensic techniques. In this case, a very thorough Phase I ESA consisting of a comprehensive historical review set the stage for a detailed Phase II subsurface investigation that resulted in the demolition of buildings and provided access to the subsurface. These actions ensured the site was adequately characterized.
- If a contaminant has a low groundwater risk factor, it suggests there will not be any synergistic effects between it and a vulnerable geologic environment. Examples of this are provided by the fate and migration of PAHs, PCBs, SVOCs, arsenic, lead, cadmium, and mercury. All of these contaminants were present, but none required groundwater remediation.
- Low areas on sites are often sources of contamination. Here the foundry portion contained sumps, pits, trenches, tanks, and floor drains.
- Outdoor storage areas offer potential sites for contaminant release.

- Successful remediation is possible using a simple and established method (excavation) for soil, accompanied by capping, deed restrictions, institutional controls, and pumping of free product.
- After effective remediation and scientifically informed planning, heavily contaminated sites can eventually turn into productive land uses.

14.8 Summary and Conclusion

This chapter presented several case studies designed to illustrate some of the principles that can form the scientific and scientifically informed planning foundations to attain sustainable urban watersheds. These principles include

- A detailed urban geological map is required to comprehensively assess a damaged site and facilitate efforts for its remediation and redevelopment.
- Geology in urban settings is dominated by unconsolidated sedimentary deposits, and these deposits are complex.
- Major urban areas are located near large surface water bodies, and the hydrologic and hydraulic connections between them often dictates the extent and severity of the contamination.
- Common contaminants exist at many large urban industrial sites, and they have similar storage methods outdoors, and common indoor or near-building release points (e.g., sumps, tanks, trenches, floor drains, and pits). This knowledge can assist source control efforts.
- Contaminant migration is closely related to the risk factors for soil and groundwater. Immobile contaminants released into vulnerable geologic environments may not create a synergistic effect because they have high soil risk factors and therefore do not reach groundwater. Mobile contaminants with low soil risk factors tend to exhibit high groundwater risk factors; they will migrate in vulnerable geologic environments and create a synergistic effect.
- Anthropogenic influences, e.g., sewer systems, can change the migration pathways for contaminants. A rigorous site investigation is required to determine if the anthropogenic affects will impede contamination or enhance it.
- Urban investigation and geologic forensic techniques are useful and necessary components of most, if not all, site investigations. Three-dimensional maps communicate information clearly and effectively.
- Remedial technologies vary and must be matched to the remediation effort based on the specific characteristics of the site. One size does not fit all here.
- The cost of environmental cleanup varies proportionately with synergistic effects and is often increased by the presence of shallow groundwater.
- Redevelopment alternatives must rely on the science used for remediation to guide the proper selection of land uses.

Many of these concepts are dramatically apparent when comparing and contrasting the case studies. The following are just a few important examples:

- Case Studies 1 (Michigan), 3 (Indiana), and 6 (New Jersey) were all foundries. All are located in areas with vulnerable geology, but only the site that used significant amounts of DNAPL VOCs (Michigan) experienced a synergistic effect between vulnerable geology and a chemical with a high contaminant risk factor. It is reasonable to conclude that the high contaminant risk factor for groundwater of DNAPL VOCs contributed to the synergistic effect.
- Case Studies 2 (Illinois) and 5 (California) are located in areas of different geological vulnerability, but both had chromium VI as the contaminant of concern. The masses of chromium VI released were nearly the same and occurred over a similar span of time (less than 40 years). The Illinois site is located in a much less geologically vulnerable area (rating of 13) compared to the California site (rating of 63). The synergistic effect occurred because the California site was located in a vulnerable geological environment.
- Case Study 4 (Wisconsin) provides an example of how anthropogenic influence represented by the presence of storm sewers and building foundations greatly influences groundwater and contaminant migration.
- Case studies 1, 2, 4, 5, and 6 demonstrate the value provided by structure demolition. This measure enables unobstructed access to the subsurface and assists with site investigation and remediation.
- Damaged sites can be effectively redeveloped into different land uses (Case studies 4 and 6).
- Case studies 1, 2, 3, 4, and 6 highlighted low areas being the sources of contamination, that is, sumps, pits, trenches, floor drains, and tanks. Outdoor storage areas are also identified as contaminant sources. Identification of these sources provides a basis for instituting pollution prevention initiatives, especially at those locations where a synergistic effect may occur.
- Remediation of soil at each of the case studies relied on soil excavation as part of the remedial process. Other remedial efforts included a variety of technologies, and in most cases a combination of options was required to successfully achieve remedial goals.
- The presence of synergistic effects is a key determinant of cost. In the six cases presented, the costs of investigation ranged from $400,000 to more than $25 million. The case studies costing the most (Michigan and California) are located where synergistic effects were present, and those costing the least (Indiana and Illinois) were located where a synergistic effect was not present.

The cases in Michigan and California (Case Study 1 and 5) were unsuccessful because the synergistic effects at these sites: (a) were extensive and created investigative and remedial work that exhausted available financial resources, (b) will require decades to be fully remediated, and (c) required time-consuming involvement from several different public and private entities. By contrast, the case studies in Illinois, Indiana, Wisconsin, and New Jersey (Case Study 2, 3, 4, and 6) were successful mainly because the full synergistic effect was not realized, and this kept the investigation, remediation, and subsequent redevelopment of these sites more manageable.

We now use the outcomes from these studies, and from the thousands not described here, to investigate ways to attain sustainable development within urban watersheds.

References

California Department of Toxic Substance and Control (DTSC). 2006. *Remedial Investigation Report and Feasibility Study*. Newport Beach, CA: Geomatrix Consultants, Inc.

Illinois Environmental Protection Agency (IEPA). 2010. *Remedial Action Plan*. Downers Grove, IL: St. John-Mittelhauser & Associates.

Indiana Department of Environmental Management (IDEM). 2005. *Remediation Completion Report*. Indianapolis, IN: Clayton Group Services.

Michigan Department of Natural Resources and Environment (MDNRE). 2010. *Environmental Investigation Report*. Novi, MI: Arcadis.

New Jersey Department of Environmental Protection (NJDEP). 2010. *Preliminary Assessment Report*. Piscataway, NU: AECOM, Inc.

Rogers, D.T., M.M. Kaufman, and K.S. Murray. 2006. Improving environmental risk management through historical impact assessments. *Journal of the Air and Waste Management Association* 56:816–823.

Wisconsin Department of Natural Resources (WDNR). 2003. *Comprehensive Site Investigation Report*. Downers Grove, IL: Clayton Group Services.

15

Science-Based Landscape Planning in Urban Watersheds

15.1 Introduction

Science and planning have much in common. Science is the process allowing us to link isolated facts into coherent and comprehensive understandings of the natural world. Planning is the process of developing a sequence of action steps to achieve a specific goal or set of goals. Besides the common ground they share as processes, there is overlap in their approach to achieving outcomes. Science attempts to predict outcomes by understanding natural processes, which requires observation, data collection, experimentation, measurement, and validation. After scientific inquiry has validated the outcome of a process, humans can design technology to take advantage of the properties that might benefit our needs to provide shelter, travel, and create. For example, through the efforts of science, we know water will boil at 100°C and generate steam, which occupies a much larger volume than liquid water. Steam turbines were then designed to take advantage of this property of fluid expansion to generate electricity.

The practice of planning tries to influence outcomes by minimizing undesired events, and in doing so often involves some of the same actions used by science: observation, data collection, measurement, and the consideration of alternatives—a form of experimentation. Consider a busy intersection without traffic lights, where the undesired event is the occurrence of accidents. Observing the traffic at the intersection, collecting traffic count data, and analyzing their patterns would start the planning process to develop a new outcome. Two of the alternatives for minimizing the undesirable events and producing a better outcome might include traffic lights or stop signs.

The relationship between science and planning goes even deeper, because science cannot be performed without some measure of planning. In fact, science embodies planning, by using a sequence of action steps to achieve a specific goal. Does planning embody science? To answer this, we need to move beyond the rhetorical and get specific.

Our concern is with contaminated urban watersheds and the sequence of actions needed to return them to sustainability. The planning concepts most relevant to this effort are urban planning/land use planning, watershed planning/management, and landscape/environmental planning. As noted in Chapter 12, there is a growing trend *in the literature* to treat urban areas as ecosystems, and a well-established linkage between planning and science exists in the literature of watershed management (Viessman and Welty 1985; Naiman 1992; NRC 1999; Loucks and van Beek 2005), and landscape/environmental planning (Ortolano 1984; Lein 2003; Marsh 2010). Literature demonstrating the relationships between geology and land use planning has also developed (Rogers 1992, 1996, 1997, 2002; Rogers and Murray 1997; Murray and Rogers 1999; Gerber and Howard 2002; Kaufman et al. 2003; Lerner 2003; Murray et al. 2004; Lamelas et al. 2007; Rogers et al. 2007).

As noted repeatedly since Chapter 2, and demonstrated with the case studies in Chapters 6, 10, and 14, the failure to incorporate geology into land use decisions and employ geologic mapping have contributed to mismatches between human activities and the land. This chapter begins by investigating why the incorporation of science from the academic/professional literature and professional practice into the community land use planning process is lagging. Our investigation includes an evaluation of the land use planning process in the United States to determine whether it is structured to accommodate science, and if not, what can be changed to accomplish its inclusion. The primary focus here is on geology, but includes other sciences. Next the concept of science-based landscape planning is defined, and applied to parcel-level sites, because at this geographic scale most of the important land use decisions are made. The chapter concludes with an overview of watershed management.

15.2 The Science–Policy Gap

The science–policy gap can be defined as "not doing what we know." In the United States, this gap was born and nurtured by many factors, including the reshuffling of educational priorities, public attitudes bred of ignorance, the differing scientific and legal conceptions of truth, and resistance to change by organized groups. With respect to education, after Sputnik was launched in 1957, the United States responded with a massive effort to upgrade math and science in the schools by passing the National Defense Education Act (Harris and Miller 2005). This level of interest and investment in science education, however, has not been sustained (Trohler 2010).

The repositioning of science in the educational curriculum may be a factor contributing to the public's ignorance about science. For instance, in a 1994 poll, almost one in ten U.S. citizens indicated they thought the moon landings were faked (Pollack 2003). As for the difference between the scientific and legal conceptions of truth, in science, only falsehoods, not truths, can be proven, so there exists some degree of uncertainty. It is this uncertainty that drives scientists to keep searching for better explanations of phenomena. In law, truth is conceived as existing "beyond a reasonable doubt." This conceptual difference partly explains why our legislation is sometimes slow to incorporate science—lawmakers want to see a smoking gun.

Resistance to change on the part of organized lobbies also fuels the science–policy gap. One example here is the resistance of the tobacco industry to antismoking legislation, despite overwhelming scientific evidence that smoking is detrimental to people's health. With other issues, however, the reasons behind the gap are cloudier. In the 1980s, science demonstrated the damages from nonpoint pollution to the nation's waterways (Humenik et al. 1987), but as of 2010 no national legislation exists for controlling agricultural runoff. Is this simply a case where the agricultural lobby stuck their money into the pockets of Congresspersons? Or, are there other considerations, such as the cost of implementing the program, and the financial burden these regulations may impose on small farmers? Could those two concerns be a smokescreen for the agricultural lobby to accomplish its objectives? Are there legislators involved who believe the moon landings were faked? Clearly, there are many factors contributing to the science–policy gap. The concern here is to find ways to minimize its impact so "what we know" can be applied more efficiently in the effort to create sustainable urban watersheds.

Fulfilling the objective of minimizing the science–policy gap in urbanized watersheds begins at the community level. This geographic scale is where the rubber really hits the road, because the designations, allocations, and arrangements of land uses are made here. Reviewing the land development planning process at the community level can help determine whether there are opportunities to narrow the science–policy gap.

15.2.1 Land Development at the Community Level

The land development process differs across regions, states, and communities but, like the planning process, has some common threads. Here is a summary of the key steps:

- *Rough plan.* A developer presents to the community what he/she wants to do with the property. This presentation is usually in the form of a conceptual diagram prepared by an architect or engineer. While acquiring the property, the developer would have completed due diligence for environmental contamination—usually a Phase I investigation.
- *Compatibility review by the community.* Professionals serving the community (e.g., planners and city managers) review the proposal to determine if it meshes with its stated goals embodied in its Master Plan. Additional review is conducted to assess compatibility with zoning regulations (e.g., structure height, parking provisions, traffic considerations, adherence to local architectural guidelines, and overall development density). There may be other ordinances developers must comply with, such as those for historic preservation, wetland protection, tree protection, soil and sedimentation control, and stormwater management. The results of these reviews are fed back to the developer, who must now develop a detailed site plan.
- *Site plan submission.* The developer produces a detailed site plan of the proposed development. Local regulations vary, but most communities require the preparation of several maps to show a plan (overhead) view of the entire development within the surveyed property boundaries, existing and proposed utilities, existing drainage and proposed stormwater management procedures, proposed landscaping, and a facade view of the structure(s).
- *Agency site plan review.* The detailed site plan is now passed on to relevant government agencies and community departments to help ensure proper site development, for example, the county road commission, the public works/engineering department, and the planning department (or planning consultant). The historic district commission may review site plans in some communities, and larger-scale projects may be reviewed at the state level by the environmental agency to assess environmental impacts. If federal funds are involved, there may be additional reviews by federal officials.
- *Public comment/Planning Commission review.* Many communities have a planning commission, whose task is to review site plans and provide a public forum for input about proposed projects. This is where a disgruntled neighbor can stop construction of the garage you want to build. At this public meeting, the planning department or planning consultant will answer questions about the proposed development from the audience and members of the Planning Commission—an elected body of laypersons from the community. If the proposal is approved by the Planning Commission, the developer has the green light to proceed with construction.

- *Issuance of building permit*. The developer can begin construction and will have to submit to local building code inspection as the project progresses. If an erosion control ordinance is present, the site may be inspected by the local enforcing agent for compliance.

This sequence of steps appears to be a logical approach to land development. You have generalists from the public who might catch problems the professionals miss during their review of the site plan. If problems arise with the proposed plan, they can be fixed by the developer provided there is not too much of a financial or scheduling burden. Substantial consideration also seems to exist for environmental concerns, like stormwater management and erosion.

There is, however, a problem of geographic scale with this approach. If the proposed development is a subdivision, the overall environmental performance of the individual parcels has already been largely compromised by how the original tract of land was divided.

Most subdivision land developed in the United States is based on the density principle. The specific lot sizes and the overall size of the entire site are predetermined. As a result, a spatial form is imposed on the landscape before development proceeds. A 20 ha site "plat" map may specify 20 single-family residential structures on lots from 0.1 to 0.2 ha. The developer gets an architect, and a site plan is developed conforming to these constraints. Any land area left over can be dedicated to greenspace, trails, parks, or detention storage, depending upon the local zoning regulations.

The problem with this approach is that it specifies what the land should hold based solely on local community goals (e.g., "low-density development"), rather than first identifying the actual physical capabilities of the landscape to support specific uses. Instead, we leap before we look, as homes are sited in flood-prone areas, native topsoil is removed indiscriminately (and often sold back to the homeowner after being adulterated to "fill"), and homes are constructed and oriented in a way that fails to maximize their energy-saving potential. These practices reflect an inefficient land use planning process and generate a haphazard development pattern.

In the United States, suburban sprawl has been the dominant pattern of metropolitan growth in the last 50 years (Downs 1998). Sprawl refers to a pattern of urban and metropolitan growth that reflects low-density, auto-dependent, and exclusionary new development on the fringe of settled areas often surrounding a deteriorating city (Squires 2002). Within many of the urbanized watersheds of the United States, a large percentage of their land area consists of suburbs comprised largely of subdivision-style developments. For instance, the heavily urbanized Rouge River watershed in southeast Michigan has an area of 1134 km^2, with the city of Detroit occupying only 370 km^2.

15.2.2 Narrowing the Science–Policy Gap

To narrow the science–policy gap, a thorough physical planning process should precede the specification of development density on parcels scheduled for subdivision. The capabilities of the landscape should dictate the uses, density, and arrangement of the built environment, with a comprehensive physical planning process conducted to assess and map the geologic environment, map (and sample) the soils, inventory and analyze vegetation, slope, drainage, and sun angles, and perform a groundwater vulnerability study. Geographic information system (GIS) overlay techniques can be used to manage and analyze the data, with field checks used to verify the GIS results. The objective is to identify those areas not suitable for development and those areas best suited to specific uses and certain development densities. In this process, the specification of density comes last.

Significant opportunities to reduce the science–policy gap also exist within the steps of the land development process outlined above. These opportunities include the elimination of ordinance cloning and the participation of environmental scientists during the reviews.

Local zoning and subdivision ordinances—some in place since the National and State Planning Enabling Acts of the 1920s—were updated to reflect environmental concerns (Hagman 1975). By the mid-1980s, most municipalities had adopted ordinances related to soil erosion and floodplains, and many were in the process of implementing ordinances for wetland protection, recycling, and stormwater management (Kaufman et al. 2002).

The differing origins of local environment-related ordinances and their application raise several questions. For instance, who developed them? How did they come to exist and evolve over time in a particular place—for example, what sort of environment and problems did the authors address? Are the ordinances equally suited to local physical conditions in other places, places far from and different from the places for which the authors designed them? Finally, if an ordinance is unsuited to local physical conditions, what are the environmental consequences?

In a review of the ordinances of the 1778 villages, townships, and cities (minor civil divisions [MCDs]) of Michigan, Kaufman et al. (2002) found that many MCDs had developed their own ordinances, others had cloned or copied ordinances, such as those regulating subdivisions, stormwater, wetlands, and floodplains/coastlines from other communities, while other MCDs had adopted (copied) these ordinances from higher levels of government "by reference" (MCL 2000).

Ordinance cloning or copying by reference can be environmentally detrimental because the process often ignores significant differences in local environmental settings and their physical processes and systems. For instance, there are profound differences in soil, topography, and surface drainage patterns between the flat, clayey lake plains of the Saginaw Valley in east-central Michigan and the hilly, morainic terrain of the Traverse Bay region in the northwest part of the lower peninsula, yet subdivision ordinances and stormwater infrastructure are the same in both (Kaufman et al. 2002). Figure 15.1 shows the different landscapes of a lake plain and moraine.

In this case, the practice of cloning or copying by reference may increase the discharge of stormwater to nearby streams and result in local flooding. Additionally, the specification of a predefined density requirement, without considering the local environmental conditions such as soil infiltration capacity, may lead to the overbuilding of drainage infrastructure. From an economic efficiency perspective, the extra infrastructure is a wasted

FIGURE 15.1
Lake plain and moraine. (Photo by Daniel T. Rogers.)

capital expenditure, which is especially critical given the current fiscal conditions of many communities.

Moreover, the development of in-house environmental ordinances is no guarantee of their satisfactory performance. Many communities use attorneys and urban planners to develop their own ordinances, yet these professionals have little or no training in environmental science. In Michigan, there have been numerous indications of substantial environmental damage resulting from the improper application of environmental ordinances (Kaufman and Marsh 1997; Kaufman 2000).

Including environmental scientists during the review process is another logical way to narrow the science–policy gap. Kaufman (1995) found that communities with more "green staff" exhibited a better response to stormwater management. Since any landscape being considered for development had its origins in geological processes, it makes sense to include a geologist to help guide the human use of landscapes. Physical geographers, ecologists, and hydrologists would also provide meaningful input to the land planning process.

15.3 Science-Based Landscape Planning

Science-based landscape planning is defined as "understanding the physical processes inherent to a natural landscape's formation and its long-term sustainability, and transforming selected components of these processes into planning principles and actions to preserve or enhance the cultural landscape." **Natural landscapes** are formed by the interactions within and between the physical systems comprising the four spheres of the geosphere: atmosphere, biosphere, hydrosphere, and lithosphere (e.g., the runoff system of the hydrosphere eroding soil and rock of the lithosphere). These types of landscapes are not planar or confined to the surface, but also contain the land below and the atmosphere above contributing to their formation and change. A **cultural landscape** is the product of human interaction with the natural landscape, with culture acting as the agent and the natural area as the medium (Sauer 1963). In effect, this interaction defines a land use system, such as urban.

Two primary objectives of science-based landscape planning are (1) to follow a scientific framework using scientific inquiry and analysis and (2) to produce results that can be directly incorporated into planning decisions. To accomplish these objectives, the following procedural sequence is employed:

- Define how a specific landscape forms and changes.
- Specify how it is spatially structured.
- Analyze the flows of matter and energy into and at the site under consideration.
- Identify the factors that stabilize this landscape, and how human activity can destabilize it.
- Implement the human activities in a way (e.g., the spatial arrangement of land uses and structures) that does not exceed the capability of the landscape to sustain itself. For example, on some landscapes, the clustering of buildings is preferable; on others, nothing should be constructed.

This general approach is adopted from Marsh (2010). Landscape formation processes are complex, and this reality creates the need for rigorous investigation, or the inputs of matter and energy from humans will render landscapes unsustainable.

15.3.1 How Landscapes Form and Change

Natural landscapes result from Earth's crust being hammered from below and above. From below, tectonic plate motion (Chapter 2) provides the tremendous force necessary to create and modify natural landscapes. At broad geographic scales including global, continental, and regional, various forms of tectonic activity (e.g., orogeny and volcanism) have created mountain ranges, ridges, valleys, trenches, and mixed terrains called mélange (Table 15.1).

After uplift, the crust is hammered by the forces of water and wind. Now, other formative processes of the landscape including waves, wind, glaciers, and runoff fed by overland flow, streamflow, soil water, and groundwater get their shot.

15.3.2 The Special Role of Water in Landscape Change

Even in glacial environments and sand deserts, the landscape is influenced heavily by running water. Flowing water is responsible for more landscape change than all other processes combined (Marsh 2010). As noted in previous chapters, special attention to water flows in all media and directions must be given during site investigations, and this principle holds true for any study of landscapes at larger geographic scales.

The properties of water account for its prominent role in landscape change. Water is heavy and relatively viscous, two properties allowing it to function as an efficient agent of erosion. Water also expands when frozen by 9%, and this increase in volume helps it break apart rock and soil. Most of the depositional processes creating sedimentary environments involve water, and these form the physical framework for urban watersheds. In addition to these properties, water exists everywhere on earth, although in varying quantities.

As noted in Chapter 12, human modifications of water resources have also had a profound impact on urban watersheds. Two of the major environmental impacts in urban

TABLE 15.1

Landforms Associated with Different Plate Boundaries

Plate Boundary Type	Associated Landforms
Convergent	
Ocean to ocean	Trench, accretionary wedge, forearc basin, magmatic mountains
Ocean to continental	Trench, mélange, magmatic mountains
Continent to continent	Fold mountains
Divergent	
Ocean	Mid-oceanic ridges
Continental	Rift valleys, shield volcanoes
Transform	
Ocean	Ridges
Continental	Varies, depends on the erosion processes operating on each side of the boundary

watersheds—stormwater and erosion—are connected by the flows of water, and land subsidence related to groundwater withdrawal has become an issue in some urban areas.

15.3.3 The Dimensions of Landscape

The formative processes of landscapes proceed within a three-tiered geometry (Marsh 2010). In the vertical direction are the atmosphere, surface, and lithosphere. The atmosphere contributes water from precipitation and wind as a result of pressure differentials. Wind velocities near the ground are dampened by friction, but between altitudes of 7.5 and 15 m (25–50 ft) they increase significantly. This increase can influence the erosion and particle transport capacity, as is seen with smokestacks.

The surface component of the vertical profile ranges from the top of tree canopies to the bottoms of root systems. Here, flows of water, soil, and rock shape and reshape the features of the landscape. Vegetation stabilizes the soil, intercepts precipitation, provides habitat for organisms, and produces oxygen for the atmosphere and organic matter returned to the soil. At the lowest level of the vertical extent, below the plant roots to the bedrock, are the aquifers which, in humid regions, accept inflows from precipitation and discharge to surface waters.

Horizontally, landscapes exhibit a length and width. This extent determines their **geographic scale**, which is defined through the observation and/or measurement of events and processes. Table 15.2 shows the different scales associated with a small sample of landscape events and processes.

The ordering of scales in Table 15.2 represents a hierarchy through which different processes operate (perform work). For example, wind operates at all scales from the global to the micro. We may also see the effects of plate tectonics at the local or microscale in the form of a hillslope, or shattered or folded rock sequences.

15.3.4 Site and Situation

For the individual land parcels within watersheds, most of the matter and energy exchanges originate "off-site." **Site** refers to a fixed place, and **situation** represents the outlying places related to the site (Figure 15.2). For example, a land parcel (the shaded rectangle) and its

TABLE 15.2

Common Associative Scales for Selected Landscape Processes

Scale	Association	Examples
Global	Entire earth	Plate tectonics, global warming
Continental	Entire continents, significant areas of multiple continents, such as Eurasia	Orogeny (mountain building), glaciations
Regional	Small parts of continents, several nations, one or more states within countries, large watersheds	Climate and weather, environmental hazards (earthquakes, floods, hurricanes)
Local	Urban metropolitan areas, neighborhoods, several small farms, small watersheds, subwatersheds	Photochemical smog, stormwater runoff, groundwater, and surface water pollution
Microscale	Areas under 0.4 ha in size	Microclimates, edge effects[a]

[a] Edge effects are located at the margins of physical entities, such as forests. Different processes occur at the edge (usually a very small area) because this location creates different relationships with other surrounding processes. For instance, the trees located at the forest edge receive more sunlight and usually exhibit larger canopies.

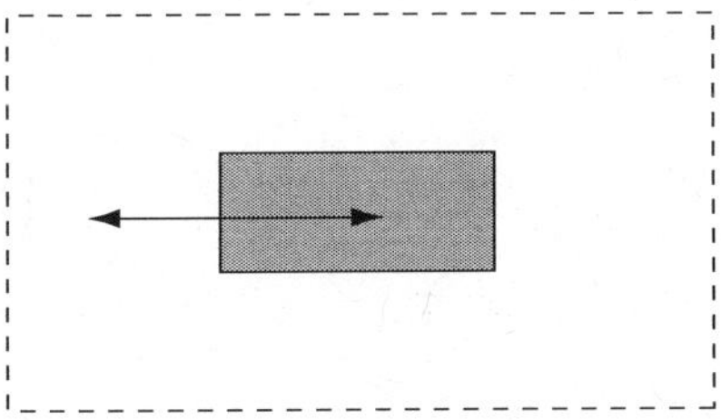

FIGURE 15.2
Site and situation.

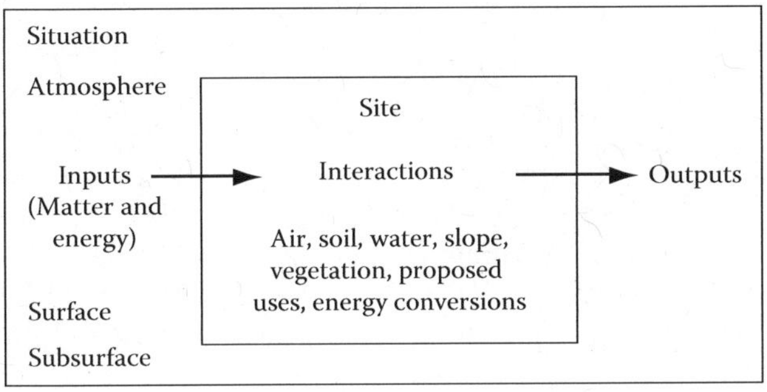

FIGURE 15.3
Inflows and interactions at a parcel-scale site.

encompassing watershed (the dashed-line outer boundary) exhibit a site and situation relationship. The bidirectional arrow indicates there are flows of matter and energy to and from the site.

What are some of these flows? From the vertical perspective, the atmosphere may deliver precipitation from the watershed area and beyond onto the parcel. At the surface level, eroded material washed down a hillslope in the headwaters region of the stream may be deposited in a sink located within the parcel. Below the surface, contaminated groundwater from outside the parcel may migrate to surface water within the parcel. Thus, the site (e.g., parcel, city, and watershed) is an intersection of matter and energy flows, where the different processes initiating and sustaining the flows interact (Figure 15.3). With land parcels, the major share of the flows originates from off-site, and the different widths of the arrows indicate the relative flow contributions of situation and site.

Although sites have readily identifiable geographic scales (e.g., a bounded city or a parcel of property), situational geographic scales often vary. For instance, the precipitation delivered to a site may have been acidified hundreds of kilometers away, while the eroded soil deposited on the site originated from the adjacent parcel. Accounting for these different situational scales allows the investigator to identify and characterize the materials and energy reaching the site and is essential to understanding the interactions on the current landscape.

15.3.5 On-Site Interactions and Thresholds

The list of landscape elements shown interacting at the site in Figure 15.3 is generalized. At a finer level of distinction, water includes precipitation, interflow, baseflow, surface runoff, streamflow, soil moisture, water vapor, and interception storage. What changes will occur to all of these water forms and flows when a new structure is built on a hillslope? Now the interactions must include a consideration of soil type (composition and texture), soil

TABLE 15.3

Matrix Showing Complex Interactions at a Site

Landscape Element	Soil Type	Soil Water	Vegetation	Slope	Precipitation	Interflow
Soil type	X					
Soil water		X				
Vegetation			X			
Slope				X		
Precipitation					X	
Interflow						X

Context: Placement of a structure on a hillslope.

water, vegetation, slope, precipitation, and interflow. In addition, these interactions need to be considered within the context of the existing amounts of shear strength and shear stress present on the slope, and the likely changes to these variables incurred by building the structure. In complex situations like this one, a matrix is a useful way to organize the interactions to help identify the data required for subsequent field and bench analyses (Table 15.3).

The landscape elements are arranged in the same sequence down the rows and across the columns. The "X" in a cell indicates no interaction, since an element does not interact with itself. We will use two interactions to demonstrate the procedure, which has four steps: (1) list an interaction, (2) identify the scientific principles governing the interaction, (3) specify the data required for collection, and (4) repeat until all interactions have been covered. Some data items listed for collection may be repeated, since different interactions may warrant a different sampling strategy.

15.3.5.1 Interaction 1: Soil Type with Soil Water

Principles: The amount of organic content influences the amount of moisture a soil is capable of holding. Soil pore space size varies by soil texture (clay, silt, and sand). Infiltration rates vary by soil type.

Data required: Soil profile, soil texture, soil organic content, soil infiltration rate (all sampled at different locations on the slope).

15.3.5.2 Interaction 2: Soil Type with Vegetation

Principles: Slope stability depends on the balance between the shear stress and shear strength forces present. Examples of shear stress forces include gravity, frost heaving, and water. Water adds weight to a soil by replacing air within the pore spaces, and the added mass makes soil more prone to downward movement. Vegetation is a cohesive force on slopes, and the soil type is one factor determining the types of vegetation present. Roots of vegetation may increase the porosity of near-surface soil and increase the infiltration capacity. Vegetation contributes organic matter to the soil, the roots help stabilize soil, and it can intercept precipitation on slopes.

Data required: Soil organic content (sampled by vegetation clusters), root depth measurements, root density measurements taken at several locations on the slope. Soil type and texture samples should be collected at locations at and near the proposed structure location. A full list of the species of vegetation present and their distributions at the site.

Vegetation canopy measurements (aerial photos can help here) to determine where vegetation removal will create the most slope instability, and where plantings can have the most stabilizing impact.

This short example demonstrates there is a significant amount of data to collect and science to know—even without considering the proposed activity. Besides sloped terrain—which we might expect to become unstable with development—other seemingly stable landscapes can be destabilized by human activity if the human activity causes a threshold present to be exceeded. For instance, wetlands can only withstand a certain amount of inundation before they become damaged. Seed banks found in freshwater wetlands play a critical role in the maintenance of many of their plant communities (Leck 1989), and their disturbance by floods can decrease seedling emergence (Peterson and Baldwin 2004). Therefore, the possibility of damaging seed banks in the wetlands of urban wetlands must be considered when routing stormwater flows.

15.4 Watershed Management

As demonstrated, significant reduction of environmental impacts at the parcel scale can be achieved with science-based landscape planning. Individual parcels, however, are situated within watersheds of larger geographic extent, and there are also many opportunities at this broader scale to apply science-based landscape planning methods. Before these efforts are attempted, there are fundamental differences between parcels and watersheds that must be considered when applying a science-based planning approach to watersheds. Some of the major differences include

- *Geographic scale.* Landscape processes operating a certain way at one scale may not mimic their behavior at another. Runoff provides an example. Since most individual parcels do not contain a stream segment, overland flow is the dominant form of runoff on the parcel. Across watersheds, however, streamflow transports the most surface water in a drainage network.
- *Landforms.* Watersheds contain entire landforms, and their distribution results in a highly varied terrain over large distances. For example, urbanized watersheds in sedimentary environments exhibit a wide range of landforms and land cover types, e.g., kames, eskers, moraines, valleys, lakes, deltas, wetlands, and forests. Parcels may exhibit significant topographic variations over a small distance, but typically do not contain a diverse set of landforms and land cover types. It is safe to assume most homeowners and factories do not have river deltas and valleys on their property.
- *Land use.* Urban watersheds contain a wide range of land uses, including agricultural, industrial, commercial, and residential. Due to zoning regulations, a singular land use almost always exists at the parcel level, and where multiple uses do exist (e.g., operating a business out of one's home) a zoning variance is required. Some urban planning strategies, such as "New Urbanism" (Duany et al. 2000) and "Smart Growth" (ICMA/USEPA 2010), advocate mixing land uses within neighborhoods to preserve the character of older cities—but the mixing occurs at scales larger than a single parcel.

- *Environmental impacts.* As a function of their larger extent, watersheds exhibit a wider variety of impacts than parcels.
- *Land tenure and institutional controls.* Most individual parcels are privately owned, but a mixture of public and private ownership exists across watersheds. The existence of hundreds, possibly thousands, of private owners having different outlooks, and with access to streams and lakes, complicates the development of a consistent management plan for water resources.
- *Levels of awareness and education.* Differences in awareness and education result largely from the scale differences. It is simply much easier to know the characteristics of an individual parcel than an entire watershed.

Despite these differences, the general processes of landscape formation, spatial structure and dynamics, stability, and human impacts also operate at the watershed scale and can be addressed by a scientific-based planning approach. A history of growth, pollution, and attempts to control water flow has created an array of environmental impacts within urbanized watersheds: buried tributaries, streambanks lined with concrete, combined sewer overflows (CSOs), degraded aquatic habitats, and soil and groundwater contamination (Figure 15.4). It will require a comprehensive approach and mammoth effort to rectify these problems. Specifying the details of this process is beyond the scope of this book. What can be accomplished here is to introduce a definition of the process of watershed management, outline its primary stages and components, and specify how the science in this book can be applied within this framework.

Watershed management is the minimization of the human impacts within and between different land use systems, with the objective of achieving a sustainable landscape. As bounded landscapes, watersheds are three-dimensional and receive inputs from outside their boundaries. They have formative processes, undergo constant change, and within urbanized watersheds, the cultural landscape often displays the outcomes of environmental impacts initiated and sustained by anthropogenic activities.

One apparent paradox of watershed management is the focus on land. To achieve a sustainable watershed, the land has to be properly managed, and since most watershed surface areas are 90%–95% land, this imperative is justified. The practical question is: How can science-based planning begin to put urbanized watersheds on the path to sustainability? The final chapter will answer this question.

FIGURE 15.4
Channelized portion of an urban river. (Photo by Kent S. Murray.)

15.5 Summary and Conclusion

Science and planning are cut from similar cloths, with both performing some of the same steps during their implementation. Yet, there is a science–policy gap existing in the United States that has prevented a full application of scientific knowledge and methods to land planning problems at different geographic scales. The reasons for this gap are complex, although opportunities exist during the land planning process to narrow it.

At the parcel-level scale, the specification of density through the uniform subdivision of a large land parcel often precedes the physical analysis of the landscape. This process compromises the ability to account for micro-scale landscape processes. Since a large percentage of the land in U.S. urbanized watersheds consists of subdivisions, ending this practice and instituting an analysis first, density last approach provides one opportunity to narrow the science–policy gap. Ending the practice of ordinance cloning and employing environmental scientists during the land development process are other ways to reduce this gap.

A science-based landscape planning approach, carried out at the parcel scale, provides additional potential to narrow the science–policy gap. Because of the site and situation relationships within watersheds, and the legal framework for land development in the United States, successful planning at the parcel scale is a prerequisite to effective watershed management.

References

Downs, A. 1998. The big picture: How America's cities are growing. *Brookings Review* 16:8–11.

Duany, A.E., Plater-Zyberk, and J. Speck. 2000. *Suburban Nation*. New York: North Point Press.

Gerber, R.E. and K. Howard. 2002. Hydrogeology of the Oak Ridges moraine aquifer system: Implications for protection and management from the Duffins Creek watershed. *Canadian Journal of Earth Sciences* 39:1333–1348.

Hagman, D.G. 1975. *Urban Planning and Land Development Control Law*. St. Paul, MN: West Publishing.

Harris, M.M. and J.R. Miller. 2005. Needed: Reincarnation of the National Defense Education Act of 1958. *Journal of Science Education and Technology* 14:157–171.

Humenik, F., M. Smolen, and S. Dressing. 1987. Pollution from nonpoint sources. *Environmental Science and Technology* 23:737–742.

ICMA/USEPA, International City/County Management Association, U.S. Environmental Protection Agency, 2010. This is smart growth. Smart Growth Network, Washington, DC. http://smartgrowthonlineaudio.org/pdf/TISG_2006_8-5x11.pdf (accessed June 20, 2010).

Kaufman, M.M. 1995. Community response to stormwater pollution in an urbanized watershed. *Water Resources Bulletin* 31:491–504.

Kaufman, M.M. 2000. Erosion control at construction sites: The science-policy gap. *Environmental Management* 26:89–97.

Kaufman, M.M. and W.M. Marsh. 1997. Hydro-ecological implications of edge cities. *Landscape and Urban Planning* 36:277–290.

Kaufman, M.M., K.S. Murray, and D.T. Rogers. 2003. Surface and subsurface geologic risk factors to groundwater affecting brownfield redevelopment potential. *Journal of Environmental Quality* 32:490–499.

Kaufman, M.M., D.L. Wigston, and E.B. Perlman. 2002. Environmental evaluation of subdivision site developments. *Environmental Management* 29:801–812.

Lamelas, M.T., O. Marinoni, A. Hoppe et al. 2007. Groundwater vulnerability map for the Ebro alluvial aquifer between Jalón and Ginel tributaries (Spain). *Environmental Geology* 53:861–878.

Leck, M.A. 1989. Wetland seed banks. In *Ecology of Soil Seed Banks*, ed. M.A. Leck, V.T. Parker, and R.L. Simpson, pp. 283–305. New York: Academic Press.

Lein, J.K. 2003. *Integrated Environmental Planning*. Oxford, U.K.: Blackwell Science.

Lerner, D. (ed.) 2003. *Urban Groundwater Pollution*. Rotterdam, the Netherland: Balkema.

Loucks, D.P. and E. van Beek. 2005. *Water Resources Systems Planning and Management: An Introduction to Methods, Models and Applications*, Paris, France: UNESCO.

Marsh, W.M. 2010. *Landscape Planning: Environmental Applications*, 5th edn. Hoboken, NJ: John Wiley & Sons.

MCL. 2000. *Michigan Compiled Laws Annotated*. St. Paul, MN: West Publishing.

Murray, K.S. and Rogers, D.T. 1999. Groundwater vulnerability, brownfield redevelopment and land use planning. *Journal of Environmental Planning and Management* 42:801–810.

Murray, K.S., D.T. Rogers, and M.M. Kaufman. 2004. Heavy metals in an urban watershed in southeastern Michigan. *Journal of Environmental Quality* 33:163–172.

National Research Council (NRC), 1999. *New Strategies for America's Watersheds*. Committee on Watershed Management, Water Science and Technology Board, Commission on Geosciences, Environment, and Resources. Washington, DC: National Research Council.

Naiman, R.J. 1992. *Watershed Management: Balancing Sustainability and Environmental Change*. New York: Springer-Verlag.

Ortolano, L. 1984. *Environmental Planning and Decision Making*. New York: John Wiley & Sons.

Peterson, J.E. and A.H. Baldwin. 2004. Seedling emergence from seed banks of tidal freshwater wetlands: Response to inundation and sedimentation. *Aquatic Botany* 78:243–254.

Pollack, H.N. 2003. *Uncertain Science…Uncertain World*. Cambridge, U.K.: Cambridge University Press.

Rogers, D.T. 1992. The importance of site observation and follow-up environmental site assessments—A case study. In *Proceedings of the National Ground Water Association Phase I ESA Conference*, pp. 218–227. Orlando, FL.

Rogers, D.T. 1996. *Environmental Geology of Metropolitan Detroit*. Novi, MI: Clayton Environmental Consultants.

Rogers, D.T. 1997. *Geologic Sensitivity Map of the Rouge River in Southeastern Michigan*. Wayne County, MI. 1:62,500. 1 Sheet.

Rogers, D.T. 2002. The development and significance of a geologic sensitivity map of the Rouge River watershed in southeastern Michigan, USA. In *Geoenvironmental Mapping: Methods, Theory, and Practice*, ed. P.T. Bobrowsky, pp. 295–319. Rotterdam, the Netherland: Balkema.

Rogers, D.T. and K.S. Murray 1997. Occurrence of groundwater in Metropolitan Detroit. In *Groundwater in the Urban Environment*, eds. J. Chilton et al., pp. 155–161. Rotterdam, the Netherland: Balkema.

Rogers, D.T., K.S. Murray, and M.M. Kaufman. 2007. Assessment of groundwater contaminant vulnerability in an urban watershed in southeast Michigan, USA. In *Urban Groundwater—Meeting the Challenge*, ed. K.W.F. Howard, pp. 129–144. London, U.K.: Taylor & Francis.

Sauer, C.O. 1963. The morphology of landscape. In *Land and Life: A Selection from the Writings of Carl Ortwin Sauer*, ed. J. Leighly, pp. 315–350. Berkeley, CA: University of California Press.

Squires, G. 2002. *Urban Sprawl: Causes, Consequences, and Policy Responses*. Washington, DC: The Urban Institute Press.

Trohler, D. 2010. Harmonizing the educational globe, world polity, cultural features, and the challenges to educational research. *Studies in Philosophy and Education* 29:5–17.

Viessman, W. Jr. and C. Welty. 1985. *Water Management: Technology and Institutions*. New York: Harper and Row.

16

Attaining Sustainable Urban Watersheds

16.1 Introduction

Geological investigations combine scientific observation, analysis, and interpretation. A by-product of these efforts is a very valuable tool—the geologic map. In many urban areas, these maps are not of sufficient detail or absent, or if available, are underutilized or ignored. This often leads to land development without sufficient knowledge of its hydrogeology, soils, and surface water features, and a diminished ability to foresee the potential vulnerabilities of the very ground where structures are erected and human activities occur. This is analogous to taking a trip without a road map—you often get lost.

Once this land has been developed, we assault it with chemicals used in manufacturing, herbicides and pesticides applied across vast tracts used for agriculture, fertilizers sprayed or spread on our lawns, and particulates and greenhouse gases emitted from our vehicles. Many of these substances are harmful to humans and the environment and interact with the geologic setting into which they are released. These events sound ominous, but with our knowledge about the properties of these substances we can determine the risks they pose based on their toxicity, mobility, and persistence. These risk assessments yield valuable knowledge about the potential environmental outcomes when chemicals are released into soil, groundwater, and air.

If we want to reduce this widespread contamination, we need to work fast because the worldwide rate of urbanization continues to accelerate. Urbanized watersheds have become the focal point for the excess matter and energy produced by the growing horde of humans. Rather than accommodating to nature, humans have tried to control it. As a result, the urban landscape is damaged and needing repair. Remnants of old factories dot the landscape, groundwater is contaminated or depleted, and any act of precipitation triggers more water pollution.

So now that we are saddled with this mess, everyone is talking about sustainability. Sustainable urban watersheds might be achieved if we could all get along and stopped emitting pollution—but how do we get there, given the mishmash of existing institutional arrangements for managing land, water, and development, and the reality that science-based planning is not performed at the watershed scale? The Rouge River watershed and its Rouge River National Wet Weather Demonstration Project (United States Environmental Protection Agency Grants #XP995743-01, -02, -03, -04, -05, -06, -07, and C-264000-01) is an example of an attempt to rectify a legacy of neglect and damage caused by 100 years of urbanization and industrialization. However, here again, the focus is on engineering a solution to the problem as opposed to reducing the source of the impacts. In this case, hundreds of millions of dollars have been spent to curb discharges of wastewater from combined sewer overflows during wet-weather events. Although the goal of the project is admirable, combined sewer overflows represent a single point source of contamination to the Rouge River, and all point sources combined represent only 20%–25% of the pollution problems facing the river.

An old Chinese proverb says a journey of a thousand miles begins with a single step. This chapter defines that first step. It begins with an overview of the concepts of sustainability and sustainable development and uses them to develop a framework for undertaking urban watershed management. Next, the principles learned from geological investigations, contaminant properties, risk assessments, and science-based planning are applied within this framework. The chapter concludes by demonstrating ways to link this first step of watershed management with the broader tasks necessary for achieving sustainable urban watersheds.

16.2 Sustainability and Sustainable Development

There is no standard definition for sustainability, and as a result the concept is open to many interpretations (Bell and Morse 2008). The working definition of sustainability used so far in this book was based on a few general principles adopted from Lynam and Herdt (1989) and Fresco and Kroonenberg (1992): (1) human activities do not inflict excessive levels of damage upon the environment, (2) society needs to allow the environment to repair itself from prior damages, and (3) these efforts will help return physical systems to dynamic equilibrium.

The term **sustainable development** originated in a 1987 report issued by the U.N. World Commission on Environment and Development: "development that meets the needs of the present without compromising the ability of future generations to meet their own needs" (WECD 1987). This definition is economic based and requires modification to accommodate the environmental imperatives for attaining sustainable urban watersheds. Specifically, in urban watersheds, sustainable development encompasses those activities designed to achieve and maintain dynamic equilibrium within the major physical systems performing their work in the atmosphere, biosphere, hydrosphere, and lithosphere. A sustainable urban watershed would have swimmable streams, swimmable lakes, a stable and regionally appropriate diversity of plant and animal populations, rehabilitated brownfields that do not pose a risk to groundwater and surface water, the absence of an urban heat island, and the absence of any identifiable linkages between its water, soil, and air qualities and the occurrence of disease in plants, animals, or humans. This is a highly idealized concept that in reality probably can never be accomplished because of the high financial cost necessary to achieve these goals. However, there is no reason not to do everything in our power, and within our financial capability, to shoot toward some compromise between the situation we face today and the idealized watershed.

16.3 A Framework for Beginning Urban Watershed Management

Land is the primary surface component of watersheds, and land-based anthropocentric activities are the major source of water contamination. In addition, urbanized watersheds exhibit a diversity of land uses and landscapes with ensuing high levels of environmental damage often present. Given these characteristics and challenges, the first step of watershed management must accomplish the following tasks: (1) prevent further land contamination and degradation, (2) preserve those land areas that are undamaged or minimally damaged, and (3) remediate and rehabilitate degraded brownfields to the fullest extent possible. These three tasks can be thought of as principles forming the basic framework for the first step of urban watershed management. Based on the definitions of sustainability and sustainable development, attainment of these tasks represents sustainable development activity.

Our definition of watershed management also views individual watersheds as sites and recognizes that impacts occur within and between multiple watersheds (Chapter 15). This reality means there are processes operating at geographic scales beyond those of a single watershed that may influence its ability to achieve sustainability. The framework developed here is upwardly compatible with the comprehensive watershed management effort required at the geographic scales of the watershed and beyond.

16.3.1 Prevent Further Land Contamination and Degradation

A sustainable outcome in urbanized watersheds is not possible without significantly preventing further land contamination and land degradation. Source control is the most effective strategy of pollution prevention in terms of cost and it has the highest environmental efficiency (Rogers et al. 2006). Source control strategies should take advantage of opportunities to integrate point and nonpoint source control measures. The identification of these locations is assisted by geology and the products of geological investigations.

Some of the integrated source control opportunities provided by the outcomes of geological investigations include

- *Geological maps.* These maps show the location of surface water, groundwater, and their flow paths. They can be used to identify areas in the watershed where potential point source releases and nonpoint source transport mechanisms (overland water flow and wind) overlap. These locations can then be prioritized for source control measures based on the contaminant risk factors (CRF) scores of the contaminants and the relative abilities of the geological environmental and the nonpoint transport mechanism to disperse them.

Three-dimensional geologic maps show the relationships between groundwater and surface water. The locations of discharge points between groundwater and surface water can help with the design of a contamination monitoring network. This information provides feedback on the efficacy of source control efforts, which is necessary for their optimization. These maps also identify locations where contamination may migrate outside of the watershed. For example, in the Rouge River watershed, the presence of a lower clay layer effectively routes groundwater horizontally into the Great Lakes (Figure 5.11).

- *Geologic vulnerability maps.* These maps indicate where stormwater flows may intersect groundwater recharge zones. As a source control measure, groundwater vulnerability assessment should be incorporated into every site planning process.
- *Stratigraphic columns and cross-sectional maps.* These outputs characterize the depositional history of the watershed and the resulting landforms. Of particular importance to source control efforts is the ability to evaluate the natural landscape's capability to store water and minimize stormwater runoff. This information also provides important parts of the science necessary to perform the sequence of steps

employed for science-based landscape planning (Section 15.3), including the flows of matter and energy onto specific sites, and landscape stability assessment.

- *Unconformities.* Geologic maps can highlight unconformities which may serve as preferred natural pathways for the migration of contaminants.
- *Anthropogenic disturbance.* These outputs often identify human-induced pathways for contaminant migration.
- *Soil characterization.* The soil samples taken at environmental investigations can characterize the soils within the watershed and update the county soil survey maps. This information can be used to implement terrain-specific erosion control measures and reduce the loss of soil and the accompanying transport of heavy metals into aquifers and streams.
- *Groundwater flow, recharge zones, and contaminants.* Combining the outcomes from geological investigations and geologic maps permits the identification of groundwater flow paths and recharge zones near high levels of contaminants in soil. This information can assist wellhead protection efforts within the watershed through the determination of appropriate capture zones computed from the hydraulic conductivities of the aquifer.

In addition to the opportunities for source control produced from parcel-scale geological investigations, there is the potential to apply integrated source control measures to atmospheric processes such as precipitation and sensible heat transfers at the watershed scale:

- *Sample the precipitation and develop a map of the pH distributions throughout the watershed.* Without cooperation across watershed boundaries, it will be impossible to restore the pH to natural levels (pH = 5.6). Nevertheless, knowing the pH distribution can identify areas where lower-pH precipitation may mobilize heavy metals in soil and release them to groundwater. Once identified, the risks of metals migrating to groundwater in these areas can be reduced by adding vegetation to increase interception and raising the organic content of the soils.
- *Develop a micro-climate classification map for the entire watershed with the objective of ameliorating the effects of heavy thunderstorms.* Begin by identifying those areas emitting excessive amounts of sensible heat. Follow this with a tree pattern analysis using aerial photos and ground surveys. Trees should not be planted until a certified arborist and a geologist have been consulted to optimize new tree planting locations. Those areas selected should have soil suited to native species and an ability to intercept significant precipitation. Other considerations include the potential to increase infiltration rates and the placement of trees at locations where they do not become conduits for contamination released at the surface. Certain trees such as willows can evapotranspirate large amounts of water per day (Minor 2009), and this capability reduces ambient air temperatures. Wherever possible, the sensible heat management strategies—such as green roofs—should be linked to stormwater management efforts.
- *Get vertical with stormwater management.* Figure 16.1 shows the paths of stormwater in an urbanized watershed.

The arrows between the paths indicate the primary direction for the water flow. With the exception of the vertical roof to downspout flow, the remaining components transfer

Roof ⟶ ↓ Downspout ⟶ Driveway ⟶ Road surface ⟶ Curbs ⟶

Storm sewer pipes ⟶ Outfall (local water body)

FIGURE 16.1
Paths of stormwater in an urbanized watershed.

water to the next component in the chain horizontally. Although precipitation reaching the ground surface can go in three possible directions (up with evaporation, down with infiltration, and horizontally with surface runoff), the human-use system has heavily favored the horizontal flow path. This routing bypasses the soil as a flow path and groundwater recharge declines. Instead, stormwater is delivered rapidly to the nearest water body where it discharges its load of contaminants.

Source control intervention for stormwater must occur before contaminants reach the environment. Since intervention in the atmosphere is impossible, the appropriate location occurs at the individual land parcel, for once water has left a lot it becomes part of the human-engineered horizontal transmission system. As discussed in Chapter 13, disconnecting downspouts is an example of source control intervention at the parcel scale. The significance of *when* to intervene with source control is illustrated by Figures 16.2 and 16.3.

Figure 16.2 shows the grid pattern typical of most urban areas. In Figure 16.3, this human-engineered pattern of development and drainage infrastructure is overlaid on the dendritic drainage patterns characteristic of the sedimentary environments found in many large urban watersheds. The result is a maze of water pathways that develops during wet weather, and these routes differ considerably from the natural paths, serving to increase the volume, velocity, and contamination of stormwater (Figure 12.8).

Source control for stormwater routes downspout flow vertically toward its natural path into the soil. In highly developed urban watersheds, this redirection of the flows minimizes the mismatch between human infrastructure and natural drainage patterns. The

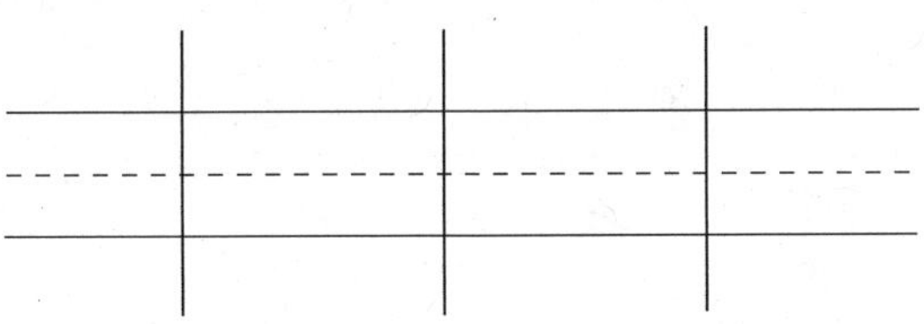

FIGURE 16.2
Grid pattern of streets, with storm sewer location shown as a dashed line.

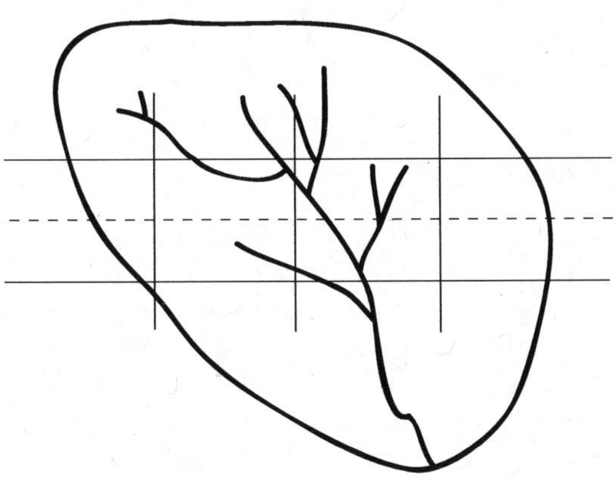

FIGURE 16.3
Overlay of infrastructure on dendritic drainage pattern (density reduced).

Roof → ↓ Downspout (↓) Driveway → Road surface → Curbs →

Storm sewer pipes → Outfall (local water body)

FIGURE 16.4
Source control for urban stormwater on the land parcel (circled).

circle in Figure 16.4 indicates the step in the flow sequence where human intervention can achieve negative feedback and help return the runoff system to equilibrium.

Retrofitting developed urban areas for stormwater source control follows a key principle of science-based landscape planning—do not do anything until you evaluate the capability of the landscape. Even in developed areas, there is substantial capability to store water before building additional infrastructure. Many of the general source control measures and micro-topographic opportunities for stormwater management were presented for this type of landscape in Chapter 13.

Vertical routing of stormwater also provides a new opportunity to design engineering approaches that use this water as an asset, instead of treating it as something to transport away quickly. The major residential uses of water are for toilet flushing and lawn watering—two uses not requiring municipally treated potable water. A combination of cisterns, rain barrels, low-cost filters, and simple plumbing modifications can make downspout water available for these uses.

- *Link performance standards to source control.* Performance standards specify an environmental condition that must be met to achieve a broader set of goals, such as reducing pollution. For example, one performance standard found in many municipalities is "post-development runoff must not exceed pre-development runoff." When planning at the site scale, proposed developments lying within zones of high water retention capacity should not be allowed to sacrifice this capability. Information from geologic investigations can identify these locations. Instead of building infrastructure such as detention ponds to meet a performance standard, the natural capacity of the landscape to hold water can be preserved. This approach is a win-win situation for environment preservation and economic development, which often conflict. Developers, instead of having to dedicate valuable property to detention storage, can take advantage of the landscape's capacity to store water and not use valuable real estate for stormwater control. Residents of the development then do not have to worry about maintaining the ponds, which are prone to sedimentation. The environment wins because detention ponds represent an end of the pipe treatment that disrupts the local hydrology by moving water out of its natural path.

16.3.2 Preserve Undamaged and Minimally Damaged Land

Any attempt to attain a sustainable urban watershed should capitalize on the assets available. If what humans have done to the land creates the most contamination in urbanized watersheds—then what we have not done to the land in these locations must become part of the solution. Land that is undamaged or minimally damaged is an asset, because it can provide a starting point for overcoming human impacts on stream channels, aquifers, and watershed ecosystems. Properly applying this principle requires a thorough knowledge of how specific landscapes were formed and how they were subsequently altered by humans. Conducting geological investigations, employing contaminant risk factors, and following a science-based landscape planning approach provide this knowledge.

Geological investigations (Chapter 4) can determine whether a landscape is damaged, and whether the source of the damage was from natural or anthropocentric origins. The inspections and data collected during Phase I and Phase II Environmental Site Assessments (ESAs) will detect the presence of human impacts, residual chemicals and their types, and the extent of their damage. Geologic maps (Chapter 5) can be used to establish the locations and cause of original landforms, and reconstruct the flows of water into, out of, and within the landscape. Since water is a key determinant of landscape change, this information will help indicate the extent of human modification. Information collated from these sources provides one of the pieces necessary to assess the degree of land degradation.

The next piece is provided by the contaminant risk factors developed for chemicals in soil, groundwater, and air. If chemicals are present at a site, the CRFs for each chemical detected can be evaluated to assess their toxicity and persistence. When this information is combined with the landforms and water flow information, the fate and transport of the contaminant can be determined (Chapters 8 and 9).

Completing the assessment of a landscape's condition is a science-based planning process (Chapter 15). An infrastructure evaluation can be used to characterize the types and levels of human modifications and assess their likelihood of destabilizing the landscape. During this process, investigators should be conscious of the potential for synergy between anthropocentric impacts and contaminant fate and transport. For example, if there are stormwater detention ponds present, these structures may serve as sinks for highly contaminated stormwater with the potential to contaminant groundwater and surface water.

Within urbanized watersheds, undamaged and relatively undamaged land may exist at some unexpected locations. In a study of brownfields in southwestern Detroit, Murray et al. (2008) found that 22 of the 87 properties (25%) were not contaminated, and another 35% were only minimally impacted. These properties are therefore available for redevelopment or may be employed for environmental purposes such as groundwater recharge, stormwater detention, and increasing the amount of tree canopy coverage.

Inputs of contaminants from adjacent and distant watersheds through acid rain and wind deposition prevent urban watersheds from being pristine. Nevertheless, large tracts of land may still possess a large share of their original capability to support sustainable development. Preservation of land as an activity thus includes the broader role of maintaining the natural landscape functions across the atmosphere, biosphere, lithosphere, and hydrosphere.

The asset-based approach presented here is derived from the main principles embodied in asset-based community planning (Kretzmann and McKnight 1993) and the watershed restoration strategy called Rapid Biotic and Ecosystem Response (Doppelt et al. 1993). Asset-based planning was developed as a way to rehabilitate damaged communities by using the assets present, rather than relying upon outside assistance and becoming "client neighborhoods." A community's main asset is its people—who have specific knowledge and skills to bring about positive change (Kretzmann and McKnight 1993).

In rapid biotic and emergency response (RBER), the initial efforts of watershed restoration do not focus on the most damaged areas. Instead, relatively undamaged areas are targeted first because they possess the ability to serve as locations for reestablishing habitat corridors, providing habitats and refugia, and rebuilding species diversity (Doppelt et al. 1993).

There is always some risk involved with taking principles from one discipline and applying them to another. The application of the evolutionary principle of "survival of the fittest" to sociology is one example (social Darwinism). Here, a spurious correlation was used to justify discriminatory immigration legislation in the early 1900s (Hofstadter 1944). Although the pitfalls of cross-disciplinary applications of principles are perhaps easier to recognize, considerable care must also be taken when principles are applied within scientific disciplines. For example, island biogeographic theory (MacArthur and Wilson 1967) can greatly assist in understanding the effects of habitat fragmentation. It does not, however, address other factors that can significantly influence which birds reside in a fragment. For example, an abundant presence of nest-robbing species could prevent certain invaders from establishing themselves (Ehrlich et al. 1988). The lesson here is that any principle should not be applied unless the specific processes involved at the relevant geographic scales are thoroughly understood.

16.3.3 Remediate and Rehabilitate Degraded Brownfields

If considered within the context of a mass balance—when the amount of contamination coming into a watershed exceeds the amount of contaminants being removed from the watershed—no progress toward sustainability can be made. The qualitative aspects of the contamination also need to be considered. Persistent and highly toxic contaminants affecting a small area may be causing more environmental damage than a less persistent and less toxic contaminant spread over a wider area. These scenarios underscore the importance and challenges related to brownfield remediation and rehabilitation in urbanized watersheds.

Geologic maps (Chapter 5) and vulnerability maps (Chapter 6) are the best places to start the process of brownfield remediation and rehabilitation. When combined with the CRFs of the chemicals at the site (Chapter 10), these tools can provide the information necessary to select the sites with the most potential for synergy between the contaminants and the geologic environment, and help prioritize the locations selected for rehabilitation. Besides the environmental aspects, other factors such as commercial marketing guidelines, financial incentives, regional infrastructure and labor resources, and local community acceptance will also affect the prioritization process (Thomas 2002).

Brownfield remediation and rehabilitation efforts are aided by thorough investigative and geologic forensic techniques (Chapter 5). A comprehensive Phase I ESA consisting of a comprehensive historical review can set the stage for a detailed Phase II subsurface investigation and result in an adequate characterization of the site. Once the site characteristics are known to their fullest extent, the selection of the remediation technology can be made (Chapter 11).

As demonstrated by the case studies in Chapter 14, the possibility exists to remediate and rehabilitate some brownfield sites to a nonpolluting and socially productive status. Redevelopment alternatives must rely on the science used for remediation to guide the proper selection of land uses. As part of the landscape, brownfields exert an influence on its environmental and social functions. If planned and restored properly, brownfield sites with large areas of undisturbed habitat may complement parks and natural reserves in the effort to increase the survival of wildlife populations and the establishment of recreational activities (Lafortezza et al. 2004). In addition, brownfields in urbanized areas have the potential to serve as important locations for stormwater management, sensible heat reduction, and groundwater recharge.

These opportunities for brownfield reuse are available in the highly industrialized city of Flint, Michigan, within the Flint River watershed. Here, brownfields comprise a

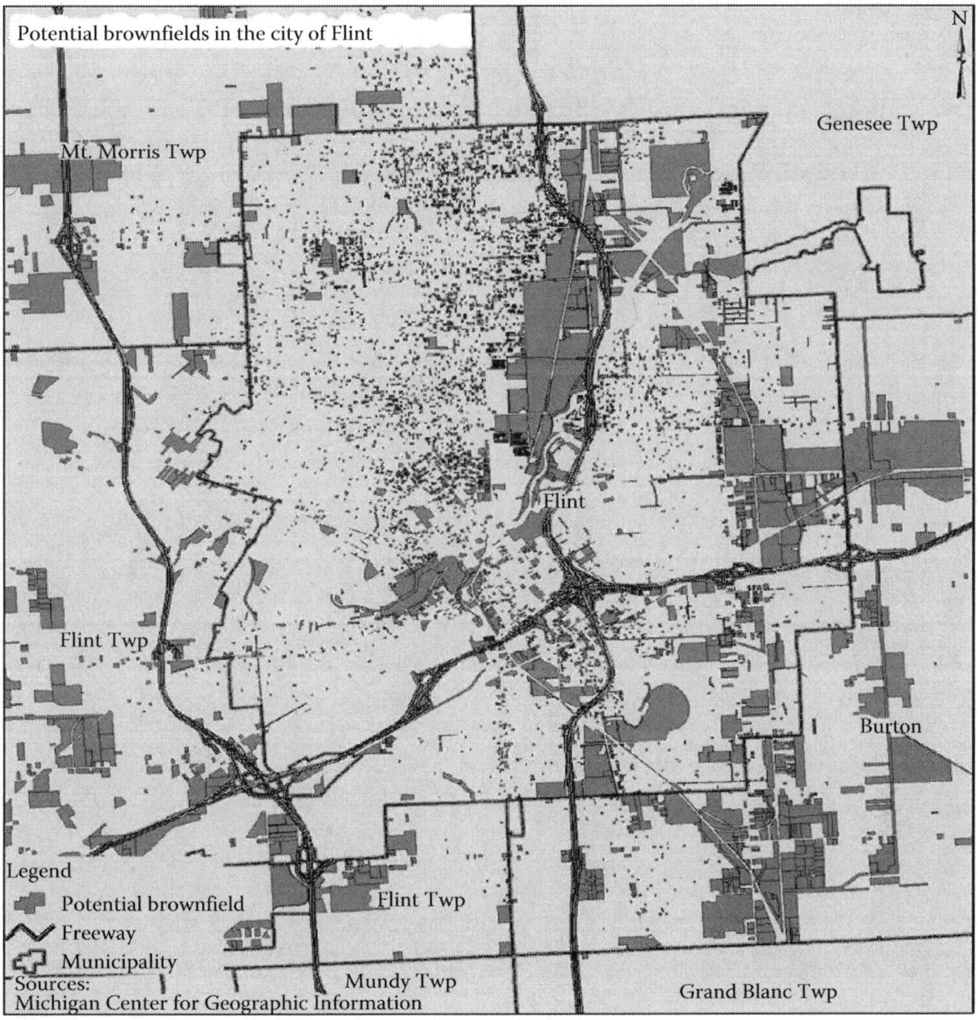

FIGURE 16.5
Brownfields may cover a significant portion of the urban area (Flint, Michigan). (Courtesy of Center for Applied Environmental Research, University of Michigan-Flint, Flint, MI, 2010.)

significant portion of the landscape, with over 7000 potential brownfields occupying 19% of the total land area (Figure 16.5).

Through the use of a geographic information system (GIS), the spatial correspondence between brownfields and geological layers of information can be derived (Chapter 5). For example, an overlay of contaminated brownfield locations on recharge zones for regional aquifers would indicate locations where aquifer vulnerability was higher. Taking this logic to the next level, the map producing this overlay could be combined with another layer of surface streams to show areas where streams might receive contaminated groundwater. This small sample out of a potentially larger set of bad outcomes associated with contaminated brownfields reinforces the need to address them as part of a first step toward sustainable watersheds.

16.4 Linking the First Step to the Big Picture

This beginning framework for attaining sustainable watersheds is like an open electrical plug—it still needs an outlet to complete the circuit. The larger picture (or outlet) is represented by the status of watershed management in the United States, so before completing the circuit, it makes sense to appraise the current state of affairs.

A large quantity of academic literature exists along with a large reservoir of technical documents detailing professional watershed management experiences. A few key themes emerge:

- Watersheds are landscape units defined by topographic boundaries; they contain multiple political units with formal boundaries not coincident with the watershed boundaries. As a result, communities with control over their land use decisions act primarily in their narrowly bounded self-interest and do not plan for the benefit of the entire watershed.
- There is not enough attention paid to or money spent for implementing nonstructural changes within watersheds. Most of the water management is performed by federally and locally contracted engineers who implement predominantly structural controls. Many of these structural controls, especially channelization measures (Chapter 12), create a variety of problems for attaining sustainable watersheds.
- Land and water management are separated in the United States. At the federal level, the National Resource Conservation Service, Bureau of Land Management, and the U.S. Forest Service (among others) manage public lands, while the Federal Energy Regulatory Commission, the Bureau of Reclamation, and the Army Corps of Engineers manage water resources (among others). Many states also have distinct agencies for addressing land and water issues. This separation impedes the progress of watershed management efforts by increasing bureaucracy.
- Response to environmental problems in the United States is reactive, rather than proactive. As a result, it takes a major disaster to catalyze change, such as the Santa Barbara, California oil spill, which helped provide the impetus for passing the National Environmental Policy Act of 1969 (Hedgpeth 1973).
- There is an urgent need for public education about all aspects of watershed management.

It would be easy to prescribe treatment for these problems. For instance, the boundary mismatch between watersheds and political units could be addressed by creating a governance system at the geographic scale of the watershed to implement watershed management priorities. This measure, however, would likely meet widespread resistance. For example, Section 208 of the Clean Water Act during the late 1970s and into the 1980s promoted voluntary regional land use planning efforts for improving water quality. Though voluntary, local communities reacted negatively to the "command and control" efforts by the federal government to impose land use decisions. These planning efforts were usually housed within regional councils of government, and most failed. Communities were eager to assume local control of their land use decisions (Malone 1990; Kaufman 2000).

Instead of specifying what *needs to be done* with respect to these broad watershed management issues, our approach is to focus on what *can be done* now with the beginning framework proposed here. Research presented so far in this book has made it clear that science—especially geology—has not been used to its full potential in urban watershed management, and without a full tank of science, attaining sustainable urban watersheds is not possible.

Our conclusion is that the best way to overcome this deficit and capitalize on the benefits generated along the way is to start at the individual parcel scale. It is at the parcel scale where geologic investigations occur, data are generated to characterize the geological environment, the major agent of landscape change—water—is given a full inventory, risk factors for chemicals are computed and applied, synergy between contaminant properties and the geological environment is evaluated, and point source controls are implemented. At the parcel scale is also where science-based landscape planning can most effectively prevent further environmental damage to land being considered for new development or redevelopment. In this strategy, contaminated sites are also an asset: the data they produce from environmental investigations benefits science-based landscape planning at uncontaminated sites and can help reduce their releases of stormwater, lower erosion rates, and protect soil, groundwater, and air through source control.

Linkages between the beginning framework presented here and the broader watershed-wide management themes are shown in Figure 16.6. Each linkage represents a way to transfer the outputs from the parcel level efforts upward to the watershed scale. Underlying this approach is the hypothesis that what is learned at the parcel level functions as a catalyst for ameliorating the environmental concerns at the watershed scale, and also provides the basis to begin aligning the out-of-whack institutional arrangements present. This process is modeled by depicting the five major themes characterizing the status of watershed management at two geographic scales. The center of the diagram identifies the linkages between the themes at the parcel and watershed scales. Discussion of Figure 16.6 begins with the boxes in the top row and proceeds downward.

16.4.1 Boundaries

The linkage preserves community control of land use decisions, but requires science-based site and watershed-scale planning (Chapter 15). The new institutional arrangement used to enable the linkage is the inclusion of watershed councils in the site plan review process. The most common criticism of community "home rule" is that this practice focuses communities' land use decisions within their own borders while ignoring the potential watershed-wide implications. Since a bad site plan has environmental implications for the entire watershed, a good site plan can help prevent further contamination and protect undamaged land. The use of a watershed-level institutional form to review site plans elevates the consideration of their outcomes to the scale of the entire watershed. Communities still retain some level of control over their land use decisions, but site plans would now promote the sustainable development of the entire watershed. The use of a formal watershed body with administrative powers might also facilitate inter-watershed collaboration on cross-boundary environmental issues, such as acid rain.

16.4.2 Structural and Nonstructural Controls

Geologic and vulnerability maps depict many of the physical processes that form landscapes, and these maps show the flows and location of surface water and groundwater (Chapters 5 and 6). Information gathered from environmental site investigations includes soil and water samples and the locations of low points in buildings where contaminant

Parcel scale	Linkage	Watershed scale
No boundary conflict between watershed and political entities	Continued community control of land use decisions, but science-based planning must be employed; Site plan reviews performed by watershed council	Watershed and political boundaries do not match
Easier to implement non-structural controls	Use geologic and vulnerability maps and the information gathered from environmental site investigations to identify the best places for nonstructural measures and source control. Incorporate parcel-level science-based planning into 319 grants	Structural controls dominate
Joint land and water management is easier to perform	Use geologic and vulnerability maps to characterize water flows and the original landscape, and with these as a guide, identify land areas within the watershed for preservation. Brownfield redevelopment must rely on the science used for the remediation; these parcels should be considered active landscape components.	Land and water management is performed separately
Site planning has proactive elements, but not enough source control is used	Use the outcomes from the science-based parcel-scale planning to guide BMP use, and as input to the watershed-wide monitoring efforts. Involve more environmental scientists in the parcel-scale planning process.	Repsonse is reactive
Education is lacking	Must be ongoing and consistent. Publish all monitoring results. Include watershed management in K-12 school curricula. Update all parcel-scale planning documents to include watershed implications.	Education is lacking

FIGURE 16.6
Linkage between themes at the parcel and watershed scales.

releases are more likely to occur (Chapters 6 and 14). Together, this body of information can help characterize areas under consideration for development, and also be used to locate common source and nonpoint source locations (16.3.1). To help increase the use of nonstructural controls at the watershed scale, science-based parcel-scale site plan implementations with nonstructural controls can be included as part of Clean Water Act Section 319 watershed planning grants.

16.4.3 Land and Water Management

More efficient and effective land preservation is achieved when land and water management are combined. The information provided by geologic and vulnerability maps permits relatively easy land and water management at the parcel scale, where the property is usually owned by one party and there is little red tape. Mimicking this decision-making process at broader scales means limiting the number of land and water management "players." Clarity of purpose is one way to streamline the management of land and water. One way to attain a clear purpose is to use an asset-based approach that views brownfields as potentially functional landscape components, and relies on remediation science to determine the ultimate use of the parcel. For instance, information produced during site remediation may indicate that some sites are best suited for stormwater management and others for new buildings. These recommendations may clash with market-driven perspectives about the reuse of a particular contaminated site. The overriding objective, however, is to strive for sustainability.

16.4.4 Proactive and Reactive Response

At the parcel scale, the very use of the site planning process to evaluate the potential uses is a proactive response. However, there needs to be more emphasis on source control, as each site should be viewed as a making a small contribution to the overall sustainability of the watershed. Site plan review at the watershed level and science-based planning can encourage and initiate proactive responses, but measurable successes are needed to sustain them. Since each site potentially contributes sediment, stormwater, and contaminants to the watershed's groundwater and streams, the conditions before and after the development can be measured. Although it would not be feasible to monitor every site, subcatchments can be monitored where multiple projects have been implemented. As the results are evaluated, the efficacy of the best management practices used within each catchment can be compared. The most effective methods can be applied to the appropriate micro-topography in other parts of the watershed. Subsequent monitoring determines the success of these efforts, and continual improvement will help sustain a proactive approach.

Involving more environmental geologists at the parcel-scale planning process will help close the science-policy gap and facilitate the transition to a more proactive response at the watershed level. This shift will occur in part because geologists understand that processes of groundwater contamination, stormwater pollution, and erosion operate at different geographic scales, and they will be able to help educate others about the need for a comprehensive approach.

16.4.5 Lack of Education

Education must be ongoing and consistent, and not limited to specific watershed management events, such as river cleanups. The lessons from the successful antismoking and antidrunk driving educational campaigns in the United States demonstrate the value of using these two attributes. In Florida, the Legacy "Truth" antismoking campaign helped cut the teen smoking rate substantially between 2001 and 2002 (Niederdeppe et al. 2004). Sustained efforts by various groups have also brought about reductions in traffic accidents related to drunk driving. For example, deaths related to drunk driving between 1982 and 2004 dropped from 59.6%–39.5% (Yi et al. 2006). These campaigns did not limit their message to specific locations associated with the behaviors—they were seen everywhere. To become successful, watershed education has to become elevated to this level. The public must know what is happening with respect to watershed management. Monitoring results should be published in local newspapers and made available on prominent websites. Watershed education should be incorporated into K-12 education. The site planning process also offers an opportunity to educate developers, land owners, and the community about watersheds during site plan reviews.

16.5 Summary and Conclusion

Sustainable outcomes in urbanized watersheds are not possible without significantly preventing further land contamination and land degradation. Knowing how the land was formed and how it functions as a landscape is critical to managing it properly, and this principle underscores the value of geology and geological investigations in watershed management. Managing the land properly will help reduce the pollution loads transported by eroded sediment and stormwater, and protect drinking water supplies in surface water and aquifers.

The effective use of science alone cannot attain sustainable urban watersheds. Comprehensive watershed management and the attainment of sustainable watersheds also depend on the success of institutional reform. The mismatches between landscape process and institutional forms will have to be resolved, and all levels of government, the private sector, and the public will need to share a common vision.

Despite the daunting challenges, we must begin the journey. The framework presented here is general and the details still need to be worked out. For instance, who comprises the "watershed group" and what are their roles? Answers to these questions must emerge through negotiations and hard work. The value of this framework lies in its specification of ways to use scientifically acquired data to enable science-based landscape planning at different geographic scales. This is the first step to achieving sustainable urban watersheds.

References

Bell, S. and S. Morse. 2008. *Sustainability Indicators: Measuring the Immeasurable?* London, U.K.: Earthscan.

Doppelt, R., M. Scurlock, C. Frissell et al. 1993. *Entering the Watershed: A New Approach to Save America's River Ecosystems*. Covela, CA: Island Press.

Ehrlich, P.R., D.S. Dobkin, and D. Wheye. 1988. *A Guide to the Natural History of the Birds of the Pribilof Islands*. Center for Conservation Biology, Stanford University, Stanford, CA.

Fresco, L.O. and S.B. Kroonenberg. 1992. Time and spatial scales in ecological sustainability. *Land Use Policy* 9:155–168.

Hedgpeth, J.W. 1973. The impact of impact studies. *Helgoland Marine Research* 24:436–445.

Hofstadter, R. 1944. *Social Darwinism in American Thought*. Philadelphia, PA: University of Pennsylvania Press.

Kaufman, M.M. 2000. Erosion control at construction sites: the science-policy gap. *Environmental Management* 26:89–97.

Kretzmann, J.P. and J.L. McKnight. 1993. *Building Communities from the Inside Out: A Path Toward Finding and Mobilizing a Community's Assets*. Evanston, IL: Institute for Policy Research.

Lafortezza, R., B. Pace, G. Sanesi et al. 2004. Planning for the rehabilitation of brownfield sites: A landscape ecological perspective. In *Brownfield Sites II: Assessment, Rehabilitation & Development*, eds. A. Donati, C. Rossi, and C. Brebbia, pp. 21–30. Ashurst, Southampton, U.K.: WIT Press.

Lynam, J.K. and R.W. Herdt. 1989. Sense and sensibility: Sustainability as an objective in international agricultural research. *Agricultural Economics* 3:381–398.

MacArthur, R.H. and E.O. Wilson. 1967. *The Theory of Island Biogeography*. Princeton, NJ: Princeton University Press.

Malone, L.A. 1990. *Environmental Regulation of Land Use*. New York: Clark Boardman.

Minor, M.A. 2009. Surface energy balance and 24-h evapotranspiration on an agricultural landscape with SRF willow in central New York, *Biomass and Bioenergy* 33:1710–1718.

Murray, K.S., D.T. Rogers, and M.M. Kaufman, 2008. *Using GIS to Map and Assess Brownfield Sites in Southwest Detroit*. Presented at the International Brownfields Conference, sponsored by the U.S. Environmental Protection Agency and the International City/County Management Association, May 5, Detroit, MI.

Niederdeppe, J., M.A. Matthew, and C. Farrelly. 2004. Confirming "truth": More evidence of a successful tobacco countermarketing campaign in Florida. *American Journal of Public Health* 94:255–257.

Rogers, D.T., M.M. Kaufman, and K.S. Murray. 2006. Improving environmental risk management through historical impact assessments. *Journal of the Air and Waste Management Association* 56:816–823.

Thomas, M.R. 2002. A weighted, multi-attribute, site prioritization and selection process for brownfield redevelopment. *Environmental Practice* 4:95–106.

WCED (World Commission on Environment and Economic Development), 1987. *Our Common Future*, Oxford, U.K.: Oxford University Press.

Yi, H., C.M. Chen, and G.D. Williams. 2006. *Surveillance Report #76: Trends in Alcohol-Related Fatal Traffic Crashes, United States, 1982–2004*; Alcohol Epidemiologic Data System. Bethesda, MD: National Institute on Alcohol Abuse and Alcoholism, Division of Epidemiology and Prevention Research.

Glossary

Abiotic degradation: degradation of a substance by processes not involving microorganisms or fungi.

Acid: any substance that when dissolved in water increases the concentration of the hydrogen ion H^+ and lowers the pH of the solution.

Acid rain: precipitation with a pH of less than 7, resulting in large part from the reaction between sulfur dioxide and water vapor in the atmosphere that produces sulfuric acid.

Acute response: characterized as a single high dose with rapid onset and disappearance of symptoms.

Absorption: a process by which a particle of gas or liquid enters a substance.

Adsorption: the adhesion of molecules of gas, liquid, or dissolved solids to a surface.

Advection: bulk transport of a liquid.

Adverse health effect: a change in body function or cell structure potentially leading to disease or health problems.

Aeolian deposits: sediments deposited by wind action. Also known as eolian deposits.

Air sparging: a groundwater remedial method that involves using injected air to volatilize contaminants in groundwater. As the injected air rises through the saturated zone and reaches the unsaturated zone, the vapors containing the contaminants are removed from the ground using a soil-vapor extraction system.

Alluvial fan: sedimentary deposit formed when a fast-flowing stream in steep terrain suddenly slows where the slope decreases; this declining velocity results in the deposition of sediments due to energy loss.

Amines: a group of organic compounds that contain nitrogen and are basic.

Angle of repose: the steepest angle a cohensionless slope can maintain without losing its stability.

Angular unconformity: an unconformity where horizontal strata or layers of sedimentary rock are deposited on tilted and eroded strata.

Anisotropic: high variability of a specific geologic feature when measured in different directions.

Anthropogenic: effects, processes, or materials derived from human activities.

Anthrosphere: human-built or developed world.

Annular seal: material of low hydraulic conductivity placed above a well screen to minimize the potential for the vertical migration of substances into the well. Also referred to as bentonite seal.

Aquifer: a mappable geologic unit composed of water-saturated porous media capable of storing and transmitting significant quantities of water under ordinary conditions.

Aquifer system: more than one generally interconnected aquifer at a given location.

Aroclor: mixture of PCB compounds distinguished by a four-digit numbering system that denotes characteristics of composition.

Artesian well: a well drilled through impermeable strata into strata receiving water from a higher elevation; this overburden of water creates pressure to force the water in the aquifer to flow upward.

Artificial groundwater recharge: artificially increasing the amount of water entering an aquifer.

Asthenosphere: upper portion of the mantle.

Asbestos: a name applied to a group of six different fibrous minerals including amosite, chrysotile, crocidolite, and the fibrous forms of tremolite, actinolite, and anthophyllite.

Atomic mass: the mass of an atom. Also referred to as atomic weight.

Atomic weight: the mass of an atom. Also referred to as atomic mass.

Bacteria: single-celled organisms.

Bar sediments: fluvial deposits along a river or stream bank, especially along the inside portions of a river or stream channel bend.

Bar-top sediments: sediments deposited on top of bar sediments; typically formed in shallow water regions in channels and abandoned channels. Also termed channel-top sediments.

Base: any substance that when dissolved in water increases the concentration of the hydroxide ion OH^- and raises the pH of the solution.

Baseflow: groundwater seepage into a stream channel.

Base-neutral-acid compounds: a group of organic compounds much less volatile than VOCs. Also referred to as semi-volatile organic compounds.

Basic: a compound with a pH over 7.

Bedding: the layering of sediments as they are formed and deposited—also referred to as stratification.

Bedrock: the continuous solid rock of the continental crust.

Bentonite seal: material of low hydraulic conductivity placed above a well screen to minimize the potential for the vertical migration of substances into the well. Also referred to as annular seal.

Benzene ring: A hexagonal ring arrangement found in benzene and other aromatic compounds, consisting of six carbon atoms with alternating single and double bonds between them. In derivatives of benzene, each carbon atom is bonded to a hydrogen atom or to other atoms or groups of atoms.

Bioaccumulation: contaminant accumulation in the body of an organism at a concentration greater than what is generally defined as background or naturally occurring.

Bioremediation: involves the introduction of microorganisms to a contaminated medium in an effort to allow natural processes to degrade the contaminants into harmless substances.

Biotic degradation: degradation of a substance by microorganisms or fungi.

Bioturbation: disturbance of sedimentary strata by living organisms such as plants and worms.

Biphenyl: organic compound composed of two benzene rings.

Boring: a hole in Earth drilled for collecting geological, chemical, physical, or other desired information about the subsurface.

Boring log: a written record of pertinent information concerning a boring.

Brownfield: an abandoned, idled, or underutilized industrial or commercial facility.

Cancer: when cells in the body become abnormal and grow or multiply out of control.

Capillary fringe: subsurface layer where groundwater seeps up from the water table by capillary action to fill pore spaces.

Capping: a remedial method where a barrier is placed over a contaminated area.

Capture zone: the areal extent where groundwater will be captured by a pumping well.

Carbon cycle: exchanges of carbon from reservoir to reservoir (e.g., oceans to atmosphere) by various chemical, physical, geological, and biological processes.

Carbon dioxide: CO_2, a common air pollutant and greenhouse gas. Anthropogenically produced through the combustion of fossil fuels.

Carbon monoxide: CO, a common pollutant toxic to humans and animal life. Anthropogenically produced through the combustion of fossil fuels.

Casing: a hollow tube placed inside a borehole providing access to the subsurface.

Cave: a natural feature beneath the surface formed by groundwater dissolving away the rock.

Cavern: an enlarged void or cavity caused by the dissolution of rock by a fluid.

Cementation: the precipitation of minerals at grain to grain contacts and within the pore spaces of sedimentary deposits; after this process is completed a sedimentary deposit becomes a sedimentary rock.

Channelization: flood control/navigation efforts for a stream or river involving methods used to change the channel geometry.

Channel-top sediments: sediments deposited on top of bar sediments; formed in shallow water regions of channels and abandoned channels. Also termed bar-top sediments.

Chemical: a substance with a specific atomic composition.

Chemical dehalogenation: a remedial method performed to remove halogens from contaminants.

Chemical oxidation: the loss of electrons.

Chemical reduction: the addition of electrons.

Chert: a sedimentary rock composed of silica (SiO_2); it may form from microscopic organisms secreting silica shells around their bodies or as a secondary precipitant.

Chlorinated solvents: a group of organic solvents containing one or more chlorine atoms within their atomic structure.

Chronic response: a stimulus lingering for a period of time after exposure to a chemical.

Clastic: consisting of fragments of rock.

Clay: a clastic sediment with a particle size of less than 1/256 of a mm (4 μm), or fine-grained minerals of hydrous aluminum-silicate composition.

Combined sewer overflow: when the wastewater volume in a combined sewer system exceeds the capacity of the sewer system or treatment plant; the excess wastewater is discharged directly to nearby streams, rivers, or other water bodies.

Compaction: the rearrangement of individual grains in a sedimentary deposit; a more tightly packed sediment forms with the mass staying the same and the volume decreases.

Completion evaluation study: a study conducted to confirm the remediation of contamination has been adequate.

Cone of depression: the three-dimensional shape of water withdrawal from a well.

Confined aquifer: a permeable geologic unit or strata bounded above and below by relatively impervious geologic material and containing water at a pressure greater than atmospheric pressure.

Confining unit: a layer of rock or unconsolidated sediment retarding the movement of water in and out of an aquifer and possessing a very low hydraulic conductivity.

Congener: refers to a specific PCB chemical family.

Conjunctive use: the planned and interchangeable use of groundwater and surface water.

Consistency: the relative ease an unconsolidated material can be deformed.

Contaminant: human introduction into the environment of substances harmful to human health or ecosystems. Also referred to as pollutant.

Contamination: the introduction of harmful substances into an ecosystem; typically anthropocentric-manufactured chemicals.
Contaminant fate and transport: a description of what happens to a contaminant after it has been released into the environment.
Contaminant risk factor: a method to calculate the potential risk of a contaminant in air, soil, or water by combining the elements of toxicity, mobility, and persistence.
Contaminant risk factor for groundwater: the environmental risk posed by specific contaminants to contaminant groundwater.
Continental drift: theory originally proposed by Alfred Wegener in 1912 that the continental land masses are not stationary and move.
Convection: vertical advection of air, water, or other fluids as a result of thermal differences.
Covalent bond: a form of chemical bonding characterized by sharing pairs of electrons between atoms.
Cross-bedding: sets of bedded sediments at different, on lapping, and inclined angles.
Cross-dating: technique used to help understand sedimentary environments and the processes leading to the origin of their sediments. Also referred to as facies analysis.
Cultural eutrophication: when anthropogenic activities such as fertilization or sewage discharges bring excess nutrients into water bodies.
Cultural landscape: the product of human interaction with the natural landscape, with culture acting as the agent, and the natural area as the medium.
Cyanide: any chemical compound containing the cyano group consisting of a carbon atom triple-bonded to a nitrogen atom.
Darcy's law: describes the flow of a fluid through a porous medium.
Debris flow: the rapid downslope plastic flow of a mass of regolith or debris. Commonly creates an apron-like or tongue-like area at its terminus with a very irregular surface.
Delta deposits: sedimentary strata originating through the deposition at the mouth of a river or stream discharging into a large body of water such as a lake or ocean.
Dense nonaqueous phase liquids: organic compounds that do not readily mix or dissolve in water and are heavier than water.
Desiccation cracks: formed when a fine-grained deposit (most often clay size particles) undergoes drying and is subsequently buried, preserving the feature. Also referred to as mudcracks.
Diagenesis: any chemical, physical, or biological change a sediment undergoes following its initial deposition.
Diffusion: the movement of a chemical from an area of higher concentration to an area of lower concentration due to the random motion of the chemical molecules. Also referred to as molecular diffusion.
Dioxins: a diverse set of halogenated substances; includes other compounds called furans.
Dip: the angle a geologic deposit, feature, or structure is tilted relative to the horizontal plane.
Disconformity: a type of unconformity existing between parallel layers of sedimentary rocks.
Dispersion: the tendency for contaminants to spread out from the path normally expected from advective flow. Also referred to as hydrodynamic dispersion.
Dissolve: to cause a substance to pass into solution.
Dolostone: a sedimentary rock composed of magnesium-calcium carbonate ($CaMg[CO_3]_2$) and associated with a marine-type environment.
Dredging: a sediment remedial method where submerged material is excavated.

Drumlins: elongated hills composed of glacial till that are much steeper on one side than the other. The steeper slope of a drumlin indicates the direction of ice movement.

Ecosystem: a self-regulating association of living plants, animals, and their nonliving physical and chemical environments.

Eddies: turbulent swirls in a fluid.

Edge effects: processes located at the margins of physical entities, such as the receipt of more light at the edge of forests.

Effective porosity: that portion of interconnected pore space permitting groundwater flow through a geologic strata. Also termed specific yield.

Effluent stream: occurs where groundwater flows into a stream. Also referred to as gaining stream.

Electrokinetics: a remedial method applied to heavy metals. Involves the imposition of an electrical field via electroosmosis designed to induce migration of heavy metals to a desired location where they are removed or encapsulated.

Elimination: a pollution prevention term used to describe the removal or storage of a certain hazardous substance.

End moraine: a moraine formed at the end of a glacier or at a location where the glacier had been stagnant during a retreating or regressive phase.

Environment: a broad term encompassing all living and nonliving things on Earth or at smaller geographic scales. The environment encompasses the natural world existing within the atmosphere, hydrosphere, lithosphere, and biosphere, and also includes the built or developed world.

Environmental concern: a condition or situation not presenting an obvious threat to human health or the environment; typically not the subject of an enforcement action.

Environmental impact statement: a study to identify and evaluate the positive and negative biophysical, social, and other environmental effects that a proposed development action may have on the environment.

Environmental risk: the probability of an event resulting in an adverse impact on the environment or humans.

Environmental risk assessment: a qualitative or quantitative assessment or investigation of the risks posed by the presence of contamination to human health or the environment.

Eolian deposits: sediments deposited by wind action. Also known as aeolian deposits.

Erosion: a complex group of related processes where rock is broken down physically and chemically and its products removed.

Esker: glacial sedimentary feature formed beneath glaciers within ice-walled tunnels of outwash that typically deposit coarse-grained gravel sediments in an irregular pattern.

ESPM: pollution prevention implemented with a stepwise evaluation process that proceeds from the most preventative measure to the least preventative measure; e = elimination, s = substitution, p = prevention, m = minimization.

Esters: a group of chemical compounds containing a modified carboxylic acid group; the acidic hydrogen atom has been replaced by a different organic functional group.

Eutrophication: the natural process of enrichment of surface waters with plant nutrients.

Evaporite: a sedimentary deposit formed by the evaporation of water.

Experimental design: the assignment of subjects to experimental groups.

Ex situ thermal treatment: a remedial method involving the excavation of contaminated soil and a subsequent thermal treatment at the surface in an effort to either destroy or transform the contaminant into a harmless substance.

Facies: characteristics of a rock or sediment mass that reflect its depositional environment.

Facies analysis: technique used to help understand sedimentary environments and the processes leading to the origin of their sediments. Also referred to as cross-dating.

Feasibility study: an evaluation of the potential remediation methods after the nature and extent of contamination have been determined.

Feedback: an outcome of a process within a system affecting the overall function of the system.

Fertilizers: chemical compounds designed to promote plant growth when applied.

Fick's law of diffusion: a relationship stating the flux of a diffusing species is proportional to the concentration gradient.

Filter pack: granular material placed around the circumference of a well screen designed to prevent suspended fine-grained sediments from entering a well.

Flaser bedding: interbedded and alternating fine and coarser-grained layers of sediment.

Floodplain deposits: sedimentary strata deposited within the floodplain of a river.

Flowing artesian well: water level of a well drilled into a confined aquifer that rises above the ground surface and flows.

Fluvial deposits: sediments deposited by a river.

Flux density: the mass of a chemical transported across an imaginary surface of unit area per unit of time.

Fossil: the remains of a once-living organism preserved in rock or sediment.

Fracturing: a remedial action involving cracking dense contaminated soil or rock so other remedial methods can work more efficiently.

Fragmentation: the division of ecosystems into spatially separated units or fragments.

Gaining stream: occurs where groundwater flows into a stream. Also referred to as effluent stream.

Geographic scale: areal extent defined through the observation and/or measurement of events and processes.

Geographic information system (GIS): computer software with the capability to input, display, analyze, and output spatial data.

Geologic formation: a fundamental unit of lithostratigraphy consisting of geologic strata with similar features in lithology, origin, or other criteria.

Geologic vulnerability: the natural properties of geologic materials acting to increase or decrease their potential exposure risk to a contaminant.

Geology: the science dedicated to the study of the history, structure, and composition of Earth.

Geomorphic system: the removal, transport, and deposition of material by wind or water.

Geophysical investigation: a type of subsurface investigation employing any number of geophysical techniques to gather geologic information in a certain area.

Glacial deposits: sedimentary strata originating from the direct or indirect action of glaciers.

Glacial lacustrine: material ranging from fine clay to sand derived from glaciers and deposited in glacial lakes mainly by glacial meltwater; many deposits are interbedded or laminated.

Glacial lake: a lake formed between the glacial front and an end moraine.

Glacial till: sedimentary material deposited directly from glacial ice.

Glacial outwash: glacial sedimentary deposits formed by the action of glacial meltwater transporting the material away from the glacial front.

Gravel: clastic sediment with a size greater than 2 mm.

Greenhouse gases: gases in the atmosphere capable of absorbing and emitting radiation within the thermal infrared range.

Ground moraine: a moraine deposit formed under the ice between lateral moraines. Ground moraine deposits may also be termed lodgment tills.

Groundwater: water beneath the surface of the ground present in pore space, fractures, or void spaces.

Groundwater recharge: a hydrologic process where water migrates downward from the surface and recharges groundwater.

Gypsum: a sulfate mineral ($CaSO_4 \cdot 2H_2O$); an evaporite deposit.

Half-life: the average amount of time required to degrade half or 50% of a specific contaminant population.

Half reaction: the loss or gain of an electron.

Halite: a mineral composed of sodium chloride (NaCl); an evaporite deposit.

Halogens: a group of elements composed of fluorine (F), chlorine (Cl), Bromine (Br), and Iodine (I).

Halogenated volatile organic compounds: a group of organic compounds with a halogen atom as part of its atomic structure.

Henry's law: a measure of the tendency for substances to volatilize.

Hepatotoxin: a chemical posing a risk of liver damage.

Herbicide: a chemical compound used to kill unwanted plants.

Heterogeneity: highly variable and poorly sorted geologic materials.

Hexavalent chromium: the metallic element chromium in its positive-6 valence (hexavalent) state; all compounds containing Cr(VI) are potential occupational carcinogens.

High resolution aquifer profiling: a technique of groundwater sampling and analysis using multiple parameters capable of creating a detailed three-dimensional depiction of the inner dynamics of groundwater contaminant plumes.

Homogeneity: well-sorted geologic materials of low variability.

Humid microthermal climate: continental climate characterized by strong seasonal variations and highly variable weather with ample precipitation throughout the year.

Hydraulic conductivity: the ability of saturated geologic media to conduct water under an induced hydraulic or pressure gradient.

Hydraulic gradient: the rate of pressure or elevation change between two or more groundwater monitoring points over the length of the flow path. Also referred to as hydraulic head.

Hydraulic head: the rate of pressure or elevation change between two or more groundwater monitoring points over the length of the flow path. Also referred to as hydraulic gradient.

Hydrodynamic dispersion: the tendency for contaminants to spread out from the path normally expected from advective flow. Also referred to as dispersion.

Hydrograph: a plot of stream discharge over a specified time interval.

Hydrologic cycle: the solar-initiated and gravity-sustained renewable flow of water between five major reservoirs: the oceans, atmosphere, ice caps, surface water, and groundwater.

Hydrolysis: cleavage of a molecule into two parts by the addition of a molecule of water.

Hygroscopic water: water tightly held onto soil particles through adsorption and not available for flow.

Igneous rocks: rocks formed by the solidification of molten material either beneath Earth's surface (termed plutonic igneous rocks) or at the surface (termed volcanic igneous rocks).

Influent stream: when a stream loses water to groundwater. Also referred to as losing stream.

In situ soil flushing: a groundwater remedial method that pumps or percolates water or chemicals into the ground. The objective is to flush or drive the contaminants present in the saturated soil to a location where they can be removed, typically by a pumping well.

In situ thermal treatment: a remedial method for soil that injects a form of heat into the affected area in an effort to mobilize the contaminants so they can be more easily removed.

Isomer: chemicals with the same molecular formula but a different molecular structure.

Isopotential level: an imaginary surface representing the static level groundwater would rise to. Also referred to as potentiometric surface, pressure level, and piezometric surface.

Isotropic: low variability of a specific geologic feature when measured in different directions.

Kame: a glacial sedimentary deposit formed as an ice contact feature by glacial meltwater; consists of a conical-shaped mass of course-grained sand and gravel with little or no fine-grained material.

Karst topography: a landscape created by groundwater slowly dissolving sedimentary rock such as limestone.

Kettle: a glacial sedimentary feature consisting of a depression formed when portions of ice from a glacial front become partially or wholly buried with glacial outwash.

Lacustrine deposits: sedimentary strata deposited in lakes.

Laminae: very thinly bedded sedimentary layers.

Land use restrictions: a remedial method of placing restrictions on land use to protect human health and environment so any contamination left in place is not disturbed.

Latency period: the duration of time without an observable effect following exposure to a chemical.

Lateral moraine: a moraine formed on the sides of glaciers.

LD_{50}: dose of a substance that is lethal to 50% of the test subjects.

Light nonaqueous phase liquids: organic compounds that do not readily mix or dissolve in water and are lighter that water.

Limestone: a sedimentary rock composed of calcium carbonate ($CaCO_3$) and formed predominantly in marine environments from the accumulation of shells, shell fragments, or from the direct precipitation from water (usually seawater).

Lithification: a complex process where unconsolidated geologic materials become rock mainly through compaction and cementation.

Lodgment till: a moraine deposit formed under the ice between lateral moraines. Lodgment till deposits may also be termed ground moraines.

Longshore current: movement of water along a shoreline caused by waves crashing along a beach at an angle.

Losing stream: when a stream loses water to groundwater. Also referred to as influent stream.

Lowest-observed-adverse-effect level: the lowest tested dose of a chemical or substance causing a harmful or adverse health effect. Also referred to as threshold effect value.

Mass wasting: the movement of rock or soil downslope by gravity without the aid of moving water, glaciers, or wind.

Maximum contaminant level: the maximum allowable amount of a contaminant in drinking water.

Mesothelioma: a type of cancer attributed to asbestos exposure.

Mechanical soil aeration: a remedial method using photolysis to remediate soil at or near the surface of the ground; typically includes a periodic tilling or mixing of soil.

Metamorphic rocks: formed by applying extreme temperature and pressure to existing rocks that rearrange their structure and mineralogy to create a new rock.

Mineral: a naturally occurring solid formed through geological processes having a characteristic chemical composition, highly ordered atomic structure, and specific physical properties.

Minimization: pollution prevention term used to describe a reduction in use of harmful chemicals or production of waste through process changes, recycling, or other methods.

Mobility: a measure of a substance's potential to migrate in the environment.

Molecular attraction: a force pulling molecules together.

Molecular diffusion: the movement of a chemical from an area of higher concentration to an area of lower concentration due to the random motion of the chemical molecules. Also referred to as diffusion.

Molecular mass: the sum of the atomic weights of all the atoms in a molecule. Also referred to as molecular weight.

Molecular weight: the sum of the atomic weights of all the atoms in a molecule. Also referred to as molecular mass.

Monitored natural attenuation: a remedial method relying on natural processes to clean up the contaminated medium.

Monitoring well: a device inserted into the subsurface that penetrates the groundwater surface for the purposes of collecting hydrologic, chemical, or other information with respect to groundwater. Also referred to as a piezometer.

Moraine deposits: sedimentary glacial deposits formed by the deposition of material from a glacier and often exposed after the glacier has receded.

Mudcracks: formed when a fine-grained deposit (most often clay size particles) undergoes drying and is subsequently buried, preserving the feature. Also referred to as desiccation cracks.

Mudflow: the rapid downslope movement of a plastic and almost fluid mass of regolith or debris.

Multiphase extraction: a groundwater remedial method designed to remove contaminants present in more than one phase or state (i.e., dissolved phase and free phase).

Mutagen: a chemical capable of causing genetic changes which could affect future generations.

Natural landscape: earth surface formations resulting from the interactions within and between the physical systems comprising the four spheres of the geosphere: atmosphere, biosphere, hydrosphere, and lithosphere.

Natural resource: air, land, water, fish, biota, wildlife, groundwater, rock, soil, aquifers, ice, or other identifiable entities on Earth considered a resource and occurring naturally.

Natural resource damage assessment: a study evaluating damages or injuries to the environment.

Negative feedback: when an output from a process in a system slows down or dampens the overall operation of the system.

Nonconformity: an unconformity where horizontal strata or layers of sedimentary rock overlie igneous or metamorphic crystalline rock.
Nonpoint source pollution: pollution from a diffuse or unknown source.
Organochlorines: an organic compound containing at least one covalently bonded chlorine atom.
Overgrowth: the importation of excess nutrients into an ecosystem.
Oxbow lake: a U-shaped body of water formed when a river meander is cut off from the main body of water flow due to erosion.
Oxidant: an atom donating an electron.
Ozone: O_3, a gas present in the upper and lower atmosphere. Ozone in the upper atmosphere absorbs ultraviolet radiation produced by the sun and is beneficial. In the lower atmosphere, ozone is considered a pollutant and is a major component of smog.
Parasite: an organism living on or within a different organism or host at the expense of the host organism.
Particulate matter: complex mixture of very small particles and liquid droplets emitted into the atmosphere. Also referred to as particle pollution.
Particle pollution: complex mixture of very small particles and liquid droplets emitted into the atmosphere. Also referred to as particulate matter.
Pebble: clastic sediment and a subset grouping of gravel with a size range from 4 to 64 mm.
Permeability: a measure of a geologic material's ability to be penetrated by water.
Permeable reactive barrier: a groundwater remedial method involving installation of a wall or fence-like structure constructed beneath the surface of the ground within the saturated zone downgradient of the contamination. The wall is composed of chemicals that degrade or destroy the contaminants of concern.
Persistence: a measure of a substance's ability to remain in the environment before being degraded, transformed, or destroyed.
Pest: living organisms occurring where they are not wanted or causing damage to crops, humans, or animals.
Pesticide: a chemical compound with the function of preventing, destroying, repelling, or mitigating any pest.
Phase I environmental site assessment: a qualitative study of the current environmental condition of a property or site.
Phase II investigation: quantitative investigation conducted at a property or site after a Phase I environmental site assessment. The objective is to confirm or refute the presence of contamination, or define the nature and extent of known contamination.
Phenol: a group of organic compounds composed of one or more hydroxyl groups attached to a carbon atom in a benzene ring.
Photochemical degradation: a chemical reaction where a substance is broken down or degraded by photons. Commonly occurs in the presence of sunlight. Also referred to as photodegradation or photolysis.
Photochemical smog: a form of air pollution formed through a reaction of air contaminants and sunlight.
Photodegradation: a chemical reaction where a substance is broken down or degraded by photons. Commonly occurs in the presence of sunlight. Also referred to as photolysis or photochemical degradation.
Photolysis: a chemical reaction where a substance is broken down or degraded by photons. Commonly occurs in the presence of sunlight. Also referred to as photodegradation or photochemical degradation.

Phreatic surface: the top of the zone of saturation. Also referred to as water table.

Phreatic zone: the region within the subsurface where all available pore space is saturated with water. Also termed zone of saturation.

Phthalate: a group of approximately 25 organic compounds consisting of esters of phthalic acid.

Phytoremediation: removal of contaminants using plants.

Piezometer: a device inserted into the subsurface that penetrates the groundwater surface for purposes of collecting hydrologic, chemical, or other information with respect to groundwater. Also referred to as monitoring well.

Piezometric surface: an imaginary surface representing the static level groundwater would rise. Also referred to as potentiometric surface, isopotential level, and pressure level.

Plate tectonics: scientific theory describing the large scale motion of Earth's lithospheric plates.

Plasticity: a property of soil or rock allowing it to be deformed beyond the point of recovery without cracking or exhibiting appreciable change in volume.

Point source pollution: pollution originating from an identifiable source such as an exhaust pipe or smokestack.

Polarity: directional charge.

Pollution: the human introduction into the environment of substances harmful to human health or ecosystems. Also referred to as contamination.

Pollution prevention: preventing contaminants from entering the environment by reducing or eliminating waste at the source. Methods include modifying production processes, promoting the use of nontoxic or less toxic substances, implementing conservation techniques, and reusing materials rather than putting them into a waste stream.

Polychlorinated biphenyl: a group of synthetically produced chemicals with 1–10 chlorine atoms attached to a biphenyl.

Polycyclic aromatic hydrocarbons: a group of more than 100 organic compounds with multiple benzene rings. Also referred to as polynuclear aromatic hydrocarbons.

Polynuclear aromatic hydrocarbons: a group of more than 100 organic compounds with multiple benzene rings. Also referred to as polycyclic aromatic hydrocarbons.

Porosity: the volume of open space within sediment, usually expressed as a percent.

Positive feedback: when the outcome of a process within a system speeds up or magnifies the system's activity or work output.

Potable water: water used for any human purpose.

Potency: the degree to which a substance or agent can cause harm to an organism when exposed. Also referred to as toxicity.

Potentiometric surface: an imaginary surface representing the static level groundwater would rise to. Also referred to as piezometric surface, isopotential level, and pressure level.

Precambrian: referring to the earliest period of geological time, between about 4600 and 543 million years ago, from when Earth was formed until the first simple forms of life appeared.

Pressure level: an imaginary surface representing the static level groundwater would rise. Also referred to as potentiometric surface, isopotential level, and piezometric surface.

Prevention: minimizing the potential of contaminants entering the environment, usually through engineering controls.

Principle of cross-cutting relationships: when a layer of sediment is cut by faulting, or the intrusion of an igneous dike or sill; the sedimentary layer is older than the event or feature cutting the layer.

Principle of faunal and floral succession: explains the relationship between biological evolution and sedimentary deposits.

Principle of original horizontality: sediments formed from sedimentary processes are generally deposited as horizontal sheets.

Principle of superposition: defines the depositional sequence of undeformed strata of rock or sediment—what is youngest is on top.

Principle of uniformity: the external and internal processes on and within Earth observed today have been operating unchanged and, at the same approximate rates, for a very long period of time throughout most of Earth's history.

Property: a parcel of land with a specific and unique legal description.

Pump and treat: a groundwater remedial method involving the extraction of groundwater by pumping it from the ground and treating the contaminated groundwater at the surface.

Radioactive decay: a spontaneous loss of energy through the emission of ionizing particles and radiation by an unstable atomic nucleus.

Rain drop prints: small circular impressions formed from raindrops hitting the soft surface of an exposed sediment; subsequent burial preserves the feature.

Receptor: the organism or ecological habitat where exposure to a substance or agent may occur.

Recognized environmental condition: the presence or likely presence of any hazardous substance or petroleum product on a property or site under conditions that may materially affect or threaten the environmental condition of the property or site, human health or the environment.

Reductant: an atom receiving an electron.

Reduction: the loss of areal coverage of an ecosystem or community.

Regolith: weathered rock devoid of organic material.

Regressive sedimentary sequence: sediment with vertically increasing coarser grains; commonly indicates decreasing water depths and increasing energy.

Remediation: meaning "remedy"—the process of cleaning up or reducing the risks posed by the presence of contamination.

Remedial action: a step or steps undertaken to remedy or lower the risks posed to human health or the environment by the presence of contamination at a property or site.

Remedial action plan: a description of the steps to be undertaken to remedy or lower the risks posed by the presence of contamination at a property or site to human health or the environment.

Remedial investigation: a comprehensive qualitative and quantitative environmental investigation conducted at large and complex contaminated properties or sites.

Retardation: the slowing of the migration of contaminants.

Ripple marks: small dune-like deposits whose long axes are perpendicular to the air or water current where they formed.

Risk assessment: a procedure used to evaluate whether there is an unacceptable risk posed to humans or the environment from natural events, human activity, or specific substances.

Rock: a solid object composed of minerals.

Rockfall or debris fall: the rapid descent of a rock mass vertically from a cliff, or by leaps down a very steep slope.

Rockslide or debris slide: rapid, sliding descent of a rock mass down a slope.
Routes of exposure: the path chemicals and substances enter the human body; includes inhalation, ingestion, and dermal absorption.
Sand: clastic sediment of a size between 1/16 and 2 mm.
Science-based landscape planning: the process of understanding the physical processes inherent to a natural landscape's formation and its long-term sustainability, and transforming selected components of these processes into planning principles and actions to preserve or enhance the cultural landscape.
Screen: a perforated pipe placed at the bottom of a well allowing water to seep into the well from an aquifer or saturated medium.
Secondary porosity: porosity formed within a geological medium after the material has been deposited.
Sediment: settled matter at the bottom of a liquid.
Sedimentary deposits: the accumulation of natural materials and sediments formed at or near the surface of Earth at ordinary temperatures and pressures.
Sedimentary rocks: rocks formed by the mechanical weathering or erosion of preexisting rock or from dissolved material precipitating from solution.
Sedimentation: the deposition of eroded material in a sink.
Semi-volatile organic compounds: a group of organic compounds much less volatile than VOCs. Also referred to as base-neutral-acid compounds.
Silt: a clastic sediment with a size range between 1/16 and 1/256 of a mm.
Simplification: the reduction in the number of species in an ecosystem or community.
Sink: location of deposition.
Sink hole: a rounded depression of the ground formed when a cave in the subsurface collapses.
Site: a parcel of land including one or more than one property or easement; a specific location.
Situation: the outlying places related to a site.
Slump: the downward slipping of a coherent body of rock or regolith along a curved surface or rupture.
Soil: the top layer of Earth's surface consisting of rock and mineral particles mixed with organic matter.
Soil creep: slow movement of a soil slope.
Soil excavation: a remedial method involving the removal of contaminated soil and its disposal at an offsite location, typically a regulated landfill.
Soil vapor extraction: a remedial method that removes contaminants from soil in the form of vapors.
Soil washing: a remedial method that "scrubs" soil to remove and separate the contaminant from the soil using detergents or a variety of chemicals depending in the type of contaminant, concentration, and soil type.
Solidification: a remedial method involving the addition and mixing of a substance designed to immobilize or entomb contamination. Also referred to as stabilization.
Solubility: a measured property of a solid, liquid, or gaseous chemical substance termed a solute to dissolve in a liquid solvent to form a homogeneous solution.
Solute: a solid, liquid, or gaseous chemical substance dissolved in anther substance.
Solvent: a solid, liquid, or gaseous chemical substance that dissolves another substance.
Sorption: refers to the action of both absorption and adsorption.
Sorting: the spatial arrangement of particles during their transport and deposition by size.
Source control: abating contamination at its source.

Spatial resolution: the smallest identifiable element in a sequence.

Speciation: a process over time where one species evolves into a different species, or one species diverges to become two or more species.

Specific heat capacity: a measure of the heat energy required to increase the temperature of a unit quantity of a substance by a unit of temperature.

Specific retention: the volume of water retained after a saturated geologic material has been drained under the force of gravity.

Specific yield: that portion of interconnected pore space permitting groundwater flow through a geologic strata. Also termed effective porosity.

Sprawl: haphazard pattern of urbanization affecting large metropolitan areas; creates more automobile dependence and longer commutes to work.

Spreading basin: a form of artificial groundwater recharge where water is spread over a large surface area composed of permeable material and permitted to seep into the ground and migrate into an aquifer below.

Stabilization: a remedial method involving the addition and mixing of a substance designed to immobilize or entomb contamination. Also referred to as solidification.

Stormwater: nonpoint pollution in urban areas initiated by wet-weather events.

Strata: layers of rock or sediment having internally consistent characteristics distinguishing them from contiguous layers.

Stratigraphic column: a vertical chronological succession of geologic units or formations specific to a given location.

Stratification: the layering of sediments as they are formed and deposited; also referred to as bedding.

Stratigraphy: the study of rock layers or unconsolidated sediment and strata; particularly their ages, composition, and relationship with other layers.

Strike: the attitude or trend of a particular deposit of geologic material.

Sulfur dioxide: SO_2; occurs naturally through volcanic eruptions and anthropogenically through the combustion of fossil fuels and some industrial processes. Sulfur dioxide is a pollutant and a component of smog.

Substitution: a pollution prevention term used to describe the use of a less toxic or potentially harmful substance instead of a more toxic or harmful substance; in ecology, when one or more organisms/species are replaced by others.

Surface risk: the probability any given site will contaminate the environment given the best available data from public sources.

Surface water: water at the surface of Earth.

Sustainability: human activities that do not inflict excessive levels of damage upon the environment; the environment is given adequate time to repair itself from prior damages; and, these efforts will help return physical systems to dynamic equilibrium.

Sustainable development: in urban watersheds this concept encompasses those activities designed to achieve and maintain dynamic equilibrium within and between the major physical systems performing their work in the atmosphere, biosphere, hydrosphere, and lithosphere.

Teratogen: a chemical capable of causing an adverse effect on a developing fetus.

Threshold: in landscapes, the point at which instability occurs.

Threshold effect value: the lowest tested dose of a chemical or substance causing a harmful or adverse health effect. Also referred to as lowest-observed-adverse-effect level.

Total porosity: the sum of effective porosity and specific retention.

Toxicity: the degree to which a substance or agent can cause harm to an organism when exposed. Also referred to as potency.

Trace fossil: preserved evidence in rock or sediment that a living organism was present (i.e., foot prints, burrows, feeding marks, etc.).

Transgressive sedimentary sequence: sediment with vertically increasing finer grains, commonly indicates increasing water depths and decreasing energy.

Transmissivity: a measure of the amount of water horizontally transmissible through a unit width by the full saturated thickness of an aquifer under a unit of hydraulic gradient per unit of time.

Transition zone: a diffuse boundary in the saturated subsurface where freshwater and saline water mix. Also referred to as zone of dispersion.

Trihalomethanes: a group of volatile organic compounds where three of the four atoms of methane are replaced by halogen atoms.

Trophic level: the position an organism occupies in a food chain.

Turbulent diffusion: random and chaotic mixing of a liquid.

Type section: a location where a particular stratigraphic column is considered complete or representative of a given geologic formation or unit.

Unconfined aquifer: a saturated geologic material where the surface of the saturated layer is equal to atmospheric pressure.

Unconformity: a buried erosional surface separating two rock masses or sedimentary strata of different ages; indicates sediment deposition was not continuous.

Unsaturated zone: that portion of Earth between the surface and the water table or zone of saturation. Also referred to as the vadose zone.

Urban geologic map: geologic map of an urban area.

Urban heat island: higher urban temperatures resulting from the excess energy input into the atmosphere from anthropogenic activities. The primary source of the additional energy comes from the replacement of vegetation with structures and pavement.

Urbanization: the processes contributing to urban growth; specifically, an increased number of people coming to the cities to live, and the subsequent city expansion through the annexation of surrounding land and adjacent communities.

Vadose zone: that portion of Earth between the surface and the water table or zone of saturation. Also referred to as the unsaturated zone.

Vapor pressure: the pressure of a vapor in thermodynamic equilibrium with its condensed phases in a closed container.

Varve: stratification formed from the seasonal fluctuation in clastic, biological, or chemical laying of sediment.

Virus: a subcellular agent capable of replicating itself inside the cells of another organism.

Vitrification: a form of solidification using electrical power to transform contaminated subsurface soil into a glass-like substance.

Volatile organic compounds: organic compounds tending to volatilize or evaporate readily under normal atmospheric pressure and temperature.

Volatilization: conversion into a vapor or gas without chemical change.

Vug: small- to medium-sized cavity inside a rock.

Watershed: the extent of land drained by a given water feature, such as a river system, lake, swamp, estuary, reservoir, wetland, bay, sea, or ocean.

Watershed management: the minimization of the human impacts within and between different land use systems, with the objective of achieving a sustainable landscape.

Water table: the top of the zone of saturation. Also referred to as phreatic surface.

Weathering: a destructive process breaking down rocks and minerals through exposure to atmospheric agents such as air, wind, water, and ice.

Wellhead protection zone: the capture zone for a protected water supply well.

Wetland: lands where water saturation is the dominant factor determining the nature of soil development and the types of plant and animal communities living in the soil and on its surface.

Zone of dispersion: a diffuse boundary in the saturated subsurface where freshwater and saline water mix. Also referred to as transition zone.

Zone of saturation: the region within the subsurface where all available pore space is saturated with water. Also termed phreatic zone.

Index

A

Abiotic degradation, 283–284
Advection, 276–277, 288
Air contaminant risk factor (CRF_{AIR})
  DNAPL compounds, 370
  environmental factor, 365
  LNAPL VOCs, 370
  PAHs and PCBs, 368, 370
  physical and chemical factor, 365–366
  pollutant factors, 365–370
Analogy, ecosystems *vs.* urban areas, 14
Angle of repose, 31
Angular unconformity, 28
Anisotropic variability, 82
Annular/bentonite seal, 153
Anthropogenic disturbance, 57–58
Aquifer, 76, 80
Artesian well, 79
Artificial groundwater recharge, 109–113
Asbestos
  contaminant fate and transport, 305
  definition, 252
  heat insulator, 252
  minerals, 252
  properties, 251
Asthenosphere, 23–24

B

Bacteria, contaminant, 263, 308
Baseflow, 96
Bedrock, 30
Benzoylmethylecgonine, 264–265
Bioremediation
  sediment remediation, 399
  soil remediation, 380–381
Biotic degradation, 282–283
Borehole geophysics, 131
Boring log, 149

C

California
  benefits, 482
  geologic cross section and soil excavation, 482, 484
  groundwater sample collection and remediation, 482–483
  groundwater, seven discrete zones treatment, 484
  site characteristics, 481
Capillary fringe, 85
Capping
  sediment remediation, 399
  soil remediation, 382
Carcinogens, 229–230
Cave, 37
Cavern, 80
Cementation, 41
Channelization, 104, 428
Chemical oxidation, 284
Chemical reduction, 284
Chlordane, 248
Combined sewer overflows, 7
Compaction, 41
Cone of depression, 95
Cone penetrometer (CPT) method, 484
Confined aquifers, 79
Confining layer, 9
Congener, 239
Conjunctive use, 111
Contaminant fate and transport
  acids and bases, 305
  asbestos, 305
  atmosphere
    air pollutants removal, 296–297
    anthropogenic sources, 294–295
    composition, 296
    deposition, 296–298
    harmful effects, 296
    layers, 293–294
    oxygen, 295
    temperature and pressure, 295
    turbulent diffusion, 293–294
  bacteria, 308
  carbon monoxide, 306
  contaminant releases, environment
    exploration and evaluation, 275
    factors, 274
    history, 274
    sources, 274–275
  cyanide, 305
  dioxins, 304

emerging contaminants, 308
fertilizers, 304–305
greenhouse gases, 306
groundwater
affinity, 291
capture zone, 292–293
contaminant factors, 288–289
dispersion, 289–290
factors affecting migration, 291–292
karst topography, 293
underground storage tank, 289–290
heavy metals, 302–303
ozone, 306–307
PAH, 301
parasites, 307
particulate matter, 308
PCB, 301–302
pesticides and herbicides, 303–304
principles
advection/convection, 276–277
degradation, 281–284
dispersion, 277
Fick's law, diffusion, 277–278
migration rate, 278–281
molecular diffusion, 277
radioactive compounds, 306
soil
definition, 284
diffusion and mass flow, 285
migration rate, 284–285
solid and liquid phase, 285
sulfur dioxide, 307
surface water
contaminant persistence and bioaccumulation, 287–289
municipal/industrial waste water plant, 286–287
turbulent flow, 286
SVOC, 302
viruses, 308
VOC
contaminant degradation rate, 300–301
groundwater, 299
photochemical smog, 298
trihalomethane, 301
Contaminant pathways, 12
Contaminant risk factor for groundwater (CRF_{GW})
calculation, 348–349
distribution, 354
examination, 350–351
LNAPL and DNAPL, 353
LNAPL VOCs, 350
mobility, 347
PAHs, 352
persistence, 348
toxicity, 347
VOC DNAPLs, 351
Contaminant risk factors (CRFs)
air, 369
anthropogenic sources, 363
calculation, 367–368
DNAPL compounds, 370
environmental factor, 365
LNAPL VOCs, 370
mobility, 366
PAHs and PCBs, 368, 370
persistence, 367
physical and chemical factor, 365–366
pollutant factors, 365–370
toxicity, 366
groundwater, 356–357
chromium VI and DNAPL, 350–351
distribution coefficient, 348–349
environmental contamination, 355
examination, CRF_{GW}, 351
Henry's law, 347
influencing factors, 346
LNAPL and DNAPL, 353
mobility, 347
PAHs, 352
persistence, 348
retardation factor, 348
toxicity, 347
volatile organic compound, 349–351
soil, 362
vs. groundwater, 364–365
LNAPL VOCs, 358–359
migration potential, 355
organic carbon partition coefficient (Foc), 357
PAHs, 361
physical/chemical factors, 362–363
retention capacity, 363
risk factors, 358–363
total organic carbon fraction (Koc), 357
surface risk
calculation, 344–345
contamination incident rates, 344
edge effects, 342
evaluation, 341–342
industrial location, 346
mean surface risk value, 346
recharge area, 342–343
Rouge watershed, 345–346
spatial resolution, 342–343

standard industrial classification code, 344
wellhead protection zones, 342
Contaminants
acids and bases, 253–254
asbestos, 251–252
bacteria, 263
benzoylmethylecgonine, 264–265
carbon monoxide, 256–258
chemicals, 225–226
cyanide, 250–251
definition, 225
dioxane isomers, 265–266
dioxins, 249–250
fertilizers, 250
greenhouse gases, 255–256
heavy metals
carcinogenicity, 246
toxicity, 246–247
uses, 245
household products, 225
organic contaminants, 226–227
ozone, 258–259
parasite, 263
particulate matter, 260, 262–263
perchlorates, 265
pesticides and herbicides
definition, 247
dichlorodiphenyltrichloroethane, 248
organochlorines, 249
pharmaceuticals and drugs, 264
polychlorinated biphenyls
chlorination, 239
physical properties, 238
structure, 240
uses, 238–239
polynuclear aromatic hydrocarbons
chemical formula and carcinogenicity, 237–238
compounds, 237
molecular structure, 237
naphthalene, 236
radioactive compounds, 254
semivolatile organic compounds
amines, 243
esters, 243–244
phenol, 242–243
phthalates, 240–242
sulfur dioxide, 260–261
toxicity
adverse health effect, 228
carcinogens, 229–230
chemical effect, 228
exposure, routes, 228–229
measurement, 227–228
types, 227
virus, 263
volatile organic compounds, 230–236
Contaminants of concern (COC), 136
Contamination, 225; *see also* Contaminants
Convection, 276–277
Covalent bond, 66
CRF_{AIR}, *see* Air contaminant risk factor (CRF_{AIR})
CRF_{GW}, *see* Contaminant risk factor for groundwater (CRF_{GW})
CRFs, *see* Contaminant risk factors
CRF_{SOIL}, *see* Soil contaminant risk factor (CRF_{SOIL})
Cultural landscape, 496
Cyanide, 250–251, 305

D

Darcy's law, 83
DDT, *see* Dichlorodiphenyltrichloroethane
Dense nonaqueous phase liquids (DNAPLs)
air contaminant risk factor, 370
contaminant risk factor for groundwater, 350–351, 353
soil contaminant risk factor, 359–360
Department of Natural Resources and Environment (DNRE), 317
Dichlorodiphenyltrichloroethane (DDT), 248
Dioxins, 249–250, 304
Disconformity, 28
Dispersion, 277–278
Dispersion/transition zone, 100, 102
DNAPLs, *see* Dense nonaqueous phase liquids
Dolostone, 39

E

Eddies, 286
Effective porosity (n_e), 87
Electrical resistivity, 131
Electromagnetics, 131
Elimination, nonpoint sources
deep-shaft combined sewer basin, 457–458
micro-topography, 455
opportunities, stormwater storage and diversion, 456–457
pollutants, 457–458
urban parcel water balance, 455–456
Emerging contaminants, 308
Environmental concern, 123
Environmental impact statement (EIS), 135–136

Environmental risk, 117
Environmental sampling methods; *see also* Subsurface environmental investigations
- air sampling, 158–161
- groundwater sampling methods, 150–156
- sediment sampling, 157
- soil sampling and description, 138–150
- surface water sampling, 157–158

Eolian, 54
Erosion rate, 34
Escherichia coli, 263
ESPM methods
- nonpoint sources
  - elimination, 455–458
  - minimization, 459
  - prevention, 458–459
  - substitution, 458
- point sources
  - elimination, 443
  - minimization, 448–449
  - prevention, 444–448
  - substitution, 443–444

Evaporite, 39–40
Excavation
- sediment remediation, 398
- soil remediation, 379

F

Fertilizers, 250, 304–305
Fick's law, diffusion, 277–278
Filter pack, 153
Flaser bedding, 46
Flowing artesian well, 80
Flux density, 277
Forensic geologist, 168

G

Gaining stream, 96
Geologic strata, 174
Geologic vulnerability
- high vulnerability site, 210–217
- low vulnerability site, 207–210
- map-building example, 202–205
- methods
  - composition, 200
  - depth to groundwater, 199
  - geologic and hydrogeologic factors, 197–198
  - geologic vulnerability matrix and scoring system, 199
  - groundwater, 201–202
- site comparison analysis, 217–218
- subsurface vulnerability and vulnerability map development, 196–197

Geologic vulnerability matrix and scoring system, 199
Glacial sediments, 54–56
Glacial till, 55
Glyphosate, 249
Gravity survey, 131
Greenhouse gases, 255–256, 306
Ground penetrating radar (GPR), 131
Groundwater; *see also* Contaminant risk factors (CRFs); Remediation; Water and hydrogeology
- contaminant risk factor, 346–355
- contaminants fate and transport, 288–293
- recharge, 70
- remediation technologies, 394–397
- sampling methods, 150–156
- watersheds, 76

H

Half-life, 281
Half reaction, 284
Hand auger, 140
Heavy metals
- contaminant fate and transport, 302–303
- soil, 315–316

Hepatotoxins, 229
Herbicide, 247–249
Heterogeneity, 80
High resolution aquifer profiling, 153
Hollow-stem auger drilling rig, 141
Home rule, 515
Homogeneity, 80
Hydraulic conductivity (K), 8–9, 83
Hydrologic cycle, 10–11
Hydrolysis, 45, 283
Hygroscopic water, 84

I

Igneous rocks, 23
Illinois
- after demolition, 470–471
- 3D cross section, Cr VI impacts, 470–471
- before demolition, 470–471
- ferrous sulfate, 473
- on-site and off-site Cr VI impacts, 470, 472
- site characteristics, 470

Indiana
facilities on-site, 474
location and surrounding region, 474
RECs and AOCs, 475–476
site characteristics, 473
Industrial point sources, pollution prevention
contaminant risk factors (CRFs), 442
ESPM methods, 443–449
financial gains and cost avoidance, 449–450
geologic vulnerability, 442
hazardous substances, 441–442
objectives, 440–441
Island biogeographic theory, 512
Isotropic variability, 82

K

Karst topography, 37

L

Lacustrine, 53–54
Laminae, 45
Light nonaqueous phase liquids (LNAPLs)
air contaminant risk factor, 370
contaminant risk factor for groundwater, 350, 353
soil contaminant risk factor, 358–359
Limestone, 39
Losing stream, 96
Lowest-observed-adverse-effect level (LOAEL), 227–228

M

Malathion, 248–249
Mass wasting process, 31–33
Maximum contaminant level (MCL), 329
Metal analytical methods, 319
Metal contamination
heavy metals, soil, 315–316
shallow groundwater and rouge river watershed, dissolved metals
contamination, 334–335
chromium detection, 335
Cu-fabricating facility, 328
heavy metals mobilization, 329, 333–334
hydrogeologic units, 327
influencing factors, 327
lead concentration, 334
mean concentration, heavy metals, 329–333
site and sample distribution, 328
soil and Rouge River watershed
analytical methods, metal, 319
copper and zinc concentration, 325
environmental contamination, 317
environmental site, 316
evaluation and screening, 317–318
glacial drift, 319
lake beds, 319–320
lead concentration, 326
lower clay unit, 325
low-occurrence metals, 320
MDEQ, 317
mercury concentration, 325
metal analysis, 319
metal concentrations, 317
metal distribution, 321–323
metal's mean concentration, 321, 323
sample location, 317–318
silty clay unit, 321, 324
Metal mobility and contamination ranking, 334
Metamorphic rocks, 23
Methyl-*tert*-butyl ether (MTBE), 233
Michigan
geologic map, 465, 468
high-resolution aquifer profiling, 467, 469
historical map, 465–466
identified recognized environmental conditions (RECs), 465, 467
pre-and post-demolition site maps, 465
site characteristics, 464
surface map, 466, 468
urban setting, aerial image, 464
Michigan Department of Environmental Quality (MDEQ), 317
Mid-level personal protective equipment, 126
Mobility, 273
Molecular attraction, 84
Molecular diffusion, 277
Mutagens, 229

N

National Pollutant Elimination System (NPDES), 399
Natural landscapes, 496
Naturally occurring disturbance, 58–59
Natural resource damage assessment (NRDA), 135
New Jersey
RECs, foundry portion, 486
remedial method, 487
site characteristics, 485
urban setting, 485–486

Nonconformity, 28
Nonpoint sources, pollution prevention
contaminant risk factors (CRFs), 454
erosion and stormwater runoff, 459
ESPM methods, 455–459
geologic vulnerability, 454–455
hazardous substances, 453–454
objectives, 451–453

O

Occupational Safety and Health Administration (OSHA), 251
Ox-bow lakes, 75
Oxidant, 284
Ozone (O_3), 258–259, 306–307

P

PAHs, *see* Polycyclic aromatic hydrocarbons
Parasites, 263, 308
Particulate matter
adverse health effects, 262–263
components, 260, 262
contaminant fate and transport, 273
human hair, 262
Perchlorates, 264
Permeability, 79
Permethrin, 249
Persistence, 273
Pesticides and herbicides, 247–249, 303–304
Pests, 248
Photochemical smog, 298
Phytoremediation, 383–384
Plasticity, 148
Plate tectonics, 23
Polarity/directional charge, 66
Pollution prevention
components, 438
definition, 437
industrial point sources
contaminant risk factors (CRFs), 442
ESPM methods, 443–449
financial gains and cost avoidance, 449–450
geologic vulnerability, 442
hazardous substances, 441–442
objectives, 440–441
nonpoint sources, urbanized watersheds
contaminant risk factors (CRFs), 454
erosion and stormwater runoff, 459
ESPM methods, 455–459
geologic vulnerability, 454–455
hazardous substances, 453–454
objectives, 451–453
United States, 438–439
Polychlorinated biphenyls (PCBs), 11, 301–302
Polycyclic aromatic hydrocarbons (PAHs)
air contaminant risk factor, 368, 370
contaminant fate and transport, 301
contaminant risk factor for groundwater, 352
soil contaminant risk factor, 358–359, 361
Porosity, 86
Potable water, 68
Potency, *see* Toxicity
Potentiometric surface, 90
Prevention, point sources
aboveground storage tank (AST), 445, 447
improper storage, drums, 444
outdoor liquid waste storage area, 445–446
proper storage, drums, 445
sealed floor drain, 445–446
spill response and training exercise, 448
spill station, 447–448
surface, outdoor storage area, 445–446
Principle of uniformity, 26
Property, 120
Pump and treat system, groundwater remediation, 390–391

R

Radioactive decay, 254
Rapid biotic and ecosystem response, 511
Recognized environmental condition (REC), 122
Redevelopment, contaminated sites
California
benefits, 482
geologic cross section and soil excavation, in situ treatment, 482, 484
groundwater sample collection and remediation., 482–483
groundwater, seven discrete zones treatment, 484
site characteristics, 481
Illinois
after demolition, 470–471
3D cross section, Cr VI impacts, 470–471
before demolition, 470–471
ferrous sulfate, 473
on-site and off-site Cr VI impacts, 470, 472
site characteristics, 470
Indiana
facilities on-site, 474
location and surrounding region, 474

RECs and AOCs, 475–476
site characteristics, 473
Michigan
geologic map, 465, 468
high-resolution aquifer profiling, 467, 469
historical map, 465–466
identified recognized environmental conditions (RECs), 465, 467
pre-and post-demolition site maps, 465
site characteristics, 464
surface map, 466, 468
urban setting, aerial image, 464
New Jersey
RECs, foundry portion, 486
remedial method, 487
site characteristics, 485
urban setting, 485–486
Wisconsin
after site demolition, ground-level views, 480
extent of impacts, 478
remediation, aerial photograph, 480
site characteristics, 477
site map, 477
west–east geologic cross section, 479
Reductant, 284
Refine, 130
Regolith/weathered rock, 30
Relative size, sand, silt, and clay particle, 145
Remedial action plan (RAP), 136
Remedial investigations, 134–135
Remediation
air pollution
dust collector., 405
regulation, 403
sources, 404
causes, 378
cost
contaminants, 406–407
contaminant soil, 409–410
controlling factors, 404–405
evaluation, 405–406
groundwater remediation, 408
remediated media, 409
site breakdown, 407–408
soil remediation, 407–408
groundwater
air sparging, 392
biological agents injection, 393
chemical agents injection, 393
geological spectrum, 395
influencing factor, 394–396
institutional controls, 394
monitored natural attenuation, 394
multiphase extraction, 394–395
permeable reactive barriers, 393
pump and treat system, 390–391
in situ soil flushing, 392
superfund sites, 396–397
technologies, 391
influencing/controlling factors, 378, 389
sediment
bioremediation, 399
capping, 399
dredging, 398–399
excavation, 398
natural attenuation, 399
source control, 398
soil
bioremediation, 380–381
capping, 382
chemical dehalogenation, 387–388
chemical oxidation, 387
electrokinetics, 386
excavation, 379
ex situ thermal treatment, 385
fracturing, 386
land use restrictions, 382–383
mechanical soil aeration, 381–382
monitored natural attenuation, 381
phytoremediation, 383–384
in situ thermal treatment, 384–385
soil vapor extraction, 383
soil washing, 384
solidification/stabilization, 386
techniques, 379
vitrification, 386–387
VOC methods, 389–390
surface water
engineered natural system, 401
spills, 401–403
wastewater treatment plant, 400
Resource Conservation and Recovery Act (RCRA), 127
Retardation, 292
Risk assessment, 131
Roto-sonic drilling rig, 142
Roto-sonic drilling rod, 142
Rouge River geologic vulnerability scoring, 205

S

Sanitary sewers, 426
Science-based landscape planning
cultural landscape, 496
definition, 496

dimensions, 498
formation and change, 497
natural landscapes, 496
objectives, 496
on-site interactions and thresholds
inflows, parcel-scale site, 499
matrix, 500
soil vegetation, 500–501
soil water, 500
role of water, 497–498
science-policy gap
definition, 492
environmental scientists, review process, 496
geographic information system (GIS), 494
lake plain and moraine, 495
land development, 492–493
site and situation, 498–499
watershed management
channelized portion, urban river, 502
vs. parcels, 501
Science–policy gap, 492–496
Secondary porosity, 88–90
Sedimentary deposits, 22
bedded/stratified depositional features, 45–48
environments
alluvial, 50
eolian, 54
fluvial, 50, 52–53
glacial, 54–56
lacustrine, 53–54
major U.S. urban centers, 52
non-clastic chemical depositional environments, 57
wetlands, 54
identification and classification methods
composition, 44–45
grain shape, 43
grain size, 43–44
sorting, 42–43
thickness, 45
remnants/identification marks, 47–49
Sedimentary rocks, 25–26
Seismic refraction and reflection, 131
Semivolatile organic compounds (SVOCs)
amines, 243
contaminant fate and transport, 302
esters, 243–244
phenol, 242–243
phthalates, 240–242
Sink, 30, 273
Sinkhole, 37
Site, 120
Soil classification by grain size, 144
Soil contaminant risk factor (CRF_{SOIL})
calculation, 357–358
DNAPL VOCs, 359–360
LNAPL VOCs, 358–359
PAHs, 358–359, 361
physical/chemical factors, 362–363
Soil flushing, remediation groundwater, 392
Soil textural chart, 145–146
Southeast Michigan Council of Governments (SEMCOG), 320
Spatial resolution, 342
Specific heat capacity, 67
Specific retention (Sr), 87
Specific yield (Sy), 87
Spreading basin, 111
Standard industrial classification (SIC), 344
Stormwater
detention basins, 431
impacts, 428
land use and quality, 429
Novi, Michigan, 430–431
regional detention, 432
relationship, 429–430
Stratigraphy
principle
crosscutting relations, 28
faunal and floral succession, 28
original horizontality, 26–27
superposition, 27–28
rock types, 8
Subsurface environmental investigations
air sampling
indoor air sampling, 158–160
outdoor air sampling, 160–161
purposes, 158
completion evaluation study, 136–137
ecological and human health risk assessment
dose–response assessment, 133
exposure assessment, 133
exposure pathways, 133
hazard identification, 133
receptors, 133
risk characterization, 133
subsurface geology, 132
toxicity, 132
EIS, 135–136
feasibility study, 134
geophysical investigation, 130–131

groundwater sampling methods
Geoprobe®, 150–151
monitoring well cover, 155
monitoring well, installation of, 150–154
steps, 155
teflon bailer containing groundwater, 156
water level indicator, 155
harmful substance, 117–118
household products, 118
NRDA, 135
phase I environmental site assessment
case-by-case basis, 120–121
de minimis, 123
health and safety, 124–126
landfilling activities, 124
REC, 123
requirements, 120
researching activities, 121
site inspection, 121–122
phase II investigation, 127–130
purposes, 137
RAP, 136
remedial investigations, 134–135
sediment sampling, 157
soil sampling and description
documentation, 143–150
drilling using a hand auger, 139–140
excavating test pits, 139
mechanical drilling methods, 139–143
sample containers, 143
surface sampling, 138
surface water sampling, 157–158
triggering mechanisms, 118
Sustainable urban watersheds
definition, 506
degraded brownfield remediation and rehabilitation
factors, 512
geographic information system (GIS), 513
land contamination and degradation prevention
grid pattern of street, 509
infrastructure, dendritic drainage pattern, 509
integrated source control opportunities, 507–508
natural drainage patterns *vs.* human infrastructure, 509
source control, urban stormwater, 510
linkages
boundaries, 515
lack of education, 517
land and water management, 516–517
proactive and reactive response, 517
structural and nonstructural controls, 515–516
principles, 506
Rouge River watershed, 505
undamaged and damaged land preservation
geologic maps, 511
island biogeographic theory, 512
overcome human impacts, 510
rapid biotic and emergency response (RBER), 511
SVOCs, *see* Semivolatile organic compounds

T

Teratogens, 229
Threshold effect value, 227–228
Total porosity, 87
Toxaphene, 249
Toxicity, 227–230

U

Unconformities, 28–29
Unified soil classification system, 147
United States, pollution prevention, 438–439
Urban geologic mapping
base map, 181–182
conceptual geologic model, 171–172
developing and constructing cross sections, 179–181
facies analysis, 177–179
fieldwork, 172–177
final draft map preparation
common methodologies, 183
GIS, 185–190
succession, 183–185
three-dimensional maps, 185, 189
information and mapping challenges, 166–167
information sources, integration, and evaluating usability, 167–171
Urban heat island (UHI), 432–433
Urbanization, watersheds
function
erosion, 423
negative feedback, 423
positive feedback, 424
matter and energy flows disruption
erosion and sedimentation, 432
stormwater, 428–432
urban heat island (UHI), 432–433

water infrastructure and ecosystems, 425–428
structure
atmosphere, 418–420
definition, 418
surface and subsurface, 420–421
three dimensions, 418
zonation, 421–422
Urban watersheds, geology
anthropogenic disturbance, 57–58
environmental problems, 21–22
geological processes
principle of uniformity, 26
principles of stratigraphy, 26–28
sediments and rocks, formation of, 22–26
unconformities, 28–29
naturally occurring disturbance, 58–59
sedimentary depositional environments
alluvial, 50
eolian, 54
fluvial, 50, 52–53
glacial, 54–56
lacustrine, 53–54
major U.S. urban centers, 52
non-clastic chemical depositional environments, 57
wetlands, 54
sedimentary deposits, identification and classification of
bedded/stratified depositional features, 45–48
bioturbation and tracks, 49
composition methods, 43–45
fossils, 49
grain shape methods, 43
grain size methods, 43–44
mudcracks, 48
raindrop prints, 49
ripple marks, 48
sorting methods, 42–43
thickness methods, 45–46
varves, 49
sedimentary processes
clastic processes, 30–38
non-clastic chemical sedimentary processes, 38–40
rock formation, diagenesis, 41
sediments formation, 40

V

Vadose zone, 84
Virus, 263, 308
Volatile organic compounds (VOCs)
air contaminant risk factor, 370
contaminant fate and transport
contaminant degradation rate, 300–301
groundwater, 299
photochemical smog, 298
trihalomethane, 301
contaminant risk factor for groundwater, 350–351
dense nonaqueous phase liquids, 233–235
light nonaqueous phase liquids, 230, 232–233
soil contaminant risk factor, 358–360
trihalomethanes, 235–236
uses, 232–233

W

Wastewater treatment plant, 400
Water and hydrogeology
anthropogenic groundwater influences, 107–112
anthropogenic surface water influences, 103–106
basic chemistry and properties
water chemistry, 66–67
water cycle, 67
water distribution, earth, 67–68
basic water chemistry, 66–67
distribution, 67–68
groundwater, 77–78
aquifer types and basic structure, 79–80
flow direction, 90–92
flow rate, 92–94
fundamental concepts, 78–79
heterogeneity and homogeneity, 80–82
hydraulic conductivity, 83
movement, 84–86
primary porosity, 86–88
secondary porosity, 88–90
water table and hydraulic gradient, 90
withdrawal, 94–96
potable water use, 68–71
surface water
drainage development and patterns, 74–76
watersheds, 71–74
surface water–groundwater interaction
lake interaction with groundwater, 98–100
ocean water, 100–103
stream interaction with groundwater, 96–98
wetland interaction with groundwater, 100

water cycle, 67
Water infrastructure and ecosystems
channelization, 428
contamination, 426
food chain and food web, 425
fragmentation, 426
navigation/flood control, 428
overgrowth, 426
overland flow, 428
reduction, 426
simplification, 426
stormwater, 427
substitution, 426
Watershed approach
contamination and risk, 11–13
human–environmental interactions, 1
hydrogeology, 8–9
industrial development, historical transformation
groundwater, 3
Rouge River, 2
industrial property abandonment, 11–13
industry evolution
Detroit's central location, 5–6
Detroit's growth, 4
Detroit's situation, 3–4
southeast Michigan, developed land, 6–7
movement of water, science and planning, 10–11
near-surface geology, 7–9
remediation of contaminated sites, 13–14
science-based landscape planning, 14–15
surface hydrology, 9–10
Watershed management
channelized portion, urban river, 502
vs. parcels, 501
Water table/phreatic surface, 79
Weathering process, 30–31
Wellhead protection zones, 342
Wetland, 54
Wisconsin
after site demolition, ground-level views, 480
extent of impacts, 478
remediation, aerial photograph, 480
site characteristics, 477
site map, 477
west–east geologic cross section, 479